Methodology of Educational Measurement and Assessment

This book series collates key contributions to a fast-developing field of education research. It is an international forum for theoretical and empirical studies exploring new and existing methods of collecting, analyzing, and reporting data from educational measurements and assessments. Covering a high-profile topic from multiple viewpoints, it aims to foster a broader understanding of fresh developments as innovative software tools and new concepts such as competency models and skills diagnosis continue to gain traction in educational institutions around the world. Methodology of Educational Measurement and Assessment offers readers reliable critical evaluations, reviews and comparisons of existing methodologies alongside authoritative analysis and commentary on new and emerging approaches. It will showcase empirical research on applications, examine issues such as reliability, validity, and comparability, and help keep readers up to speed on developments in statistical modeling approaches. The fully peer-reviewed publications in the series cover measurement and assessment at all levels of education and feature work by academics and education professionals from around the world. Providing an authoritative central clearing-house for research in a core sector in education, the series forms a major contribution to the international literature.

Lale Khorramdel • Matthias von Davier
Kentaro Yamamoto
Editors

Innovative Digital-Based International Large-Scale Assessments

Foundations, Methodologies, and Quality Assurance

Springer

Editors
Lale Khorramdel
Lynch School of Education and Human Development
TIMSS & PIRLS International Study Center at Boston College
Chestnut Hill, MA, USA

Matthias von Davier
Lynch School of Education and Human Development
TIMSS & PIRLS International Study Center at Boston College
Chestnut Hill, MA, USA

Kentaro Yamamoto
Independent Consultant for Psychometric Methodology and Large-Scale Assessment (formerly ETS)
Princeton, NJ, USA

ISSN 2367-170X ISSN 2367-1718 (electronic)
Methodology of Educational Measurement and Assessment
ISBN 978-3-031-90950-4 ISBN 978-3-031-90951-1 (eBook)
https://doi.org/10.1007/978-3-031-90951-1

This Springer imprint is published by the registered company Springer Nature Switzerland AG
The registered company address is: Gewerbestrasse 11, 6330 Cham, Switzerland

Acknowledgements

This volume is the result of a collaborative effort, and we extend our heartfelt gratitude to everyone who contributed to its realization.

First and foremost, we thank all the authors whose expertise and insightful contributions have brought this volume to life. Their dedication and rigorous scholarship have been essential in creating a comprehensive volume on digital-based innovations in international large-scale assessments.

We also wish to express our special appreciation for the exceptional work of Emily Kerzabi, our Editorial Manager and Copyeditor. Emily's meticulous attention to detail, unwavering commitment, and expert guidance have been instrumental in shaping this volume and ensuring the highest quality.

Finally, we are grateful to ETS for their invaluable support.

Thank you all for your remarkable contributions and support!

Contents

Part III Innovations in Translation, Data Processing, and Scoring

Part IV New and Advanced Methodologies for Analyzing Response and Process Data

Part I
Introduction: Technology as a Driver of the Evolution of International Large-Scale Assessments

Chapter 1
Technology as a Driver of the Evolution of International Large-Scale Assessments: Constructs, Frameworks, and the Psychometric Qualities of Test Results

Lale Khorramdel, Matthias von Davier, and Kentaro Yamamoto

Abstract This chapter outlines the motivation behind the creation of this book on foundations, innovations, and quality assurance in digital-based international large-scale assessments (ILSAs). It examines the critical role of technology as a driving force in the evolution of ILSAs, highlighting how advancements in digital tools and methodologies have transformed assessment practices in the past decades. The chapter also provides a comprehensive overview of the book's content, detailing each part's key themes and contributions. Additionally, it presents a perspective on the future of digital-based ILSAs, discussing emerging trends and potential directions for research and practice in this rapidly changing field.

1.1 Technology Is Changing Everything

Technology is transforming our lives at an unprecedented pace. While some of us still remember when owning a computer or mobile phone was rare and how the internet revolutionized how we access and share information, others have never known a world without laptops, tablets, smartphones, and smart homes. Today,

L. Khorramdel (✉) · M. von Davier
Lynch School of Education and Human Development, TIMSS & PIRLS International Study Center at Boston College, Chestnut Hill, MA, USA
e-mail: lale.khorramdel@bc.edu

K. Yamamoto
Independent Consultant for Psychometric Methodology and Large-Scale Assessment (formerly ETS), Princeton, NJ, USA

L. Khorramdel et al. (eds.), *Innovative Digital-Based International Large-Scale Assessments*, Methodology of Educational Measurement and Assessment,
https://doi.org/10.1007/978-3-031-90951-1_1

some children learn to operate a touchscreen before they can even speak, and discussions about screen time are common in many households. Developments in technology have been a driver in globalization for decades, increasing collaborations and opportunities across borders, all while presenting new challenges. The COVID-19 pandemic has accelerated the broad use of technology even further. Video calls are now a staple not only for personal communication via Skype, Messenger, FaceTime, and WhatsApp, and many others but also for professional and learning interactions through platforms like Zoom and Microsoft Teams. More and more schools and universities offer online programs accessible from anywhere, and employers provide remote job opportunities.

The most recent wave of technological excitement—or shock for some—has come with the broad public availability of artificial intelligence (AI) tools, such as ChatGPT, Llama, and CoPilot. These tools leverage large language models and machine learning to combine the capabilities of search engines with generative AI. With simple prompts, users can generate various texts (e.g., poems, emails, essays, even entire book chapters), summarize or combine existing texts, create fake images, audio, and videos, produce designs and art, and can generate test and survey questions. Virtual realities now appear convincingly real, enabling experiences like flying over the Alaskan landscape, joining a past family celebration through a lifelike video, or attending a private concert by Alicia Keys—all brought into our homes through devices like Apple's Vision Pro.

These rapid developments are profoundly impacting and changing how we think, learn, work, collaborate, and communicate. Developing new skills to navigate through these technology-rich environments is inevitable, and education systems must adapt (OECD, 2018; Redecker & Johannessen, 2013). School curricula need updating, new resources must be provided, and teacher training programs require modernization. Learning environments are changing as well and are becoming increasingly gamified, using advanced AI and machine learning techniques to personalize the users' learning experience. One example is the popular Duolingo app which has provided an interactive platform for learning foreign languages since 2011 and that also offers tools for learning mathematics since 2022. Guided through different learning levels by a diverse set of avatars and virtual characters, learners respond to various types of tasks and questions. They can collect points and compete with friends, family members, and strangers internationally in a learning environment that automatically adapts to individuals' proficiency levels, providing a tailored and motivating experience (Bilham, 2021; Portnoff et al., 2021).

1.2 Technology Is Revolutionizing ILSAs

Considering these new technological developments and learning environments, traditional paper-and-pencil tests are becoming less relevant for assessing skills and proficiencies and may fail to engage test-takers adequately. Therefore, tests and assessments must adapt as well and utilize digital forms of administration such as computers, laptops, tablets, or mobile devices. This also applies to national large-scale assessments (NLSAs) and international large-scale assessments (ILSAs) as driving forces of changes and innovation in education worldwide. *This edited volume* offers a comprehensive overview of the technology-based innovations that have transformed NLSAs and ILSAs over the past decade, along with the research and considerations accompanying these changes. It documents the groundwork and foundations for ongoing innovations and explores potential future trajectories of large-scale assessments as they continue to evolve and advance.

This section begins with an overview of existing ILSAs and their primary objectives, followed by an examination of the technological innovations that have already reshaped the landscape of ILSAs.

1.2.1 Existing ILSAs and Their Main Goals

In exploring prominent ILSAs, it is essential to begin with the National Assessment of Educational Progress (NAEP, launched in 1969), which monitors and compares average student achievement across U.S. states. Although NAEP is not an international assessment, it shares significant similarities and serves a comparable purpose by measuring educational outcomes across diverse educational systems, demographics, and socioeconomic factors within the United States. As a pioneer in large-scale assessment methodologies (e.g., Mislevy et al. 1992), the statistical models and insights developed through NAEP informed many ILSAs (e.g., von Davier et al., 2006).

Following NAEP, the International Association for the Evaluation of Educational Achievement (IEA) launched the Trends in International Mathematics and Science Study (TIMSS) in 1995, adopting many of the methods and procedures from NAEP. The Progress in International Reading Literacy Study (PIRLS), the International Computer and Information Literacy Study (ICILS), and the International Civic and Citizenship Education Study (ICCS) were all launched by IEA. IEA's studies are focused on education, are rooted in academic achievement, and are based on a comprehensive survey of school curricula internationally.

The other player is the Organisation for Economic Co-operation and Development (OECD). Their most well-known assessment is the Programme for the International Student Assessment (PISA, launched in 2000), but there are other programs as well, such as the Programme for the International Assessment of Adult Competencies (PIAAC), which followed the International Adult Literacy Survey (IALS, launched in 1994) and Adult Literacy and Lifeskills Survey (ALL, launched in 2003) preceded by the National Adult Literacy Survey (NALS, launched in 1992). In contrast to IEA's studies, these OECD studies focus less on monitoring educational systems, and more on skills vital for labor market success and economic growth.

Moreover, regional large-scale assessments provide comparisons across different (mostly low- and middle-income) countries belonging to specific regions that share the same language and a similar culture. They are an important tool for providing data that help measure progress towards the United Nations' (UN) Sustainable

Development Goal 4[1] of the 2030 Agenda. Without claiming to provide a complete list, examples are the Programme for the Analysis of Education Systems (PASEC), the Regional Comparative and Explanatory Study (ERCE), the Southern and Eastern African Consortium for Monitoring Educational Quality (SACMEQ), the Southeast Asia Primary Learning Metrics (SEA-PLM), and the Pacific Island Literacy and Numeracy Assessment (PILNA). For a comprehensive overview of educational surveys and large-scale assessments, see Khorramdel et al. (2023).

ILSAs provide cross-country comparisons of students' average achievement and achievement distributions in different domains of interest, especially reading, mathematics, and science. Moreover, they provide trend measures for each participating country over time by administering their assessments in recurring cycles. For examples, PIRLS is administered every 5 years, TIMSS every 4 years, PISA used to be administered every 3 years but switched to a 4-year cycle starting in 2021, and PIAAC is administered every 10–12 years. ILSAs also provide key information about different contexts of student learning such as school resources, home environments, teaching practices, and student's confidence, attitudes, and well-being, which can provide further insight into the characteristics of successful students and schools. Each new assessment cycle relies on assessment content from the previous cycle to enable a link to the past (trend) as well as newly developed materials to address framework updates and changes in education systems and societies. Blocks of existing trend items are combined with new items in a matrix sampling design where each respondent only receives a subset of the items (see more information about ILSA designs in Chap. 5). The development and piloting of ILSAs is a complex, iterative process involving multi-disciplinary teams, all designed to ensure that the instruments, test delivery platforms, and test designs meet the highest standards of quality. This process aims to provide reliable, valid, and comparable data. To achieve this, each assessment cycle is comprised of a field test and a main data collection or main survey. The field test is conducted prior to the main survey to evaluate the quality of the instruments, administration procedures, and data processes across various stakeholders and countries. Figure 1.1 illustrates the considerations that inform the development and design of ILSAs, as well as the key outcomes of this rigorous process.

ILSAs have become crucial tools for the evaluation of educational systems regarding quality, equity, and efficiency in learning outcomes across countries and economies, influencing education reforms (OECD, 2023). Therefore, important goals for ILSA data are to provide information about the current state of education across countries and to link to the past through trend measures that allow monitoring changes over time. At the same time, data are meant to allow for initiating meaningful changes that prepare education systems and students for the future. Hence, ILSAs need to account for updates and innovations regarding their frameworks, the measured content and constructs, and questions around inclusiveness and fairness that require innovating test designs and analysis methods, as well.

[1] https://sdgs.un.org/goals/goal4

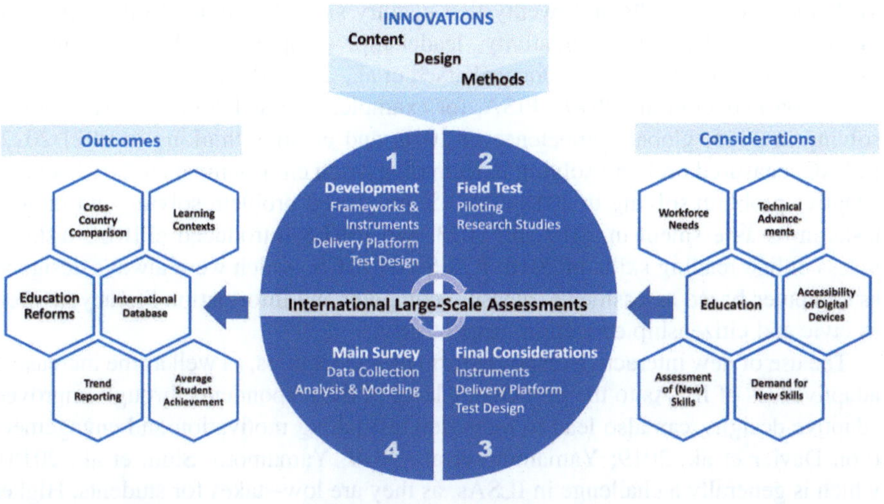

Fig. 1.1 Considerations, innovations, and outcomes of ILSAs

1.2.2 Technological Innovations Introduced in ILSAs in The Past Decade

In recent years, ILSAs (and NAEP) have transitioned from paper-based to computer-based or digital-based test administrations, reflecting the need for more innovative and motivating assessment methods. This shift is essential to accurately measure and enhance students' capabilities in an increasingly digital world. The PIAAC 2012 cycle was largely computer-based with a paper-based option for respondents with insufficient computer experience as part of its adaptive design (Yamamoto, Khorramdel, et al., 2018; Yamamoto, Shin, et al., 2018). PISA transitioned to a fully computer-based administration in 2015, followed by TIMSS in 2019 and PIRLS in 2021. NAEP started transitioning to a computer-based administration in 2009 and to a digital administration utilizing touch screens in 2024.

Transitioning from paper-based to digital-based administration modes offers numerous benefits for ILSAs, enabling the implementation of innovations that improve measurement precision and validity and extend their impact and reach. These include improved standardization of testing procedures, more efficient data collection and processing, the increased use of automated scoring to benefit scoring accuracy, and computer-adaptive testing to enable a wider range of countries, with different cultures, languages, and average proficiency levels, to participate (Redecker & Johannessen, 2013; Tyack et al., 2024; von Davier et al., 2019, 2022, 2024; Yamamoto et al., 2019; Yamamoto, Shin, et al., 2018). Moreover, digital platforms allow for the use of new and interactive item types (e.g., drag-and-drop, highlighting text, hotspot, drop-down, free drawing, bar chart tool) and interactive features (such as videos, gifs, popup windows). They also enable the measurement of innovative constructs including Information and Communication

Technology (ICT) skills and twenty-first-century skills like critical thinking, communication, collaboration, creativity, leadership, complex problem solving, and information and technology literacy (Kirsch et al., 2013; Redecker & Johannessen, 2013; von Davier et al., 2019). PISA, for example, assessed collaborative problem solving in 2015, global competence in 2018, and creative thinking in 2021/2022. PIAAC measured problem solving in technology-rich environments in 2012, adding adaptive problem solving in 2018. TIMSS introduced problem solving and inquiry tasks in its assessment in 2019 and 2023, and PIRLS introduced ePIRLS tasks to assess online reading skills in 2016. ICILS and ICCS, which were always designed as computer-based assessments, measure computer and information literacy, as well as civic and citizenship education, respectively.

The use of new interactive response formats and features, as well as the increasing adaptiveness of ILSAs to the proficiency level of the respondents through improved adaptive designs, can also lead to increased test-taking motivation and engagement (von Davier et al., 2019; Yamamoto et al., 2019; Yamamoto, Shin, et al., 2018), which is generally a challenge in ILSAs, as they are low-stakes for students. Higher test-taking motivation in digital-based assessments is indicated by, for example, lower omission rates and higher reported enjoyment and test-taking experience (Hosseini et al., 2014; Li et al., 2017). Digital administration also allows for the collection of log file and process data, which offer additional information about how students navigate through the assessment, that can be utilized to examine data quality and validity and to improve scoring and proficiency estimates, ultimately leading to more accurate and fair test scores (e.g., Anghel et al., 2024; Provasnik, 2021; von Davier et al., 2019). Figure 1.2 provides an overview of some of the technology-based innovations introduced in ILSAs, which are usually piloted in field tests before being implemented in main surveys.

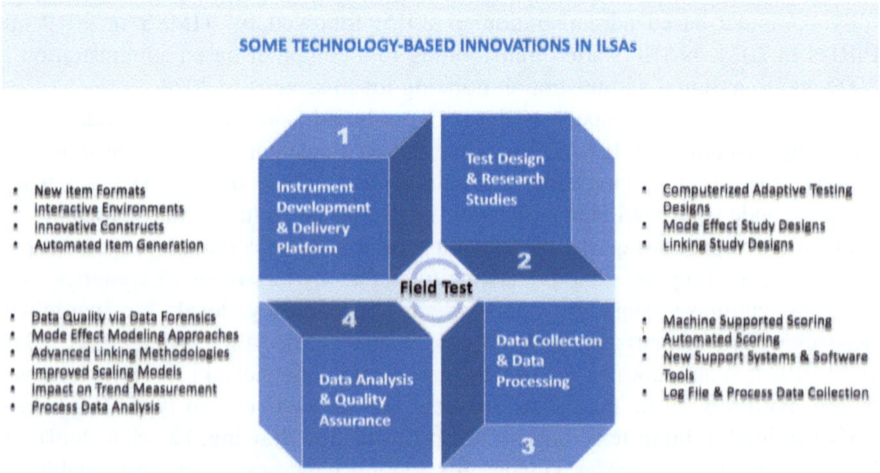

Fig. 1.2 Examples of digital innovations in ILSAs (piloted in field tests)

Despite the numerous advantages of digitizing ILSAs, there are challenges in implementing these changes internationally (standardized across different cultures and languages) and in maintaining comparability to past assessment cycles, therefore, potentially impacting the reporting of trends (Beller, 2013; Kirsch et al., 2013; von Davier et al., 2019). To report test scores from separate assessment cycles on a common scale, assessments are linked through a common set of trend items administered in each cycle. To link paper-based and digital-based cycles, sophisticated assessment designs are used to ensure trend results can be reported even if trend items administered in both assessment modes are not completely equivalent (e.g., von Davier et al., 2023). Item characteristics, such as difficulty, may vary across modes, exhibiting so called mode effects. These differences can depend on the item type, facet, or content area to varying extents (Harrison et al., 2023; Lowrie & Logan, 2015). Mode effects occur not only between paper-based and digital-based assessments but also between different types of digital devices, which may not be directly comparable, such as computers and laptops versus touchscreen-based tablets and smartphones (Nikou & Economides, 2013). Due to rapid technological development, predicting how devices will function in the future and how they will compare when used for administering assessments is challenging.

Potential mode effects are not the only threat for reporting trends. Changes in school curricula, constructs (e.g., reading on paper versus digital reading), instruments, and frameworks are competing with the goal of reporting trend. In addition, global events, such as the recent COVID-19 pandemic, may lead to changes in education systems, making comparisons with past assessments more difficult. Moreover, item drift over time, differential digital emphasis in school curricula, or the digital divide (i.e., different availability of technology leading to differences in ICT skills) can challenge the comparability of measurements. Innovations in assessment methodologies, such as test designs (e.g., linear versus adaptive testing) and scaling models, add to this complexity. Some of these changes and differences cannot be fully controlled or assessed, while others are purposefully implemented to advance, improve, and innovate ILSAs. However, innovations need to be carefully and properly implemented and accompanied by state-of-the-art research studies (as part of a field test or a large-scale linking study; see Fig. 1.2) to ensure both goals are met: a robust link to past assessment cycles for a continuous reporting of trends and enhancing and modernizing the measurement properties of ILSAs to facilitate their transition into the future and the digital age.

1.3 This Book: Foundations, Emerging Principles, and Guidelines for Innovating ILSAs

While the rapid evolution of technology and its impact on education and assessments is challenging to predict, we aim at describing current best practices and emerging principles for transitioning large-scale assessments from traditional paper-based formats to a digital-based administration and for using technology to further innovate and advance them. *This edited volume* illustrates these by providing guidelines and operational examples from the leaders and measurement experts who have transitioned large-scale assessments, such as PISA, PIAAC, TIMSS, PIRLS, and NAEP into the digital era, redefining how we think about, design, implement, and use these assessments. It also includes contributions from scientists who have conducted extensive research on data from these digital-based assessments, showcasing new opportunities and insights provided by log file and process data, as well as data from innovative domains.

This book is structured in six parts that address various areas of technology-based innovations in ILSAs, with the current chapter presenting.

Part I: Introduction: Technology as a Driver of the Evolution of International Large-Scale Assessments

Part II: Advances in Test Development, Design, and Data Collection (Chaps. 2–7)

This part covers the creation of new digital-based item types and interactive environments, addresses the importance and complexity of multistage adaptive test designs, and explores the essential functions of digital testing platforms. The presented advancements underscore the continuous need for adaptation and innovation in assessment development and practices to meet evolving educational demands.

Lennon and Kirsch (Chap. 2) discuss the development of new computer-based item types and interactive environments for assessments like PISA and PIAAC. They highlight evolving frameworks, platform capabilities, and scoring methods that enhance relevance by measuring skills essential for success in a digital world. The chapter details the complex development process, requiring collaboration among multidisciplinary teams with specialized expertise.

Mullis and Martin (Chap. 3) focus on PIRLS 2021, detailing the shift to a digital multilingual reading assessment for fourth-grade students to align with their evolving reading experiences. The chapter details the integration of digitalPIRLS and ePIRLS, and the use of log data for examining reading strategies and introducing operational efficiencies. It also addresses the challenges of implementing digital assessments across diverse schools, languages, and countries.

Ou et al. (Chap. 4) explore new constructs and item types introduced by digital technologies, focusing on scenario-based item types across PISA and PIAAC core domains (reading and mathematics, and in PISA also science) and innovative domains like problem solving in technology-rich environments, collaborative problem solving, global competence, and creative thinking. The chapter also addresses challenges and opportunities in analyzing process data for inferential and predictive purposes.

Yamamoto et al. (Chap. 5) discuss the advantage of using multistage adaptive testing (MST) designs in digital-based ILSAs on the example of PISA and PIAAC. They illustrate psychometric considerations for the design and data analysis complexities. The authors demonstrate how MST can balance assessment efficiency with estimation accuracy (especially for heterogeneous groups of countries), enable comprehensive constructs representation, and prevent participant fatigue.

Frey and Fink (Chap. 6) address bias in computerized adaptive testing (CAT) due to item position effects in ILSAs. They introduce a balanced statistical design and an extended continuous calibration strategy (CCS) to control these effects and show with simulation studies that this method produces virtually unbiased ability estimates and achieves a lower mean square error (MSE) compared to an unbalanced design. The authors offer guidelines for using CCS in ILSAs to ensure accurate ability estimates and minimize bias.

Wagner and Kerzabi (Chap. 7) review the digital testing platforms for PIAAC and PISA, detailing the necessary functions for accurate assessment creation, dissemination, and data collection. They discuss innovations like interactive item types, log and process data collection, new test designs, automatic scoring, and the possibilities and limitations of web-based test administration, offering general considerations for designing digital platforms for ILSAs.

Part III: Innovations in Translation, Data Processing, and Scoring (Chaps. 8–13)

This part explores the impact of a digital-based administration on translation and linguistic quality assurance, discusses the use of psychometric forensics to detect anomalies and ensure data integrity, and highlights automated scoring methods that improve the accuracy of scoring constructed response items. Collectively, these innovations illustrate ongoing advancements that enhance assessment processes for improving data quality and reliability.

Dept et al. (Chap. 8) explore the impact of transitioning from paper-based to computer-based assessments on translation and linguistic quality assurance (LQA), emphasizing the need for updated models and technologies to capitalize on advances in translation methodologies. They discuss the prerequisites and implications of this paradigm shift, which paves the way for the professionalization of survey instrument translation.

Yamamoto et al. (Chap. 9) focus on psychometric forensics, utilizing process data, automated procedures, and contextual information to identify anomalies and potential sources of error. Drawing on examples from PIAAC and PISA, the authors outline various data quality procedures and monitoring techniques designed to safeguard data integrity and enhance the reliability, validity, and comparability of survey outcomes.

Kerzabi et al. (Chap. 10) outline procedures like coder training and coding designs to enhance the accuracy of human scoring for constructed response items. They describe the Open-Ended Response System (OERS) and Open-Ended Coding System (OECS) used in PISA to improve coder reliability and efficiently manage large volumes of constructed responses.

Flor and Cahill (Chap. 11) explore the relevance of automated scoring for open-ended constructed responses in ILSAs. They provide an overview of the history and current state of technologies for both English and other languages, differentiating between essay and content scoring. The chapter highlights the benefits and critiques of automated scoring, while also discussing crucial considerations such as validity, fairness, and the handling of atypical responses.

Shin et al. (Chap. 12) present the Machine-Supported Coding System (MSCS) implemented in PISA 2018 for the automated scoring of short constructed response items across multiple languages. The authors illustrate that the MSCS is designed as a supervised learning model, which utilizes an exact-matching method, and demonstrate that it leads to a significant reduction in scoring burden while achieving high prediction accuracy.

Zehner and Goldhammer (Chap. 13) discuss ReCo, software designed for the automatic coding of short text responses. They demonstrate its effectiveness and explore its potential integration with systems like PISA's Machine-Supported Coding System (described in Chap. 12) to harmonize human coding across various languages.

Part IV: New and Advanced Methodologies for Analyzing Response and Process Data from Digital-Based International Large-Scale Assessments (Chaps. 14–23)

This part presents new and advanced methodologies for linking assessments across modes, groups, and over time, and for analyzing data from new item types and innovative domains. Moreover, it emphasizes the role of research studies for properly implementing innovations, and the use of process data indicators for modeling unplanned missing responses, guessing behavior and response style bias. Altogether, the presented modeling approaches highlight significant advancements in data analysis for improving measurement accuracy and validity in both trend and new measures.

von Davier et al. (Chap. 14) discuss new approaches for scale linking and modeling mode effects introduced in PISA 2015 during the transition to computer-based assessments. They illustrate how these methods enable more comprehensive measurement invariance testing by examining item-by-group interactions and exploring the impact of different assessment modes on parameter estimates. Additionally, they demonstrate that the new scale linking, based on concurrent calibration and more general IRT models, provides a better model-data fit and more robust item parameter estimates while preserving trend measure integrity.

Jewsbury et al. (Chap. 15) address mode effects in the transition from paper-based to digital-based administration in NAEP 2017. They find that item difficulty increases in the digital assessment, especially for constructed response items, with this trend being consistent across subjects and grades. A common population linking approach was applied to address these discrepancies, resulting in few significant differences in mean estimates for states, districts, and subpopulations between digital and paper results,

Mazzeo et al. (Chap. 16) describe a method for incorporating linking error into the approximation of NAEP standard errors when linking assessments. Their aim is to balance the implementation of digital-based innovations in assessment content, test delivery, and analysis methods while ensuring valid trend results. The chapter also outlines ongoing theoretical and practical efforts to refine and enhance the methods employed in 2017.

von Davier et al. (Chap. 17) address scoring and scaling challenges for innovative, interactive tasks in ILSAs, emphasizing the need to interpret results in terms of reliability, validity, and comparability. The authors propose a general approach for analyzing data from innovative domains and illustrate automated scoring of text sequence responses and feature generation for PISA 2015 collaborative problem-solving items as specific examples.

Keskpaik et al. (Chap. 18) demonstrate how to effectively manage the complexity and volume of process data captured in technology-enhanced assessments by utilizing both data-driven approaches (based on educational data mining) and theory-driven methods (based on content expert knowledge). The authors discuss the integration of didactics, technology, and analytics to develop meaningful process data indicators and hypotheses, enabling valid interpretations of the data.

Ulitzsch et al. (Chap. 19) highlight the value of process data indicators in analyzing large-scale assessment data by incorporating response time into the modeling of omitted responses and unreached items, thereby reducing bias in the estimation of item and person parameters, as well as group statistics. The authors review four contemporary model-based approaches, each addressing different missingness mechanisms, and illustrate their effectiveness using empirical data from PISA 2018.

Shin et al. (Chap. 20) showcase the usefulness of response time data when incorporated into population modeling to improve measurement accuracy in ILSAs. They describe the effective handling of item-level response time data and illustrate how including this information improves proficiency score reporting. Using data from PISA 2015, the authors present findings that informed the decisions for the PISA 2018 population modeling approach.

Molenaar and van Rijn (Chap. 21) explore various psychometric models for simultaneously analyzing responses and response times, focusing on using response times as collateral information about individual differences in latent ability within assessed domains. They discuss current challenges and future directions in response time research in large-scale educational assessments and its potential to enhance reliability.

Kroehne et al. (Chap. 22) investigate the relationship between extreme response style and rapid responding in non-cognitive measures, examining whether these behaviors are similar or distinct phenomena. The authors propose a joint modeling approach that utilizes item-level response time thresholds and IRTree models, illustrating its application with data from PISA 2015.

Pokropek and Khorramdel (Chap. 23) introduce the IRT Explanatory Model for Guessing (IRT-EMG), which incorporates response time and other process data into mixture models to detect and account for various types of guessing behavior, thereby enhancing assessment reliability and validity. The chapter illustrates how the model can identify items particularly prone to guessing and outlines its features and advantages, demonstrating its application using PIAAC data.

Part V: Software Tools for Analyzing Large-Scale Assessment Data (Chaps. 24–28)
This part provides an overview of various software tools designed for analyzing complex data sets that are based on sophisticated test designs and include large proportions of missing data, multiple imputations (plausible values), sampling weights, and newly available log file and process data. It highlights some of the essential tools available for effectively analyzing digital-based ILSA data.

Gonzalez (Chap. 24) details the IEA's IDB Analyzer, a tool configured for processing and analyzing most ILSA databases as well as a few national assessments, highlighting its main features and capabilities. The chapter demonstrates the two modules of the IDB Analyzer and its application based on empirical data.

Barone and Rogers (Chap. 25) describe ETS's Data Explorer (DX), available on a web-based platform, for analyzing national and international large-scale assessment data, emphasizing its ease of use and computational capabilities for primary and secondary analysis of complex databases.

Pokropek and Żółtak (Chap. 26) focus on statistical tools and methods for analyzing ILSA datasets using Stata and R: *p*isatools, piaactools, and repest packages from Stata and mitools, survey, and svyPVpm libraries from R. The authors present analyses involving descriptive statistics, frequency tables, OLS regression, and logistic regression. They show the proper estimation of sampling variance and use of plausible values, in a short and accessible way.

Kroehne and Goldhammer (Chap. 27) discuss software tools for log data analysis from computer-based assessments, comparing the R environment, the LogFSM package, and the PIAAC Log Data Analyzer based on several key requirements. The chapter outlines a workflow for preparing log data by aggregating low-level features into high-level process indicators which can be used in data analysis.

Zhang et al. (Chap. 28) introduce EdSurvey, an R package designed for handling and analyzing complex large-scale survey response data, including plausible values. Using empirical examples, the authors illustrate how to use EdSurvey to perform data processing, exploration, manipulation, and some statistical analyses, including tables, gap analysis, and linear and logistic regressions.

Part VI: The Future of International Large-Scale Assessments (Chap. 29)
This part explores the future of ILSAs, emphasizing the need for adaptation to rapid technological advancements and shifting societal dynamics. Predictions for future innovations underscore the critical role of ILSAs in addressing future educational needs.

Avvisati and Borgonovi (Chap. 29) explore the evolving role of ILSAs in the context of rapid technological change, highlighting how these assessments must adapt to remain relevant amid shifts in work, learning, and social interaction influenced by information and communication technology. They discuss future transformations, such as technological innovations, population aging, and international migration, and propose strategies to ensure ILSAs effectively measure both foundational and emerging skills while maintaining trust in their findings.

1.4 Outlook

As we look to the future, it becomes clear that adaptation is crucial for maintaining ILSA relevance in an ever-changing world. Striking a balance between implementing necessary innovations and preserving trend measures is increasingly challenging, particularly as ILSAs strive to keep pace with rapid technological advancements. The traditional emphasis on trend measurement—comparing results across cycles—needs to be reexamined. While trend data has long been a cornerstone of ILSAs, we must acknowledge that different cycles may not be directly comparable if we prioritize making future assessments truly relevant. This shift implies that preserving trends and designs is no longer the primary goal. Instead, the focus should be on assessing the skills that matter most for individual and societal well-being, such as those that contribute to income, longevity, and overall quality of life.

However, trend measures still hold value if changes across assessment cycles are not too drastic. Maybe shorter trend comparisons across two consecutive cycles only or comparing results within a single cycle across different countries and subgroups may offer sufficient insight without the need for extensive cross-time comparisons. Ultimately, ILSAs must evolve to serve the most relevant measurement objectives, ensuring they continue to provide meaningful data in a rapidly evolving landscape.

Looking ahead, further innovations will be instrumental in shaping the future of ILSAs beginning with enhancements in test development, data processing techniques, and design innovations. Machine learning and AI is set to enhance automated scoring, particularly for scoring image (e.g., Tyack et al., 2024) and multilingual text responses with greater precision and consistency (e.g., Jung et al., 2024; Shin et al., 2024). Additionally, generative AI and large language models are opening exciting possibilities for automated item generation (e.g., Bezirhan & von Davier, 2023; Yaneva & von Davier, 2023; von Davier, 2018), offering the potential to not only streamline the creation of assessment content, but also to introduce a broader variety of tasks and formats that better capture diverse competencies. These advancements in automated scoring will enhance adaptive testing designs in ILSAs, enabling proficiency estimates that guide adaptive pathways through assessments to be based not only on responses to selected and multiple-choice items but also on answers to constructed response items.

Regarding test and content development, we may also witness the development of more gamified and highly interactive tasks, potentially utilizing features of virtual realities and advanced adaptive technologies to deliver a more personalized and engaging test-taking experience. Such advancements could greatly enhance the relevance of ILSAs, ensuring they align more closely with the evolving needs of learners and educators in a rapidly changing digital landscape. However, a significant challenge will be effectively integrating and implementing these innovations into the complex test designs and within the multilingual and multicultural context of ILSAs, where technology access may greatly vary, and maintaining the comparability of data and test results is essential for ensuring valid and reliable outcomes.

As you explore this edited volume, you will encounter fundamental and groundbreaking developments that enabled the transition from paper-based to digital-based international large-scale assessments (ILSAs). The chapters were written and submitted between 2019 and 2023. Although the publishing process took time, the content remains highly relevant, with methods and innovations still actively used and further refined in operational test development, design, and data analysis for ILSAs. Most, if not all, of these innovations continue to inform the design of current ILSA cycles.

Many innovations, such as AI-supported automated item generation and automated scoring of text and image responses—already applied in studies like TIMSS, PIRLS, and PISA—continue to evolve through active research groups, with additional initiatives (e.g., SIGMIE at NCME) driving further methodological and AI-based advancements. We encourage you to follow the chapter authors and their ongoing groundbreaking research, as the insights presented here provide a foundation for current and future progress in digital-based ILSAs.

Acknowledgements The initial idea for this book was developed while the editors of this volume were part of the Center for Global Assessment at ETS.[2] Several chapters are based on their work in the Center on projects, such as PISA and PIAAC.

We extend our heartfelt gratitude to all the authors and colleagues whose contributions and support made this extraordinary collection of research, methods, and procedures possible.

We also acknowledge the challenges posed by the COVID-19 pandemic and the numerous changes in jobs and affiliations that were navigated while working on these book chapters. The author's resilience and collaboration have been invaluable throughout this process.

References

Anghel, E., Khorramdel, L., & von Davier, M. (2024). The use of process data in large-scale assessments: a literature review. *Large-Scale Assessment in Education, 12*, 13. https://doi.org/10.1186/s40536-024-00202-1

Beller, M. (2013). Technologies in large-scale assessments: New directions, challenges, and opportunities. In M. von Davier, E. Gonzalez, I. Kirsch, & K. Yamamoto (Eds.), *The role of international large-scale assessments: Perspectives from technology, economy, and educational research* (pp. 25–45). Springer. https://doi.org/10.1007/978-94-007-4629-9_3

Bezirhan, U., & von Davier, M. (2023). Automated reading passage generation with OpenAI's large language model. *Computers and Education: Artificial Intelligence, 5*. https://doi.org/10.1016/j.caeai.2023.100161

Bilham, J. (2021). Case study: How Duolingo utilises gamification to increase user interest. https://raw.studio/blog/how-duolingo-utilises-gamification/

Harrison, S., Kroehne, U., Goldhammer, F., Luedke, O., & Robitzsch, A. (2023). Comparing the score interpretation across modes in PISA: an investigation of how item facets affect difficulty. *Large-Scale Assessment in Education, 11*, 8. https://doi.org/10.1186/s40536-023-00157-9

Hosseini, M., Abidin, M. J. Z., & Baghdarnia, M. (2014). Comparability of test results of computer based tests (CBT) and paper and pencil tests (PPT) among English language learners in Iran.

[2] https://www.ets.org

Procedia-Social and Behavioral Sciences, 98, 659–667. https://doi.org/10.1016/j.sbspro.2014.03.465
Jung, J. Y., Tyack, L., & von Davier, M. (2024). Combining machine translation and automated scoring in international large-scale assessments. *Large-Scale Assessment in Education, 12*, 10. https://doi.org/10.1186/s40536-024-00199-7
Khorramdel, L., von Davier, M., Yamamoto, K., & Kirsch, I. (2023). Educational surveys: Conceptual overview. In R. J. Tierney, F. Rizvi, & K. Erkican (Eds.), *International encyclopedia of education* (Vol. 14, 4th ed., pp. 347–358). Elsevier. https://doi.org/10.1016/B978-0-12-818630-5.10044-2
Kirsch, I., Lennon, M., von Davier, M., Gonzalez, E., & Yamamoto, K. (2013). On the growing importance of international large-scale assessments. In M. von Davier, E. Gonzalez, I. Kirsch, & K. Yamamoto (Eds.), *The role of international large-scale assessments: Perspectives from technology, economy, and educational research* (pp. 1–11). Springer. https://doi.org/10.1007/978-94-007-4629-9_1
Li, D., Yi, Q., & Harris, D. (2017). *Evidence for paper and online ACT comparability: Spring 2014 and 2015 mode comparability studies* (ACT Working Paper 2016-2). https://www.act.org/content/act/en/research/pdfs/R1616-paper-and-online-testing-2017-04.html
Lowrie, T., & Logan, T. (2015). The role of test-mode effect: Implications for assessment practices and item design. In *In pursuit of quality mathematics for all. Proceedings of the 7th ICMI-East Asia regional conference on mathematics education* (pp. 649–656). http://mathted.weebly.com/uploads/7/8/5/0/7850000/pp_649_tom_lowrie_final_paper_edited.pdf
Mislevy, R. J., Beaton, A. E., Kaplan, B., & Sheehan, K. M. (1992). Estimating population characteristics from sparse matrix samples of item responses. *Journal of Educational Measurement, 29*(2), 133–161. http://www.jstor.org/stable/1434599
Nikou, S. A., & Economides, A. A. (2013). Student achievement in paper, computer/web and mobile based assessment. In *BCI (Local)* (p. 107). https://doi.org/10.1016/j.chb.2015.09.025
OECD. (2018). *The future of education and skills: Education 2030* (OECD Education Policy Perspectives, No. 98). OECD. https://doi.org/10.1787/54ac7020-en.
OECD. (2023). *PISA 2022 results (volume I): The state of learning and equity in education*. PISA, OECD. https://doi.org/10.1787/53f23881-en
Portnoff, L., Gustafson, E.N., Bicknell, K., & Rollinson, J. (2021). *Methods for language learning assessment at scale: Duolingo case study*. Educational Data Mining. https://educationaldatamining.org/EDM2021/virtual/static/pdf/EDM21_paper_230.pdf
Provasnik, S. (2021). Process data, the new frontier for assessment development: rich new soil or a quixotic quest? *Large-Scale Assessment in Education, 9*, 1. https://doi.org/10.1186/s40536-020-00092-z
Redecker, C., & Johannessen, Ø. (2013). Changing assessment — Towards a new assessment paradigm using ICT. *European Journal of Education, 48*(1), 79–96. https://doi.org/10.1111/ejed.12018
Shin, H. J., Andersen, N., Horbach, A., Kim, E., Baik, J., & Zehner, F. (2024). Operational automatic scoring of text responses in 2016 ePIRLS: Performance and linguistic variance. *IEA Research Report*. IEA. Obtained from: https://www.iea.nl/sites/default/files/2024-04/Operational-Automatic-Scoring-of-Text-Responses-ePIRLS.pdf
Tyack, L., Khorramdel, L., & von Davier, M. (2024). Using convolutional neural networks to automatically score eight TIMSS 2019 graphical response items. *Computers and Education: Artificial Intelligence, 6*, 100249., ISSN 2666-920X. https://doi.org/10.1016/j.caeai.2024.100249
von Davier, M. (2018). Automated item generation with recurrent neural networks. *Psychometrika, 83*, 847–857. https://doi.org/10.1007/s11336-018-9608-y
von Davier, M., Khorramdel, L., He, Q., Shin, H. J., & Chen, H. (2019). Developments in psychometric population models for technology-based large-scale assessments: An overview of challenges and opportunities. *Journal of Educational and Behavioral Statistics, 44*(6), 671–705. https://doi.org/10.3102/1076998619881789

von Davier, M., Khorramdel, L., & Yamamoto, K. (2024). Linking international large-scale assessments over time and administration modes—Scaling, linking, and mode effects in PISA 2015. In L. Khorramdel, M. von Davier, & K. Yamamoto (Eds.), *Innovative digital-based international large-scale assessments—Foundations, methodologies and quality assurance procedures*. Springer.

von Davier, M., Mullis, I. V. S., Fishbein, B., & Foy, P. (Eds.). (2023). *Methods and procedures: PIRLS 2021 Technical report*. Boston College, TIMSS & PIRLS International Study Center. https://pirls2021.org/methods

von Davier, M., Sinharay, S., Oranje, A., & Beaton, A. (2006). The statistical procedures used in national assessment of educational progress: Recent developments and future directions. In C. R. Rao & S. Sinharay (Eds.), *Handbook of statistics* (Vol. 26, pp. 1039–1055). Elsevier. https://doi.org/10.1016/S0169-7161(06)26032-2

von Davier, M., Tyack, L., & Khorramdel, L. (2022). Scoring graphical responses in TIMSS 2019 using artificial neural networks. *Educational and Psychological Measurement*. https://doi.org/10.1177/00131644221098021

Yamamoto, K., Khorramdel, L., & Shin, H. J. (2018). Introducing multistage adaptive testing into international large-scale assessments designs using the example of PIAAC. *Psychological Test and Assessment Modeling, 60*, 347–368. https://www.psychologie-aktuell.com/fileadmin/Redaktion/Journale/ptam_3-2018_347-368.pdf

Yamamoto, K., Shin, H. J., & Khorramdel, L. (2018). Multistage adaptive testing design in international large-scale assessments. *Educational Measurement: Issues and Practice, 37*, 16–27. https://doi.org/10.1111/emip.12226

Yamamoto, K., Shin, H. J., & Khorramdel, L. (2019). *Introduction of multistage adaptive testing design in PISA 2018* (OECD Working Paper). https://doi.org/10.1787/b9435d4b-en

Yaneva, V., & von Davier, M. (Eds.) (2023). *Advancing natural language processing in educational assessment*. Routledge. https://doi.org/10.4324/9781003278658

Part II
Advances in Test Development, Design, and Data Collection

Chapter 2
Innovations in Item Development for Computer-Based Assessments

Mary Louise Lennon and Irwin Kirsch

Abstract This chapter focuses on innovations in test development that have made it possible to increase the relevance and utility of international large-scale surveys. In response to the rapid evolution of digital technologies, policy makers, educators, and other stakeholders are increasingly interested in better understanding the extent to which student and adult populations possess the skills that are required to succeed in changing times. Experience with the Program for International Student Assessment (PISA) and the Program for the International Assessment of Adult Competencies (PIAAC) is presented, focusing on the development of computer-based items in the legacy constructs of reading, mathematical, and scientific literacy for PISA and literacy and numeracy for PIAAC. Innovative work around these cognitive domains has been driven by three factors: (1) assessment frameworks that have evolved to reflect the growing importance of digital information, (2) platform capabilities that support both interactive environments and new item types that can be used to assess the range of skills required to navigate, critically analyze, and problem solve in complex digital environments, and (3) the implementation of innovative scoring and data capture capabilities. Collaborative efforts among domain experts, psychometricians, test developers, platform specialists, data analysts, and translation specialists, have led to new capabilities that make it possible to enhance the relevance of large-scale international assessments by expanding both the range of skills that can be measured and the means by which those skills are assessed.

The utility of large-scale assessments relies on their ability to answer key questions about the skills possessed by populations of interest and the ways in which those skills are related to educational, economic, and social outcomes (Kirsch & Braun, 2020). As the demands of technologically infused economies and global competition have interacted to change the ways in which we study, work, and live, the questions being asked about skills have changed as well. Policy makers, educators and other

M. L. Lennon (✉) · I. Kirsch
ETS, Princeton, NJ, USA

L. Khorramdel et al. (eds.), *Innovative Digital-Based International Large-Scale Assessments*, Methodology of Educational Measurement and Assessment,
https://doi.org/10.1007/978-3-031-90951-1_2

stakeholders want to better understand how prepared their populations are to succeed in jobs that increasingly require "twenty-first century skills" such as problem solving, critical thinking, collaboration, and digital literacy. And they need to understand whether, in times of rapid change, individuals have the skills and initiative needed to learn on their own and continuously upgrade what they know and can do.

In line with what Messick and others have referred to as construct-based or evidence-centered assessment, developing instruments to address these issues requires defining the skills and knowledge associated with constructs of interest, identifying the behaviors that reveal that set of skills and knowledge, and then describing the types of tasks or situations that elicit such behaviors (Messick, 1994; Mislevy et al., 1999). The constructs that underlie many national and international large-scale assessments have been expanding over time to include the skills required to navigate, critically analyze, and problem solve in complex and data-intensive digital environments. And as these assessments transition to technology-based platforms, the opportunities to measure those skills have expanded. It is now possible to leverage the tools and capabilities these platforms provide to improve measurement, as well as to develop and deliver items that reflect the kinds of technology-based tasks of increasing importance today and into the future.

This chapter focuses on innovations in the measurement of legacy constructs for the cognitive domains in two major international large-scale surveys[1]: The Program for International Student Assessment (PISA), which is a survey of the performance of 15-year-old students in some 88 countries, and the Program for the International Assessment of Adult Competencies (PIAAC), which is a household survey that measures the proficiencies of adults ages 16–65 in over 30 countries.[2] The legacy constructs in these surveys include those cognitive domains that were measured via paper-and-pencil assessments in the past: reading, mathematical, and scientific literacy in PISA, and literacy and numeracy in PIAAC. While a paper-and-pencil option has continued to be available in both assessments, in order to measure the kinds of skills of increasing interest to policy makers and other stakeholders, new development has focused on items designed to be delivered on a variety of technology-based platforms, including desktops, laptops, and tablets.[3]

In each cycle of PISA, one of the legacy constructs functions as the major domain. The major domain receives a new framework, and new items are developed

[1] See Ou et al. (2025) of this volume for a discussion of new constructs and item types in large-scale assessments.

[2] For more information about PIAAC see https://www.oecd.org/skills/piaac/. For more information about PISA see https://www.oecd.org/pisa/.

[3] Throughout this chapter, the term "computer-based" is used to refer PISA cycles 2015–2022 and PIAAC Cycle 1, which are the focus of the discussion. Starting in 2015, PISA has been administered on school computers and PIAAC Cycle 1 was administered on laptops. Because large-scale assessments are likely to move to a wider variety of delivery platforms (including tablet-based delivery for PIAAC Cycle 2) the term "technology-based" is used to refer more broadly to future assessments and assessment tasks.

Table 2.1 New and trend items across computer-based assessment (CBA) cycles in the PISA and PIAAC legacy domains

	Cycle	Legacy domain	New computer-based items developed	Trend items adapted from paper	Trend items selected from previous CBA cycle
PISA Computer-based assessment of students, age 15	2015	Scientific Literacy	x	x	
		Reading Literacy		x	
		Mathematical Literacy		x	
	2018	Scientific Literacy			x
		Reading Literacy	x		x
		Mathematical Literacy			x
	2022	Scientific Literacy			x
		Reading Literacy			x
		Mathematical Literacy	x		x
PIAAC Computer-based assessment of adults, ages 16–64	Cycle 1 (administered on laptops)	Literacy	x	x	
		Numeracy	x	x	
	Cycle 2 (administered on tablets)	Literacy	x		x
		Numeracy	x		x

Note. In addition to the legacy domains, each assessment included an innovative domain in each cycle: Collaborative Problem Solving (PISA 2015), Global Competence (PISA 2018), Creative Thinking (PISA 2022), Problem Solving in Technology-Rich Environments (PIAAC Cycle 1) and Adaptive Problem Solving (PIAAC Cycle 2)

to reflect that framework. In 2015, scientific literacy was the major domain, followed by reading literacy in 2018, and mathematical literacy in 2022. Thus, as shown in Table 2.1, when each legacy construct was a major domain, all new item development focused on computer-delivered items. In the case of PIAAC, both the literacy and numeracy domains received a new or updated framework in each of its two cycles. New literacy and numeracy items were developed for delivery on laptops in the first cycle, conducted in three rounds from 2012 to 2017, and for delivery on tablets in the second cycle, which was conducted in 2022 and 2023.[4]

[4]Because tablet-based technologies are becoming increasingly familiar and require less advanced ICT skills, the expectation is that use of tablets in PIAAC Cycle 2 will significantly reduce the percentage of individuals who report having computer experience but refuse to take the survey on

In the initial computer-based cycles of each assessment (2015 for PISA and Cycle 1 for PIAAC), existing paper-and-pencil items were redeveloped for computer delivery to maintain trend measurement. Mode effect studies were conducted in both surveys to establish comparability across assessment modes.[5] Beginning with the 2018 cycle of PISA and PIAAC Cycle 2, trend items were selected from the previous cycle, meaning that they included both some of the items originally adapted from paper and some of the computer-based items that had been newly developed for the previous cycles.

This chapter focuses on how technology has allowed us to broaden both what we measure and how we measure it. The chapter is organized as follows. First, the role that frameworks and expert groups have played in expanding legacy constructs is discussed. This is followed by a description of a series of innovations introduced for PISA and PIAAC, including the development of simulated technology-based environments (such as websites, email applications, and spreadsheets), new item types and response modes, and innovative scoring and data capture capabilities enabled by technology-based platforms. Additionally, the chapter describes the ways in which the item development process has changed in the context of developing computer-based assessments. Finally, the chapter concludes with a general discussion of trend measurement and the ways in which issues related to screen layout, as well as response modes and scoring, were considered as PISA and PIAAC transitioned from their paper-based origins to computer-based assessments.[6]

2.1 The Development of Assessment Frameworks

Large-scale surveys are typically based on frameworks that provide a detailed plan for what is to be measured and how results will be interpreted and reported. While the frameworks for legacy constructs have evolved to reflect a growing interest in measuring the range of knowledge and skills needed in today's increasingly technology-driven societies, they share common characteristics both across domains and over time. As implemented in PISA and PIAAC, the assessment frameworks provide a blueprint for the development of the cognitive instruments and interpretation of the results. Each framework is developed by a panel of experts with relevant knowledge and experience in the domain of interest. The expert groups work in close collaboration with the teams responsible for the assessment design, development and

---

the computer, which was the case for an average of 10% of the population across the participating countries in Cycle 1.

[5]For more information about the mode effect studies see the Technical Report of the Survey of Adult Skills (PIAAC) (OECD, 2013) and PISA 2015 Technical Report (OECD, 2017).

[6]Note that PIAAC has been a technology-based assessment since its inception. As discussed later, in PIAAC Cycle 1, trend items were taken from two previous paper-based large-scale assessments of adult skills: The International Adult Literacy Survey (IALS) and the Adult Literacy and Lifeskills Survey (ALL).

authoring of the items, and delivery of the assessment and interpretation of the results. This collaboration, which allows domain experts, psychometricians, test developers, platform specialists, data analysts, and translation specialists to share their perspectives, is key to the successful design, development and delivery of the assessment.

2.1.1 *The Role of Assessment Frameworks*

The framework for any evidence-centered assessment performs several key functions, including the following.

- **Define the domain.** The definition sets the boundaries for what will and will not be measured in an assessment and makes the assumptions underlying the design and development of the assessment explicit.
- **Organize the domain.** Once the definition is developed, it is important to think about the kinds of tasks that represent the skills and abilities included in that definition. Those tasks must then be categorized in relation to the construct definition to inform test design and result in meaningful score reporting. This step makes it possible to move beyond a laundry list of tasks or skills to a coherent representation of the domain that enables policy makers and others to summarize and report information in more useful ways.
- **Identify and specify key task characteristics that should be used to build the assessment and report results.** Task characteristics are the key components or features of the domain that will be used to guide the construction of the assessment tasks and characterize performance. In the PIAAC literacy domain, for instance, these features have been defined to include cognitive task demands, text characteristics and social contexts. Each feature is then further defined in the framework. For example, the literacy framework specifies that the cognitive task demands include accessing (identifying a relevant text in a set or locating information within a text), understanding (literal, inferential and multiple text comprehension), and evaluating (evaluating accuracy, soundness and task relevance). Text characteristics and social contexts are similarly specified in the framework. Test developers use these task characteristics to guide the construction of test items. Thus, each item developed for the literacy assessment is situated in a particular social context, involves one or more texts of a specific type and format, and is associated with cognitive skills judged by the domain experts to be necessary to arrive at a correct response. Together, these characteristics contribute to the overall difficulty and discrimination of each item.
- **Provide the foundation for the interpretation of what drives difficulty and explains performance.** The framework presents hypotheses about factors that are expected to make an item easier or more difficult. These hypotheses may be informed by existing theory and research in a domain, as well as experience based on other assessments. Factors such as the explicitness of the problem statement,

the complexity and number of texts associated with an item, or the number of calculations required to compute an answer, are all examples of the types of features that may impact difficulty. These factors are manipulated by test developers to develop items that are expected to range in difficulty. Following data collection, as part of the scaling and analysis process, the relative location of each item along the domain scale is determined based on how easy or difficult it proved to be for respondents. Analyses of these assessment data are then subsequently used to validate and adjust the interpretative scheme in order to more accurately reflect task difficulty and respondent performance.

As an example, the framework for the PIAAC Cycle 1 literacy domain defined a number of features that were anticipated to impact the difficulty of the assessment items. These included: The level of cognitive processing required, the amount of text that must be processed, the transparency of information in the text as it related to the presented task or question, and whether or not there was competing information in the text.[7] Items were developed that reflected these features of the construct. Based on analyses of the main study data, the location of each item along the literacy scale was determined. Items were then reviewed by the expert group and test developers to evaluate the extent to which these features could, in fact, be used to explain performance. The validated features were then used to characterize tasks at various points along the described proficiency scale, which ranges from the lowest level of Below Level 1–Level 5. The connection between the hypothesized difficulty drivers and the ways in which they explained performance can be seen in the descriptors for each level. For example, the description of tasks at literacy Level 1 includes statements such as the following, with the associated difficulty driver included in parentheses for reference:

- "Most of the tasks at this level require the respondent to read relatively short … texts" [*amount of text that must be processed*]
- "to locate a single piece of information," [*level of cognitive processing*]
- "which is identical to or synonymous with the information given in the question or directive." [*relationship between information in the text and the presented question*]
- "Little, if any, competing information is present." [*extent of competing information in the text*]

By fulfilling the functions described above, a framework goes far beyond providing a basic list of test specifications. Rather, the framework contributes to the development of a coherent assessment. As shown in Fig. 2.1, a coherent assessment is one in which the framework, instrumentation and interpretation are integrated.

In a coherent assessment:

---

[7] For a complete discussion of the difficulty factors and descriptions of each level in the described proficiency scale for PIAAC Cycle 1 Literacy, see the Technical Report of the Survey of Adult Skills (OECD, 2013)

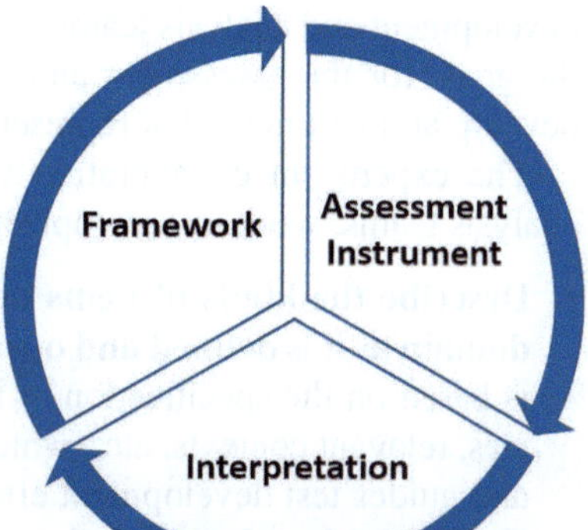

Fig. 2.1 The three components of a coherent assessment

1. The framework, as explained above, defines the construct measured in the assessment, explains how the scale or scales for the construct will be reported, and defines the characteristics of items that will measure the construct. Thus, the framework sets forth a blueprint for the types and numbers of items needed to measure the construct as it has been defined, as well as the plan for how the results will be reported.
2. Assessment items are developed based on the task characteristics and distribution of items specified in the framework.
3. An interpretative scheme is developed that describes task difficulty and examinee performance. That scheme is based on the foundation established by the framework and the scaling and analyses of the survey data that is based on the items built to reflect that framework.

2.1.2 *The Role of Expert Groups*

The development of each domain framework is guided by an advisory group with expertise in that domain. The expert group is responsible for developing a framework that reflects current thinking in the field and provides the guidance needed to develop and interpret the assessment. To support the development of new items, the expert group must work collaboratively with test developers, psychometricians, and members of the platform development team from the outset of the project. This collaboration is designed to balance the opportunities and constraints presented by an assessment. While frameworks may, and often do, include aspirational goals for how a construct might be measured, the experts need to understand what can be developed and delivered given the operational constraints of a large-scale assessment. These constraints include budgets, timelines, available testing time, the need to successfully deliver a survey to hundreds of thousands of respondents in countries around the world, and the ability to capture all the necessary data. Equally important, by working collaboratively with the expert groups, members of the design, development and analysis teams can better understand the experts' broader vision of the goals for the assessment and the underlying theoretical constructs in order to develop an instrument that represents those measurement goals.

The experts, in collaboration with members of the design, development and analysis teams, work to accomplish the following tasks.

- **Describe the kinds of items needed to assess the skills and abilities in each domain as it is defined and operationalized in the framework.** This description is based on the specification of task characteristics (for example, key competencies, relevant contexts, etc.) which, as previously described, organizes the domain and guides test development efforts.
- **Review and understand the proposed assessment design in order to define the number and types of items that will be needed for the domain.** Where a domain needs to include trend measurement, this includes agreeing on the numbers of new and trend items. For each of the cognitive domains, the experts define the different task types for new items and the range of difficulty needed to fulfil the assessment design. Particularly where the assessment design includes multistage adaptive testing, there is a need to focus on defining tasks that allow for improved measurement at both the lower and upper ends of the scales. Such a design allows for an expansion of the range of skills that are assessed in order to better represent the full range of skills in the populations of interest. This is especially important as new countries and economies look to participate in international large-scale assessments and, therefore, require the measurement of a broader range of skills.
- **Define the behaviors of interest for technology-based tasks and begin to specify how the considerable amount of log file data might be organized in ways that will result in meaningful analyses and reporting.** These specifications, which reflect a new and expanded role for the experts in the context of technology-based assessments, serves to not only inform data analysis following test administration, but guides test development to help ensure that the item pool includes items that elicit the behaviors of interest.
- **Define the range of functionality to be used for new items.** When thinking about the design of new tasks, both the experts and test developers will have questions about the kinds of tools and displays that can be supported by the platform. Similarly, the platform specialists will need to understand the kinds of functionality that would best support tasks of interest. The goal of discussions among the experts, test developers, and the platform development team is to define the intersection between the kinds of functionality that might be desirable for measuring the constructs and the functionality that is practicable to implement and deliver.

Additional responsibilities of the expert groups include: Reviewing the assessment tasks as they are developed, confirming the framework classifications, or mapping, that developers have assigned to each item based on the task characteristics defined in the framework, approving the selection of the item pool for the field trial to ensure adequate coverage of the construct or domain, reviewing analyses of the field trial data and supporting the selection of items for the main study, and reviewing the results of the main study in order to develop a described proficiency scale for the domain.[8]

2.2 Development of New Computer-Based Items

The heart of any large-scale assessment is the instrumentation developed to yield data of importance to policy makers and others with an interest in the outcomes of the survey. New technology-based items are of particular interest as they are designed to reflect an evolving understanding of the constructs being measured, as well as the range of skills required in today's changing times. The more recent frameworks for the legacy domains in both PISA and PIAAC reflect this interest, emphasizing the growing importance of digital information, representations, devices and applications across educational, personal, work and social contexts, as the following examples show.

The PISA 2018 reading literacy framework emphasizes the evolving nature of reading literacy.

> In the past, the primary interest when evaluating student reading literacy proficiency was the ability to understand, interpret and reflect upon single texts. While these skills remain important, greater emphasis on the integration of information technologies into citizens' social and work lives requires that the definition of reading literacy be updated and extended. It must reflect the broad range of newer skills associated with literacy tasks required in the 21st century (OECD, 2019).

Similarly, the PIAAC Cycle 2 literacy framework focuses on the importance of digital reading in the lives of adults.

> The current framework aims at describing reading literacy in the present-day context, in which digital reading is a central aspect of active participation in society. Three core sets of abilities are required for skillful reading in the complex information environments readers interact with: 1) ability to navigate within and between networked documents, 2) ability to comprehend and integrate multiple and sometimes disparate sources of information, and 3) ability to critically evaluate the information presented (OECD, unpublished draft).

In both cases, these frameworks emphasize the importance of including innovative tasks focused on digital and interactive texts, and on skills such as comparing,

[8]Both PISA and PIAAC employ a design where a set of items is administered in the field trial (or field study) to collect data about how those items function within and across all participating countries. The field trial includes both trend and new items to establish the stability of the trend item parameters and the overall quality of the new items. Based on the field trial data, a subset of items reflecting both the construct and the assessment design in terms of numbers and types of items is selected for the main study (or main survey). Scaling and analysis of the main study data are used to report on performance within and across countries and to develop a described proficiency scale that characterizes performance along the scale for each domain. For more detailed information about described proficiency scales, see Kirsch (2001, 2003).

integrating, and synthesizing information across multiple sources.[9] Likewise, the most recent frameworks in mathematics and numeracy have argued that these assessments must also reflect changes in the demand for skills in these domains.

The PISA 2022 mathematical literacy framework describes the ways in which evolving technologies have impacted the need for new and expanded skills in this domain.

> PISA [2022] aims to consider mathematics in a rapidly changing world driven by new technologies and trends in which citizens are creative and engaged, making non-routine judgements for themselves and the society in which they live.... The definition of mathematical literacy explicitly includes the use of mathematical tools. These tools include a variety of physical and digital equipment, software and calculation devices. Computer-based mathematical tools are in common use in workplaces of the 21st century, and will be increasingly more prevalent as the century progresses both in the workplace and in society generally. The nature of day-to-day and work-related problems and the demands on individuals to be able to employ mathematical reasoning (both deductive and inductive) in situations where computational tools are present has expanded with these new opportunities—creating enhanced expectations for mathematical literacy (OECD, 2018).

This perspective is echoed in the PIAAC Cycle 2 numeracy framework:

> … adults are presented with ever-increasing amounts of information of a quantitative or mathematical nature through Internet-based or technology-based resources. New means of communication and types of services have changed the way individuals interact with governments, institutions, services and each other, and social and economic transformations have, in turn, changed the demand for skills as well. More so than in prior decades, a wider range of quantitative and mathematical information is more readily available, but this information has to be located, selected or filtered, interpreted, at times questioned and doubted, and analyzed for its relevance to the responses needed (OECD, unpublished draft).

As these examples show, the frameworks for these legacy constructs increasingly emphasize the importance of extending what is measured and delivering authentic tasks that reflect evolving technologies and the skills associated with their use. These same evolving technologies have made it possible to develop platforms that support the delivery of assessment tasks to measure these critical skills. Measuring skills of interest and making use of the tools and capabilities afforded in these platforms has required the development of innovative environments and item designs, innovative scoring and data capture capabilities. and new development processes.

2.2.1 *Innovative Environments*

As noted, the skills associated with the use of tools and capabilities within technology-based environments are increasingly becoming integrated into the legacy

---

[9]Note that each of these assessments include what are called 'non-continuous' texts, where information is arranged not in sentences or paragraphs but in a matrix format. Examples include tables, graphs, charts and forms.

constructs so that they more accurately reflect the ways in which individuals access, use, and communicate information. The transition to computer-based assessments has made it possible to develop and implement simulated environments that support these types of technology-based tasks and to design interfaces that allow respondents to operate within those environments. For example, simulated websites have been included in the PISA reading literacy and PIAAC literacy and numeracy assessments to expand the measurement of digital reading and numeracy skills. In a typical simulated website, respondents are able to move freely among the available web pages, access information via links, and scroll or page through sources with longer content, much as they would in real-life contexts.[10] Other environments that have been included in the delivery platforms for PISA and PIAAC include simulated email and spreadsheet environments and simulated applications, such as route planners and expense trackers. In the design and development of such environments, test developers have remained cognizant of the fact that PISA and PIAAC are not assessments of ICT literacy. Simulated, interactive environments have therefore been designed to be simple and intuitive, and tutorials have been developed to familiarize respondents with all of the tools and functionality used in the assessments.

Another feature of technology-based environments—the ability to dynamically view information in different representations—has been incorporated into the designs for some new items. For example, in some units, respondents can opt to view data in a variety of formats including bar graphs, pie charts, line graphs and scatter plots.[11] Other units allow respondents to change the display of a graph by selecting the information to display on each axis. Capabilities such as these make it possible to assess how respondents reason with different representations of information, a skill that is of particular interest in the science and math/numeracy domains.

The authentic and interactive environments that have been developed and implemented in the last several cycles of PISA and PIAAC have broadened what can be measured both in terms of the nature of the tasks that can be included, the types of stimulus materials that can be used, and the ways in which the presentation of these materials can be staged. The scenario-based units designed for the PISA 2018 reading literacy and PIAAC Cycle 2 literacy assessments are examples of staging that would not be possible in a paper-based assessment. This staging is critical, as the updated frameworks for reading and literacy have emphasized the importance of including units that contain multiple texts or sources of information in

[10] It should be noted that the simulated websites developed for these assessments include a limited number of pages and links in order to reduce the respondent burden and, from an operational perspective, the translation burden. The range of available functionality is limited for that same reason, as well as to provide a more generic interface that is less likely to become dated as technologies change.

[11] In the context of these assessments, a "unit" is defined as one or more stimulus materials focused on a particular topic and the set of questions (also referred to as items) associated with those materials.

order to reflect the demands of reading in online and technology-based environments The scenario-based approach also reflects the understanding expressed in each framework that the way in which readers approach texts is primarily determined by their reading goals, which are, in turn, informed by their understanding of the context and task demands (Kendeou, 2020; Sabatini et al., 2014, 2019). Traditionally, assessments of reading literacy have focused on a series of single texts, each associated with a number of discrete questions. In contrast, and aligned with the view that reading is a purpose-driven activity, in this scenario-based design each unit begins with an introduction that sets the context and lays out the general purpose for reading the text, or set of texts, it includes. For example, an introduction might say that you want to purchase a new bicycle and are looking at a series of online reviews. Or you need to make a home repair and are checking instructions from a number of sources. In each case, the context is set in the introduction and all of stimulus materials and items in the unit are focused on this general context as it has been defined for the respondent. As new texts or stimulus materials are presented throughout the unit, the purpose for each is introduced in turn.

One goal for each of these large-scale assessments is to provide information about individuals with a range of skills and knowledge. The staging that can be implemented when units contain multiple texts makes it possible to include easier questions at the start, often focusing on just a single text, and then build to more challenging items that involve multiple stimuli. This allows respondents with a range of skills to demonstrate what they know and can do. Examples from both PISA and from PIAAC follow to illustrate the scenario-based model.

2.2.2 PISA 2018 Reading Literacy: Scenario-Based Environment

This first example is from a released PISA 2018 reading literacy unit that focuses on the island of Rapa Nui (Easter Island) and several theories about how and why the large trees that once covered the island disappeared. The unit includes three texts: A web page from a blog by a professor doing field work on the island, a review of a book about the collapse of several civilizations including the one on Rapa Nui, and an article from an online science magazine that presents an alternative theory about why the trees disappeared. The texts are presented in a staged manner with some questions focused on an individual text and others requiring students to integrate information across two or more sources. As shown in Fig. 2.2, the interface designed for these scenario-based units includes tabs that allow respondents to select the text or stimulus material they wish to view and to toggle freely among the available materials, similar to the way readers use tabs within an internet browser to view multiple pages.

Figure 2.3 shows the introductory screen for the Rapa Nui unit. The general purpose for reading is defined: You will be attending a lecture and have been asked

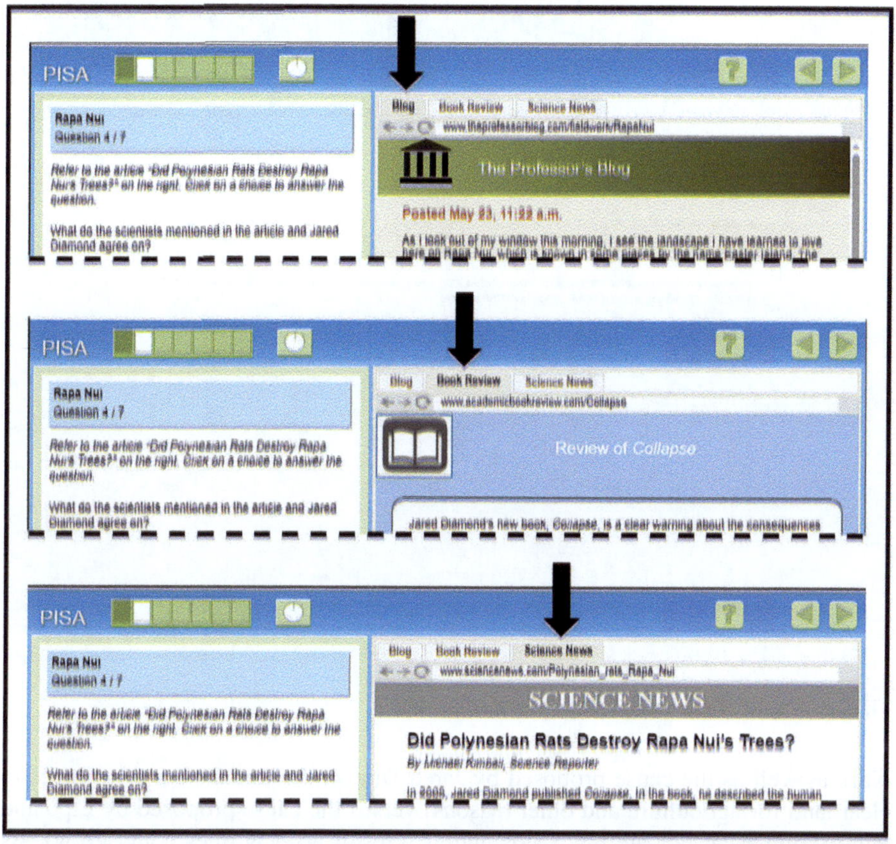

Fig. 2.2 Tabbed interface used for scenario-based units, where each tab displays a different text

to read some material to learn more about the topic. In addition, the first text is introduced.

The first question in the unit is shown in Fig. 2.4. To answer the question ("… when did the professor start her field work?"), the student must locate specific information in the blog post. Only one text is available at this point and the correct answer, "nine months ago", is in the text (i.e., "Later today, I will take a walk through the hills and say good-bye to the moai that I have been studying for the past nine months.")

Each of the two remaining texts is introduced in turn with a number of questions focusing on each. By the end of the unit, all three texts are available. In the question shown in Fig. 2.5, students are asked to integrate information across the texts with respect to the theories put forth, on the one hand, by Jared Diamond (mentioned in both the book review and Science News article) and, on the other hand, by Carl Lipo and Terry Hunt (mentioned in the Science News article). Using the tabs, student can freely toggle among the stimulus materials in order to locate the information required to answer the question. To answer correctly, students must drag and drop the provided options to identify the effect (The large trees disappeared from Rapa

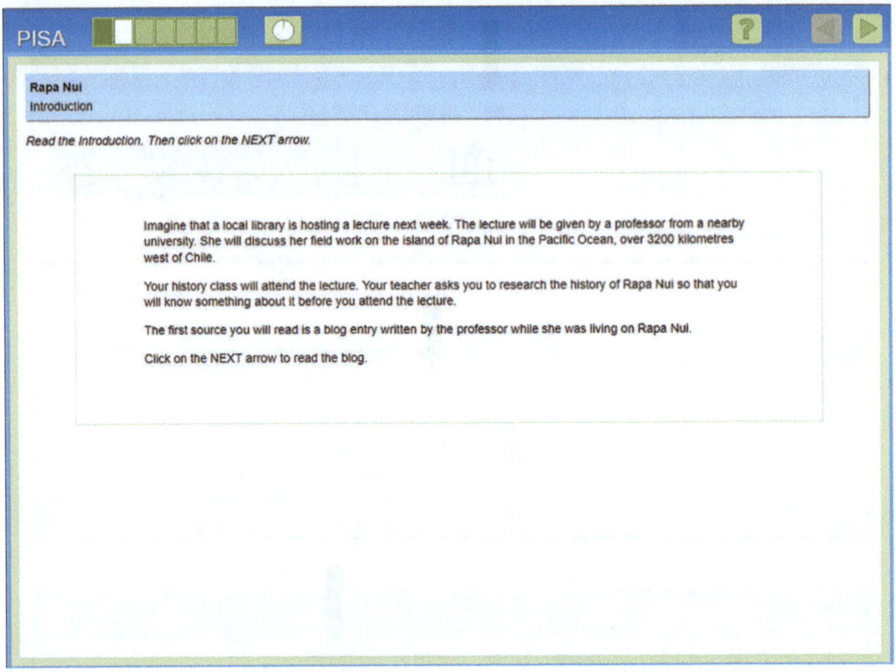

Fig. 2.3 Rapa Nui—introduction

Nui), as well as the cause proposed by Jared Diamond (Humans cut down trees to clear land for agriculture and other reasons) versus the cause proposed by Lipo and Hunt (Polynesian rats ate tree seeds and as a result no new trees could grow). As expected, because this item required integrating information across multiple sources, and generating inferences, and creating a representation of causes and effects using a subset of the available options, this was one of the more difficult items in the assessment.

2.2.3 PIAAC Cycle 2 Literacy: Scenario-Based Environment

An example of a similar scenario-based unit in PIAAC Cycle 2 is one that focuses on job training in the context of a fictional local company that is incorporating new technology into the manufacturing process at its plant. This unit includes three texts that are linked via the premise that a local newspaper has devoted a special section to the issue of worker training that includes articles by employees, advertisements from training companies, and editorials related to the topic.

The unit begins with an article written by an employee about his choice to participate in an online training course rather than enroll in a classroom-based course at a community center. Respondents are asked two questions that focus on the information provided in this article. The second text is then presented—an

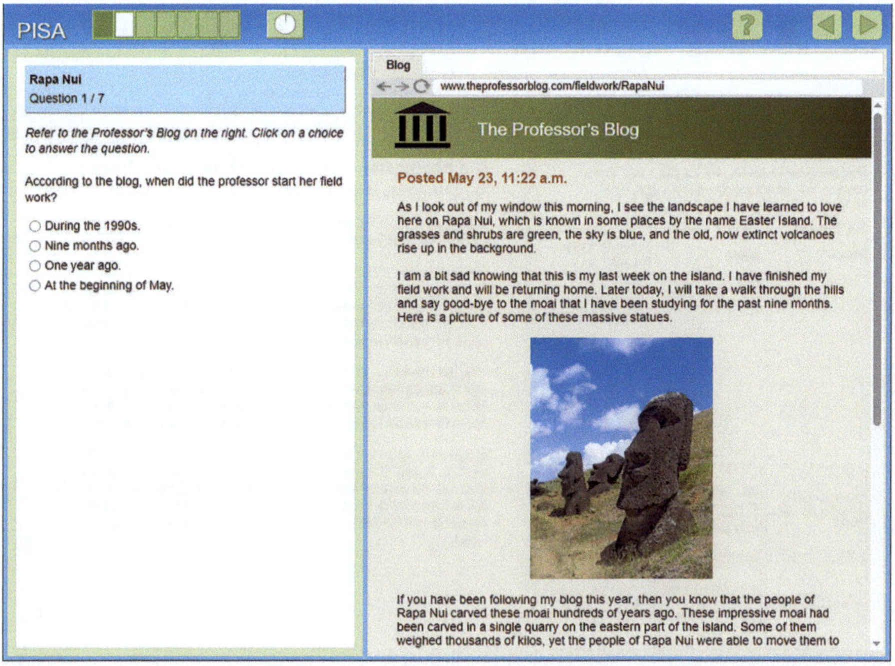

Fig. 2.4 Rapa Nui—question 1

advertisement from a company that provides online training programs for workers. Respondents are asked questions that require only the use of the advertisement, as well as several that require them to compare the information in the advertisement with that in the employee article. By using the same interface shown in Fig. 2.2 above, respondents are able to view each of the two sources. Finally, an editorial is presented that supports a blended learning approach, where online and classroom instruction, as well as hands-on, on-the-job training, are integrated. Again, respondents are asked a series of questions focused on a single text—in this case, the editorial. The last question in the unit requires respondents to integrate information across all three sources in order to identify which would support the viewpoints expressed in a number of statements.

The scenario-based tasks in PISA and PIAAC make it possible for respondents to demonstrate skills related to understanding and integrating information from multiple sources. In addition to the samples shown, other units in both assessments included simulated websites with hyperlinks and multiple web pages. The interface and functionality used in the design of these tasks reflect the real-life contexts from which they were taken. The kinds of tasks that can be presented using multiple texts and interactive digital texts such as these make it possible to both reflect the definitions and explanations of the domains that are provided in the frameworks for these assessments and extend what can be measured in important ways.

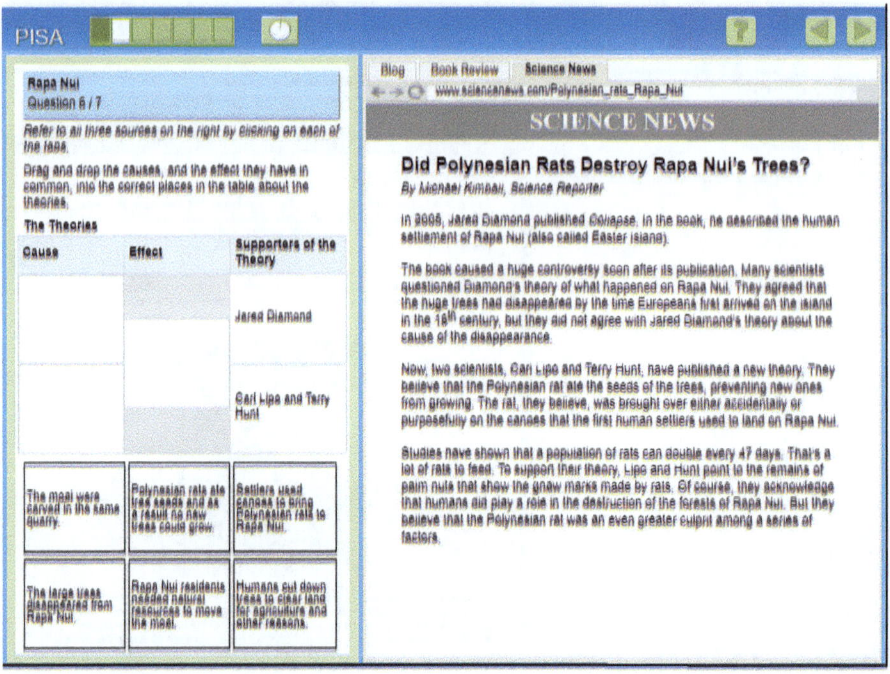

Fig. 2.5 Rapa Nui—question 6

A second example of an innovative environment is the simulation environment used in the PISA 2015 assessment of scientific literacy, a design that was replicated in some units included in the 2022 assessment of mathematical literacy.

2.2.4 PISA 2015 Scientific Literacy: Simulation Environment

Scientific literacy was the major domain in the 2015 cycle, meaning that a new framework and new items were developed for this legacy construct. As previously noted, the 2015 cycle of PISA marked a shift in this survey from a primarily paper-based delivery mode that included optional computer-based modules to a fully computer-delivered assessment. As a result, all new items for scientific literacy were designed as computer-based items. The 2015 framework for this domain defined a scientifically literate person as one who both understands, and is capable of undertaking, a basic set of competencies. These competencies included being able to: (1) Explain phenomena scientifically, (2) evaluate and design scientific inquiry, and (3) interpret data and evidence scientifically. The 2015 cycle presented the opportunity to improve the measurement of scientific literacy across all these competencies but, in particular, the use of simulations allowed for a more authentic assessment of the ability to design scientific inquiry. The simulations developed for

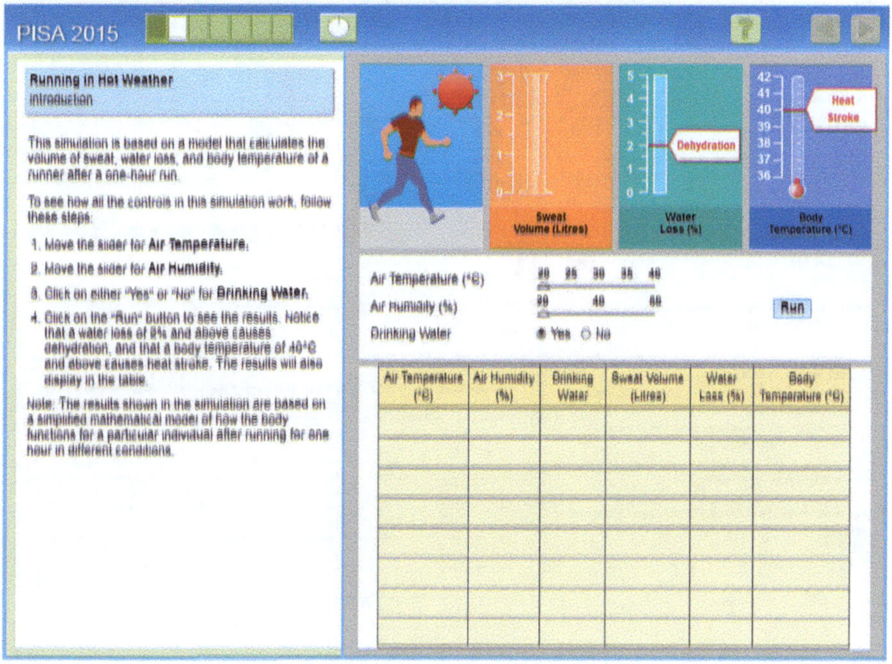

Fig. 2.6 Running in hot weather simulation—introduction

this purpose allowed students to set up and run virtual experiments, selecting variables and values for those variables and using the resulting data to respond to presented questions. The simulations were open-ended in nature and allowed students to run multiple trials, evaluate results that displayed in both animations and data tables on the screen, and adjust their approach as needed.

One unit focused on how a runner's body temperature is affected by air temperature, humidity levels, and whether or not the runner drinks water over the course of a long-distance event. As shown in Fig. 2.6, for each trial, respondents were asked to set the air temperature and humidity and select "Yes" or "No" for drinking water. When the "Run" button was selected, the data associated with the selected variables displayed in the table on the screen and the runner's sweat volume, water loss and body temperature were also shown in the visual display at top panel in the simulation pane. When the selected conditions triggered dehydration or heat stroke, those health dangers were highlighted with red flags.

As was the case with all the interactive science units, before the first question was presented, students were introduced to the controls in the simulation and allowed to practice setting each, as shown in the set of instructions in the left pane of the screen in Fig. 2.6. The interactions themselves were designed to be simple to use and consistent across units, but it was important to be sure that students understood how to manipulate the controls before beginning the items. The interface was designed so that students could not skip this introductory task in any of the interactive units. If

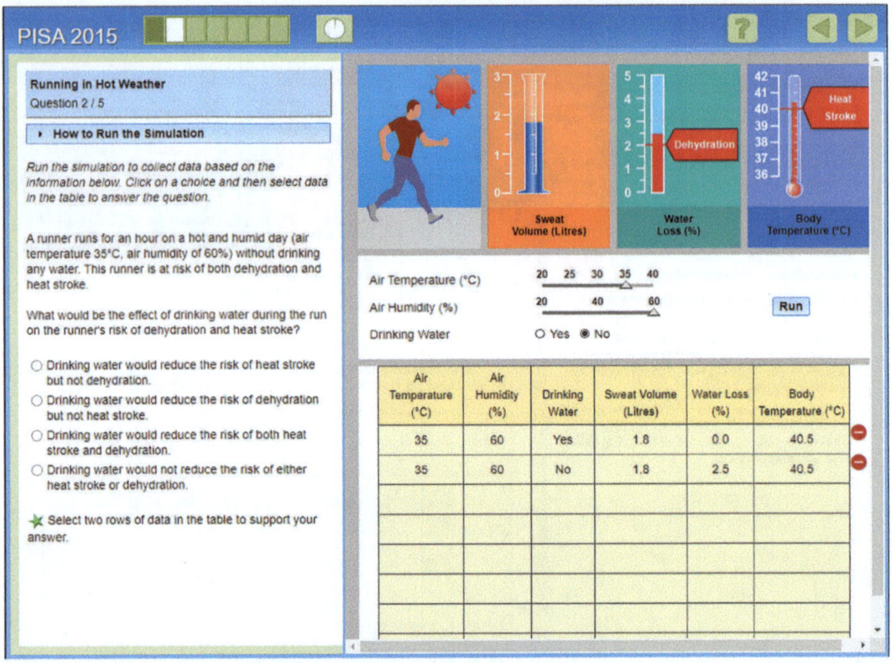

Fig. 2.7 Running in hot weather simulation—question 2

students did not take the requested actions after a specified period of time, a message displayed to remind them of the steps to follow, and, if they still did not respond, those steps were demonstrated via an animation.

The item shown in Fig. 2.7 illustrates one model used for these simulation tasks: Students must run a series of trials based on the provided information and then both answer a multiple-choice question and identify the data from one or more trials that supports their answer. In this particular case, students were asked to run the simulation holding the air temperature and humidity constant using specified values. They needed to manipulate just one variable, i.e., whether or not the runner drank water. The simulation showed that running under the specified conditions without drinking water led to both dehydration and heat stroke. In contrast, drinking water reduced the risk of dehydration but not the risk of heat stroke. Students needed to run the simulation twice—once with drinking water set to "No" and once with it set to "Yes"—in order to collect the data that supported the correct answer. Figure 2.7 shows the results after running the simulation under both conditions.

This was one of the more difficult items in the science assessment. To receive full credit, students had to select the correct multiple-choice option (B) AND select the following two rows in the data table based on the experiments they ran:

- Air temperature set to 35 °C, 60% air humidity and 'No' for drinking water
- Air temperature set to 35 °C, 60% air humidity and 'Yes' for drinking water

Units such as this one typically included a set of questions that ranged from easy to very difficult. Easier questions typically provided specific values for variables, requiring students to carry out a straightforward procedure and interpret the resulting display correctly, as per the example in Fig. 2.7. Questions increased in difficulty based on the number of variables that needed to be manipulated and the extent to which students needed to determine the values for those variables on their own. The most difficult questions typically required students to extrapolate beyond the data they had collected through the stimulation to draw more general conclusions about the relationships between or among variables.

These variations in how the items were constructed and how the difficulty drivers were manipulated by test developers directly reflected the task characteristics defined in the scientific literacy framework. Data from the assessment were then used to evaluate the extent to which these variables accounted for variance in the difficulty distributions of the items and could be used to describe performance along the scientific literacy scale and, in this case, along the "Evaluate and Design Scientific Inquiry" subscale. Results showed that the hypotheses about what might drive difficulty were supported by the assessment data, so these features were used to describe points along the proficiency subscale, which was defined by seven levels ranging from the lowest level of Level 1b–Level 6. Students at Level 1b were described as being able to "follow simple instructions to carry out a scientific procedure, [running] a simulation to extract a single data point." In contrast, students at Level 4 could "conduct experiments involving two or more independent variables in a constrained context and justify aspects of their experimental design, drawing on procedural and epistemic knowledge." Thus, these interactive science tasks are not only examples of innovative environments that extended what could be measured in this domain, but they also illustrate how a coherent assessment of scientific literacy was developed, integrating the framework, instrumentation and interpretation by design.

2.2.5 Innovative Item Types and Response Modes

The range of item types and response modes in the computer-based versions of the PISA and PIAAC legacy domains have included the following.

- **Drag and drop:** Moving images or text boxes around the screen to indicate sequences, associations, categorization or matching.
- **Drop-down menus:** Choosing from a list of words or symbols to complete a sentence or indicate a selection.
- **Hot spots:** Clicking or tapping on a sentence in a text, a cell in a table, or a portion of an image to indicate a selection.
- **Text entry:** Typing an answer with either limited or unlimited response lengths based on the item design.

- **Numeric entry:** Typing a number (for delivery platforms that include a keyboard) or using a provided keypad to indicate a numeric response (for the tablet-based delivery in PIAAC Cycle 2 where respondents do not have access to a keyboard).
- **Radio buttons:** Indicating one or more selections or values (typically used in simulation tasks).
- **Highlighting:** Selecting and highlighting words and sentences (used only in PIAAC Cycle 1).
- **Selected response:** Indicating an answer in either a single selection or multiple selection multiple-choice item.

A few specific examples of how these response modes were implemented are included below, focusing on the PISA 2022 mathematical literacy items and the PIAAC Cycle 2 numeracy items.[12]

Figure 2.8 shows a PISA item where respondents were asked to use drag and drop to complete a pattern. Multiple instances of each of the two tile patterns were available and students could drop them into any available location and move them at will until they were satisfied with their response. The computer platform makes it possible to automatically score responses to items such as this one based on the final state of the selections.

In PIAAC, the drag and drop functionality was used to allow respondents to "draw" a graph by dragging a bar upward to a selected point on the Y-axis based on data provided in a scenario. In another item, respondents were able to drag an onscreen ruler to measure several objects shown on the testing screen in order to compare their lengths.

Figure 2.9 is an example of sorting functionality that allows respondents to click on a column in a spreadsheet so that the data in that column is sorted. This format was used in both PISA and PIAAC and allows respondents to freely select and change the column to be sorted. An additional functionality that was included for some items was the ability to specify that the arithmetic mean for the numbers in a selected column be displayed. Tools such as these allowed respondents to view the display of data in interactive ways to support their evaluation of that information.

The sample item types and response modes included here could only be implemented in an interactive computer-based assessment environment, and, thus, they represent innovations in how skills can be measured. It should be noted that other item types included in these computer-based assessments—such as multiple-choice and open-ended text entry items—could just as easily be used in a paper-and-pencil assessment. However, for all item types, the computer platform allowed for innovative scoring and data capture capabilities. These capabilities expanded the information that could be collected about both the answers respondents provided and the processes they engaged in when interacting with the assessment materials, enriching the data collected about what individuals know and can do.

---

[12]In both cases in this section, screen shots are taken from the sample items in the PISA 2022 Mathematics Framework (OECD, 2018).

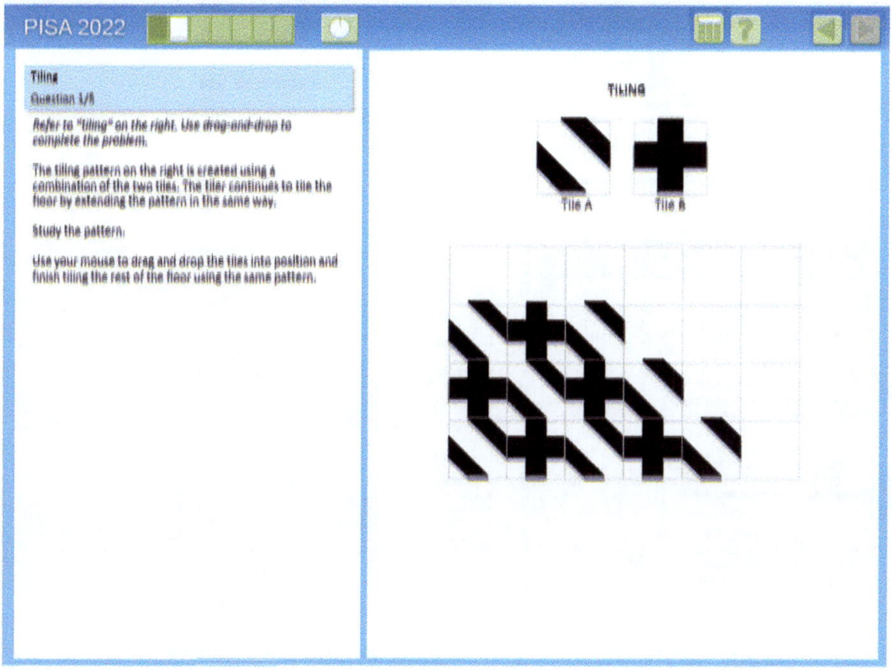

Fig. 2.8 Drag and drop to complete a pattern

2.2.6 Innovative Scoring and Data Capture Capabilities

For any given task, the PISA and PIAAC computer platform collects a variety of information including time to first action, total time, number of actions, and specific actions taken, such as clicks, taps, etc. This information provides direct evidence of the processes and strategies that respondents use and, therefore, broadens the type of inferences we can make about their knowledge and skills. In order to make the best use of the capability to capture process and strategy information, it is important to use the framework variables and hypotheses about the task characteristics that will impact performance to define those behaviors that provide the best evidence about performance and, therefore, are important to capture and score.

The collection of timing information, while a seemingly simple benefit of computer-based assessments, has been enormously helpful in supporting both data analysis and test development efforts. For example, non-response data can be treated in a more sophisticated way by taking timing data into account. While the problem of rapid guessing has been described in high-stakes assessments (Wise & DeMars, 2005), the nature of large-scale surveys such as PISA and PIAAC does not compel respondents to guess, but rather to skip items for reasons that may be unrelated to skills, for example due to perceived time pressure or a lack of engagement. In the PIAAC Cycle 1 data analyses, if very little time was spent on an item (5 seconds or

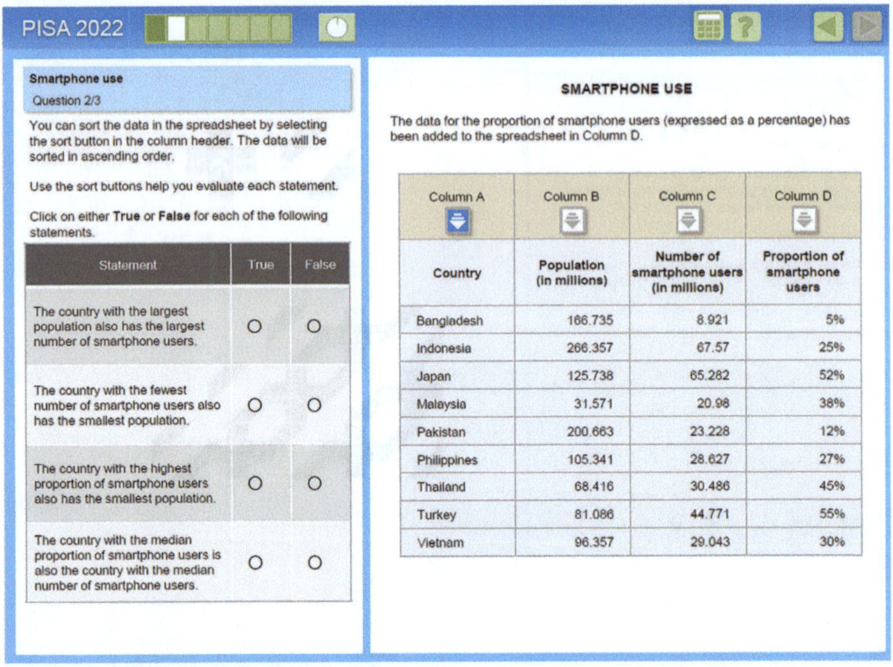

Fig. 2.9 Spreadsheet with sorting functionality

less) the item was judged to be skipped and was coded as 'not reached/not administered' (OECD, 2013; Weeks et al., 2014). If, however, a respondent spent more time on an item, or performed more than the minimum number of actions, a missing response would be assumed to be an intentional omission and the item was scored as incorrect. This approach was found to provide more accurate information about responses that increased the proficiency estimation. The use of timing information has also been helpful in detecting and documenting instances of data fabrication (Yamamoto & Lennon, 2018). Finally, timing data is useful to test developers in informing the selection of main study items based on the field trial data and in assembling forms for the main study. Test developers can use the data regarding the mean time spent on each item during the field trial to inform the construction of parallel blocks of items or the implementation of an adaptive design for the main study.

The most basic information that is collected by the computer platform and included in the performance data is a respondent's answer and the associated score. A large proportion of items in PISA, and all of the items developed for PIAAC, have been designed to be scored automatically. Test developers define the correct response(s) for each item and this information is implemented into the system by the programmers. One component of the data analysis process is a review of any items that are flagged as not performing as expected, either internationally or in specific countries. In these cases, the scoring rules and their implementation are checked. Where errors are found, those can be centrally corrected, which is another advantage of the computer-based platform.

The ability to automatically score all participant responses in real time is a key benefit of a computer-delivered assessment. Automatic scoring makes it possible to implement adaptive testing designs in large-scale surveys. Both PIAAC and PISA have implemented multistage adaptive testing designs, which have improved the efficiency of these assessments, allowed for more complex assessment designs, and provided more reliable information about respondents' skills (Yamamoto et al., 2019; Yamamoto, Shin et al., 2018).[13] Automated scoring also increases the reliability of scores, both within and across countries, and reduces country burden in terms of having to recruit, train, and supervise scorers. In addition, data entry demands are eliminated, as responses are automatically captured. Data files for computer-scored items are cleaner, with fewer errors and inconsistencies compared to a paper-based test. Finally, the data can be tracked automatically to determine if parts are missing or corrupted.

Beyond the specification of correct and incorrect scores, test developers also have increasingly focused on predefining additional variables of interest. While research efforts that focus on data mining once an assessment has been administered have contributed to our understanding of the responses provided and processes in which respondents engage, efforts to predefine variables are construct driven from the outset and are designed to support the understanding and interpretation of respondent behaviors. As will be discussed next, as part of the PIAAC Cycle 2 test development process, two types of variables have been defined: Navigation and scoring variables.

2.2.7 *Navigation Variables*

As noted previously, many PIAAC Cycle 2 literacy units use a scenario-based approach with multiple stimuli. Additionally, several units include simulated websites. In each of these cases, respondents have the opportunity to view multiple screens of information and navigate freely among them. It was hypothesized that the extent to which respondents navigate through information and the efficiency with which they are able to locate relevant information would be factors that might explain performance on some of these items. Therefore, variables were predefined to capture the pages viewed for each item in a unit. In addition to assigning a variable

[13]Multistage adaptive testing is a variant of item-level adaptive testing. In item-level adaptive testing, a response to a single item determines the next item presented. The multistage design algorithms work on a testlet, or cluster, level, in which responses to a number of items determine the next stage or testlet presented to the test taker. This design makes it possible to collect more performance information and, therefore, increases the selection accuracy for the next testlet. Data from the field test provide the initial IRT parameters that are used to construct the adaptive testing algorithm implemented in the main study.

to each possible page that the respondent might view, test developers identified the page(s) with relevant information for each item so that these could be easily identified in the data. This information will be reviewed by the expert group following the collection of the PIAAC's main study data to investigate the extent to which these variables can be used to help explain item performance.

The collection of navigation data goes beyond investigating actions for single items. Because these actions of interest are common across many of the literacy units, it is possible to collect information that can be integrated and generalized to reflect key skills associated with digital texts. To date, most of the work involved with defining such variables has been conducted by test developers with input from data analysts. While the chairs of both the PIAAC literacy and numeracy expert groups provided input into this process, this work points to a potential expanded role for the expert groups as a whole. In the future, members of the expert groups could be asked to help develop hypotheses about the types of variables that might be expected to be of interest in identifying similarities or differences in response patterns or styles across various subgroups as part of the framework development work.

2.2.8 Scoring Variables

While the identification of navigation variables is expected to inform analyses of item performance and the development of the PIAAC literacy proficiency scale, a second set of variables was defined to provide data to inform examination of field trial responses. Following field trial data collection, test developers meet with the expert groups to review item performance and select items for the main study. Part of that work sometimes includes rechecking the assigned scores for items. It is typical for the experts to ask whether respondents selected alternative or additional answers when discussing item performance. Because the schedule dictates that the experts meet immediately following the conclusion of the field trial data analysis process, there is little to no time to delve into the log files to evaluate other responses for some of the more complex item types. However, predefining additional scoring variables makes it possible to have such data readily on hand. Therefore, for both the literacy and numeracy domains, a set of variables was defined during item development to capture additional responses of interest. For example, one new item type in the literacy assessment uses a 'tap on a sentence' response mode. In the course of country review and translation, questions were sometimes raised about whether a sentence that was not initially included as a correct response should perhaps also be credited. Where developers felt that the inclusion of an additional sentence might, indeed, be warranted, that selection was assigned a specific variable. This will make it possible to discuss and evaluate alternative item scoring with the expert group, where it is supported by the data, and get input from the experts prior to finalizing the scoring rules for the main study.

2.2.9 Scoring Open-Ended Items

While PIAAC was designed from its first cycle as a fully computer-scored assessment in order to support its adaptive design, PISA has retained a proportion of open-ended human-scored items (referred to as human-coded) in each of its legacy constructs. These item types are judged to be important to include in this assessment, but they do pose a number of operational and measurement challenges. From an operational perspective, human-scored items are both labor and time intensive. Test developers must develop detailed scoring guides that must be translated and verified. National centers must hire a team of scorers and those scorers must be trained and supervised. And the reliability of the scoring process, both within and across countries, must be monitored. Another challenge related to human-scored items results from the increased interest in implementing adaptive testing in large-scale assessments in order to better adjust the set of items administered in an assessment session to the abilities of particular respondents. Because the results for human-scored items are not available during the testing session, these items cannot contribute to the implementation of adaptive assessment designs. Issues such as these led to a series of studies investigating the feasibility of implementing a machine-supported coding system for some open-ended items. A pilot study was conducted using PISA 2015 data and the system became fully operationalized for the 2018 field trial and main study.

In brief, the machine-supported coding system (MSCS) makes it possible for frequently observed, identical responses to receive the same appropriate score automatically so that the scoring effort for these responses can be minimized (Yamamoto et al., 2017; Yamamoto, He, et al., 2018). As discussed in more detail in Chap. 12, the system does not 'interpret' responses per se, but relies solely on matching responses that have been human-scored in previous PISA cycles. That is, existing raw responses specific to a national language are collected. In cases where a common response (say, for example, "30") is repeatedly seen for an item, the correct scoring for that response is confirmed. Then, in future test administrations, that score is assigned to the same responses (e.g., all instances of "30"). Using the system, all instances of nonresponse (such as items that are left blank) and all responses with existing, verified scoring can be automatically scored. Only those responses that cannot be matched to an identical response stored in the database are assigned to human scorers. For open-ended items that have simple, straightforward responses, the use of the MSCS has demonstrated important efficiency gains. In the PISA 2018 main study, the coding burden was reduced, on average, by 33% in mathematical literacy, 29% in trend reading literacy items, 23% in new reading items, and 25% in scientific literacy (OECD, 2020). Improved scoring reliability and the efficiency gains realized by reducing the scoring burden support the use and continued development of this innovative system.

2.2.10 The Development Process for New CBA Items

Creating items for computer-based surveys requires a development process that differs from paper-based development in several significant ways. As discussed earlier, one of the most important differences is the expanded importance of early collaboration between key members of the development and delivery teams. From the start, members of the expert groups, test developers, platform specialists, and psychometricians need to be in agreement about the types of items that will be developed, how they will display and function, and what behaviors will be captured and scored. As noted earlier, to achieve this understanding, it is important for lead test developers and platform specialists to attend meetings of the expert groups.

In addition to establishing procedures to ensure shared agreement about the goals and possibilities of the assessment instruments, a set of more specific processes is important to support the development and delivery of items. These include developing storyboards, creating programmed prototypes for usability testing, and defining rules for scoring and capturing a range of response data.

2.2.11 Role of Storyboards in the Design, Development and Review Processes

For PISA and PIAAC, PowerPoint storyboard templates are created that reflect the screen design for each survey. Using storyboards to draft items provides several key advantages.

- The process of developing storyboard templates prompts detailed discussions between test developers and programmers about the display and functionality of different item types, stimulus materials and tools. Separate templates are typically developed for the various environments to be included in the assessment (for example, simulated websites, emails, and blogs). The templates are also used to document designs for the display of multiple texts and tools such as keypads for entering numbers.
- Using the same set of templates during the development process helps ensure consistent display and design across items. This is helpful for test developers and limits the tendency to create unique displays for specific units. Reducing the number of "one-offs" streamlines programming and decreases respondent burden, as test takers only need to learn about a few environments that use consistent functionality.
- Graphic artists can refine the rough storyboards created by test developers to improve the layout, document specifications for fonts and colors to facilitate programming, and include any needed graphics.
- Linguistic considerations, such as the space required to accommodate translations of button labels, etc., can be addressed at the beginning of the development

process, thereby limiting the number of language-specific layout adaptations that might be required. Additionally, developers can mock up screens using quick translations of the content to identify if the layout needs adjusting to accommodate differences across languages (e.g., a French translation of an English stimulus can be, on average, 15–25% longer, Japanese and Korean characters require more vertical space, etc.). The goal is to have a consistent display across languages with the same amount of information on screen and to avoid what might be called "differential scrolling"—e.g., a text that fits on a single screen in some languages but requires scrolling in other. If a stimulus or question can reasonably be adjusted from the outset that saves time and effort later in the process.

- The storyboards can function as a communication tool between developers and programmers. Particularly for interactive stimuli and items, test developers need to think through and document how the system should respond to particular behaviors. For example, what should happen when a respondent taps on a link on a web page? Or, equally important, how should the system respond if a respondent does not take an expected action? Adding notes to the storyboards that specify system responses documents the developers' expectations and facilitates the work of programmers.
- Having items in storyboard format facilitates the review process as reviewers can more easily envision what the items will look like when delivered in the assessment. Early feedback from both subject matter experts and national centers is particularly important in a computer-based environment because of the time and cost associated with authoring items. Working from storyboards typically helps reviewers provide more detailed and useful feedback.
- Finally, the notes pane in the storyboards has been used to document how each item is classified according to the framework task characteristics and to record the scoring rules for the item. The completed storyboards can, thus, provide a permanent record of each item and its associated metadata.

Figure 2.10 is an example of a PIAAC Cycle 2 template for a literacy unit that includes two texts, one of which is a simulated web page. The sample shows how the test developers and platform specialists worked with interface designers and graphic artists to create the template for units that included more than one stimulus, with tabs at the top of the right pane that function as navigation devices. The template for the simulated browser included a toolbar with navigation elements and an address bar. Once these designs were developed and agreed upon, test developers could use them for all applicable units and these templates could be included in the domain tutorial developed to introduce respondents to the environments used in the assessment.

2.2.12 Prototypes for Usability Testing

In cases where new displays and/or functionality are proposed by test developers, prototypes are developed to support usability testing. These prototypes may be

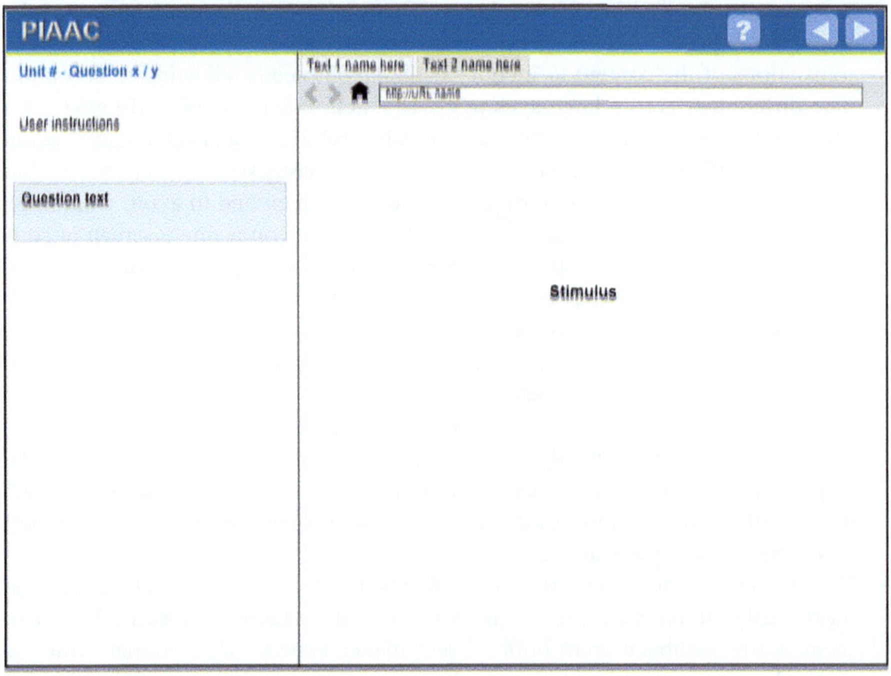

Fig. 2.10 Sample PIAAC storyboard template

developed as storyboards or low-fidelity authored items (e.g., items without final graphics or full functionality). This testing is conducted to inform item development and provide evidence that new formats are not expected to impact performance. The usability testing also informs the types of information and practice to include in the tutorials that respondents receive prior to beginning the assessment. Testing is conducted with a limited number of individuals who represent the range of ages and computer experience expected in the testing population.

The results and insights from the first round of usability testing directly guide further development and re-engineering of the instruments. Best practice dictates that several cycles of studies should be conducted to build the optimized assessment instrument. Ideally, the development cycle includes at least one final round of testing to validate changes implemented in the previous round, but this must realistically be dictated by timeline and budgetary constraints.

2.2.13 New Test Assembly Tools

Since the first cycle of PIAAC, test developers have used a proprietary tool developed at Educational Testing Service (ETS) to support the selection of items and assembly of forms or adaptive clusters for the PISA and PIAAC main study instruments. This interactive tool uses the estimated item parameters from the field trial to dynamically generate information curves for selected sets of items, making it possible to quickly evaluate the impact of including or excluding particular items on the test as a whole. For example, in cases where the assessment design specifies multiple blocks of items that should be of approximately equal difficulty, the test characteristic curves (TCC) for the blocks can be easily compared. Figure 2.11 shows the curves for seven trend clusters in the PISA 2015 mathematical literacy assessment, which confirm that the item selection matches the design. If the curves showed that one or more clusters was much harder or easier than the others, test developers could make adjustments to the selection to try and improve the comparability of the clusters.

As shown in Fig. 2.12, the selection tool also generates the test information function (TIF) for a set of items. In the TIF, the height of the curve corresponds to the measurement accuracy and the highest point indicates the point on the difficulty scale where measurement is most accurate. This particular graph was generated to check that item selection met the goals of an adaptive design. This TIF illustrates that, as specified in the assessment design for this instrument, the easiest testlets maximize information at the lower range of the scale, while the hardest testlets maximize information at the upper range.

The interactive test assembly tool has also been used to evaluate the selection of trend items. One task for test developers in PIAAC Cycle 2 was to select a subset of items from Cycle 1 that would function as trend items. The goal was to have the Cycle 2 items, as a set, both reflect the framework and assessment design for the new cycle and mirror the measurement characteristics demonstrated in Cycle 1. The two test information functions shown in Fig. 2.13 show that the TIF for the proposed Cycle 2 trend items closely matches that for the items in first cycle in terms of their shapes not in the sizes due to different number of items. The illustration in Fig. 2.13 was used in discussions with the PIAAC Cycle 2 literacy expert group to inform their

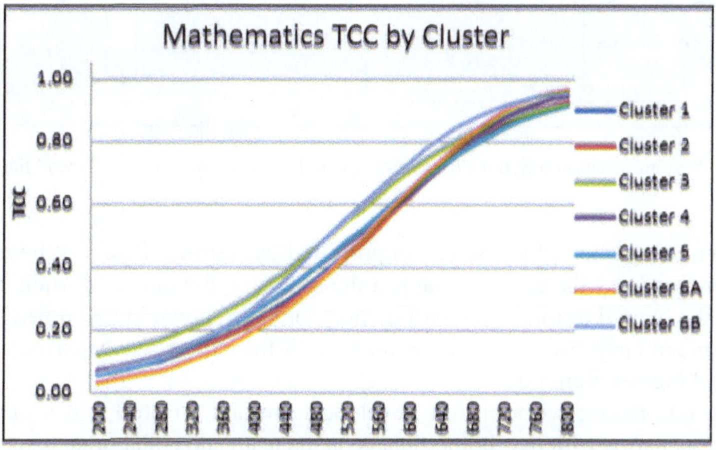

Fig. 2.11 Test characteristic curves (TCC) for seven clusters in PISA 2015 mathematics

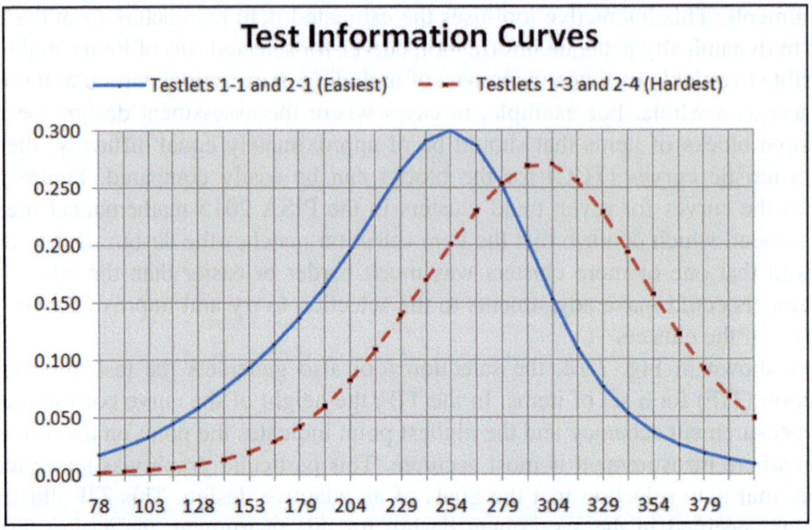

Fig. 2.12 Test information functions for the easiest and the hardest testlets in an adaptive design

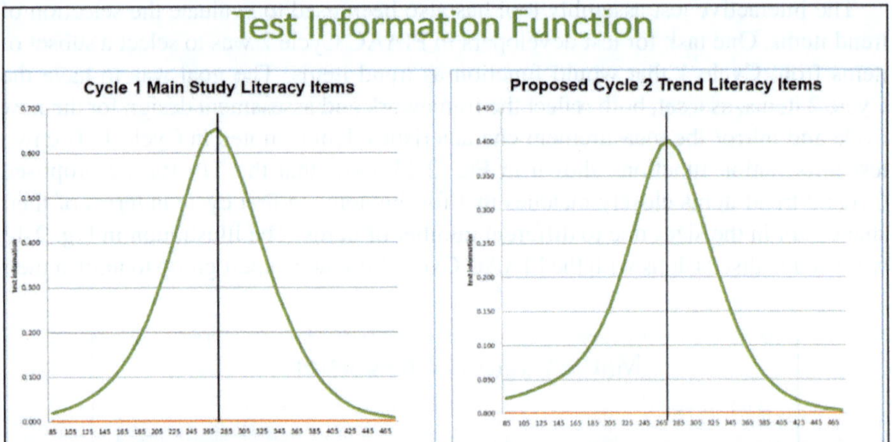

Fig. 2.13 Test information functions for PIAAC cycle 1 main study and cycle 2 trend literacy items

review and discussion of the selection proposed by test developers. Where suggestions were made by the experts to adjust the selection, the impact of such modifications on the overall test information function could be shown in real time during the discussion, and adjustments could be made until the experts were satisfied with the final trend literacy item pool.

Using this interactive tool, test developers are able to draft and share various selection scenarios with psychometricians on the team and collaborate to agree on an optimal selection of items that balances content considerations, framework coverage, and item performance. The tool is useful not only for item selection purposes, but the graphs that are generated provide a clear visual representation of item selections that can be shared with governing boards, national centers, and other stakeholders. The information functions that are used to support the item selection process are subsequently validated using the final item parameters established in the main study. Those final representations are typically used in reports and presentations about the study.

2.3 Maintaining Trend Measurement Across Modes

In addition to extending measurement to reflect evolving constructs, international large-scale assessments, such as PIAAC and PISA, are designed to measure changes in the proficiency of populations over time. To link cycles of these assessments psychometrically, trend items must be included. As the PISA and PIAAC assessments moved from paper-and-pencil to computer- and tablet-based delivery, this process presented new challenges. The key when moving paper-based trend items to a technology-based platform is to ensure that the cognitive demands of the tasks do not change and that the performance of the items remains stable. Therefore, in the case of both PISA 2015 and PIAAC Cycle 1, when paper-and-pencil items were first moved to the computer, considerable efforts were made in terms of the display and functionality of items to ensure the performance of those items would not change. Additionally, the assessment design was developed to support mode effect studies that were conducted in order to confirm that the overwhelming majority of the item parameters would remain invariant and, thus, ensure comparability with performance in previous paper-based cycles.

It should be noted that the following discussion is historical in nature as it focuses on decisions made in 2012 for the PISA items that were being transitioned from paper to computer delivery in the 2015 cycle. Because these items originated in a paper environment, they did not require the kinds of skills that are necessary in interactive online contexts. Such skills include being able to locate information using search engines, evaluating the credibility of sources, and synthesizing information across multiple sources.[14] In the context of PISA 2015, these were static texts being displayed on a computer screen. As such, the focus at the time was on issues around screen layout, response modes, and scoring with the goal of maintaining trend measurement across modes.

[14]See a discussion of these "new literacies," many of which were the focus of new items development in PISA 2018 and in PIAAC, in Leu et al. (2020).

2.3.1 General Screen Layout

PISA's 2015 cycle serves as an example of the kinds of considerations that were required when moving from a primarily paper-based delivery mode that included optional computer-based modules to a fully computer-delivered assessment. Recall that, in the 2015 cycle, paper-and-pencil items in all three domains of scientific, reading, and mathematical literacy were moved to computer delivery, and new computer-based items were developed for the domain of scientific literacy.

The first challenge was to develop the overall screen design and general interface. This work was done at the beginning of the development cycle in 2012. The goal was to create a design that was forward looking, while still ensuring that PISA could continue to provide reliable trend data. This meant that the design needed to support the range of display options and interaction modes required by new, innovative items, while also facilitating the display of paper-and-pencil trend items being moved onto the computer.

The screen design for PISA 2015 needed to be forward thinking in the short term, given that items authored in 2012 would not be delivered until 3 years later, as well as in the longer term, as some items would be moved forward as trend items in the 2018 cycle and beyond. While rapidly changing technology environments can be difficult to predict, it was clear at the time that computer displays were moving towards wider screen formats. From a design perspective, this meant that horizontal space was becoming increasingly more plentiful relative to the available vertical space.

Moving from a paper display to the computer is akin to moving from a portrait display, which is taller than it is wide, to a landscape display, which is wider than it is tall. As shown in Fig. 2.14, text taken from a paper-and-pencil format and presented unchanged on the computer screen would not fit in the available vertical space and would not take advantage of the available horizontal space. This would result in an inefficient use of screen space, with blank borders on the left and right, and the need for scrolling to view all the text.

2.3.2 Display of Longer Texts

One of the factors considered in the design for PISA 2015 was the extent to which scrolling might be required to view some of the longer texts that had been used in the paper-based assessment. Reviews of research at that time suggested that a scrolling format had some negative impact on reading comprehension. For example, in three studies comparing reading comprehension between a scrolled and paged format, results indicated that a scrolling format reduced comprehension, especially for readers with lower working memory capacity (Higgins et al., 2005; Pommerich, 2004; Sanchez & Wiley, 2006). Another earlier study comparing paper-and-pencil to a computer-based display suggested that scrolling interfered with students' ability to

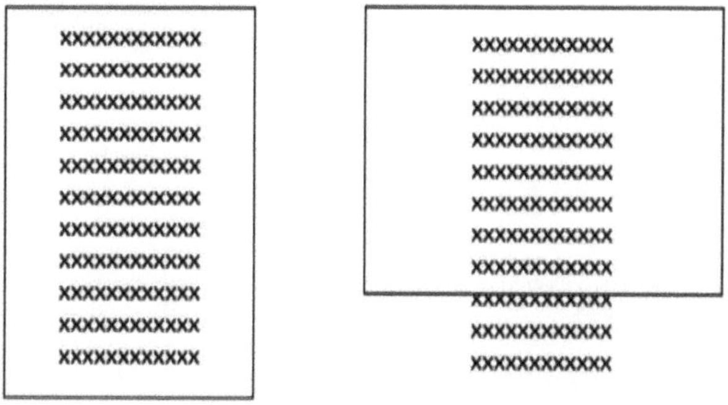

Fig. 2.14 Text on paper (left) versus the same text on a computer display (right)

scan for key words and cued phrases (Choi & Tinkler, 2002). Finally, a literature review that synthesized the results of 81 studies comparing computer-based versus paper-and-pencil multiple-choice tests for primary and secondary school students suggested that scrolling does not allow a student to use spatial memory clues when reading, which can negatively impact performance (Kingston, 2009).

One way to display text on a computer screen and limit the amount of scrolling required would have been to create a screen design that employed longer line lengths. While this was possible, given the width of computer screens, research from that time showed that, in general, longer line lengths impact the readability of text. For example, a study by Ling and van Schaik (2005) showed that line length can make a difference in a student's ability to retrieve information. Dyson and Haselgrove (2001) found that medium line lengths tend to support effective reading as measured by responses to reading comprehension questions. While specific recommendations in terms of the number of characters per line tended to vary across studies, in general, line lengths on the computer that were consistent with those on paper tended to be most readable and most preferred by study participants.

In the PISA 2015 cycle, all of the reading literacy items consisted of materials that had previously been administered in a paper-and-pencil format. Because reading was a minor domain in 2015, these items functioned as trend items and maintaining trend measurement across modes was of utmost importance. A number of stimulus materials were quite long and sometimes presented on more than a single page in the paper-based booklets. Based on relevant research at the time and, following consultations with members of the reading expert group, the decision was made to present longer texts on static screens with a paging interface that allowed students to move from page to page throughout the text. It was felt that this would be more similar to the original paper format and eliminate the need to scroll through longer texts. Decisions about where to split the text across pages in the computer versions were driven by the need to keep the presentation as similar as possible to the paper-based display and to ensure that the translations across all languages would have the same information displayed on each page. Figure 2.15 shows the paging display that was used. The assessment data ultimately showed that this design did not impact

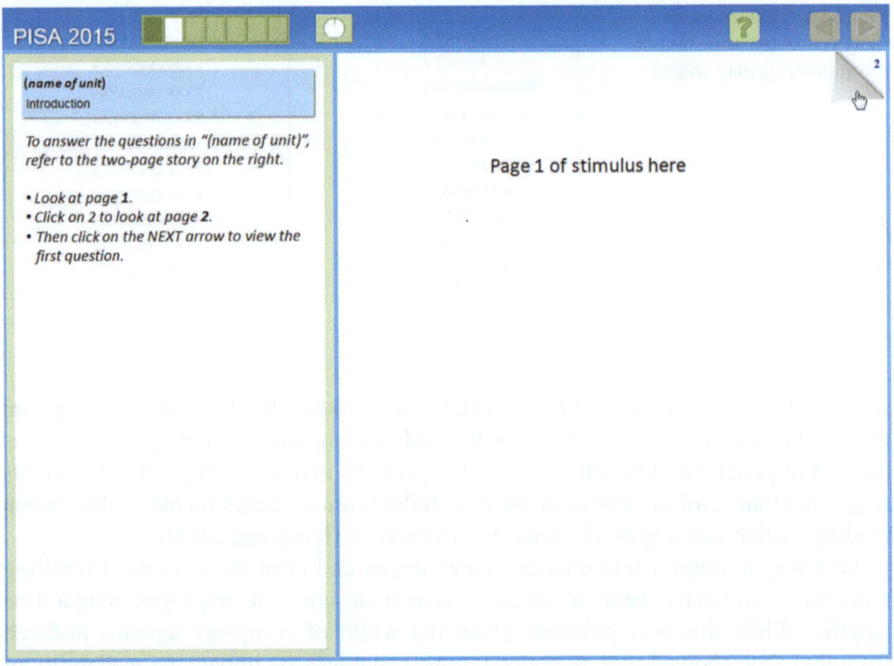

Fig. 2.15 PISA 2015 paging navigation

item performance, supporting the psychometric link for the trend items in the 2015 cycle.[15]

When designing this interface, a number of safeguards were included to ensure that students saw all the pages in each unit and understood how to navigate among them.

- Students were introduced to the paging interface in the tutorial administered prior to the assessment.
- Prior to encountering the first question for any stimulus that spanned more than a single screen, students were instructed to click on each page of the stimulus, as shown in the directions on the left pane in Fig. 2.15.

[15]Note that the 2018 cycle of PISA included a new framework for reading literacy and new item development. Given the evolving focus on reading in digital environments, the 2018 reading expert group believed that it was important to represent digital forms of reading in authentic formats. These experts, therefore, supported a display that included scrolling for new items, where maintaining trend measurement was not an issue. Of the units selected for the main study, the texts that required scrolling averaged just over 400 words and, at most, only required scrolling to view a paragraph or two. Usability studies were conducted to investigate whether students had difficulty switching between the paging interface for the trend items that continued to be used in 2018 and the new items introduced in this cycle. Those studies found that, with the support of the domain tutorials, students were able to successfully use both displays.

- The "NEXT" button did not become active until students had clicked on each page. Thus, students could not proceed to the first question in the unit until they had viewed each page in the stimulus.
- Each turned down page corner was animated so it moved a bit when students hovered the cursor over it. This animation was included to further draw students' attention to the paging display.

2.3.3 Vertically Split Screen Design

In recognition of changing displays and the interest in reducing both scrolling and long line lengths, a vertically split screen design, similar to the one used in the first cycle of PIAAC in 2012, was adopted for PISA. As shown in Fig. 2.16, the directions and question were displayed in the left pane while the stimulus was on the right. This display was reversed for countries with right-to-left languages. The design was flexible in that the relative width of the two panes could be adjusted across units to optimally display unit-specific content.

Splitting the screen vertically resulted in a display area for stimulus materials that was taller than it was wide. Because this more closely mirrored the width-to-height ratio of paper, this was an advantage for displaying the paper-based trend items

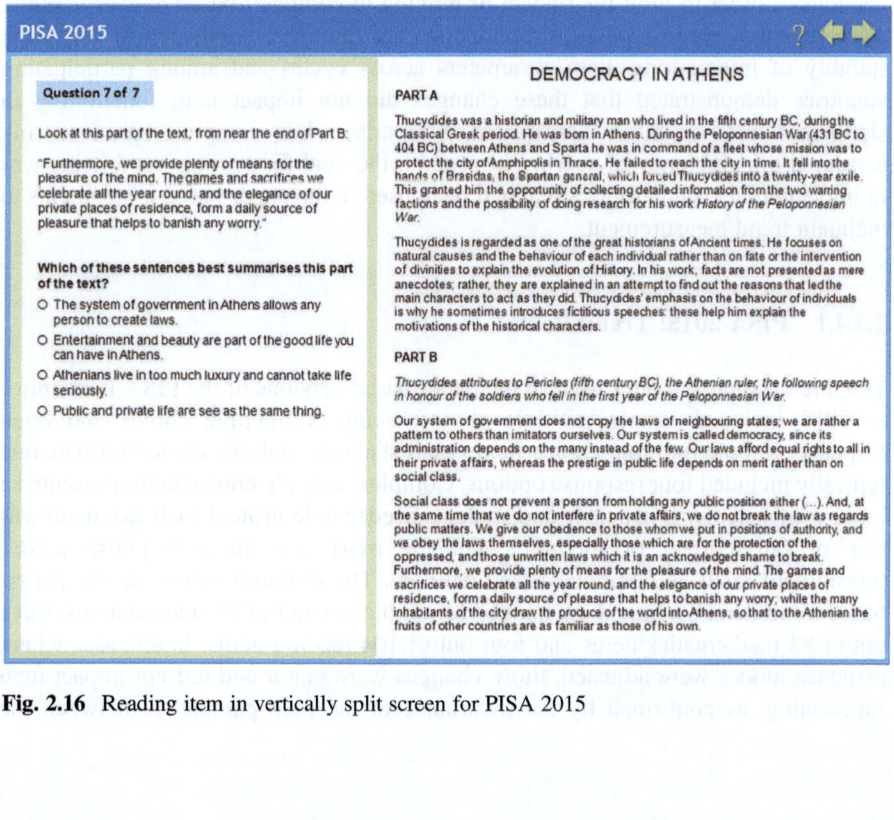

Fig. 2.16 Reading item in vertically split screen for PISA 2015

included in all the PISA 2015 legacy domains. Figures 2.16 and 2.17 show the same item in both a vertically and horizontally split screen, illustrating that the vertically split design eliminates the need for scrolling in this case.

In summary, a challenge for the PISA 2015 computer-based assessment was to develop an interface and screen design that would allow trend items to be displayed in ways that were similar to those used in the existing paper-based assessment. The paging interface that was used for longer texts was one way that challenge was addressed. Additionally, the vertically split screen design achieved a number of important goals and has been used in subsequent PISA cycles as well as both cycles of PIAAC. The design allowed text to be formatted using more natural line lengths, resulting in improved readability. It also accommodated displays across the range of languages used in these international assessments. Finally, in the case of trend items, the design facilitated the display of paper-and-pencil items being moved onto the computer and, based on the mode effect studies and assessment data, resulted in stable item parameters across modes.

2.3.4 Maintaining Trend Response Modes and Scoring

The response modes for many of the trend items in PISA 2015 and PIAAC Cycle 1 were, quite intentionally, kept as similar as possible to their paper-and-pencil versions in order to limit the impact of moving to computer-based delivery. While some revisions were required, the mode studies that were conducted to verify the stability of international item parameters across cycles and among participating countries demonstrated that these changes did not impact item functioning in significant ways in either assessment. Similarly, while computer-based scoring increased the efficiency of scoring in general, the goal for scoring trend items was to implement the same rubrics that had been used in the paper-based assessments to maintain trend measurement.

2.3.4.1 PISA 2015: Trend Items

The three response modes used in the paper-based versions of the PISA items prior to 2015 included simple multiple choice, complex multiple choice and open response. Simple multiple-choice items used standard multiple-choice formats that typically included four response options. Complex multiple-choice items presented a number of statements in a table and students needed to indicate if each statement was true or false/correct or incorrect. For the most part, these response modes transitioned easily to computer-based delivery. The response modes for the paper-based trend items remained unchanged for all but one out of 85 science items, three out of 83 mathematics items and four out of 104 reading items. In all cases where response modes were adjusted, those changes were minor and did not impact item functioning, as confirmed by the invariance of the item parameters between the

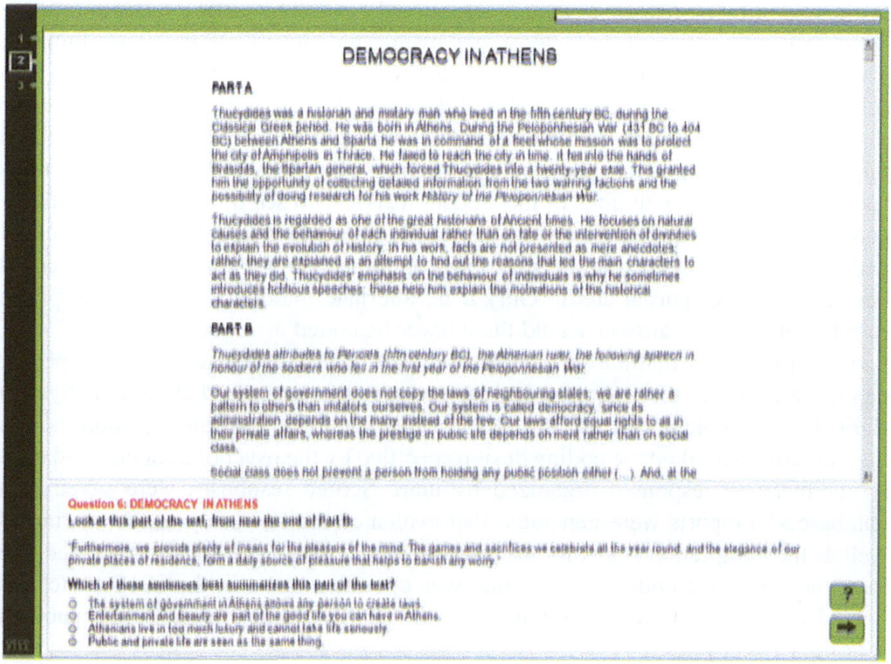

Fig. 2.17 Reading item in horizontally split screen

paper- and computer-based versions. Examples of these adjustments included changing a paper item that asked students to circle one of eight provided choices to a computer item where students needed to click on one of the eight choices presented as a multiple-choice options; revising a paper item where students were asked to write down a day of the week to one where all 7 days of the week were listed and students were asked to click on their selection; and updating a paper items where students colored in a bar on a graph to one where they clicked on a bar and it highlighted.

When implemented in the computer platform, a majority of PISA items, including multiple-choice, drag and drop, and numeric entry items, could be automatically scored. For standard response modes, such as multiple choice or numeric entry, automated scoring was performed using a rule-based system. The correct answer (or partially correct answers in the case of partial credit items) were defined based on Boolean rules defined in a custom syntax. Simple conditionals were possible, e.g., to support different combinations of checkboxes in a multiple selection item where two out of three correct options should be selected. For numeric response items, the rules could check for string matches, which required an exact match against a known correct answer, or numeric matches, which used numeric equivalence to check an answer. For example, 34.0 would match 34 if the correct response was defined as a numeric match, but not if it were defined as a string match. A challenging part of evaluating numeric responses in an international context like PISA was determining how to parse the string of characters typed by the student and interpret it as a number. There are differences in decimal and thousands separators that must be taken into account based on conventions used within countries and local usage. Use of these separators is not always consistent within a country, especially with increased migration and the pervasiveness of the internet. For PISA 2015, the coding rules tested multiple interpretations of the student response to see if one of them could be coded as correct. The numbers were parsed in different ways, changing the decimal and thousands separators, and testing each option to see if a correct response could be granted full or partial credit. Only if no alternate interpretation of the response resulted in a correct answer would the answer be coded as incorrect.

For PISA items that allowed an open-ended response, where students typed or entered an answer that required human scoring, the Open-Ended Coding System (OECS) was implemented for the 2015 cycle. The OECS assigned responses to a human scorer based on the coding design specified by the psychometricians and each scorer received responses organized by item. Scored responses were saved in a database and reports were generated that evaluated the reliability of the scoring as well as the completion status for each scorer and item. As previously discussed, the machine-supported coding system that was operationalized in the 2018 cycle, has helped reduce the burden for human scoring, while maintaining consistency of scoring across cycles.

2.3.4.2 PIAAC Cycle 1: Trend Items

In contrast to PISA, an important requirement in PIAAC Cycle 1 was that all items be computer scored. This was a desired feature in order to implement adaptive testing but had important implications for the response modes implemented for trend items and how they could be scored. The PIAAC trend items were taken from previous paper-and- pencil adult assessments including the International Adult Literacy Survey (IALS) and the Adult Literacy and Lifeskills Survey (ALL). Each of the items in these surveys was open-ended and the goal for the computer-based instruments was to retain this response mode both to maintain the real-life focus of the assessment and to maintain the psychometric link between PIAAC and prior surveys. While those prior paper-and-pencil surveys allowed respondents to write responses ranging from a word or two to several sentences, the use of automated scoring for such responses was not possible for PIAAC given that the assessment was to be delivered in over 30 languages.

Previous ETS work on a derivative computer-based test for individuals had defined a set of computer-scoreable, open-ended response modes for items that were taken from, or similar to, those used in IALS and ALL. This work had shown that item parameters for the paper-and-pencil items were not impacted when those items were adapted to allow respondents to indicate their answers by clicking on the stimulus, typing numeric answers, and highlighting responses in text. Additionally, each of these three response modes required only basic computer skills—an important consideration given that the test needed to be accessible to adults with a range of computer experience. Development for PIAAC therefore proceeded on the assumption that trend items could be adapted to employ these three response modes and still maintain item parameters from previous assessments, an assumption that was ultimately supported by the field trial data.

Both from a test development and national implementation perspective, the highlighting items were most challenging as responses were language dependent. These items allowed respondents to highlight one or more words, phrases and sentences in a text to answer questions. For each response, developers defined a minimum correct response, as well as a maximum correct response. They based those judgments on ETS's previous work to develop open-ended, computer-scoreable items, as well as experience in scoring paper-based responses. In the paper-based IALS and ALL assessments, respondents were given credit for correct answers when they underlined or circled information in the stimulus instead of writing an answer on a response line. Existing rules for what constituted a correct response in those situations thus helped guide the development of rules for highlighted responses in PIAAC. As this was the most language-dependent response mode, countries were actively involved in implementing and testing the minimum/maximum rules for their national versions of these item types. It should be noted that, while the highlighting items served an important purpose in linking the first cycle of PIAAC with the IALS and ALL surveys, this response mode was not used for any of the new items developed in the second cycle given the time required to implement and test the scoring for this language-dependent response mode. PIAAC is, nevertheless, an example of how trend measurement can be enhanced via the move to computer-based implementation of automatic scoring. Such scoring supports the use of adaptive testing algorithms, improves scoring reliability and reduces country burden.

2.3.5 The Continuing Role of Paper-Based Items in PISA and PIAAC

It should be noted that while all new item development for PIAAC and PISA has focused on computer-based items, PISA and PIAAC Cycle 1 maintained a paper-based component. In PISA, the paper-and-pencil booklets are available for countries that are not able or willing to test their students by computer. In PIAAC Cycle 1, booklets were available for adults who were unable or unwilling to take a computer-based test.

The paper booklets differ from the computer- and tablet-based versions of the assessments in a number of ways, including differences in the total number of items and the number of booklet versions or forms, with the computer version have many more in both instances. More important, however, is the fact that the paper-based items do not reflect the new aspects of the legacy constructs as those have been identified in more recent frameworks. In both PIAAC Cycle 1 and PISA, the paper booklets include only items that had been developed for paper-and-pencil administration in earlier cycles. In PISA, the paper booklets that were developed for the 2015 cycle, and have been used since, include items from the 2003, 2006, 2009 and 2012 cycles. In PIAAC Cycle 1, the trend items came from the IALS and ALLS surveys that had been developed in mid 1990s and early 2000s. Given their age and mode of administration, these items necessarily do not reflect the types of interactive, technology-based tasks that have been increasingly developed for, and included in, the computer-based assessments. They also do not reflect other changes in the constructs as those have evolved over time. Thus, while these items can still be put on the same scale as those administered on the computer, the results of the paper-based assessment are less directly linked to the computer-based assessment and more model based. This divergence will only increase over time, making the interpretation of results increasingly challenging.

The interest of policy makers, educators, and other stakeholders in assessing skills that are of growing importance in educational, workplace and community contexts continues to drive the focus on developing and delivering assessments that employ technology-based platforms to assess technology-based skills. This creates a healthy tension between measuring such skills and maintaining trends, which is an understandable interest of participating countries. As previously discussed, this interest led to the adaptation and inclusion of older paper-based items in the 2015 cycle of PISA and in PIAAC Cycle 1. While some of these items continue to be included in these surveys, their relative contribution to overall measurement diminishes with each cycle as they are replaced by technology-based trend items that have been developed to reflect the evolution of the legacy constructs. For example, in PIAAC Cycle 1, all 31 of the literacy trend items came from the paper-based IALS and ALL sources. In contrast, in Cycle 2, only 11 of those paper-based trend items were included and the remaining trend items were those that had been developed for the computer-based version of the literacy assessment in Cycle 1. The greater tension is for countries that use only the paper-based booklets, where the link is only to the past. Currently, there has been a slow but steady move away from relying only paper-based instruments. For the PISA 2022 cycle, there were some 88 participating countries/economies, but only five used paper only. In order to reduce the ICT demands for PIAAC, Cycle 2 used a tablet-based delivery mode. The introduction of this tablet-based version of the assessment eliminated the need for paper booklets. The field test results from that cycle showed that adults were both willing and able to complete the assessment on a tablet and participating countries, therefore, agreed that a paper-based version was not needed (OECD, Forthcoming). However, if, in the end, the use of paper-based instruments is important to include in international large-scale surveys, particularly to ensure the participation of developing countries, there may come a time when it becomes difficult to establish comparability from both a statistical and interpretive point of view and organizations may choose to report findings on separate scales.

2.4 Conclusion

Large-scale surveys must be responsive to changing times if they are to yield relevant information about the skills of the populations being assessed. The rapid evolution of digital technologies has driven the need to understand more about the extent to which individuals have the skills required to access, use and communicate information in these new environments. The frameworks developed for each of the legacy domains in PIAAC and PISA over the past decade have emphasized the importance of defining and assessing those skills. And the move to technology-based platforms has enabled a range of innovations that have helped realize that goal.

The development of innovative interfaces and environments has made it possible to situate assessment tasks in authentic contexts. By utilizing simulated websites, apps, and emails, stimulus materials can reflect real-life digital environments. The inclusion of interactive simulations and scenario-based environments that require the use of multiple sources of information support the introduction of new item types that can be used to assess the range of skills required to navigate, critically analyze and problem solve in complex digital environments.

Technology-based assessments have also made it possible to introduce innovative scoring and data capture capabilities. The implementation of real-time automatic scoring has not only contributed to less labor intensive and more reliable data collection, but it has made it possible to expand the use of adaptive testing. Capturing process information, including timing and respondent actions, helps both in evaluating the quality and fairness of the items and in increasing our understanding of the strategies respondents use, enhancing what we can learn about their knowledge and skills. Pre-defining actions of interest that are hypothesized to contribute to our understanding of how respondents with differing skill levels interact with stimulus materials and complete presented tasks has the potential to further increase the utility of these data.

Designing, developing and delivering any technology-based assessment is a complex endeavor involving multiple teams of people, each with specific areas of expertise. The complexity of the process highlights the importance of collaboration among those individuals and teams, which include domain experts, psychometricians, test developers, platform specialists, data analysts, translation specialists, and those responsible for operational aspects of a survey, as well as governing boards for the programs and the national centers of participating countries. This collaboration is all the more important when the goal is to introduce the kinds of innovation seen in assessments such as PIAAC and PISA. A tension always exists between what is desired or hoped for in an assessment versus what is possible under constraints that exist at that time. But in the best of collaborations, as has been the case in these assessments, each of the teams is committed to working towards improving the measurement of key core domains.

We have focused on the ways in which the innovations in item development that were introduced in the PISA and PIAAC legacy constructs have utilized the capabilities afforded by technology-based platforms to expand the range of skills that can be measured. In addition, these innovations have made it possible to overcome some of the challenges inherent in the transition from a paper-based to a technology-based platform. For example, it was necessary to develop a screen design and interface that allowed the first trend items to display and function in ways that reflected their paper-based origins, while also making it possible to display new items that reflect the demands of online and technology-enabled environments. Additionally, these designs needed to be sufficiently forward thinking to reduce the possibility of becoming quickly outdated, which remains a challenge in times of rapid technological change. It is undoubtedly the case that a range of technical, methodological and political challenges will continue as international large-scale assessments continue to evolve. In a recent paper, Kirsch and Braun (2020) discuss a number of such challenges. From an item development perspective, some of those important challenges include the following. The expanded use of adaptive testing will require the development of larger item pools with increased real-time scoring, which presents development, translation and operational challenges. An interest in providing assessments on a wider range of devices presents challenges related to delivery, data capture and storage, and, most importantly, comparability of measurement. Realizing the full potential of the information that technology-based assessments yield requires more extensive use of logfile data. Engaging expert groups and test developers in the process of predefining variables of interest within that data is one innovative strategy that is being explored, but more work in this area will be needed. And finally, the desire by donor organizations and others to increase the participation of low- to middle-income countries will likely increase the range of skills and abilities that need to be assessed, which will require more complex designs and more challenges associated with maintaining comparability across modes and cycles.

At the outset of this chapter, we noted that technology has made it possible to broaden both what we measure and how we measure it. It has been critical to broaden what is measured in order to be responsive to questions about whether student and adult populations have the skills they need to succeed in changing times. While the introduction of new constructs has been one way that programs such as PISA and PIAAC have responded to such questions, as has been shown, the legacy constructs have met this challenge as well. With each new cycle of development, the ways in which skills within the legacy constructs are measured have evolved to take full advantage of (and often drive) the development of new methodologies and platform capabilities. The innovations that have been introduced have broadened the relevance and utility of these surveys and increased the quality and validity of the data, strengthening the foundation upon which future surveys can continue to build.

References

Choi, S. W., & Tinkler, T. (2002). *Evaluating comparability of paper and computer-based assessment in a K-12 setting*. Paper presented at the Annual Meeting of the National Council on Measurement in Education, New Orleans, LA.

Dyson, J., & Haselgrove, M. (2001). The influence of reading speed and line length on the effectiveness of reading from screen. *International Journal of Human-Computer Studies, 54*, 585–612. https://doi.org/10.1006/ijhc.2001.0458

Higgins, J., Russell, M., & Hoffman, T. (2005). Examining the effect of computer-based passage presentation on reading test performance. *The Journal of Technology, Learning, and Assessment, 3*(4) https://ejournals.bc.edu/index.php/jtla/article/view/1657

Kendeou, P. (2020). The assessment of reading for understanding. In P. D. Pearson, A. S. Palincsar, G. Biancarosa, & A. I. Berman (Eds.), *Reaping the rewards of the reading for understanding initiative*. National Academy of Education.

Kingston, N. M. (2009). Comparability of computer- and paper-administered multiple-choice tests for K-12 populations: A synthesis. *Applied Measurement in Education, 22*(1), 22–37.

Kirsch, I. S. (2001). *The International Adult Literacy Survey (IALS): Understanding what was measured* (Research Report No. RR-01-25). Educational Testing Service. https://doi.org/10.1002/j.2333-8504.2001.tb01867.x

Kirsch, I. S. (2003). Measuring literacy in IALS: A construct centered approach. *International Journal of Educational Research, 39*, 181–190. https://doi.org/10.1016/j.ijer.2004.04.002

Kirsch, I., & Braun, H. (2020). Changing times, changing needs: Enhancing the utility of international large-scale assessments. *Large-Scale Assessments in Education*. https://doi.org/10.1186/s40536-020-00088-9

Leu, D. J., Kiili, C., Forzani, E., Zawilinski, L., McVerry, J. G., & O'Byrne, W. I. (2020). The new literacies of online research and comprehension. In C. A. Chapelle (Ed.), *The concise encyclopedia of applied linguistics* (pp. 844–852). Wiley-Blackwell.

Ling, J., & van Schaik, P. (2005). The influence of font type and line length on visual search and information retrieval in web pages. *International Journal of Human-Computer Studies, 64*(5), 395–404. https://doi.org/10.1016/j.ijhcs.2005.08.015

Messick, S. J. (1994). The interplay of evidence and consequences in the validation of performance assessments. *Educational Researcher, 23*, 13–23. https://doi.org/10.3102/0013189X023002013

Mislevy, R., Steinberg, L., & Almond, R. (1999). *Evidence-centered assessment design*. Educational Testing Service. https://doi.org/10.1016/j.pse.2014.11.003

OECD. (2013). *Technical report of the survey of adult skills (PIAAC)*. OECD Library. https://www.oecd.org/skills/piaac/_Technical%20Report_17OCT13.pdf

OECD. (2017). *PISA 2015 technical report*. OECD Library. https://www.oecd.org/pisa/data/2015-technical-report/PISA2015_TechRep_Final.pdf

OECD. (2018). PISA 2021 mathematics framework (draft). https://www.oecd.org/pisa/sitedocument/PISA-2021-mathematics-framework.pdf

OECD. (2019). PISA 2018 reading framework. In *PISA 2018 assessment and analytical framework*. OECD. https://doi.org/10.1787/5c07e4f1-en.

OECD. (2020). *PISA 2018 technical report*. OECD. https://www.oecd.org/pisa/data/pisa2018technicalreport/

OECD. (Forthcoming). *Technical report of the survey of adult skills 2023 (PIAAC)*. OECD Library. https://www.oecd.org/en/about/programmes/piaac/piaac-publications.html

Ou, L., Kerzabi, E., Rosen, Y., & von Davier, A. (2025). New constructs and item types in large-scale assessments. In L. Khorramdel, M. von Davier, & K. Yamamoto (Eds.), *Computer-based international large-scale assessments: Concepts, methodologies and quality assurance procedures*. Springer.

Pommerich, M. (2004). Developing computerized versions of paper tests: Mode effects for passage-based tests. *The Journal of Technology, Learning, and Assessment, 2*(6), 1–44. https://www.learntechlib.org/p/106318/

Sabatini, J., O'Reilly, T., Halderman, L., & Bruce, K. (2014). Broadening the scope of reading comprehension using scenario-based assessments: Preliminary findings and challenges. *International Journal Topics in Cognitive Psychology, 114*, 693–723. https://doi.org/10.3917/anpsy.144.0693

Sabatini, J., O'Reilly, T., Weeks, J., & Wang, Z. (2019). Engineering a 21st century reading comprehension assessment system utilizing scenario-based assessment techniques. *International Journal of Testing*. https://doi.org/10.1080/15305058.2018.1551224

Sanchez, C. A., & Wiley, J. (2006). An examination of the seductive details effect in terms of working memory capacity. *Memory & Cognition, 34*(2), 344–355. https://doi.org/10.3758/bf03193412

Weeks, J., von Davier, M., & Yamamoto, K. (2014). Design considerations for the programme for international student assessment. In L. Rutkowski, M. von Davier, & D. Rutkowski (Eds.), *Handbook of international large-scale assessment* (pp. 259–276). Chapman & Hall.

Wise, S. L., & DeMars, C. E. (2005). Low examinee effort in low-stakes assessment: Problems and potential solutions. *Educational Assessment, 10*(1), 1–17. https://doi.org/10.1207/s15326977ea1001_1

Yamamoto, K., He, Q., Shin, H. J., & von Davier, M. (2017). *Developing a machine-supported coding system for constructed-response items in PISA* (ETS Research Report No. RR17-47). Educational Testing Service. https://doi.org/10.1002/ets2.12169

Yamamoto, K., He, Q., Shin, H. J., & von Davier, M. (2018). Development and implementation of a machine-supported coding system for constructed-response items in PISA. *Psychological Test and Assessment Modeling, 60*(2), 145–164. https://www.psychologie-aktuell.com/fileadmin/download/ptam/2-2018_20180627/02_PTAM-2-2018_Yamamoto_v2.pdf

Yamamoto, K., & Lennon, M. (2018). Understanding and detecting data fabrication in large-scale assessments. *Quality Assurance in Education: An International Perspective, 26*(2), 196–212. https://doi.org/10.1108/QAE-07-2017-0038

Yamamoto, K., Shin, H. J., & Khorramdel, L. (2018). Multistage adaptive testing design in international large-scale assessments. *Educational Measurement: Issues and Practice, 37*(4), 16–27. https://doi.org/10.1111/emip.12226

Yamamoto, K., Shin, H. J., & Khorramdel, L. (2019). *Introduction of multistage adaptive testing design in PISA 2018* (OECD Education Working Papers 209). OECD. https://doi.org/10.1787/19939019

Chapter 3
Innovative Digital Reading Assessment for Fourth Grade Students: IEA's PIRLS 2021

Ina V. S. Mullis and Michael O. Martin

Abstract IEA's PIRLS (Progress in International Reading Literacy Study) provides trends in reading achievement at the fourth grade for 60-plus countries around the world. Assessed every 5 years since 2001 in a paper-and-pencil format, PIRLS in 2021 transitioned to a digital assessment in about half of the countries. These countries will participate in a unified digital assessment that integrates digitalPIRLS, a digital version of the PIRLS 2021 assessment presented through an innovative user interface, and ePIRLS, an assessment of online informational reading first administered in 14 countries on a voluntary basis in 2016. The remaining countries will administer PIRLS 2021 in paper-and-pencil format, as they did in previous assessment cycles. PIRLS 2021 will keep pace with the changing nature of fourth graders' reading experiences, capture information (log data) about students' reading strategies, and introduce operational efficiencies into the complex endeavor of an assessment conducted across thousands of schools internationally and in multiple languages. This chapter addresses the development of digitalPIRLS and ePIRLS undertaken by the TIMSS & PIRLS International Study Center at Boston College, including a discussion of the challenges associated with a transition to a digital assessment—including the development of items and a digital testing platform—and the opportunities a digital reading assessment presents.

IEA's PIRLS (Progress in International Reading Literacy Study) provides trends in reading achievement at the fourth grade for 60-plus countries around the world. First administered in 2001 and conducted every 5 years since, PIRLS 2021 is the fifth in this series of international reading assessments, providing 20 years of trends in

This chapter was submitted in May, 2021, before the international report and results for PIRLS 2021 were published.

I. V. S. Mullis (✉) · M. O. Martin
IEA's TIMSS & PIRLS International Study Center at Boston College, Chestnut Hill, MA, USA
e-mail: ina.mullis@bc.edu

L. Khorramdel et al. (eds.), *Innovative Digital-Based International Large-Scale Assessments*, Methodology of Educational Measurement and Assessment,
https://doi.org/10.1007/978-3-031-90951-1_3

comparative reading achievement across countries. For PIRLS 2021, the TIMSS & PIRLS International Study Center at Boston College, which directs PIRLS, is focused on the transition to a digital assessment system. digitalPIRLS will include a variety of reading texts presented in an engaging and visually attractive format that will motivate students to read and interact with the texts and improve student response rates. Countries that administer digitalPIRLS can take advantage of the benefits of a computer-based assessment, including greater operational efficiency in translation and translation verification, data entry, and scoring, without the need for printing or shipping.

The PIRLS 2021 computer-based assessment features two approaches to assessing reading at the fourth grade: digitalPIRLS and ePIRLS. digitalPIRLS uses a new user interface to assess reading comprehension across a wide variety of reading materials that students are likely to encounter in and out of school. ePIRLS assesses how well students can read online information: Guided by a teacher avatar, students navigate through a simulated internet environment to obtain information and answer questions.

This chapter describes the challenges and processes in evolving from a paper-and-pencil to a computer-based international large-scale assessment. This involved developing a new digital user interface, conducting an international search for appropriate reading passages, and converting the selected stories and articles, many beautifully illustrated, to a digital format. New systems were developed to import, translate and deliver the passages and items to the students, and collect and score their responses. The chapter also describes how paper-based PIRLS 2016 pioneered the ePIRLS assessment of online reading, including the assessment approach and a summary of results. This was an important stepping stone to a completely digital PIRLS 2021.

3.1 The PIRLS 2021 Reading Assessment Framework

The PIRLS reading assessment framework (Mullis & Martin, 2019) has been updated for each assessment cycle to keep PIRLS on the cutting edge of reading assessment. This has included incorporating measurement of reading processes; extending the range of difficulty to address a wider breadth of student achievement; addressing the increasing role of online reading; and now, in a completely digital assessment maintaining relevance with students' reading experiences in- and out-of-school, relating reading strategies to reading achievement, and increasing operational efficiency.

In the PIRLS reading assessment framework, reading is viewed as a constructive and interactive process, whereby meaning is constructed through the interaction between reader and text in the context of a particular reading experience (Kintsch, 2012). To reflect fourth grade students' reading experiences in and out of school, PIRLS makes every effort to select universally applicable reading materials from a variety of print and online sources. The transition to digital assessment provides a

Table 3.1 Percentages of the PIRLS and ePIRLS reading assessments devoted to each reading purpose and comprehension process

	PIRLS (%)	ePIRLS (%)
Purposes for reading		
Literary experience	50	0
Acquire and use information	50	100
Processes of comprehension		
Focus on and retrieve explicitly stated information	20	20
Make straightforward inferences	30	30
Interpret and integrate ideas and information	30	30
Evaluate and critique content and textual elements	20	20

Note. ePIRLS is denoted separately in the framework because as an assessment of online reading, it only is assessed as part of digitalPIRLS and does not have a paper counterpart

further opportunity for PIRLS to present students with reading material that is consistent with the kind that young readers may encounter in their everyday lives. Students are exposed to a wide variety of multimodal reading materials: picture books that contain elaborate images and informational articles and websites that have photos, diagrams, and maps, as well as dynamic features such as animations, videos, hyperlinks, and pop-up boxes. The PIRLS reading materials and experiences are designed to give students opportunities to demonstrate a repertoire of linguistic skills and cognitive and metacognitive strategies and use background knowledge to actively construct meaning from the texts, along with associated visual images and to provide responses to the assessment questions.

As shown in Table 3.1, the PIRLS 2021 framework focuses on the two overarching purposes for reading that account for most of the reading done by young students in and out of school: reading for literary experience (narrative texts that tell a story) and to acquire and use information (texts that convey facts, concepts, processes, and discoveries). In addition, PIRLS integrates four broad-based reading comprehension processes within each of the two reading purposes: focus on and retrieve explicitly stated information, make straightforward inferences, interpret and integrate ideas and information, and evaluate and critique content and textual elements.

For each assessment cycle, the PIRLS framework, reading materials, and assessment questions are products of a lengthy, collaborative effort between the TIMSS & PIRLS International Study Center and the participating countries. The countries' National Research Coordinators (NRCs) participate in updating and reviewing the framework, identifying and extensively reviewing the reading passages, writing and reviewing the assessment items, translating the assessments into the languages in their countries, and adhering to rigorous data collection standards and quality control procedures.

3.2 digitalPIRLS

Maximizing the assessment opportunities provided by a digital environment requires capitalizing on its benefits and addressing its challenges. A digital reading assessment can motivate students with engaging, interactive materials, collect data about the reading comprehension skills and strategies student use, and when and how effectively they use them, and provide some operational efficiencies. However, the underlying software systems are very complex and resource-intensive to develop and document. PIRLS also has additional challenges associated with conducting digital fourth grade reading assessments in an international setting: accommodating the alphabets of many written languages, some of which use right-to-left formats, and the lack of standard internet and digital devices across—and even within—countries.

3.2.1 Adapting to the Constraints of Digital Devices

The PIRLS assessments are administered to students in school settings, so development of digitalPIRLS had to assume that countries would utilize equipment and resources available in schools as much as possible. If countries use devices that schools already have, this increases the likelihood that students will be familiar with the devices, but it also may increase the diversity of devices from school to school and result in operational problems. Countries can buy or rent equipment to use to administer PIRLS. This improves standardization but increases costs and perhaps requires training to familiarize students with the device. Regardless of the approaches adopted by the countries, digitalPIRLS 2021 runs on a wide variety of computers and tablets, with smaller devices becoming more popular.

Because the PIRLS 2021 assessment framework requires presenting students with a wide variety of materials typical of those read in their everyday experiences in and out of school, the reading passages are drawn from existing books, articles, and websites and are written by successful authors who understand writing for a young audience. The reading passages are provided and reviewed by the participating countries to ensure they are representative of the literary and informational materials their students read. To ensure a range of difficulty, the reading passages span approximately 500–800 words, with the stories supported by colorful illustrations and the informational articles supported by visuals, such as photos, diagrams, charts, and graphs.

PIRLS has included such authentic materials since the initial 2001 assessment, but books for young students increasingly have a higher proportion of pages devoted to beautiful and engaging illustrations than to text. This change could be due to publishers' attempts to bridge the gap between traditionally text-heavy classroom materials and the internet's multimodal feast of visuals (Anstey & Bull, 2006).

For digitalPIRLS, this posed the challenge of presenting reading passages with more pictures and graphics together with the assessment questions on the small screen of a digital device. The digitalPIRLS goal of making reading passages as inviting and engaging as the materials in students' books, magazines, and digital devices meant devoting as much of the available screen space as possible to the text and accompanying images in a reading passage.

But what about the assessment questions? The PIRLS framework specifies that each student be given two reading passages, each with approximately 15 questions in a variety of formats: about half selection-type and half constructed-response (at least one requiring a longer answer). Having approximately 30 items per student provides a solid basis for the psychometric achievement scaling methods used to calculate national estimates of student achievement in the participating countries (Foy & Yin, 2017).

In the paper-and-pencil PIRLS assessments, typically each literary story or informational article is presented and then followed by its questions; another format interleaving the questions on the pages facing the relevant text was developed for and used in countries with less proficient readers. However, neither of these formats was suitable for digitalPIRLS. The small screen size would mean more screen pages to see an entire passage compared to the paper version, and then the questions would be placed much further from the text in the digital version than in the paper format. Also, squeezing an authentically reproduced reading passage and the questions side-by-side onto a single screen was impossible.

digitalPIRLS needed a new way for students to navigate between a reading passage and questions, and that method needed to be at least as easy as paging back and forth through a paper booklet.

3.2.2 Developing the digitalPIRLS User Interface

According to Constantine and Lockwood's (1999) usage-centered design published two decades ago, a good design of a user interface (UI) provides all options and materials necessary to accomplish the given task visible in the simplest way possible, without distracting the user. This advice, excellent but difficult to follow, is especially important in developing a user interface for fourth grade students with a broad range of experience in using computers and tablets: from extensive to hardly any at all. The digitalPIRLS UI needed to facilitate achievement on the reading assessment, not hamper it.

Throughout its development, the digitalPIRLS UI was subjected to a series of external reviews, beginning with the first meeting of the PIRLS 2021 National Research Coordinators (NRCs) from the participating countries. Figure 3.1 shows the initial prototype digitalPIRLS UI proposed to the NRCs, using a passage from PIRLS 2016 about sea turtles as an example. On the left there is a timer and a way for students to monitor their progress on the items via a column of numbers that change color when a question is completed. Students can navigate through the reading

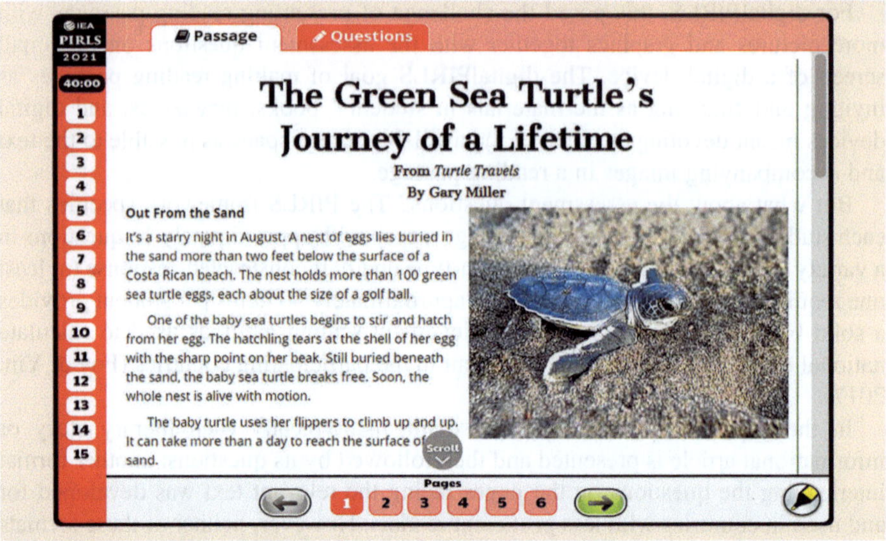

Fig. 3.1 Initial prototype for the digitalPIRLS UI

passage using the scroll bar on the right, or the arrows and page numbers at the bottom of the screen. A highlighter tool is available in the lower right corner should students want to use it as they read. The tabs at the top of the screen enable toggling between the reading passage and the questions. However, accessing the questions makes the passage disappear and vice versa.

In retrospect, not surprisingly, the PIRLS NRCs gave this proposal a lukewarm reception. At the most negative, one NRC hypothesized that the somewhat cluttered screen and cumbersome navigation between the passage and questions would make the digitalPIRLS assessment much more difficult for fourth grade students to manage than the existing paper format.

Fortunately, the software design staff at the TIMSS & PIRLS International Study Center eventually met the challenge of developing a more streamlined and easier-to-navigate user interface. After several more iterations, including a pilot in two English-speaking countries and cognitive interviews with fourth grade students, digitalPIRLS had a new interface that was well received by the various PIRLS constituencies: reading specialists, measurement specialists, and the PIRLS NRCs.

Figure 3.2 shows the new digitalPIRLS UI applied to the same PIRLS 2016 passage about sea turtles. The new interface exemplifies the simplicity principle and is well-suited for an international population of fourth grade students, some of whom may have little familiarity with digital devices. This streamlined design makes it possible to have both the reading passage and the questions readily available, without distracting students with extra information.

Most of the screen is devoted to presenting the passage text together with its supporting images, in this case a map. The scroll bar on the right can be used to scroll or click through pages using the darker rectangles, and there are up and down

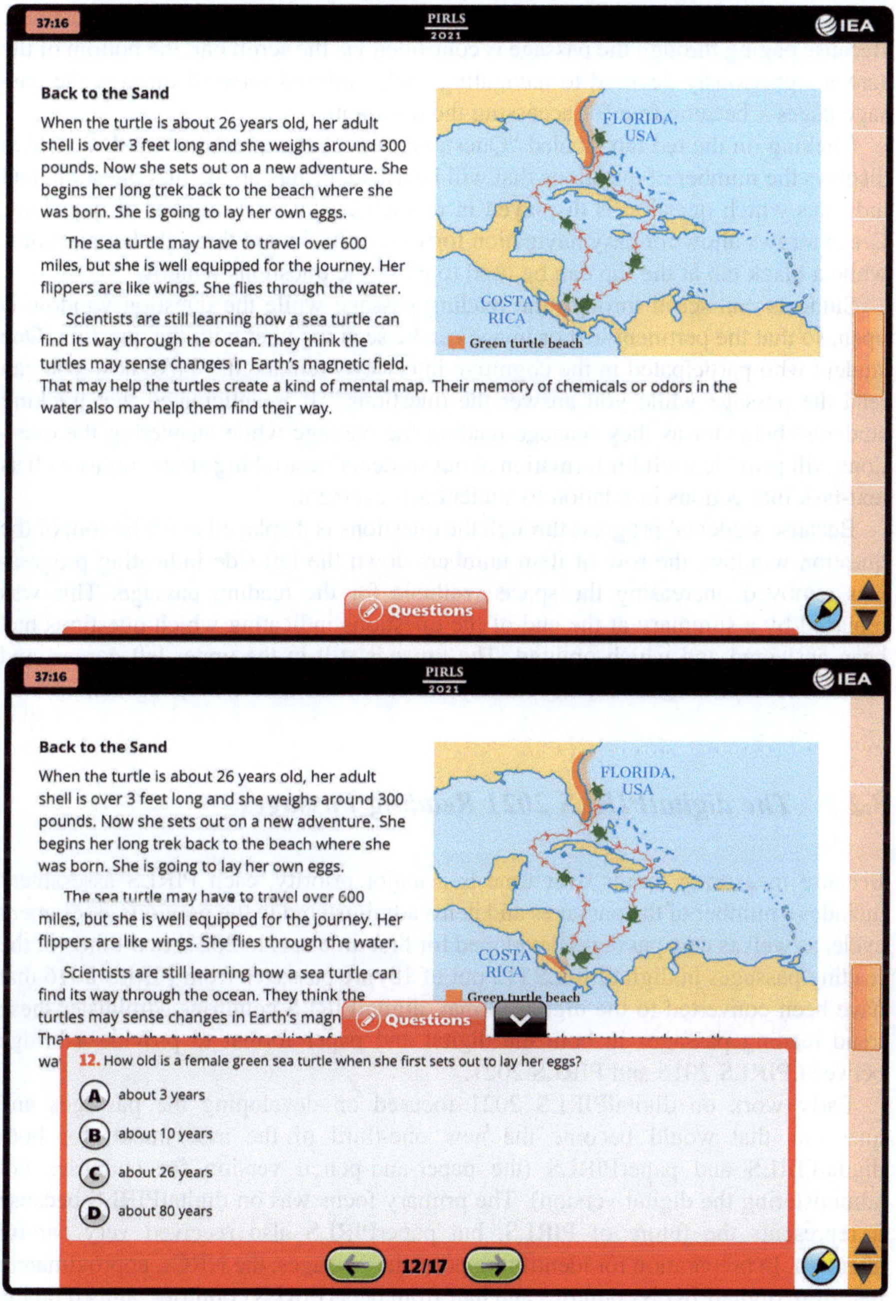

Fig. 3.2 Improved digitalPIRLS UI

navigation arrows. The highlighter tool from the initial prototype was also retained. Because paging through the passage is controlled via the scroll bar, the bottom of the screen—previously devoted to navigating backward and forward through the passage pages—became free for accessing the questions.

Clicking on the red tab labeled "Questions" opens the questions window, which displays the number of questions that will be asked, brings up the first question, and indicates which question is displayed in relation to the total number of questions. Green arrows allow for easy navigation forward or backward through the questions, while a black tab at the top can be used to close the questions window.

Students can scroll through the reading passage while the question window is open, so that the pertinent text or image can be seen together with the question. One student who participated in the cognitive interviews remarked, "I like how you can read the passage while you answer the questions." It is anticipated that tracking students' behavior as they manage reading the passage while answering the questions will provide useful information about students' test-taking strategies as well as text-task interactions in relation to student achievement.

Because students' progress through the questions is displayed at the bottom of the question window, the row of item numbers down the left side indicating progress was removed, increasing the space available for the reading passage. This was replaced by a summary at the end of the questions indicating which questions had been answered and which omitted. The timer is still in the upper left corner, and students receive a notification about any omitted items before they log out.

3.2.3 The digitalPIRLS 2021 Reading Passages

Because measuring trends over time is a major priority, each PIRLS assessment includes a number of the passages and items administered in the previous assessment cycle, as well as new passages developed for first time use. In 2021, two-thirds of the reading passages in digitalPIRLS (12 out of 18) are passages from PIRLS 2016 that have been converted to the digital format. digitalPIRLS countries administer these trend reading passages in both the digital and paper format to provide a bridge between PIRLS 2016 and PIRLS 2021.

Early work on digitalPIRLS 2021 focused on developing the passages and questions that would become the new one-third of the assessment, for both digitalPIRLS and paperPIRLS (the paper-and-pencil version for countries not administering the digital version). The primary focus was on digitalPIRLS because it represents the future of PIRLS, but paperPIRLS also received very careful attention. In preparation for identifying potential passages, the NRCs, approximately half from digitalPIRLS countries and half from paperPIRLS countries, agreed on the passage characteristics necessary for PIRLS 2021. To be suitable for use in PIRLS, passages are required to meet criteria for coherence, have a reading difficulty level appropriate for fourth grade students, and contain suitable, interesting content, free of cultural bias. The passages also need to be colorful and visually engaging for digitalPIRLS, while remaining suitable for printing in booklets for countries using the paper format. Because the PIRLS 2021 digital and paper results are reported on the same PIRLS reading achievement scale, the passages in both formats have essentially the same text and supporting graphics and very similar layouts. In several cases, the digital version of an informational passage includes an interactive feature: The graphic has some animation, the diagram is presented via a pop-up rather than appearing on the next page, or vocabulary definitions are accessed through hyperlinks instead of appearing in text boxes.

The development plan called for 12 new reading passages and items—twice as many as needed for the actual PIRLS 2021 assessment. These would be validated by an international field test with large, nationally-representative samples of students from each country to ensure having six passages of very high quality for inclusion in the 2021 assessment. Identifying, reviewing, revising, and formatting and developing items for 12 new digitalPIRLS passages for the field test was a long process.

The search for passages began with an online discussion board where the NRCs could post and discuss prospective passages during April–September 2018. The TIMSS & PIRLS International Study Center identified the passages that were well-liked by most of the NRCs. Then, together with the PIRLS 2021 Reading Development Group (RDG) composed of international experts nominated by the participating countries, staff sorted out the reading passages that were engaging and culturally appropriate across countries and provided good coverage of the assessment framework. The NRCs further reviewed these passages, eliminating several of them, and then used the remaining reading passages as the basis of an item-writing workshop. Following the workshop, the reading passages and items were further reviewed by the RDG and the NRCs, and the required 12 passage/item sets were selected for field testing. The reading passages developed for field testing, all produced in visually attractive and engaging and matching digital and paper formats, represent a variety of stories (i.e., fables, historical events, contemporary), as well as a variety of informational content and ideas (i.e., science, environmental issues, personal experiences).

3.3 ePIRLS Assessment of Online Reading

At an ever-increasing pace, the internet is changing what it means to be a good reader. Students need the skills, strategies, and practices necessary to learn from reading information on the internet (Leu et al., 2013). New educational policies and instruction are being implemented to meet this need, and new assessment methods are necessary to monitor their effectiveness. In response, the ePIRLS assessment of online reading will be an integral part of PIRLS 2021.

The ePIRLS assessment of online reading began as an optional, computer-based extension to PIRLS 2016 to assess students' achievement in reading on the internet. Fourteen pioneering PIRLS 2016 countries administered this new computer-based assessment of online reading. ePIRLS assesses online reading for informational purposes in a school setting: Guided by a teacher avatar, students work through a series of webpages in a simulated internet environment, obtaining information and answering questions for school projects.

3.3.1 ePIRLS Assessment Framework

Based on current research about online reading (Goldman, 2014; Leu et al., 2013), the *PIRLS 2016 Assessment Framework* adopted the position that reading for informational purposes on the internet in a non-linear environment with much more information requires new reading comprehension skills and strategies (Mullis & Martin, 2015). In this respect, online reading in an internet environment is a similar construct to traditional reading, but with a number of different features. Internet reading consists of searching through a network of texts where readers are responsible for creating their own paths. Readers first must access the appropriate website, and then use navigation strategies (e.g., menus, tabs, and links) to move efficiently within and across one website or webpage to the next. Beyond the reading comprehension processes assessed by PIRLS, ePIRLS assesses two internet-specific processes: selecting a website that meets a particular information need and using online interactive features to locate information within websites.

3.3.2 Assessing Online Reading at the Fourth Grade

As early as PIRLS 2011, a subset of the countries suggested extending PIRLS to include students' achievement in online reading, and several small meetings were held. Initially, countries envisioned a web-based assessment where students accessed the actual internet. However, this was not feasible for a number of reasons, primarily because the PIRLS assessment was exclusively administered in paper booklets, a decade from being web-based. Moreover, websites appropriate for fourth grade students do not exist in many of the PIRLS languages. More fundamentally, at its core, PIRLS is an assessment that needs to collect a certain amount of data to estimate students' reading achievement. The assessment of internet reading had to be more than an invitation to surf the web—fourth grade students could not be left on their own to browse for interesting information; they needed to answer questions about their internet reading as required by the PIRLS framework.

Fortunately, in time for PIRLS 2016, research by the foremost online reading authority in the United States, Professor Donald Leu at the University of Connecticut, provided a foundation for addressing the major assessment issues. Leu and his colleagues had obtained a grant from the U.S. National Science Foundation to develop an Online Reading Comprehension Assessment (ORCA), where eighth grade students researched science issues in an internet environment (Leu et al., 2008). The ORCA model had eighth grade students read and conduct science research in a simulated web environment led by a teacher avatar, an approach that was deemed viable for PIRLS. The simulated web environment could provide the internet reading content and the teacher avatar could ensure that students would be given an opportunity to read and answer the questions.

3.3.3 Developing ePIRLS 2016

The TIMSS & PIRLS International Study Center, in consultation with Donald Leu and with assistance from the Australian Council for Educational Research (ACER), began the arduous process of adapting the ORCA model to create an international assessment of online reading for fourth grade students. To be consistent with the PIRLS framework, the assessment needed to address one or both of the PIRLS reading purposes: literary or informational. To increase ePIRLS' relevance to school instruction and learning, the idea was to situate ePIRLS in a classroom environment. In collaboration with the interested PIRLS 2016 countries, it was generally agreed that consulting the internet to complete school reports or projects was common at the fourth grade, much more than reading stories online. To reflect students' internet reading experiences, ePIRLS would assess only the informational purpose for reading (at least to begin). More specifically, the teacher avatar would guide the fourth grade students in completing authentic, school-like assignments involving social studies and science topics (aligned with topics in the TIMSS science assessment framework at fourth grade; Jones et al., 2013).

Developing ePIRLS 2016 focused on maintaining the authenticity of the internet reading experience as much as possible. A simulated Google search window (used with permission) lists simulated websites containing information about the social studies or science topics. To read about a topic, students navigate through the websites and webpages, each with a variety of features such as photos, graphics, pop-ups, videos, multiple tabs, and hyperlinks as well as advertisements. In the assessment window, to the right of the browser window, the teacher avatar guides students through the ePIRLS school project, prompting them to read particular webpages and then asking questions about the texts. To ensure that students are given the opportunity to answer all the questions, they are permitted to navigate back after being guided through the entire set of websites. Students can navigate back and forth until the assessment time expires. The teacher avatar may not be as representative of students' real-life experiences in searching for particular information on the internet, but the teacher avatar keeps the fourth-grade students focused on reading the webpages and answering the assessment questions.

Creating websites that are as engaging and captivating as possible further motivates students to keep responding to the assessment questions. Concerted effort was given to choosing interesting topics that can be equally enjoyed by boys and girls and that would also contribute to learning social science and science content (e.g., impact of key historical events, environmental issues, space exploration). In accordance with the PIRLS framework, ePIRLS websites were adapted from existing websites and made as colorful, animated, and interactive as possible in an assessment situation.

Each ePIRLS task typically incorporates three websites, each with sub-navigation, totaling about 6–12 webpages (1000–1200 words per task). The ePIRLS simulated websites provide more information than necessary to answer the questions, reflecting the environment of actual online reading, which often involves sorting through a considerable amount of information to achieve one's goals. Consistent with the PIRLS framework and design, each task has at least 15 questions to be answered in up to 40 min; each student completes two tasks according to a specific rotation plan.

To provide countries with an example of an ePIRLS task, a prototype task about how changes in polar bears' environment, including melting of the polar ice caps, were impacting their lives, was developed. The task contained photographs of polar bears from the internet (including showing the polar bears stranded on ice floes), as well as graphical presentations from the internet to illustrate how much the size of the polar ice cap is shrinking. The prototype ePIRLS task was reviewed, revised, programmed, and eventually provided to the NRCs to share within their countries to encourage participation in the first ePIRLS 2016 assessment. Subsequently, 14 countries with the interest and necessary computer infrastructure joined the ePIRLS 2016 venture, most of them among the higher-performing PIRLS countries. The majority of the PIRLS 2016 countries had little or no experience in any type of computer-based testing at that time (2013–2014 for a field test in 2015), so there was a sense of stepping into the future. The ePIRLS 2016 countries had a desire to partake in computer-based testing to learn more about it, as well as an interest in assessing internet reading.

The goal for the first ePIRLS assessment was to have at least three internet reading tasks. Following standard practice at the TIMSS & PIRLS International Study Center, this meant developing six tasks (twice as many) for the field test. Developing the six tasks was considerably more time-consuming than anticipated, with each step being extremely challenging: identifying the topic; creating the simulated websites; developing the assessment items, including items in new formats that took advantage of the digital environment; and designing the software. Feedback from cognitive interviews with fourth grade students, pilots in several countries, and iterative reviews by the ePIRLS NRCs provided crucial input that was used to improve the tasks as they were being developed.

Throughout the development process, anecdotal evidence indicated that students liked the colorful and engaging internet-based materials. Even more satisfying, the first computer-based field test conducted by the TIMSS & PIRLS International Study Center went relatively smoothly. The assessment materials appeared appropriate for the fourth-grade students: low percentages of item non-response, a medium level of difficulty, and high percentages of students liking the tasks. After discussion with the ePIRLS 2016 NRCs, five of the six field-tested tasks were retained for inclusion in the assessment (with a total of 91 items).

In the 14 participating countries, ePIRLS 2016 was administered to the same students as PIRLS 2016, but separately—on the day after the PIRLS assessment, in a setting with the appropriate digital devices. Assessing the same PIRLS sample of students for ePIRLS was efficient and enabled comparing the results of the two assessments. However, the administration burden was substantial, involving finding equipment and arranging facilities for computer-based testing and attempting to ensure the students returned the second day (not entirely achieved in some countries).

Everything considered, ePIRLS 2016 was successful in the 14 participating countries. The timing data revealed that, on average, 91% of the students reached all of the ePIRLS items. In general, the students liked working on the ePIRLS tasks (89% on average). And, the students had a satisfactory measure of success (57% correct, on average, across all items).

After the ePIRLS 2016 assessment, the TIMSS & PIRLS International Study Center released two of the five tasks: "Mars," about space exploration, and "Elizabeth Blackwell," about the first woman doctor. To illustrate the interactive nature of ePIRLS, the two tasks were made available to the public on the TIMSS & PIRLS International Study Center website in the same way they were given to students. Visitors to the "Take the ePIRLS" webpage can work through the tasks, enter their answers, and see how they performed using the scoring key provided (TIMSS & PIRLS International Study Center, 2017).

Figure 3.3 shows two screenshots from the "Mars" task, which began with a website about the characteristics of Mars and the early fascination with its "canals" seen though telescopes, followed by a website about advances in space exploration to learn more about Mars, including the ongoing on-site data being collected by Mars rovers, and concluded with an online newspaper editorial called the "Gift of Curiosity," about the rover named Curiosity finding evidence of water on Mars. The two screenshots illustrate how the browser window showed information about exploring Mars and the assessment window contained the teacher avatar (Mr. Webster) providing guidance and the assessment questions.

3.3.4 ePIRLS2016 Results

The ePIRLS 2016 results were reported in *ePIRLS 2016 International Results in Online Informational Reading* (Mullis et al., 2017). The ePIRLS results were reported on the PIRLS reading achievement scale, so that relative achievement on the two assessments could be compared. A few highlights from the report follow.

Figure 3.4 shows average student achievement in internet reading for the ePIRLS 2016 countries. As previously mentioned, most of the ePIRLS countries were high-achieving PIRLS countries, and, consistent with that, these countries also had relatively high achievement on ePIRLS. Singapore was the top-performing country, followed by Norway and Ireland.

Figure 3.5 compares relative average achievement on ePIRLS to PIRLS average achievement in the paper-and-pencil format for only the reading passages assessing

Fig. 3.3 Screenshots from the ePIRLS Task "Mars"

reading for information. Most of the 14 countries performed relatively higher on one or the other of the two assessments of informational reading.

According to an informal exchange among the National Research Coordinators from the 14 ePIRLS 2016 countries, a relative advantage in ePIRLS may have been related to how familiar students were with using computers in school contexts, especially as part of assessments or classroom instructional activities. However,

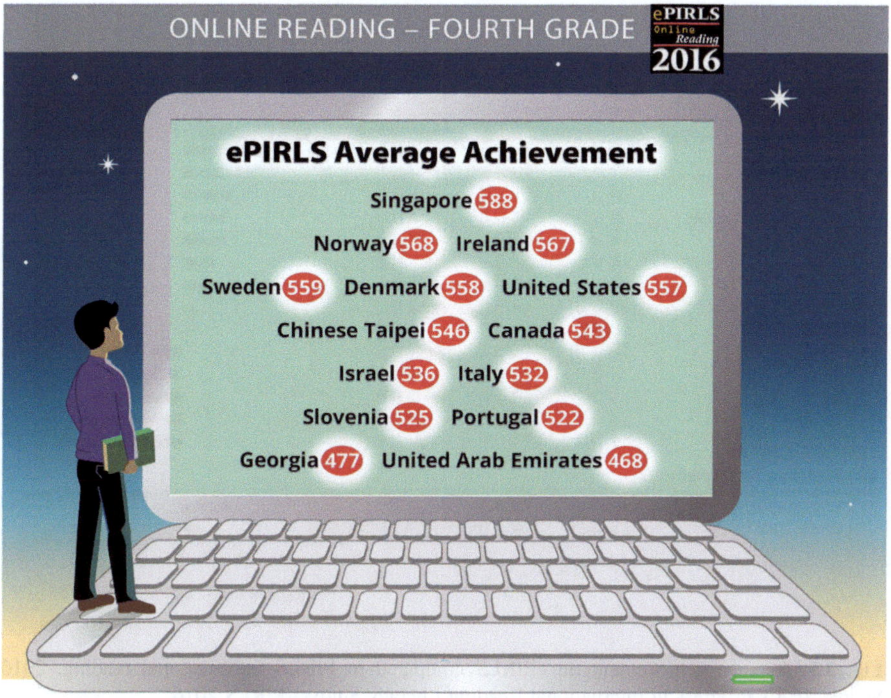

Fig. 3.4 ePIRLS average achievement in online reading at the fourth grade, 2016

the PIRLS 2016 data does not indicate whether this familiarity led to a mode effect between the "e" and paper modes or higher levels of achievement in online reading. More research about this question will be conducted after PIRLS 2021, which will have informational reading in the digital and paper formats, as well as ePIRLS. PIRLS 2021 also will collect more context questionnaire data about students' digital learning experiences.

ePIRLS process data about number of actions and action sequences indicate that students had little difficulty navigating using the ePIRLS tabs and links; 93% of students navigated to all required web pages. Students who did not (7% on average) were automatically directed to the correct webpage so they could continue the assessment. Interestingly, only 8% of the students, on average, were distracted by the advertisements, which were attractive but not nearly as intrusive as typical advertisements on the actual internet.

To analyze students' strategies for integrating information across websites, six questions required students to combine information from websites they previously viewed in the task, together with information from the current website. Most of the students (83%) did not navigate back to previously viewed pages, even though many of them should have; on average, 55% of the students did not navigate back for another look and answered incorrectly, and those students had the lowest ePIRLS achievement (517). A substantial percentage of the students (26%) remembered the

Country	ePIRLS Online Informational Average Scale Score	PIRLS Informational Average Scale Score	Difference
‡ Denmark	558 (2.2)	544 (2.7)	15 (1.9)
† United States	557 (2.6)	543 (2.9)	13 (1.4)
[3] Singapore	588 (3.0)	579 (3.3)	9 (1.1)
United Arab Emirates	468 (2.2)	460 (2.7)	8 (1.3)
Norway (5)	568 (2.2)	560 (2.4)	8 (1.5)
[3] Israel	536 (2.3)	530 (2.4)	6 (1.4)
Sweden	559 (2.3)	555 (2.6)	5 (1.4)
Ireland	567 (2.5)	564 (3.0)	3 (1.5)
[1 2] Canada	543 (3.2)	540 (3.2)	3 (1.4)
[2] Portugal	522 (2.2)	528 (2.4)	-6 (1.1)
[1] Georgia	477 (3.3)	487 (3.4)	-10 (2.4)
Italy	532 (2.1)	549 (2.4)	-17 (1.9)
Slovenia	525 (1.9)	544 (2.1)	-19 (1.1)
Chinese Taipei	546 (2.0)	569 (2.2)	-24 (1.5)
Benchmarking Participants			
Abu Dhabi, UAE	431 (4.1)	422 (4.9)	9 (2.6)
Dubai, UAE	528 (1.6)	524 (2.1)	4 (1.0)

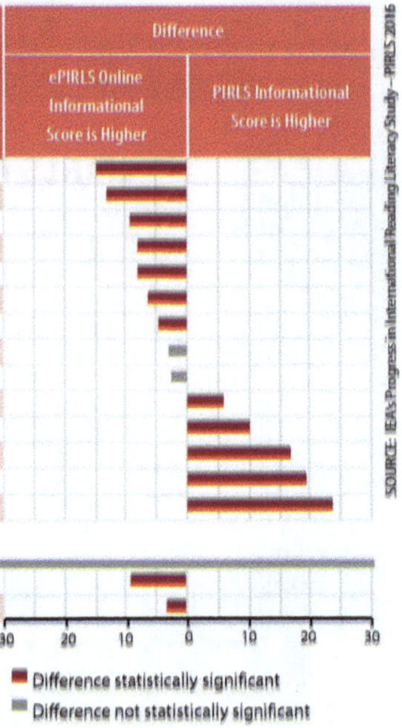

Fig. 3.5 Average achievement on ePIRLS compared to PIRLS informational reading, 2016. Source: IEA's Progress in International Reading Literacy Study—PIRLS 2016

information for the correct answers without needing to look back, and they had relatively high achievement (572). Interestingly, the small percentage of students (7%) who both looked back and answered correctly achieved the highest score (589), even slightly higher than the students who remembered the information for the answers. Finally, the 10% of students who did look back but answered incorrectly had comparatively midlevel ePIRLS achievement (540). Apparently, better readers recognize they should navigate back to view the texts again to answer the questions, even if they cannot do it successfully.

3.3.5 *ePIRLS in 2021*

Building on the success of ePIRLS 2016, the development of ePIRLS 2021 is well underway. For the 30-plus countries transitioning to a digital assessment environment, ePIRLS will have full status together with digitalPIRLS in one unified assessment. Integrating ePIRLS and making it an essential component of PIRLS is consistent with a growing global dependence on the internet as the primary source for distributing all types of information, and the increasing amount of digitally based reading instruction in schools.

In tandem with developing digitalPIRLS 2021 reading passages for field testing, the TIMSS & PIRLS International Study Center developed two new ePIRLS 2021 tasks for field testing, to replace the two tasks from ePIRLS 2016 that were released on the study center's website. Of course, the ePIRLS tasks newly developed for 2021 retain the core assessment method: a browser window with the look and feel of the internet to present the informational reading content and an assessment window where a teacher avatar guides the students through the websites and presents the assessment questions.

The two tasks created for ePIRLS 2021 reflect content about contemporary issues as well as the increasingly multimodal and interactive aspects of the internet, setting the stage for examining students' behavior regarding text-task interactions. They include some more difficult types of questions, with an eye to increasing the range in difficulty across the ePIRLS 2021 tasks. As part of developing the materials for PIRLS 2021, the two ePIRLS tasks have been field tested and undergone extensive reviews and revisions by the PIRLS 2021 Reading Development Group and the National Research Coordinators.

In the 30-plus countries transitioning to digitalPIRLS 2021, the ePIRLS tasks will be integrated with the digitalPIRLS assessment according to the assessment design. Students will be given two digitalPIRLS reading passages, two ePIRLS tasks, or a hybrid assessment of one digitalPIRLS passage, and one ePIRLS task. This design for PIRLS 2021 positions PIRLS to complete the transition to a fully digital environment in 2026:

- The results for digitalPIRLS 2021 will provide a foundation for trends into the future.
- The digitalPIRLS 2021 countries also will administer the paperPIRLS trend booklets to bridge back to PIRLS 2016, giving all 2021 countries (digital and paper formats) that have participated in earlier PIRLS assessments trends back to PIRLS 2001.
- The results for ePIRLS 2021 will provide a foundation for trends into the future in online informational reading, including trends for the 14 countries that participated in ePIRLS 2016.
- The hybrid booklet equivalents, containing both digitalPIRLS and ePIRLS, provide a basis for research about the relationship in achievement between informational reading assessed in the two digital formats. Countries also will have an estimate of their informational reading achievement from the paper bridge booklets.

To help provide improved data about digital contexts for teaching and learning, the PIRLS 2021 teachers will be asked a broader range of questions about how digital devices are being used in classrooms. Students respond to an enhanced PIRLS 2021 scale about their self confidence in using digital devices.

3.4 PIRLS 2021 International Data Collection: digitalPIRLS and ePIRLS

In digitalPIRLS countries, the assessment instruments are delivered to countries via the Assessment Master designer, translation management system, and delivery systems developed by SoNET Systems—a part of RM Results, a world-leading provider of eAssessment services. The assessment content in each country's own language(s) may be delivered online via the web through individual USB drives that connect to each student's digital device or by means of PC that acts as a local server. As described previously, each student receives two reading passages/tasks and items, with 40 min available for reading each one and answering the 15 or so questions about it. Students' responses are uploaded and stored on IEA Hamburg's servers, which provide a system that supports scoring teams from countries in reliably scoring the students' answers to the constructed response questions. After scoring, IEA Hamburg will review countries' digital data, as well as data from the countries that used the paper format, for accuracy and consistency within as well as across countries. Then, IEA Hamburg will send the data to the TIMSS & PIRLS International Study Center for analysis and reporting.

Prior to the main PIRLS 2021 data collection, there was an international field test in March–April 2020, including digitalPIRLS, paperPIRLS, and ePIRLS, after which the TIMSS & PIRLS International Study Center, the Reading Development Group, and the participating countries reviewed the results. Data collection for PIRLS 2021, in digital and paper formats, with digitalPIRLS 2021 incorporating ePIRLS 2021, began in the Southern Hemisphere in October–December 2020, and in the Northern Hemisphere in March–June 2021. Due to the COVID-19 pandemic, data collection was extended until March–June of 2022. Students' achievement results are scheduled to be reported in December 2022.

3.5 Conclusion

Over the two decades spanned by PIRLS 2001 through 2021, reading instruction has changed in response to new research about effective reading strategies—and so has the nature of reading—with students' need to read and learn online. In response, the PIRLS assessments have evolved with the times: creating and implementing innovative assessment methods that reflect the new literacies. PIRLS in 2021 is transitioning to a digital environment; the reading passages reflect the colorful and engaging images and broad array of photos, diagrams, and charts found in the materials typically read by students, and a new streamlined user interface enables students to manage their own assessment strategies and read the passages and questions at the same time. Integrated into the same digital assessment, ePIRLS provides a simulated internet environment and requires students to demonstrate the skills, strategies, and practices that enable them to read and learn online.

References

Anstey, M., & Bull, G. (2006). *Teaching and learning multiliteracies: Changing times, changing literacies*. International Reading Association.

Constantine, L., & Lockwood, L. (1999). *Software for use: A practical guide to the models and methods of usage-centered design*. Addison-Wesley.

Foy, P., & Yin, L. (2017). Scaling the PIRLS 2016 achievement data. In M. O. Martin, I. V. S. Mullis, & M. Hooper (Eds.), *Methods and procedures in PIRLS 2016* (pp. 12.1–12.38). https://timssandpirls.bc.edu/publications/pirls/2016-methods/chapter-12.html

Goldman, S. R. (2014). Reading and the web: Broadening the need for complex comprehension. In R. J. Spiro, M. DeSchryver, P. Morsink, M. S. Hagerman, & I. P. Thompson (Eds.), *Reading at a crossroads? Disjunctures and continuities in current conceptions and practices*. Routledge.

Jones, L., Wheeler, G., & Centurino, V. (2013). The TIMSS 2015 science assessment framework. In I. V. S. Mullis & M. O. Martin (Eds.), *TIMSS 2015 assessment frameworks*. TIMSS & PIRLS International Study Center, Boston College.

Kintsch, W. (2012). Psychological models of reading comprehension and their implications for assessment. In J. P. Sabatini, E. R. Albro, & T. O'Reilly (Eds.), *Measuring up: Advances in how to assess reading ability* (pp. 21–37). Rowman & Littlefield.

Leu, D., Kinzer, C., Coiro, J., Castek, J., & Henry, L. (2013). New literacies: A dual-level theory of the changing nature of literacy, instruction, and assessment. In D. E. Alvermann, N. J. Unrau, & R. B. Ruddell (Eds.), *Theoretical models and processes of reading* (6th ed., pp. 1150–1181). International Reading Association.

Leu, D., Kulikowich, J., Sedansk, N., & Coiro, J. (2008). Framework document: The new literacies of online research and comprehension. In *Assessing online reading comprehension: The ORCA project*. A grant proposal to the US Department of Education, Institute for Education Sciences.

Mullis, I. V. S., & Martin, M. O. (Eds.). (2015). *PIRLS 2016 assessment framework*. TIMSS & PIRLS International Study Center, Boston College.

Mullis, I. V. S., & Martin, M. O. (2019). PIRLS 2021 reading assessment framework. In I. V. S. Mullis & M. O. Martin (Eds.), *PIRLS 2021 assessment frameworks*. http://timssandpirls.bc.edu/pirls2016/framework.html

Mullis, I. V. S., Martin, M. O., Foy, P., & Hooper, M. (2017). *ePIRLS 2016 international results in online informational reading*. http://timssandpirls.bc.edu/pirls2016/international-results/

TIMSS & PIRLS International Study Center at Boston College. (2017). Take the ePIRLS assessment. http://timssandpirls.bc.edu/pirls2016/international-results/epirls/take-the-epirls-assessment/

Chapter 4
New Constructs and Item Types in Large-Scale Assessments

Lu Ou, Emily Kerzabi, Yigal Rosen, and Alina A. von Davier

Abstract Faced with the evolving environments of work and life in the new century, new technologies have introduced new occupational, social, and educational constructs in need of measurement, such as problem-solving in technology-rich environments and digital reading literacy. Through the ubiquity of digital technologies, large-scale, computer-based assessments are uniquely able to assess such constructs within the general population through innovative item types, such as scenario-based tasks. In this chapter, we review the new constructs in large-scale assessments that have sprung from rapid societal and technological change and the new item types that make it possible to assess such constructs. We provide examples from two OECD international, large-scale assessments and discuss the related challenges posed by new constructs and instruments, as well as consider future directions.

Technology drives radical changes in modern work and life. In only a few decades, computers have gone from being large, bulky equipment, requiring additional connectivity infrastructure and long processing time, to being small enough to fit in the palm of a hand, with enough processing power, connectivity, and available data to bring information instantaneously to any user. Access to such technology is becoming fairly commonplace throughout the globe, changing the way people work, educate, and socialize. The ubiquity of computers—particularly the high volume of

---

L. Ou (✉)
San Jose, CA, USA

E. Kerzabi
ETS, Princeton, NJ, USA

Y. Rosen
BrainPOP, New York, NY, USA

A. A. von Davier
Duolingo, Pittsburgh, PA, USA

L. Khorramdel et al. (eds.), *Innovative Digital-Based International Large-Scale Assessments*, Methodology of Educational Measurement and Assessment,
https://doi.org/10.1007/978-3-031-90951-1_4

computers in schools and the mobility offered by laptops, tablets, and mobiles—is what makes it possible to conduct large-scale assessments digitally. And in turn, because computers have become so common, the ways in which we use computer technology have become meaningful and measurable competencies.

With such dramatic changes in technology, education must adapt to the evolving technology-rich environment to prepare individuals to interact with and through technology and offer new solutions and tools to advance learning and teaching (Khorramdel et al., 2024; von Davier et al., 2013). To accommodate for such technological changes, assessments ought to be updated with evolving constructs of interest and innovative item types. By definition, constructs are the substantive psychological traits or processes of interest to be measured by assessments, while items are the components of an assessment designed to gauge participants' responses and infer unobserved psychological states associated with a construct. Updated assessments with new constructs and item types still need to maintain high standards of validity, reliability, fairness, and generalizability (Beller, 2013).

Relevant continued efforts are particularly evident in low-stakes national (e.g., the National Assessment of Educational Progress [NAEP]) and international large-scale assessments (e.g., the Organisation for Economic Co-operation and Development [OECD]'s Programme for International Student Assessment [PISA] and Programme for the International Assessment of Adult Competencies [PIAAC]). As large-scale assessments share many similarities in their design and administration, this chapter will review the development of new constructs and item types in the OECD's international large-scale assessments. The main examples provided in this chapter are taken from these two latter assessments, PIAAC and PISA. The OECD commissioned PIAAC, the OECD's survey of adult skills, to assess modern workplace skills, with an entire domain dedicated to technology-use skills. PISA, the OECD's assessment of 15-year-old students, calls for a new innovative domain at each assessment cycle, providing the opportunity to identify and measure new constructs of interest amidst ongoing global change. After discussing the design considerations that go into the development of new constructs and item types in large-scale assessment, we look at the new constructs and scenario-based item examples from the OECD's PIAAC and PISA assessments. We then present some of the challenges inherent in the measurement of new constructs in large-scale assessments and conclude the chapter by looking ahead to the possibilities for future research, such as the use and interpretation of collected process data, new construct domains of interest, and some of the measurement technologies that could be used in such assessments.

4.1 Developing New Constructs and Item Types in Large-Scale Assessments

In order to be successful in twenty-first-century educational and workforce settings, one needs skills that are mediated by technology (Griffin et al., 2012). Advances in technology shape the possibilities for assessment, allowing for the measurement of new constructs and by enabling better measurement of existing constructs. Newly assessed constructs are often technology-mediated and evaluate skills that may not have been commonplace in populations when previous assessments were administered. As the world becomes more digital and interconnected, it has become important to also measure performance and proficiency in digital and interconnected environments. For example, digital reading literacy is a new subscale of the Reading domain in PISA that has been introduced by technological developments in reading platforms (e.g., on a computer, tablet, mobile) and needs to be measured with corresponding technology. Similarly, problem-solving in technology-rich environments in PIAAC is a somewhat recent construct domain that is of interest due to the predominance and importance of technology in modern life and is now measurable with advances in computer-based stimulus presentation and data recording (Kirsch & Lennon, 2017). Both of these are examples of digital scenario-based assessments, the focus of this chapter. Scenario-based assessments embed tasks in contexts that usually (partially) represent real-world situations (such as those that might occur in a common setting, like in school, the workplace, or a social environment) and record various forms of process data as participants navigate their way towards a solution. They are designed to evoke natural responses to real-life problems and collect evidence of capabilities beyond the selected or constructed response to problem-solving tasks. Technology further drives the design, development, presentation, and scoring of such tasks. As computer technologies develop, the ability to deliver a wide range of scenario designs and task functions similarly develops. It is possible to capture the necessary process and outcome data across broad populations, both geographic and demographic, as these technologies have become more prolific. With secondary analyses of process data from large-scale assessments, more is learned about how to interpret human behavior on such items and draw valid inferences. From design, through test administration, data collection, and scoring, digitally-mediated constructs and corresponding item types in scenario-based assessments have rapidly become a major field of educational research.

Currently, scenario-based assessments are generally administered through simulations and gamified tasks (which should be considered separately from game-based assessments). Here the gamification refers primarily to enhanced engagement, while game-based assessment may include additional features, such as competitiveness. Driven by contextual learning and assessment, gamified and simulation-based item types are designed to provide participants with an engaging experience in an authentic interactive environment, such that their responses better reflect their capabilities. However, there are many practical and psychometric considerations in designing these assessments (Mislevy et al., 2014; Zapata-Rivera & Bauer, 2012).

For instance, the assessment needs to evoke and record relevant evidence to support reasoning about competencies based on the assessment scores, while balancing the collection of appropriate evidence with enhanced engagement. Simulations are used in both PIAAC and PISA, as simulated settings are often fairly standard, such as in a typical scientific experiment or web interface. The actions a participant can make in both simulations and gamified tasks are generally limited and controllable; therefore, gamified tasks could also be a possibility in large-scale assessments but would require slightly greater interactivity between the participant and the programmed task. For international assessments like PISA, this would likely require increased technological capabilities in certain countries or school districts in order to be feasibly administered. Whether one could demonstrate or learn important knowledge or skills from a game as an assessment still requires validation (Shute et al., 2009). For example, Dicerbo (2017) presented a four-iteration development process of refining an evidence model in a game-based assessment of mathematical understanding, where small-scale validation studies were first conducted to generate data that update tentative hypotheses about participants' actions and uncover unknown behavior before a larger-scale study was implemented to confirm stronger hypotheses. The feedback loop between assessment design, hypothesis generation, and validation results in clearer arguments about the links between constructs, tasks, and evidence, and increases the validity of the assessment. Thus, one has more guidance in collecting and analyzing evidence, as well as making inferences from the scores. Including gamified tasks may be beneficial for both summative and formative assessments (Zapata-Rivera & Bauer, 2012); however, game-based assessments as large-scale assessments are still challenging to develop. Games are also associated with the high cost, and the complexity of game designs makes it hard to scale up for broad coverage of topics or a wide implementation across states and countries. Therefore, the examples in this chapter are all simulation tasks, such as immersive digital environments, that have been or are currently in use in the OECD's international large-scale assessments.

Scenario-based assessments are designed to meet the need of measuring new constructs and offer an insight into what traditional assessment measures could not. For example, it is expected that performance and proficiency in simulated settings are better estimated than in previous, paper-and-pencil type tests; e.g., asking a participant to solve a problem in a simulated computer program, such as a word processor, would yield higher-quality data about the participant's abilities than would a multiple-choice test on the functions of that computer program. In order to measure such skill constructs with high validity, reliability, fairness, and generalizability, new item types are developed in digital environments with more engaging context and more functionality to seamlessly track a participant's response and behavior. To this point, the Evidence-Centered Design framework (ECD; Mislevy et al., 2003) has been established specifically for use in designing scenario-based assessments. The core idea behind ECD is that the following three components need to be developed during the design process: (1) a student model that defines constructs and underlying processes to be measured; (2) a task model that elucidates the activities participants undertake and that links the task design with the evidence needed; and (3) an evidence model that links potential observables from the tasks to the constructs of interest in the form of a measurement model and scoring rules. This ensures a tight alignment between the measurement model, the task design characteristics, and the produced observable outcomes. However, the deliberate definition of each of the three framework components requires careful consideration and iterative updates (Rupp et al., 2010), which may pose challenges to longitudinal and large-scale, international assessments. Changes in the task model or how the task is evaluated through iterative updates may, and should, change over time or in different contexts. While some changes in the task, such as updating a simulated interface (e.g., a web browser) to be more modern, make considerable sense (presenting participants with an outdated interface may even introduce a source of error not present in earlier test administrations; Wagner & Kerzabi, 2024), other changes, like how an item is scored, may be an improvement, but would not necessarily be useful in measuring trend over time. PIAAC and PISA use outcome variables (e.g., whether the task was solved partially correctly, correctly, or incorrectly) to score scenario-based tasks, but the test delivery systems do the capture and record the participants' actions through engagement with such tasks, and these actions may inevitably become artifacts of patterns of behavior and the subject of investigative research. Thereby, the ECD framework or the extended ECD—the ECD extended to include process data, data-driven algorithms, and eventually, learning and changes in the ability during the learning process (Arieli-Attali et al., 2019) are important to consider when designing scenario-based assessments, specifically the task build and the action and other process data to be captured.

Computer-based assessments can easily collect a plethora of process data, of which, only some or none may be used in scoring. This is because process data pose challenges in associating quantitative evidence with qualitative constructs, as well as utilizing or developing modeling and estimation approaches to analyze and interpret the data (see Part IV of this book). Traditional psychometric models, such as item response theory, may not be readily applicable to the process data collected in the scenario-based assessments. Exploratory data analyses and data mining tools are often needed to find patterns in complicated data and help generate hypotheses to inform our understanding and reasoning of our actions. Further, such as computational psychometrics, which concerns the assimilation of data-driven algorithms into the psychometrics frameworks, and the development of proper measurement models that accommodate data structure and synthesize evidence into meaningful scores or parameter values needs investigation and validation (Mislevy et al., 2014; von Davier, 2017; von Davier et al., 2021).

The opportunities and challenges introduced by new technologies further complicate the design and implementation of scenario-based tasks in large-scale assessments, which are constrained by the technologies available in classrooms across the country and around the world. Unlike small-scale assessments in a classroom or a school, large-scale assessments aim at providing comparable indicators on participant performance and education quality across schools, regions, states, and countries (von Davier et al., 2013). Large-scale assessments tend to report on group-level characteristics that are valid for an education system, rather than individual results.

They are supposed to inform policy, research, and practice in education and cover a broad range of domains, despite limited testing time and budget. During implementation, they must often use complex sampling designs to assemble representative samples and apply complex assessment designs, in which each individual only takes a block of items, and each item is only given to certain groups of individuals. Therefore, unique challenges in large-scale assessments include not only defining constructs that are important for assessments and designing items to measure the constructs in a sensible way, but also summarizing and comparing performance on the items and computing estimates with both sample design and measurement uncertainty taken into account.

In the next section, we provide examples of scenario-based tasks that have been used in the core and innovative domains of two OECD assessments: PIAAC and PISA.

4.2 Measuring New Constructs with Scenario-Based Assessments

There are many examples of scenario-based tasks used in assessment, but as a model paradigm, we present the new, technology-based constructs in PISA and PIAAC. We summarize the definitions and the organizations of these construct examples in Table 4.1. Beginning with PIAAC, which was developed as a computer-based assessment from its inception, through PISA's new innovative domains in every cycle of assessment, the fast-paced generation of new constructs, and the item types that facilitate them, illustrate how technology and global change drives the evolution of modern educational measurement.

4.2.1 Problem Solving in Technology-Rich Environments (PSTRE) in PIAAC

The Programme for the International Assessment of Adult Competencies (PIAAC) is the OECD's survey of adult skills. The large-scale international survey targets adults aged 16–65 in the domains of literacy, numeracy, and problem-solving in technology-rich environments (PSTRE) in roughly 40 countries worldwide. Primarily a work-skills survey, PIAAC collects and reports information on the cognitive abilities and skills of the general adult population needed to operate in modern workplace settings. PIAAC has primarily administered as a computerized-adaptive assessment since its inception in 2010 (OECD, 2013) via laptop (Cycle 1) or tablet (Cycle 2) by interviewers, one-on-one with adults. The goal of PIAAC is to shed light on the adult skills that are necessary for an economically robust workforce and to understand more fully how training and education can further adult skills across

Table 4.1 Chapter construct examples overview

Construct	Year	Domain	Definition	Organization
Problem Solving in Technology-Rich Environments	2010	PIAAC Core	Using digital technology, communication tools, and networks to acquire and evaluate information, communicate with others, and perform practical tasks	Cognitive (goal setting, monitoring, planning and organization, acquiring, evaluating, and using information) Technology (hardware, software, commands, functions, representations) Task (contexts, complexity, explicitness of problem statement)
Problem Solving	2012	PISA Innovative	The capacity to engage in cognitive processing to understand and resolve problem situations where a method of solution is not immediately obvious, including the willingness to engage with such situations to achieve one's potential as a constructive and reflective citizen	Problem-solving processes (exploring and understanding; representing and formulating; planning and executing; and monitoring and reflecting) Nature of the problem situation (interactive, static) Problem Context (technology, personal, social)
Science	2015	PISA Core	The ability to engage with science-related issues, and with the ideas of science, as a reflective citizen	Competencies (explain, evaluate & design, interpret) Knowledge (content, procedural, epistemic) Context (personal, local/national, global)
Reading	2018	PISA Core	Understanding, using, evaluating, reflecting on and engaging with texts in order to achieve one's goals, to develop one's knowledge and potential and to participate in society	Processes (read, locate, understand, evaluate and reflect) Texts (source, structure, format, type) Scenarios (personal, public, educational, occupational)
Mathematics	2021	PISA Core	The capacity to reason mathematically and formulate, employ, and interpret mathematics to solve problems in a variety of real-world contexts	Processes and capabilities Content Context
Collaborative Problem Solving	2015	PISA Innovative	The capacity of an individual to effectively engage in a process whereby two or more agents attempt to solve a problem by sharing the	Competencies (shared understanding, take action, team organization) Student background

(continued)

Table 4.1 (continued)

Construct	Year	Domain	Definition	Organization
			understanding and effort required to come to a solution and pooling their knowledge, skills and efforts to reach that solution	Context (problem scenarios, team composition, task characteristics, medium)
Global Competence	2018	PISA Innovative	The capability to examine local, global and intercultural issues, understand and appreciate different perspectives and worldviews, interact successfully and respectfully with others, and take responsible action toward sustainability and collective well-being	Global competence: Knowledge, skills, Attitudes, and Values Global understanding Cognitive processes: Content (global or intercultural issues) scenarios (students as researchers, reporters, mediators or team-members, debaters)
Creative Thinking	2022	PISA Innovative	The competence to engage productively in the generation, evaluation and improvement of ideas, that can result in original and effective solutions, advances in knowledge and impactful expressions of imagination	Competences and processes: generate diverse ideas, generate creative ideas, evaluate and improve ideas Domains (written expression, visual expression, social problem solving, and scientific problem solving)

different age ranges. The PSTRE framework notes that two conditions for the advancement of technology-rich societies involve the use of specific digital problem-solving skills and the distribution of those skills across the general population (OECD, 2009). The questions PSTRE tries to answer include, whether individuals are able to interact with digital interfaces, like a computer screen to send an email or use a word processor, and what proportion of the population, and with what demographic characteristics, are able to perform such tasks.

While PSTRE was cutting-edge for an international large-scale assessment at the time, smaller exploratory studies laid the foundation for such an undertaking, determining how such a domain might function and how it could be validly scored. In 2003, a study on problem-solving in technology-rich environments was carried out as an exploration of the innovative use of computers in administering the NAEP assessments (Bennett et al., 2007). In the study, problem-solving was measured with technology in two scenarios: the Search scenario, in which participants needed to locate and synthesize scientific information from a simulated web environment, and the Simulation scenario, in which participants conducted a physics experiment by watching animated displays after manipulating factors such as mass. The intended construct of interest was problem-solving proficiency, which was conceptualized as a combination of scientific inquiry and computer skills in the Search scenario and further composed of the scientific inquiry synthesis skill, scientific inquiry exploration skill, and computer skills in the Simulation scenario. Observables, such as participants' choices in their experimental design, served as evidence for the constructs and were converted to scores according to expert rubrics. The multivariate observed scores were fitted by Bayesian item response theory and graded response models (Samejima, 1969) with prior distributional assumptions on the context effect. Parameter estimates were obtained by Monte Carlo Markov Chain techniques and used to construct overall and subskill scores of problem-solving. This showed that actions taken during problem solving could not only be scored, but also be placed on a scale.

Years later, the PIAAC design was made possible by the mobility of laptops and other digital devices, and the ubiquity of such devices drove the development of the PSTRE domain as a construct urgently in need of measurement. How adults interact with and successfully use technology at home and in the workforce had yet to be investigated in a large-scale assessment, nor at an international level. While having cognitive subscales, the PSTRE domain evaluates adult proficiency in four technology subscale dimensions (e.g., use of hardware devices, software applications, commands and functions, and interaction with digital representations, such as text, sound, graphics, and video) as well as three task dimensions (e.g., the purpose of the task, such as personal or for work, the explicitness of the task to be completed, and the intrinsic complexity of the task). Task complexity may include the number of steps to solve the problem, the diversity of the operations to perform, the likelihood of reaching an impasse or an unexpected outcome, and the number of constraints to be satisfied in the task. To thoroughly and validly assess problem-solving skills in digital environments, traditional item types, such as multiple choice and constructed response, are insufficient measures. Thus, the PSTRE domain aligns perfectly with the capabilities of scenario-based assessment methods and immersive item types, in which participants may interact with a simulated word processor to complete a task or search the web on a simulated web browser. An example of such an item is shown in Fig. 4.1 (NCES, n.d.).

As a forerunner in large-scale scenario-based assessment design, the PSTRE domain set a standard for what an immersive digital assessment can look like and can measure on a global level. The test delivery software captures a variety of process data, such as action sequences and actions taken, as well as basic response time data. Further, proficiency estimates are able to be placed on a single composite scale. Following suit, similarly immersive environments later appeared in the core domains of PISA, such as Reading, in which a new construct of digital literacy could be assessed (see the section on the PISA Reading domain later in this chapter).

4.2.2 Problem Solving in PISA

Every 3 years, PISA assesses 15-year-olds who are about to finish compulsory education across countries, to gauge if they are prepared to use what they have

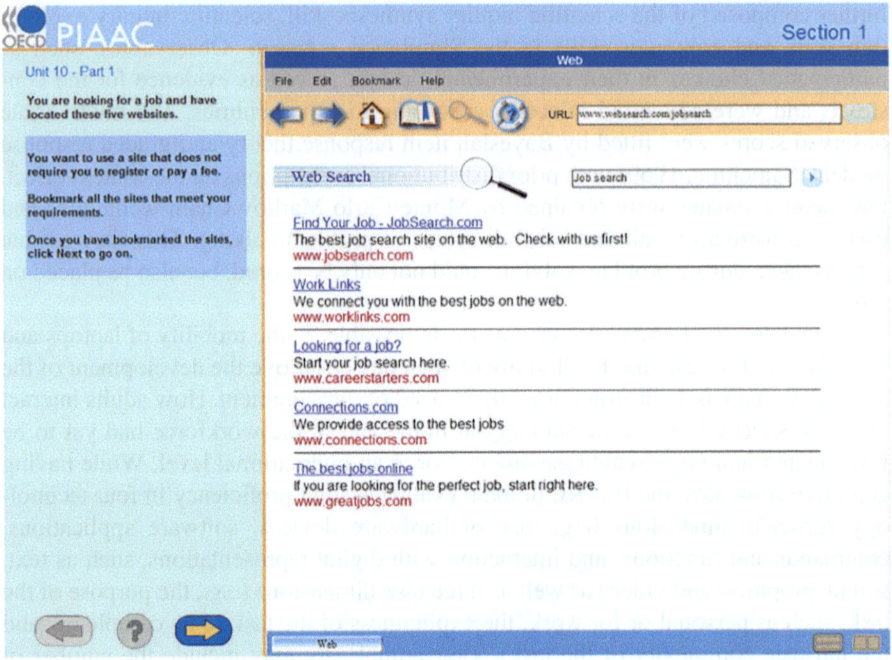

Fig. 4.1 Example PSTRE Item in a simulated web browser

learned in school and apply that knowledge to real-life situations and problems. A year after the assessment, PISA publishes group-level results to provide comparable indicators on participant performance and schooling practices across countries, inform effective educational systems, and suggest potential improvements. As entailed by the goal in assessing 15-year-olds' ability to participate in societies, the assessments have evolved to include new item types and subscale constructs for the traditional, *core domains* of reading, mathematics, and science, as well as new constructs beyond these cognitive competencies in new, *innovative domains*.

One such innovative domain was Problem Solving in PISA, administered as a paper-and-pencil test in 2003, and delivered again via computer in 2012 when advances in technology allowed for more opportunities to increase efficiency and effectiveness of measurement. Problem-solving competency is central to one's ability of applying what has been learned to real-world problems and is defined as one's "capability to engage in cognitive processing to understand and resolve problem situations where a method of solution is not immediately obvious" (OECD, 2013). It requires coordination of cognitive, motivational, and affective factors to achieve a goal that is often blocked by lack of knowledge or obvious strategies and potentially needs to be achieved by creative and critical thinking. The computer-based assessment of individual problem-solving competency made it possible to present interactive and dynamic problems, engage students with the question, and record data of various types in real-time during the assessment.

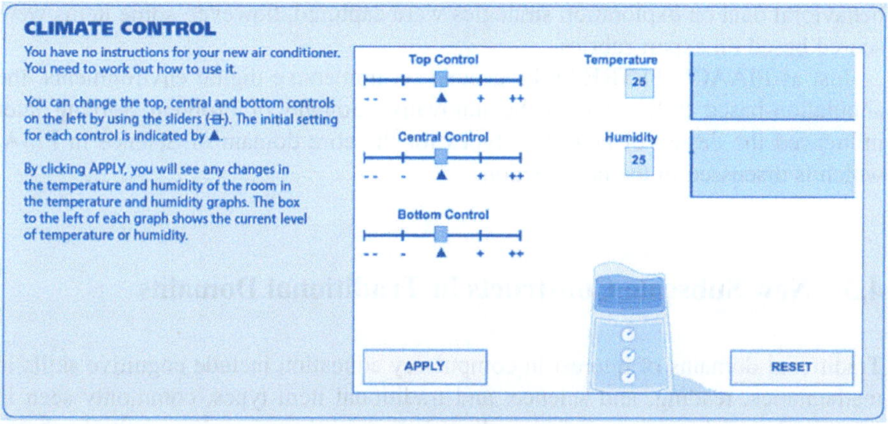

Fig. 4.2 Example PISA problem solving item: climate control

In the PISA 2012 assessment framework, the problem-solving processes include exploring and understanding, representing and formulating, planning and executing, and monitoring and reflecting. As the processes are considered as internal to one's cognitive and emotional system, they need to be inferred by one's behavior and outputs, such as exploration strategies and answers to assessment items. To make these inferences, test developers drew from previous research and incorporated two main types of interactive tasks in the PISA 2012 problem-solving assessments: "MicroDYN" systems (Greiff & Funke, 2009) and "finite-state automata" (Buchner & Funke, 1993). A "MicroDYN" task is based on a small system of causal relations and involves manipulating and controlling input variables to reach assigned states of output variables. A "finite-state automata" is a system with a finite number of states, inputs, and outputs, where the system's current state and inputs uniquely determine its next state and outputs (Buchner & Funke, 1993). In both kinds of tasks, participants need to explore the system to understand its functioning, provide interventions to achieve a goal, and articulate explanations or proposals for improvement. For example, in an interactive scenario, such as the "Climate Control" task shown in Fig. 4.2 (OECD, 2014), participants must troubleshoot a malfunction in a device by conducting some experimentation to collect data and uncover additional relevant knowledge to control the device. As acquisition and application of knowledge in an interactive environment are different from that in static problems, the inclusion of both interactive and static problems allowed for the measurement of a broader set of problem-solving competencies than had been possible with paper-based instruments. Automatic coding of certain actions and products was also possible via computer administration. Real-time capture of behavior data provided reliable information, including action sequences and response times, and contributed to the measurement of competencies, such as exploration strategies that would otherwise depend on participants' retrospective reports. As with PSTRE in PIAAC, the result of the problem-solving assessment was reported on a single composite scale. Although behavioral data on exploration strategies were captured, however, some items were scored based on expert rubrics.

Just as PIAAC's PSTRE broke ground on immersive digital environments, the simulation-based tasks seen in the innovative domain of Problem Solving later influenced the design of new item types for the core domain of Science in PISA, which is discussed in the next section.

4.3 New Subscale Constructs in Traditional Domains

Traditional domains of interest in compulsory education include cognitive skills in mathematics, reading, and science, and traditional item types, commonly seen in paper-and-pencil tests, range from selected-response items, such as multiple-choice questions, to constructed-response items in which participants construct short or long answers to an open-ended question. The development of new subscales within these domains is made possible with the development of new item types, which is an important adaptation of large-scale assessments to societal changes. Now having the technological capacity and capability to administer scenario-based items, traditional domains should see new item types, as well as new subscales, to accurately assess skill proficiencies among more modern skill-use situations. From PISA 2015 onward, all domains were delivered via computer (OECD, 2017), and the use of computer technology enabled the presentation of non-traditional and non-static stimulus material, such as animations and interactive simulations.[1] Starting with science in 2015, followed by reading in 2018, and mathematics in 2022, each domain saw new interactive, immersive, and scenario-based item types and the assessment of new subscales. The following sections review the general modifications made to each domain in turn.

4.3.1 Science

Scientific literacy was the core domain in PISA 2015 and refers to knowledge of science or of science-based technology. PISA 2015 refined and extended the previous, paper-based construct of scientific literacy to target three new subscale competencies: (1) explain phenomena scientifically, (2) evaluate and design scientific enquiry, and (3) interpret data and evidence scientifically. "Scientific knowledge" had evolved to include content, procedural, and epistemic knowledge, which underlaid the development of items newly developed for PISA 2015.

Interactive tasks based on simulations of scientific enquiry were added to gauge individuals' scientific knowledge and competencies. For example, the interactive

[1] A paper-based version was always made available although in the traditional domains only.

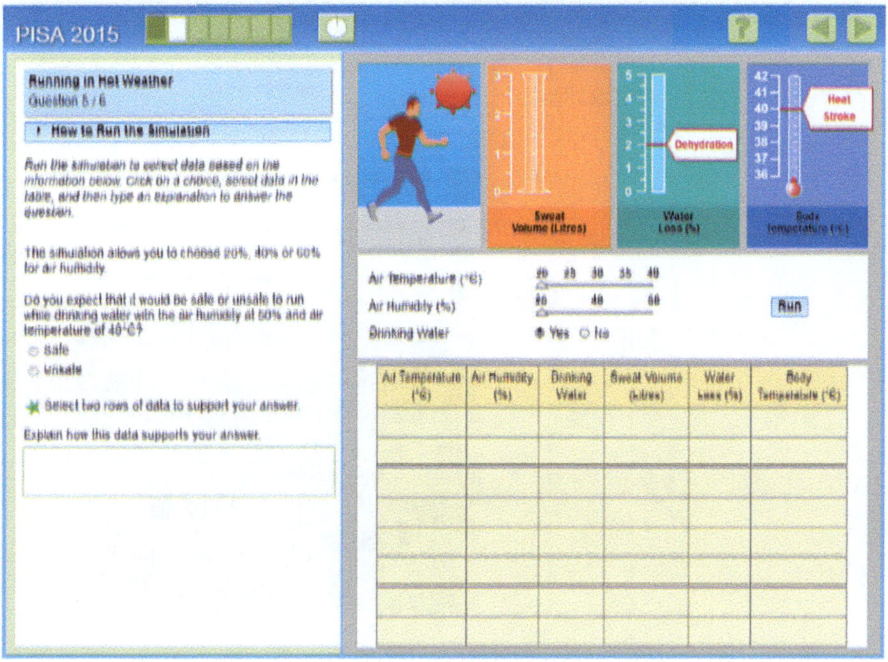

Fig. 4.3 Example PISA science item: running in hot weather

simulation-based task shown in Fig. 4.3 (OECD, 2015) examines the relationship between weather (e.g., air temperature and humidity) and body condition (e.g., sweat, water loss, body temperature) after a one-hour run. Participants are able to manipulate the weather variables and record data on body conditions to answer subsequent questions about potential health dangers a runner might encounter under certain conditions. To support their answer, students would need to select multiple rows of their collected data. In another simulation-based item, shown in Fig. 4.4 (OECD, 2015), participants were asked to investigate environmental factors on two slopes; they needed to run experiments and take measurements, which permitted the assessment of their procedural knowledge, as well as their competency to design and evaluate scientific enquiry. Responses to tasks in these categories were scored as multiple-choice items or constructed responses if they were considered as open-ended and, therefore, could be analyzed as traditional items.

The main theme behind the new item types in PISA 2015 Science is their interactivity, requiring the student to engage in the scientific process in order to be able to answer a scientific question appropriately. Such items are frequently included amongst many of the science subscales, such as interpreting data and evidence scientifically and procedural knowledge. Further, the process data collected on these tasks, such as action sequences, may be useful in determining whether students are employing efficient problem-solving strategies, with analyses only made possible through interactive item interfaces.

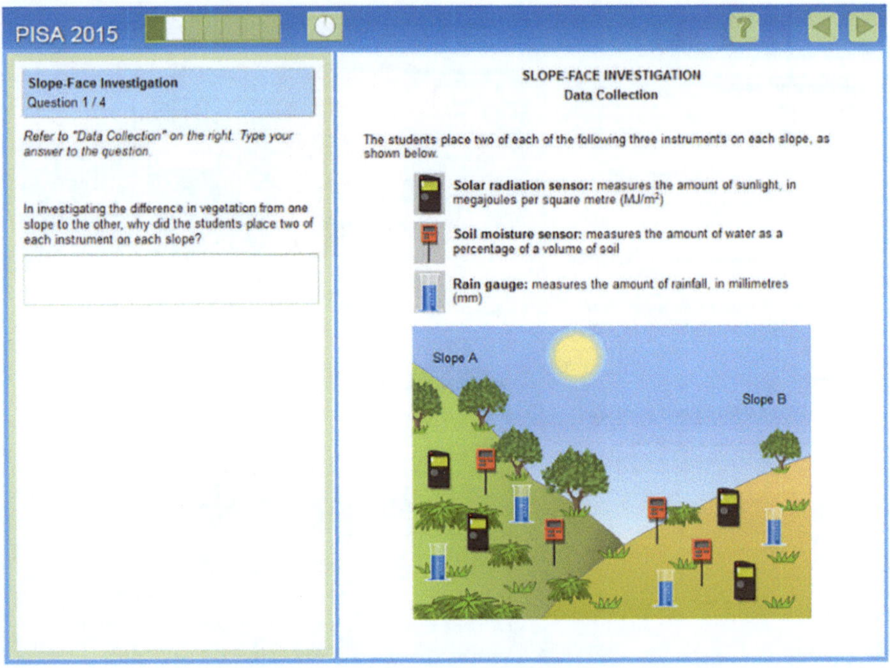

Fig. 4.4 Example PISA science item: slope-face investigation

4.3.2 Reading

Interactive tasks tended to be a good match in the domain of science but were less applicable in the domain of reading, in which navigation through information and conceptual understanding takes precedence over evaluation through trial and error. Reading was the core domain in PISA 2018, and its content capitalized on new mediums for reading and sharing information. As information technologies increasingly pervade modern life and work, digital screens have replaced printed documents, which in turn have increased the amount and altered the structure of published information, as well as the ways in which individuals interact with written content. While the ability to read and comprehend a single text remains important, there has been a shift in the competencies required for a new, more relevant construct of reading literacy toward analysis and integration of relevant information from multiple sources. Hence, in PISA 2018's revised definition of reading literacy, new aspects of digital literacy, including the evaluation of all forms of handwritten, printed, or screen-based texts, were added as an integral part of the literacy construct, in addition to understanding, using, and reflecting on the texts. Text processing, including reading fluency, locating information, understanding, as well as evaluating and reflecting on the content, and task management processes were identified as underlying the reading processes. Texts and scenarios were also updated and refined in the design of the PISA assessment (OECD, 2019a).

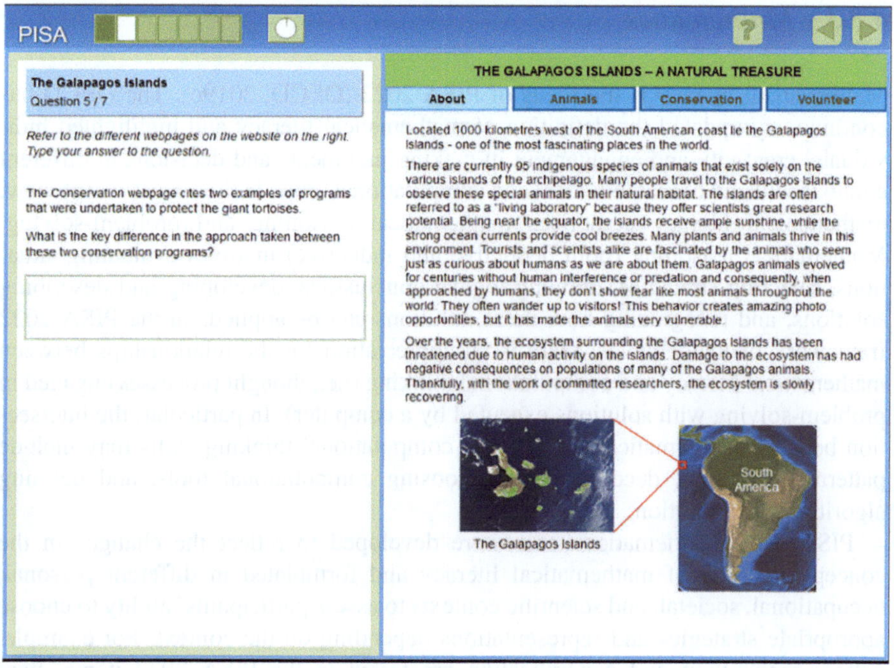

Fig. 4.5 Example PISA reading item: the Galapagos Islands

In particular, PISA 2018 included scenario-based assessments to provide overarching goals of one's reading activity and to enhance engagement and validity of the large-scale international assessments. The scenarios were designed to be immersive and mimic the authentic personal, public, educational, or occupational situations, and elicit one's natural interactions with the source material, such as researching information on an organization's website. Figure 4.5 (OECD, 2019d) shows an example of one PISA 2018 reading item in which the student must navigate across different web pages to answer the item question.

As a final note, new item types and constructs do not always need to be high-tech. In a large-scale international assessment like PISA, both floor and ceiling effects (e.g., making the test too difficult or too easy for any individual student) are worrisome possibilities that need to be addressed in computer-based assessment. To increase measurement precision at the lower end of the proficiency distribution, the subscale of reading fluency, an assessment of basic literacy, was given to all students prior to taking any of the main reading domain items. Reading fluency items are sentence-level, efficiency tasks, in which students must determine whether the material presented is a sentence or not. There is a speeded component to this subscale as well, in which a series of eleven "sentences" of 5–7 words must be responded to within three minutes. Both accuracy of the student response and time it took to respond to these generally low-demand conditions were recorded by the assessment delivery software.

4.3.3 Mathematics

Mathematical literacy is the focus of PISA 2022 (OECD, 2019c). The OECD has continuously updated the definition of mathematical literacy and highlighted individuals' creativity and engagement in making judgments and decisions in different contexts with aid of mathematical and computational tools. In their most recent view, mathematical literacy comprises mathematical reasoning and problem-solving. Mathematical reasoning (both deductive and inductive) involves evaluating situations, selecting strategies, drawing logical conclusions, developing and describing solutions, and recognizing how those solutions can be applied. In the PISA 2022 framework of mathematics, there is an appreciation for the relationships between mathematical literacy and computational thinking (i.e., thought processes engaged in problem-solving with solutions executed by a computer). In particular, the intersection between mathematical literacy and computational thinking skills may include pattern recognition, decomposition, choosing computational tools, and defining algorithms in a solution.

PISA 2022 mathematics items were developed to reflect the changes in the conceptualization of mathematical literacy and formulated in different personal, occupational, societal, and scientific contexts to assess participants' ability to choose appropriate strategies and representations depending on the context. For example, pattern recognition and generalization are tested in the PISA 2022 items. Item formats include selected-response items and open and closed constructed-response items. Participants may be asked to show their work or provide constructed response rationale to support their computations. An on-screen text editor tool allows students to incorporate common mathematical operators and functions into constructed responses, such as square root, exponent, and pi. Further, a computer-based assessment may enable the use of hints, statistical software, geometric construction, visualization utilities (e.g., use of a spreadsheet with sorting), and virtual measuring instruments beyond traditional calculators, reflecting the increasingly popular digital tools with which individuals solve real-world problems. Simulation-based assessments are included in PISA 2022, in which participants use a simulator to collect data and identify changes and relations. For example, the savings simulator, shown in Fig. 4.6, requires the student to manipulate a set of variables (e.g., savings period, monthly deposit, annual interest rate, and total savings) to collect data in order to answer the questions (OECD, 2019c).

With a growing interdependency between mathematical literacy and the use of computer technology to reach conclusions mathematically, the PISA 2022 mathematics assessment is keeping pace with modern realities in education and the workplace. However, the mathematics framework developers identify the distinction between mathematical demands and computer-based item demands when developing new items and new item types. Specifically, it is important that the interface of new items not be difficult for the student to use; otherwise, the proficiency in using the digital interface may be what is measured by a particular item, not the desired construct of mathematical proficiency. It remains important to be cautious and

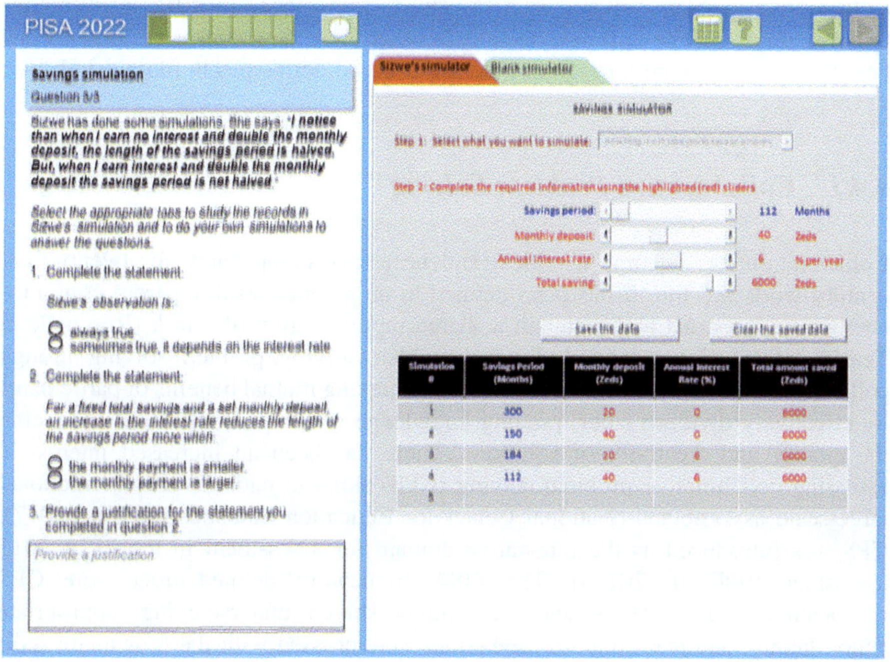

Fig. 4.6 Example PISA mathematics item: savings simulation

cognizant in the development of new items and constructs such that the collected data are interpretable and free of unnecessary noise.

As a final note, an accompanying background questionnaire is administered for each PISA domain; in 2022, data on mathematical reasoning, computational thinking, and four focal content areas (i.e., growth phenomena, geometric approximation, computer simulations, and conditional decision making) will be collected.

4.4 Innovative Domains and New Constructs for a Global Environment

Turning from technology-driven constructs, the next sections look at the innovative constructs in PISA, commonly assessing twenty-first-century skills. Collaboration and measuring individuals' contributions to collaborative efforts is a major field of research in twenty-first-century skills. Similarly, global and intercultural competencies have become a greater concern as college students from different cultures and business professionals from different countries come together both in-person and online. Finally, modern problems, such as climate change or dealing with the fallout of a global pandemic, may need creative solutions. With sights set on the future, PISA's innovative domains for the past three cycles were created to measure collaborative problem-solving in 2015, global competence in 2018, and creative thinking in 2022. The next few sections look at these new construct domains and how they are measured through new scenario-based item types in more detail.

4.4.1 Collaborative Problem Solving

Collaboration has become a critical and necessary competency for twenty-first-century work and life. It has been defined as an activity resulting from continued attempts to construct and maintain a shared representation of a task (Roschelle & Teasley, 1995). Compared to individual problem-solving, problem-solving through collaboration has the potential advantage of achieving mutual benefits of participants by effectively dividing labor, pooling knowledge and experiences, and enhancing the quality and creativity of solutions. There has been an increased interest in assessing collaborative problem solving (CPS) skills in national and international large-scale assessments (National Center for Education Statistics [NCES], 2017). CPS was introduced as the innovative domain for assessment in the PISA 2015 assessment (OECD, 2019a). The PISA framework defined three core CPS competencies—establishing and maintaining shared understanding, appropriate action taking, and team organization—that were crossed with the four major individual problem-solving processes to be assessed in various contexts.

As depicted in Fig. 4.7 (OECD, 2015), much of the collaborative portion of CPS scenario-based tasks was handled through a chat interface, in which students selected from a set of responses on how best to collaborate to solve the problem. This, essentially, made each step in the collaboration process a multiple-choice item in terms of scoring. Although these are scenario-based items, outcome data presented in a multiple-choice item format made scoring and scaling proficiency estimates in the CPS domain similar to the approach used in the other PISA domains. Students did not collaborate with one another but, rather, with virtual agents to collaborate and work through the tasks. These choices in the interface design and scoring were made in part because of the unique sampling and reporting design of PISA, in which estimates of country-level competencies are of interest. The choice to use an agent rather than pairing students with other students was for consistency and comparability of results. In order for proficiency estimates to be comparable across countries, students must take the same assessment under the same testing conditions, which includes the same scenarios in which they are assessed on collaborative skills. Human to agent collaboration is a much more controlled environment than human to human collaboration; team composition and individual experience and skills of the other team members, need to be controlled for in the measurement in order for one student's responses to the item to be comparable to that of others. In order to vary the CPS context, agents were programmed to assume different roles, attitudes, and levels of competency across different assessment parts. The types of interactions between the agents and the participants were pre-defined based on a previous state of the participants' actions or the degree to which the problem had been solved. This design

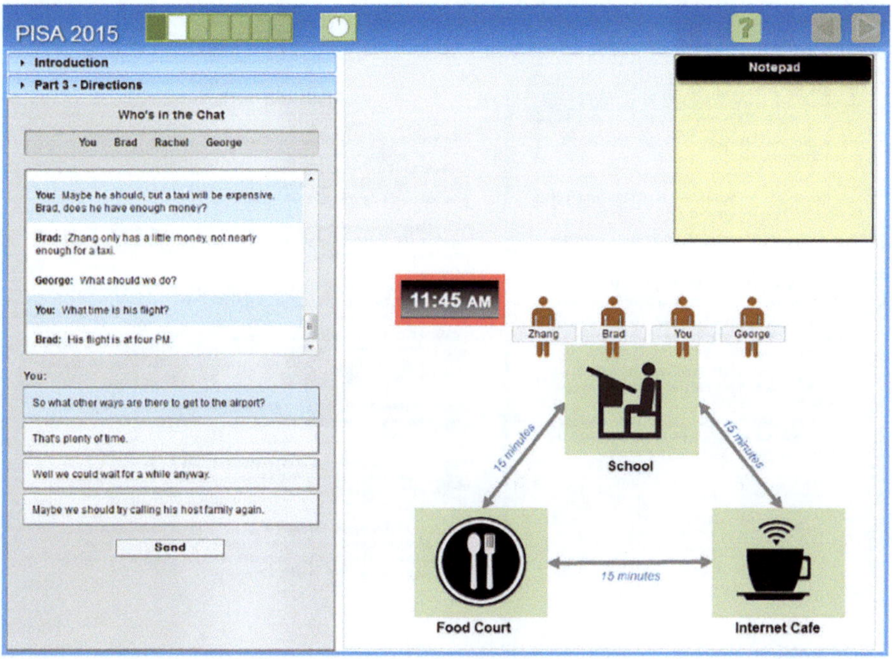

Fig. 4.7 Example PISA collaborative problem solving item: the visit

also enabled the administration of PISA under a variety of logistic constraints, such as a 30-min time block, limited network capacity for real-time communications, and cultural and language differences across countries.

4.4.2 Global Competence

In order to prepare young people to become global citizens in the era of economic, political, and cultural globalization, PISA 2018 took up the challenge to assess global competence as an innovative domain and defined it as the capability to "examine local, global and intercultural issues, understand and appreciate different perspectives and worldviews, interact successfully and respectfully with others, and take responsible action toward sustainability and collective well-being" (OECD, 2019a). Operationally, PISA 2018 assessed a construct of "global understanding," which was identified as supported by four cognitive processes, namely capacities: (1) to evaluate information, formulate arguments, and explain complex situations; (2) to identify and analyze multiple perspectives and worldviews; (3) to understand differences in communication in diverse social and cultural contexts; and (4) to evaluate actions and short and long-term consequences. When combined with knowledge of global issues and intercultural knowledge, these four cognitive

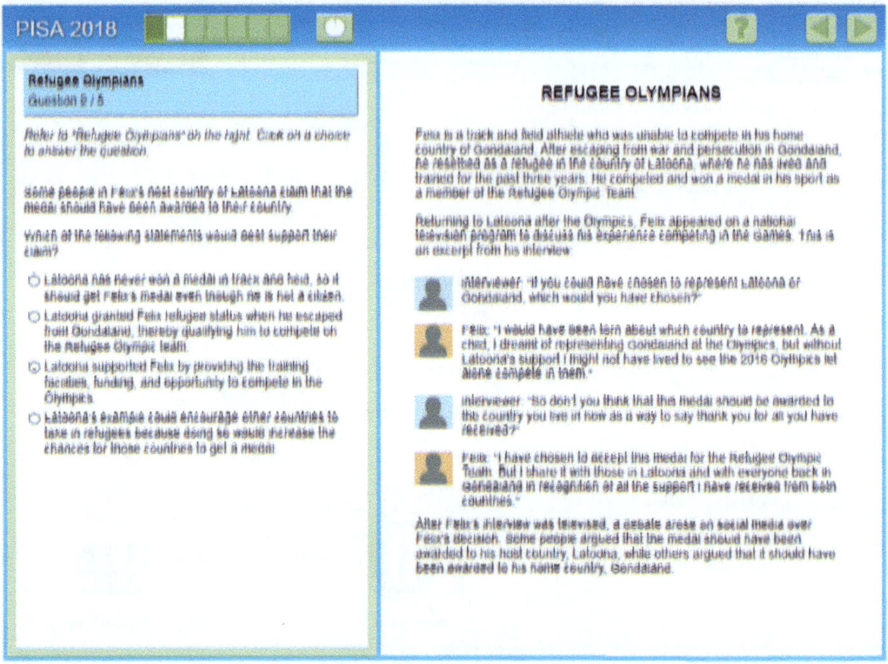

Fig. 4.8 Example PISA global competence item: refugee Olympians

processes respectively correspond to the four dimensions in the definition of global competence mentioned earlier, which are information processing, perspective taking, communication, and action.

The PISA assessment of global competence consisted of a cognitive test on the construct of "global understanding," and a questionnaire to assess one's knowledge of global and cultural issues, social and cognitive skills, and attitudes of openness. The cognitive tests included scenario-based items that covered a wide range of content domains, contexts, and complexity levels. The presentation of the stimulus materials (text and image) was tailored to engage 15-year-olds and reduce the reading load. In those tasks, participants were asked to assume roles such as researchers, reporters, mediators, team members, and debaters. An example item is shown in Fig. 4.8, in which students are asked to take on multiple perspectives (OECD, 2020). Both open-ended and closed responses were collected, with open-ended responses scored according to expert rubrics.

4.4.3 Creative Thinking

Creative thinking is another important competency needed for modern society. While the development of artificial intelligence allowed machines to conduct repetitive work with high efficiency, creative thinking is a quality that enables an individual to come up with solutions outside the box and take on jobs that machines cannot do. PISA 2022 (OECD, 2019b) will assess creative thinking as an innovative domain and has defined it as "the competence to engage productively in the generation, evaluation and improvement of ideas, that can result in original and effective solutions, advances in knowledge and impactful expressions of imagination." It focuses on the creative thinking skills that can be facilitated and developed at school with a goal to inform relevant teaching activities and education systems, minimizes the impact of intelligence, artistic talent, and domain knowledge, and highlights the engagement processes of generating, evaluating, and iterating upon ideas. Because standardized tests have different constraints related to test time and platforms, PISA 2022 focused on creative written and visual expression and creative scientific and social problem solving to provide broad coverage. In terms of measurement, competencies are expected to be evidenced through generating diverse and creative ideas and evaluating and improving ideas. Items and the associated coding schema for assessing creative thinking have been developed based on the evidence-centered design framework. A typical item might ask participants to freely write down ideas, an evaluation of a solution, or improvements for a specific task within a suggested time limit (Rosen et al., 2020). For example, an item may ask students to generate ideas for what all people can do to save water or ideas for how to build a bicycle of the future. Items may also have a more generally creative aim, such as those requiring visual expression, as shown in Fig. 4.9; such items will also incorporate a drawing tool and toolbox, often with available stamps, colors, and shapes (OECD, 2019b). Indicators on task motivation and features that facilitate exploration are also considered during test development in order to record relevant process data and encourage research.

To sum up, over the years, the PISA assessments have continuously updated the constructs in the core domains and introduced new constructs in the innovative domains. New item types in a typical form of scenario-based assessments have been designed and developed in pursuit of better measurement quality. As we mentioned above in separate domains, similar innovations with potential differences in construct definition and implementation can also be seen in other large-scale assessments such as NAEP and PIAAC. For instance, the assessment teams have shared knowledge and experience in assessing new constructs such as problem solving and collaborative problem-solving skills for better practices and measurement qualities. In particular, as we will notice in the next section, researchers working on different assessments have been developing analytical methods that address the same set of data issues and are applicable across similar assessments. See, for example, Bergner and von Davier (2019) for a review of the innovative constructs and methods in NAEP.

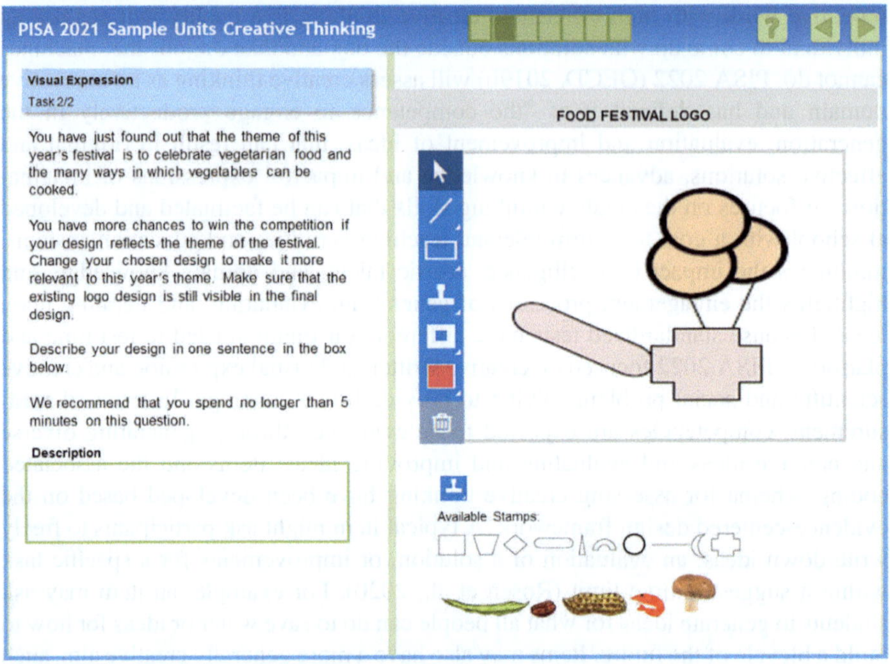

Fig. 4.9 Example PISA creative thinking item: visual expression

4.5 Developments and Challenges in Analyzing New Data

Scenario-based assessment items, such as simulated experiments, immersive digital environments, and collaborative tasks, generally have been scored with expert rubrics and may be modeled using item response theory, as with the examples of PISA and PIAAC. However, with the collection of both process data and simultaneous outcome data, computational psychometrics emerges as an active research field in support of such assessments, particularly with collaborative problem solving (CPS) tasks (Polyak et al., 2017; von Davier, 2017). Data from CPS tasks include both individual and team outcomes and process data. Outcome data are the individual or team's correct/incorrect/partial credit responses to a task and can often be analyzed with traditional psychometric models. With collaborative tasks, actions, communications, products (potentially including answers to probes that explicitly assess one's cognitive state), and response-time can be used to infer participants' states and proficiency in the CPS cognitive and social processes. Some constructs, such as turn-taking and participation, are easy to measure from the process data, while other constructs, such as action-taking can be harder to measure from the sheer volume and complexity of potential process data space. Process data that are considered as independent outcome data, for example, response times or a selection of action/chat, can be analyzed with traditional modeling approaches. When the time dependence between the process data is of interest, time-series methods and data mining tools are often needed to analyze the complex data in collaboration (NCES, 2017).

As mentioned before, types of process data range from simple measures, such as response time, use of hints, and multiple attempts, to time series data of actions and keystroke logs (Zhang et al., 2016, see also Part IV of this book). In scenario-based assessments, typical process data are action sequences, with or without timestamps, in which each action is considered as a categorical value. Researchers have used different statistical modeling and data mining tools to explore the data patterns, represent the data, and extract measures or features from the data that are associated with success or failure outcomes. These insights are not only useful for understanding participants' behavior but also instructional for designing iterated assessments.

One example use of process data is with the NAEP Technology and Engineering Literacy (TEL) scenario-based assessments, in which action sequences of eighth graders in troubleshooting and repairing a broken hand pump were collected. They were further rated using expert rubrics on two dimensions of problem-solving in technology-rich environments—systematicity and efficiency. With this data, researchers have developed methods to analyze the categorical action sequences and discover the psychometric properties that are related to human-rated scores (Hao et al., 2015; Shu et al., 2017; Zhu et al., 2016; Bergner et al. 2014).

Another example is a recent study on strengthening the validity of NAEP Science technology-enhanced tasks through the analysis of process data and testing the impact of various design features on student' engagement (Rosen et al., 2021). In the study, the criteria for identifying low-engagement tasks focused on events observed in the process data including missing responses, the correctness of the responses, time spent on each item and entire task, etc. The findings indicate that reducing cognitive load by changing item types, removing dependencies, breaking grouped items into separate items, and modifying the mode of data generation had a positive impact on performance as well as student preference in terms of making the task seem more engaging and accessible.

In international large-scale assessments, such as PISA and PIAAC, statistical and mining approaches have also been utilized to represent and analyze action sequences in problem-solving tasks (e.g., Liu et al., 2018; Qiao & Jiao, 2018). Xu et al. (2018) applied a latent class analysis approach to analyze recurrent events in a problem-solving item in PISA 2012. Chen et al. (2019) used an event history analysis model to predict the final outcome and duration of problem-solving from its patterns. Ren et al. (2019) extracted and clustered goal pursuit behavior from process data to investigate goal balancing in problem-solving tasks. He and von Davier (2015) extracted n-gram adjacent substrings from action sequences as features and explored their relations with successful and unsuccessful problem-solving behaviors. More recently, Liao et al. (2019) identified typical patterns in action sequences between subgroups characterized by background variables that significantly predicted general problem-solving skills in PIAAC. Moreover, latent features could also be extracted from action sequences, for instance, in the multidimensional scaling framework, and were found to be predictive of different background variables and assessment scores in PIAAC (Tang et al., 2020).

Scoring and making inferences about collaborative actions may pose further challenges. As touched on in the example of the Collaborative Problem Solving domain in PISA, collaboration with a virtual agent (human-to-agent, H-A, assessments) tend to have a finite set of agent responses and/or a finite set of possible human inputs, such as with the collaborative problem-solving domain in PISA 2015, in which the majority of CPS items could be scored as simple or complex multiple choice. Human-to-human (H-H) assessment, however, or even a hybrid of the two, can be expected to have far more response variability. Yuan et al. (2019) proposed a paradigm to extract indicators from H-H process data and modeling the dyad data with a within-item multidimensional Rasch model, which could be feasible for CPS assessment. Actually, in studies comparing the H-H and H-A assessments, similarities and differences in participants' performance were found in the H-H and H-A modes. In one study, while there was no significant difference in other performance measures, participants assessed in the H-A mode had higher CPS scores than those assessed in the H-H mode, probably because CPS with a computer agent involved more shared understanding, progress monitoring, and feedback (Rosen & Foltz, 2014). In another study, although H-H collaborations facilitated the production of more problem solutions, the participants in the H-A mode were found to encounter significantly more conflict situations that were necessary to form shared understanding than in the H-H mode (Rosen, 2014). These results suggested the potential of using H-A interactions to assess a wide range of CPS skills despite its constraint on the freedom of communication.

Nevertheless, modeling and validation of scoring of more complex process data, such as those with time-dependent structures, needs further development and examination. Such modeling and validation work can be especially useful in open-ended scenarios in which no optimal action sequence exists.

4.6 Future Directions

Technology has facilitated promising developments of new constructs and item types in large-scale assessments. Because large-scale assessments often need to cover a large population and a variety of items in wide-ranging contexts, technology will play an increasingly important role in scaling up the assessments and shaping future learning and teaching.

While it is not clear exactly how technology or global change may influence new constructs of interest, it has become clear over the last few years that the future of at least some industries may become remote. Working or learning online, either alone or through online collaboration, has raised concerns over social and emotional well-being and posed the question of how well students learn remotely and through online tasks. New constructs may include measurements of emotional intelligence or actual human-to-human interactions, rather than human-to-agent. Advances are being made in methods of measuring how individuals actually relate to one another through digital interfaces. For example, Educational Testing Service's (ETS) prototype for science collaborative assessment, the Tetralogue system (von Davier et al., 2017), has several versions of collaboration modes in which H-H conversations are allowed, and computerized agents can serve as a moderator and/or a collaborator. However, it is uncertain how such approaches would scale to large-scale assessments.

Regarding new item types, the OECD's PISA and PIAAC assessments, so far, have met the need of assessing new technology-based or global-change constructs using scenario-based items, such as immersive digital environments, simulated experiments with data collection, and simulated social interactions through digital interfaces, like chat. While these are all examples of simulated settings, adding gamified elements to item tasks could be a next step in new item types for measuring new constructs. Gamified elements would likely require some adaptive algorithms to respond to different actions taken by the respondent, and measurement of such items would likely require some verification.

Scenario-based assessments can further integrate virtual reality (VR) and augmented reality (AR) tools to create engaging scenarios. These technologies can utilize high-dimensional simulations to evoke and capture participants' natural reactions that are not currently captured by basic computer systems, while maintaining the assessments in a standardized way. In these scenarios, participants' experiences of the environment and the feedback they receive as a result of their behavior can be enhanced. The simulations can also create new environments where physical rules can change, enabling assessments of constructs, such as creativity, and transferability or generalization of skills. While the VR and AR tools hold a lot of promise in assessments, the current cost of setting up the tools and training test administrators can be high, preventing large-scale implementation at present. However, just as computers in schools were once few and limited, new technologies may become more common and cost-effective, enabling their use in large-scale assessment. Likewise, with advances in sensory technologies and their more widespread use, the collection of electrophysiological, eye-tracking, and physiological data that give real-time measurements directly about the cognitive and affective states is also promising in future assessments. However, as mentioned before, there can be many challenges in understanding, representing, and analyzing new types of data for inferential and predictive purposes. In addition, the collection of behavioral and process data requires efficient data storage and data management capabilities.

As a closing remark, while we embrace that technology drives and supports the development of new constructs and item types in international large-scale assessments, it is important to be mindful about the ultimate goal of these assessments, to inform education across populations (von Davier et al., 2013). The processes from defining a construct, developing valid and reliable items and scoring rules, to collecting data and interpreting analytic results, will take time and thoughtful consideration, as they will have an impact on educational systems and practices, aiming to improve teaching and learning.

References

Arieli-Attali, M., Ward, S., Thomas, J., Deonovic, B., & von Davier, A. A. (2019). The expanded evidence-centered design (e-ECD) for learning and assessment systems: A framework for incorporating learning goals and processes within assessment design. *Frontiers in Psychology, 10*, 853. https://doi.org/10.3389/fpsyg.2019.00853. PMID: 31105616; PMCID: PMC6498139.

Beller, M. (2013). Technologies in large-scale assessments: New directions, challenges, and opportunities. In M. von Davier, E. Gonzalez, I. Kirsch, & K. Yamamoto (Eds.), *The role of international large-scale assessments: Perspectives from technology, economy, and educational research* (pp. 25–45). Springer. https://doi.org/10.1007/978-94-007-4629-9

Bennett, R. E., Persky, H., Weiss, A. R., & Jenkins, F. (2007). *Problem solving in technology-rich environments. A report from the NAEP technology-based assessment project* (NCES 2007-466). National Center for Education Statistics, U.S. Department of Education. http://nces.ed.gov/nationsreportcard/pdf/studies/2007466.pdf

Bergner, Y., Shu, Z., & von Davier, A. A. (2014). Visualization and confirmatory clustering of sequence data from a simulation-based assessment task. In *Proceedings of 7th international conference on educational data mining* (pp. 177–184). http://educationaldatamining.org/EDM2014/uploads/procs2014/long%20papers/177_EDM-2014-Full.pdf

Bergner, Y., & von Davier, A. A. (2019). Process data in NAEP: Past, present, and future. *Journal of Educational and Behavioral Statistics, 44*(6), 706–732. https://doi.org/10.3102/1076998618784700

Buchner, A., & Funke, J. (1993). Finite-state automata: Dynamic task environments in problem-solving research. *The Quarterly Journal of Experimental Psychology, 46*(1), 83–118.

Chen, Y., Li, X., Liu, J., & Ying, Z. (2019). Statistical analysis of complex problem-solving process data: An event history analysis approach. *Frontiers in Psychology, 10*, 486. https://www.frontiersin.org/article/10.3389/fpsyg.2019.00486

Dicerbo, K. E. (2017). Building the evidentiary argument in game-based assessment. *Journal of Applied Testing Technology, 18*(S1), 7–18.

Greiff, S., & Funke, J. (2009). Measuring complex problem solving: {T}he {Micro-DYN} approach. In F. Scheuermann & J. Björnsson (Eds.), *The transition to computer-based assessment: New approaches to skills assessment and implications for large-scale testing* (pp. 157–163). https://doi.org/10.2788/60083

Griffin, P., McGaw, B., & Care, E. (2012). *Assessment and teaching of 21st century skills*. Springer.

Hao, J., Shu, Z., & von Davier, A. A. (2015). Analyzing process data from game/scenario-based tasks: An edit distance approach. *Journal of Educational Data Mining, 7*(1), 33–50. http://www.educationaldatamining.org/JEDM/index.php/JEDM/article/view/JEDM072

He, Q., & von Davier, M. (2015). Analyzing process data from problem-solving items with N-grams. https://doi.org/10.4018/978-1-4666-9441-5.ch029

Khorramdel, L., von Davier, M., & Yamamoto, K. (2024). Technology as a driver of the evolution of international large-scale assessments: Constructs, frameworks, and the psychometric qualities of test results. In L. Khorramdel, M. von Davier, & K. Yamamoto (Eds.), *Innovative digital-based international large-scale assessments—Foundations, methodologies and quality assurance*. Springer.

Kirsch, I., & Lennon, M. L. (2017). PIAAC: a new design for a new era. *Large-Scale Assessments in Education, 5*(1). https://doi.org/10.1186/s40536-017-0046-6

Liao, D., He, Q., & Jiao, H. (2019). Mapping background variables with sequential patterns in problem-solving environments: An investigation of United States adults' employment status in PIAAC. *Frontiers in Psychology, 10*, 646. https://doi.org/10.3389/fpsyg.2019.00646

Liu, H., Liu, Y., & Li, M. (2018). Analysis of process data of PISA 2012 computer-based problem solving: Application of the modified multilevel mixture IRT model. *Frontiers in Psychology, 9*, 1372. https://www.frontiersin.org/article/10.3389/fpsyg.2018.01372

Mislevy, R. J., Oranje, A., Bauer, M. I., von Davier, A. A., Hao, J., Corrigan, S., Hoffman, E., DiCerbo, K., & John, M. (2014). *Psychometric considerations in game-based assessment.* https://web.archive.org/web/20160320151604/http://www.instituteofplay.org/wp-content/uploads/2014/02/GlassLab_GBA1_WhitePaperFull.pdf

Mislevy, R. J., Steinberg, L. S., & Almond, R. G. (2003). Focus article: On the structure of educational assessments. *Measurement: Interdisciplinary Research and Perspectives, 1*(1), 3–62. https://doi.org/10.1207/S15366359MEA0101_02

National Center for Education Statistics (NCES). (2017). Collaborative problem solving: Considerations for the National Assessment of Educational Progress. https://hdl.handle.net/10993/31897

National Center for Education Statistics (NCES). (n.d.). *Sample problem-solving in technology-rich environments items*. Programme for the International Assessment of Adult Competencies (PIAAC). Institute for Education Sciences. https://nces.ed.gov/surveys/piaac/sample_pstre.asp

OECD. (2009). *PIAAC problem solving in technology-rich environments: A conceptual framework* (OECD Education Working Papers, No. 36). https://doi.org/10.1787/220262483674.

OECD. (2013). *PISA 2012 assessment and analytical framework: Mathematics, reading, science, problem solving and financial literacy*. https://doi.org/10.1787/9789264190511-en

OECD. (2014). *PISA 2012 results: Creative problem solving, students skills in tackling real-life problems (volume V)*. https://www.oecd.org/pisa/keyfindings/pisa-2012-results-volume-V.pdf

OECD. (2015). *PISA 2015 released field trial cognitive items.* https://www.oecd.org/pisa/pisaproducts/PISA2015-Released-FT-Cognitive-Items.pdf

OECD. (2017). *PISA 2015 assessment and analytical framework: Science, reading, mathematic, financial literacy and collaborative problem solving.* https://doi.org/10.1787/9789264281820-en

OECD (2019a). *PISA 2018 assessment and analytical framework.* https://doi.org/10.1787/b25efab8-en

OECD. (2019b). *PISA 2022 creative thinking framework (third draft).* https://www.oecd.org/pisa/publications/PISA-2021-creative-thinking-framework.pdf

OECD. (2019c). *PISA 2022 mathematics framework.* https://pisa2022-maths.oecd.org/ca/index.html

OECD. (2019d). *PISA 2018 released field trial new reading items.* https://www.oecd.org/pisa/test/PISA-2018-Released-New-REA-Items.pdf

OECD. (2020). *PISA 2018 results (volume VI).* https://doi.org/10.1787/d5f68679-en

Polyak, S. T., von Davier, A. A., & Peterschmidt, K. (2017). Computational psychometrics for the measurement of collaborative problem solving skills. *Frontiers in Psychology*. https://doi.org/10.3389/fpsyg.2017.02029

Qiao, X., & Jiao, H. (2018). Data mining techniques in analyzing process data: A didactic. *Frontiers in Psychology, 9*, 2231. Retrieved from https://www.frontiersin.org/article/10.3389/fpsyg.2018.02231

Ren, Y., Luo, F., Ren, P., Bai, D., Li, X., & Liu, H. (2019). Exploring multiple goals balancing in complex problem solving based on log data. *Frontiers in Psychology, 10*, 1975. Retrieved from https://www.frontiersin.org/article/10.3389/fpsyg.2019.01975

Roschelle, J., & Teasley, S. D. (1995). The construction of shared knowledge in collaborative problem solving. In C. O'Malley (Ed.), *Computer supported collaborative learning* (pp. 69–97). Springer. https://doi.org/10.1007/978-3-642-85098-1_5

Rosen, Y. (2014). Comparability of conflict opportunities in human-to-human and human-to-agent online collaborative problem solving. *Technology, Knowledge, and Learning, 19*(1-2), 147–164. https://doi.org/10.1007/s10758-014-9229-1

Rosen, Y., Stoeffler, K., San Pedro, S., & Schneiders, J. (2021). *Strengthening the validity of NAEP through increased student engagement in technology-enhanced assessments: Research report.*

Rosen, Y., Stoeffler, K., & Simmering, V. (2020). Imagine: Design for creative thinking, learning, and assessment in schools. *Journal of Intelligence, 8*(2), 16. https://www.mdpi.com/2079-3200/8/2/16

Rosen, Y. V., & Foltz, P. W. (2014). Assessing collaborative problem solving through automated technologies. *Research and Practice in Technology Enhanced Learning, 9*(3), 389–410.

Rupp, A. A., Gushta, M., Mislevy, R. J., & Shaffer, D. W. (2010). Evidence-centered design of epistemic games: Measurement principles for complex learning environments. *Journal of Technology, Learning, and Assessment, 8*(4), 1–48. https://ejournals.bc.edu/index.php/jtla/article/download/1623/1467/0

Samejima, F. (1969). Estimation of latent ability using a response pattern of graded scores. *Psychometrika Monograph Supplement, 34*(4, Pt. 2) https://psycnet.apa.org/record/1972-04809-001

Shu, Z., Bergner, Y., Zhu, M., Hao, J., & von Davier, A. A. (2017). An item response theory analysis of problem-solving processes in scenario-based tasks. *Psychological Test and Assessment Modeling, 59*(1), 109–131. http://www.psychologie-aktuell.com/fileadmin/download/ptam/1-2017_20170323/07_Shu.pdf

Shute, V. J., Ventura, M., Bauer, M., & Zapata-Rivera, D. (2009). Melding the power of serious games and embedded assessment to monitor and foster learning: Flow and grow. *The Social Science of Serious Games: Theories and Applications, 1*(643), 295–321.

Tang, X., Wang, Z., He, Q., Liu, J., & Ying, Z. (2020). Latent feature extraction for process data via multidimensional scaling. *Psychometrika, 85*(2), 378–397. https://doi.org/10.1007/s11336-020-09708-3

von Davier, A. A. (2017). Computational psychometrics in support of collaborative educational assessments. *Journal of Educational Measurement, 54*(1), 3–11. https://doi.org/10.1111/jedm.12129

von Davier, A. A., Hao, J., Liu, L., & Kyllonen, P. (2017). Interdisciplinary research agenda in support of assessment of collaborative problem solving: Lessons learned from developing a collaborative science assessment prototype. *Computers in Human Behavior*. https://doi.org/10.1016/j.chb.2017.04.059

von Davier, A. A., Misley, R., & Hao, J. (2021). *Computational psychometrics. New methodologies for a new generation of digital learning and assessment*. Springer.

von Davier, M., Gonzalez, E., Kirsch, I., & Yamamoto, K. (2013). *The role of international large-scale assessments: Perspectives from technology, economy, and educational research*. Springer. https://doi.org/10.1007/978-94-007-4629-9

Wagner, M., & Kerzabi, E. (2024). The digital testing platform for PIAAC and PISA. In L. Khorramdel, M. von Davier, & K. Yamamoto (Eds.), *Innovative digital-based international large-scale assessments—Foundations, methodologies and quality assurance*. Springer.

Xu, H., Fang, G., Chen, Y., Liu, J., & Ying, Z. (2018). Latent class analysis of recurrent events in problem-solving items. *Applied Psychological Measurement, 42*(6), 478–498. https://doi.org/10.1177/0146621617748325

Yuan, J., Xiao, Y., & Liu, H. (2019). Assessment of collaborative problem solving based on process stream data: A new paradigm for extracting indicators and modeling dyad data. *Frontiers in Psychology, 10*, 369. Retrieved from https://www.frontiersin.org/article/10.3389/fpsyg.2019.00369

Zapata-Rivera, D., & Bauer, M. (2012). Exploring the role of games in educational assessment. In J. Clarke-Midura, M. Mayrath, & D. Robinson (Eds.), *Technology-based assessments for 21st century skills: Theoretical and practical implication from modern research* (pp. 147–169). Information Age.

Zhang, M., Hao, J., Li, C., & Deane, P. (2016). Classification of writing patterns using keystroke logs. In L. A. van der Ark, D. M. Bolt, W.-C. Wang, J. A. Douglas, & M. Wiberg (Eds.), *Quantitative psychology research* (pp. 299–314). Springer. https://doi.org/10.1007/978-3-319-38759-8

Zhu, M., Shu, Z., & von Davier, A. A. (2016). Using networks to visualize and analyze process data for educational assessment. *Journal of Educational Measurement, 53*(2), 190–211. https://doi.org/10.1111/jedm.12107

Chapter 5
Improved Test Designs and Multistage Adaptive Testing in Large-Scale Assessments

Kentaro Yamamoto, Hyo Jeong Shin, Frederic Robin, Lale Khorramdel, and Laura Halderman

Abstract Multistage adaptive testing (MST) designs are becoming common in international large-scale assessments (ILSAs). For example, an MST was implemented for the Programme for the International Assessment of Adult Competencies (PIAAC) in 2012 for about 40 countries and in the 2018 and 2022 cycles of the Programme for International Student Assessment (PISA) for more than 70 countries. Using the example of PISA 2018 in the domain of Reading, this chapter addresses some important psychometric considerations in designing and implementing MST and analyzing the collected data in the context of ILSAs. Specifically, this chapter presents 1) the conditions considered when introducing MST, 2) how the items and units (i.e., a set of items) were developed and assigned, and 3) the aspects that were monitored and examined, including the performance of the MST delivery platform, decisions for scaling procedures in analyzing the data, and alternative ways of determining the proportion correct under the MST design. The chapter concludes with a summary on utilizing and monitoring the MST in ILSAs.

K. Yamamoto
Independent Consultant for Psychometric Methodology and Large-Scale Assessment (formerly ETS), Princeton, NJ, USA

H. J. Shin (✉)
Sogang University, Seoul, South Korea
e-mail: hshinedu@sogang.ac.kr

F. Robin · L. Halderman
ETS, Princeton, NJ, USA

L. Khorramdel
Lynch School of Education and Human Development, TIMSS & PIRLS International Study Center at Boston College, Chestnut Hill, MA, USA

L. Khorramdel et al. (eds.), *Innovative Digital-Based International Large-Scale Assessments*, Methodology of Educational Measurement and Assessment,
https://doi.org/10.1007/978-3-031-90951-1_5

5.1 Test Designs in ILSAs

International large-scale assessments (ILSAs) assess the skills, knowledge, and attributes of various populations across countries and economies, focusing on group-level scores (Kirsch et al., 2013). ILSAs are becoming more important and popular among researchers and policymakers to make inferences about performance of a broad range of populations across the world. Because the reporting goal of ILSAs is not to assess individual examinees but to describe populations and subpopulations worldwide (Rutkowski et al., 2010; von Davier et al., 2006), the test design in an ILSA needs to account for widely heterogeneous performance across students, within and across participating countries. Historically, large-scale surveys have often used some form of a balanced incomplete block (BIB) design, either in paper-based assessment (PBA) or computer-based assessment (CBA). A BIB spiraling design for large-scale surveys was developed by Messick et al. (1983) and was introduced in the National Assessment of Educational Progress (NAEP), in which each student was administered only a subset of the total item pool. The use of this BIB design in ILSAs allows a large number of items to be used to fully represent comprehensive assessment frameworks (Organisation for Economic Co-Operation and Development [OECD], 2016a, 2016b) and describe the relevant features of the proficiency of subpopulations.

Recently, with technology becoming an essential part of learning, problem solving, and daily communication, many ILSAs have transitioned or are transitioning to computer to remain relevant. For example, the Programme for International Student Assessment (PISA) switched from a PBA to a primarily CBA in the 2015 cycle, with about 90% of countries choosing CBA in 2018, and almost all selecting CBA in 2022. The Programme for the International Assessment of Adult Competencies (PIAAC) administered the CBA to 74% of test takers on average in its first cycle (beginning in 2012), and the Trends in International Mathematics and Science Study (TIMSS) began transitioning to CBA in 2019. Switching to CBA allows content experts to implement new, innovative, and more realistic item types (e.g., simulation-based) that incorporate features and functions only afforded by computer to assess new domains (e.g., collaborative problem solving in PISA 2015) and to collect process data (e.g., response times, number of actions).

The mode change of ILSAs from PBA to CBA also allows for the implementation of adaptive testing. All nonadaptive tests, including the BIB design, administer a set of items to students independent of their proficiencies. This approach measures examinees with average proficiency within a group quite well, but it measures less precisely those at the high and low ends of the proficiency scale (Hambleton & Swaminathan, 1985; Lord, 1980; Weiss, 1974). Nonadaptive designs cannot offer uniform measurement precision along the proficiency distribution of a population, which is a problem for obtaining accurate group measurement. As observed in PISA, characteristics of student populations, as well as their distribution of proficiencies, vary a great deal within a country as well as across countries. While nonadaptive designs (fixed form clusters) are not optimized for widely heterogeneous groups of students, adaptive testing is considered more promising. It is because adaptive testing has proven advantageous for obtaining relatively consistent, efficient, and precise measurement of examinees across the entire proficiency distribution (Lord, 1980; Oranje et al., 2014; Wainer, 1990).

Adaptive tests generally belong to one of two groups: item-level adaptive tests or multistage adaptive tests (MSTs). In item-level adaptive testing, adaptation occurs after every item, which is attractive for testing programs that focus on individuals in which a relatively narrow construct is assessed using a large item pool of independent items.[1] The ILSAs mentioned here aim to measure highly complex constructs that require contextual cognitive processing under realistic circumstances or in situations that are as close to the real world as possible (OECD, 2016b). The recent advent of interactive assessments has brought ILSAs closer to this goal. Scenario-based items/item sets can provide rich environments in which students manipulate item features and engage with the assessment, making items that are related by particular contexts and stimuli quite useful and applicable in ILSAs. For example, PISA 2018 includes scenarios in which students are provided an overarching purpose for reading a collection of thematically related texts in order to complete a high-level task, such as responding to some larger integrative question or writing a recommendation based on a set of texts.

Aligned with the requirements of ILSAs, MST adaptation occurs after a prescribed set of items, referred to as an *item set* or *unit*. Item sets are treated as the smallest unit in an MST and often comprise a single set of related items (i.e., the items may share the same item stimulus, such as a reading passage). MST designs "[strike] a balance among adaptability, practicality, measurement accuracy, and control over test forms" (Zenisky et al., 2010) and have several practical advantages. MST can be considered particularly useful when various constraints need to be taken into account. In ILSAs, measurement is based on a broad framework of numerous items, items are translated across multiple languages, and group-level scores of heterogeneous populations must be reported; with these constraints to consider, an MST design is a good choice. In brief, the primary goals of implementing MSTs in ILSAs are to reduce measurement error for heterogeneous populations without overburdening individual respondents, controlling the content composition of each test form, and facilitating different item types associated with a specific unit to best measure the construct. More details about advantages and expected gains of MST designs in PISA and PIAAC are explained in Yamamoto et al. (2018), and the implemented MST design in PISA 2018 is described in Yamamoto et al. (2019).

The following terms are used throughout this chapter:

- *item*: a task to which an examinee is directed to provide a response that receives a score, including machine- and human-coded items

[1] See van der Linden and Glas (2000, 2010) and Yan et al. (2014) for the most recent and comprehensive reviews about this topic.

- *unit* or *item set*: a set of items (usually ranging from two to eight items) that are designed to share similar or identical content, stimuli, or reading passage[2]
- *block*: a set of multiple units that constitutes each stage that are often called as 'module' in other literature
- *cluster* is used as a corresponding term for *block* in nonadaptive tests
- *stage*: a singular phase of an MST, after which, test adaptation occurs; hence, multi-*stage* adaptive assessment

In the following sections, we use the example of PISA 2018, with the major domain of Reading, to address some considerations in designing and implementing an MST and analyzing the collected data. More specifically, this chapter presents 1) what psychometric properties were considered in designing an MST, 2) how the items and units were assigned to each stage, 3) what aspects can be monitored to evaluate the administration of an MST, and 4) what scaling procedures are recommended for analyzing collected data. The chapter will conclude with practical guidance and implications for utilizing MSTs in international and national large-scale assessments.

5.2 A Prerequisite for Introducing the MST in PISA: Unit Order Effects

When implementing an MST in an ILSA, which inherently leads to different item or unit placement, it should be verified that psychometric properties of items hold, regardless of item or unit position across different blocks (i.e., absence of item or unit position effects). For example, suppose that one item is slightly harder when it appears in the second position than in the first. When an item response theory (IRT) model is applied, ignoring item order within the block, the item parameters represent the average of the item difficulty across the two positions (Mazzeo & von Davier, 2008). But proficiencies are overestimated relative to this average for the group of students taking the item in the first position and are underestimated for the group taking it in the second. In other words, when the item position effect is present, item difficulty would be dependent on the position and would translate to biases in proficiency estimation if not modelled (Mazzeo & von Davier, 2008). Therefore, item position effects are an important concern for assuring the quality of inferences in ILSAs (e.g., Le, 2007; Nagy et al., 2019; Wu et al., 2019).

In PISA, blocks consist of multiple units, and the unit-order positioning in each block can be different. Because the order of items is maintained within the unit, unit order effects can be investigated. For example, a certain unit can be presented as the first unit in some blocks, but as the second unit in others. Without a thorough examination of item or unit position effects, biases in the parameter estimation

[2]Note that the term "unit" used in this paper is closer to what is called "testlet" in the measurement literature (e.g., Wainer et al., 2007). Wainer and Kiely (1987) described "testlet" as a group of items that relate to a single topic, such as a reading passage.

may occur (Leary & Dorans, 1985). Hence, possible item position effects have to be examined in order to proceed with the MST, such as whether the magnitude of item or unit position effects is ignorable or not. Also, note that reliable item parameters can be estimated only in the main survey with proper sampling weights. The field trial is conducted in preparation for the main survey; thus, the data are limited in terms of sampling and accuracy of statistical inferences. In this section, we focus on the unit order effect as one of the important psychometric considerations for designing and implementing the MST in the PISA main survey.

5.2.1 Invariance of Item Parameters by Unit Order

To investigate the feasibility of an MST in the PISA 2018 main survey for the major domain of Reading, the PISA 2018 field trial was designed for CBA countries to investigate possible unit-order effects. While the order of items within a unit does not change, the position of a unit across blocks does change. That is, it is ideal for item parameters to be invariant across all positions of unit order. If the unit order is shown to impact item parameters and proficiency estimates, an MST design should not be implemented (e.g., a cluster design should be retained). To that end, field trial data was collected to study unit order effects by manipulating fixed and varying positions within 30-min (intact) blocks, and students were randomly assigned to three groups with different unit orders.

The study design used for the field trial was a type of randomized control trial in a sense that the treatment (unit order) was randomly manipulated between groups of students. The assumption of randomly equivalent groups allows us to examine whether the same administered units behave differently given unit order. The details of the design are shown in Fig. 5.1: In each CBA country, students were randomly assigned to one of these three groups, and the designated test forms in each group were randomly assigned. The unit order design is based on three different groups in which test forms are presented with different unit orders: in Group 1, trend Reading items are presented in a fixed unit order (FUO); in Group 2, trend and new Reading items are presented in a variable unit order (VUO); in Group 3, new Reading items are presented in an FUO. Each cluster consisted of multiple units, and the ordering of the units was always fixed and consistent in FUO forms in Groups 1 (forms 1–18) and 3 (forms 43–54). In contrast, unit order varied across the VUO forms in Group 2 (forms 19–42). For example, with three units (A, B, C in the order of easy to hard) per cluster, the order of units in FUO forms was always consistent as ABC, while the ordering of units in VUO forms was one of the two alternate orders, either ACB or BAC. More comprehensive sets of ordering were possible (e.g., BCA, CAB, CBA); however, these conditions were not considered. This decision was made because it was expected that test-taking motivation could be negatively impacted by having the most difficult unit (C) appear before relatively easier units (A or B). Note that each test form contains four clusters, with two alternate unit orders (ACB or BAC). When the test was initialized for each student, one of the permutations from among the

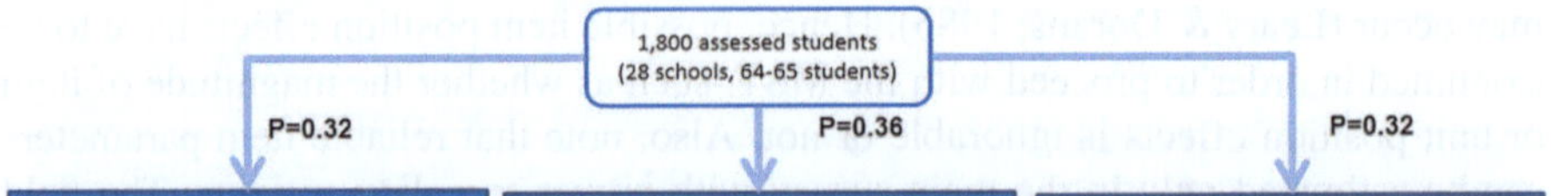

GROUP 1 – CBA Trend
FUO (Forms 01-18)

Forms	Cluster 1	Cluster 2	Cluster 3	Cluster 4
01	S1	S4	M1	M2
02	S3	S6	M3	M4
03	S5	S2	M5	M6ab
04	M2	M3	S2	S5
05	M4	M5	S4	S1
06	M6ab	M1	S6	S3
07	M1	M4	R1	R2
08	M3	M6ab	R3	R4
09	M5	M2	R5	R6ab
10	R2	R3	M2	M5
11	R4	R5	M4	M1
12	R6ab	R1	M6ab	M3
13	R1	R4	S1	S2
14	R3	R6ab	S3	S4
15	R5	R2	S5	S6
16	S2	S3	R2	R5
17	S4	S5	R4	R1
18	S6	S1	R6ab	R3

GROUP 2 – CBA Trend/New R
VUO (Forms 19-42)

Forms	Cluster 1	Cluster 2	Cluster 3	Cluster 4
19	R1	R14	R12	R7
20	R2	R16	R14	R9
21	R3	R18	R16	R11
22	R4	R8	R18	R13
23	R5	R10	R8	R15
24	R6ab	R12	R10	R17
25	R13	R1	R10	R9
26	R15	R2	R12	R11
27	R17	R3	R14	R13
28	R7	R4	R16	R15
29	R9	R5	R18	R17
30	R11	R6ab	R8	R7
31	R11	R18	R1	R8
32	R13	R8	R2	R10
33	R15	R10	R3	R12
34	R17	R12	R4	R14
35	R7	R14	R5	R16
36	R9	R16	R6ab	R18
37	R16	R17	R15	R1
38	R18	R7	R17	R2
39	R8	R9	R7	R3
40	R10	R11	R9	R4
41	R12	R13	R11	R5
42	R14	R15	R13	R6ab

GROUP 3 – CBA New R/GC
FUO (Forms 43-54)

Forms	Cluster 1	Cluster 2	Cluster 3	Cluster 4
43	R7	R8	R10	R14
44	R8	R9	R11	R15
45	R9	R10	R12	R16
46	R10	R11	R13	R17
47	R11	R12	R14	R18
48	R12	R13	R15	R7
49	R13	R14	R16	R8
50	R14	R15	R17	R9
51	R15	R16	R18	R10
52	R16	R17	R7	R11
53	R17	R18	R8	R12
54	R18	R7	R9	R13

Fig. 5.1 PISA 2018 field trial computer-based assessment design. Note. R1-R6ab represent CBA trend reading clusters and R7-R18 represent CBA new reading clusters. M1-M6ab represent CBA trend mathematics clusters and S1-S6 represent CBA trend science clusters (assembled from 2015 trend and new items). GC1-GC4 represent CBA new global competence clusters. Note: FUO = fixed unit order; VUO = variable unit order

16 different permutations of the four clusters (4^2; e.g., 1111, 1112, …, 2222) was randomly assigned.

Since all Reading items were used in both FUO and VUO forms, item performance could be compared between FUO and VUO. More specifically, trend Reading items could be compared between Group 1 (FUO) and Group 2 (VUO), and new Reading items could be compared between Group 3 (FUO) and Group 2 (VUO). Performance of items was investigated based on the proportion correct and response times, as well as the invariance of item parameters across different groups using multiple-group IRT models. These analyses were conducted using the data from 42 out of the 53 countries[3] that were administered VUO (Group 2).

Table 5.1 and Fig. 5.2 show the comparison of the average percent correct between FUO and VUO per cluster. At the cluster level, the differences in the average percentage correct between FUO and VUO ranged from 0.17 to 2.24 for trend items and 0.09 to 1.16 for new items. Across clusters, the difference between VUO and FUO was 0.62 for trend items and 0.12 for new items on average. Statistically, these differences were not significant: $z = 0.18$ for the first five clusters,

[3]These 53 countries, referred to as *Batch 1* countries, submitted field trial data relatively early. Although Batch 1 was not a complete PISA dataset, major psychometric decisions needed to be made in a timely manner in preparation for the main survey; therefore, Batch 1 did include data from the majority of PISA participating countries.

Table 5.1 Average percent correct by groups per reading cluster

	Cluster	Group 1 (FUO)	Group 2 (VUO)	Group 3 (FUO)
Trend reading items	R01	60.05	57.81	
	R02	50.83	51.00	
	R03	56.82	56.28	
	R04	64.47	63.80	
	R05	51.18	51.89	
	R6A	61.33	60.09	
	R6B	60.08	59.24	
	Average	57.40	56.78	
New reading items	R07		64.66	64.80
	R08		65.75	66.35
	R09		64.52	64.91
	R10		64.69	65.23
	R11		60.95	61.34
	R12		62.27	61.58
	R13		52.50	52.59
	R14		56.16	55.22
	R15		64.42	63.26
	R16		51.28	50.67
	R17		47.59	48.08
	R18		55.80	55.18
	Average		59.22	59.10

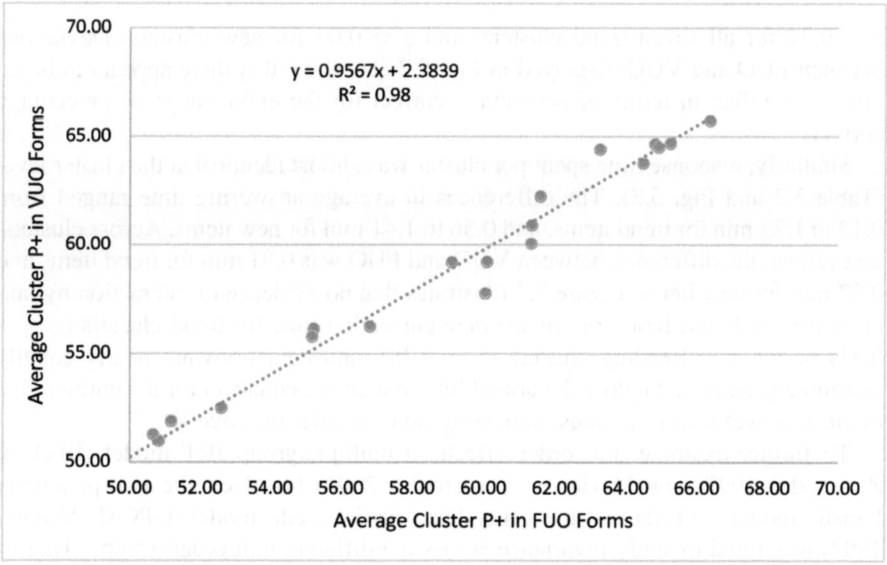

Fig. 5.2 Comparison of percentage correct between FUO and VUO in PISA 2018 field trial

Table 5.2 Average time spent by groups per reading cluster (in minutes)

	Cluster	Group 1 (FUO)	Group 2 (VUO)	Group 3 (FUO)
Trend reading items	R01	19.36	18.03	
	R02	18.97	18.75	
	R03	17.88	17.67	
	R04	18.30	18.43	
	R05	17.84	17.71	
	R6A	15.90	15.85	
	R6B	13.87	13.65	
	Average	17.97	17.66	
New reading items	R07		17.23	16.58
	R08		21.01	21.56
	R09		20.02	20.22
	R10		26.15	26.71
	R11		23.19	23.50
	R12		21.88	21.42
	R13		22.04	21.52
	R14		20.56	19.15
	R15		21.50	20.55
	R16		19.40	19.02
	R17		21.95	21.86
	R18		21.69	21.28
	Average		21.38	21.11

$z = 0.24$ for all seven trend clusters, and $z = 0.04$ for new clusters. Deviations between FUO and VUO displayed in Fig. 5.2 also show that there appears to be no unit order effect in terms of percentage correct for the entire range of percentage correct.

Similarly, response time spent per cluster was almost identical at the cluster level (Table 5.2 and Fig. 5.3). The differences in average answering time ranged from 0.13 to 1.33 min for trend items, and 0.56 to 1.41 min for new items. Across clusters, on average, the difference between VUO and FUO was 0.31 min for trend items and 0.27 min for new items. Figure 5.3 illustrates that no evidence of interaction by unit order can be found based on the average answering time for trend clusters ($z = -0.31$) or for new Reading clusters ($z = 0.28$), and deviations are nearly equally distributed above and below the dotted line (indicating equality) for the entire range of cluster-level response times, indicating no unit order bias overall.

To further examine unit order effects, a multiple-group IRT model (Bock & Zimowski, 1997; von Davier & Yamamoto, 2004) based on the two-parameter logistic model (2PLM) and the generalized partial credit model (GPCM; Muraki, 1992) was fitted to study invariance across the different unit order groups. That is, the IRT model was characterized with three different unit-order groups (Groups 1–3), and the measurement invariance of item parameters in each group was evaluated. If there is any significant item-by-unit order group interaction, it indicates

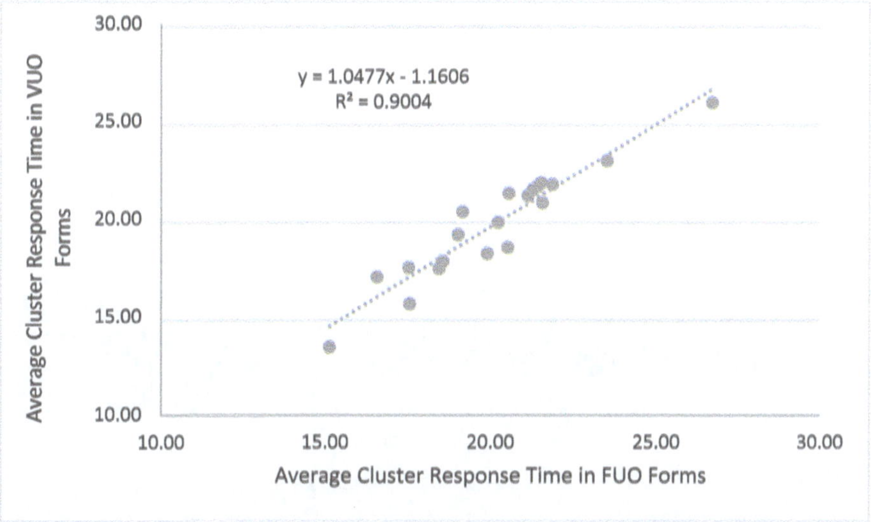

Fig. 5.3 Comparison of time spent per cluster between FUO and VUO in PISA 2018 field trial (in minutes)

that item parameter estimation is affected by the unit order. During the IRT analyses, all item parameters were constrained to be equal across the three groups: The trend item parameters were fixed to those from PISA 2015, and the (common) item parameters for new items were estimated. Then, the measurement invariance of item parameters across the three groups was evaluated using item fit statistics, specifically, root mean squared deviation (RMSD; OECD, 2017; Joo et al., 2021). The same operational IRT scaling threshold for the PISA field trial cognitive domains, RMSD >0.15, was used to evaluate item misfits.

In this context, item-by-unit order interactions are the main source of item misfits, demonstrating that, if present, the common item parameters do not work for a certain group (either VUO or FUO). Results showed that there are no item-by-unit order interactions between Group 1 and Group 2 for trend items or between Group 2 and Group 3 for new items with RMSD ≥0.15. Figures 5.4 and 5.5 provide the distribution of RMSD values for trend items taken by Group 1 and Group 2 and for new items taken by Group 2 and Group 3 respectively. The red line in both figures indicates a threshold of RMSD = 0.15. Figures 5.4 and 5.5 show that the RMSD values for all items (displayed on the x-axis) far below this threshold, indicating very good item fit within all unit order groups.

In summary, comparison of the percentage correct and response time per cluster between FUO and VUO, as well as evaluation of the measurement invariance of item parameters between FUO and VUO, suggests that bias on item parameter estimation due to differential unit order in the clusters was negligible for the major domain. Therefore, the field trial results confirmed the feasibility of introducing an MST into the main survey, as unit order effects were found to be negligible.

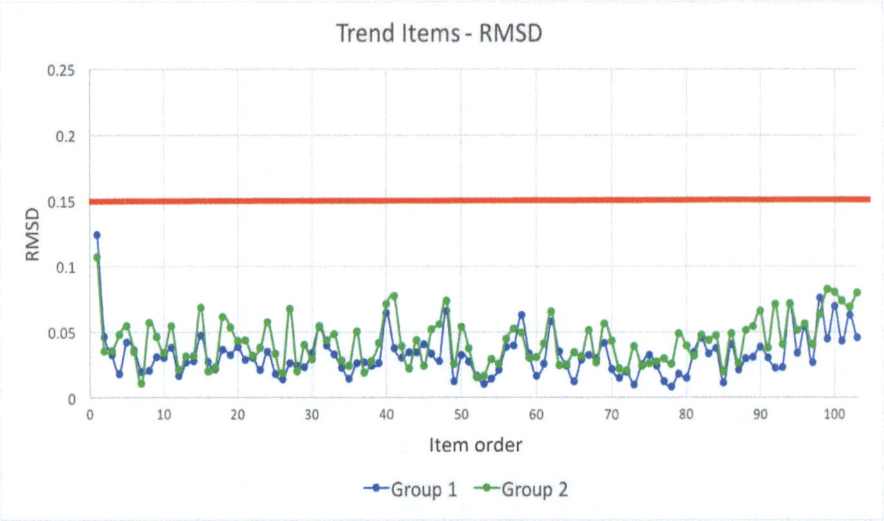

Fig. 5.4 Distribution of RMSD values for trend items administered in Group 1 and Group 2

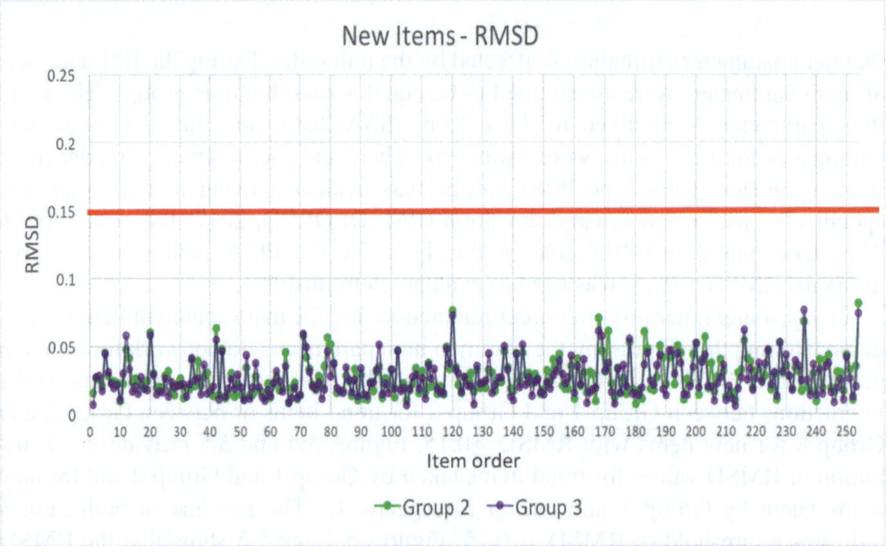

Fig. 5.5 Distribution of RMSD values for new items administered in Group 2 and Group 3

5.3 Assignment of Units and Items in Designing the MST in PISA

Given that unit order effects were negligible, the MST design was introduced in the PISA 2018 main survey for the major domain of Reading. In ILSAs, a group of content experts define the frameworks that map out the assessment instrument in each domain, which specify the task characteristics to be manipulated by test developers and outline the relative proportion of items to be developed based on the key variables associated with those task characteristics. First, we focus the next section on the newly developed units and items and how they were selected for the main survey after being administered in the field trial. Then, we discuss how all trend and new Reading items were assigned according to the MST design in PISA 2018 Reading.

5.3.1 Selection of Items and Units for the Main Survey

Test developers made the initial selection of new items for the main survey based on item statistics from the field trial data. Operationally, analyses of the first batch of field trial data are important for making timely psychometric decisions and for investigating if there are any unexpected issues with data collection or data management processes. In the first stage of item selection, the functioning of complex multiple-choice items (that have a table format; i.e., rows of short prompts with binary response options, such as Yes/No and True/False) was carefully monitored. Final scores on complex multiple-choice items were determined based on the combinations of multiple responses provided in each row. Thereby, very difficult rows (i.e., proportion correct close to or less than 50%) could be eliminated. Rows with a negative or nearly zero item-total correlations were also eliminated. Then, scoring rules were developed for complex multiple-choice items by content experts based on performance across all rows in a table. In some cases, several items were combined into polytomous items with partial credit. In other cases, if all rows in a table item were easy, items were combined and scored dichotomously, with only full credit and no credit scoring (OECD, 2020).

Once scoring models were created for the complex multiple-choice items, all field trial items were scaled using multiple-group IRT models by country and language groups (Yamamoto & Mazzeo, 1992; Bock & Zimowski, 1997). In the second stage of item selection, several criteria were used to monitor item functioning using the field trial data. More specifically, items that showed any of the following were carefully examined:

1. extreme values of proportion correct (<0.20 or >0.90);
2. low item-total correlations (<0.20);
3. low estimated item slopes (<0.40);

4. many item-by-country interactions or *misfit* in many countries (>10% of countries with misfit);
5. low coder reliability in many countries (>10% of countries).

Nineteen of the 254 new field trial items were eliminated based on these criteria.

Finally, in the third stage of item selection, modifications were made at the unit level; specifically, units were shortened, such that they would fit into the MST design for Reading. Shorter units had approximately five items, while longer units had seven. In addition to their statistics, items were selected in the final stage such that unit stimuli (e.g., texts) were effectively used by the final set of items, and, ideally, complete units were able to trigger varying cognitive processes and provide varying pieces of data relating to different cognitive scales of interest. Considering that subscales are reported for the major PISA domain, an effort was made to balance those subscales. For example, the number of items categorized with cognitive processes already represented in the trend units were reduced, and items that represented the new cognitive processes were preferentially selected. When all selection criteria were met, items that had better statistics were retained over those with worse statistics. At this final selection stage, 50 items were eliminated, resulting in a total of 245 items (45 units) total for the main survey.

5.3.2 Assignment of Units to the MST Design

The Reading MST design imposed several constraints that had to be balanced with the goals of the framework during unit assignment for the main survey. These constraints are described here to provide a context for the unit assignment conducted by test developers. A simplified flow of the MST design is summarized in Fig. 5.6: it comprised three stages (Core, Stage 1, and Stage 2), and an individual respondent took one or two blocks at each stage.

5.3.2.1 Core

The core stage was the introductory stage with eight blocks that each consisted of only two, average difficulty units. Five units were used for eight different core blocks: One unit appeared in four core blocks while the other four units appeared in three core blocks, and hence, all core blocks were deliberately linked across different core blocks. The goal was to have approximately 10 items (80%) machine-coded items in each block, which meant each unit had to have approximately 4–5 items. Within each unit, there needed to be as many machine-scored (i.e., simple or complex multiple-choice) items as possible so that there was enough information to make routing decisions into Stage 1. Human-coded (i.e., constructed-response) items could not be scored during test administration and, thus, were not used for routing decisions. Thus, the five units assigned to the Core

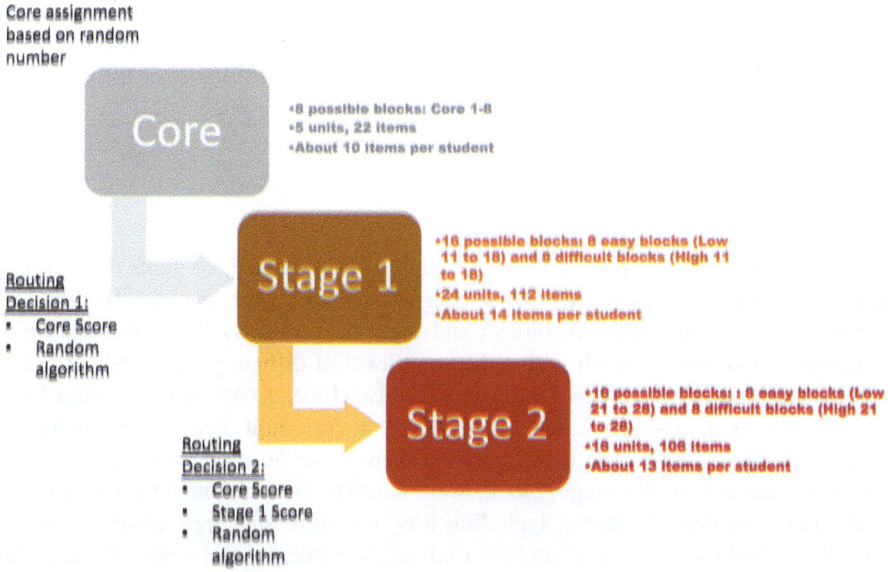

Fig. 5.6 Simplified flow of PISA 2018 MST design for reading

blocks had to have a large proportion of machine-scored items of average difficulty. Representing both trend and new units as equally as possible across the core blocks was also considered important for the trend. Fifty percent of all trend items in blocks 1–5 were human-coded, which limited the number of units with a majority of machine-scored items. Nevertheless, a good balance of trend and new units was created that maximized the number of machine-scored items.

5.3.2.2 Stage 1

After the Core stage, students were routed to either an easier (lower difficulty, labelled as "L") or more difficult (higher difficulty, labelled as "H") block in Stage 1. In total, 24 units were used in 16 different Stage 1 blocks (eight difficult blocks and eight easy blocks); Each unit appeared in two different blocks with a similar level of difficulty (e.g., H-H or L-L). Each block in Stage 1 consisted of three units with approximately 15 items per block, and, like the Core stage, a high proportion (60%) of machine-scored items were required so that most items could contribute to the routing decisions for Stage 2. As with the Core, it was important to balance the representation of trend and new units, so at least one trend and one new unit was assigned to each block. In Stage 1, the test developers felt it was important to represent the new multiple source and/or digital reading framework elements within each block. This would ensure that every student, no matter their skill level, would have an opportunity to interact with the new aspects of the framework, primarily multiple sources of information. Thus, unit assignment within Stage 1 required test developers to balance the number of items in a unit, the number of machine-scored items in a unit, the trend or new status of a unit, whether the unit contained a single source or multiple sources of information, and the average difficulty of the unit (low or high average difficulty).

5.3.2.3 Stage 2

Stage 2 included the longest units, both in terms of the number of items in each unit and the length of stimulus texts. As in Stage 1, 16 units were used for 16 different Stage 2 blocks (eight difficult blocks and eight easy blocks): Half of those units appeared in two blocks, each with a different level of difficulty (e.g., H-L or L-H), and the other half of the Stage 2 units appeared two blocks, each with a similar level of difficulty (e.g., H-H or L-L). Units in this stage could also contain a higher proportion of human-coded items because there were no more routing decisions following Stage 2. In this stage, blocks were constructed such that each block had a wider range of item difficulty, including low, medium, and high difficulty. Here, each block had only two units and approximately 14 items total across those units. Every effort was made to include at least one new unit that represented the new aspects of the framework, as was done in the Stage 1 blocks. This would further ensure that every student encountered the new content, multiple sources, and digital aspects of the construct.

5.3.2.4 MST Stage Routing

The described design was initially developed as the standard design and is referred to as *Design A* (Core > Stage 1 > Stage 2). A supplementary component, *Design B*, was introduced to further balance the position effects by switching the order of Stage 1 and Stage 2. That is, students who took Design B were administered Stage 2 items first and then moved to Stage 1 items (Core > Stage 2 > Stage 1). Design A was administered to 75% of students and Design B was administered to 25% of students per country.

5.3.3 Evaluation of the Unit Assignment

Once the item selection was complete, and the units were assigned to the MST stages, simulation studies were conducted to assess the performance of the design using the preliminary item parameters obtained from the field trial. The details of the simulation studies are described in Yamamoto et al. (2018, 2019). In short, the simulation studies suggested that the item parameters could be recovered with minimal errors and that the MST design would improve measurement precision for the entire range of the skill distribution, particularly at the extreme lower and higher ends. More specifically, the simulation study showed about a 5% reduction of measurement error on average over the range of typical skill distributions (Yamamoto et al., 2019), and 11% more test information was gained compared to a similar number of randomly selected units (OECD, 2020).

Given that the MST design generated 192 paths (combination of Core, Stage 1, and Stage 2 blocks), it was not possible for the reading experts to review all those combinations of item sets and make recommendations for the selection. Instead, at a meeting of the PISA Reading domain experts, a thorough explanation of the item selection process and the unit assignment was presented, as well as the characteristics of the final item pool. The item pool was evaluated on a holistic level, considering the representation of the cognitive processes across the entire item pool and the distribution of difficulty and construct representation within each stage of the multistage adaptive design. At the end of the meeting, the experts signed off on the multistage adaptive design and the unit assignment. In the future, developments of an automated algorithm to assign units to the MST design would be very useful to reduce the risk of misplacement of units in certain paths and to improve the balance of unit selection across different paths, considering various constraints that should be met in PISA (e.g., subscale reporting, sufficient time to take the test).

5.4 Psychometric Considerations in Analyzing MST Data in PISA

The MST for the Reading domain in PISA 2018 was designed to improve measurement precision across a wider range of proficiencies, and at the same time, collect the data needed for item analyses and IRT scaling. Therefore, it was important to verify that the MST was implemented as intended (see, *Monitoring the Performance of MST Delivery Platform*). In addition, two important psychometric aspects were also speculated and developed. One is related to the classical item analysis perspective, particularly the proportion correct (*Development of equated proportion correct*). Another is related with the decisions for IRT-based scaling procedure, particularly, how to incorporate MST information in the scaling procedure (*Scaling Procedures in Analyzing the MST Data in PISA*).

5.4.1 Monitoring the Performance of the MST Delivery Platform

The MST designs and platforms for PISA 2018 and PIAAC were both developed at Educational Testing Service (ETS), utilizing the accumulation of intellectual assets and knowledge from ETS's previous experience with ILSAs. Three aspects of MST delivery were closely monitored to evaluate whether the MST design was

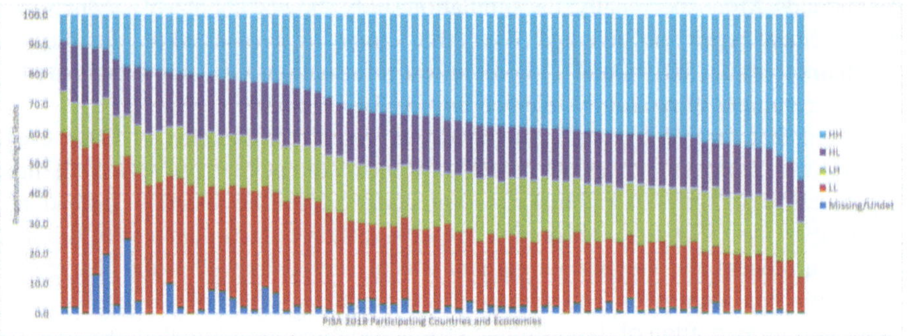

Fig. 5.7 Reading MST proportions of students routed to higher and lower difficulty Stage 1 and Stage 2 blocks

implemented to students as designed and intended. First, the proportions of students routed across stages was verified. Next, it was checked that the proficiency distributions of students routed through Design A and Design B were equivalent. Finally, the design called for some proportion of students to be intentionally misrouted given their estimated proficiencies (i.e., high ability students routed to an easier block), so the equality of proficiency distributions between the intentionally misrouted and the well-routed students was reviewed. Assuring this data yield is the first, essential step in verifying the quality of the data collected through the MST design. In the next sections, we explain how all three data review aspects can confirm that the delivery and performance of an MST platform was as intended, such that the quality of the data collected through the MST design is high.

Regarding the first aspect, proportions of students routed across stages, three criteria were used to monitor the PISA Reading MST delivery platform within each stage and across stages. Within the stage, random assignment of students was closely monitored: first, if students were randomly assigned to each MST core block and 2) if students were randomly assigned either to Design A (75%) or to Design B (25%). Results confirmed the random assignment to Core blocks worked well, with a uniform distribution of about 12.5% across eight different blocks, and further routing to Designs A and B had the desired proportions (75% and 25%, respectively) within and across all participating countries. Next, across the stages, the more complex routing of students from Core to Stage 1 and from Stage 1 to Stage 2 blocks were evaluated in detail. Figure 5.7 shows the proportions of students routed to difficult and easy blocks across Stage 1 and Stage 2.[4] Some students were categorized as missing/undetermined when they did not complete certain stages and, thus, were not assigned a Stage 1 or both Stage 1 and Stage 2 blocks.

[4]Difficult blocks in both Stage 1 and Stage 2 blocks are labelled as "HH" (high options in both Stage 1 and Stage 2), the difficult Stage 1 and easy Stage 2 blocks are labelled as "HL" (high option in Stage 1 and low option in Stage 2), the easy block in Stage 1 and the difficult block in Stage 2 block are labelled as "LH", and easy blocks for both Stage 1 and Stage 2 are labelled as "LL".

Table 5.3 Average percent correct, not-reached, and omitted for MST reading by stage position

	% Correct		% Not-Reached		% Omit	
	Design A	Design B	Design A	Design B	Design A	Design B
Core Items	57.73	57.85	0.09	0.27	3.7	3.57
Stage 1 Items	**55.15**	47.51	**1.84**	12.57	**5.49**	7.42
Stage 2 Items	47.8	**54.83**	12.29	**1.73**	6.19	**5.17**

Note. Comparable cells between designs A and B are greyed

In Fig. 5.7, the lowest to highest performing countries are shown from left to right. As intended, in lower-performing countries, smaller proportions of students were assigned to more difficult blocks (light blue and purple) while in higher-performing countries, smaller proportions of students were assigned to easier blocks (red and green). However, as intended, every block was assigned to at least 25% of the total country sample. As shown in the figure, adding up light blue and purple (HH and HL) bars for the lowest performing countries, and adding up red and green (LL and LH) bars for the highest performing countries, exceeded 25% each. This confirmed that, regardless of countries' proficiency distributions, the MST design consistently provided the minimum number of responses per item needed for IRT scaling and appropriate item coverage across the whole range of item difficulty.

Second, the equality of proficiencies between Design A and Design B was monitored. Note that Design B was created by reversing the presentation of Stage 1 and Stage 2 to reduce the risk of bias in item parameter estimation in case Stage 2 items show higher omission rates, not-reached ratio, or lower proportion correct due to the time limit (a.k.a., item position effects), and Design A and Design B were randomly assigned to 75% and 25% of students per country. Therefore, proficiency distributions between Design A and Design B were expected to be equal. Table 5.3 shows the average Core, Stage 1 and Stage 2 block percent correct, not-reached and omission rates across all countries by stage position (Designs A and B). Note that the position of Stage 1 items and Stage 2 items was swapped between Designs A and B, and exactly the same set of items was used. When comparing the same blocks across Designs A and B, the results are very similar in terms of percent correct, not-reached, and omitted—for example, percent not-reached was 1.84 for Stage 1 in Design A, and the corresponding value was 1.73 for Stage 2 in Design B. Comparable cells between designs A and B are greyed.

In addition, group-level statistics (group means and group standard deviations) were compared between Design A and Design B to evaluate the equality of proficiencies more directly. Separate multiple-group IRT analyses and group statistics for Design A datasets and Design B datasets were estimated separately. As Fig. 5.8 shows, group means and group SDs estimated for each design resulted in very similar and comparable results. Taken together, the equality of proficiencies between Design A and Design B was confirmed, as the evidence for quality assurance of the PISA data collected through this MST design.

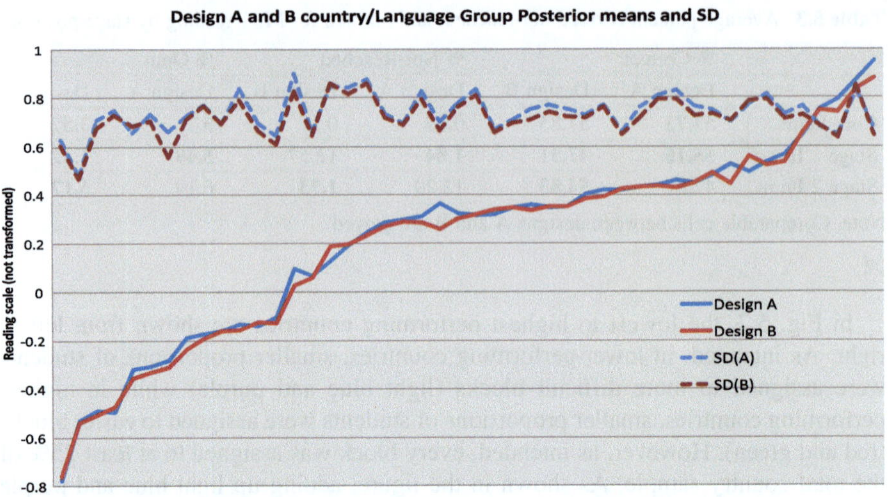

Fig. 5.8 Comparison of group statistics between Design A and Design B. Note. X-axis indicates individual countries, and they are sorted by the group means of Designs A and B

Finally, the equality of proficiencies between intentionally misrouted and well-routed students was reviewed. According to the PISA MST design (Yamamoto et al., 2019), a pre-defined portion of students are intentionally misrouted to the next stage: 10% of high-performing students were routed to a low block and 10% of low-performing students were routed to a high block. The remaining 90% of students were routed to a block appropriately matching their performance, including many in the middle range of performance whose subsequent block was randomly assigned. This intentional misrouting was introduced mainly to control the item exposure rate that is critical for item parameter estimation (Chen et al., 2014), but it would not be applicable to the individual-level reporting testing programs that utilize MST with known item parameters. If intentional misrouting was not adapted, but rather, all students were routed according to their performance deterministically, item parameters would be biased (Shin et al., 2021; Svetina et al., 2019); thus, such MST designs would not be applicable for the purpose of estimating item parameters. Although misrouted students contribute to more accurate item parameter estimation, it should still be confirmed that the final performance estimation was not overly affected by the set of assigned items that were mismatched to their performance levels. Because the performance of intentionally misrouted and well-routed students is expected to be equal, student performance between the two routing designs was used as an additional verification of MST quality. More specifically, students were grouped according to the blocks taken at Stage 1 (either L for easy blocks or H for difficult blocks) and their performance level (L for low-performing and H for high-performing students). Across 73 countries, the average performance of students in each category is summarized in Table 5.4. As expected, the absolute differences of

Table 5.4 Average PV1 of students by Stage 1 blocks and their performance levels

	90% well-routed	10% Misrouted	Absolute difference
L(S1)-L(Perf)	344.36	345.54	1.18
L(S1)-H(Perf)	536.52	535.74	0.78
H(S1)-L(Perf)	361.48	363.77	2.29
H(S1)-H(Perf)	562.80	559.68	3.12

average performance between well-routed students and misrouted students were very small or negligible in each case, resulting in a difference of 0.78 to 3.12 points on PISA Reading scale scores (OECD, 2020).

5.4.2 Scaling Procedures in Analyzing the MST Data in PISA

In most testing situations where MST is administered, item parameters are pre-calibrated and treated as known; then, item fits are evaluated to examine the parameter drift (Glas, 2010). However, in the context of ILSAs, including PISA, item parameters are estimated using the sample of nationally representative data collected during the main survey with proper sampling weights. More specifically, during the first run of multi-group concurrent calibrations, the item parameters are constrained so that only one set of common or international parameters is estimated per item to model the data for all the country-by-language groups (OECD, 2020). As part of the calibration process, the fit (or misfit) of the common item parameters to the data for each pre-defined group is evaluated. Then, item-by-group interactions are identified when the fit to the data was found to be poor (the fit statistic was worse than a chosen threshold value). From the second calibration run, new unique or group-specific item parameters are estimated in the group or groups in which misfit was found in the first round of calibration. In subsequent calibrations, the item fit threshold is gradually lowered until the target threshold is reached by allowing additional group-specific item parameters to be estimated. This stepwise procedure is fundamental for optimizing the comparability of item parameters by keeping the common item parameters as much as possible. By allowing unique item parameters for items that show item-by-group interactions—in contrast to excluding such items or accepting poor common item parameter fit—measurement error is reduced without introducing bias. The research base for this approach can be found in Meredith (1993), Glas and Verhelst (1995), Yamamoto (1997), Glas and Jehangir (2014), Meredith and Teresi (2006), Oliveri and von Davier (2011, 2014), as well as in von Davier et al., 2019).

Therefore, in the context of ILSAs, it is very important to balance the need for accurate measurement across a heterogeneous proficiency range with the need for sufficient response data across all items in all countries for item parameters to be accurately estimated. Given the understanding that item calibration and scaling procedures can be different by MST designs (Jewsbury & van Rijn, 2020), a series of simulation studies were conducted in advance to verify if the implemented MST designs would be able to recover the item parameters. The final PISA MST design was determined based on the key findings from simulation studies, which showed the acceptable level of item parameter recovery (Yamamoto et al., 2019).

In addition, scaling procedures for the MST data collected during the PISA 2018 main survey required that data be carefully selected and evaluated, particularly with respect to the item parameter estimation. In this regard, the use of appropriate IRT models and marginal maximum likelihood (MML) estimation methods are important in the analysis procedure under MST designs. The multiple-group IRT model (Yamamoto & Mazzeo, 1992; Bock & Zimowski, 1997) is beneficial in ILSAs because it is suitable for analyzing incomplete item administration designs, such as MST and BIB, yet it still correctly reflects the uniqueness of each sub-population through a unique mean and variance. The work of Glas (1988) is important here because it shows that no bias is introduced for computer adapted tests or MST when the adaptive decision is based solely on the cognitive item responses. Glas (1988) has shown that the specification of the probability of a response pattern under MST is identical to the specification in latent trait models with a fixed design (where the desirable feature called "ignorability" holds). This rule can be applicable not only for the Rasch model (Rasch, 1960) but also for the 2PLM and for both dichotomous and polytomous items. Additionally, Glas et al. (2000) showed that ignorability holds when the MML estimation is used, in both models in MST designs. A recent study (Shin et al., 2022) also provides theoretical justification for the use of a unidimensional IRT scaling method for MST designs when the routing decision is based on the vector of item responses, such as in the case of PISA MST. Thus, unique means and variances were estimated for each group, and the MML, with an expectation-maximization method, was used via mdltm software (von Davier, 2005; Khorramdel et al., 2019) for scaling the PISA MST data.

One further consideration is related to the inclusion of MST-specific information in the group definition for the multiple-group IRT model. Because high-performing students and low-performing students are likely to take different sets of items that are matched to their abilities under the MST design, it could be reasonable to designate unique skill distributions by performance levels given that it does not affect item parameter estimation. For example, based on the routing rules of PISA 2018 MST design, country-by-language group definitions can be further broken down by the student performance levels, resulting in groups tripled by recognizing 1) high performing, 2) middle performing, and 3) low performing students in each specific country-by-language group. In this regard, a recent study (Shin et al., 2022) provides theoretical justification for excluding the MST-specific information (i.e., performance levels) because routing decisions are not independent of item responses conditional on latent ability, violating the typical assumptions of multi-group IRT models. To verify the approach for the specific PISA 2018 MST design, such possibilities were also examined in terms of the item parameter recovery and item fit statistics using the simulated data presented in Yamamoto et al. (2019).

Figure 5.9 compares the item parameter estimates when the MST-specific information was included (empty dots) and excluded (solid dots) against the generating

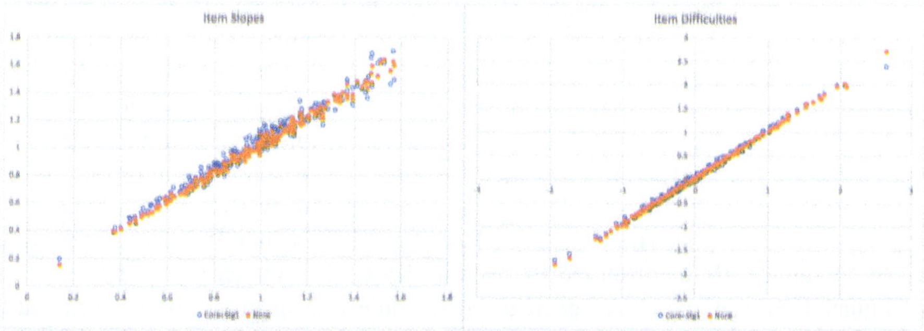

Fig. 5.9 Scatterplot of item parameter estimates (y-axis) against the generating values (x-axis). Note: The performance variable was included in the left panel and excluded in the right

values on the x-axis for one simulated dataset. The calibration sample here combines well-routed and intentionally misrouted samples. As can be seen from the figure that, although very similar, when MST-specific information was included, item parameter recovery was less satisfactory, particularly for item slopes. Biases in slopes ranged from −0.10 to 0.06 when the performance levels were not included (labelled as "None"), while the corresponding range, with MST information included, was from −0.19 to 0.12 (labelled as Core+Stg1). Therefore, this small simulation study confirmed that MST-specific information does not help item parameter recovery at the IRT-based scaling procedure, as to not introduce biases in item parameter estimation.

5.4.3 Development of Equated Proportion Correct

With nonadaptive assessments, in which each block of items is administered to randomly equivalent respondents through a BIB design, standard item statistics (such as percentage correct, omitted, and not-reached) are comparable across items, as well as within and across countries. Here, item statistics represent complete samples, because randomly equivalent groups respond to every item. For adaptive tests, however, standard item statistics are not similarly comparable because item sets are not randomly assigned. To equate statistics between adaptive and nonadaptive assessments, the differential proportion of routing paths taken within the MST, across countries and subgroups within the country, must first be considered. By selecting item sets to match the student's ability, higher performing students in the Reading MST were likely to be assigned to more difficult items, and lower performing students were likely to be assigned to easier items. As a result, it cannot be assumed that students' abilities are randomly equivalent across items in the MST. For example, the subsamples of students routed to the more difficult MST blocks are expected to be more proficient than the total sample of students; therefore, a higher proportion of correct responses would be expected from this subsample than would be expected from the total sample. Therefore, while classical observed proportion correct (P+ or *observed P+*) and other statistics (e.g., not-reached, omitted) are still helpful for identifying items with potential scoring or other issues, classical observed proportion correct is no longer comparable between adaptive and non-adaptive designs (i.e., across cycles within countries/economies) nor across countries/economies within a cycle. Therefore, to facilitate the comparison of proportion correct across countries and across cycles, an *Equated Proportion Correct* (*Equated P+*) statistic was developed. The PISA 2018 Reading MST Equated P+ was developed to be comparable with the observed P+ used in PISA 2015 and earlier cycles.

Equated P+ statistics have been used by some testing programs to verify and ensure the comparability and quality of operational items and provide the information needed to assemble new test forms (e.g., Ali & Walker, 2014). In general, the Equated P+ is equivalent to the proportion correct that would have been obtained from the non-adaptive designs used in previous PISA cycles. It accounts for the differences in proficiencies between the sample who responded to the item and the total sample. More specifically, the statistic is computed based on the item parameter estimates calibrated based on the multiple-group IRT model and the mean deviation (MD) statistics. For the dichotomously scored and polytomously scored items used in PISA, and modelled using the 2PLM and GPCM described above, the Equated P+ can then be computed as follows (Ali & Walker, 2014):

$$\textit{Equated } P+_g = \int wp_g(X|\theta)f_g(\theta)d\theta, g = 1, \ldots, G. \tag{5.1}$$

As noted by the subscript g, the Equated P+ is computed for each country-by-language group (g = 1, …,G) with each student having a sample senate weight w; $p_g(X|\theta)$ is the probability of correct response given the proficiency θ; and $f_g(\theta)$ is the group-specific density distribution on the students' ability scale. To facilitate computations, the integral in the equation can be approximated using individual's proficiency point estimate $\widehat{\theta}$ using the weighted likelihood estimates (WLE; Warm, 1989):

$$\int wp_g(X|\theta)f_g(\theta)d\theta \approx \frac{1}{\sum_{s=1}^{N_g} w_s} \sum_{s=1}^{N_g} w_s p_g(X|\theta). \tag{5.2}$$

Note that PISA aims to measure the constructs over a huge heterogeneous population from many countries together on the same common scale. Thus, the above statistics based on the IRT models still would result in some degree of deviation from the observed data. The MD statistics are useful for quantifying such deviations, such as the differences in the proportion correct between the model prediction and the observed responses. Thus, the MD statistic was added to the computation of Equated P+ to account for the deviations and provide more accurate Equated P+ values:

$$\text{Equated } P+_g = \frac{1}{\sum_{s=1}^{N_g} w_s} \sum_{s=1}^{N_g} w_s p_g(X|\theta) + MD_g. \tag{5.3}$$

MD_g quantifies the deviation between group-specific observed and expected ICCs (von Davier, 2005). The integral over students per group was approximated with Gaussian quadrature points q ranging from -5 to 5 with $f_g(\theta_q)$ in the next equation representing weights at each quadrature point:

$$\int \left[p_g^{obs}(X|\theta) - p_g^{exp}(X|\theta)\right] f_g(\theta) d\theta \approx \sum_{q=1}^{Q} \left[p_g^{obs}(X|\theta_q) - p_g^{exp}(X|\theta_q)\right] f_g(\theta_q). \tag{5.4}$$

The Equated P+ values were reported to be between 0 and 1, regardless of response type (dichotomous or polytomous). For polytomous items, this was done by averaging the probability of correct response $p_g(X|\theta)$ over the partial credit categories:

$$p_g(X|\theta) = \frac{1}{k} \sum_{k=0}^{m} k p_g(X = k|\theta), k = 0, 1, \ldots, mi \tag{5.5}$$

where $p_g(X = k|\theta)$ is the probability of score of k (0, 1, 2,..., m_i) for item i and is computed from the generalized partial credit model.

Figures 5.10 and 5.11 show the comparison of observed P+ versus Equated P+ for one country participating in the PISA 2018 main survey. Items are identified by the MST stage (Core, Stage 1 and Stage 2) and the difficulty of the blocks to which they belong (Low-Low, Low-High, High-Low, or High-High).[5] Results in Fig. 5.10 were obtained without adding MD in the equating, and the results in Fig. 5.11 were with adding MD. The figures show the quality of the newly developed Equated P+ statistics and the benefits of adjusting for deviations by adding MD in the computations. Importantly, observed P+ and Equated P+ for Core items were nearly identical (more so in Fig. 5.11), as they should be, since no adaptation had yet taken place at the Core stage. Results for Stage 1 and Stage 2 items showed the expected pattern of adjustment. More specifically, Fig. 5.11 shows that as Stage 1 and Stage 2 blocks are adaptively selected for more (or less) proficient subsamples of students, the items in High (difficult) MST blocks have lower Equated P+ than observed P+, and items in Low (easy) blocks have higher Equated P+ than observed P+. These patterns, as well as the Low-High and High-Low patterns that show much smaller differences between observed P+ and Equated P+, were expected according

[5] Note that PISA 2018 MST design for reading is based on the deliberate linkage of units across blocks. For example, each unit appears in multiple blocks for a given stage, not only just one block. All Stage 1 units appear exclusively either low (easy) or high (difficult) blocks multiple times, but half of Stage 2 items appear across low (easy) and high (difficult) blocks twice. Those units that appear across different difficulties of blocks are labeled as “low-high” or “high-low”.

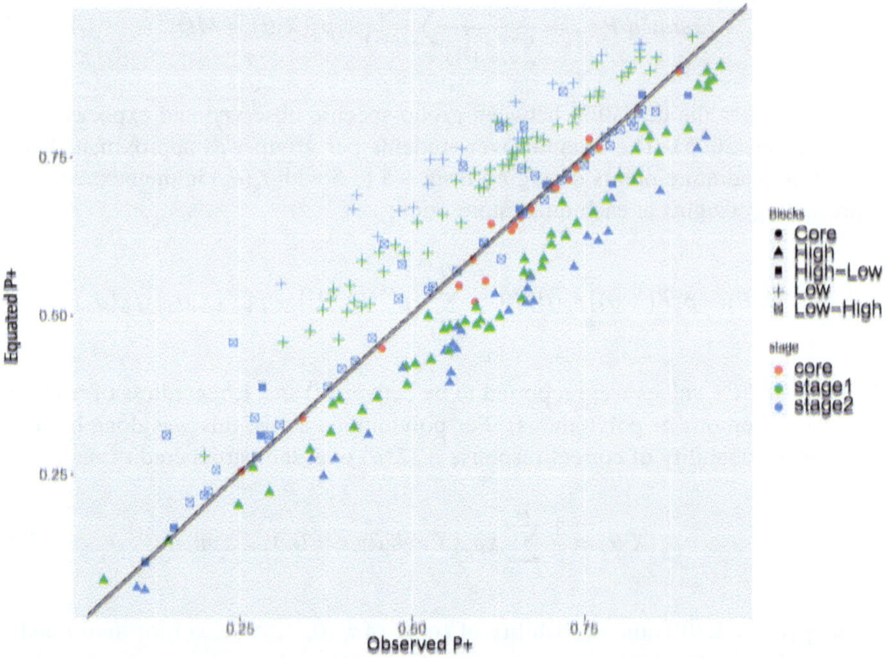

Fig. 5.10 Scatterplot of Observed P+ and Equated P+ from one example country when Equated P+ was computed without adjusting for item deviations (MD)

to the PISA MST design. In sum, this confirmed the quality of the reported Equated P+ values and the choice of including MD in the computations.

Overall, Table 5.5 summarizes the average differences between observed P+ and Equated P+ by MST stage and block type across countries. On average, for Stage 1 and Stage 2 items that link across blocks within the same difficulty (High or Low for Stage 1 and High-High or Low-Low for Stage 2), the Equated P+ is further from the grand mean of observed P+ by approximately 8–12%. For items that link across blocks with different difficulties (Stage 2 High-Low or Low-High), the difference was less than 3% on average.

5.5 Conclusion

For any assessment, it is difficult to find an ideal balance between the accuracy of estimation and delivering the assessment in an efficient way, as well as sufficiently representing the assessment constructs and avoiding fatigue effects or a potential loss of motivation. MST designs in ILSAs were introduced in this context with the aim to provide more accurate and efficient measures, especially given that participating groups and individuals are extremely heterogeneous and that a large number

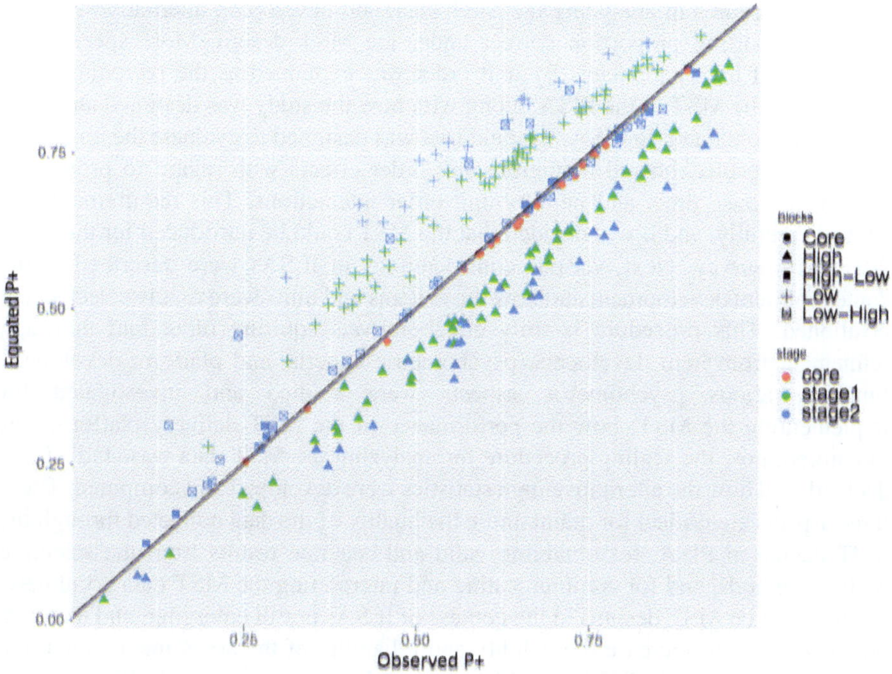

Fig. 5.11 Scatterplot of observed P+ Equated P+ from one example country when Equated P+ was adjusted by adding the item deviations (MD)

Table 5.5 PISA 2018 reading average observed P+ and equated P+ across countries by MST stage and blocks in which the items are linked

Stage	Testlet	Observed P+	Equated P+
Core		57.7	58.0
Stage 1	High	53.1	46.3
	Low	53.7	62.0
Stage 2	High and high	51.8	41.9
	High and low	47.7	48.5
	Low and high	48.0	50.7
	Low and low	52.6	66.5
All		52.2	53.2

of items need to be developed and implemented to represent the skills of interest. With these constraints, the MST was designed and thoroughly studied based on simulated and empirical data from the field trial and discussed with the technical advisory boards, which include international scientists and experts in the fields of statistics and psychometrics.

This chapter presents 1) what conditions were considered in introducing the MST, 2) how the items and units (i.e., a set of items) were developed and assigned, and 3) what psychometric aspects were speculated and considered, such as monitoring the performance of the MST delivery platform, making decisions for the scaling procedures in analyzing the MST data, and developing alternative ways of equating reporting proportion correct under the MST design. More specifically, invariance of item parameters by unit order was explained as the prerequisite for introducing the MST in the PISA, along with how the study was designed and what results were obtained. In PISA, the field trial was designed to evaluate the unit-order effect, and results showed negligible unit order effects with regard to proportion correct, response time, and item-by-unit order interactions. This requirement was studied carefully, and it was decided that the MST could be introduced for the PISA 2018 main survey. Next, various considerations in ILSAs were described in the context of item development and how those items and units were finally selected and positioned. This procedure is truly collaborative, requiring input and thorough refinement from item developers, psychometric experts, and platform developers. Finally, various psychometric aspects were studied and investigated for implementing the MST: how the performance of the MST delivery platform was monitored, how the scaling procedure for analyzing the MST data was studied and decided, and how the alternative item statistics were developed and computed. These three aspects are critical for maintaining the quality of the data collected through the MST design in PISA, for obtaining valid and accurate results from the scientific scaling methods, and for communicating and interpreting the MST data accurately.

Research on MST designs in the context of ILSAs is still emerging, and there are ongoing works to increase the validity and reliability of the assessments. First, the current MST design in PISA could be improved further in terms of efficiency and accuracy. Although item position effects were partially mitigated by reversing the presentation of Stage 1 and Stage 2 blocks in Design B, more improvements can be made, such as alternative designs with a fully balanced design of items and unit positions under the MST designs. Also, an automatized algorithm to assign units and items considering various constraints imposed in the ILSA context may prove to be very useful to employ the MST designs more efficiently if construct representation can be controlled. Furthermore, in PISA, threshold values that classify respondents into three groups (i.e., high, medium, low)[6] and the proportions of partial randomization (e.g., currently 10% of students are directed to a routing opposite their performance) could be optimized through further simulation studies, which could lead to an adjustment to the randomization method and threshold values. In the future, more simulation studies that examine the robustness of MST designs in ILSAs could be computed—for example, effects of the item-by-country interactions and the omission rates per item—to establish safeguards against potential threats to validity.

[6]Note that PISA 2018 MST reading blocks have two difficulty levels, either easy (low) or difficult (high), while students' performances are classified into three performance groups, high, medium, and low. For each performance group, probability of being routed to either easy or difficult block is different (Yamamoto et al., 2019).

References

Ali, U. S., & Walker, M. E. (2014). *Enhancing the equating of item difficulty metrics: Estimation of reference distribution (RR–14-07)*. Educational Testing Service. https://doi.org/10.1002/ets2.12006

Bock, R. D., & Zimowski, M. F. (1997). Multiple group IRT. In W. J. van der Linden & R. K. Hambleton (Eds.), *Handbook of modern item response theory* (pp. 433–448). Springer.

Chen, H., Yamamoto, K., & von Davier, M. (2014). Controlling MST exposure rates in international large-scale assessments. In D. Yan, A. A. von Davier, & C. Lewis (Eds.), *Computerized multistage testing: Theory and applications* (pp. 391–409). Chapman and Hall/CRC.

Glas, C. A. W. (1988). The Rasch model and multistage testing. *Journal of Educational Statistics, 13*, 45–52. https://doi.org/10.3102/10769986013001045

Glas, C. A. W., & Jehangir, K. (2014). Modeling country specific differential item functioning. In L. Rutkowski, M. von Davier, & D. Rutkowski (Eds.), *Handbook of international large-scale assessment*. CRC Press.

Glas, C. A. W., & Verhelst, N. D. (1995). Testing the Rasch model. In G. H. Fischer & I. W. Molenaar (Eds.), *Rasch models: Foundations, recent developments, and applications* (pp. 69–95). Springer.

Glas, C. A. W., Wainer, H., & Bradlow, E. (2000). MML and EAP estimation in block-based adaptive testing. In W. J. Van der Linden & C. A. W. Glas (Eds.), *Computerized adaptive testing: Theory and practice* (pp. 271–287). Kluwer Academic Publishers.

Glas, C. A. W. (2010). Item parameter estimation and item fit analysis. In W. J. van der Linden & C. A. W. Glas (Eds.), *Elements of adaptive testing* (pp. 269–288). Springer.

Hambleton, R. K., & Swaminathan, H. (1985). *Item response theory: Principles and applications*. Kluwer-Nijhoff.

Jewsbury, P. A., & van Rijn, P. W. (2020). RT and MIRT models for item parameter estimation with multidimensional multistage tests. *Journal of Educational and Behavioral Statistics, 45*(4). https://doi.org/10.3102/1076998619881790

Joo, S. H., Khorramdel, L., Yamamoto, K., Shin, H. J., & Robin, F., (2021). Evaluating item fit statistic thresholds in PISA: The analysis of cross-country comparability of cognitive items.

Khorramdel, L., Shin, H., & von Davier, M. (2019). GDM software mdltm including parallel EM algorithm. In M. von Davier & Y.-S. Lee (Eds.), *Handbook of psychometric models for cognitive diagnosis* (pp. 603–628). Springer.

Kirsch, I., Lennon, M., von Davier, M., Gonzalez, E., & Yamamoto, K. (2013). On the growing importance of international large-scale assessments. In M. von Davier, E. Gonzalez, I. Kirsch, & K. Yamamoto (Eds.), *The role of international large-scale assessments: Perspectives from technology, economy, and educational research* (pp. 1–11). Springer.

Le, L.T. (2007). Effects of item positions on their difficulty and discrimination: A study in PISA science data across test language and countries. Paper presented at the 72nd annual meeting of the psychometric society, Tokyo. https://research.acer.edu.au/cgi/viewcontent.cgi?article=1001&context=pisa.

Leary, L. F., & Dorans, N. J. (1985). Implications for altering the context in which test items appear: A historical perspective on an immediate concern. *Review of Educational Research, 55*, 387–413. https://doi.org/10.3102/00346543055003387

Lord, F. M. (1980). *Applications of item response theory to practical testing problems*. Lawrence Erlbaum Associates.

Mazzeo, J., & von Davier, M. (2008). *Review of the programme for international student assessment (PISA) test design: Recommendations for fostering stability in assessment results (OECD working paper: (EDU/PISA/GB(2008)28))*. Organisation for Economic Co-operation and Development.

Meredith, W. (1993). Measurement invariance, factor analysis and factorial invariance. *Psychometrika, 58*, 525–543.

Meredith, W., & Teresi, J. A. (2006). An essay on measurement and factorial invariance. *Medical Care, 44*(11), S69-S77. Chicago.

Messick, S. J., Beaton, A. E., & Lord, F. M. (1983). *National Assessment of Educational Progress Reconsidered: A New Design for a New Era.* (NCES 15-TR-20). Washington, DC: National Center for Education Statistics.

Muraki, E. (1992). A generalized partial credit model: Application of an EM algorithm. *Applied Psychological Measurement, 16*(2), 159–177.

Nagy, G., Nagengast, B., Frey, A., Becker, M., & Rose, N. (2019). A multilevel study of position effects in PISA achievement tests: Student- and school-level predictors in the German tracked school system. *Assessment in Education: Principles, Policy & Practice, 26*(4), 422–443. https://doi.org/10.1080/0969594X.2018.1449100

Oliveri, M. E., & von Davier, M. (2011). Investigation of model fit and score scale comparability in international assessments. *Psychological Test and Assessment Modeling, 53*(3), 315–333. http://www.psychologie-aktuell.com/fileadmin/download/ptam/3-2011_20110927/04_Oliveri.pdf

Oliveri, M. E., & von Davier, M. (2014). Toward increasing fairness in score scale calibrations employed in international large-scale assessments. *International Journal of Testing, 14*(1), 1–21. https://doi.org/10.1080/15305058.2013.825265

Oranje, A., Mazzeo, J., Xu, X., & Kulick, E. (2014). A multistage testing approach to group-score assessments. In D. Yan, A. A. von Davier, & C. Lewis (Eds.), *Computerized multistage testing: Theory and applications* (pp. 371–390). Chapman and Hall/CRC.

Organisation for Economic Co-operation and Development. (2016a). *PISA 2015 assessment and analytical framework: Science, reading, math, and financial literacy.* OECD Publishing. http://www.oecd.org/publications/pisa-2015-assessment-and-analytical-framework-9789264281820-en.htm

Organisation for Economic Co-operation and Development. (2016b). *PISA 2018 draft analytical framework.* https://www.oecd.org/pisa/data/PISA-2018-draft-frameworks.pdf.

Organisation for Economic Co-operation and Development. (2017). *PISA 2015 Technical Report.* http://www.oecd.org/pisa/data/2015-technical-report/.

Organisation for Economic Co-operation and Development. (2020). *PISA 2018 Technical Report.* https://www.oecd.org/pisa/data/pisa2018technicalreport/.

Rasch, G. (1960). *Probabilistic models for some intelligence and attainment tests, expanded edition.* University of Chicago Press/Nielsen & Lydiche.

Rutkowski, L., Gonzalez, E., Joncas, M., & von Davier, M. (2010). International large-scale assessment data: Issues in secondary analysis and reporting. *Educational Researcher, 39*(2), 142–151. https://doi.org/10.3102/0013189X10363170

Shin, H. J., Yamamoto, K., Khorramdel, L., & Robin, F. (2021). Increasing measurement precision of PISA through multistage adaptive testing. In *Quantitative psychology: The 85th annual meeting of the Psychometric Society, Virtual* (pp. 325-334). Springer International Publishing.

Shin, H., Jewsbury, P., & van Rijn, P. (2022). Generating group-level scores under response accuracy-time conditional dependence. *Large-Scale Assessments in Education, 10*(4). https://doi.org/10.1186/s40536-022-00122-y

Svetina, D., Liaw, Y. L., Rutkowski, L., & Rutkowski, D. (2019). Routing strategies and optimizing design for multistage testing in international large-scale assessments. *Journal of Educational Measurement, 56*(1), 192-213.

van der Linden, W. J., & Glas, C. A. W. (2000). *Computerized adaptive testing: Theory and practice.* Kluwer Academic Publishers.

van der Linden, W. J., & Glas, C. A. W. (2010). *Elements of adaptive testing.* Springer.

von Davier, M. (2005). *A general diagnostic model applied to language testing data (RR-05-16).* Educational Testing Service. https://doi.org/10.1002/j.2333-8504.2005.tb01993.x

von Davier, M., Sinharay, S., Oranje, A., & Beaton, A. (2006). Statistical procedures used in the National Assessment of Educational Progress (NAEP): Recent developments and future directions. In C. R. Rao & S. Sinharay (Eds.), *Handbook of statistics, 26: Psychometrics.* Elsevier.

von Davier, M., & Yamamoto, K. (2004). Partially observed mixtures of IRT models: An extension of the generalized partial-credit model. *Applied Psychological Measurement, 28*(6), 389–406. https://doi.org/10.1177/0146621604268734

von Davier, M., Yamamoto, K., Shin, H., Chen, H., Khorramdel, L., Weeks, J., Davis, S., Kong, N., & Kandathil, M. (2019). Evaluating item response theory linking and model fit for data from PISA 2000–2012. *Assessment in Education: Principles, Policy & Practice, 26*(4), 466–488. https://doi.org/10.1080/0969594X.2019.1586642

Wainer, H. (1990). *Computerized adaptive testing: A primer*. Lawrence Erlbaum Associates.

Wainer, H., Bradlow, E. T., & Wang, X. (2007). *Testlet response theory and its applications*. Cambridge University Press.

Wainer, H., & Kiely, G. L. (1987). Item clusters and computerized adaptive testing: A case for testlets. *Journal of Educational Measurement, 24*(3), 185–201. https://doi.org/10.1111/j.1745-3984.1987.tb00274.x

Warm, T. A. (1989). Weighted likelihood estimation of ability in item response theory. *Psychometrika, 54*, 427–450. https://doi.org/10.1007/BF02294627

Weiss, D. J. (1974). *Strategies of adaptive ability measurement (RR 74-5)*. University of Minnesota, Psychometric Methods Program. https://files.eric.ed.gov/fulltext/ED104930.pdf

Wu, Q., Debeer, D., Buchholz, J., Hartig, J., & Janssen, R. (2019). Predictors of individual performance changes related to item positions in PISA assessments. *Large-Scale Assessments in Education, 7*(5). https://doi.org/10.1186/s40536-019-0073-6

Yamamoto, K. (1997). A chapter: Scaling and scale linking. In *International adult literacy survey technical report*. Statistics Canada.

Yamamoto, K., & Mazzeo, J. (1992). Item response theory scale linking in NAEP. *Journal of Educational Statistics, 17*(2), 155–174. https://www.jstor.org/stable/1165167?seq=1#page_scan_tab_contents

Yamamoto, K., Shin, H., & Khorramdel, L. (2018). Multistage adaptive testing design in international large-scale assessments. *Educational Measurement: Issues and Practice, 37*, 16–27. https://doi.org/10.1111/emip.12226

Yamamoto, K., Shin, H., & Khorramdel, L. (2019). *Introduction of multistage adaptive testing design in PISA 2018, OECD education working papers*. OECD Publishing. https://doi.org/10.1787/19939019

Yan, D., von Davier, A. A., & Lewis, C. (2014). *Computerized multistage testing: Theory and applications*. Chapman and Hall/CRC.

Zenisky, A. L., Hambleton, R. K., & Luecht, R. M. (2010). Multistage testing: Issues, designs, and research. In W. J. van der Linden & C. A. W. Glas (Eds.), *Elements of adaptive testing* (pp. 355–372). Springer.

Chapter 6
Controlling for Item Position Effects When Adaptive Testing Is Used in Large-Scale Assessments

Andreas Frey and Aron Fink

Abstract Item position effects are typically present in international large-scale assessments (ILSAs) and can cause bias when computerized adaptive testing (CAT) is used. This study introduces an extension of the continuous calibration strategy (CCS) that aims to control for item position effects in CAT by incorporating a balanced statistical design to rotate clusters. The proposed method performed well in a simulation study with ILSA conditions ($N = 2000$). The CCS led to virtually unbiased ability estimates independent of the size of the item position effects across five cycles of a hypothetical ILSA test. Beginning with the second test cycle, the mean square error (*MSE*) of the CCS was smaller than its unbalanced version and smaller than random item selection if item position effects were present. With regard to the item response theory model used in conjunction with the CCS, the *MSE* of the ability estimates for the 2PL was lower from Test Cycle 3 on, but for Test Cycles 1 and 2, the *MSE* for the 1PL was lower. The results on the bias and *MSE* of the ability estimates also show that considering three clusters in the CCS is sufficient to control for item position effects in ILSA conditions. To conclude, the CCS is capable of avoiding bias in ability estimates due to item position effects, which would otherwise result when CAT is used. Additionally, due to its capability of calibrating new items during operational testing, the CCS can help to reduce the effort connected with the field trials conducted in ILSAs.

A. Frey (✉)
Goethe University Frankfurt, Frankfurt am Main, Germany

Centre for Educational Measurement (CEMO) at the University of Oslo, Oslo, Norway
e-mail: frey@psych.uni-frankfurt.de

A. Fink
Goethe University Frankfurt, Frankfurt am Main, Germany

L. Khorramdel et al. (eds.), *Innovative Digital-Based International Large-Scale Assessments*, Methodology of Educational Measurement and Assessment,
https://doi.org/10.1007/978-3-031-90951-1_6

One of the major current challenges of international large-scale assessments (ILSAs) is the appropriate consideration of student performance heterogeneity (Frey & Hartig, 2020). ILSAs typically account for this heterogeneity by aligning the difficulty of the presented test material to the performance level of the tested students. This alignment promises, on the one hand, an increase in the statistical efficiency of the derived latent proficiency distributions and, on the other hand, a reduction of unwanted effects on emotional states, such as boredom and daydreaming, on the side of the test takers (Asseburg & Frey, 2013). In order to maximize these advantages, ILSAs are incorporating more and more adaptivity into their assessments.

In 2009, the Programme for International Student Assessment (PISA), for example, introduced the option for low-performing countries to include easy item clusters in the assessment (OECD, 2012). However, this coarse alignment of the average item difficulty on the level of countries showed only small advantages regarding bias and root mean square error (RMSE) of ability estimation (Rutkowski et al., 2018). The Progress in International Reading Literacy Study (PIRLS) uses a similar approach. Here, a less difficult version of the regular assessment, called PIRLS Literacy, is available for low-performing countries (Mullis & Martin, 2015).

After these adaptations at the country level, the next step was to adapt the difficulty level of the presented test items to the observed performance level of the students with multistage testing (MST; e.g., Zenisky et al., 2010). Thereby, high-performing students are routed to more difficult modules (each containing a set of items) and low-performing students to easier modules. Thus, the adaptation takes place on the more fine-grained level of students within countries and not on the level of countries. The number of stages and the number of alternative modules at each stage, nevertheless, are typically small in MST, resulting in a medium level of adaptation. With the Programme for the International Assessment of Adult Competencies (PIAAC; OECD, 2013) and PISA 2018 (Yamamoto et al., 2018b), two major ILSAs recently adopted MST. The efficiency gains (comparison of test information between MST and nonadaptive testing) achieved with MST were somewhat higher than those achieved with the adaptation on the country level, with 4–31% for PIAAC (Yamamoto et al., 2018a) and 4–7% for PISA (Yamamoto et al., 2018b).

Additional efficiency gains can be expected if even more fine-grained adaptations are implemented. Thus, the obvious next step is the application of computerized adaptive testing (CAT; e.g., van der Linden & Glas, 2010) on the most fine-grained level of adaptation possible. For many ILSAs, this level is testlets (set of items linked to a common stimulus) and, for some ILSAs, it is even at the level of single items. Incorporating CAT into ILSAs harvests the efficiency gains due to optimally aligning the test difficulty to the student performance while being able to account for the restrictions imposed by the individual ILSA. For the optimal case of item-level CAT, a frequently mentioned rule of thumb states that unrestricted CAT needs only about half the items to reach the same measurement precision as sequential testing. Even in nonoptimal and heavily constrained cases, such as using CAT with the actual item pool of PISA 2000 to 2006 and taking the PISA-specific restrictions into consideration, content-balancing, fixed linking items, no instant scoring of open-format items, and selection at the testlet level, CAT still results in a 73% increase in measurement efficiency compared to sequential testing (Frey & Seitz, 2011). In direct comparison with the MST design used for reading in PISA 2018 with a test design that was as adaptive as possible but at the same time met all PISA restrictions, the obtained test information increased by 22–27% (Frey et al., 2023). Note that CAT maximizes the measurement efficiency on the level of individual students. Thereby, it leads to the collection of highly informative responses to test items. In ILSAs, these are not used to derive point estimates of student abilities. Rather, the responses are scaled with an item response theory model (IRT; e.g., van der Linden, 2016), which is combined with responses given to background questions to derive multivariate proficiency distributions on the population level. From these proficiency distributions, plausible values (PVs; e.g., Mislevy, 1991) are drawn and used to calculate the reported statistics as estimations on the population level (group means, variances, percentages, etc.). Point estimates of student abilities are not used to calculate the reported ILSA results. Nevertheless, they are very useful when it comes to examining how the efficiency of the IRT model-part of the described methodology can be fostered. For example, Frey et al. (2013) reported that multidimensional adaptive testing leads to an increase in the precision of all 10 unconditioned literacy subdimensions of PISA to such an extent that they can be used for reporting, while taking PISA-specific constraints into account. Usually, the subdimensions are only reported for the respective major domain in PISA (e.g., OECD, 2017).

However, despite its advantageous measurement efficiency, CAT is based on more assumptions that can be violated than is the case in sequential testing. One assumption that has not yet gained much attention is the assumption that items function in the same way regardless of the position in the test in which they are presented (i.e., item position effects). The neglect of this aspect is problematic because item position effects (IPEs) must be expected in ILSAs and because each individual item can be presented in many different positions in CAT while the same fixed item parameters are used for all positions. We will show that ability estimation can be biased in CAT if IPEs are present. With the present chapter, we want to (a) increase the awareness of CAT-specific challenges for ILSAs due to IPEs, (b) describe a method that makes it possible to control for IPEs in CAT, and (c) report results from a simulation study that show how successful this method actually is for IPEs with a magnitude typical for ILSAs.

The rest of the chapter is organized as follows. In the next section we summarize the literature on IPEs and describe IPE-induced challenges for sequential and adaptive testing. These challenges do not mean that CAT should be avoided in ILSAs, but that methods need to be established that control for the unwanted effects of IPEs. We will elaborate on such methods and discuss their usefulness for ILSAs in the following section. Among the available methods, we regard the continuous calibration strategy (CCS; Fink et al., 2018) as the most promising method. Therefore, the subsequent sections are devoted to the description of the CCS and the methods and results of a Monte Carlo simulation examining it under ILSA conditions. The chapter closes by discussing the implication of the results for future ILSAs.

6.1 Item Position Effects in Computerized Adaptive Testing

A typical CAT construction process (e.g., Thompson & Weiss, 2011) includes an extensive calibration study to obtain accurate and precise item parameter estimates. These item parameter estimates are treated as known and are used for item selection and ability estimation in the operational CAT phase. This procedure draws from the assumption that item parameters are invariant with respect to (a) test takers, (b) time, (c) testing situations, and (d) the position in which items are presented in the test. In a thorough CAT construction process, the appropriateness of the first three assumptions is examined with (a) differential item functioning (DIF; e.g., Holland & Wainer, 1993), (b) item drift analyses (e.g., Donoghue & Isham, 1998), and (c) standardization of the testing situation, respectively. The question of whether the assumption that item parameters are invariant across positions holds is (d) typically not yet addressed.

6.1.1 Empirical Evidence for Item Position Effects

Not checking the IPE assumption is inappropriate because the literature on IPEs provides strong empirical evidence that variations in item parameter estimates across positions in a test are more the rule than the exception. Most studies report a decreasing proportion of correct answers, and thus, an increase in item difficulties, towards the end of a test (Albano, 2013; Debeer et al., 2014; Debeer & Janssen, 2013; Frey et al., 2017; Hartig & Buchholz, 2012; Hecht et al., 2015; Hohensinn et al., 2008; Le, 2007; Meyers et al., 2008; Nagy et al., 2016, 2019; Qian, 2014; Weirich et al., 2014; Wu et al., 2019; Yen, 1980). For PISA 2009, Debeer et al. (2014) reported an average standardized difference in difficulty estimates between the first and the last cluster in a booklet of $d = 0.51$. In other words, the difficulty of an item increases by 0.51 standard deviations of the reading ability when it is moved from cluster position 1 to cluster position 4. After the switch from paper-and-pencil testing (PPT) to computer-based testing (CBT) in PISA 2015, smaller IPEs were reported. There was a substantial difference in the average proportion of correct answers (P+) between the fourth and the first cluster position; for CBT in PISA 2015, it was 50% (mathematics), 22% (reading), and 19% (science) smaller than for PPT in PISA 2012 (OECD, 2017). In PISA 2018, which was conducted with CBT, the P+ differences in mathematics remained at this low level but, for science, they returned to a value slightly higher than that found in PISA 2012 (OECD, 2020). Some studies also reported the contradictory result of decreasing item difficulties towards the end of a test (Davis & Ferdous, 2005; Schweizer et al., 2009, 2011). Furthermore, other studies indicated that IPEs may be of a general or an item-specific nature. For example, Kingston and Dorans (1984) found negative effects of the item position on the item difficulty estimates for the subscale “reading comprehension” and positive effects on the item difficulty estimates for the subscale “analysis of explanations" in a nonadaptive section of the Graduate Record Examination (GRE) General Test. In addition to these general item position effects, Albano (2013) determined that the magnitude and direction of IPEs may differ between items. Lastly, some individual studies reported only small or even no IPEs (Hohensinn et al., 2011; Li et al., 2012; Shin et al., 2021; Zwick, 1991). To summarize the empirical results on IPEs: It can be concluded that variations in item difficulty estimates across positions in a test exist in many types of assessment, including major ILSAs, and that these IPEs can be of a general (the same for all items) or an item-specific nature.

6.1.2 Problems Due to Item Position Effects in Sequential Testing

In traditional tests with one test form presenting the same set of items in the same order to each test taker, IPEs are typically not examined. Actually, the examination of IPEs is not possible with data stemming from such traditional tests because item position is completely confounded with items. Nevertheless, this does not mean that IPEs are unproblematic for traditional tests. Just as in studies using several test forms (such as ILSAs), the existence of IPEs causes two problems for traditional tests. First, they can spuriously increase the precision of a test and all of the derived measures, such as the reliability. This is because items are completely confounded with positions so that any systematic variance caused by positions adds to the systematic variance caused by the construct that should be measured. On the item side, this means that construct irrelevant variance is incorporated into item parameter estimates. This is inappropriate for all constructs whose definition does not include systematic effects that occur over the course of the test (e.g., fatigue). Thus, this is a validity issue, which typically cannot be detected (because items are presented in fixed positions) and can lead to an overestimation of the precision of a test. Second, IPEs can jeopardize criterion-referenced test-score interpretations, which are frequently used for reporting ILSA results. Often, criterion-referenced test-score interpretations are based on the relative standing of individuals with respect to the items on the latent scale of interest. This makes it possible to formulate statements about whether an individual can solve one or more items with at least a certain probability. The definition of proficiency levels in ILSAs is typically based on such considerations (e.g., OECD, 2017). In the presence of IPEs, such inferences are no longer straightforward. Note that this is the case for both, tests using only one fixed test form and assessments with balanced booklet designs (Frey et al., 2009), such as ILSAs. For tests with one fixed test form, the location of the items on the scale is shifted by the effect (typically an increase in difficulty) with respect to the position in which it is presented in the test. This results in varying item difficulty shifts for different items. For tests using a balanced booklet design, all items are presented in all possible positions (typically blocks of items) to the same number of test takers.

Thus, all items are subject to the average effect across all possible positions. If the IPE is the same for all items (general IPEs), the complete set of items is shifted with respect to the test takers by the average IPE. In the case of item-specific IPEs, the items are shifted differently. In all cases and for both tests with one fixed test form and tests relying on a balanced booklet design, inferences incorporating the probabilities of individuals to solve certain items are incorrect. Note that we are only discussing IPEs with regard to item difficulty estimates here. Things get more complicated if IRT models including more parameters than the difficulty parameter are considered. Some empirical results showed that relations can exist between the position in the test and the item discrimination parameter estimates (Bäckström & Björklund, 2016; Le, 2007; Schweizer et al., 2009, 2011). To avoid making this chapter too complicated, we limit our explanation to the case of IPEs with regard to difficulty estimates, on which most of the research on IPEs focusses.

6.1.3 Problems Due to Item Position Effects in CAT

The two problems of overestimating test precision and jeopardizing criterion-referenced test-score interpretations discussed above concern both sequential and adaptive tests. For CAT, however, IPEs are even more problematic for three additional reasons. First, comparisons on the individual and the group levels are likely to have limited precision if item-specific IPEs are present; that is, the items in the item pool of the adaptive test vary with respect to the degree to which their item parameter estimates are influenced by IPEs. In this case, the item sets selected for some test takers can be more strongly influenced by IPEs than the item sets selected for other test takers. The average ability differences between these two groups of test takers, thus, represent not only differences in the construct of interest but also differences in the average position-specific effect. Second, if IPEs and the operational usage frequency of items are stochastically related, ability estimations on the individual and the group levels will be biased. This is the case if an item set in which IPEs differ from the average IPE (of all items in the pool) is presented more often than others in the operational CAT phase. If items with a large IPE regarding item difficulties are presented very often, the average difficulty of the test will be artificially high. Because item parameters will not be estimated anew in CAT, this will transform directly into biased ability estimations. Third, in the presence of IPEs, bias also results if the operational CAT test length is not equal to the calibration test length. This is because the IPE that was present in the calibration study is integrated into the item difficulty estimates. In a fully position-balanced design, where the same number of responses results for each item in each possible position, this is just a simple addition of the average IPE per item across positions to the individual item difficulty parameter.

We will now exemplify the effect of the difference in test length between calibration and operational CAT with a small simulation-based demonstration also used in a conference presentation by Frey et al. (2015; see also Born, 2019). For this simulation, we used the item parameter estimates from an existing CAT item pool that measures student competences in science (for more information on this test, see Spoden et al., 2018). In the calibration study of this test, a balanced incomplete booklet design with a test length of 33 items was applied. This made sure that all items were presented in each possible position equally often. The gathered responses of $N = 1632$ students were scaled using the one parameter logistic model (1PL). The resulting item parameter estimates were then used to simulate CAT with maximum Fisher information as item selection criterion and a test length varying from one to 33 items. To generate data, standard normally distributed abilities for $N = 1000$ simulees and general IPEs (same for all items) were used with 100 replications. These can be interpreted as additional difficulties that were added to each position. The IPEs were derived from the empirical results of the calibration study. To obtain stable IPE estimates, they were estimated for groups of five or four items. The resulting IPEs (−0.174, −0.267, −0.118, 0.072, 0.076, 0.252, 0.160) were transformed to a mean of zero. The typical overall pattern that the proportion of correct responses was larger at the beginning of the test than at the end of the test can be seen even though the trend was not monotone. This simulation setup resembles a typical procedure in a CAT construction process in which the item parameter estimates derived from a calibration study are considered as known in the operational phase of the test, although the test itself might have a different length than the calibration study.

Figure 6.1 shows the average bias of the person parameter estimates (maximum a posteriori with standard normal prior; Mislevy, 1986) for the simulated adaptive test depending on the number of presented items. Bias is clearly related to the test length. With increasing test length, the bias initially gets larger, then reaches its maximum at 10 items, and then declines after a short plateau of about five items. When the CAT test length is 33 and, thus, equal to the calibration test length, virtually unbiased

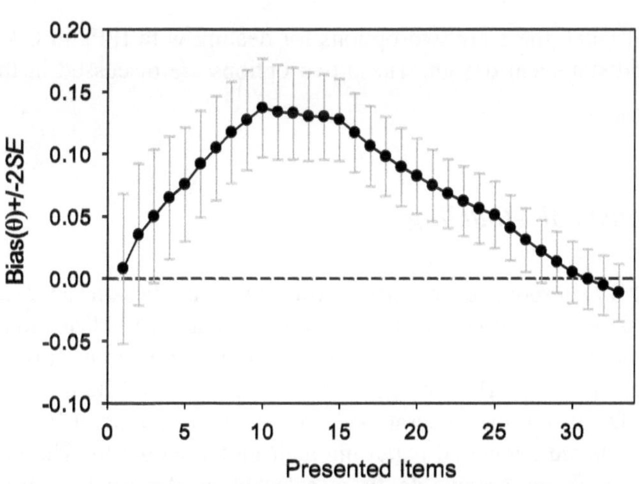

Fig. 6.1 Average bias of ability estimates as a function of number of presented items

ability estimates result. This pattern becomes immediately clear when taking into account the fact that the specific difficulty estimate obtained for each item in the calibration study was used, regardless of the position in which the item was presented. Each of these item difficulty estimates is the sum of the true item difficulty (item difficulty without any IPE) and the average of the 33 position-specific effects. For an item presented in the first position, this means that the difficulty estimate derived from the calibration is used (which is equal to the true item difficulty because the IPEs are standardized to have a mean of zero in this example), which is slightly too high. The appropriate item difficulty for this position would be the true item difficulty plus the IPE of position one, which is $-0.174/5 = -0.038$. This value does not transform directly into bias because the information obtained from just one response is so sparse that the prior with a mean of zero has a large influence on the ability estimate. For the second item, the difference between the item difficulty used (true item difficulty) and the appropriate item difficulty for this position (true item difficulty $+2 \cdot (-0.038)$) is even larger. It follows that the bias is a function of the average of these differences for the presented items. The bias starts to decrease at a test length in which the IPE becomes smaller than the previous IPE, because the average IPE has decreased. In the example, this is the case from a test length of 11 items. Note that, depending on the IPEs, other bias curves result. Only for the test length of 33 items (i.e., the test length of the calibration) are the used item parameters appropriate, which leads to unbiasedness. In this example, the relationships described lead to an unneglectable bias for most test lengths ranging up to $d = 0.14$ (14% of the *SD* of student ability).

6.2 Methods to Overcome Item Position Effects in Computerized Adaptive Testing

On a general level, there are two options for dealing with IPEs in CAT: statistical modeling and statistical design. These two options are discussed in the following two sections.

6.2.1 Statistical Modeling

Frey et al. (2017) introduced a multistep procedure that relies on statistical modeling. In Step 1, a series of multifaceted Rasch models (Linacre, 1994) is initially fitted. In the basic setup, three models are compared: the Rasch model, the Rasch model with an additional facet capturing the general IPE (same for all items), and the Rasch model with facets for the general IPE and an item-specific IPE. In Step 2, the different models are compared regarding their global model fit. The most parsimonious model with an acceptable fit is selected as the final calibration model.

Subsequently, the parameter estimates of the selected model are used in the operational CAT phase for item selection and ability estimation. Thereby, IPEs are only incorporated into CAT if there is empirical evidence for their existence. If this is the case, the method is not restricted by assumptions about the functional form of the IPEs. The separately parameterized IPEs can represent any functional form found in the empirical data at hand. The downside of this flexibility is that the models can comprise large numbers of parameters, which require large samples for their precise estimation. Because the requirements regarding sample sizes can be reduced by merging adjacent positions without substantially restricting the interpretability of the results, this does not seem to be a serious problem for ILSAs, which rely on large samples anyway. For studies already in progress, the applicability of Frey et al.'s (2017) method may, nevertheless, be limited for another reason: It might not be feasible for ILSAs already in progress, because it may be necessary to change the scaling model, and this is accompanied by numerous methodological challenges (e.g., linking, comparability across assessment cycles).

6.2.2 Statistical Design

As an alternative to statistical modeling, statistical design can be used to control for IPEs in CAT. Such a possibility was outlined in the initial publication on the CCS by Fink et al. (2018). Below, we will elaborate further on this approach and then analyze its capabilities to control for unwanted IPEs by means of a Monte Carlo simulation. Fink et al. (2018) argued that typical reasons for why CAT is not used even though it would be advantageous to do so are that a separate large calibration study and/or the construction of a very large number of items for the calibration study are not feasible. To extend the application area of CAT to such situations, they suggested adopting the CCS for tests that are used repeatedly and that strive to measure the same construct each time. The main objectives of the CCS are (a) to orient the item calibration towards the capacities available for test development, (b) to optimize item parameter estimation by using item responses across test cycles, (c) to maintain the reporting scale across time, (d) to continuously increase the adaptivity of the test during its operational use, and (e) to reach robustness by checking for item parameter drift. A test following the principles of the CCS starts in a nonadaptive manner and becomes more and more adaptive across cycles. A separate calibration study is not needed. The CCS can be combined with the shadow testing approach (van der Linden & Reese, 1998) for constraint management.

In this context a test cycle is regarded as the whole test procedure, including steps such as test assembly, test administration, and the analysis of the test results. The CCS distinguishes between three types of item clusters, *calibration clusters, adaptive clusters,* and *linking clusters*. Each of these cluster types contains several items with specific features and has a specific goal. The calibration clusters contain uncalibrated items and make it possible to establish the item pool in the first test cycle and to enlarge it by seeding in new items in the subsequent test cycles. The adaptive clusters contain items that are administered adaptively using the item parameter estimates from the previous test cycle in order to enhance the measurement precision. Finally, the linking clusters utilize common items to establish a common scale across test cycles using a common-item nonequivalent group design (Kolen & Brennan, 2014). Items in the calibration and the linking clusters are administered to every test taker of one test cycle. If the CCS is used for situations where no item parameter estimates are available at the start of the procedure, the first test cycle is comprised of calibration clusters only. From the second test cycle onwards, tests are composed of all three types of clusters. If increasing the item-pool size is not of interest, the calibration clusters can become additional adaptive clusters to further increase the adaptivity of the test. There is no restriction to the size of the item pool in the CCS.

The algorithm of each test cycle in the CCS following the first test cycle has seven steps (see Fig. 6.2 for a flow chart of the CCS):

1. Common item selection for the linking clusters. Items are selected so that they represent the construct to be measured well and provide high test-information across a broad ability range.
2. Test assembly and test administration.
3. Temporary item parameter estimation based on the responses of the current test cycle.
4. Scale transformation of the common items. Due to the inequivalence of the test takers in the two consecutive test cycles, their parameter estimates are not on the same scale and, thus, the scale has to be transformed before testing for item parameter drift (IPD).
5. Detection of IPD (Battauz, 2019; König et al., 2021). Common items showing significant IPD are excluded from the fixed parameter calibration (FPC; e.g. Kim, 2006) in Step 6 and estimated freely.
6. FPC. For all items with no significant IPD, the available responses from all test cycles are used in a concurrent scaling procedure for item parameter estimation. In this scaling, the parameters of the final common items are fixed at their estimated values from the previous test cycle, whereas all the other items are estimated freely.
7. Person parameter estimation using the item parameters that resulted from Step 6.

The cluster structure of the CCS can be used to control for IPEs. One option is to use a Latin square design to specify the order of the clusters (e.g., Frey et al., 2009; Williams, 1949), which results in a test design that is balanced for both cluster positions and first-order carryover effects (effect of a preceding cluster on the next cluster; e.g., Giesbrecht & Gumpertz, 2004). Thus, asymptotically, the impact of potential IPEs is the same for all clusters. Because the CCS uses three cluster types, the smallest Latin square design is of size 3 × 3. In order to achieve a higher resolution of item positions, the three item clusters could be divided into more fine-grained, equally sized subclusters. Table 6.1 shows an example of such a test design with six clusters.

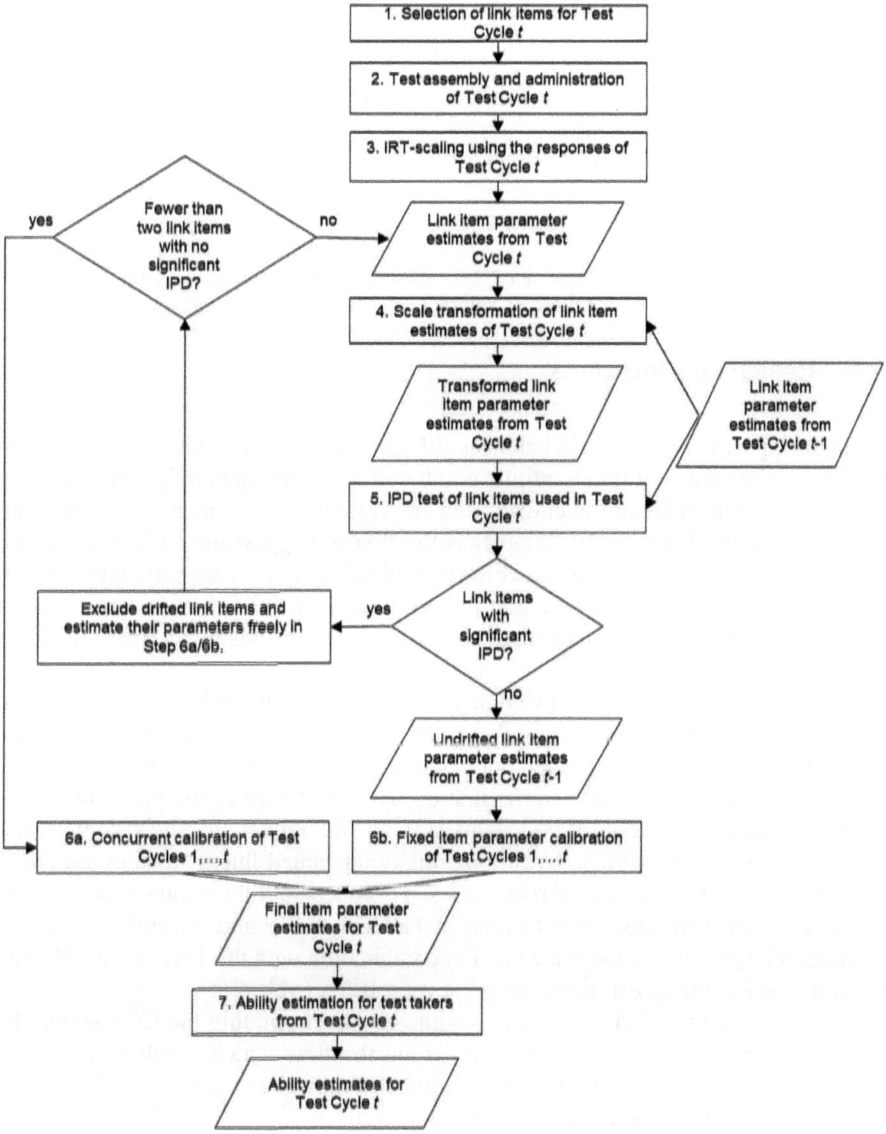

Fig. 6.2 Flowchart of the continuous calibration strategy

Each test taker receives a randomly chosen test form. As in many ILSAs, spiraling can be used for this purpose. As a result of this, on the sample level, each cluster is presented in each of the six positions to a nearly equal number of test takers, and each cluster precedes each of the other clusters exactly once. Therefore, IPEs and first-order carryover effects cannot lead to systematic bias.

Table 6.1 Test design in the continuous calibration strategy based on latin square design for six clusters

	Cluster position					
Test form	1	2	3	4	5	6
1	L1	L2	C2	A1	C1	A2
2	L2	A1	L1	A2	C2	C1
3	A1	A2	L2	C1	L1	C2
4	A2	C1	A1	C2	L2	L1
5	C1	C2	A2	L1	A1	L2
6	C2	L1	C1	L2	A2	A1

Note. L1-L2 = linking clusters; A1-A2 = adaptive clusters; C1-C2 = calibration clusters

6.3 Research Questions

It can be expected that the CCS balances IPEs asymptotically. That is, for an infinite number of test takers, or for an infinite number of items arranged in large numbers of clusters, possible IPEs are averaged, and no systematic bias in person parameter estimates results. Because this chapter is the first to suggest the CCS with cluster balancing, it is yet unknown whether the typical ILSA settings are sufficient to reach unbiasedness. This motivates the first research question:

Research Question 1: Can the CCS control for item position effects in ILSA settings?

The data requirements are different for the 1PL and the 2PL. Even though the number of available responses and the number of items grow across test cycles due to the calibration procedure carried out in the CCS, noteworthy differences can result between the 1PL and the 2PL in the first cycles with regard to the precision of the ability estimation. If such differences exist, then IRT model-specific recommendations are necessary on how to avoid potentially unwanted fluctuations in the precision of the ability estimation across cycles. To understand the entanglement of the IRT model and the number of test cycles in the CCS, the second research question is:

Research Question 2: Does the CCS in combination with the 1PL or the 2PL lead to more precise ability estimates across several ILSA cycles?

The complexity added to ILSA procedures by incorporating the CCS should be kept to a minimum. Therefore, it is important to know which number of clusters (used for balancing) is sufficient to adequately control for IPEs in ILSAs. This stimulates the last research question of this study:

Research Question 3: What is the minimum number of clusters required in the CCS in order to adequately control for IPEs typical in ILSAs?

6.4 Method

6.4.1 Study Design

To answer the research questions stated above, a comprehensive Monte Carlo simulation based on a full factorial design with three between- and one within-factor was conducted. The first between-factor, *IRT model* (1PL, 2PL), included two IRT models frequently used in ILSAs. Under the 2PL, the probability of a correct response $u_{ij} = 1$ of person $j = 1, \ldots, N$ with the latent ability level θ_j to an item i is defined as:

$$P\left(u_{ij} = 1 | \theta_j, a_i, b_i\right) = \frac{\exp\left[a_i(\theta_j - b_i)\right]}{1 + \exp\left[a_i(\theta_j - b_i)\right]}, \tag{6.1}$$

where a_i is the discrimination parameter and b_i is the difficulty parameter of item i. Eq. 6.1 reduces to the 1PL when the discrimination parameter a_i is set to a constant value for all items (Birnbaum, 1968). For the 1PL in this study, the discrimination parameters were set to $a_i = 1$. The second between-factor, *IPE* ($d_{IPE} = 0.00$, $d_{IPE} = 0.26$, $d_{IPE} = 0.51$, $d_{IPE} = 1.02$), introduces a variation of the IPEs. The medium IPE level was chosen based on the findings from Debeer et al. (2014), who reported an average IPE of $d_{IPE} = 0.51$ from the first to the last cluster position across the 65 participating countries in PISA 2009. On the basis of this the low ($d_{IPE} = 0.26$) and the high level ($d_{IPE} = 1.02$) of the factor IPE were calculated by halving and doubling the medium level, respectively. With the third between-factor, *item selection* (CCS3, CCS6, CCS12, UCCS, random), different types of item selection and test assembly were varied. The first three levels of this factor represent the CCS as described above. The numbers 3, 6, and 12 after CCS stand for the number of clusters into which the test was divided. Item clusters were presented using a Latin square design and were randomly assigned to the test takers. In addition, two baseline conditions were simulated. The UCCS condition represents an unbalanced variant of the CCS, in which the order of clusters was the same for every test taker. Starting from Test Cycle 2 of the UCCS condition, the cluster order was linking cluster, calibration cluster, and adaptive cluster. In the random condition, items were chosen randomly from an item pool whose size increased by $\frac{1}{3} \cdot T$ in each test cycle, where T is the test length. For each condition, the algorithm ran for five test cycles. Therefore, the last factor was the within-factor *test cycle* (1, 2, 3, 4, 5). The fully crossed design resulted in 2 x 4 x 5 x 5 = 200 conditions. For each condition, $R = 100$ replications were analyzed with regard to the evaluation criteria as defined below.

The simulation was carried out in R (R Core Team, 2019) using the "mirtCAT" package (Chalmers, 2016) to simulate the tests and the "mirt" package (Chalmers, 2012) for item and person parameter estimation. Scale transformation and tests for IPD were conducted with the "equateIRT" package (Battauz, 2015). The corresponding functions were called from R code, which was specifically written to carry out the CCS.

6.4.2 Data Generation

In each replication, the difficulty parameters b_{i1} for item i in position $p = 1$ were randomly drawn from a normal distribution, $b_{i1} \sim N\left(-\frac{1}{2} \cdot d_{IPE}, 1.25\right)$. Assuming a linear increase across positions, $p = 1, \ldots, T$, the position-specific difficulty parameter b_{ip} of item i in position p was generated by:

$$b_{ip} = b_{i1} + \frac{(p-1) \cdot d_{IPE}}{(T-1)}. \tag{6.2}$$

Discrimination parameters a_i were randomly drawn from a lognormal distribution with a_i~lognormal $(0, 0.25)$ and kept constant across positions. The ability parameters of $N = 2000$ test takers in each test cycle were drawn from a standard normal distribution, $\theta \sim N(0, 1)$. A test length of $T = 24$ was chosen to mimic the test length for one domain in PISA. The test takers' responses were generated in line with either the 1PL or the 2PL. In order to simulate a response pattern that is realistic for ILSAs, missing values (omitted responses and not-reached items) were incorporated into the generated data.

6.4.2.1 Omitted Responses

Omitted responses were simulated based on the position-specific omission rates (ORs) observed in PISA 2015. Due to the fact that only cluster-position-specific ORs were reported in PISA 2015, the ORs for the cluster positions 1, 2, 3, and 4 were assigned to the first, second, third, and fourth item-position quartiles, respectively. A correlation of −.35 was assumed between ability and OR. This is an average value for this correlation and is typically observed in ILSAs (Frey et al., 2018; Robitzsch, 2016; Weeks et al., 2016). To integrate omissions, after generating ability parameters, a second set of parameters (omission parameters), which were negatively correlated with ability, was drawn from a standard normal distribution for each position $p = 1, \ldots, T$, resulting in an $N \times T$-matrix of position-specific omission parameters. For every test taker whose position-specific omission parameter was greater than the critical z-value z_{1-OR}, the response to the item in this position was recoded into 0 (incorrect). Table 6.2 shows the final cluster-position-specific ORs in comparison to the ORs in PISA 2015 (OECD, 2017).

Table 6.2 Omission rates per cluster position for the present study and PISA 2015

	Cluster position			
Study	1	2	3	4
This study	2.80	4.22	3.63	4.41
PISA 2015	2.80	4.25	3.65	4.85

Table 6.3 Percentages of not-reached items for the present study and PISA 2012

	Number of not-reached items									
Study	0	1	2	3	4	5	6	7	8	>8
This study	88.62	2.15	1.17	1.48	0.65	0.47	0.70	0.82	0.27	3.67
PISA 2012	88.73	2.37	1.17	1.31	0.61	0.42	0.66	0.78	0.26	3.69

6.4.2.2 Not-Reached Items

The number of not-reached items, defined as the series of missing values at the end of the test, was oriented towards PISA 2012 and treated as not administered. In order to calculate the percentages of not-reached items for PISA 2012, the mean of the booklet-specific percentages was taken. Table 6.3 shows the percentages of not-reached items in this study compared to PISA 2012 (OECD, 2014).

6.4.3 Specification of the CCS

The numbers of items in the calibration clusters, the adaptive clusters, and the linking clusters were set to the equal size of $\frac{1}{3} \cdot T$. As the item pool increased by $\frac{1}{3} \cdot T$ starting from Test Cycle 2, the resulting item-pool size I_t after Test Cycle t was:

$$I_t = T + \frac{(t-1) \cdot T}{3}. \tag{6.3}$$

Therefore, after Test Cycle 5, the final item pool was comprised of $I = 56$ items. For the conditions CCS6 and CCS12, the three types of clusters were subdivided into two and four subclusters, respectively. During the test, within the adaptive and the linking clusters, maximum a posteriori (Mislevy, 1986) was used to derive the provisional ability estimates, which serve as the basis for adaptive item selection. The items in the calibration clusters were not scored during the simulated testing sessions. For the adaptive item selection, the maximum information criterion was applied (Lord, 1980). Items in the calibration and the linking clusters were selected randomly.

For the link item selection, only items that had already been administered in one of the previous test cycles and did not serve as link items in the preceding test cycle were eligible. The selection procedure resembled the procedure used by Fink et al. (2018). To ensure that the link items represented the item-difficulty distribution in the pool, in each test cycle, $t > 1$ items were first assigned to five difficulty categories (very low, low, medium, high, and very high) based on their estimated difficulty parameters from the previous test cycle. Two items each from the medium, the low, and the high category and one item each from the remaining categories were chosen. To carry out the scale transformation in Step 4 of the CCS, the Stocking-Lord method (Stocking & Lord, 1983) as recommended by Born et al. (2019) was used. To test IPD, a Wald-type test, as implemented in the "equateIRT" package, was used (Battauz, 2019) and was conducted iteratively (Candell & Drasgow, 1988). Drifted link items were excluded from FPC and estimated freely without imposing constraints.

In each test cycle, marginal maximum likelihood (MML; Bock & Aitkin, 1981) was used to estimate the item parameters for both the temporary item parameter estimation and the FPC. Subsequent to the FPC, final person parameters were estimated using weighted likelihood estimation (WLE; Warm, 1989).

6.4.4 Evaluation Criteria

The accuracy and the measurement precision were evaluated by examining the bias and the mean square error (*MSE*) of the ability estimates. Let $\hat{\theta}_{jtr}$ and θ_{jtr} denote the estimated and true abilities of the jth test taker in Test Cycle t of the rth replication, $j = 1,\ldots,N$ and $r = 1,\ldots R$. Then,

$$Bias_t = \frac{1}{R \cdot N} \sum_{r=1}^{R} \sum_{j=1}^{N} \left(\hat{\theta}_{jtr} - \theta_{jtr} \right), \tag{6.4}$$

and

$$MSE_t = \frac{1}{R \cdot N} \sum_{r=1}^{R} \sum_{j=1}^{N} \left(\hat{\theta}_{jtr} - \theta_{jtr} \right)^2. \tag{6.5}$$

6.5 Results

In order to answer the question of whether the CCS can control for IPEs in ILSA settings (Research Question 1), as a first step, the accuracy of the ability estimates was analyzed for the individual test cycles. Figure 6.3 illustrates the bias of the ability estimates per test cycle across the different simulation conditions. Each of the three balanced CCS conditions, as well as the random condition, led to virtually unbiased ability estimates independent of the magnitude of the IPEs. Only the UCCS

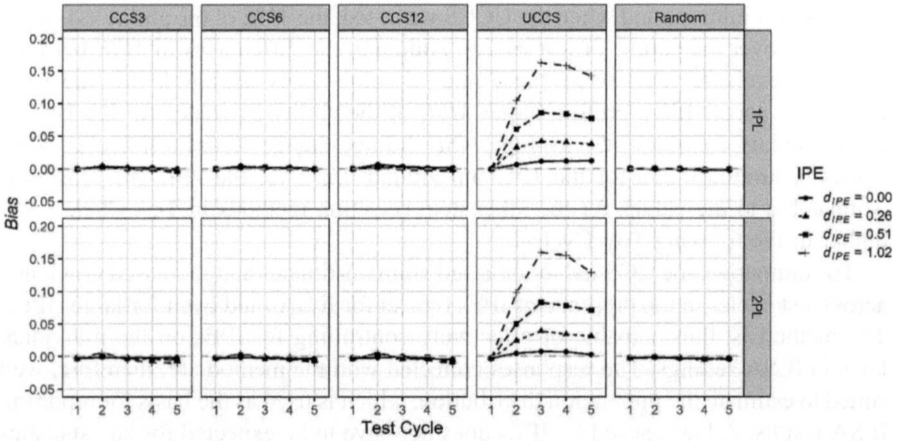

Fig. 6.3 Bias of the ability estimates for five item-selection methods. *Note*. CCS3, CCS6, CCS12 = balanced continuous calibration strategy with 3, 6, and 12 clusters, respectively; UCCS = unbalanced continuous calibration strategy; random = random item selection, five test cycles, and four levels of item position effects (IPEs) under either the 1PL or the 2PL for a test of 24 items ($N = 2000$)

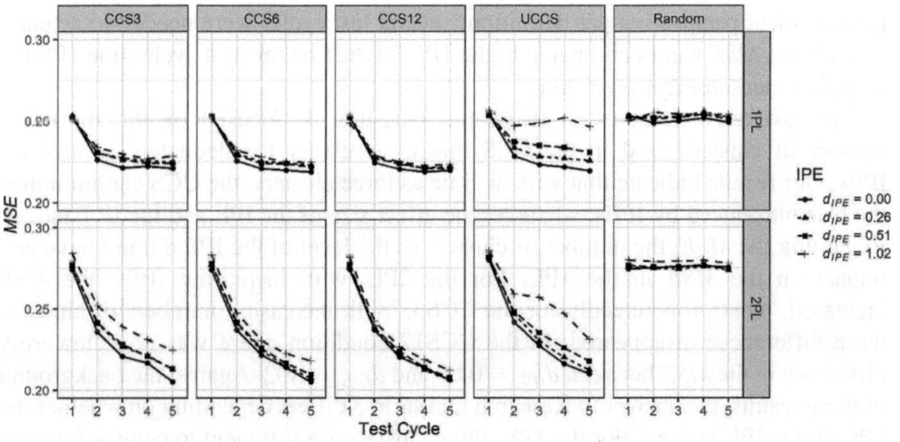

Fig. 6.4 Mean square error (*MSE*) of the ability estimates for five item-selection methods. *Note*. CCS3, CCS6, CCS12 = balanced continuous calibration strategy with 3, 6, and 12 clusters, respectively; UCCS = unbalanced continuous calibration strategy; random = random item selection, five test cycles, and four levels of item position effects (IPEs) under either the 1PL or the 2PL for a test of 24 items ($N = 2000$)

condition resulted in systematic bias when IPEs were present, with larger IPEs resulting in larger bias. The bias increased until Test Cycle 3 and then decreased slightly.

As a second step, measurement precision was analyzed by comparing the *MSE* between the simulated conditions. Figure 6.4 summarizes the results regarding *MSE*.

In all CCS conditions and when the UCCS was used, the *MSE* of the ability estimates decreased over the course of test cycles. Although the differences in *MSE* between the four levels of IPEs were small and in some cases negligible for the CCS conditions, larger IPEs led to a larger *MSE* of the ability estimates for the UCCS. In the random condition, the average *MSE* did not change across test cycles. With values of around 0.250 for the 1PL and around 0.275 for the 2PL, the *MSE* was comparable to the values for the other item selection methods in Test Cycle 1 but higher fir the follwing Test Cycles.

To summarize, the CCS led to unbiased ability estimates and increased efficiency across test cycles in the presence of IPEs typical for ILSAs and even for larger IPEs. The method is, thus, capable of effectively controlling for IPEs on the individual level in ILSA settings. The responses collected with the method are, therefore, well suited to estimate the population distribution, which is used as the basis for reporting ILSA results. A bias caused by IPEs does not have to be expected for any statistics derived from this distribution (e.g., by using plausible values).

The second research question focused on the differences between the 1PL and the 2PL across test cycles. As can be seen in Fig. 6.3, both IRT models led to virtually unbiased ability estimates for all five test cycles. Comparing differences in *MSE* between the 1PL and the 2PL model (Fig. 6.4), however, reveals that even though the 2PL is capable of achieving higher test-information than the 1PL from a mathematical perspective per definition, several test cycles were needed to actually reach lower *MSE* values compared to the 1PL. In the first few test cycles, the 1PL led to higher measurement precision.

The last research question (Research Question 3) focused on the minimum number of clusters used in the CCS that is necessary to adequately control for IPEs. Our results indicate that with as little as three clusters, the CCS can minimize the bias introduced by IPEs, whatever the effect size of the IPE and the IRT model. Regarding the *MSE*, the number of clusters or the level of the IPE did not have any impact on the *MSE* in the 1PL. For the 2PL, with increasing IPEs, the *MSE* increased, most pronouncedly for the CCS3. With increasing numbers of clusters, these differences disappeared. In the CCS12 condition, there was no noteworthy difference in the *MSE* between $d_{IPE} = 0.00$ and $d_{IPE} = 1.02$. Against the background of these results, the answer to Research Question 3 differs depending on whether the 1PL or the 2PL is used. For the 1PL, three clusters are sufficient to control for even very large IPEs. When using the 2PL for IPEs of a magnitude typical for ILSAs ($d_{IPE} = 0.51$), three clusters are also sufficient. If very large IPEs are to be expected, however, 12 clusters are recommended to ensure that the systematic variance introduced by the IPEs is fully controlled for.

6.6 Discussion and Outlook

IPEs are typically present in ILSAs. They can cause additional problems when CAT is used. On a general level, the problems of using CAT when IPEs are present can be solved by statistical modeling or by statistical design. The latter is more suitable for ILSAs because it does not necessitate changes in the scaling model. In this chapter, we introduced an extension of the CCS of Fink et al. (2018), which applies statistical design principles to control for IPEs across several CAT cycles.

The proposed method performed well in a simulation study that resembled ILSA settings. Beginning with the second test cycle, the CCS showed a smaller *MSE* in comparison to its unbalanced counterpart, to the UCCS, and to random item selection in the presence of IPEs. Furthermore, the CCS led to virtually unbiased ability estimates. The UCCS, however, resulted in substantial bias in Test Cycles 2 to 5. With regard to the results obtained for the UCCS, two points will be discussed here. First, it was observed that the *MSE* of the UCCS was smaller than the *MSE* in the random condition, even for relatively large IPEs of up to $d_{IPE} = 0.51$ and even though substantial bias was only found for the UCCS. For very large IPEs of $d_{IPE} = 1.02$, the *MSE* was comparable between the UCCS and the random condition. This result pattern indicates that the ability estimation with the UCCS is off target (i.e., is biased) but, nevertheless, very precise. Because the differences from the true ability value (whose average is the bias) are squared in the calculation of the *MSE* (Eq. 6.5), relatively large shifts due to IPEs are necessary to lower the *MSE* below the results found for random item selection. Second, the bias observed in the UCCS condition decreased slightly after Test Cycle 3. This can be explained by the fixed cluster order in the UCCS condition. Starting from Test Cycle 2 of the UCCS, the linking cluster was always presented in the first cluster position, and only one third of the items initially calibrated in the first test cycle were presented in this cluster. Therefore, in the second test cycle, more items could be selected as common items, which included an IPE associated with cluster positions 2 or 3 and were, therefore, too difficult for the first cluster position than items that were presented in the first cluster position in the first test cycle. The problem repeated itself in the third test cycle: The items available for selection as common items were items from the first test cycle that did not serve as items in the second test cycle and new items that were initially calibrated in the second test cycle. As new items are always administered in the middle cluster position, their position-specific item difficulties are composed of the item difficulties in cluster position 1 and the mean IPE of cluster position $\frac{T}{3} + 1$ to $\frac{2 \cdot T}{3}$, whose expected value is $\frac{d_{IPE}}{2}$. The item difficulty parameters of these items are too high when used in the linking cluster, because the position-specific item difficulties for the first cluster position are composed of item difficulties in cluster position 1 and the mean IPE of cluster position 1 to $\frac{T}{3}$, whose expected value is $\frac{d_{IPE}}{3}$. Due to the fact that, in the FPC, link-item difficulty parameters are fixed at their values from the previous test cycles and their difficulty parameters are systematically too high for the cluster position in which they are presented when IPEs are present, the scale is anchored too far to the right, leading to the observed overestimation of ability. Furthermore, an additional effect works in the opposite direction. This goes back to the fact that initially calibrated items are only administered in the first or the third cluster position. Because item-difficulty parameters are updated after every test cycle (or after every second test cycle for link items), the UCCS also kind of balances itself for item positions, and, therefore, bias decreases across test cycles. However, because the resulting decrease in bias is relatively slow, the CCS should be preferred to the UCCS if IPEs are present.

The simulation also revealed systematic differences in the average precision of the ability estimates that can be expected if the CCS is used with the 1PL rather than the 2PL. The 2PL started to have a smaller *MSE* of the ability estimates from Test Cycle 3 on; prior to that, the *MSE* of the 1PL was smaller. From a practical point of view, this means that, for situations in which decisions regarding the IRT model are possible, the CCS with the 2PL should be preferred (apart from fit-related considerations) if achieving a very high level of efficiency in the long run is important, and the CCS with the 1PL should be preferred if a high efficiency in the first two cycles is important. For applications where the IRT model is not under discussion, the results mean that a relatively low precision has to be expected, especially when the 2PL is used in the first cycles. This can be compensated for by extending the tests with some items in the first cycles.

Lastly, the simulation results provide guidance on the number of clusters that the CCS should include. When using the CCS with the 1PL, three clusters are sufficient to control for even large IPEs. These three clusters are also sufficient when applying the CCS with the 2PL (in terms of bias and *MSE*). However, for large IPEs, 12 clusters are recommended to fully control for the systematic variance caused by IPEs.

For ILSAs, the capability of the CCS to calibrate new items on the fly during operational testing might also be an interesting feature. Because one important objective when carrying out field trials in ILSAs is trying out new items, the CCS might be an option to reduce or avoid the effort connected with this additional study. The capacity saved can be used, for example, to configure the CCS in the best possible way. The empirically observed IPEs, as well as the number of items and the test length for the specific ILSA, can then be taken into account. However, it has to be noted that ILSAs contain more sources of statistical error than those that were considered in the simulation study presented here. Besides the sampling error, which is unaffected by the test design, variations of item parameter estimates between countries do exist. The CCS is a good method to deal with these item-by-country interactions because the parameters of the link items can be country-specific if necessary. This makes it possible to distinguish between common and unique item parameters, as was done in PISA 2018, for example (OECD, 2020). Moreover, nonstatistical requirements can be accounted for by using constraint management methods in combination with the CCS. This was not done in the simulation study presented here but is possible and necessary for ILSAs. In addition to the minimization of the measurement error based on the responses to the cognitive items, ILSAs face numerous nonstatistical requirements that are accounted for with the study design. For example, for domains with rich and complex specifications, many items are needed to measure them. Constraint management methods can be used to achieve an appropriate construct coverage at the population level (e.g., by constraining the proportions of items presented for subdomains). Heuristic constraint management methods (e.g., Born & Frey, 2017) or very powerful optimal design techniques, such as the shadow testing approach of van der Linden and Reese (1998), can be used for this purpose. With these methods, all typical ILSA-specific design requirements can be implemented directly into the adaptive algorithm.

A comparison in terms of a simulation study with the procedure with the highest degree of adaptivity in ILSAs so far would be particularly interesting; this procedure was first adopted for the Reading domain in PISA 2018 (see Yamamoto et al., 2025 of this volume) and then extended to the Mathematics domain in PISA 2022 (OECD, 2024). As shown in Frey et al. (2023), all constraints that are accounted for in PISA with the multistage testing design used can be met with constraint management in the context of a highly adaptive testing design. Furthermore, IPEs can be easily controlled for with the CCS, as shown in this chapter. What is not yet known is the expected precision and bias of the statistics on the population level that are reported in PISA. A future simulation study should also include constraint management to guaranty appropriate content coverage and to show how item-by-country interactions can be handled. Besides that, it is important to reproduce the special sample structure of the respective ILSA, because these studies do not draw a simple random sample but, instead, complex samples (e.g., two-stage stratified sample in PISA). Based on the results of such a simulation study, it can be considered whether the next step towards more adaptivity is worthwhile or not. Summing up, the proposed extension of the CCS proved capable of resolving the problematic bias issues in ILSAs introduced by IPEs. As incorporating the cluster balancing in the CCS does not have disadvantages compared to the unbalanced UCCS, the balanced version is recommended for all applications, even those in which it is not sure that IPEs will be an issue. If IPEs are an issue, the CCS will control for them; if not, no disadvantages result.

Conflict of interest On behalf of all authors, the corresponding author states that there is no conflict of interest.

This work was supported by the German Federal Ministry of Education and Research (BMBF) [grant number 16DHL1005].

References

Albano, A. D. (2013). Multilevel modeling of item position effects. *Journal of Educational Measurement, 50*, 408–426. https://doi.org/10.1111/jedm.12026

Asseburg, R., & Frey, A. (2013). Too hard, too easy, or just right? The relationship between effort or boredom and ability-difficulty fit. *Psychological Test and Assessment Modeling, 55*, 92–104.

Bäckström, M., & Björklund, F. (2016). Is reliability compromised towards the end of long personality inventories? *European Journal of Psychological Assessment, 35*, 1–8. https://doi.org/10.1027/1015-5759/a000363

Battauz, M. (2015). equateIRT: An R package for IRT test equating. *Journal of Statistical Software, 68*, 1–22. https://doi.org/10.18637/jss.v068.i07

Battauz, M. (2019). On wald tests for differential item functioning detection. *Statistical Methods & Applications, 28*, 121–130. https://doi.org/10.1007/s10260-018-00442-w

Birnbaum, A. (1968). Some latent trait models and their use in inferring an examinee's ability. In F. M. Lord & M. R. Novick (Eds.), *Statistical theories of mental test scores* (pp. 395–479). Addison-Wesley.

Bock, R. D., & Aitkin, M. (1981). Marginal maximum likelihood estimation of item parameters: An application of an EM algorithm. *Psychometrika, 46*, 443–459. https://doi.org/10.1007/BF02293801

Born, S. (2019). *Umgang mit Kontexteffekten beim computerisierten adaptiven Testen [consideration of context effects in computerized adaptive testing]*. Friedrich-Schiller-Universität.

Born, S., Fink, A., Spoden, C., & Frey, A. (2019). Evaluating different equating setups in the continuous item pool calibration for computerized adaptive testing. *Frontiers in Psychology, 10*, 1277. https://doi.org/10.3389/fpsyg.2019.01277

Born, S., & Frey, A. (2017). Heuristic constraint management methods in multidimensional adaptive testing. *Educational and Psychological Measurement, 77*, 241–262. https://doi.org/10.1177/0013164416643744

Candell, G. L., & Drasgow, F. (1988). An iterative procedure for linking metrics and assessing item bias in item response theory. *Applied Psychological Measurement, 12*, 253–260. https://doi.org/10.1177/014662168801200304

Chalmers, R. P. (2012). Mirt: A multidimensional item response theory package for the R environment. *Journal of Statistical Software, 48*, 1–29. https://doi.org/10.18637/jss.v071.i05

Chalmers, R. P. (2016). Generating adaptive and non-adaptive test interfaces for multidimensional item response theory applications. *Journal of Statistical Software, 71*, 1–39. https://doi.org/10.18637/jss.v071.i05

Davis, J., & Ferdous, A. (2005). *Using item difficulty and item position to measure test fatigue*. American Institutes for Research. https://www.air.org/sites/default/files/downloads/report/AERA2005Test_Fatigue11_0.pdf

Debeer, D., Buchholz, J., Hartig, J., & Janssen, R. (2014). Student, school, and country differences in sustained test-taking effort in the 2009 PISA reading assessment. *Journal of Educational and Behavioral Statistics, 39*, 502–523. https://doi.org/10.3102/1076998614558485

Debeer, D., & Janssen, R. (2013). Modeling item-position effects within an IRT framework. *Journal of Educational Measurement, 50*, 164–185. https://doi.org/10.1111/jedm.12009

Donoghue, J. R., & Isham, S. P. (1998). A comparison of procedures to detect item parameter drift. *Applied Psychological Measurement, 22*, 33–51. https://doi.org/10.1177/01466216980221002

Fink, A., Born, S., Spoden, C., & Frey, A. (2018). A continuous calibration strategy for computerized adaptive testing. *Psychological Test and Assessment Modelling, 60*, 327–346.

Frey, A., Bernhardt, R., & Born, S. (2017). Umgang mit Itempositionseffekten bei der Entwicklung computerisierter adaptiver tests [handling of item positions effects in the development of computerized adaptive tests]. *Diagnostica, 63*, 167–178. https://doi.org/10.1026/0012-1924/a000173

Frey, A., Born, S., & Bernhardt, R. (2015, September). *Umgang mit Itempositionseffekten beim computerisierten adaptiven Testen* [accounting for item position effects in computerized adaptive testing]. Presentation at the 12th meeting of the section methods and evaluation of the German psychological society (DGPs), Jena, Germany.

Frey, A., & Hartig, J. (2020). Methodological challenges of international student assessment. In H. Harju-Luukkainen, N. McElvany, & J. Stang (Eds.), *Monitoring of student achievement in the 21st century* (pp. 39–49). Springer. https://doi.org/10.1007/978-3-030-38969-7_4

Frey, A., Hartig, J., & Rupp, A. (2009). Booklet designs in large-scale assessments of student achievement: Theory and practice. *Educational Measurement: Issues and Practice, 28*(3), 39–53. https://doi.org/10.1111/j.1745-3992.2009.00154.x

Frey, A., & Seitz, N. N. (2011). Hypothetical use of multidimensional adaptive testing for the assessment of student achievement in PISA. *Educational and Psychological Measurement, 71*, 503–522. https://doi.org/10.1177/0013164410381521

Frey, A., Seitz, N. N., & Kroehne, U. (2013). Reporting differentiated literacy results in PISA by using multidimensional adaptive testing. In M. Prenzel, M. Kobarg, K. Schöps, & S. Rönnebeck (Eds.), *Research on PISA* (pp. 103–120). Springer. https://doi.org/10.1007/978-94-007-4458-5_7

Frey, A., Spoden, C., Goldhammer, F., & Wenzel, S. F. C. (2018). Response time-based treatment of omitted responses in computer-based testing. *Behaviormetrika, 45*, 505–526. https://doi.org/10.1007/s41237-018-0073-9

Frey, A., König, C., & Fink, A. (2023). A highly adaptive testing design for PISA. *Journal of Educational Measurement*. Advance online publication. https://doi.org/10.1111/jedm.12382.

Giesbrecht, F. G., & Gumpertz, M. L. (2004). *Planning, construction, and statistical analysis of comparative experiments*. Wiley.

Hartig, J., & Buchholz, J. (2012). A multilevel item response model for item position effects and individual persistence. *Psychological Test and Assessment Modeling, 54*, 418–431.

Hecht, M., Weirich, S., Siegle, T., & Frey, A. (2015). Effects of design properties on parameter estimation in large-scale assessments. *Educational and Psychological Measurement, 75*, 1021–1044. https://doi.org/10.1177/0013164415573311

Hohensinn, C., Kubinger, K. D., Reif, M., Holocher-Ertl, S., Khorramdel, L., & Frebort, M. (2008). Examining item-position effects in large-scale assessment using the linear logistic test model. *Psychology Science Quarterly, 50*, 391–402.

Hohensinn, C., Kubinger, K. D., Reif, M., Schleicher, E., & Khorramdel, L. (2011). Analysing item position effects due to test booklet design within large-scale assessment. *Educational Research and Evaluation, 17*, 497–509. https://doi.org/10.1080/13803611.2011.632668

Holland, P. W., & Wainer, H. (1993). *Differential item functioning*. Erlbaum.

Kim, S. (2006). A comparative study of IRT fixed parameters calibration methods. *Journal of Educational Measurement, 43*, 355–381. https://doi.org/10.1111/j.1745-3984.2006.00021.x

Kingston, N. M., & Dorans, N. J. (1984). Item location effects and their implications for IRT equating and adaptive testing. *Applied Psychological Measurement, 8*, 147–154. https://doi.org/10.1177/014662168400800202

Kolen, M. J., & Brennan, R. L. (2014). *Test equating, scaling and linking: Methods and practices* (3rd ed.). Springer. https://doi.org/10.1007/978-1-4939-0317-7

König, C., Khorramdel, L., Yamamoto, K., & Frey, A. (2021). The benefits of fixed item parameter calibration for parameter accuracy in small sample situations in large-scale assessments. *ducational Measurement: Issues and Practice, 40*, 17–27. https://doi.org/10.1111/emip.12381.

Le, L. T. (2007, July). Effects of item positions on their difficulty and discrimination: A study in PISA science data across test language and countries. Paper presented at the 72nd annual meeting of the psychometric society, Tokyo.

Li, F., Cohen, A., & Shen, L. (2012). Investigating the effect of item position in computer-based tests. *Journal of Educational Measurement, 49*, 362–379. https://doi.org/10.1111/j.1745-3984.2012.00181.x

Linacre, J. M. (1994). *Many-facet Rasch measurement* (2nd ed.). MESA Press.

Lord, F. M. (1980). *Applications of item response theory to practical testing problems*. Lawrence Erlbaum.

Meyers, J. L., Miller, G. E., & Way, W. D. (2008). Item position and item difficulty change in an IRT-based common item equating design. *Applied Measurement in Education, 22*, 38–60. https://doi.org/10.1080/08957340802558342

Mislevy, R. J. (1986). Bayes modal estimation in item response models. *Psychometrika, 51*, 177–195. https://doi.org/10.1007/BF02293979

Mislevy, R. J. (1991). Randomization-based inference about latent variables from complex samples. *Psychometrika, 56*, 177–196. https://doi.org/10.1007/BF02294457

Mullis, I. V. S., & Martin, M. O. (Eds.). (2015). *PIRLS 2016 assessment framework*. TIMSS & PIRLS International Study Center, Boston College.

Nagy, G., Lüdtke, O., & Köller, O. (2016). Modeling test context effects in longitudinal achievement data: Examining position effects in the longitudinal German PISA 2012 assessment. *Psychological Tests and Assessment Modeling, 58*, 641–670.

Nagy, G., Nagengast, B., Frey, A., Becker, M., & Rose, N. (2019). A multilevel study of position effects in PISA achievement tests: Student- and school-level predictors in the German tracked school system. *Assessment in Education: Principles, Policy & Practice, 26*, 422–443. https://doi.org/10.1080/0969594X.2018.1449100

OECD. (2012). *PISA 2009 technical report*. OECD Publishing. https://www.oecd.org/pisa/pisaproducts/50036771.pdf

OECD. (2013). *Technical report of the survey of adult skills (PIAAC)*. OECD Publishing. https://www.oecd.org/skills/piaac/_Technical%20Report_17OCT13.pdf

OECD. (2014). *PISA 2012 technical report*. OECD Publishing. https://www.oecd.org/pisa/pisaproducts/PISA-2012-technical-report-final.pdf

OECD. (2017). *PISA 2015 technical report*. OECD Publishing. https://www.oecd.org/pisa/data/2015-technical-report/

OECD. (2020). *PISA 2018 technical report*. OECD Publishing. https://www.oecd.org/pisa/data/pisa2018technicalreport/

OECD. (2024). PISA 2022 technical report. OECD Publishing. https://doi.org/10.1787/01820d6d-en

Qian, J. (2014). An investigation of position effects in large-scale writing assessments. *Applied Psychological Measurement, 38*, 518–534. https://doi.org/10.1177/0146621614534312

R Core Team. (2019). *R: A language and environment for statistical computing [software]*. R Foundation for Statistical Computing.

Robitzsch, A. (2016). Zu nichtignorierbaren Konsequenzen des (partiellen) Ignorierens fehlender item responses im large-scale assessment [on non-negligible consequences of (partially) ignoring missing item responses in large- scale assessments]. In B. Suchań, C. Wallner-Paschon, & C. Schreiner (Eds.), *PIRLS & TIMSS 2011—die Kompetenzen in Lesen, Mathematik und Naturwissenschaft am Ende der Volksschule: Österreichischer Expertenbericht* (pp. 55–64). Leykam.

Rutkowski, D., Rutkowski, L., & Liaw. (2018). Measuring widening proficiency differences in international assessments: Are current approaches enough? *Educational Measurement: Issues and Practice, 37*(4), 40–48.

Schweizer, K., Schreiner, M., & Gold, A. (2009). The confirmatory investigation of APM items with loadings as a function of the position and easiness of items: A two-dimensional model of APM. *Psychology Science Quarterly, 51*, 47–64.

Schweizer, K., Troche, S. J., & Rammsayer, T. H. (2011). On the special relationship between fluid and general intelligence: New evidence obtained by considering the position effect. *Personality and Individual Differences, 50*, 1249–1254. https://doi.org/10.1016/j.paid.2011.02.019

Shin, H. J., Yamamoto, K., Khorramdel, L., & Robin, F. (2021). Increasing measurement precision of PISA through a multistage adaptive testing. In M. Wiberg, J. González, D. Molenaar, U. Böckenholt, & S. J. Kim (Eds.), *Quantitative psychology. 85th annual meeting of the psychometric society, virtual, 2020*. Springer.

Spoden, C., Frey, A., & Bernhardt, R. (2018). Implementing three CATs within eighteen months. *Journal of Computerized Adaptive Testing, 60*, 38–55. https://doi.org/10.7333/1809-060338

Stocking, M. L., & Lord, F. M. (1983). Developing a common metric in item response theory. *Applied Psychological Measurement, 7*, 201–210. https://doi.org/10.1177/014662168300700208

Thompson, N. A., & Weiss, D. J. (2011). A framework for the development of computerized adaptive tests. *Practical Assessment, Research & Evaluation, 16*. Retrieved from http://pareonline.net/getvn.asp?v=16&n=1

van der Linden, W. J. (Ed.). (2016). *Handbook of item response theory. Volume one: Models*. Chapman & Hall/CRC.

van der Linden, W. J., & Glas, C. A. W. (Eds.). (2010). *Elements of adaptive testing*. Springer.

van der Linden, W. J., & Reese, L. M. (1998). A model for optimal constrained adaptive testing. *Applied Psychological Measurement, 22*, 259–270. https://doi.org/10.1177/01466216980223006

Warm, T. A. (1989). Weighted likelihood estimation of ability in item response theory. *Psychometrika, 54*, 427–450. https://doi.org/10.1007/bf02294627

Weeks, J. P., von Davier, M., & Yamamoto, K. (2016). Using response time data to inform the coding of omitted responses. *Psychological Test and Assessment Modeling, 58*, 671–701.

Weirich, S., Hecht, M., & Böhme, K. (2014). Modeling item position effects using generalized linear mixed models. *Applied Psychological Measurement, 38*, 535–548. https://doi.org/10.1177/0146621614534955

Williams, E. J. (1949). Experimental designs balanced for the estimation of residual effects of treatments. *Australian Journal of Chemistry, 2*(2), 149–168.

Wu, Q., Debeer, D., Buchholz, J., Hartig, J., & Janssen, R. (2019). Predictors of individual performance changes related to item positions in PISA assessments. *Large-Scale Assessments in Education, 7*(1), 5. https://doi.org/10.1186/s40536-019-0073-6

Yamamoto, K., Khorramdel, L., & Shin, H. J. (2018a). Introducing multistage adaptive testing into international large-scale assessments designs using the example of PIAAC. *Psychological Test and Assessment Modeling, 61*, 347–368.

Yamamoto, K., Shin, H. J., & Khorramdel, L. (2018b). Multistage adaptive testing design in international large-scale assessments. *Educational Measurement: Issues and Practice, 37*(4), 6–27. https://doi.org/10.1111/emip.12226

Yamamoto, K., Shin, H. J., Robin, F., Khorramdel, L., & Halderman, L. (2025). Improved test designs and multistage adaptive testing in large-scale assessments. In L. Khorramdel, M. von Davier, & K. Yamamoto (Eds.), *Innovative digital-based international large-scale assessments—foundations*. Springer.

Yen, W. M. (1980). The extent, cause and importance of context effects on item parameters for two latent trait models. *Journal of Educational Measurement, 17*, 297–311. https://doi.org/10.1111/j.1745-3984.1980.tb00833.x

Zenisky, A., Hambleton, R. K., & Luecht, R. M. (2010). Multistage testing: Issues, designs, and research. In W. J. van der Linden & C. A. W. Glas (Eds.), *Elements of adaptive testing*. Springer.

Zwick, R. (1991). Effects of item order and context on estimation of NAEP reading proficiency. *Educational Measurement: Issues and Practice, 10*(3), 10–16. https://doi.org/10.1111/j.1745-3992.1991.tb00198.x

Chapter 7
The Digital Testing Platforms for PIAAC and PISA

Michael Wagner and Emily Kerzabi

Abstract This chapter reviews the digital testing platforms for PIAAC and PISA as a means to provide general considerations when working with or designing a testing platform for an international or large-scale assessment. Such a platform needs to act as a hub for various functional purposes of the assessment and the communication between them. The test delivery system and questionnaire authoring tool are the primary focus, as they are the two most common and necessary components of digital testing platforms. In conjunction, these tools allow for accurate and efficient assessment creation, dissemination, and data collection, and may further provide users with the means to collaborate when the authoring tool is incorporated into a larger repository or communication system. Future innovations of large-scale assessments are also discussed, specifically in regard to new interactive item types and interfaces, the collection of log and process data, the incorporation of new test designs and automatic scoring, and the possibilities and limitations of web-based test administration.

The digital testing platform for an international large-scale assessment (ILSA) should be able to do far more than accomplish the primary purpose of test delivery and data collection. It should also be able to manage test creation, revision, translation, adaptation, and the assemblage of a large number of test forms among multiple languages with minimal human intervention across each function. In a nutshell, the platform should act as a hub for each functional purpose of the assessment and the communication between them.

When designing a successful digital testing platform for an ILSA, there are several aspects that should be considered; for example: What types of items will

M. Wagner
Independent Researcher (formerly ETS), Princeton, NJ, USA

E. Kerzabi (✉)
ETS, Princeton, NJ, USA
e-mail: ekerzabi@ets.org

L. Khorramdel et al. (eds.), *Innovative Digital-Based International Large-Scale Assessments*, Methodology of Educational Measurement and Assessment,
https://doi.org/10.1007/978-3-031-90951-1_7

be designed and delivered (e.g., trend items; multiple-choice, interactive items, collaborative items, essay)? What is the test design, and does it require automatic scoring and designated routing (e.g., are forms fixed or adaptive)? What response and process data will be collected (e.g., response time, number of actions, keystrokes)? What types of users will be using the platform (e.g., assessment designers, school administrators, national centers), and how will they use the platform (e.g., to collaborate, to communicate, as a reference)? Beyond the test itself, what other data will be securely and centrally stored (e.g., user information, translation details, coder training and handbook materials, assessment design information)? These questions and several more guided the design of the digital testing platform for the Programme for the International Assessment of Adult Competencies (PIAAC), the first major international assessment to be delivered primarily via computer. Educational Testing Service (ETS) later built on this knowledge to develop testing platforms for the Programme for International Student Assessment (PISA). While testing platforms may include several possible components, two main features can be expected from these digital testing platforms: a test delivery system and a questionnaire authoring tool.

The test delivery system is used to administer the test and record response data. Testing is usually conducted via computer or tablet, and the testing system may be permanently installed on the computer hardware, as in PIAAC where tablets are retained by test administrators; or only temporarily installed via a software package, as in PISA where local school computers are used for test administration. In ILSAs, the system must be able to support multiple languages and unique modifications for each country and language group administering the assessment. Beyond presenting items, the test delivery system is further responsible for data capture and secure data transmission following completion of testing. It must adhere to a specified test design, specifically presenting the correct items to the correct participants following a fixed, non-adaptive layout, or it must include a scoring function and follow an algorithm to select test content based on previous responses in an adaptive layout (discussed more in the section on *Assessment Designs*). The delivery system should also log system actions, such as what process was initiated, by which user, and at what time. Such information is useful for data quality procedures when reviewing the collected data.

The questionnaire authoring tool is used by test designers to create the assessment as it will be delivered (such as authoring items the background questionnaire and the cognitive assessment). Through this system, trend/linking items and new items can be authored and reviewed. *Trend* or *linking items* are those that have appeared in, and are invariant across, previous cycles of assessment and measure trends over time; as such, they are generally re-authored for computer as close to previous paper or other digital version of the item as possible (such that the item continues to link between cycles and modes of assessment). *New items* have not appeared in previous cycles of assessment and, in the cases of PIAAC and PISA, often incorporate interactive features to more accurately measure constructs of interest; thus, new items may need to emulate an immersive digital environment (e.g., the functionalities of web browsers, word processors, messaging software, or email interfaces), and relational database programs need to be built directly into the items to accommodate all possible interactions. Item authoring and item review features may support collaborative test design, such as allowing test administrators to review item layout, content, and translation accuracy ahead of test administration. Additionally, as international assessments, PIAAC and PISA are administered in multiple languages and across different cultures with unique nuances that may be pertinent during testing. It is particularly useful for each country and language group to be able to review items in the authoring tool to verify that all translations and minor item adaptations result in a modified item that is qualitatively the same as the source item.

In this chapter, we describe how the systems built for PISA and PIAAC were architected, specifically for use by diverse groups, each with broad functionality and interface requirements. We start by outlining the unique survey conditions of each assessment and then turn to the platform capabilities that enable a wide set of users to collaborate and develop the assessment instrument in each case. We conclude with remarks on the flexibility of assessment platforms to accommodate future changes in large-scale assessment.

7.1 Test Delivery Systems for PIAAC and PISA

Testing conditions are an important factor for how the assessment can be delivered and how assessment data are collected and stored. PIAAC and PISA are administered under very different circumstances, and two possible approaches for test delivery and collection can be illustrated from their examples. PIAAC is administered through a one-on-one interview via computer (laptop or tablet) brought to the respondent's household by the interviewer and is administered to one person at a time. PISA is typically administered by local proctors and school authorities to several students at a time, in school settings, and using school computers. In each case, a unique test delivery system was created to administer the assessment and record the collected data.

7.1.1 PIAAC Test Administration

With PIAAC, the computer is dedicated solely for assessment delivery and data collection. This allows for the assessment software to be permanently installed on the interviewer's laptop, referred to as the *host system*. The software on the host system consists of three parts:

- the country's case management system (CMS), which manages the sample and interview status;
- a delivery platform virtual machine or *guest system*, which includes the test interface environment;

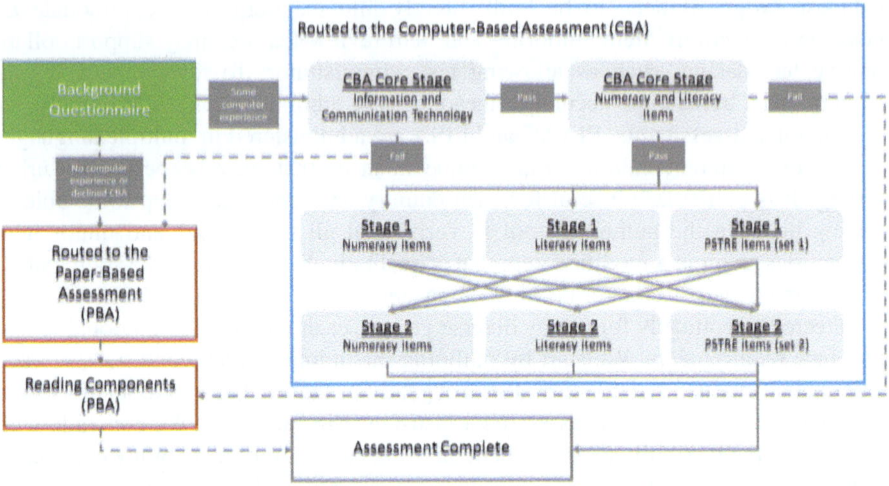

Fig. 7.1 PIAAC cycle 1 testflow routing

- the background questions and cognitive items administered within the guest system (see Upsing et al., 2013).

In PIAAC, the choice of operating system (OS) for the guest system is a Debian Linux system (Aoki, 2019), which ensures a controlled environment and has the benefit of being open-source. For security, the guest system and the laptop host system are insulated from one another, with restricted communications. Provided scripts launch the instrument and upload response data from the host system only. Testing itself is conducted offline to prevent any external communications.

The test is administered through the guest system, which accommodates an adaptive test design; specifically, adaptive tests do not conform to a pre-determined set of items, but rather, a respondent's answers to previous items influence what future items or sets of items will be presented. The testflow of PIAAC is shown in Fig. 7.1. For PIAAC, responses to the background questionnaire (BQ) and performance on subsequent stages determines the assessment routing. If the participant indicates that they have no computer experience on the BQ or declines the computer-based assessment (CBA), they are routed to the paper-based assessment (PBA). If the participant indicates having some computer experience but fails the first core stage dealing with information and communication technology, they are routed to PBA. In the second core stage, a mix or numeracy and literacy items are presented; if this stage is failed, a basic reading components assessment is administered via PBA. If this stage is passed, the participant is routed to CBA where they receive a full set of numeracy, literacy, or problem solving in technology rich environments (PSTRE) items in two different test stages. In total, each participant receives items from two out of three domains or PSTRE only: literacy and numeracy, literacy and PSTRE, numeracy and PSTRE, or two sets of PSTRE (see Fig. 7.1). Note that the literacy and numeracy items in Stage 1 are identical to the items in

Stage 2, while the items in the two stages of PSTRE are mutually exclusive. The interview is complete after Stage 2, and the data are retained on the host system.

7.1.2 PISA Test Administration

While the system used by PIAAC is on a dedicated computer, schools participating in PISA run the testing software temporarily on whatever computers the school may have available. From a delivery standpoint, a web-based, online system is more efficient, as results can be automatically uploaded to a central server; however, as with PIAAC, an offline version of the assessment is more reliable for administration and security purposes, especially considering the wide variation of Internet infrastructure in schools around the world.

The PISA assessment software is securely downloaded and then run from a USB drive, a local hard drive, or a network drive; the software does not need to be installed on the school computers. Because of the wide variety of OS and hardware capabilities of school computers globally, compatibility diagnostics are included in the software package, which verify that the computer meets the specifications of the PISA assessment. An example output of the compatibility diagnostic is shown in Fig. 7.2. This diagnostic is run by schools in advance of the testing day and then again immediately prior to testing (on the day of the test administration). This software verifies that no alterations to the computer systems have been made in the interim that may affect testing, that no background processes are running that might interfere with test administration, and that there are no other issues that could present problems or compromise testing security during test administration.

During the PISA assessment, the screen is locked to prevent test takers from leaving the test or viewing non-test applications. To begin the assessment, from the main screen, the student selects one of two testing sessions: the cognitive assessment or the questionnaire; the test moderator will prompt students with the appropriate option, depending on which point students are in the testflow. Once logged in, the

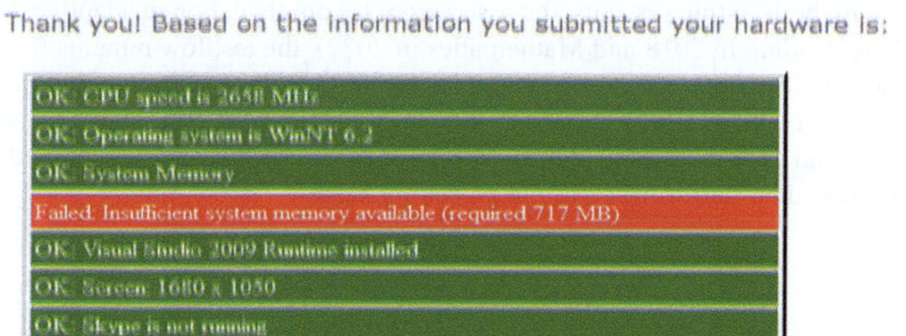

Fig. 7.2 Example compatibility diagnostic output report

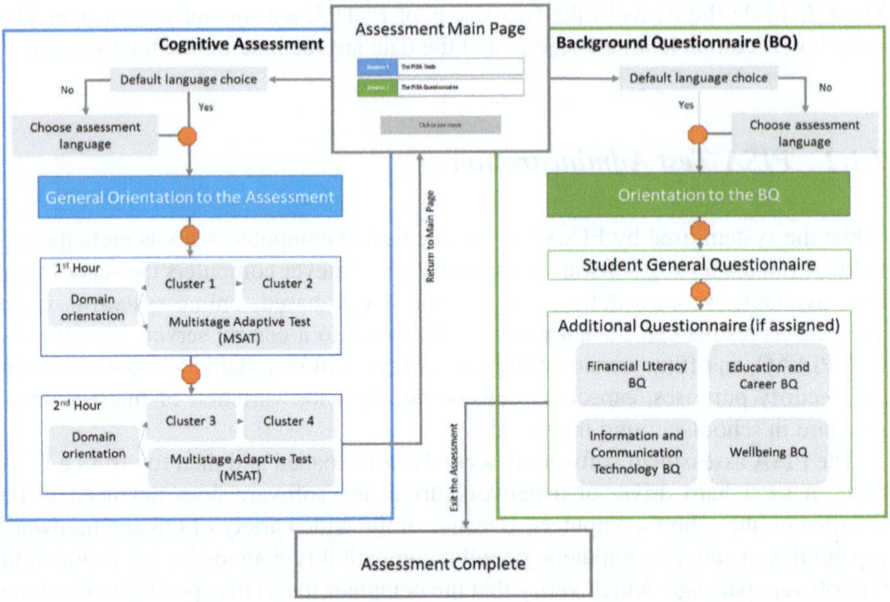

Fig. 7.3 PISA 2018 testflow for the nonadaptive domains

assessment will guide the student through the appropriate screens. The PISA 2018 testflow for nonadaptive domains is shown in Fig. 7.3. Because students work independently, mandatory pauses (indicated by an octagon in the workflow) are implemented between major sections of the assessment, and students only move forward following prompting from the test moderator. Each session has an orientation followed by assessment content. In Session 1 (the cognitive assessment), students will have an additional orientation for the domain being administered, followed by two clusters of items each hour. Following this session, students are routed back to the assessment main page. In the background questionnaire session, students receive a general questionnaire before moving through specific sub-questionnaires, and not all sub-questionnaires are assigned to all students. Following both testing sessions, the assessment is complete. For an adaptive domain (such as Reading in 2018 and Mathematics in 2022), the testflow remains the same, except that items are not administered in clusters. Within the hour in which the adaptive domain is assessed, participants will begin at a core stage (similar to PIAAC) and then be routed through difficult or easy items in stage 1 and stage 2, based on their performance in the preceding stage.

7.1.3 Data Collection

In both assessments, data quality, data retention, and data security (both item security and participant security) are of the highest concern. In PIAAC, response data are first handled by the CMS, implemented by each country to manage the work of the interviewers. Response data files are automatically copied from the guest system and sent back to the national center for further processing. The files remaining in the guest system are automatically archived to prevent data duplication but can be recovered if necessary. In PISA, results, by default, are written to the USB drive (or computer hard drive or network drive) used for running the delivery system software. Countries have a range of options for managing these files. Many return the USB drives to the national center where the data files are extracted using custom software. Alternatively, if an Internet connection is available at the end of the testing session, the data files can be automatically uploaded to an Internet server from which countries can retrieve them. To prevent inadvertent data loss for countries that leave the data on USB drives, it is recommended that they enable the included automatic upload feature as a backup mechanism, so that copies of the data would be available if USB drives are lost or damaged.

7.1.4 Data Cleaning and Data Quality

Computer delivered assessment data are generally much better structured and more regular in how they are captured compared to PBA data. Paper assessments tend to have issues with missing booklets, unreadable pages, and unreadable responses, which are not an issue with CBA data that are directly input through a digital interface. Using CBA does not eliminate the need for data cleaning, but the issues of concern are different. CBA data issues may include multiple occurrences of the same student ID, platform or code bugs, and use of the platform in ways that are not anticipated. These issues are normal, but data consistency checks should be performed to eliminate or correct faulty data.

Both external and internal data consistency checks are recommended. *External consistency* refers to verifying that the expected external data, like specific respondents or schools, are found within the collected data. For example, sampling data shows which respondents should have collected data, and it can be verified that each respondent from the sample does have collected data, and only one set of collected data. *Internal consistency* refers to verifying that internal expectations, like test design and routing, were correctly met. For example, it can be verified that respondents' assignments to fixed test forms were random, which can usually be checked by reviewing a distribution of respondent proficiency, or that the proportion of respondents routed to a specific test form or stage in an adaptive test design was as expected (for more on adaptive testing, see the section on *Test Designs*, as well as Yamamoto, Shin, et al., 2025, of this volume).

Psychometricians may also identify specific variables of interest that can be built into the system that will allow for further checks to verify expectations; this is especially important when implementing a new test design and for being able to detect anomalous data with precision (see Yamamoto, Ali, et al., 2025, for more on data quality procedures). For example, PISA captures four pieces of data for each administered item: 1) the raw response, such as option "A" or a typed answer; 2) the scored response for computer scored items, such as "correct", or the coded response for human coded items, such as code "12," identifying the type of written response given; 3) the item timing, which is the amount of time spent on the item and is commonly referred to as *response time*; and 4) the number of actions, or number of clicks on the screen that a respondent made while on the item. Aggregates of these collected data are used by data analysts and psychometricians as measures of quality assurance. The time spent on an item, for example, can be aggregated for each respondent at the level of the unit, the domain, or the entire test, allowing the total test time by country, language, and domain to be checked for abnormalities or to provide descriptive statistics of countries with generally faster or slower responding students. Both response time and the percentage of correct responses can be used to identify effects of an item appearing in different test positions. In general, it is best for an item to be invariant (the item should not be more or less difficult or require more or less time) across different positions in a test. Because most ILSAs and adaptive tests vary the position of items and units in a test, it is vital that the item remain invariant, such that valid inferences can be drawn from assessment results (see Frey & Fink, 2025, for more on item position effects).

For both PIAAC and PISA, a great deal of work goes into making sure that everything functions as intended at the time of testing and that the collected data are as expected. In the next section, we discuss how technical platform development should be guided by the needs of the assessment and organized around the users designing it.

7.2 Questionnaire Authoring Tool

Beyond test delivery and collection, a wide range of users must collaborate to develop and modify items, review item translations, check item display and layout, and communicate important changes and procedures. While these operations can be conducted on an ad-hoc basis, as was sufficient in PIAAC, a centralized platform can also be created to accommodate multiple user roles and facilitate remote collaboration. Such a platform was developed for PISA, called the *PISA Portal.* To make such global collaboration possible, the PISA Portal platform utilizes standard web technologies (HTML, CSS, and JavaScript).

7.2.1 PISA Portal Organization

The PISA Portal interface is specific to the various roles of project participants, structured by the functions and deliverables required for each of those roles, and provides the tools needed to deliver those functions. The structure of the portal is outlined in Fig. 7.4. While all users have this same basic layout, user views are tailored to each user's role; for example, users from a country's national center are only able to access information pertinent to their country, including the tasks assigned to the country and country-specific item adaptations.

The "Home" screen automatically loads when a user logs-in. On this screen, there is a panel displaying a welcome message, any news and announcements, and a series of quick links. These links are also available under the "Links" tab as a dropdown menu. The majority of project work is completed through the "Tasks," "Translation," and "Documents" tabs.

The portal's "Tasks" tab is the main area for tracking the workflow of the assessment. Here, a listing of all current tasks are stored in a data frame, which note the country, language, the title of the task, the responsible party (such as PISA reviewer or project manager), the workflow type (such as a document requiring review or software package to be downloaded), the workflow step (such as "translation," "edit," or "final review"), the date due, and the status of the task (such as "new," "ongoing," or "overdue"). Each country's users will have access only to their

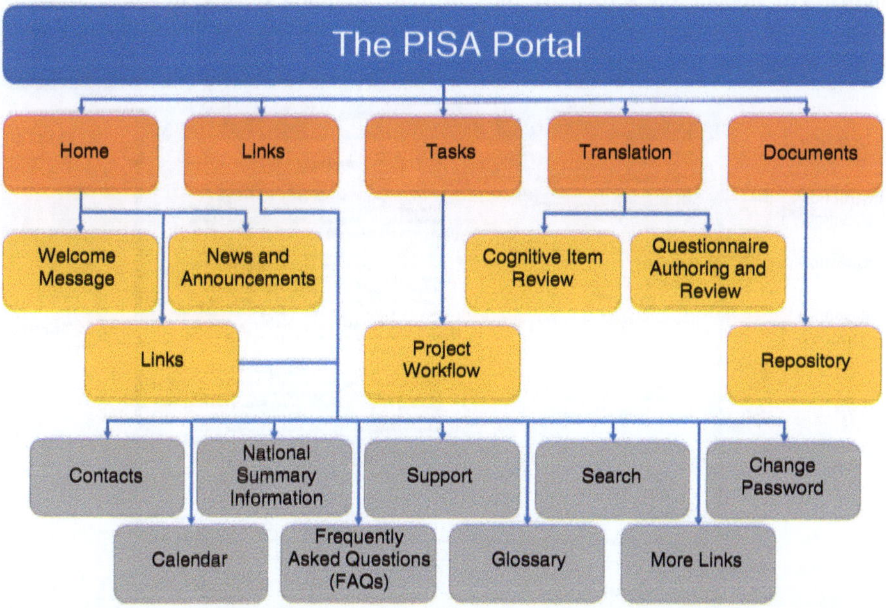

Fig. 7.4 Overview of the PISA portal

own project workflow tasks; however, test administrators will be able to see the status of open tasks for all countries.

7.2.2 *Item Authoring, Translation, and Review*

Because PISA is administered in several languages and in varying cultural and national contexts, items must be translated into each assessment language at the national level, and countries may further elect to make slight modifications to retain the qualitative meaning of the item within the national context. PISA "source" items are available in French and English, and national users may use either source to make appropriate national adaptations. Because national users are not expected to have the technical background necessary to author complex items directly, national users input the desired adaptations into a template, and PISA technical experts carry out the desired modifications to both standard items (multiple-choice and constructed-response formats) and innovative items (with interactive interfaces, designed to expand the measurement of the PISA domains) for each national assessment. An example of the text fields that national users may modify are shown directly on an example item in Fig. 7.5. While the unit title is modified at the unit level, and the instructions should remain consistent across similar types of items, the question, the response options, and stimulus will receive the most adaptation and careful review. Nationally adapted cognitive items (saved as XILFF files) can be reviewed under the "Translation" tab of the PISA Portal from a cognitive item preview panel; these items will appear on screen as they will to students during the test. Because background questionnaire items are more straightforward, typically just a line of

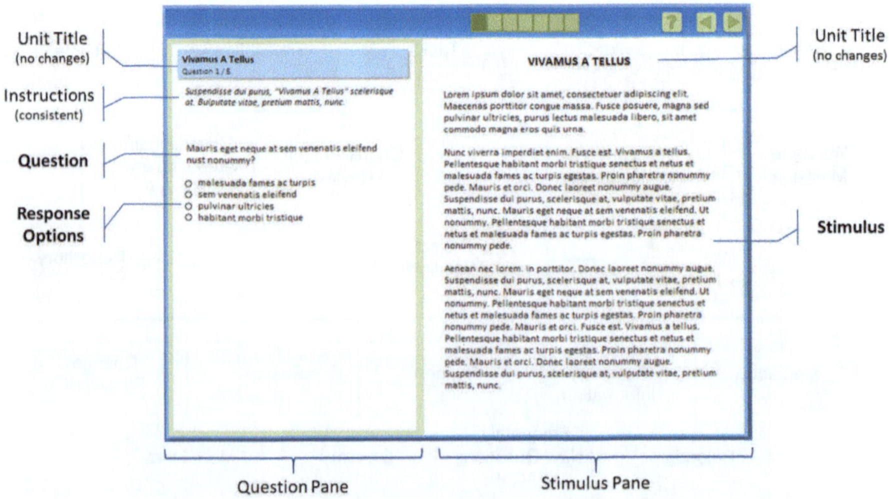

Fig. 7.5 Standard text fields used in national, item-level adaptations

text, these items can be authored directly by national users. The questionnaire authoring panel launches an authoring tool, and adaptations can be reviewed from the questionnaire review panel. All item adaptations must undergo thorough translation, compatibility, and equivalency reviews, and both cognitive and background questionnaire items will be reviewed frequently by national users.

With both the tasks and translation panels, the PISA Portal offers test developers tools to efficiently monitor workflows and publish item content while providing national users with clear process steps, as well as the opportunity to review items and make modifications. To this end, the portal supports unique layout changes on a per country and per language basis (to account for variations in language length and characteristics); however, it is recommended that layout changes be avoided if not absolutely necessary, as items should remain as close to the source version as possible, and needless layout changes can quickly increase the workload of the development experts responsible for executing each modification. As a technical design consideration, it is most useful to create digital source items with enough space to accommodate longer translations and to implement layouts in such a way that text can expand predictably. When the national assessment items are finalized, the item translation and display for each country and language are available for review on the PISA Portal. This is particularly helpful for psychometricians and data analysts during the scaling phase of the assessment if an item is found to function differently (differential item functioning or DIF; see Glas & Jehangir, 2014) in a particular country-language group (e.g., it can be investigated whether an adaptation could be responsible for poor model fit for the item).

7.2.3 Repository and Communication Hub

A secondary function of the PISA Portal is to act as a central repository of assessment information. In the "Documents" tab, the portal follows a hierarchical folder structure and includes a record of meeting notes, coding guidelines for human coders in the field, and assessment design reference material, among other related documents. Centralized repositories are useful for securely sharing files in an organized way and are recommended for any project requiring many isolated but compounding tasks, a high volume of remote work, or a diverse user group (e.g., varying native languages, work cultures, or time zones). Through the user-based design, the portal facilitates simple and straightforward access to the key resources needed by the various stakeholders involved in the development and delivery of PISA.

7.3 Adaptability for Future Assessments

Innovative, computer-based assessments tend to expand assessment measurement capabilities, and, therefore, require that their technology platforms adapt alongside advances in the assessment. There are several examples from both PISA and PIAAC in which innovation in either the domain content, item content, or assessment design drove technical changes in item presentation, in interface functionality, and in the types of data being collected.

7.3.1 New and Interactive Item Types

PIAAC was the first ILSA to devote an entire domain to assessing the cognitive skills required for making use of computer-based information to solve problems. The PIAAC PSTRE domain measures the ability to analyze various requirements of a task, define goals and plans, and monitor progress (Lennon & Tamassia, 2013) within a digital environment. An important aspect of the PSTRE domain is to provide user interfaces that are as realistic as possible to provide an accurate assessment of respondent ability. For example, participants work in standard spreadsheet and word processing interfaces, navigate through a web browser with simulated features (forward, back, home, and bookmark), and compose email communications and other tasks in a simulated email client that allows respondents to reply, reply to all, forward, send, and move email to folders. These simulated interfaces provide a wide array of process data, including time spent, actions taken (clicking and typing responses or selections from drop-down menus, such as "file" and "edit" in the simulated word processor), and the sequence in which actions were completed. This information provides direct evidence of the processes and strategies respondents use to complete assigned tasks and allows for better inferences about their knowledge and skills related to PSTRE.

In 2015, PISA introduced interactive items in the major domain of Science and in the innovative domain of Collaborative Problem Solving (CPS), with the goal of more accurately measuring constructs of interest (i.e., problem solving, evaluating, and designing scientific enquiry). In Science, interactive items may require that students manipulate item features, such as running a simulation to gather data and then responding to items based on the observed results, as shown in Fig. 7.6. In this science item, the student is meant to find at what point a person running outdoors will suffer either dehydration or heat stroke under different air temperature and air humidity conditions, also considering whether the runner is drinking water. When the simulation is run, the data are collected in the table below; the student must answer the question in the item and include data that support the answer (Organisation for Economic Co-operation and Development [OECD], 2015).

In CPS, the platform was designed to support common communication interfaces, such as simulated email, internet, and chat, similar to the PSTRE domain in PIAAC.

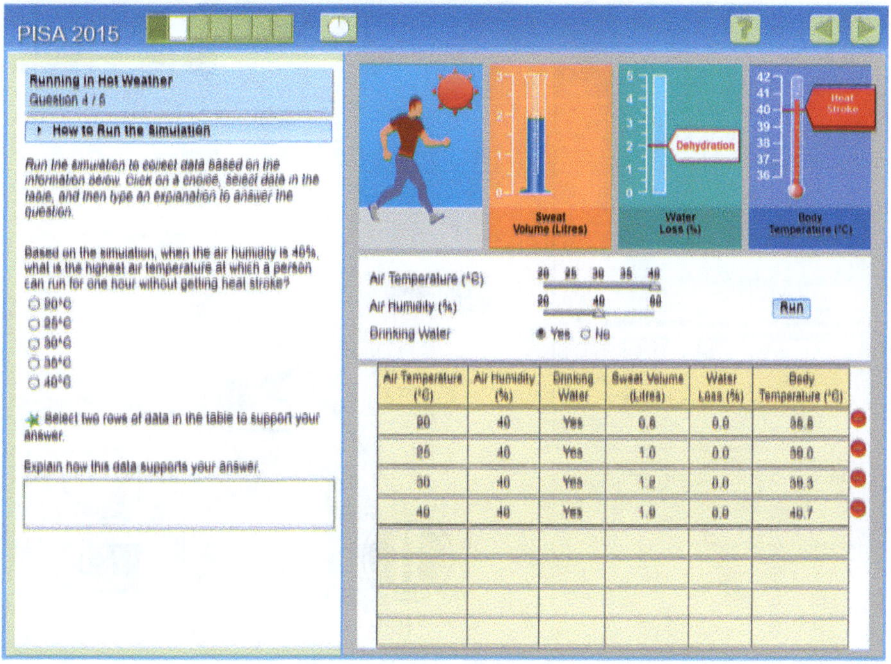

Fig. 7.6 Example interface of a PISA science item with the simulation feature

Through these interfaces, test participants collaborate with virtual agents or other participants through predefined options; an example item is shown in Fig. 7.7, in which the student is in a group chat with virtual agents (in roles as other students), and they must arrange to meet at a certain place at a certain time, although one member of the chat is not responding. In the chat interface, the participant will receive a message or messages from a virtual agent(s) and respond by selecting a response from a dropdown list, such as requesting that a task be performed or asking if a previously requested task has been completed. Data collection for this type of interaction is rather straightforward, as predefined text selection is fairly analogous to conventional multiple-choice items in assessments. Other types of actions can be performed in other interfaces, such as checking the environment to see if an action has been performed by another agent or completing an action that another agent failed to perform. Consequently, the platform collects the participant's message communications, actions, and reflections throughout the process of collaborative problem solving with other agents. These process data are saved to a log file and later uploaded. In this innovative domain, most sent messages and performed actions can be automatically scored.

However, utilizing complex interfaces, such as those of the PIAAC PSTRE and the PISA Science and CPS scenario-based tasks, poses several challenges. One challenge is to keep the user interfaces up to date and consistent with current technology. These assessment tasks are typically reused over multiple years, for

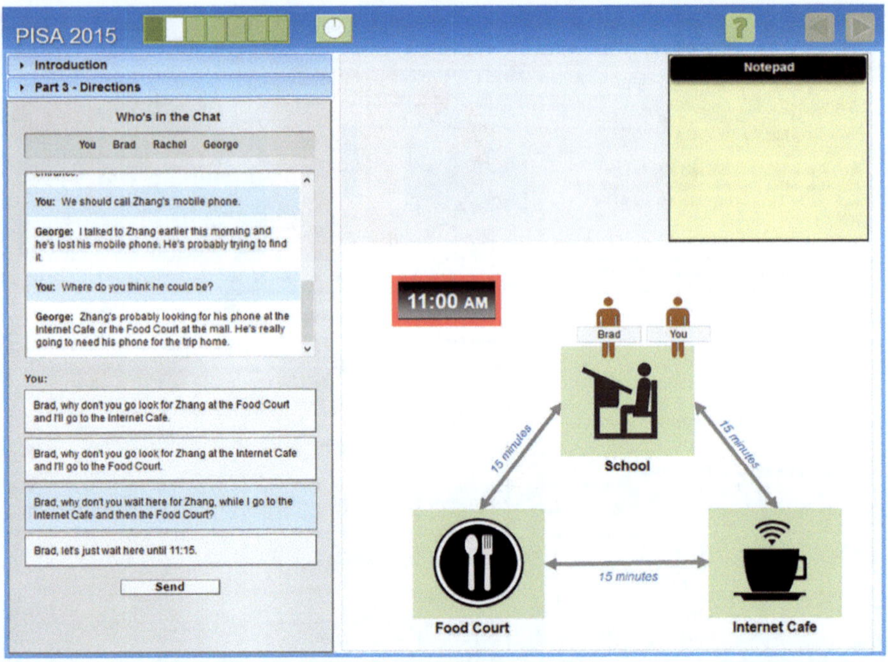

Fig. 7.7 Example interface of a PISA collaborative problem solving item

example, a cycle of PIAAC can be up to 10 years in length. When user interfaces do not match what users are accustomed to in their everyday lives, there is a risk of introducing non-construct related bias into the measurement. An example of this is with email interfaces. While most people once used a dedicated email program like Microsoft Outlook or Apple Mail, today, many people use alternate means to read and send email, such as web-based applications like Google Mail or smartphone-based interfaces. If presented with an old, desktop application-based interface, it may be unfamiliar and interfere with the respondent's performance on the task. This leads to another challenge with the use of complex environments: In order to be assured that all users have equal understanding of the environment they are using, an orientation or tutorial module is typically used, which allows the user to learn about the interface functionality and practice using it. This can seem burdensome for knowledgeable users and can take up valuable testing time. These and other challenges should be considered carefully before including such complex functionality in computer-based assessments.

7.3.2 Assessment Designs

Another innovative aspect of CBAs is the multistage-adaptive test design (MSAT; see Yamamoto, Shin, et al., 2025 for an overview in PISA). Adaptive testing systems are capable of presenting each test taker with items or units of appropriate difficulty, rather than with items that may be too difficult or too easy. PIAAC has been an MSAT since its inception, with participants taking one stage of the assessment at a time before moving onto the next. Test takers are routed based on responses to items in the background questionnaire and performance on each stage of the assessment. In PISA, the reading domain in 2018 also utilized a stage-level adaptive design and is based solely on stage-level performance for routing decisions. Adaptive designs allow for more accurate ability measurement at the highest and lowest ends of the proficiency distribution, which is particularly important in ILSAs because they are expected to sample from widely diverse population subgroups.

One challenge of MSAT designs is the scoring of constructed-responses (open-ended responses). From a technical standpoint, scoring constructed-responses is not problematic, but only if test designers can provide a finite list of correct responses. In PIAAC, developers identified a mostly-exhaustive list of possible constructed responses for machine scoring in the adaptive design. PISA, on the other hand, does not automatically score constructed responses in real-time; rather, it relies on the automatic scoring of the multiple-choice responses for routing decisions, while constructed responses are scored by human coders following assessment completion and are not used for routing decisions during adaptive test administration.

Alternative to real-time automatic scoring, testing platforms may also be developed to support constructed-response items that do not utilize an open-ended text field. For example, for items in which the correct response is located in the item stimulus (such as a reading passage or figure), item design options, such as highlighting the correct response in the passage or clicking on an image, a cell in a table, or a location on a graph to indicate the correct response, may allow for machine scoring more easily than is possible with a text field.

7.3.3 Log and Process Data Collection

The types of data to be collected are an important consideration of any large-scale CBA. In a CBA environment, the collection of an array of process data, such as response time, number of actions, and other auxiliary data, is not only possible but also beneficial. Such data can be used to detect data fabrication (Yamamoto, Ali, et al., 2025) and guessing (Pokropek & Khorramdel, 2025), inform research on cognitive processes (Kespaik et al., 2025) and response styles (Kroehne et al., 2025), identify disengaged responding (Ulitzsch et al., 2019), and has even shown promise in improving the accuracy of and population modeling (Shin et al., 2025; Molenaar & van Rijn, 2025). The flexibility of a testing platform to support the collection of new types of data increases the usability of the assessment for a variety of research purposes and, thereby, may also increase the longevity and utility of the assessment.

7.3.4 Web-Based Test Administration

Following the fallout of the COVID-19 pandemic, a question frequently being asked of test developers was whether or not certain large-scale tests can be administered online. The answer to that question depends on the technical capabilities of the users (students and schools) taking the test and the quantity and quality of data required by test developers and psychometricians. First, if the assessment needs to support interactive item types with rich media interfaces and collect complex process data, the requirements for the users are much higher, such as internet speed and latency. If the internet were to disconnect midway through the assessment, safeguards would need to be in place to make sure that data is frequently backed up, both so that it is not lost and that the user's experience is unaffected by the loss of internet (which would add a confounding factor influencing the ability to make valid inferences about the student's assessment results). Further, if an adaptive testing design is being used, in which item responses are scored in real-time, internet speed and a consistent connection are vital. Connection and presentation speed influence process data collection as well, like number of actions and response time. If an item were to load slowly or be disrupted by a dropped connection, the amount of time spent on the item and the number of actions, particularly on interactive items in which multiple actions are expected, may not be accurately recorded. However, if only the response need be recorded, and interaction with the item is minimal or not necessary, online delivery is much more feasible. A rich survey can be administered over the internet, but it is unlikely that it will reach every part of the sample of interest. In that case, an alternative or backup plan, such as in-person testing for those without sufficient internet access, would need to be in place. Regardless of the complexity of the data to be collected, online testing is not likely to affect the types of test designs that can be administered.

Test security is another concern of web-based testing. The lack of a human proctor and secure testing location requires that additional measures to be taken to ensure that cheating does not occur. Measures to prevent individual cheating may include installing temporary software to block other programs from running while the test is underway or the use of a webcam and remote proctor to monitor the test taker during the assessment. Similarly, software to prevent the theft of confidential testing information by a third party must also be considered. The unauthorized release of test information, such as item content, can be costly to replace and would compromise the validity of inferences that can be drawn from that testing information. However, requiring installation of local software to manage these test security issues obviates many of the advantages of online testing, while adding the additional requirement of a reliable internet connection.

7.4 Discussion

Digital platforms for ILSAs must be developed considering a multitude of factors, including the conditions of and requirements for test administration, the types of items that will be administered, and what tools will be needed by the many different users it will support (e.g., from assessment developers and translators to national project managers and local test administrators). The test delivery system and questionnaire authoring tool are the primary components of digital testing platforms. In PIAAC, the delivery system is permanently installed on the host system, and the assessment is run in an isolated guest system environment. In PISA, the delivery system is temporarily downloaded and installed on local systems, usually school computers, or run off a USB. In both assessments, data are frequently backed up, and retained on local drives until upload to a central server or direct delivery via USB is confirmed following test completion. These test delivery systems must accommodate multiple languages and be able to support a wide set of non-adaptive and adaptive test designs. Further, the platforms must be flexible enough to adapt to new learning-assessment environments and accommodate new item types, new domains, and new assessment strategies. The questionnaire authoring tool can either be a stand-alone component or integrated into a larger network of systems, connecting test developers, test programmers, and test administrators. In PISA, the authoring tool is incorporated into the PISA Portal, where users are able to collaborate and track item progression. The Portal is organized by tasks, tools, and roles and is a clear method of disseminating necessary information and allowing for communication between centralized assessment developers and national assessment teams around the world. These tools allow for accurate and efficient assessment creation and dissemination.

Considering the future of large-scale assessments, digital testing platforms, like those used by PIAAC and PISA, must balance a very wide and often conflicting set of requirements. For example, the measurement of new constructs requires new and more complex item types, often providing immersive, simulated contexts and the necessary capture of complex user actions. However, in an ILSA environment, in which testing time is constrained and respondents have a wide range of proficiency that the test content must capture, it is critical that these item types be simple to use, require little or no prior content knowledge, and do not introduce construct irrelevant bias, such as unfamiliarity with or difficulty using an immersive item interface affecting a person's observed proficiency in the domain. New items are frequently being added to the item bank, aiming to measure new constructs or new aspects of existing constructs, and must be incorporated into the test delivery platform and authoring tool. Such item types include the basic item types used in most ILSA cognitive domains (e.g., multiple-choice and constructed response), as well as innovative item types, such as those that mimic digital environments in the PIAAC PSTRE domain and several of the new interactive items introduced in PISA since 2015 when the assessment began computer administration. The collection of log and process data from both trend and new items, including response time and number of actions or action sequences, need to be accommodated, as they provide important information for both test quality assurance and secondary research. For example, such data can be used by researchers to study cognitive processes and response behavior in ILSAs and are expected to improve methods for verifying assessment reliability and validity (Yamamoto, Ali, et al., 2025). Test designs are also frequently updated in ILSAs; the adaptive testing implemented in both PISA and PIAAC has demanded real-time response scoring and has required specific routing guidelines to be programmed and accurately applied. Finally, web-based test administration is another possibility that may be pursued, and while it is expected that little of the assessment need be changed or sacrificed in terms of quality and data collection, much is dependent on the capabilities of the user. In circumstances where internet connection and speed do not allow for the test to be implemented as designed, or where test security cannot be guaranteed within a reasonable degree, alternative test administration options must be available.

The possibilities of technical innovations are exciting and numerous, while simultaneously challenging. It is expected that the quality, quantity, and types of data being collected through digital platforms will continue to have a meaningful impact on the field of large-scale and educational research.

References

Aoki, O. (2019). Debian Linux system [computer software]. https://www.debian.org/doc/manuals/debian-reference/.

Frey, A., & Fink, A. (2025). Controlling for item position effects when adaptive testing is used in large-scale assessments. In L. Khorramdel, M. von Davier, & K. Yamamoto (Eds.), *Innovative digital-based international large-scale assessments—foundations, methodologies and quality assurance*. Springer.

Glas, C., & Jehangir, K. (2014). Modeling country-specific differential item functioning. In L. Rutkowski, M. von Davier, & D. Rutkowski (Eds.), *Handbook of international large-scale assessment* (pp. 97–115). https://doi.org/10.1201/b16061

Kespaik, S., dos Santos, R., & Salles, F. (2025). Preparing and analyzing log and process data in large-scale assessments. In L. Khorramdel, M. von Davier, & K. Yamamoto (Eds.), *Innovative digital-based international large-scale assessments—foundations, methodologies and quality assurance*. Springer.

Kroehne, U., Khorramdel, L., Goldhammer, F., & von Davier, M. (2025). Extreme response styles and rapid responding: Two sides of the same medallion? A joint modeling approach using IRTree and mixture IRT models. In L. Khorramdel, M. von Davier, & K. Yamamoto (Eds.), *Innovative digital-based international large-scale assessments—foundations, methodologies and quality assurance*. Springer.

Lennon, M. L., & Tamassia, C. (2013). *Chapter 2: The development of the PIAAC cognitive instruments* (Survey of Adult Skills Technical Report). https://www.oecd.org/skills/piaac/_Technical%20Report_17OCT13.pdf.

Organisation for Economic Co-operation and Development. (2015). *PISA 2015 Released Field Trial Cognitive Items* (OECD publication). https://www.oecd.org/pisa/test/PISA2015-Released-FT-Cognitive-Items.pdf.

Pokropek, A., & Khorramdel, L. (2025). Analyzing test-taking behaviors through process data using the IRT explanatory model for guessing. In L. Khorramdel, M. von Davier, &

K. Yamamoto (Eds.), *Innovative digital-based international large-scale assessments—foundations, methodologies and quality assurance*. Springer.

Shin, H. J., von Davier, M., & Yamamoto, K. (2025). Incorporating timing data into PISA population modeling. In L. Khorramdel, M. von Davier, & K. Yamamoto (Eds.), *Innovative digital-based international large-scale assessments—foundations, methodologies and quality assurance*. Springer.

Ulitzsch, E., von Davier, M., & Pohl, S. (2019). A hierarchical latent response model for inferences about examinee engagement in terms of guessing and item-level nonresponse. *British Journal of Mathematical and Statistical Psychology, 73*, 83–112. https://doi.org/10.1111/bmsp.12188

Upsing, B., Goldhammer, F., Schnitzler, M., Baumann, R., Johannes, R., Barkow, I., Rölke, H., Jars, I., Latour, T., Rlichart, P., Jadoul, R., Henry, C., & Wagner, M. (2013). *Chapter 8: Development of the integrated computer platform* (Survey of Adult Skills Technical Report). https://www.oecd.org/skills/piaac/_Technical%20Report_17OCT13.pdf

Molenaar, D., & van Rijn, P. (2025). Modeling item responses with response times as collateral information in large-scale educational assessments. In L. Khorramdel, M. von Davier, & K. Yamamoto (Eds.), *Innovative digital-based international large-scale assessments—foundations, methodologies and quality assurance*. Springer.

Yamamoto, K., Ali, U., Robin, F., Kapur, L., & Kandathil, M. (2025). Psychometric forensics to maintain and improve data quality in large-scale assessments. In L. Khorramdel, M. von Davier, & K. Yamamoto (Eds.), *Innovative digital-based international large-scale assessments—foundations, methodologies and quality assurance*. Springer.

Yamamoto, K., Shin, H. J., Robin, F., Khorramdel, L., & Halderman, L. (2025). Improved test designs and multistage adaptive testing in large-scale assessments. In L. Khorramdel, M. von Davier, & K. Yamamoto (Eds.), *Innovative digital-based international large-scale assessments—foundations, methodologies and quality assurance*. Springer.

Part III
Innovations in Translation, Data Processing, and Scoring

Chapter 8
Translation in Computer-Based International Large-Scale Assessment: Not a Stand-Alone Component

Steve Dept, Andrea Ferrari, Manuel Souto Pico, and Danina Lupsa

Abstract From the perspective of translation and linguistic quality assurance (LQA), the evolution of international large-scale surveys from paper-based to computer-based assessments has had far-reaching implications: Elaborate processes that apply to one testing mode may be inadequate or ill-suited for another. This chapter briefly describes the state of the art of translation and LQA in computer-delivered assessments, and endeavours to identify opportunities in this field that may have been overlooked: As international large-scale assessments (ILSAs) migrated to technology-rich environments, existing translation models have often been replicated without being challenged. The focus has been on maintaining the same time frame and containing budget growth rather than on investigating efficient ways to leverage advances in translation technology and methodology. Paradoxically, current translation and verification processes for computer-based items fail to harness the full potential of translation technology. This chapter will examine prerequisites and implications of a shift in paradigm leading to professionalization of survey instrument translation.

Definitions of translation and adaptation abound in the literature on test and survey methodology. Authoritative authors such as Hambleton et al. (2005), Harkness (2003, 2007), Harkness et al. (2003, 2010), Iliescu (2017), Joldersma (2004) and Van de Vijver (2003) commonly refer to *translation* as a somewhat narrow term in view of the breadth of expertise required to produce multiple versions of assessment instruments or in view of the underlying objectives of functional equivalence and cross-linguistic comparability. These authors suggest that the term *adaptation* is either to be preferred (Hambleton et al., 2005; Joldersma, 2004) or that it encompasses test translation as a subset (Harkness et al., 2010; Iliescu, 2017). Iliescu explains that test translation is linguistically driven (content over intent), while test

S. Dept (✉) · A. Ferrari · M. Souto Pico · D. Lupsa
cApStAn, Brussels, Belgium
e-mail: steve.dept@capstan.be

L. Khorramdel et al. (eds.), *Innovative Digital-Based International Large-Scale Assessments*, Methodology of Educational Measurement and Assessment,
https://doi.org/10.1007/978-3-031-90951-1_8

adaptation is validity-driven (intent over content), and this is certainly a workable distinction.

However, in this chapter, we shall mostly focus on technical aspects of translation and use terminology that is more prevalent in the language service industry than in test and survey methodology. Professional translation includes processes to make good use of a variety of resources, which we shall refer to as *language assets*, and a set of linguistic quality assurance processes. A holistic approach to test translation is not limited to capturing content from the source text and conveying it in a suitably fluent and intelligible manner in the target language. Rather, the source version and related documentation are used to form a broad understanding of the author's intent and idea, and, subsequently, the translator attempts to reproduce that idea in the target language in a manner that is most likely to remain true to the original intent. This may require intentional deviations from the source text—which, incidentally, are often referred to as *adaptations*—to ensure that the target version maintains the intended properties and characteristics.

With a view to avoiding confusion, we shall consistently use the term *translation* in a very broad sense to refer to a set of processes making it possible to produce comparable target-language versions. This set of processes may include—without being limited to—steps to optimise the source version, the creation of workflows, leveraging language assets (glossaries, style guides, translation memories), the actual work done by translators and language experts, consultation of domain experts, and linguistic quality assurance (LQA). The latter is an umbrella term for a sequence of processes designed to make it possible to meet a pre-determined set of translation quality standards. Translation verification is a subset of LQA and consists of checking to what extent these standards are met. Test *adaptation*, in a sense that "includes decisions about whether a test can measure the same construct in the new language and culture" (Iliescu, 2017, p. 21), is a research topic in its own right and is not specific to computer-based assessments. It is not the focus of this chapter, and this explains our choice to use *translation* to refer to the translation, adaptation and LQA processes described here.[1]

In the transition from pencil-and-paper to computer-based international large-scale assessments (CB-ILSAs), several approaches to translation and LQA were explored. However, the most common starting point has been to reproduce in a computer-based environment what had previously been successful for pencil-and-paper instruments. In PISA, PIAAC, e-TIMSS, and e-PIRLS, for example, attempts were made to replicate the translation models used in pencil-and-paper tests in the new computer-based mode, adapting and tweaking them as needed (see Sect. 8.2.5). In this process, workarounds were identified to circumvent obstacles, and fixes were created to patch up flaws. Only a few basic functionalities from word processors were emulated, and, failing better methods, abundant use was made of manually

[1] In other contexts, this set of processes is often referred to as *localisation*. However, this term is also used in different ways depending on the type of content that is being localised and, considering that the term *translation* is known by all survey methodologists, it was deemed the most suitable choice.

copy-pasting content from one mode to the other. Three salient observations about the transition from paper-based assessments (PBA) to computer-based assessments (CBA) were selected to inform this chapter.

First, envisaging the computer-based environment merely as a different medium, a different setting in which one can replicate and adapt translation procedures previously used in pencil-and-paper tests, is, at the very least, inadequate and inefficient. It seems preferable to design a new paradigm that takes into account the limitations, requirements, and, of course, the full range of opportunities connected with the computer-based assessment mode.

This leads to the second observation: Designing such a paradigm, i.e., creating a sophisticated translation workflow for assessment instruments, greatly benefits from a multidisciplinary approach. Ideally, testing organisations and their test developers, platform engineers, and translation technologists should work together from the earliest planning stages.

The third observation is that, because testing organisations are responsible for assembling and delivering the assessment, their preference is to gain or maintain some level of control over all components of the testing environment, including the translation workflows. Time constraints often make it difficult to test additional tools, and a number of organisations have shown a tendency to reinvent the wheel when designing an assessment platform. In cases where an understanding of complex translation processes is lacking, this approach has led to investing in the development of relatively primitive translation interfaces and workflows, with a view to integrating those into their platform rather than resorting to existing state-of-the art translation technology outside their platform.

A testing platform that accommodates multiple language versions but does not feature translation technology at all is a substandard translation environment. Non-professional translators may find it practical to enter translations in free text fields in a platform, but trained translators find it unprofessional to work without adequate tools. The approach of an all-in-one platform with limited translation technology is at odds with the technology-driven need to professionalise translation in CB-ILSAs: While such an approach may be understandable from the perspective of platform engineers, it fails to leverage the language assets and mature translation technology that professional translators are accustomed to and that support the quality of their work.

As a result, one needs to be aware that switching to a computer-based testing environment can generate adverse effects, such as additional work for translators, a surge in manual operations, and a decrease in quality and consistency of translated versions of assessment instruments.

In this chapter, we shall first set the stage by examining the requirements and reviewing the state of the art and variants currently implemented in major assessments. This aspect has gained importance in computer-based assessments, as interoperability, interdependency, and traceability require a multidisciplinary approach. This will naturally lead us to describe the translation data exchange standards as well as the implementation of standards, i.e., know-how and best practices. An introduction to the translation productivity tools known as computer-assisted translation tools, or CAT tools, is provided in Sect. 8.3.3, which will help the reader understand the stakes of a trade-off between centralising all translation-related functionality inside the testing platform or models that involve more cross-functional collaboration between platform designers and translation technologists, so as to integrate existing tools with the platform. Considering that the portfolio of skills required from assessment translators has evolved together with technology, it will be useful to supplement this section with information on the profile of professional users (and how to recruit them to translate CB-ILSA). To put this into perspective, examples of cooperation between platform engineers and translation technologists and of academic exchanges (research institutes and field practitioners) are provided. To conclude this chapter, the discussion and outlook will take the form of a plea to embed translation into the assessment design rather than to treat it as a stand-alone component that follows after the assessment is developed and piloted.

8.1 The Requirements

In an effort to keep this chapter focused, we shall refrain from reviewing needs and requirements that apply to assessment instruments in general. For this purpose, *Methodological Advances in Cross-National Surveys of Educational Achievement* (Porter & Gamoran, Eds, 2002); *Adapting Tests in Linguistic and Cultural Situations* (Iliescu, 2017) and *Implementation of Large-Scale Education Assessments* (Lietz, Creswell, Rust & Adams, Eds, 2017) are examples of widely recognized reference works. Translation of any data collection instrument is geared towards providing comparable measurement across languages and cultures and, in ILSAs, across modes and over time. To achieve this for computer-based assessments, it is necessary that the teams in charge of item development, platform development and translation/LQA, respectively, communicate and agree on principles, procedures, workflows, cut-off points and coordination before the items are developed, before workflows are defined, and before the translation process begins. This approach follows the trend to increase upstream LQA efforts to reduce downstream disparities and related corrective action (Grisay, 2002; Hambleton & Zenisky, 2011; Harkness & Behr, 2008).

Regardless of the translation approach, the following prerequisites can be regarded as essential for translation of CB-ILSA:

1. A *source version* needs to exist, i.e., a measurement instrument and data collection instrument crafted by trained test developers/questionnaire authors/psychologists who have determined what they wanted to measure and designed an instrument to capture relevant data. This implies the existence of a conceptual framework to which the questions are mapped.
2. As far as possible, the measurement characteristics of that source version should be documented, i.e., the way a stimulus is expected to prompt a response strategy

should be described; universality or at least generalisability of the constructs should be established.

3. Once a consensus has been reached on a translation approach, the source version needs to be made available in a format that translators can use. This should include an extraction mechanism, i.e., a way to break the source down into translatable and untranslatable components—such as structural markup—and, if possible, hide or lock untranslatable inline components. The process of deciding which parts of the document are translatable and which parts are not is referred to as *extraction*.
4. When preparing the source version for translation, attention should be given to facilitating the translation task. One noteworthy way to achieve this is to split the translatable parts of source version into smaller units, usually referred to as segments. The process of splitting a text into small, manageable parts—usually sentences—is referred to as *segmentation*.
5. There needs to be a simple process to preview both the source version and the translated version of the assessment or questionnaire at any stage of the process, preferably without file manipulation. Translators and reviewers need to preview the assessment items or questions in context because they need to understand the task that respondents are asked to perform or the question they are asked to answer.
6. Although this is not a prerequisite in a strict sense, there should be a structured repository for translation notes, linguists' comments, and other meta-data, which should either travel together with the translation package—if the translation is exported out of the platform and imported back—or inside the platform, preferably at the segment level.

These prerequisites may seem technical and not directly linked with translation needs and requirements of computer-delivered assessments. However, in a complex environment in which the main objective is to maximize comparability across language versions of data collection instruments, it is reasonable to assume that one needs to facilitate the linguists' task. This can be achieved by bringing together the language assets they can leverage and by fostering work conditions in which they can confidently rely on their translation expertise and subject matter expertise to generate good translations, rather than depend on their detective skills.

It is paradoxical that a technology-rich environment should have adverse effects for any of the players involved: The new opportunities for test authors and for data analysts can go hand in hand with streamlined translation processes using the power of state-of-the-art translation technology, more secure workflows with more transparency, and better traceability (see Sects. 8.3.1 and 8.3.2).

8.2 The State of the Art and Variants

8.2.1 The Translation Culture in Large-Scale Assessments

A number of major ILSAs migrated gradually from the PBA mode to the CBA mode (e.g., OECD/PISA as of PISA 2006 for some optional components, with a full transition in PISA 2015; IEA/PIRLS in 2016 with ePIRLS, IEA/TIMSS in 2019 with eTIMSS) or started out in mixed PBA-CBA mode (OECD/PIAAC in 2011). In all these cases, there was no fresh start regarding the translation model. Rather, the starting point for translation processes in CB-ILSA has always been an ingrained legacy of methods and processes from the PBA era.

Our contention is that major opportunities are lost if the translation processes for CB-ILSAs are mere adjustments, tweaks and fixes to routines inherited from PB-ILSAs, without regard for the specificities of CBA or for recent developments in translation technology. These processes need to be redesigned and re-engineered, and it is crucial that they be embedded in the instrument development process from the outset, which calls for early and close collaboration between the people in charge of item development, platform development, and translation.

In the light of the above statement, we shall examine translation processes that are currently used in CB-ILSAs and are regarded as good practices. Let us look at these processes in chronological order in terms of the development process.

8.2.2 Source Preparation and Optimization: At Content Level

By definition, source preparation and optimisation consist of measures taken before the source version is disseminated for translation.

A preparation step with a high return on investment is to submit assessment and questionnaire items for a translatability assessment. This methodology (described more fully in Dept, 2013) has been applied in PISA 2015, PISA 2018, and PISA 2022, in PIAAC Cycle 2, and in a large number of European surveys, multilingual polls and attitudinal surveys. The test development team submits draft versions of new items to a pool of experienced linguists who, together, cover a broad range of target language groups. These linguists produce draft translations of those items. Their translations are not intended for further use but to help them identify and describe the problems translators would face if they had to translate/adapt the questions in their current form. A set of 'translatability categories' (e.g., 'potentially ambiguous,' 'unnecessarily complex,' 'potential cultural issue') is used to report on the potential translation, adaptation, and cultural issues identified.

The feedback is then collated by a senior linguist, with a special focus on issues that are raised by more than one linguist. In the resulting translatability report, the identified problems are listed along with either (1) a suggested translation/adaptation note to clarify a given term or expression, or to indicate the type of adaptation that may be necessary; or (2) a proposal for alternative wording (a new formulation that circumvents the problem without loss of meaning); or (3) a combination of both. The translatability report is sent to the test developers or questionnaire authors, who ultimately decide which elements of this feedback should be taken into account. They, thus, have the opportunity to eliminate (e.g.) ambiguities, Anglo-Saxon idiosyncrasies that may be difficult to render in certain languages, double-barrelled questions, or unnecessary complexity. They are also made aware of potential cultural issues. In short, an attempt is made to fine-tune the initial version of the items to achieve a more translatable source version.

Outside the realm of assessments, at least one large organisation is experimenting with the technique of 'Amplified Source Content' for its surveys on customer experience/satisfaction, which exploits the outcome of a translatability assessment to a fuller extent: Each item is sent for translation with a complement of structured meta-data that includes explanations and paraphrases, alternative wording, adaptation notes, and acceptable and unacceptable variants in translation. Similarly, other organisations outside the domain of assessment complement the 'raw' source with meta-data based on the DITA standard (described further in Sect. 8.3.1.6, including how the standard could be applied in ILSAs). It would certainly be worthwhile to investigate the feasibility/benefits of bringing such techniques across into item development for CB-ILSAs.

Also, new and innovative item types that are or are likely to be used in CB-ILSAs should receive extra scrutiny at this early stage to prevent or mitigate unforeseen problems.

For example, PIAAC Cycle 1 employed a 'highlight in the stimulus' response mode (Upsing et al., 2011). While this was necessary in order to provide the required psychometric link between existing paper-and-pencil items that allowed respondents to underline or circle responses in the text and the new CBA versions of these items, it gave rise to the difficult issue of defining the minimum acceptable response in translated versions of the texts. A word in English could correspond to more than one word in some languages or to part of a (compound or inflected) word in other languages; or, a 'continuous' phrase in English could correspond to separate pieces of text in another language. This problem had to be resolved "on the go" during execution of the project. New CB-specific item types could involve audio or video content, simulations, gaming, avatars, interactions with virtual agents, and other techniques yet to come. The translatability assessment may well need to be re-engineered to cater effectively to these novelties.

As another example, in an innovative assessment domain such as Collaborative Problem Solving—assessed in PISA 2015 (see OECD, 2017)—chat boxes were used to allow for interaction between virtual agents and respondents. Chatting is a form of communication that carries its own specific challenges. The informal language of a chat discussion cannot easily be rendered in a diglossic language such as Arabic, where the gap between standard written language and a casual chat is much wider than in English. Also, in a number of languages, it is necessary to have at least two gender versions of the chat supposedly representing the respondent's possible collaborative steps: a female respondent may not relate to a segment occurring in the male form. This is a technical challenge that can be solved with a dynamic text approach: Chat segments representing the respondents' possible utterances are duplicated in the authoring platform. The duplicated segments are identical in English but translated into different gender forms in languages where this is necessary. The system needs to know upfront whether the respondent is a male or a female, and only the corresponding gender form of the translation will be displayed. This can be taken into account during source optimisation exchanges involving test developers, platform engineers, and translation technologists.

We would complete this overview of source preparation processes at content level with a glimpse into a possible future. Developments in the rapidly evolving field of translation technology, particularly machine translation (see Sect. 8.3.3 for further information and discussion), could call for significant changes in good practice in a not distant future.

8.2.3 Source Preparation and Optimization: At Technical Level

Technical preparation steps are facilitated by early and close collaboration between the item development, platform development, and translation specialists. In PIAAC and PISA, for example, the tagged XML Localisation Interchange File Format (XLIFF) was used. The XLIFF is an international standard and attention needs to be given to how the XLIFF files are generated. Ideally, these XLIFFs can be processed with the professional translators' preferred CAT tool, and they export/import seamlessly from and into a platform providing fast previews and reliable version management. Where applicable, this configuration will support a single-source/multi-channel publishing approach, i.e., a source that presents only the content that is common for all implementations that might be planned (e.g., computer, tablet, paper booklet) and indicates the variants specific to each implementation (e.g., "click" for computer, "tap" for tablet, "circle" for paper booklet), with the goal of separating publishing modes from source format and language tasks.

Once again, new item types and evolving translation technology need to be watched. In the previously mentioned example of Collaborative Problem Solving in PISA 2015, chats between the respondent and virtual agents implied that non-linear text needed to be translated and validated (the sequence of texts shown to the respondent depends on their responses). What does 'preview' mean in this case? Linguists (translators, reviewers) need to navigate through a test unit making different choices so as to 'hit' all the possibilities—and to be complete, to do so twice, impersonating both a male and a female respondent. They could use help from the platform designers, as the preparation of files at the technical level must be thought out in a way to support an efficient (focused, selective) translation process, specifically designed for non-linear text and text variants.

8.2.4 Preparation for Translation

Once the source version is finalized, the usual practice in CB-ILSAs is to release files for translation along with requirements/recommendations, training sessions and/or kits, and language assets.

The requirements/recommendations concern:

- The minimum acceptable translation design, most often a variant of the TRAPD approach (Harkness, 2007; Harkness et al., 2004), i.e., double translation followed by reconciliation or adjudication, followed by expert or committee review, with documentation of the entire process.
- Selection criteria for translators/reconcilers—see Sect. 8.4.1.

Training sessions, face-to-face or online, are organized for all linguists who will intervene in the translation/validation processes (in a first stage translators and reconcilers/adjudicators, then verifiers and reviewers). The trainings cover the specifics of localizing ILSA instruments and instructions for using the related software and platform. If direct training sessions are not feasible, training kits are provided to those who will train the linguists.

Language assets made available to the translation and verification teams include:

- Glossaries, style guides, translation memories from previous rounds, rules for automated quality assurance checks—which, in robust CAT tools, are included in the package that contains the content to be translated (see Sect. 8.3.3).
- General translation and adaptation guidelines and segment-specific translation and adaptation notes—the latter may travel together with the content in the same XLIFF or can be entered in separate monitoring files.

The above-listed practices may not have been challenged during the transition from PBA to CBA.

In particular, the TRAPD translation model has practically acquired the status of dogma, but it existed before the switch to CBA and the sweeping changes of the last few years in translation technology. It is time to re-evaluate it and confront it with other translation approaches that include a combination of centralised translation management, machine translation engines trained with assessment-specific corpora, cognitive pre-testing of translated item pools with an immediate feedback loop to translation teams, and online adjudication meetings with linguists and subject matter experts. A specific definition of *translation quality* for CB-ILSAs could focus on functional equivalence of source and target versions, which implies that *translation quality* needs to include an analysis of the results of the data collection.

8.2.5 Translation

In regard to the translation phase sensu stricto, computer-based testing environments have added a new dimension to a dilemma that was already prevalent for PB-ILSAs. When ILSAs were predominantly pencil-and-paper tests, decentralised translation models were applied in a majority of studies, partly because participating entities are mostly keen to retain ownership of their version of the data collection instruments and partly because the direct cost of a centralised translation model may have seemed prohibitive.

- In a decentralised model, participating entities (countries or regions) produce or adapt a version, whereas guidance and quality assurance (before, during, and after) is the remit of the contractors implementing the project.
- In a centralised model, the contractors are responsible for the production of the national or regional versions, whereby participating entities provide input and possibly review and sign off on the centrally produced materials.

The advantages of a centralised model are (1) less dispersion of resources; (2) better control over compliance with guidelines, particularly regarding the professionalism of translation teams and their use of state-of-the-art tools; and (3) less risk of disclosure of confidential materials.

With some transparency/clarity concerning costs, the translation model could become a hybrid one where entities are given the choice. In the case of CB-ILSAs there are further reasons to favour the centralised model insofar as possible, as will be discussed when examining field trial to main survey processes: updates to source and focused verification.

8.2.6 Post-Translation Processes: Verification, In-Country Review, Final Checks

In the prevailing decentralised model, in all major ILSAs, a *verification*, i.e., an external review of materials translated by participating entities, is carried out as a key and incontrovertible quality assurance procedure, carried out by experienced linguists trained to identify survey-specific equivalence issues. This predominantly 'manual' operation is well documented (Dept et al., 2010) and has undoubtedly proven its worth in the era of PB-ILSAs. Thanks to progress in translation technology, in recent years it is increasingly complemented by quality-assurance-oriented technology, which can automate repetitive tasks and consistency checks, freeing verifiers' time and energy for higher-level issues: checking semantic equivalence, balancing the requirement of comparability against the requirements of fluidity and naturalness of the target language.

A closer collaboration between platform developers and linguists, as advocated in this chapter, could nevertheless further improve the effectiveness and efficiency of the 'manual' verification procedure in the context of CB-ILSAs. Verifiers work by comparing a target version to a source version, segment by segment in full context, using previews of how the units will be displayed to test takers on screen. An optimal preview system would guide verifiers visually through mechanisms linked to the segment-specific translation notes and other meta-information. For example, key correspondences between stimulus and questions (literal and synonymous matches to be respected for psychometric equivalence) could be shown via connecting lines and/or colour schemes. As another example, a note advising about a needed or recommended cultural adaptation could pop up when the cursor moves over the term concerned. One could also imagine a system in which the linguist needs to check a box next to each segment-specific guideline to confirm that it was addressed.

More help from platform designers might be needed in unforeseeable ways, as new CB-specific item types could involve audio or video content, simulations, gaming, avatars, interactions with virtual agents, and other techniques yet to come.

Technological support has also been harnessed for the post-translation phases that follow verification: the 'in-country review' (review of verification suggestions by participating entities, including negotiation and resolution of any controversies) and the 'final check' on correct implementation of agreed edits. For example, convenient 'difference reports' are generated so that each successive player can visualise what has been changed in the version they are reviewing versus the previous one.

It is currently possible to program alerts based on expansion rate (e.g., if the German version of assessment materials is on average 1.17 the length of the English source, one could program alerts for any translated segment that is not between 1.07 and 1.27 times the length of the source segment). One could program alerts for translated versions in which the proportional length of correct and incorrect responses, known as 'key' and 'distractors,' respectively, deviates from the source version by more than 5%. The relative length of key and distractors is an important psychometric characteristic than can be lost even in a semantically correct translation.

8.2.7 Field Trial to Main Survey Processes: Updates to Source and Focused Verification

All major ILSAs are planned with a field trial phase using smaller samples of respondents before the main survey phase when the 'real' data collection on full samples takes place. The translation work occurs prior to the field trial, which is where all new items (but not necessarily 'trend items' from previous rounds) are tested to see how they work in all participating entities/regions. Linguistic quality assurance continues after the field trial.

After the field trial follows the analysis of assessment results, which can reveal entity-item or language-item interactions indicating possible comparability problems and has an impact in the selection of items retained for the main data collection.

These items may be revised, so that a number of edits ('updates to the source' in the assessment parlance) need to be reflected in the translated target versions. The selection could include items that seemed to work well for a majority of participating entities, while behaving differently for some entities—who are then invited to review them and to propose edits if the suspected cause of the problem is a translation issue.

A challenge related to the switch from PBA to CBA in the major ILSAs—while keeping the same cycle length, e.g., 3 years for PISA—is the reduced timeline between the field trial and main survey for the above-described processes, owing to the need to lock, earlier, the computer-based delivery systems so that they can be adequately tested. This has meant less opportunity to make updates to source versions (more often, a decision has to be made whether to retain an item as is or discard it) and to revise translations in light of data analysis (so that either more items need to be discarded for all, or some entities will need to go to the main survey with items known to work less well in their version). In the authors' opinion, much of the lost opportunity here is due to an unquestioned adherence to the decentralized model that prevailed in the PBA era. Opening up validated materials to the scrutiny of participating entities is time-consuming and means a higher risk of unnecessary edits ("fixing what isn't broken"). A well-designed field trial to main survey translation review process that allows participating entities to review and sign off on centrally revised materials—with revisions strictly limited to what is necessary—could alleviate the timeline pressure and allow for better use of field trial data to optimize source and translations for the main survey.

8.3 The Technology Challenge: Harnessing the Power of Translation Technology

If one worked in autarky, in a closed, self-sufficient environment, one could design, program, and author without giving interoperability standards much attention. In ILSAs, just as in the translation industry at large, it is a given that many organisations will be involved and will need to work together. The contractors implementing the studies may change over time, and new players will need to take over tools, data, and environments from their predecessors. In this context, documentation, transparency, adaptability, and interoperability are four key priorities, which also apply to translation, processes and formats. The selection of standards, and the procedures that need to be followed to ensure compliance with these standards, are crucial components to address the four priorities.

8.3.1 Translation Data Exchange Standards: Addressing the Interoperability Challenge

8.3.1.1 Overview

A standard can be defined as a universally agreed upon and commonly used fashion of doing something, which becomes registered or codified in such way that all players concerned can easily refer to it. In the translation industry, we use the term "standards" to refer to format specifications with which translation data exchange must comply.[2]

Translation data exchange standards are important for two reasons: They allow for interoperability and they condense knowledge.

Interoperability is the capability to smoothly exchange files between tools. In the context of this chapter, we focus on translation data exchange standards because they determine how engineers should create translation files so that translation tools can handle them without issues or data loss. If this interoperability did not exist, it would be extremely challenging to share content among different parties, and one would need to spend money, effort, and time on engineering solutions instead of using tools that simply work out of the box when they exchange files. Melby and Zetzsche (2010) eloquently explain the importance of standards using the example of the Trans-Siberian Railway between China and Moscow, on which trains had to be raised and have their wheels changed before they could cross the border, until the standard rail gauge was defined in 1846.

Interoperability means, for example, that my translation tool will be able to open a translation kit that someone else has created with another tool, and that a third party should be able to edit my translation in yet another tool, and that the creator of the original kit will be able to retrieve the edited translation from the kit when it is handed back to them. It also means that users (e.g., linguists) can be free to use the translation tool of their choice (typically, the tool they are most proficient with). Translators working with their preferred tools has an impact on their satisfaction at work and on the quality of their work. By the same token, it means that clients are not bound to a specific provider or system.

In a way, Excel and text files could be considered de facto data standards too, and they are interoperable to some extent. However, they are not *translation data exchange* standards (e.g., they do not foresee the representation of language codes, or understand rules to present the text in segments, etc.), and that brings us to our second point: condensing knowledge. Standards embody the know-how and expertise of people who know what pitfalls must be avoided and what users need. As such, data standards are meant not only to provide different parties with a common format but also to help them to work more efficiently.

[2]To avoid confusion with ISO translation services standard (ISO 17100:2015), we will call them "translation data exchange standards," meaning format specifications for translation data exchange.

The translation data exchange standards described here are XML vocabularies, which means that the vocabularies can be validated by means of other XML utilities: One can check that a file is what it is supposed to be. If we used Excel, there would be no way to check that the files comply with the standard (about XML standards in translation technology, see Roturier, 2019 and Savourel, 2001).

8.3.1.2 XML Localisation Interchange File Format (XLIFF)

XLIFF is a format created to standardize the interchange of translation data across the different steps of the translation process, while allowing interoperability between tools (Savourel et al., 2008). It is designed in consensus with prominent software vendors and translation providers, taking into account the needs of the industry (from the vantage point of translators, customers and manufacturers). The translation data it may contain includes core data, such as the extracted translatable content in the source language and its translation, as well as optional linguistic support content, e.g., alternative translations, or meta-data, such as administrative annotations or validation status of the translation.

In 2014, OASIS[3] approved version XLIFF 2.0 (Comerford et al., 2014; Savourel, 2014; Schnabel et al., 2015), which offers technical benefits over its predecessor (version XLIFF 1.2, currently the most widely used version in the translation industry). One interesting novelty is the Change Tracking Module, which can be used to perform version control and store revision information.

How is XLIFF used? Whatever the format of the source content is (XML, HTML, InDesign, Word, Excel, JSON, etc.), the engineer extracts the translatable text from that source content and puts it in a source XLIFF file, which at this point is untranslated and contains only text in the source language. The part of the document that is not extracted is called the skeleton, and it consists of structural markup, which serves to define the formatting of the text. The source text in the XLIFF file is segmented (see Sect. 8.3.2.1), and the translator must add the translation of each segment in this file using a translation editor (see Sect. 8.3.3). By the end of the translation task, the XLIFF is bilingual: Each segment in the source language is linked to its translation.

Once the translation exists, the file can go through different levels of review (revision, verification, proofreading, etc.) before being handed back to the engineer, who will then retrieve the target version and merge it with the skeleton from the original document, creating a new document that is structurally identical to the original document but contains the text in the target language. The outcome is equivalent to replacing the translatable text in the source-language document with

[3] OASIS (Organization for the Advancement of Structured Information Standards) is a non-profit consortium that drives the development, convergence and adoption of open standards for the global information society. (See https://www.oasis-open.org/).

the translation while preserving all structural or inline codes that must not be translated or altered.

In an XLIFF file, translatable content can be separated from formatting markup. Regardless of the original file format, the interface and the form in which text is displayed in the translation editor remains constant, i.e., the source text to be translated, and the translation that is progressively added. By hiding the complexities of the original formatting, XLIFF simplifies the task and allows the translator to focus on translating the translatable text.

The alternative to using XLIFF would be to send the source document to the translators, and then receive an equivalent document containing the translation, in which way two parallel but independent documents would need to be handled. Instead, XLIFF allows one bilingual format where the two parallel versions (source text and translation) can travel together in a linked form (where every source segment is linked to its translation). XLIFF also allows the engineers to have control over leverage of previous translations to make sure that they are reused efficiently and consistently across versions and across cycles (see Sect. 8.3.1.3).

Open-source Java libraries, e.g., Okapi XLIFF Toolkit for XLIFF 2.0 (Okapi Framework, 2016a) and Okapi Rainbow for XLIFF 1.2 (Okapi Framework, 2016b) and Maxprogram's OpenXLIFF Filters (Raya, 2021a) and XLIFF Manager (Raya, 2021b) allow platform engineers to create, read, write and manipulate XLIFF, which should facilitate adoption of the standard for any organisation.

8.3.1.3 Translation Memory Exchange (TMX)

A translation memory (TM) is a central concept in translation technology. It allows for the storage of segment pairs from the current or previous projects for the purpose of re-using them. If a segment of text has been translated already, it should not (need to) be re-translated from scratch the next time it appears. The TMX format allows the export/import of translation memories for the purposes of exchange, but a TMX file can also be used as a translation memory itself where the translations can be stored and re-used directly. A TMX file must be compliant with the TMX specification (Translation Memory eXchange, 2021). The bilingual content stored in a translation memory corresponds to what was in the relevant XLIFF files and reflects the structure and segmentation of those XLIFF files. See Savourel (2001) for information about the TMX standard.

8.3.1.4 Segmentation Rules Exchange (SRX)

An SRX file defines segmentation rules (see Sect. 8.3.2.1 below) and can be shared among different parties, so that segmentation can be applied consistently in the creation of different translation kits by different people and at different points in time. SRX files are validated against the SRX specification (Segmentation Rules eXchange, 2021).

8.3.1.5 Internationalization Tag Set (ITS)

ITS is a W3C recommendation that defines a set of data categories that can be used to specify different translation-related aspects of an XML or HTML5 document, e.g., what must be translated or displayed depending on certain criteria, using the data categories as selectors. This rule-based approach makes is possible to filter and extract only content that is necessary to translate for a certain locale[4] (e.g., some content may not apply to a given version but does apply to others) or that is necessary to display for a certain administration mode (e.g., the final presentation must include the text "Click on the link" if the mode is "screen" but exclude if mode is "print," or vice versa), while having one single source content container rather than different containers. See Sasaki et al. (2006) for further insight into the role of ITS in translation workflows.

8.3.1.6 Darwin Information Typing Architecture (DITA)

DITA is an XML-based, tool-independent standard to author, organize, and manage topic-based content in a minimal and structured way, with a focus on reusability of content (OASIS, 2018). DITA includes three main topic types: task, concept, and reference, which can be further specialized for different goals. Some specializations are already built in, such as the Learning and Training specialization, which includes topics for learning assessments. A learning assessment topic presents questions or interactions that use a sub-set of the Question-Test Interoperability (QTI) specification, implemented as a DITA domain specialization [DITA TC].

Also, similarly to ITS's data categories, DITA offers support for "conditional text" through profiling attributes and conditions, so that certain parts of the document are selected for display in the final preview if certain conditions are met. This would have multiple applications in the world of assessment and questionnaires: e.g., display one sentence according to the respondent's gender (male vs. female), or qualifications (ISCED levels, etc.), or marital status (single vs. married vs. divorced vs. widow/er etc.), or administration mode (print vs. screen). There are libraries that can help easily convert DITA to XLIFF (Raya et al., 2013), making it worthwhile for item authors and platform developers to explore the potential of DITA.

[4] A locale is a language-region combination, e.g., English for Ireland, Hausa for Nigeria or French for Luxembourg.

#	Source version	Translation
1	<opens some-attribute="some-long-value" id="1938472938473.129374329874" rid="UNIT.2.0" fontsize="11"/>You will now learn about the next exercises in the survey.<closes id="1938472938473.129374329874" rid="UNIT.2.0"/><opens id="2393743293498734923.23974392" rid="UNIT.4.0" fontsize="11"/>Please click <b>Next</b> to continue.<closes id="2393743293498734923.23974392" rid="UNIT.4.0"/>	<opens some-attribute="some-long-value" id="1938472938473.129374329874" rid="UNIT.2.0" fontsize="11"/>Ýôú ωïℓℓ ñôω ℓèářñ áβôúƭ ƭλè ñèжƭ èжèřçïƨèƨ ïñ ƭλè ƨúřƲè¥.<closes id="1938472938473.129374329874" rid="UNIT.2.0"/><opens id="2393743293498734923.23974392" rid="UNIT.4.0" fontsize="11"/>Þℓèáƨè çℓïçk <b>Nèжƭ</b> ƭô çôñƭïñúè.<closes id="2393743293498734923.23974392" rid="UNIT.4.0"/>

Fig. 8.1 Example of an XLIFF segment without codes conversion and without segmentation

8.3.2 Implementation of Standards: Know-How and Best Practices

We have seen that XLIFF and the other standards have the *potential* of streamlining the translation process. However, if these standards are not used properly, they can constrain translators in predictable and unpredictable ways.

Compliance with data standards is only part of the picture. Standards are a first step to allow engineers make the most of translation technology. However, since data standards apply to a wide range of situations, they must remain flexible enough to accommodate different needs. A standard can be used in counter-productive ways and create nightmare situations where translators find it extremely challenging to do their job, thus, jeopardizing the quality of the translation and, with it, the comparability of the data to be collected.

For example, the underlying principle of the relationship between XLIFF (containing the text to be translated) and TMX (containing previous translations that could be reused) is that the text is split in small chunks that are more manageable both for brains and for algorithms. The standard supports this splitting but does not require it. That opens the door to generating unsegmented texts that are unmanageable both for translators and for software tools. For example, the text in the XLIFF files might still be left unsegmented and include long chunks of texts full of long non-translatable codes.

Figure 8.1 shows a real example (with obfuscated parts, though) of a moderately unwieldy segment pseudo-translated into a fictitious language in OmegaT, coming from an XLIFF file that has been created without converting non-translatable codes to inline elements and without using segmentation:

Users (translators, reviewers, etc.) often work in XLIFF files containing many segments like this, struggling to stay focused on the text they must read in the source and/or review in the target, perhaps unaware that XLIFF files do not need to be so difficult to handle (Morado Vázquez & Torres del Rey, 2015). Indeed, Fig. 8.2 shows the same segment prepared with inline elements and segmentation (where

#	Source version	Translation
1	<x0/>You will now learn about the next exercises in the survey.	<x0/>Ýôú-MASC ωïℓℓ ňôω ℓèářň áβôúŧ ŧλè ňèжŧ èжèřçïƨèƨ ïň ŧλè ƨúřՍè¥.
2	<x0/><x1/>Please click <x2/>Next<x3/> to continue.<x4/>	<x0/><x1/>Þℓèáƨè çℓïçk <x2/>Nèжŧ<x3/> ŧô çôňŧïňúè.<x4/>

Fig. 8.2 Example of an XLIFF segment with codes conversion and with segmentation

#	Source version	Translation
1	You will now learn about the next exercises in the survey.	Ýôú ωïℓℓ ňôω ℓèářň áβôúŧ ŧλè ňèжŧ èжèřçïƨèƨ ïň ŧλè ƨúřՍè¥.
2	Please click <x0/>Next<x1/> to continue.	Þℓèáƨè çℓïçk <x0/>Nèжŧ<x1/> ŧô çôňŧïňúè.

Fig. 8.3 Example of an optimally prepared XLIFF segment

non-translatable codes are represented as inline elements and the text has been split by sentence in two segments):

The result above is a huge improvement compared to the original preparation, but it can be further improved. Codes that embed the whole paragraph or a whole sentence do not need to be placed in a different position in the translation, and therefore it serves no purpose to include them in the translatable text. The file preparation shown in Fig. 8.3 would be optimal in terms of translation-friendliness, leaving those codes unextracted as part of the skeleton, and keeping as inline elements the only codes that the translator must handle—the formatting tag for bold type:

If one uses existing tools that help achieve this out of the box (see the last paragraph in Sect. 8.3.1.2), achieving the optimal result shown in Fig. 8.3 does not require much effort.

The example above is not extreme. In fact, it has been the case that some XLIFF files contain segments so long that they do not fit in the translation editor and with more non-translatable codes than translatable text. Since these working conditions for the translators and reviewers highly increase the chances to make mistakes, the impact of good preparation on translation quality is obvious.

In a nutshell, a standard alone is not enough to guarantee the implementation of an efficient process. The examples above show that a valid, standard-compliant XLIFF with too many untranslatable codes and poor segmentation can be challenging to work with. Translation know-how and translation technology expertise are indispensable to handle the level of detail that the actual data standard does not cover.

8.3.2.1 Segmentation

When prepared for translation (e.g., when converted to XLIFF format or when imported directly into a translation tool), a text is divided, by default, into its structural units (i.e., paragraphs, text boxes, cells, etc.). However, those units can be large, containing several sentences. In the translation industry, the standard practice in most cases is to use the sentence as the processing unit, and that is the purpose of segmentation.

Segmentation is necessary for two reasons:

1. It makes it easier to leverage translation memories (i.e., to compare new content with previously translated content for the purpose of re-using parts that appear in different materials within the same project or have not changed between different cycles of a project).
2. It makes the translation task cognitively more manageable, helping the linguist focus on one piece of text at a time while keeping an eye on the whole document.

One does not try to chew a whole steak at once. Much rather, one uses utensils to cut the steak into small pieces, so that one can manage these easily, one piece at a time. Similarly, the scope of the translation is a complete document, but from the cognitive point of view, it is much more manageable to consider one small chunk of text at a time than the whole document. This might seem obvious, yet segmentation is often missing when an organisation prepares a text for translation in an ILSA project.

Using sentence segmentation is of great help for all parties involved in the translation process, but there is one more thing that needs to be considered: Segmentation must be consistent across the organisation (among members of a team who must prepare similar content for translation) and over time (within the lifecycle of a translation process or across different cycles of a project that involves the same kind of content). When different people apply segmentation single-handedly, they might take divergent decisions, and the result will be inconsistent segmentation and lack of leverage. The Segmentation Rules Exchange (SRX) standard can help achieve this consistency if all engineers concerned within the organisation(s) have access to the same SRX ruleset (for example, shared in a common repository of configuration files) and use it consistently for all language engineering tasks.

8.3.2.2 Inline Codes

Linguists should focus on the text, rather than worry about formatting issues and try to figure out unfathomable programming codes. A file for translation should include the translatable sentences extracted from the source document but not the structural codes (e.g., the skeleton) between sentences.

On the other hand, some words or expressions within sentences might have formatting (e.g., bold, italics, invisible markup that is meant to register mouse clicking actions, etc.), or sentences might embed some elements that are not translatable but must be replicated in the translations (e.g., buttons, images, symbols). Those extra-linguistic elements are encoded as inline codes and markup, and XLIFF provides a mechanism to abbreviate each code block in the form of less obtrusive tags. These tags are recognized and protected by CAT tools and allow translators to apply the original formatting in the translation without cluttering the view.

8.3.3 CAT Tools and Other Translation Technology

We have introduced the technology and best practices available to the engineers to prepare the files for translation. Let us take a closer look at the technology and best practices that are available to linguists to perform their task (See Zetzsche, 2018, for a compendium with reviews and inventories). For the purpose of this chapter, we provide the reader with (1) a working definition of a CAT tool; (2) an overview of functionalities that could and should be used when translating computer-based assessment instruments; and (3) illustrative examples of CAT tool implementation in ILSAs.

8.3.3.1 Definition of a CAT Tool

'Computer-assisted translation tool' is a broad term that can encompass any computer software used by human translators during the translation process to improve their working conditions and increase translation quality (Bowker, 2002). In this chapter, the term 'computer-assisted translation tool,' or CAT tool, will be understood as a computer environment which (1) supports the translation of different file formats ranging from simple text documents (.txt, .docx, etc.) to complex file formats (HTML, IDML, XLIFF); and (2) allows the user to use and create language assets (e.g., terminology databases, translation memories) and to perform spellchecks, concordance searches, and basic quality assurance checks.

One should be aware of the fundamental distinction between CAT tools and machine translation (MT): A CAT tool cannot operate without a human linguist who uses the technology available to translate, whereas in MT, the MT engine is responsible for producing the output (Bowker, 2002). The website of a free and open-source tool that meets our CAT tool definition, OmegaT, expresses this unambiguously: "It is a tool intended for professional translators. It does not translate for you!" (OmegaT, 2019).

At the time when this chapter was written, it was premature to include MT in CB-ILSA translation and adaptation workflows for a number of reasons: MT output varies significantly across language combinations and across domains; measures of translation quality applied to MT output do not include notions such as psychometric properties of items, level of difficulty of questions, attractiveness of distracting information, or suitable adaptation of authentic context. However, it is not unlikely that MT will play a predominant role in this rapidly evolving technological environment.

8.3.3.2 Functionalities of CAT Tools

Before going through examples of implementation of open source technology in ILSAs, both in centralised and decentralised workflows (see Sect. 8.2.5), let us first examine the main functionalities of CAT tools and their advantages for translation.

There is a set of functionalities that are considered a "must have" for every computer assisted translation tool. The most important one is supporting the translation process itself. This goes from being able to handle different file formats, such as standard-compliant XLIFFs through the appropriate filters to supporting segmentation, to being able to lock and protect content.

When linguists translate in a CAT tool, their translations are saved and stored in a translation memory (TM) for later retrieval. This saves time and effort for linguists and increases the consistency within a project and across project cycles. If a sentence occurs several times in the source document and the document is correctly segmented, the translator will only need to translate the first occurrence of that sentence. At the next occurrence, the translation of the same sentence will be automatically pulled from the TM. Translation memories can also be useful across cycles if, for example, a TM from the previous cycle of an ILSA is used as reference; even if the source sentence is not identical, the linguist will see suggestions from the TM and can also perform concordance searches in a TM on specific terms, structures, parts of phrases and find the results in their target language.

Terminology lookup is another important functionality of a CAT tool. Proficient CAT tools can import reference terminology, either as simple bilingual glossaries containing term pairs, or as more complex termbases with definitions, context, examples, and documented sources. Termbase results are displayed in the CAT tool interface, and most CAT tools feature predictive typing: The translator types the first letter of the target term in their language, and the tool suggests inserting the full term by pressing a key on the keyboard or performing a mouse click. Using termbases increases consistency and offers the possibility of performing adherence checks from inside the CAT tool. All instances of non-compliance with the termbase are flagged, and the linguist can go through them one by one to check whether the deviation from terminology is justified or an oversight.

Basic quality assurance checks can be performed from inside most CAT tools: completeness (to verify if the entire document was translated), spelling, spacing, numeral consistency (to check if the same numerals appear in the source and target segments), identical source and target segment (to flag if a target segment was accidentally left in the source language), inconsistent target (to check if a repeated source segment was translated in different ways), inconsistent source (to check if two different source segments have the same translation), etc. These checks improve the quality of the translated materials, and, if they are performed at different stages of the translation workflow, they ensure that the next person in the workflow can focus on linguistic and content issues rather than minor details.

We have already established that, by definition, CAT tools do not automatically translate for you. We shall not examine the topic of machine translation (MT) in this section. Let us nevertheless mention that a machine translation engine can be called up by means of an API from most CAT tools, but the human translator decides when and which engine to call. A short history and a classification of the different types of MT are available for professionals (Slator Language Industry Intelligence, 2018) and non-professionals (Bessenyei, 2017). Several approaches have been adopted by linguists calling up MT engines: allowing one MT engine to fully translate the document, and the human linguist later reviews it ("post-editing"); leveraging TM content before calling up the MT; or the method applied by Zetzsche, which lets the MT engine make a suggestion only if it is a good match with what the human translator has already started typing.

8.3.3.3 Putting CAT Tools to Work in Translation of CB-ILSA

When the power of CAT technology is harnessed, a broad range of non-trivial issues are caught and resolved on the fly without the need of additional resources. It has been demonstrated in a production environment that ILSAs can benefit from the advantages offered by CAT technology, which most professional translators are familiar with. The examples below show how open-source CAT tools were used in ILSAs for a centralised and a decentralised translation, respectively.

In 2013, the translation process of the Education & Skills Online assessment [OECD: Education & Skills Online] followed a centralised model, in which the contractors were responsible for the production of nine versions for seven participating countries. The language service provider commissioned for this task, cApStAn, worked with a basic open source XLIFF editor to translate the files following the double translation and reconciliation approach. All language experts received a training on how to use the basic XLIFF editor which, while it fell short of the definition of a CAT tool used in this section, still became a proof of concept for successful use of translation technology in a complex ILSA translation workflow (OECD, 2018).

In 2016, a group of PISA 2018 participating countries and economies agreed to use a different open source CAT tool, OmegaT, to produce their field trial version of data collection instruments. They received training and support from cApStAn, the PISA Contractor in charge of linguistic quality assurance, and the pilot was satisfactory: a mix of professional and non-professional players successfully edited their XLIFF files. As a result, the use of OmegaT has been extended to all participating countries and economies in OECD/PIAAC Cycle 2 and OECD/PISA 2022.

In both examples above, the source XLIFF files were not fully compliant with the XLIFF standard, and using an open-source tool was, in fact, the only possibility because it allowed some customization to accept non-compliant XLIFF files.

8.4 The People Challenge: Professionalization and Cooperation

8.4.1 *Profile of Professional Users (and How to Recruit Them to Translate CB-ILSA)*

As discussed earlier in this chapter (see Sect. 8.2.5), there may be good reasons to use professional translators if one wishes to make good use of translation technology. Whether the centralised or the decentralised approach is preferred, who should be commissioned to use the tools and produce the translation? It is a deep-rooted tradition in ILSAs to have domain experts actively contribute to translating the assessments into their mother tongue, even though they do not have any training in translation. Their input is precious. For example, it is desirable to have a mathematics subject matter expert ascertain that the translation of a mathematics assessment will be understood by respondents in the target language the same way the English source version is understood by English-speaking respondents. However, it is a risk to replace translation expertise with subject matter expertise, e.g., to entrust the translation to the domain expert. It is often thought that translation is an ancillary task and that anyone who has some knowledge of the source language (often English) is fit to perform that task. This is one of the myths denounced by Hambleton et al. (2005).

In the context of PB-ILSAs, the combination of skills required did not explicitly include mastery of translation technology, although a majority of professionals already used CAT tools since the late nineties. In the context of CB-ILSAs and the automation of the translation process, a different set of skills is necessary in addition to linguistic skills and subject matter expertise. If the translator has not received advanced training in translation and is not aware of the difficulties that loom ahead, tools that are designed to help may elicit a counter-productive trial and error process, which could affect the quality of the translated version of the assessment. For example, sentence-level segmentation may result in 'chopped' sentences when a non-professional prudently adheres to the syntactic structure of the source version and ends each target sentence where the source segment ends.

How can suitable professional translators be identified when any bilingual person can decide to become a freelance translator? There are a few things one should keep in mind when going through applications or browsing the profiles directories of language service providers. First of all, according to UNESCO's recommendations , translators should translate into their native language variant (UNESCO, 1976). Education should be part of the selection criteria, keeping in mind that formal translation training is not available for all language combinations. However, just like a law degree does not guarantee a good law practitioner, a translation degree does not guarantee that its holder is a good translator: A difference should be made between the study of translation and the practical skill of translation (García de Toro, 2008). There are translators for any type of content and their experience and specialization should be in line with your needs. If one needs to translate a mathematics assessment, it is better to choose a translator who has studied mathematics and has experience in test item translation than a literary translator.

Since using the appropriate technology is key to a quality translation of CB-ILSAs, technical skills are also important. The professional translator should have experience with CAT tools.

The most reliable selection criterion is assessing the quality of the translator's work. It is good practice to devise a short test for the translator, with content similar to that of the assessment, prepared using the appropriate tool and adequate resources. The translator should receive instructions similar to what he or she will be confronted with during the translation process. The translation should then be evaluated by a senior linguist and, if possible, by a bilingual domain expert, who will be able to determine if the work meets the most important quality standard for ILSAs: comparability.

Working with professional translators does not reduce the importance of domain experts and specialists in the translation process. They can, for example, intervene once the translation is done and act as reviewers, suggesting changes and implementing them before the materials go through a final proofreading.

8.4.2 Cooperation Between Platform Engineers and Translation Technologists

The configuration between platform engineers and translation technologists may vary from one ILSA to another: In a centralised approach, the translation technologist may prepare translation files for linguists that belong to the same organisation, while in a decentralised approach, the translation technologist (or, occasionally, the platform engineer) may have to prepare files for a variety of players from different organisations. The translation technologist's role may be limited to providing linguistic and technical advice to platform engineers who are responsible for preparing the source content for translation into all the target languages. Technical issues may be related to encoding of Unicode characters, CAT tool configuration, preparation of XLIFF or TMX files, etc.

In recent years, multilateral or bilateral preliminary meetings were organised, involving translation technologists and platform engineers from organisations responsible for managing ILSA content. Some have been more fruitful than others, and among the key factors contributing to success we could list (1) task distribution; (2) engineers' flexibility and capacity to onboard suggestions; and (3) follow-up after the preliminary discussions.

An example of a success story involved one face-to-face meeting with the translation technologists and the platform engineers, which paved the way for an efficient translation workflow. Several work setups were discussed until consensus was reached: The translation technologists agreed to take care of XLIFF file preparation and translation kit creation; the platform engineers produced XML exports with Internationalization Tag Set (ITS) data categories (that would allow to filter content and produce different target versions from one single master version that accommodates local adaptations that apply to one version but not to the others). Follow-up was done by e-mail and calls afterwards to conduct some testing and better define the initial requirements and constraints.

Suggestions from both sides were understood and accepted, decisions were made quickly from the start, even though some testing and negotiation was necessary along the process, and the chosen distribution of tasks was beneficial for the project and for all parties: Engineers were happy to be discharged of the intricacies of fine-tuning XML file filters, and translators received optimal translation kits that they could handle as expected. Each group of players did what they knew how to do best, the project ran smoothly, and the final result was very satisfactory.

In some of the major ILSAs, there is an ongoing cooperation between platform engineers and translation technologists: Fruitful meetings are held, and continuous follow-up happens all along the project, even across project cycles. However, it remains difficult to achieve consensus and systematically apply good practices. Certain requirements that are basic and standard practice for translation specialists seem to be challenging for platform engineers (e.g., converting formatting codes as protected inline tags) and, in some cases, implementation of a basic requirement, such as segmentation, is not regarded as a priority.

Reasons for difficulties in implementing best practices in file preparation might include tight timelines, but we could also mention a lack of awareness of the real impact that these best practices have on the ability of translators to do their work and, eventually, on the quality of the ILSA project (i.e., comparability of results). Platform engineers are not involved in translation tasks, so perhaps it should be envisaged to organise workshops to acquaint them with the types of technical problems translators routinely confront. There is a considerable amount of work involved in re-thinking testing platforms and content management systems to integrate translation data exchange standards. However, the workload can be contained by using existing libraries that have already solved the same problem years before rather than writing utilities from scratch (see the last paragraph in Sect. 8.3.1.2).

8.4.3 Academic Exchanges (Research Institutes and Field Practitioners)

To the best of our knowledge, there are few formal opportunities for CB-ILSA researchers to meet translation experts and practitioners and to be exposed to ideas coming from them. One notable exception was the OECD-GESIS Seminar on Translating and Adapting Instruments in Large Scale Assessments, held 7–8 June 2018, where cApStAn was invited to present some basic ideas about translation best practices and some examples of the negative impact that poor file preparation has on the working conditions of translators and the quality of their product (Lupsa, 2018).

Based on the experience of multidisciplinary workshops and seminars, we plead that more initiatives of this kind are necessary. They can have a very positive impact on the evolution of paradigms of CB-ILSA translation. In this kind of encounter, survey researchers and testing organisations gain insight into the vast realm of possibilities that technology offers to improve translation quality and comparability across languages and cultures.

8.5 Discussion and Outlook

The transition from PBA to CBA mode for ILSAs has occurred at a time when new advances in survey methodology have promoted translation of assessment instruments to a research field in its own right. Good practices were described in the literature and, rather logically, testing organisations and contractors tasked with administering ILSAs sought to replicate these good practices in the technology-rich environment of CB-ILSA platforms. Translation technology evolved in parallel, in the translation industry rather than in the testing industry. Specialized computer-assisted translation tools that help translators produce consistent, high-quality translations have become the norm, but they are mostly difficult to integrate with all-in-one assessment delivery platforms in which translation is only one step in a complex workflow. Multi-purpose environments that aim to attend to many aspects of the survey development are prone to fall short when it comes to supporting the features and functionality professional translators routinely rely on.

A most fruitful experience can be drawn from the interaction and symbiosis of two types of tools, such as a CB-ILSA platform and a free and open-source CAT tool, where the former exports and imports files that can be translated in the latter. In other words, it is recommended to use or write an interface that makes it possible to export content for translation and re-import the final translated and validated content in XLIFF format or in a format that can be easily prepared as XLIFF (e.g., XML, QTI, etc.). This content can, thus, be handled in mature translation tools that already exist and have been evolving for decades. It is in those tools that new developments in machine translation are tested in experimental conditions, with appropriate quality assurance checks built in.

In this context, another factor comes into play: professionalisation of translation in CB-ILSA. As discussed in Sect. 8.4.1, the tradition of commissioning subject matter experts to play an eminent role in the actual translation is entrenched in ILSA culture, and it has led to unnecessary variability in translation quality. While subject matter experts are undoubtedly required to determine appropriateness of the language for the target audience, content-related accuracy, relevance, and equivalence in the level of difficulty, it is hardly their profession to craft a suitably fluent translation that also remains semantically equivalent to the source. Decentralised translation models have given participating entities great latitude in selecting the players working on the local versions of the assessment instruments. This brings a varying mix of professional and non-professional players into the arena, which in turn restricts the use of state-of-the-art translation technology. If the power of the mature tools that have evolved for decades and continuously evolve today is to be harnessed, professionalisation is inevitable.

A centralised approach to translation, whereby professional linguists produce a version that is then reviewed by the participating entities, including subject matter experts, should be considered as a viable alternative to decentralised, multidisciplinary TRAPD approaches that are often incompatible with effective and efficient use of technology. Multidisciplinary review and adjudication meetings can still be embedded in the workflows, and participating entities would retain ownership of their target version. However, the starting point would be a higher overall standard due to supervised use of translation technology to its full extent.

References

Bessenyei, G. (2017). Neural machine translation: The rising star. *memsource*. https://www.memsource.com/blog/neural-machine-translation-the-rising-star/

Bowker, L. (2002). *Computer-aided translation technology: A practical introduction*. University of Ottawa Press. https://doi.org/10.5007/7005

Comerford, T., Flip, D., Raya, R., & Savourel, Y. (2014). *XLIFF version 2.0. OASIS standard* [online]. Available at: https://docs.oasis-open.org/xliff/xliff-core/v2.0/xliff-core-v2.0.html

Dept, S. (2013). *Translatability assessment of draft questionnaire items*. Paper presented at the 5th Conference of the European Survey Research Association (ESRA), Ljubljana, 2013. https://www.europeansurveyresearch.org/conf/uploads/93/403/56/130716_ESRA2013_Dept_TranslatabilityAssessment.pdf?

Dept, S., Ferrari, A., & Wäyrynen, L. (2010). Developments in translation verification procedures in three multilingual assessments: A plea for an integrated translation and adaptation monitoring tool. In J. A. Harkness, M. Braun, B. Edwards, T. P. Johnson, L. Lyberg, P. P. Mohler, B.-E. Pennell, & T. W. Smith (Eds.), *Survey methods in multinational, multiregional and multicultural contexts* (pp. 157–173). Wiley. https://doi.org/10.1002/9780470609927

García de Toro, C. (2008). *Translation studies: An overview*. Cadernos de Tradução. https://doi.org/10.5007/2175-7968.2019v39n3p291

Grisay, A. (2002). Translation and cultural appropriateness of the test and survey material. In R. Adams & M. Wu (Eds.), *PISA 2000 technical report* (pp. 57–70). Organisation for Economic Co-operation and Development.

Hambleton, R. K., Merenda, P. F., & Spielberger, C. D. (2005). *Adapting educational and psychological tests for cross-cultural assessment*. Lawrence Erlbaum.

Hambleton, R. K., & Zenisky, A. L. (2011). Translating and adapting tests for cross-cultural assessments. In D. Matsumoto & F. J. R. van de Vijver (Eds.), *Cross-cultural research methods in psychology* (pp. 46–74). Cambridge University Press. https://doi.org/10.1017/CBO9780511779381.004

Harkness, J. A. (2003). Questionnaire translation. In J. A. Harkness, F. J. R. van de Vijver, & P. P. Mohler (Eds.), *Cross-cultural survey methods* (pp. 33–56). Wiley.

Harkness, J. A. (2007). Improving the comparability of translations. In R. Jowell, C. Roberts, R. Fitzgerald, & G. Eva (Eds.), *Measuring attitudes cross-nationally: Lessons from the European social survey* (pp. 79–93). Sage.

Harkness, J. A., & Behr, D. (2008). *How instrument quality affects translation: Insights from a film of a team translation review*. International Sociological Association Research Committee 33.

Harkness, J. A., Pennell, B.-E., & Schoua-Glusberg, A. (2004). Survey questionnaire translation and assessment. In S. Presser, J. Rothgeb, M. Couper, J. Lessler, E. Martin, & E. Singer (Eds.), *Methods for testing and evaluating survey questionnaires*. Wiley. https://doi.org/10.1002/0471654728

Harkness, J. A., van de Vijver, F. J. R., & Mohler, P. (2003). *Cross-cultural survey methods*. Wiley-Interscience/Wiley.

Harkness, J. A., Villar, A., & Edwards, B. (2010). Translation, adaptation, and design. In J. A. Harkness, M. Braun, B. Edwards, T. P. Johnson, L. Lyberg, P. P. Mohler, B.-E. Pennell, & T. W. Smith (Eds.), *Survey methods in multinational, multiregional and multicultural contexts* (pp. 117–140). Wiley.

Iliescu, D. (2017). *Adapting tests in linguistic and cultural situations*. Cambridge University Press. https://doi.org/10.1017/9781316273203

Joldersma, K. J. (2004). *Cross-linguistic instrument comparability*. Michigan State University: Unpublished manuscript.

Lietz, P., Cresswell, J., Rust, K., & Adams, R. (Eds.). (2017). *Implementation of large-scale education assessments*. Wiley.

Lupsa, D. (2018). Translation technology in ILSAs: Striving to combine apps, discernment and capacity-building. In *OECD-GESIS seminar on translating and adapting instruments in large scale assessments, Paris, 7–8 June 2018*. https://www.oecd.org/skills/piaac/events/Lupsa_for%20website_180606_OECD-GESIS_Lupsa_Translation_technology_in_ILSAs_180606.pdf

Melby A. & Zetzsche, J. (2010). *Data standards: Can they help us and can we help them?* In The ATA Chronicle, October 2010. www.internationalwriters.com/toolkit/10_October_Standards.pdf

Morado Vázquez, L., & Torres del Rey, J. (2015). Teaching XLIFF to translators and localisers. Localization focus. *The International Journal of Localisation, 14*(1). https://doi.org/10.1075/jial.19002.tor

OASIS. (2018). *Darwin information typing architecture (DITA) version 1.3*. http://docs.oasis-open.org/dita/dita/v1.3/dita-v1.3-part0-overview.html

OECD. (2017). *PISA 2015 results (volume V): Collaborative problem solving*. https://doi.org/10.1787/9789264285521-en

OECD. (2018). *Education & skills online technical documentation*. http://www.oecd.org/skills/ESonlineassessment/assessmentdesign/technicaldocumentation/

Okapi Framework. (2016a). *Okapi XLIFF Toolkit for XLIFF 2.0* [Computer software]. https://bitbucket.org/okapiframework/xliff-toolkit/wiki/Home

Okapi Framework. (2016b). *Okapi Rainbow for XLIFF 1.2* [Computer software]. https://okapiframework.org/wiki/index.php?title=Rainbow

OmegaT. (2019). OmegaT the free translation memory tool. https://omegat.org/

Raya, R. (2021a). *OpenXLIFF Filters* [Computer software]. Github. https://github.com/rmraya/XLIFFManager

Raya, R. (2021b). *XLIFF Manager* [Computer software]. Github. https://github.com/rmraya/OpenXLIFF

Raya, R., Schnabel, B., & Hackos, J. (2013). *DITA 1.2 feature article: Using XLIFF to translate DITA projects*. OASIS DITA Adoption Technical Committee. https://www.oasis-open.org/committees/download.php/48340/DITA12XLIFFArticle.pdf

Roturier, J. (2019). XML for translation technology. In M. O'Hagan (Ed.), *The Routledge handbook of translation and technology*. Routledge.

Sasaki, F., Lieske C., & Rahtz, S. (2006). Internationalization and localization of XML: Introducing ITS. In *XTech 2006, Amsterdam*.

Savourel, Y. (2001). *XML internationalization and localization*. Sams.

Savourel, Y. (2014, *June*). An introduction to XLIFF 2.0. *Multilingual*. Available at https://www.multilingual.com/article/201406-42.pdf

Savourel, Y., Reid, J., Jewtushenko, T., & Raya, R. M. (2008). *XLIFF version 1.2. OASIS standard* [online]. http://docs.oasis-open.org/xliff/v1.2/os/xliff-core.html

Schnabel, B., Hackos, J. T., & Raya, R. (2015). *A practical guide to XLIFF 2.0*. XML Press.
Segmentation Rules eXchange. (2021). Wikipedia. https://en.wikipedia.org/wiki/Segmentation_Rules_eXchange
Slator Language Industry Intelligence. (2018). Slator 2018 neural machine translation report.
Translation Memory eXchange. (2021). Wikipedia. https://en.wikipedia.org/wiki/Translation_Memory_eXchange
UNESCO. (1976, November 22). *Recommendation on the legal protection of translators and translations and the practical means to improve the status of translators*. UNESCO, Legal Instruments. http://portal.unesco.org/en/ev.php-URL_ID=13089&URL_DO=DO_TOPIC&URL_SECTION=201.html
Upsing, B., Gissler, G., Goldhammer, F., Rölke, H., & Ferrari, A. (2011). Localisation in international large-scale assessments of competencies: Challenges and solutions. Localisation focus. *The International Journal of Localisation, 10*(1), 44–58. https://www.localisation.ie/resources/publications/2011/166
Van de Vijver, F. J. R. (2003). Bias and equivalence: Cross-cultural perspectives. In J. A. Harkness, F. J. R. van de Vijver, & P. P. Mohler (Eds.), *Cross-cultural survey methods* (pp. 143–155). Wiley.
Zetzsche, J. (2018). *The translator's tool box: A computer primer for translators, version 13.5*. International Writers' Group.

Chapter 9
Psychometric Forensics to Maintain and Improve Data Quality in Large-Scale Assessments

Kentaro Yamamoto, Usama S. Ali, Frederic Robin, Lokesh Kapur, and Mathew M. Kandathil

Abstract Data quality is of utmost importance in large-scale assessments, as it directly impacts the reliability, validity, and comparability of survey outcomes. It is the role of psychometric data forensics to identify poor-quality or invalid data, enabling the detection of anomalies and potential error sources. Such forensics include harnessing process data to extract meaningful insights into test-taker behavior, leveraging cumulative databases to compare data congruency across multiple reference points, automating procedures to expedite anomaly detection and optimize resources, and collecting information about testing conditions and environmental contexts to better illuminate observable patterns in the data. Using examples from the OECD's Programme for the International Assessment of Adult Competencies (PIAAC) and the Programme for International Student Assessment (PISA), this paper outlines these various data quality procedures and monitoring approaches to safeguard data integrity throughout the assessment lifecycle, while recognizing the dynamic and time-sensitive nature of data validation.

Data quality that represents accuracy, completeness, and integrity is critical for any skill assessment including the large-scale assessment. Without it, the interpretations and conclusions drawn from the collected data are undermined. Ensuring data

---

K. Yamamoto
Independent Consultant for Psychometric Methodology and Large-Scale Assessment (formerly ETS), Princeton, NJ, USA

U. S. Ali (✉)
ETS, Princeton, NJ, USA

South Valley University, Qena, Egypt
e-mail: uali@ets.org

F. Robin · L. Kapur · M. M. Kandathil
ETS, Princeton, NJ, USA

L. Khorramdel et al. (eds.), *Innovative Digital-Based International Large-Scale Assessments*, Methodology of Educational Measurement and Assessment,
https://doi.org/10.1007/978-3-031-90951-1_9

quality in the context of a large-scale assessment is a complex effort, as these surveys involve detailed and interconnected processes and procedures carried out for multiple populations and by multiple agencies, including assessment design, instrument development, sampling, data collection, data transfer, item scoring and coding, data analysis, scaling, and reporting. Fortunately, the move to computer administration has facilitated this effort, in that the type and volume of information collected has expanded. This additional information makes it possible to improve quality control procedures with the goal of increasing the reliability and validity of surveys.

Data quality validation is a dynamic, multi-phase, and comprehensive process that requires implementing policies and procedures across all components of assessment delivery and data collection, as every component is interconnected, and each step contributes uniquely to overall quality. Any errors introduced at any step can have consequences for all subsequent steps. This process should be proactive in collecting evidence verifying that each step has been accurately implemented and certain aspects are maintained and carried over to merit data integrity. Data forensics in the psychometric context will play this role and employs statistical and non-statistical techniques to maintain the level of quality in the assessment and to detect any issues that would allow for timely action to be taken to help improve the quality. Searching for psychometric evidence to support the integrity of data and any subsequent psychometric analyses and inferences is referred to as *psychometric data forensics* or *psychometric forensics*. For international surveys, where multiple agents and stakeholders each with unique strengths, weaknesses and sometimes contextual limitations across numerous countries are involved in data collection and data processing, establishing processes that focus on the reduction and detection of errors is essential to ensure comparable, valid, and reliable measurement. Maintaining and improving quality in large-scale assessments begins with producing and communicating standards and guidelines. The importance and utility of such documents is recognized in the Standards for Educational and Psychological Testing (2014), published collaboratively by the American Educational Research Association (AERA), American Psychological Association (APA), and National Council on Measurement in Education (NCME), as well as in the quality control guidelines issued by the International Test Commission (ITC, 2014).

Due to the need for coordination among different agents and stakeholders, but on a slightly different scope and intention, the Organisation for Economic Co-Operation and Development (OECD) has established technical standards and guidelines to ensure that the design and implementation processes of its surveys yield high-quality and internationally comparable data (see OECD, 2014, 2018a, 2018b). These standards and guidelines continue to develop as we accumulate experiences in surveys conducted with features and in new contexts to help ensure that the sources of systematic survey variability are kept to a minimum and that the collected and reported assessment data are reliable, valid, and comparable across countries and over time. The standards include a quality assurance and quality control program for survey operations that involves collecting a range of information via written reports, telephone conferences, and in-person meetings. For more detailed descriptions of the quality assurance and quality control procedures implemented in more recent developments of PISA and PIAAC, see, for example, *Cycle 2 PIAAC Technical Standards and Guidelines* (OECD, 2018a) and *PISA 2021 Technical Standards* (OECD, 2018b).

Seeking quality in assessments leads to two types of procedures: (a) quality monitoring procedures for survey operations, and (b) data quality procedures conducted during data processing. This chapter describes the key data quality procedures implemented in large-scale assessments that are critical to the accurate use of assessment data to inform organizations and individuals responsible for the design and delivery of such surveys. The focus is on psychometric forensics techniques that can lead to detection of poor quality or invalid data in computer-based large-scale assessments. The primary emphasis of this chapter is on evaluating data quality as part of the data processing and analysis phase. Techniques used to evaluate data quality in previous cycles of the OECD's Programme for the International Assessment of Adult Competencies (PIAAC) and Programme for International Student Assessment (PISA) are included as examples throughout the chapter.

9.1 Data Quality in Large-Scale Assessments

Under data quality frameworks, numerous potential sources of error exist, as well as opportunities for data falsifications during data collection that may lead to problems with assessment data validity. The sources of error may be related or specific to the nature of the assessment, the contributors to the data collection process, or the stakeholders. For example, item preknowledge and other forms of cheating would be among the main threats to data quality in a high-stakes college admission test (e.g., Haberman & Lee, 2017) and both test takers and institutions receiving the scores would be impacted.

International large-scale assessments produce group scores and, therefore, are low stakes for individual respondents who will not receive a personal score; however, these assessments are commonly considered high-stakes for the country administering the assessment (and receiving the group score) and may be high-stakes for other groups, such as for policymakers and researchers who use the data and for interviewers and contractors who contribute to the assessment process. Under these conditions, quality control procedures need to focus on many potential sources of error and resulting data fabrication issues. To be able to properly detect fabricated data and identify their source, one must understand the specific nature of the assessment. For example, PIAAC is a survey of adults in which interviewers are paid to administer the assessment in respondents' homes. Given this financial incentive, it is imperative to check that responses (and respondents) have not been fabricated by the interviewer (Yamamoto & Lennon, 2018). In this case, monitoring tools such as data dashboards and post hoc analyses of response patterns can be used to detect such problems and better verify the quality of the collected data.

9.1.1 Data Quality as an Outcome of a Multi-Faceted Process

Statistical techniques along with other problem-solving tools can provide the technical basis for quality control and improvement throughout data processing. The OECD quality framework has four elements (OECD, 2011, p. 6):

- A definition of quality and its dimensions
- A procedure for assuring the quality of proposed new statistical activities
- A procedure for regularly evaluating the quality of existing statistical activities
- A set of broad principles and quality guidelines covering all phases of the statistical production process

There are many steps focused on corresponding aspects of data quality that occur even before data collection. Plans for monitoring data collection should be developed during the survey design phase, and their implementation can include many aspects, such as the validation of interviews (a portion), monitoring of data accumulation, and confirming the location and time of the data collection. Such features allow for information about data collection to be captured for every respondent. Coming from the idea of data forensics, additional procedures allow assessment developers to keep track of every piece of data, including processes, transmission, and possession. Producing daily outputs of the collected data is an example of one quality procedure to monitor the progression of data collection and make it possible to later check for any aberrant structures in the data. In large-scale assessments, quality control cannot be treated as a separate process from the design nor an afterthought procedure that can be implemented independently. It is important to emphasize that data quality cannot fully be assured if quality procedures are applied only after the data are collected. All data-handling participants need to share the same goal and be aware of what should be monitored throughout the data collection process. Specialists in large-scale assessments should try their best to construct and perform data quality control actions before, during, and after data collection efforts for high-quality data and valid inferences.

Following the sampling standards and guidelines, each participating country should carry out a data verification in the field by asking the interviewees to verify that interviews did take place, with whom, and when. In one example case following these procedures, the national team of one PIAAC participating country found that of 1961 interviews selected for verification, 239 (about 12%) were falsified, and that 70 different interviewers were responsible for at least one of those falsified interviews. These falsified cases in addition to others were deleted from the final sample after a thorough investigation (OECD, 2019). This is an example of applying a data quality feature by the countries that revealed an issue in the assessment data and the decision taken regarding such invalid data.

9.1.2 *Data Quality Attributes*

It is important to define the dimensions of quality. Following data collection, and during the data processing and analysis phases of a large-scale assessment, psychometric data forensics should be applied to evaluate data quality. Coherent data structure representing the design specifications is essential evidence to validate and comes first. Verification of the completeness of the data is another essential step to check if quantity and pattern of missingness is acceptable and within the norms. The data quality checks also include examinations of response rates, item completion rates, and functioning of the delivery platform in terms of workflow, random assignment, data capture, and timing information.

The following sections focus on five dimensions of data quality that can be checked as part of post-data collection procedures. These include:

- The structure of data reflects the assessment design
- The completeness of data (e.g., data quantity and the pattern of missing data)
- The reliability of scoring
- The credibility of responses based on timing data
- The consistency and measurement invariance or the conformance of item parameters over time

Each section can then explain (1) how the researchers can determine whether the data meet the requirement stated in the section title and (2) what they can do if it does not. These explanations are associated with empirical real-data examples that provide instances of data that meet the quality standards and in other cases show a deficiency or an issue in the data. The examples are taken from the data analysis employed in PIAAC cycle 1 and PISA previous cycles.

9.1.2.1 Structure of Data Reflects the Assessment Design

Before data are used for any analyses, the data should meet the requirements stated in the bulleted list below that reflect the assessment design features. Assessment design features may cover everything from sampling design to item delivery. The design aspects that need to be evaluated may include whether or not the collected data reflect:

- The targeted number of sampled schools, households, and/or test takers
- The specified number of respondents across test modes, where applicable
- The specified number of respondents per booklet, unit, and item
- The specified number of items per respondent

First, the collected data should reflect the sampling design. This may include reviewing that the data includes responses from the expected number and proportion of sampled target groups (i.e., schools, regions, demographics) as the sampling protocol intended. For example, countries participating in PISA are required to sample a minimum of 150 schools to obtain a representative sample of their national population of 15-year-old students, with a targeted a total sample size of 6300 students (OECD, 2020). Assessment delivery can then be reviewed. If the assessment is offered in different modes of administration (e.g., paper and computer), an expected percentage of test takers will be assigned to each assessment mode according to the design, and the data should reflect this expected distribution. In the context of computer-based large-scale assessments, the assessment platform should reliably assign a pre-defined form to each sampled study participant. Additionally, if the distribution of booklets or forms is expected to be uniform or adhere to a designed distribution across respondents, the data should reflect this as well. If the assessment is adaptive (i.e., item-level or multistage adaptive), it is important to verify that each item or item set is exposed to a reasonable distribution of respondent proficiencies. This is to ensure valid estimation of the item characteristics, such as estimated item parameters in a given population. Further, it is also important to verify that each respondent was administered their assigned set of items according to the assessment design. With computer-based assessment (CBA), this may mean checking the data for aberrant breaks in individual response patterns or exceptionally long response times, indicative of technical difficulties. Ultimately, the statistical equivalence of the samples collected in assessment needs to be established by verifying that the intended random assignment of respondents to the necessary assessment components was successful. Meeting the specified sample sizes or their ratios is a sign of a successful random assignment. If the assessment design is not implemented as planned (e.g., inaccurate assignment), subsequent analyses may not be as useful in determining, for example, the equivalency of scales across the different modes of administration. In general, how well the data collection follows the assessment design and sampling strategy is an important indicator of the quality.

Example 1 (Evaluating the Structure of Data) Figure 9.1 is a flowchart representing possible paths for a respondent taking the PIAAC assessment, and the percentages shown are percentages of all participants. The purpose is to show the actual percentages in the data, to be compared with the percentages that were anticipated by the assessment design. In PIAAC, each respondent is expected to respond first to a background questionnaire (BQ) and, given their responses, are then routed to the cognitive assessment administered on either computer (laptop or tablet) or paper. Such validation should take place at the level of each group of interest, as well as at the aggregate level. For PIAAC, this quality control procedure is conducted both within and across participating countries, with Fig. 9.1 showing aggregated results from 22 countries. While PIAAC was designed primarily to be delivered on a computer, a paper-based assessment (PBA) was included for those who were either unable or unwilling to use a computer. After accounting for small percentages of respondents who did not complete the BQ, the cognitive assessment, or both, the routing to PBA and CBA was assessed. It was anticipated that some respondents with no computer experience, along with those refusing to take the test on the computer, would be routed to the paper assessment. All other respondents

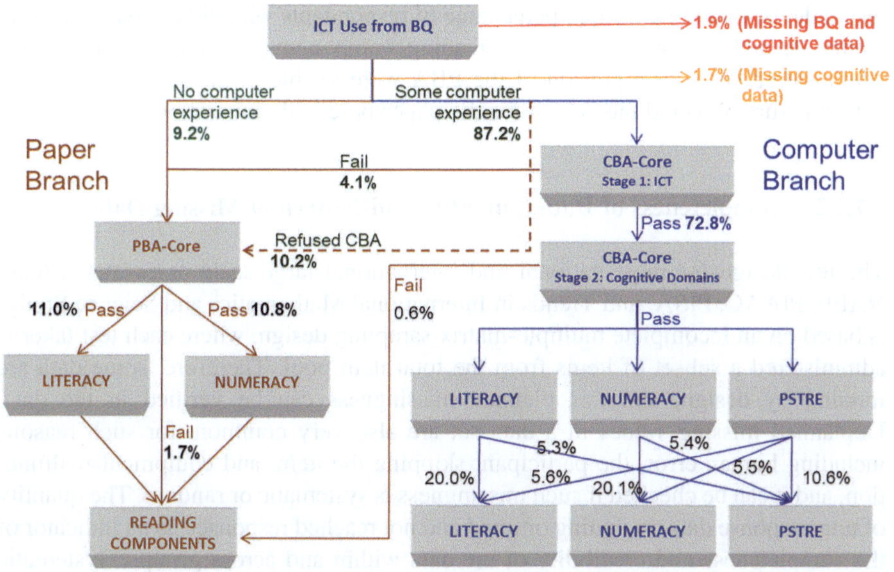

Fig. 9.1 PIAAC assessment design and international routing data. *Note. ICT* information and communications technology; *CBA* computer-based assessment; *PBA* paper-based assessment; *PSTRE* problem solving in technology rich environments

would be routed to the computer assessment. At this point, 100% of respondent paths can be verified.

Respondents directed to the CBA were administered a two-stage core assessment, first of basic computer skills, then of basic literacy and numeracy skills. Poor performance on the first stage routed the participant to the PBA, and poor performance at the second stage routed them to a paper-based assessment of basic reading skills, reading components. By design, respondents who passed both CBA core sections were routed to one of three possible paths: 50% of respondents were expected to receive a combination of literacy and numeracy tasks, 33% were to receive PSTRE combined with either literacy or numeracy, and 17% were to receive only PSTRE. After accounting for those routed to PBA in Fig. 9.1, about 55% of the participants passing both core stages were assigned only literacy or numeracy assessments, 30% were assigned a mix of literacy or numeracy and PSTRE, and 15% were assigned only PSTRE. The structure of data as represented by the proportions of paths reflects accurate execution of the assessment design across countries.

For participants assigned the PBA, a core assessment of basic literacy and numeracy skills was administered first, followed by a more thorough assessment of one domain: either literacy or numeracy. Ideally, the literacy and numeracy assessments would be assigned in equal proportion. As shown, randomly equivalent samples were assigned to literacy and numeracy items representing 11.0 and 10.8% of the total sample of those taking the assessment, respectively. Finally, all respondents, including a small percentage of respondents who did not pass the core, would be administered a block of reading components. Again, all participants administered some or a portion of the PBA were visible in the routing paths; the data structure reflected the design for the paper-based administration.

9.1.2.2 Completeness of Data (Quantity and Pattern of Missing Data)

The test design for most national and international large-scale assessments (e.g., NAEP, PIAAC, PISA, and Trends in International Mathematics and Science Study) is based on an incomplete multiple-matrix sampling design, where each test taker is administered a subset of items from the total item pool. Therefore, some data are missing by design, and this planned missingness can be verified in the data. Unplanned missing values in a data set are also very common, for such reasons including human error, the participant skipping the item, and equipment malfunction, and it can be checked if such missingness is systematic or random. The quantity of non-response data, including omitted and not reached responses, is an indicator of the completeness and credibility of the data within and across groups. Systematic missingness, such as a higher portion of missing data for a specific item may indicate issues in item translation in a given country or a system delivery issue. It is also important to review the frequencies of scores (e.g., proportion of correct or incorrect responses for dichotomous items) for every item to make sure that all item score categories are present within each assessed group. Missing one or more score categories, especially for polytomously scored items (e.g., items with a partial credit score), would impact the item parameter estimation and item parameter congruence, described later.

At the respondent level, the amount of missing data beyond what is expected in the design is a source of error that affects the overall quality of data. The respondent's missing data can be in the form of omitted responses (e.g., item skips) or early termination of the survey. Further, certain amounts or patterns of missing data between respondents can be associated with data collection issues, including the fabrication of data, particularly when similar patterns are seen originating from the same interviewer or subgroup sample (e.g., a particular school or region). For example, the pattern of booklets assigned can be a valuable insight. For example, under the PIAAC CBA design, each respondent is administered three booklets, one core booklet (includes items from two domains) and two booklets in the second stage (one for each domain); the completeness pattern for these booklets, with "1" indicating a completed booklet and "0" indicating a not-completed booklet would be: 111. Any pattern of incompleteness, such as 101 (i.e., missing the second booklet), can be suspect. This incompleteness of data raises a red flag that the observed data may be of poor quality. The proportion of respondents routed to each domain can also have an impact on data missingness. As in both the PIAAC PBA and CBA, each respondent has a specific probability of being assigned to a specific domain (literacy, numeracy, and, for CBA, PSTRE) upon completion of the core. It is expected that the core section will have a complete data set within each

Table 9.1 Data available for analysis and path of assessment (dataset 1)

Index	Country ID					Total
	1	2	3	4	5	
Total records	6104	5702	5034	6306	7289	32,498
% Completed BQ	99.9	99.5	99.3	99.6	99.5	99.5
% PBA	25.3	49.9	27.1	52.9	50.3	50.8
% CBA	74.2	48.4	72.0	42.3	44.4	44.8
% Reading components	26.9	52.6	27.9	54.3	53.5	52.6
% Completed BQ & cognitive	99.4	98.3	99.1	95.2	94.6	95.6

PBA and CBA samples, while the subsequent domains will have an incomplete data matrix. If the distribution of the data across the three domains is not consistent with the expected proportion of respondents routed to each domain, there will be a lot of missing data for one domain in comparison to the other domains. Such missingness cannot support subsequent analyses for reaching valid inferences, such as the item parameter estimation for those with low response rates (i.e., high omit rates). In order to check for completeness of data, both the pattern and proportion of missingness need to be verified.

Example 2 (Completeness of Data per Country) The aggregate data across and within all PIAAC participating countries were evaluated to verify that it matched expectations based on the assessment design. Across countries (i.e., the international sample shown in Fig. 9.1), the overall level of missingness or incomplete data was low. The percentage of respondents who did not complete BQ and cognitive items was 1.9% and the percentage of the respondents with no data on cognitive items was 1.7%. Within countries, routing percentages for five example cases (Country 1–Country 5) are shown in Table 9.1. In general, available data indicated that the great majority of participants responded to the BQ and core cognitive items in at least the minimum volume expected for each country (i.e., completeness of data per country ranged from 95.2 to 99.4%, with an average of 95.6% across these five countries).

Example 3 (Variability in Number of Interviews by Interviewers) In PIAAC, interviewers are considered a potential source of error in data collection for large-scale assessments as the BQ is administered by the interviewer, and the cognitive assessment is taken by most respondents in the CBA format under the supervision of the interviewer. The number of interviewers employed by each country may result in large variability in the average number of interviews per interviewer among countries. The ratio of standard deviation over mean (i.e., coefficient of variation) provides comparable proportional variability, in which large variability may indicate anomalous data collection.

In Table 9.2, the variability in the number of interviews conducted by each interviewer within each country is presented for five countries. Within two countries, 2 and 5, the number of interviews across interviewers is nearly uniform. The coefficients of variation (i.e., ratio of standard deviation to the mean respondents

Table 9.2 Summary statistics of interviews per interviewer within each country for two datasets

Dataset 1				Dataset 2			
Country	N	Mean	CV	Country	N	Mean	CV
1	228	27.8	0.81	A	248	21.4	0.99
2	63	90.5	0.36	B	122	40.4	1.63
3	189	26.6	0.62	C	108	51.3	0.85
4	191	33.0	0.71	D	97	52.5	0.78
5	206	35.4	0.37	E	123	50.2	0.77
				F	111	49.2	0.81
				G	64	83.3	0.82
				H	297	17.5	0.27

Note. *N* number of interviewers; *CV* coefficient of variation

per interviewer) were much higher for the other three countries (1, 3, and 4 ranged between 0.62 and 0.81).

Applying the same analysis of evaluating the variability in the of number of interviews per interviewer is shown across eight participating PIAAC countries in Dataset of Table 9.2. In review of the coefficient of variation, the variability in the number of interviews conducted by each interviewer varies widely across countries. Variation is somewhat similar amongst most of countries except two: Country H (lowest variation) and Country B (highest variation). In Country H, the number of interviews across interviewers is nearly uniform, and the collected interviews were conducted by a large number of interviewers ($N = 297$). Alternatively, a large proportion of interviews in Country B were conducted by a relatively small proportion of interviewers (i.e., 10 interviewers were responsible for 40% of interviews, see Fig. 9.2). There is nothing unusual about having interviewers who are more productive than others but evaluating the data from another perspective is a further check of data credibility.

9.1.2.3 Accuracy and Reliability of Scoring

Scoring responses is an essential task with major consequences if it is inaccurately performed. Some assessments may employ an automated scoring algorithm to code (i.e., apply a value indicating the type of a given response) or score (i.e., apply a value indicating the correctness of response), such as the use of the longest common subsequence algorithm. This algorithm, while automated, emulates the leniency shown by human scorers in cases where underlining or highlighting responses would typically be evaluated. To make such an algorithm viable, machine scored items in international surveys need to adhere to certain guidelines such that they can be accurately scored. For example, in cycle 2 of PIAAC, countries are asked to carefully check that they do not split or combine sentences during translation for sentence item type, which would make it easier for selection by tapping and avoid any scoring errors. In other computer-based large-scale assessments, such as PISA,

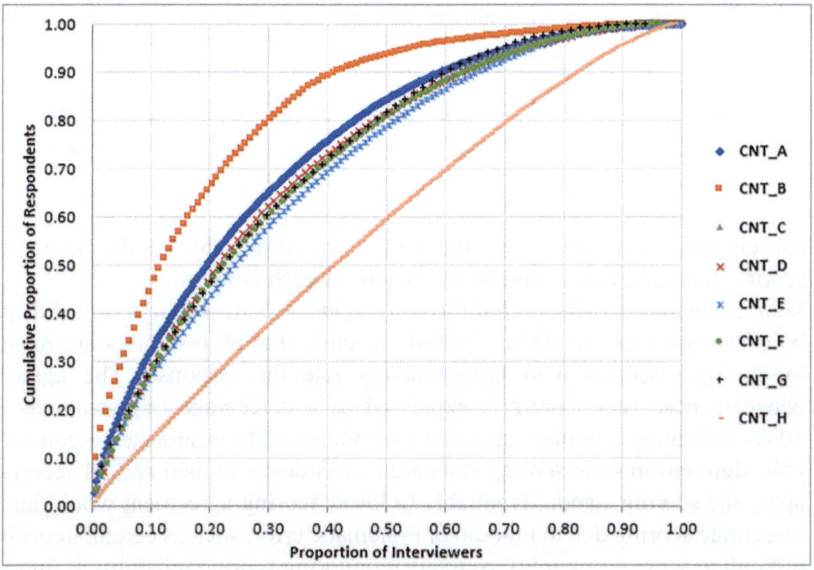

Fig. 9.2 Cumulative proportions of respondents by interviewers

machine scoring has not totally replaced human scoring. Humans recognize with ease if a respondent highlights or underlines the correct phrase even if they carelessly error omit one or two characters at the end of the line, at the beginning, or somewhere in the middle of the text (OECD, 2019). Regardless of the mechanism, obtaining an accurate score for a response is vital for valid inferences about participant ability and item properties.

Scoring reliability, such as the agreement between two scorers on human scored items, is often viewed as a measure of score validity both within and across countries in an international large-scale assessment. According to *Standard 11.3* of the *PIAAC Technical Standards and Guidelines* (OECD, 2018a, pp. 157–159) associated with scoring paper-and-pencil instruments, two quality check procedures are established:

(a) *Across-country scoring reliability—agreement with anchor booklets*[1]: While all responses are reviewed most of the time by two scorers within a country to ensure that all responses within a country are consistently scored, *anchor booklets*, a set of responses in English seen by multiple scorers across countries, ensures that each national scoring team applies the scoring standards consistently. Anchor booklets in English are scored by two bilingual scorers in each country to verify that scoring rules are applied in the same way across all

[1] This common set of booklets was prepared by test developers and distributed to all countries. Item responses in these booklets were based on actual responses collected in the field, as well as responses that reflected key points on which scorers were trained. Because responses were provided in English, scoring teams in each country designated two bilingual scorers responsible for the double-scoring process.

Table 9.3 Percentage of agreement with anchor booklets and between scorers across countries

	Country							
Scoring reliability	A	B	C	D	E	F	G	H
Anchor booklet agreement	97.0	97.5	97.7	96.9	97.5	96.0	97.4	96.2
Between-scorer agreement	99.2	99.8	99.0	99.3	99.3	98.7	98.6	98.0

participating countries. Thus, the scores are comparable to the international scoring standard, and would be evaluated for each country.

(b) *Within-country scoring reliability—agreement between scorers:* All paper booklet responses are 100% verified, meaning that all responses are reviewed by multiple scorers who independently rate the response. The agreement between these two scorers is expressed as a percentage of agreement, with 100% indicating complete agreement on the score to award the response. This procedure ensures the scoring standard is accurately applied and all scorers can apply the scoring standards reliably (a lower scoring agreement would indicate inaccurate scoring due to random or systematic error, such as certain items being difficult to score accurately). Actively monitoring scoring reliability is important for identifying scoring inconsistencies or scorer misapplication of the scoring standards early in the scoring process.

Example 4 (Accurate and Reliable Across- and Within-Country Scoring) It is expected that the scoring reliability to be very high with good training, but some errors are expected due to human judgement. Therefore, low or perfect reliabilities can indicate misunderstanding of the scoring standards of response categories or deviation from the scoring procedures (e.g., group scoring to get an agreed-upon score instead of independent scoring). The scoring agreement for eight PIAAC countries is shown in Table 9.3. Agreement with the anchor booklet to assess across country reliability is above 96% in all countries, indicating that countries applied the international scoring standard in the same way. Similarly, the agreement between scorers within a country is also high, above 98% in each country. Taken together, these two agreement measures indicate that within each country, scores applied by human scorers can be considered reliable and accurate. Both measures ensure the validity of collected responses.

On the other hand, there were cases in which poor quality of scoring was present (e.g., low or perfect scoring reliability). Further investigation found that collaborative scoring and scoring resolution were carried out to attain the perfect scoring reliability for all items for some countries and the data no longer can represent accuracy of scoring. Low reliability among multiple scorers on a particular set of items can indicate either inherently difficult items to score or scoring instructions may not be clear and not well understood. Low reliability of a scorer on an item or a set of items need to be monitored and corrected through various ways. Scoring reliability of all items and by all scorers were monitored throughout the scoring following a particular design that focused on identifying inaccurate scoring and retrain inaccurate scorer or replace.

9.1.2.4 Credibility of Responses Implied by Timing Data

The use of computer-based assessments allows for the collection of data reflecting how test takers progress through the assessment, such as time each test taker takes to respond to an item and the number of actions (i.e., mouse clicks) before submitting an answer. The data on progression known as "process data," are available in log files and can provide beneficial performance data in addition to item response accuracy (Lee & Haberman, 2016). For example, timing data can be informative for evaluating the level of participant engagement over the testing period. Very little time spent on the assessment usually indicates low effort. Too much time spent could be an indication of technical problems, most often associated with the test delivery platform, although it could simply indicate a participant with low ability. The availability of response times in computer-based large-scale assessments, such as PIAAC and PISA, makes it possible to evaluate administration problems, level of engagement, position effects, relationships of items with proficiency, and between-country differences. In PIAAC, timing data, in addition to other process data, such as the number of actions, are used in reasonably scoring omitted responses (OECD, 2019).

Example 5 (Rapid Responses) Interviewers, as part of data collection efforts, can affect the quality of data, as interviewers have the power to shorten the data collection period and replicate data. In this example, we evaluated the total response time of CBA respondents, in which the proportion of CBA respondents ranges from 42 to 74% across countries. We focus on two countries (shown in Table 9.1 as Country 1 and Country 3) with similar proportions of CBA data: 74.2% and 72.0%, respectively. The cognitive assessment was designed to take about 60 min. Nearly 90% of respondents in both countries completed the assessments within 60 min (i.e., the expected testing time for this assessment). However, as shown in Fig. 9.3, these two countries differ greatly in the proportion of respondents completing the assessment in less than 30 min (i.e., half of the allocated assessment time): 26% in Country 1 and 2% in Country 3. Note that each point in any curve presented in Fig. 9.3 represents the proportion of respondents that completed the assessment in less than or equal to a given time for that participating country.

In addition to the total response time analysis, the evaluation of item-level response time provides another indication of data quality. In an assessment administered by interview, such as PIAAC, these data can be used to flag possible data fabrication. Given the number of items attempted by each respondent, the average item response time for each respondent is calculated per interviewer to check for rapid responses. The number of rapid responses (and rapid respondents) per interviewer was calculated using data that were filtered according to the following criteria:

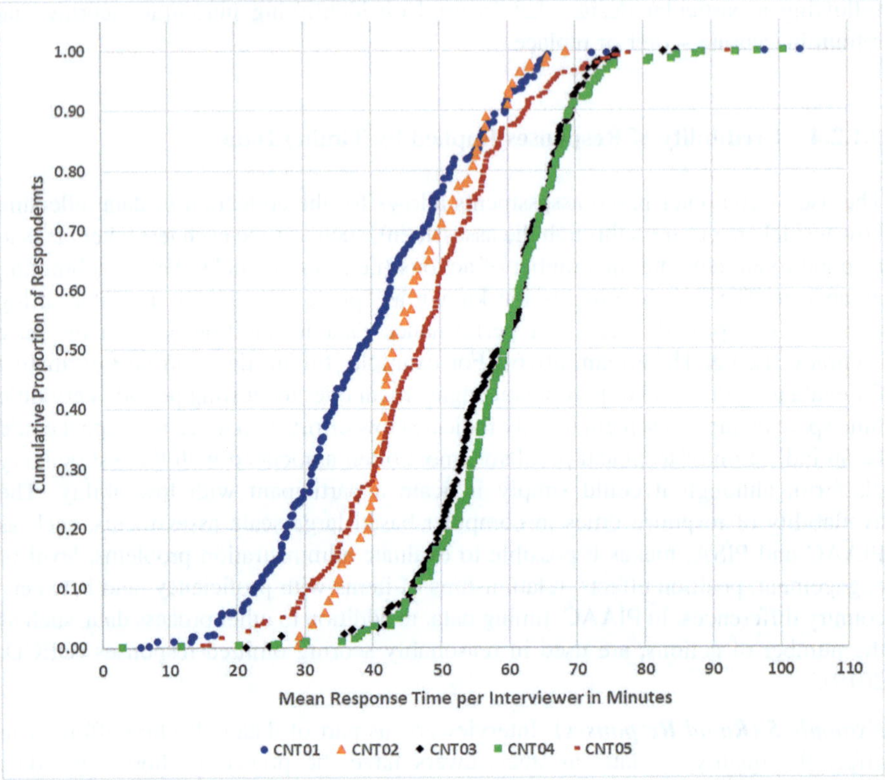

Fig. 9.3 Cumulative distribution by mean response time per interviewer across five countries

(a) A CBA respondent for whom total response time is available.
(b) A minimum number of attempted items is at least ten to assure the cognitive section was started.
(c) A minimum number of respondents per interviewer was at least 20 to obtain a stable statistic.
(d) An average item response time for an interview is less than or equal to 15 s (i.e., equivalent to an overall assessment time of ten minutes for a planned hour-long assessment).

This analysis provided 17 interviewers; each has multiple participants flagged with rapid responses, see Table 9.4.

In Fig. 9.4, three graphs on the left display distributions of respondents based on the total time (in minutes) on CBA cognitive items for three countries: B, E, and H. Two countries, E and H, show the range of distributions, with respondents in Country H taking shorter time compared to respondents in Country E. Country B is the outlier on these graphs and calls for further investigation. Three graphs on the right display cumulative distributions. It is notable that nearly 30% of respondents used less than 20 min on cognitive items in two domains designed for 60 min in Country B.

Table 9.4 Interviewers with cases of rapid responding (mean item response time <15 s) for country 1

Interviewer ID	Total # of cases	# of flagged cases	% of flagged cases
1	34	15	44.1
2	90	9	10.0
3	62	7	11.3
4	45	7	15.6
5	37	7	18.9
6	58	6	10.3
7	21	5	23.8
8	38	5	13.2
9	26	4	15.4
10	24	4	16.7
11	21	4	19.0
12	31	3	9.7
13	67	3	4.5
14	48	3	6.3
15	70	2	2.9
16	51	2	3.9
17	23	2	8.7
Total	**751**	**88**	**11.0**

Example 6 (Response Time and Interviewer Interaction) Timing data carry essential information that can be used along with other data (e.g., response accuracy, rate of interviews) to verify abnormal behaviors from various key players (e.g., respondents, interviewers). While analyzing the Country B data, the average time to complete the cognitive portion of the assessment was significantly less than other groups (e.g., 30% of respondents finished in under 20 min, which is half the time of other groups). Therefore, the interaction between response time and interviewer was further examined using analysis of variance. The assessments with shortest total time were not equally distributed among all interviewers, with this variance most pronounced in the Country B data. The variance component of interviewer on the response time, as shown in Table 9.5, indicates significant impact of the interviewer on the response time. Country B has the largest interviewer effect by far among all studied groups.

Interviewers interact with data quality in terms of accuracy and completeness. For example, Country B had a relatively large group of interviewers (122 in total), seven interviewers accounted for about 20% of the completed cases, and the mean assessment time for these interviews was less than 20 min. One interviewer among them assessed 105 cases with an average answering time of 5.8 min. The actual responses of 1032 respondents interviewed by these interviewers were, therefore, investigated very closely, looking at response patterns in addition to timing data. This analysis showed that the short assessment time was largely attributable to an unusually large proportion of non-response (equivalent to skipping) over most cognitive items after

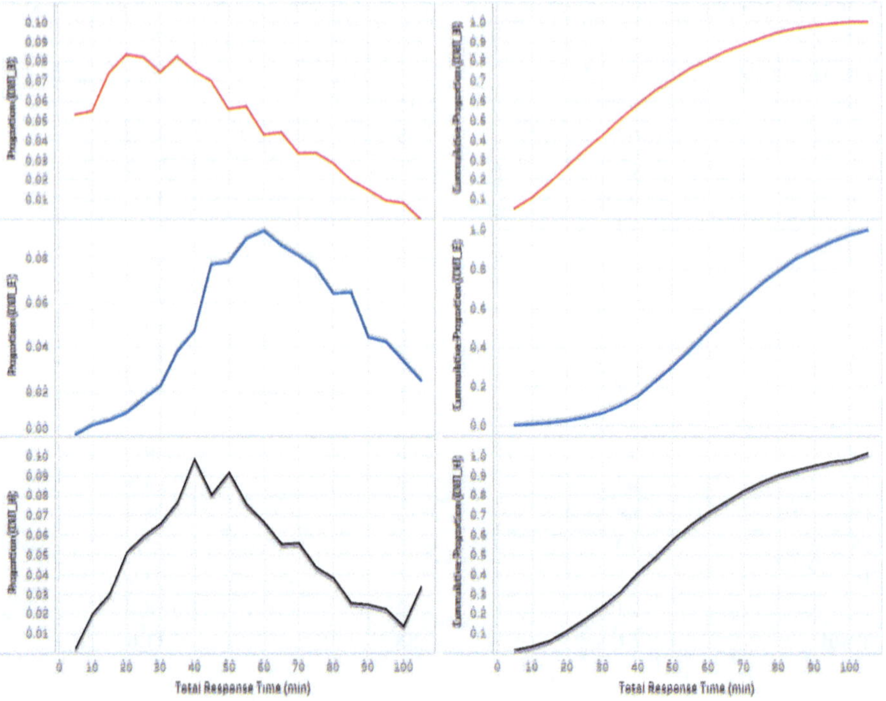

Fig. 9.4 Distribution (left panels) and cumulative distribution (right panels) of respondents based on the total time in minutes across three country groups

Table 9.5 Summary of analysis of variance for interviewer effect on response time

	Country							
	A	B	C	D	E	F	G	H
MS_{Intv}	2202.3	9189.7	4621.0	7085.4	1920.9	2322.1	6049.4	1517.4
df_{Intv}	245	115	107	96	122	106	59	286
MS_{error}	911.0	422.8	699.0	629.0	642.1	685.1	591.2	517.2
df_{error}	3178	3051	3489	3580	5387	3964	3935	1979
F	2.417	21.73	6.61	11.26	2.99	3.39	10.23	2.93

correctly responding to the first six items in a set. In addition, there were some cases where the literacy items were nearly all non-response, but almost all numeracy items were completed with unusual regularities in the provided responses.

Figure 9.5 shows that a relatively limited number of interviewers in Country B accounted for a large proportion of respondents with short response times. The strong interviewer-response time interaction for this set of interviewers in Country B implies an interviewer's impact on interviewee's response time resulting in degradation in data. This is an indication of the interviewer's interference with respondents, where no effect of interviewers on the data should be present.

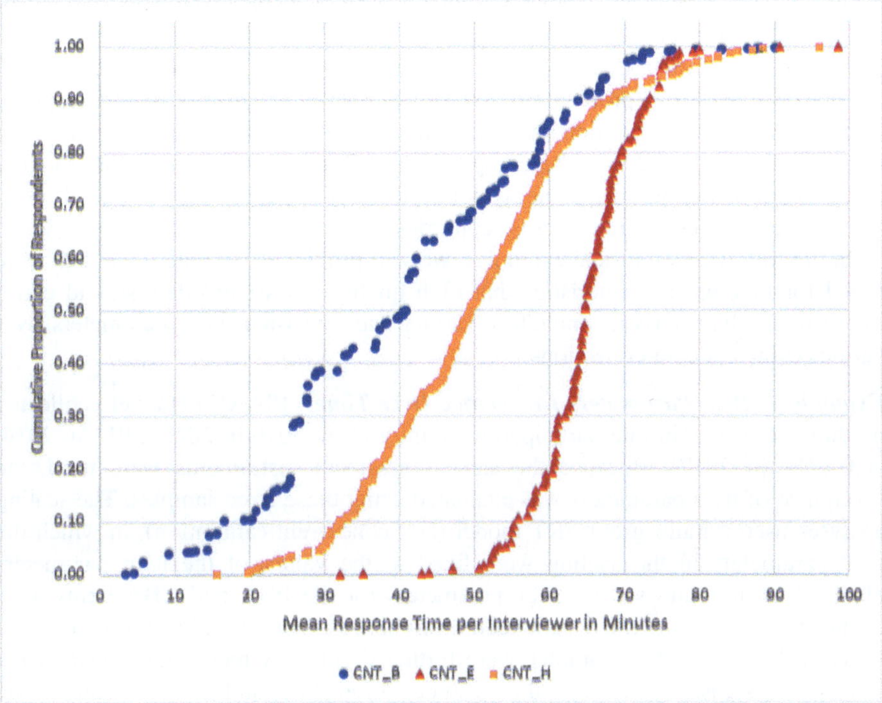

Fig. 9.5 Proportion of respondents by mean response time per interviewer across three groups

9.1.2.5 Consistency and Measurement Invariance

In the context of international large-scale assessments, establishing measurement invariance across countries is essential for valid between-country comparisons (Tijmstra et al., 2019). The assessment design and associated scaling methodology stand on the well-established phenomena of consistent psychometric properties between the item responses and latent ability (i.e., invariant item parameters). This consistency of the item parameters has been regularly observed in the National Adult Literacy Survey, the International Adult Literacy Survey, Adult Literacy and Lifeskills, three rounds of PIAAC cycle 1, and eight PISA cycles from 2000 to 2022 in both field trial and main survey (e.g., OECD, 2019, 2020). To evaluate the nature and the regularity of invariance, the size of deviations of item-by-mode-by-country over time are routinely examined. Generally, the invariance of item parameters has been observed for a large portion of those items that have been administered across modes (PBA and CBA).[2] This substantial invariance of item parameters across modes, countries, cycles, and multiple assessments over time

[2]Note that it is not possible for all items to be used in both modes. For example, some paper-based items that require drawing could not be implemented on computer. Similarly, some items developed

assures the comparability of inferences that can be made based on the surveys, and hence, it is used as a measure for data quality. In other words, if the data do not meet such invariance, further quality checks may be necessary to ensure that this violation did not originate from problems introduced in data-processing (e.g., data capture, response coding, data fabrication, etc.). Having only a few country-specific item parameters slightly reduces the comparability of scales across countries roughly proportional to the number of items with unique parameters, but the overall measurement error is comparable across countries.

In the following empirical examples, we will provide cases where the data were tested for inherent characteristics that high quality assessment data should carry. These main characteristics include invariant and consistent item parameters over various cycles of data collections.

Example 7 (Item Parameters Invariance Over Time) PIAAC data were collected on three occasions in one participating country ($N = 5010$ in 2012, 2911 in 2014, and 4719 in 2017). Because the same assessment instruments were used, the invariance of item parameters was evaluated using these three samples. The scaling analyses used a multi-group IRT model (i.e., concurrent calibration), in which the item parameters in the scaling were fixed to the values of the item parameters obtained in the main survey. IRT parameters for the PBA and CBA items were evaluated to determine if the same (common) item parameter could be used for these samples. Note that these samples were further divided by mode (CBA and PBA). The item fit statistics used were the mean deviation (MD) and the root mean square deviation (RMSD). Using these measures, we quantify the magnitude and direction of deviations in the observed data from the estimated item characteristic curve for each item. Items that showed deviations from the common item parameters were assumed to work differently across samples and to, therefore, harm the link between modes or between groups. Poorly fitting items were identified using an RMSD $>$ 0.15, and a MD > 0.15 or < -0.15 field trial criterion where a value of 0 indicates no discrepancy (i.e., a perfect fit of the model). It was assumed for such items that the common item parameters were not appropriate, and group-specific unique item parameters were estimated in a second step.

In the analyses, the responses to the common item across samples in assessment occasions were accurately described by the identical item parameters. No unique item parameters needed to be estimated, as these analyses supported measurement invariance and the basis of comparability over time. Therefore, the comparability of scales across the three data collection periods was established successfully.

Example 8 (Internal Inconsistency of Within-Country Data Across Cycles) In international large-scale assessments, IRT scaling was closely monitored for item-by-country interactions. In the context of PISA, item-by-country-by-mode pairs that were identified as being inappropriately modeled by the international (common) item

to leverage capabilities present in the computer environment, such as conducting simulated experiments or using text with hyperlinks, could not be administered on paper.

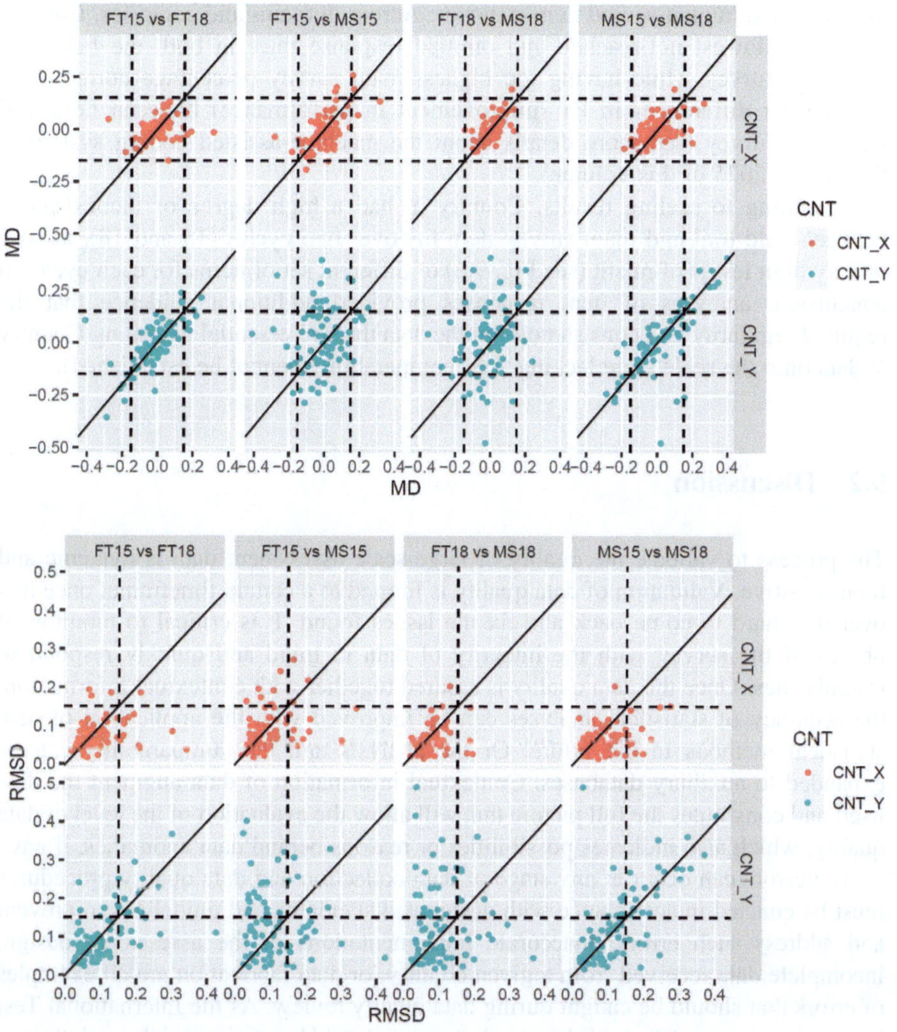

Fig. 9.6 Scatter plots of fit statistics in field trial and main survey

parameters were released (freely estimated) and received a unique set of item parameters. The number of required releases was monitored to evaluate whether it was in the range seen in prior assessments. In the following case, a very large number of items required group-specific item parameters.

To evaluate the extent to which scaling requirements are met for a given country, PISA data from two cycles over field trials and main surveys were scaled following operational scaling procedures, and fit statistics, common and unique parameters, and proficiency estimates were produced. Figure 9.6 shows these results for Country Y, using Country X as a reference for what is typically observed in the data. First, preliminary item analyses indicated some level of inconsistency in machine- and human-scored item statistics: Across domains and countries, Country Y had the lowest not-reached and omitted response rates in both the field trail and main survey while having the highest main survey percentage correct (P+) and highest differences in P+ (improvement in performance) between field trail and main survey (see more details about the machine-assisted coding system in Shin et al., 2025 of this volume).

According to scaling results, Country Y had a high degree of inconsistency between field trail and main survey item fits over two cycles (15 and 18), as well as very high levels of misfit (see Fig. 9.6) of different set of items for each cycle. In conclusion, analyses of item responses provided additional evidence that the required regularity and consistencies in the data that are essential to scaling Country Y data onto the scale were lacking (i.e., comparability cannot be established).

9.2 Discussion

The process to validate the quality of large-scale assessment data is dynamic and time sensitive. Validation of data quality is limited to a certain timeframe; once it is over it is hard to come back and fix the issues found. It is critical to monitor all phases of the survey with the integrity of data in mind and quickly respond to irregularities. Once the data quality is assured together with contextual information, the accuracy of statistical analyses can be improved with the application of new statistical methods in the future. Or the analysis and the comparability can be expanded to auxiliary databases. Contextual information of data augment the data itself and constitutes the full picture that will allow the evaluation of the level of data quality, which also indicates possibilities or reasons behind data anomalies, if any.

The errors can occur at any stage of data collection and data quality procedures must be enacted in accordance with designated standards and guidelines to prevent and address such errors. Inaccurate implementations of the assessment design, incomplete data received from a given country, or data fabrication are all examples of errors that should be caught during data quality review. As the International Test Commission notes, "To err is human, but errors should be minimized through the use of adequate quality control measures" (ITC, 2014). Poor quality, invalid, or fabricated data are highly consequential, such that they undermine the reliability, validity, and comparability of the survey results among subpopulations and across administration cycles. Invalid data destroy the linkage between assessment objectives and results and reduce the credibility and validity of a survey altogether. Poor data quality is every stakeholder's concern and should be eliminated to the furthest extent possible (Yamamoto & Lennon, 2018). Accordingly, five dimensions of data quality were considered in this chapter, including adherence to the design, level of missing data, reliability of scoring, credibility of responses based on timing data, and consistency and comparability of assessment results. Psychometric data forensics to collect evidence seeking maintenance and improvement of data quality is relatively new in the context of educational measurement and developing and adapting to the new features and design of the assessment. More methods and practices will be developed and standardized that could improve data quality and yield a better understanding of data issues and sources of poor-quality data. These practices include using process data that will allow psychometric teams to define and extract more meaningful information about test taker behaviors. Specifying and collecting useful process data variables is an important step, which helps to ensure data quality, among other various objectives. Accordingly, extracted process data would be used for elevating the operational practices (e.g., verify if the test was administered as intended, identify disengaged responses). Using cumulative database to provide a sound basis to compare congruency of data in multiple references, automating the procedures to enhance detection speed and reduce efforts and costs, and capturing more data about testing conditions and environmental context to detect any administration issues are among the recommended practices that would potentially improve the quality of assessment data.

References

American Educational Research Association, American Psychological Association, & National Council on Measurement in Education. (2014). *Standards for educational and psychological testing*. AERA.

Haberman, S. J., & Lee, Y.-H. (2017). *A statistical procedure for testing unusually frequent exactly matching responses and nearly matching responses* (Research Report No. RR-17-23). Educational Testing Service. https://doi.org/10.1002/ets2.12150

International Testing Commission (ITC). (2014). ITC guidelines on quality control in scoring, test analysis, and reporting of test scores. *International Journal of Testing, 14*, 195–217. https://doi.org/10.1080/15305058.2014.918040

Lee, Y.-H., & Haberman, S. J. (2016). Investigating test-taking behaviors using timing and process data. *International Journal of Testing, 16*, 240–267.

Organisation for Economic Co-Operation and Development (OECD). (2011). Quality framework for OECD statistical activities. http://www.oecd.org/officialdocuments/publicdisplaydocumentpdf/?cote=std/qfs(2011)1&doclanguage=en

Organisation for Economic Co-Operation and Development (OECD). (2014). PIAAC technical standards and guidelines. https://www.oecd.org/skills/piaac/PIAAC-NPM(2014_06)PIAAC_Technical_Standards_and_Guidelines.pdf

Organisation for Economic Co-Operation and Development (OECD). (2018a). Cycle 2 PIAAC technical standards and guidelines.

Organisation for Economic Co-Operation and Development (OECD). (2018b). PISA 2021 technical standards. https://www.oecd.org/pisa/pisaproducts/PISA2021_TechnicalStandards.pdf

Organisation for Economic Co-Operation and Development (OECD). (2019). Technical report of the survey of adult skills (PIAAC) (3rd ed.). https://www.oecd.org/skills/piaac/publications/PIAAC_Technical_Report_2019.pdf

Organisation for Economic Co-Operation and Development (OECD). (2020). PISA 2018 technical report. https://www.oecd.org/pisa/data/pisa2018technicalreport/

Shin, H. J., Yamamoto, K., He, Q., & von Davier, M. (2025). Automatic scoring of constructed responses in PISA using the machine-supported coding system. In L. Khorramdel, M. von Davier, & K. Yamamoto (Eds.), *Computer-based international large-scale assessments: Concepts, methodologies and quality assurance procedures*. Springer.

Tijmstra, J., Bolsinova, M., Liaw, Y.-L., Rutkowski, L., & Rutkowski, D. (2019). Sensitivity of the RMSD for detecting item-level misfit in low-performing countries. *Journal of Educational Measurement*. https://doi.org/10.1111/jedm.12263

Yamamoto, K., & Lennon, M. L. (2018). Understanding and detecting data fabrication in large-scale assessments. *Quality Assurance in Education: An International Perspective, 26*(2), 196–212. https://doi.org/10.1108/QAE-07-2017-0038

Chapter 10
Technical Platforms for the Improvement of Coding Reliability: OECS and OERS

Emily Kerzabi, Kentaro Yamamoto, and Chentong Chen

Abstract In the 2018 administration of the Programme for International Student Assessment (PISA), there were over 7-million constructed responses collected from students on items in the major domain of Reading alone, the majority of which were assigned to 1300 human coders, across 139 unique country-by-language groups. This sheer volume of data requiring human coding necessitates computer-based coding systems to facilitate independent coding and support coding teams. This chapter focuses on PISA's international coder training, efficient coding designs, and most importantly, the coding systems employed to manage the vast number of responses assigned to coding teams working across the globe during test administration. The Open-Ended Response System (OERS) and Open-Ended Coding System (OECS) provide national coding teams with formative statistics on coder agreement and overall coding reliability throughout the coding process, giving them insight and awareness of how accurately codes are being applied. These systems automatically alert lead coders to low levels of coding agreement and flag patterns of inaccurate coding. With such coding systems in-place, not only can large amounts of constructed response data be effectively managed in international large-scale assessments, but measures of coding accuracy and reliability can be calculated and reviewed for areas of future system and assessment improvement.

International large-scale assessments (ILSAs), such as the Programme for the International Assessment of Adult Competencies (PIAAC) and the Programme for International Student Assessment, emphasize measuring the fidelity of information

---

E. Kerzabi (✉)
ETS, Princeton, NJ, USA
e-mail: ekerzabi@ets.org

K. Yamamoto
Independent Consultant for Psychometric Methodology and Large-Scale Assessment (formerly ETS), Princeton, NJ, USA

C. Chen
Independent Researchers (former ETS), Princeton, NJ, USA

L. Khorramdel et al. (eds.), *Innovative Digital-Based International Large-Scale Assessments*, Methodology of Educational Measurement and Assessment,
https://doi.org/10.1007/978-3-031-90951-1_10

processing skills rather than presence and quantity of static knowledge. Hence, since the inception of both of these surveys, it was decided that not all items could be strictly a multiple-choice format. Issuing a test of all simple multiple-choice items may be easy to score, but a variety of item types (such as multiple-answer multiple-choice, matching, short and long constructed response, true/false, and fill-in-the-blank), referred to as a *mixed format* assessment, can better trigger varying cognitive response processes and incorporate a wide array of item difficulty (OECD, 2019, p. 49). In other words, mixed format tests generally allow for a more precise and a broader measurement of skills than single format tests. Since the introduction of computer-based assessment (CBA), a variety of new item types, such as drag-and-drop and data-collection items (the test taker collects simulated data needed to answer the question), have been introduced; however, nearly one-third of the test items in both PISA and PIAAC remain constructed response (or open-ended) items that require human scoring. However, scoring different types of items, especially when the item pool or the number of test takers is large, presents certain challenges.

To illustrate these challenges and how to overcome them, this chapter will focus on the coding of constructed response items in PISA and the systems employed to manage the vast number of responses, in support of coding teams across languages and countries, and to provide an overview of coding reliability statistics in the field. While items such as simple and complex multiple-choice items can be computer-scored with accuracy, constructed-response items often require a human coder to code the response. To provide a general idea of why systems for managing coding are necessary, consider that in 2018, PISA administered 300 constructed response items across five domains and two modes, spread over 81 countries and economies and across 57 languages, for a total of 139 unique country-by-language groups, and 206 coding teams. In the major domain of Reading, alone, there were over 7,000,000 constructed responses across 59 paper-based and 82 computer-based items, 80% of which required human coding and were assigned to nearly 1300 coders. With this much volume across coding teams, and the vital need for comparability of coding (i.e., codes need to be applied in the same way so that the coding results are comparable across countries and languages), software to monitor and report coding statistics throughout the coding process can be invaluable for maintaining high data quality. PISA uses the Open-Ended Coding System (OECS) in countries/economies administering the CBA mode and uses the Open-Ended Recording System (OERS) in countries/economies administering the paper-based assessment (PBA). The introduction of these systems to support real-time coding checks was necessary to support coding teams in correcting coding irregularities and to improve coding itself.

These coding systems, however, can only support coding. Coding quality will only be as good as the training coders have received and only as reliable as the coding design established for evaluating the codes applied. Before going into the specifics of the OECS and OERS platforms, we provide an overview of the training PISA administrators provide and the type of coding design used in the next section. Because every assessment is different, we also detail some points for assessment designers to consider when implementing a coding design and preparing a coding software to support coding.

10.1 Training and Coding Designs to Minimize Error

Ahead of coding student responses, it is vital that coders receive specialized training on all items they will be coding. In PISA, for example, all national coding team leaders (or other coding team representative) must participate in a series of two, centralized, international trainings, and they are responsible for relaying information from the international trainings to the rest of their coding team. All material at international trainings is in English, so those in attendance must be bilingual in English and the language of the assessment. For consistency, training materials are not provided in all languages of assessment but, rather, in a single language from which national-level training materials can be adapted, as all members of a coding team may not be sufficiently bilingual to understand the international coding material without support from their lead coder or other training attendee. It is important that international trainings work through any confusing aspects of the assessment material, and it has been observed in PISA that discussing item-level issues as a team helps to avoid coding errors (Organisation for Economic Co-operation and Development [OECD], 2020). International and national trainings take place ahead of the field trial and again before the main survey of the assessment, and they are both cumulative (they build on one another) and summative (cover the full range of content).

10.1.1 Coding Guidelines

PISA's coder trainings generally revolve around the *coding guidelines*, which comprise a series of manuals that detail, item-by-item, the types of responses coders can expect to receive and provide a code associated with each response type. Such guidelines are possible because short constructed-responses, as administered in PISA, tend to have a finite set of commonly given answers, for which, the coding guidelines can describe and provide an appropriate code for each type of commonly observed response. In this way, the PISA coding guidelines for short answer responses are carefully written to avoid ambiguity while broad enough to cover all or most possible responses to the item.

Coders in PISA are not asked to *score* responses directly (e.g., denote a level of correctness, such as full, partial, or no credit), but to *code* responses with the appropriate category associated with the type of response noted in the coding guidelines (e.g., in an item asking about how it rains, a code of "21" could be assigned to a response referencing the density of air vapor and a code of "22" could be assigned to a response referencing the gathering of clouds in low pressure zones; both answers could be considered correct, but each student has provided slightly different reasoning). This is not a rubric-based method of evaluation but, rather, a matching-based method, in which the observed response is matched with a type of response in the coding guidelines, and the associated code is applied. However, in an

ILSA being administered across so many different languages and cultures, the coding guidelines may not be as all-encompassing as necessary to account for the types of student responses received world-wide. Therefore, PISA also provides coding teams with a "hotline" for coding questions, so that when a coding team observes responses not reflected in the coding guide, the item developers themselves are able to offer support on how best to code such responses. Although the coding guidelines are generally referenced and updated by coding teams throughout coding, PISA developers also encourage lead coders to conduct item-specific training just ahead of coding to make sure that the coding guidelines are fresh in the minds of coders, and, if coding errors are discovered, item-specific retraining should be conducted.

As a general note on coding guidelines, the length and complexity of a coding guide for an item may depend on the cognitive process elicited by the item, such as the cognitive subscales of "locate information," "understand," or "evaluate and reflect" in the reading domain. For items with a higher level cognitive subscale or cognitive difficulty, in particular, it may not be sufficient to only provide examples of what constitutes a full/partial/no credit response, but also define a minimum of what is acceptable for credit. This minimum may be helpful when the item is assessed across languages with varying linguistic nuances.

10.1.2 Coding Design Considerations

10.1.2.1 Number of Coders per Coding Team

There are two primary goals regarding coding design: efficiency (completing coding as quickly and as cost effectively as possible) and accuracy (ensuring that all applied codes are free of error). Coding efficiency may be obtained by employing a large number of coders, each coding a response one time. This would result in completing coding quickly, but without verifying that the coders have correctly and consistently applied the coding schema, it would also likely result in unaddressed coding inaccuracies and unverifiable estimates of the reliability of the applied codes. On the other hand, perfect coding accuracy could be achieved by having every response coded by multiple, very well-trained coders, to ensure the comparability and reliability of all applied codes across all responses. However, this would take a long time, to extensively train, review, and correct coding, with a high amount of redundant coding, so it would not be cost effective. Therefore, a good coding design must balance the goals of efficiency and accuracy, such that coding teams do not get bogged down in verifying potentially thousands of codes, but coding accuracy, while not perfect, still remains high.

In ILSAs, this means that the coding designs of each country/economy may differ, but each design is grounded in a set of common considerations that help to balance the demands of efficiency and accuracy of coding:

Table 10.1 Determining the number of expected constructed responses

	Language (s)	Sample size (%)	Domains	% taking domain	Average constructed-responses administered	Expected responses
Country A	Language 1	6900 (100%)	Reading	100%	18	124,200
			Mathematics	50%	8	27,600
			Science	50%	12	41,400
			Financial	30%	6	12,420
Country B	Language 1	6200 (90%)	Reading	100%	18	111,780
			Mathematics	50%	8	24,840
			Science	50%	12	37,260
	Language 2	700 (10%)	Reading	100%	18	12,420
			Mathematics	50%	8	2760
			Science	50%	12	4140

- Sample size: How large is the sample taking the assessment?
- Languages being assessed: How many languages are being assessed, and what proportion (or percentage) of the sample will be taking the assessment in each language?
- What proportion (or percentage) of that sample will be taking each domain, and how many responses in each domain are expected to need human coding?
- Timeframe and workload: How many days of active coding are expected, and how many responses per day are coders expected to code?
- Minimum coders and coder attrition: If the assessment is given in multiple languages across countries, how many coders are required to be bilingual and how many coders are expected to drop out during the coding process? In general, coding agreement calculation will require a minimum of two coders.

How the answers to these questions can be used to determine the appropriate number of coders per coding team is illustrated in Tables 10.1 and 10.2. Note that these tables are not specific to the PISA assessment, but, rather, serve as an example only. In Table 10.1, the number of responses expected is determined by multiplying the sample size by the percentage of that sample taking the assessment in either language, then multiplying that result by the percentage of that sample taking each domain and by how many constructed-response items each student will take in that domain. Table 10.2 takes the expected number of responses and divides them by a timeframe of 10 working days of active coding (roughly 2 weeks) with an expectation that coders will be able to code about 1000 responses per day. This results in the minimum number of coders needed to code the expected responses in that timeframe. For coding agreement calculation, there should be a minimum of two coders on a coding team (see Sect. 10.1.2.2), and if the assessment is assessed in multiple languages across countries, two coders should be bilingual in the language of assessment and in the international training language (for example, English in PISA). It is okay if the minimum number of two coders are also the bilingual coders.

Table 10.2 Determining the number of coders per coding team

	Language(s)	Domains	Expected responses	Minimum coders	Bilingual coders	Expected attrition	Total coders
Country A	Language 1	Reading	124,200	13	2	2	15
		Mathematics	27,600	3	2	1	4
		Science	41,400	5	2	1	6
		Financial	12,420	2	2	1	3
Country B	Language 1	Reading	111,780	12	2	2	14
		Mathematics	24,840	3	2	1	4
		Science	37,260	4	2	1	5
	Language 2	Reading	12,420	2	2	2	4
		Mathematics	2760	1	2	0	2
		Science	4140	1	2	0	2

As shown in Tables 10.1 and 10.2, different coding designs should be selected for each coding team by language and domain (OECD, 2020). Here, the number of recommended coders on a coding team ranges from 2 to 15, which is primarily dependent on the size of the sample being administered in each language in each domain. As illustrated in Table 10.1, the reading domain was administered to 100% of the sample, whereas mathematics and science were only administered to 50% each; in PISA, this distinction is referred to as major and minor domains, respectively. It is also possible administer optional domains, such as financial literacy, which, in the example, was administered in Country A but not Country B. The number of coders on a coding team is also influenced by the expected number of constructed-response items in each domain. In PISA 2018, in which reading was the major domain and contained the greatest number of constructed-response items administered to each student, anywhere between 2 and 32 coders were recommended. In the minor domains (science, mathematics, financial literacy, and global competence), between 2 and 12 coders were recommended.

10.1.2.2 Determining the Reliability of Coding

Knowing how many coders are necessary is only the first part of the design; how coding reliability, or the consistency of coding, will be calculated is the second part. As with any human coding, there are common coding irregularities that need to be mitigated: Coders may apply codes with systematic bias (more harshly or more leniently score responses) or code inconsistently; coders may also mistakenly interpret how to code responses to a particular item. In ILSAs, additional issues arise: Entire coding teams may code with systematic bias or mistakenly interpret coding guidelines for specific items. It is extremely important that codes be consistently applied by each coder, by each coding team, and across coding teams so that inconsistent coding does not affect how the item functions during scaling (for more on item scaling in PISA, see Chapter 9 of the Technical Report; OECD, 2019). To verify the consistency and reliability of coding, some sort of coding agreement statistic is used to determine coding reliability, and coding agreement can be calculated within coding teams, between coding teams, and within coders.

To calculate reliability of coding within a coding team, or between coders, the codes of at least two independent coders must be compared across a set of selected responses. We refer to the coding of the same response by multiple independent coders as *multiple-coding*. Multiple-coding can be conducted across the full set of responses if the total number of responses is low, otherwise a proportion of the total (varying by the total number of responses) or a specific number of the total (fixed to a specific number of responses, invariant of the total) can be multiple-coded. In PISA, because the number of responses varies widely by coding team, a fixed set of 100 responses for each item are multiple-coded, allowing agreement statistics to be calculated between coders of the same coding team. Multiple coding is discussed further in Sect. 10.4.

Because PISA is administered across several countries, languages, and cycles of assessment, it is also important to evaluate agreement between coding teams against a fixed coding standard. PISA uses *anchor-coding* as a fixed coding standard, which comprises a series of 10–30 responses per item in English that the two bilingual coders of all coding teams must code, across assessment languages and cycles of assessment. Because the correct code for each anchor response is known, between-group agreement statistics can be calculated. For more on anchor coding, see Sect. 10.3.

Finally, the remainder of the responses that are neither anchor responses nor multiple-coded responses will be seen by only one coder, referred to as *single-coding*. True single-coding agreement could be calculated by having the same coder multiple-code the same response and comparing the same coder's code on the same response over multiple occurrences; however, in ILSAs, this is not a realistic option (e.g., it is not efficient). However, because responses are randomly assigned to coders in PISA, within-coder reliability on single-coded responses can be determined by looking at the distribution of assigned codes across coding categories to see if a single coder is consistently applying codes similar to other coders. For more on single-coding, see Sect. 10.5.

To calculate coding reliability, or the expected accuracy of an assigned code, there are multiple statistics that can be used. PISA uses direct percentage agreement, Cohen's kappa (Cohen, 1960), and Chi square coefficient (Pearson, 1900). The percentage agreement ranges from 0% (no agreement) to 100% (complete agreement). If there are two coders, and one assigns a code of 0 and the other assigns a code of 1, there is 0% agreement; if they both assign a code of 0, there is 100% agreement. The formula for calculating the average agreement for coder *i* on item *j* is shown below:

$$\mathit{Average\ Agreement}_{ij} = \frac{\mathit{Number\ of\ Agreements\ on\ Multiple\ Coded}_{ij}}{\mathit{Number\ Multiple\ Coded}_{ij} \times \left(\mathit{Number\ of\ Coders}_{j} - 1\right)} \tag{10.1}$$

where the *number of agreements on multiple coded* and the *number multiple coded* is reflective of the number of responses that have been coded thus far (e.g., if 50 out of 100 responses have been multiple coded at a given time point, the agreement would reflect only those 50 responses, and the number multiple coded would equal 50), and would increase as coding progresses. At the item level, overall agreement across coders on item *j* is a simple average of item agreement for coders $i_1, i_2 \ldots i_N$ on item *j* from Eq. (10.1), assuming that all coders multiple coded the same number of responses (otherwise, it would be more accurate to weight each coder's average agreement by the number of responses they multiple-coded).

The Cohen's kappa coefficient is another statistic that may be used alongside percentage agreement and is often recommended for reporting. The kappa statistic is considered more robust than percentage agreement because it provides a correction for agreement by chance; it ranges from 0 (no agreement) to 1 (complete agreement).

To monitor the distribution of responses to coding categories by each coder, the Chi square statistic can be used to compare the equivalency of code proportions between coders. PISA uses all three statistics for reporting coding reliability.

While these designs and statistics are useful, they do little to alert coding teams to inadequate coding reliability if they are only reviewed in a summative step (e.g., when coding is complete and no longer able to be fixed). An ongoing, formative assessment of coding reliability is greatly preferred, such that low coding reliability can be addressed while coding is still taking place. Of course, in an ILSA, this can be incredibly difficult without systems in place to support coding teams in their efforts to monitor coding reliability. Because unreliable coding introduces excessive noise into the data and reduces the amount and quality of information related to test-takers' responses and proficiency, two software systems were developed in PISA to monitor ongoing coding reliability and use these statistics to alert coding managers to coding issues.

10.2 Systems to Minimize Error

As previously stressed, any inconsistency in coding needs to be identified and corrected as soon as it is discovered in order to prevent further data corruption (e.g., incorrect codes from being applied on many more responses). To address this issue, the OERS for PBA and OECS for CBA were developed in PISA to support coding teams in real-time. With coding consistency and quick correction paramount, the OERS and OECS report coding consistency at three crucial levels during coding:

1. Between-groups of coders (across countries, languages, and cycles of assessment): All coding teams of different countries/economies and languages must apply codes consistently, and trend[1] items must be coded consistently with previous cycles of the assessment.
2. Between-coder/within-group: All coders must apply codes consistently across all responses for each item.
3. Within-coder: Coders must apply codes consistently across all responses for each item.

These three levels are crucial points at which coding must be verified. When coders differ in their coding, the accuracy of the assigned scores for human-coded items is thrown into question, which can jeopardize the reliability of the resulting scale and the overall validity of the assessment. In this respect, the OERS and OECS were developed to increase the quality of human-coded data, and the remainder of this chapter details the operations and functionalities of these systems, providing both an overview of the technical aspects of the OERS and OECS and outlining how each

[1] Trend items are those that are administered across multiple cycles of assessment and allow for the measurement of change over time.

system supports coding teams in mitigating error at each of the three levels described in the list above. We conclude with a short discussion.

10.2.1 The Open-Ended Response System (OERS) for Paper-Based Coding

The OERS is the software for monitoring coding reliability in the paper-based PISA. The OERS is an offline application and does not require a steady internet connection to operate. The OERS works in conjunction with the main reporting system of PISA, the data management expert (DME), to produce coding reports. These two software packages must be installed on the same computer and in the same directory. Because coders code responses on paper, the OERS relies on the manual input of data into the DME from coding sheets used by coders (OECD, 2014). Once data have been input, the OERS produces reports in a macro-enabled Microsoft Excel® spreadsheet. Because data are manually input, coding evaluations occur at three points, each corresponding to a different level of required coding consistency:

- *Evaluation Point 1*: Anchor coding review (between-groups of coders)
- *Evaluation Point 2*: Multiple-coding review (between-coders of one group)
- *Evaluation Point 3*: Single-coding review (within-coder)

At each evaluation point, coding agreement is closely monitored and, if necessary, corrected before proceeding to subsequent phases of coding. Items with less than 92% average agreement across coders on multiple-coded or anchor responses are flagged by the system, and any coder with less than 85% agreement with other coders on a single item (Eq. 10.1) is flagged. Further, coders with a coding category distribution (i.e. percentage of codes in each coding category for the item) disproportionate to the average for the item are also flagged. Once coding is complete, all data are manually uploaded to a secure central server.

10.2.2 The Open-Ended Coding System (OECS) for Computer-Based Coding

The OECS is the system for coding computer-based responses in PISA. National coding teams connect to the OECS via the web, code responses in a web browser, and coded data are automatically saved to a secure central database. Because responses are already digital, the OECS can offer more streamlined approaches to coding than can the OERS. For example, instead of following structured data checkpoints (as is done in the OERS), coding overviews can be generated at-will, and ongoing coding reports are systematically populated and saved every 24-h. The OECS is also set up so that coders are able to code responses on a single item or a set

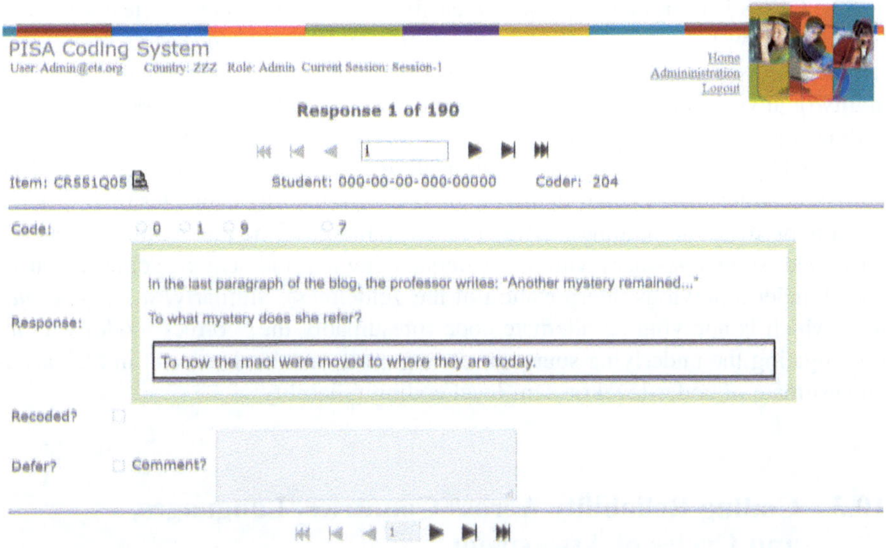

Fig. 10.1 Coder-view of a response in OECS

of related items all at once, rather than coding an entire booklet of mixed items (as in the PBA), which allows coders to focus on one item context at a time.

Each coder logs in to the OECS with his or her specified coder ID, which corresponds to an ID in the coding design. Item responses assigned to the coder are accessible from the main page. When an item is selected, the OECS will load the first available response. The coding screen (shown in Fig. 10.1) is segmented into four sections: "Code" with a selection of possible codes for the item; "Response," which contains the item stem and the response input by the student; a "Recoded?" checkbox, which (if checked) indicates that the selected code is not the original value; and a "Defer?" checkbox with a "Comment?" section, in which the coder and lead coder can communicate if the coder is uncertain how to code the response. When the coder selects the appropriate code at the top, the "Record Code" button saves the code, and OECS automatically navigates the coder to the next, uncoded response. Lead coders are able to view any deferred responses from the OECS main page, and lead coders can either code these responses themselves, contact the coding "hotline" for guidance, and consider additional training for the coder(s) if similar types of responses are commonly deferred.

There are additional functions also contained on the coding screen shown in Fig. 10.1. A unit preview button (magnifying glass next to the item ID) launches a preview window showing the set of items (unit) in which the current response occurs. Navigational buttons allow the coder to move forward 1-item, forward 20-items, back 1-item, back 20-items, go to the first response in the coding batch, and go to the last response. The coder can navigate to a different item by clicking the "Question Selection" link or by completing all of the responses for the current item.

The OECS is designed for coders to easily code and managers to clearly review; however, browsing navigation is limited. For example, there is no search option, and list creation is minimal. The exclusion of these features was intentional to limit the ability of anyone other than the coder from changing coding values. When errors in coding are discovered, specific procedures to correct coding are in place that are supported by the system (such as recoding all responses for a specific item or coder). Partial recoding would result in inaccurate coding reliability estimates (e.g., when a coder approaches an item new/different interpretation of the coding guidelines, but only codes some responses with this understanding in mind, leaving some responses coded under a previous interpretation of the guidelines). Similarly, *scoring resolution*, which is applying an alternate code (presumably the "correct" code) without investigating the underlying source of coding error, would also result in an inaccurate estimate of coder-level or item-level coding reliability.

10.3 Coding Reliability Across Countries, Languages, and Cycles of Assessment

In general, human coding must be consistent across all groups of coders, coding teams, and cycles of assessment to allow comparable results between these groups and over time; in a trend assessment, this principle of comparability must also be applied across cycles of assessment (Mazzeo & von Davier, 2014). To facilitate such a requirement in PISA, coding guidelines and trainings are implemented in the same way in all participating countries and economies and across all languages; for items that have been issued in previous cycles, coding is completed according to the same base framework as previous cycles of the assessment. To verify that all coding teams are applying coding guidelines correctly (i.e. according to the same set of coding rules), an immutable coding reference is required for each item in a single common language to which all group-level coding can be compared. In PISA, anchor coding is the process of coding a common set of English responses for each item and comparing agreement against an anchor key (similar to an answer key), containing the correct code for all anchor responses.

By design, both CBA and PBA coding teams conduct anchor coding. Anchor responses are a set of responses in English that reflect a range of possible responses likely to be seen in the assessment. At least two coders in each country-language group are required to be bilingual in English and the language of the assessment, so in addition to coding in the language of the assessment, they can also code these anchor responses. By assessing the agreement between all bilingual coders with a single anchor key (containing the correct code for the English anchor responses), it can be determined if the coding guides have been applied consistently across countries/economies, languages, and cycles of assessment.

10.3.1 Anchor Coding in the OERS and OECS

In OERS, anchor coding is conducted at Evaluation Point 1 (anchor coding review). The purpose of Evaluation Point 1 is to check for the consistency of coders across countries by trying to ensure that they apply the coding guidelines in the same way. PBA bilingual coders code ten anchor response for each human-coded item. The codes applied by the bilingual coders are checked against the common anchor key to verify that the coders, and thereby the coding teams, have understood and can apply the coding guidelines correctly. The codes are input into the OERS via the DME, and coding teams evaluate the anchor coding reliability report (shown in Fig. 10.2) and address any coding discrepancies. This anchor coding report shows the number of anchor responses coded by each coder (n), the percentage agreement each coder has with the key, and the average percentage agreement with the key for each item. Because there are only twenty comparisons to the anchor key (ten for each coder), two or more coding errors per item or coder should be investigated (such as ItemB in Fig. 10.2, with an overall item-level agreement of 90%, which is below the desired 92% agreement). Following this review, anchor coding output is submitted to PISA data management teams, noting questions, concerns, and issues that arose during coding.

In the OECS, instead of ten anchor responses for each item, thirty anchor responses are used. While the same ten responses per item used in the PBA remain in the set of CBA anchor responses, it was decided that using thirty responses, with sixty code comparisons to the key, is a more robust measure of between-group reliability, an improvement on the PBA system. As suggested above, in the OERS with only ten anchor responses, coders are flagged for incorrectly coding a single item, which may only be useful for identifying systematic irregularities if both bilingual coders miscode the same response (the coding issue can be identified as

Cluster	Item	Unit Name	Average (1)	Coders 201(3)	Coders 203(3)	Notes to Contractor
R1	ItemA	Unit1	1.00	1.00 (n=10)	1.00 (n=10)	
R1	ItemB	Unit1	0.90	0.90 (n=10)	0.90 (n=10)	
R1	ItemC	Unit1	1.00	1.00 (n=10)	1.00 (n=10)	
R1	ItemD	Unit1	0.95	1.00 (n=10)	0.90 (n=10)	
R1	ItemE	Unit1	1.00	1.00 (n=10)	1.00 (n=10)	

Fig. 10.2 OERS example anchor coding report

a misapplication of coding guidelines in this particular response). With thirty responses for each coder, it is easier to discern the cause of coding issues (linked to the coder, the item, or the type of response). The decision to increase the number of anchor responses was made in the PISA 2018 cycle and is an example of a minor adjustment that can be made to improve the accuracy of reliability estimates without disrupting trend linkage. The other difference between the OERS and the OECS is that the OECS does not separate anchor coding from the assessment coding (no Evaluation Points). When a bilingual coder selects an item to begin coding, the OECS presents anchor responses first (if the coder is a designated bilingual coder), then subsequently begins presenting test taker responses. This positioning is meant to be a smooth transition for bilingual coders, and a separate holding point for coding anchor responses is unnecessary.

10.3.2 Administrative Considerations for Between-Group Coding Reliability

After receiving anchor coding data from all coding teams, assessment designers should review coding agreement statistics and use it to make informed decisions about future rounds or cycles of assessment. When low anchor coding agreement is observed across several countries/economies on a single item, it indicates that coders ubiquitously struggled to correctly apply the coding criteria. In this case, the training for this item should be improved before the next assessment cycle. If the item is still problematic in the following field trial, even after the coding guides for the item have been improved, it should be considered for exclusion ahead of the main survey.

It is also important to consider the number of anchor responses and number of comparisons to the anchor key. Too few anchor responses may not provide enough information to make informed decisions about coding guides or items; too many anchor responses and coding efficiency decreases (coders are over-burdened with coding anchor responses). The PISA administrators have learned from past experience how to balance accuracy of reliability estimates with coding burden efficiency. Prior to the PISA 2015 cycle, coding and scoring reliabilities were not calculated for all items, rather for only a subset of items. This provided domain-level reliability estimates, and coding burden was minimal; however, because cognitive items may widely vary in content, elicited cognitive process, and other subscales, domain-level coding and scoring reliabilities are not sufficient.

Producing item-level scoring and coding reliabilities is also useful during item scaling. In ILSAs, the equivalency of item performance across different populations is evaluated. More precisely, it is examined whether the item functions in the same way across different languages, countries, and assessment cycles (OECD, 2017; Oliveri & von Davier, 2011). In the case of human-coded items, errors introduced by coders' misinterpretation of coding guidelines may lead to item-by-language, item-by-country, or item-by-cycle interactions. This can be examined through coding and scoring reliability estimates. Ideally, coding teams will correct such coding errors ahead of data submission. However, item-level coding statistics provide evidence to what extent this was successful and can help in making decisions about the treatment of items during scaling in cases where errors or other issues are observed.

10.4 Within-Group, Between-Coder Reliability

As with between-group coding reliability, ILSAs should also evaluate between-coder reliability, making sure that codes are consistently applied (in the same way) by all coders across item responses. In PISA, the vast majority of data quality checks focus on between-coder agreement, which is evaluated by comparing the coding decisions of two or more coders on the same response. The more responses that are compared across coders, the more robust a statistic of item-level coding agreement can be provided. In the PISA coding design, 100 responses are coded by multiple coders (multiple-coding) in both the OERS and OECS.

10.4.1 Multiple-Coding in the OERS and OECS

In the OERS, multiple-coding is completed at Evaluation Point 2 (multiple-coding review). Because the assessment responses are contained in paper booklets rather than in a digital file, coders use a coding sheet (shown in Fig. 10.3) to indicate the appropriate response code for multiple-coding. In this example, all possible codes are listed on the coding sheet: "9" and "7" generally indicate omitted responses, "0" indicates no credit, and "1" and "2" indicate both full and partial credit codes. Regarding scoring, items QA, QC, and QD in Fig. 10.3 are all dichotomous items (no credit or full credit) while QB is polytomous (has a third code for partial credit). By recording codes on sheets separate from the test booklet, each coder independently codes student responses (no coder should be able to see the codes awarded by previous coders). Because the OECS works with digital responses, no coding sheets

READING CLUSTER											
Question Number	Unit Name	Item Label	Response Codes (Circle the appropriate code)							Missing	N/A
QA	Unit 1	Unit1Q0A	0	1						9	7
QB	Unit 1	Unit1Q0B	0	1	2					9	7
QC	Unit 1	Unit1Q0C	0	1						9	7
QD	Unit 1	Unit1Q0D	0	1						9	7

Fig. 10.3 Example coding sheet for PBA coding

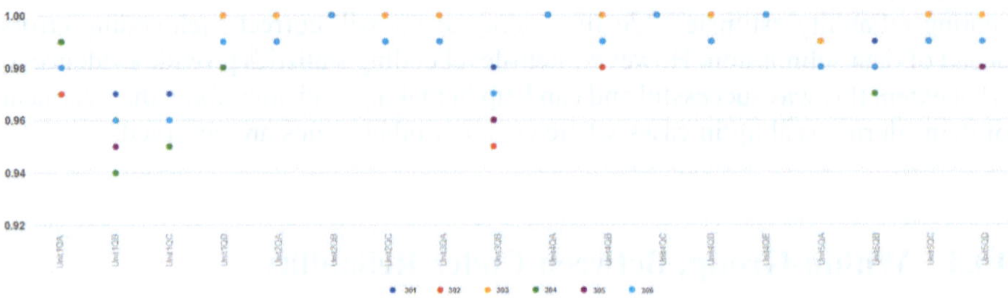

Fig. 10.4 Example proportion agreement graph in the OECS

				Coders					
Cluster	Item	Unit Name	Average	301	302	303	304	305	306
M1	Unit1QA	Unit1	95% (n=3840, g=255, t=1993)	97% (n=640 a=619)	97% (n=640 a=621)	96% (n=640 a=611)	95% (n=640 a=607)	95% (n=640 a=608)	93% (n=640 a=594)
M1	Unit1QB	Unit1	88% (n=3840, g=420, t=1965)	90% (n=640 a=560)	90% (n=640 a=565)	86% (n=640 a=529)	91% (n=640 a=570)	90% (n=640 a=561)	84% (n=640 a=513)
M1	Unit1QC	Unit1	88% (n=3840, g=609, t=1993)	88% (n=640 a=536)	89% (n=640 a=540)	86% (n=640 a=519)	89% (n=640 a=546)	89% (n=640 a=543)	86% (n=640 a=518)
M2	Unit1QD	Unit1	99% (n=3840, g=290, t=1959)	98% (n=640 a=629)	99% (n=640 a=637)	99% (n=640 a=637)	99% (n=640 a=637)	99% (n=640 a=633)	99% (n=640 a=637)

Fig. 10.5 Example proportion agreement table in the OECS

are necessary, and coders work independently from computer terminals, possibly even from remote locations. Coding reliability statistics can be generated throughout coding in the OECS rather than only at a final checkpoint.

The OERS and the OECS produce similar reports on coding reliability and proportion coding agreement, primarily:

- *Proportion Agreement*, which indicates the proportion and percentage of agreement among coders for each item and is presented in two forms, a graph and table
- *Coding Category Distributions*, which provide a detailed breakdown of proportion and percentage agreement on each coding category for an item

The OECS provides both an overview graph (Fig. 10.4) and a breakdown in tabular view (Fig. 10.5) of proportion agreement. In Fig. 10.4, note that the x-axis identifies the item and the y-axis identifies the level of between-coder agreement; the color of the data point corresponds to the coder ID. In this example, coding is incomplete, indicated by only one or two coders appearing with codes for an item. In this coding design, responses for each item will be multiple-coded by four of the six coders. When only one coder has coded an item, the proportion of coding categories will show 100% (1.00 as a decimal) agreement.

A minimum of 92% agreement across all responses for each item is expected in PISA. It was observed that, while employing an efficient coding design, this was an attainable minimum level of accuracy across coding teams (countries/economies, languages, and domains), so it became the conventional level of acceptable coding agreement in PISA. At the item-level, an average range of about 92–98% between-coder agreement is normally observed. The details of item-level agreement are reported in the proportion agreement table (Fig. 10.3), and any item-coder pair with agreement below 85% is flagged by the OECS. Generally, a coder with coding agreement below 85% across all items indicates coding irregularity or coding bias (OECD, 2017), specifically that the coder is scoring too harshly or too leniently compared to other coders. In Fig. 10.5, any item-coder pair below 85% agreement is highlighted in yellow and indicates the need for further investigation (for example, coder 306 on the second item). In the "Average" column, the number of comparisons (n), the number of machine coded responses (g) that were not seen by coders but were used to calculate averages (see Shin et al., 2025 of this volume for more on machine coding in PISA), and the total number of responses (t) are reported. In the columns noted by coder ID, the number of comparisons for the coder (n) and the number of agreements for the coder (a) are also shown.

10.4.2 Coding Team Interventions for Issues with Between-Coder Agreement

Coding reports can provide insight into several potential sources of error, such as coding confusion, coder bias, and item issues. When a coder has low between-coder agreement on one (or two) items (flagged by the coding system), it is possible that that the coder misunderstood the coding guide. The coder can be re-trained on how to code the item, and then he or she can recode all responses for that item with a correct interpretation of the coding guide in mind. When a coder has low between-coder agreement across several items, this is another indication of coder bias or inconsistent coding. In this case, the coder should be re-trained on all items for which he or she shows poor between-coder agreement, and again, item responses should be recoded. In the worst case, if the coder is unable to correctly interpret and apply the coding guidelines and retraining is insufficient, the coder may need to be replaced. Systematic bias should be identified and corrected as early as possible, since the workload to correct coding across items increases over time.

The coding reports can also alert coding teams to issues with items. When several coders disagree on how to code a single item, the cause may be related to the adaptation of the item or the item's coding guide. In this case, it is useful to check the text and interpretation of the item: Does the translation faithfully match that of the original item? Are there any typos? Could the item be ambiguous for test takers (test taker responses are outside of what is covered in the coding guide)? Issues with the item adaptation or coding guidelines should be reported so that they can be addressed in subsequent cycles. If the problem is identified during the field trial, it can be corrected ahead of the main survey. If the problem is identified during the main survey, the item may also show misfit during scaling; the item will either need unique item parameters (the information from the item is still useful for the country/economy, but functions differently in other groups, so the results are not comparable) or the item will need to be to be dropped (if the problem has affected item coherency).

It is also useful to consider properties inherent to the item construct; for example, some items have more than one type of correct answer (often those intended to invoke higher-order cognitive processes) or additional partial-credit coding. Compared to an item with a single correct answer found in the text, items with multiple codes for full-credit, partial-credit, or no-credit are likely to be more challenging for coders to code with a high degree of certainty and consistency. Coding managers should check if the item construct could be reflected in the coding reliability statistics by looking at the coding category distributions to see where coders most commonly disagree. For items with multiple codes for full-credit, partial-credit, or no-credit, it can be expected that between-coder agreement will be lower than on items with only one code for those score categories. Overall, coding teams should thoroughly investigate coding issues, take appropriate steps to correct coding, and report their actions to assessment administrators.

10.5 Within-Coder Coding Reliability

While between-coder agreement is a straightforward comparison of one coder to another, calculating true within-coder agreement would require that a coder recode the same response on multiple occasions. For coders already coding a thousand ILSA responses, this additional workload is likely not cost-effective nor efficient. Another solution is to monitor the coding category distributions of coders as they single-code item responses. Because single-coded items are randomly distributed, each coder's coding category distribution on single-coded responses for an item is expected to be near the item-level average. For coders with coding proportions that differ from the average, it is likely that they are coding irregularly or with systematic bias (rather than that they were assigned systematically different responses than other coders, for example).

In the OERS, single coding is conducted at Evaluation Point 3 (single coding review). At this point, no recoding or retraining can take place to ensure that coders maintain a consistent interpretation of the coding criteria from Evaluation Point 2. In this final stage, single-coding is completed directly in the test booklet, rather than on scoring sheets, and is the final code that will be entered into the international database.

In the OECS, there is no finite separation of single- and multiple-coding, which is another solution to make sure that the statistical reliability of multiple-coding remains the same for single-coding. The drawback, however, is that if an issue becomes apparent from the multiple-coding, between-coder agreement statistics, and

			Coders									Notes to Contractor
Unit Name	Category	Average	201	202	203	204	205	206	207	208	209	
Unit1 - ItemA	0	0.13	0.13	0.14	0.10	0.11	0.15	0.15	0.11	0.11	0.17	
	1	0.70	0.74	0.71	0.75	0.71	0.66	0.65	0.74	0.75	0.64	
	7	.	:	.	.	.	.	.	.	.	.	
	8	.	:	.	.	.	.	.	.	.	.	
	9	0.17	0.13	0.16	0.14	0.17	0.19	0.20	0.15	0.14	0.19	
		(n=1500)	(n=150)	(n=200)	(n=200)	(n=150)	(n=200)	(n=150)	(n=150)	(n=150)	(n=150)	
Unit1 - ItemB	0	0.13	0.13	0.13	0.12	0.11	0.14	0.13	0.10	0.13	0.16	
	1	0.70	0.71	0.70	0.74	0.71	0.66	0.66	0.75	0.73	0.64	
	7	.	:	.	.	.	.	.	.	.	.	
	8	.	:	.	.	.	.	.	.	.	.	
	9	0.18	0.16	0.17	0.15	0.18	0.20	0.21	0.15	0.14	0.20	
		(n=1500)	(n=150)	(n=200)	(n=200)	(n=150)	(n=200)	(n=150)	(n=150)	(n=150)	(n=150)	

Fig. 10.6 Example coding category distribution sheet in OERS

Coding Category Distributions

			Coders		
Cluster	Item	Unit Name	Average	302	304
M1	Unit1QA	Unit1	00 15% | 11 6% | 12 6% | 13 1% | 21 68% | 99 0% (n=579)		
M1	Unit1QB	Unit1	01 48% | 02 28% | 11 3% | 12 10% | 21 4% | 99 4% (n=514)		
M1	Unit1QC	Unit1	00 57% | 11 5% | 12 8% | 13 5% | 21 0% | 22 19% | 23 3% | 99 0% (n=462)	00 52% | 11 5% | 12 10% | 13 6% | 21 0% | 22 19% | 23 5% | 99 0% (n=231) (chi^2=5.836)	
M2	Unit1QD	Unit1	0 69% | 1 11% | 9 19% (n=556)		

Fig. 10.7 Example coding category distribution table in OECS

a coder's responses must be recoded, all single-coded responses for that item and coder would also need to be recoded.

Both the OERS and OECS produce coding category distribution reports (Fig. 10.6 for OERS and Fig. 10.7 for OECS) where each coder's coding category distribution is compared to the average coding category distribution at the item level, based on the Chi square statistic. The coding systems will automatically flag the item-coder pair when the Chi square test indicates significantly different coding. In the example shown in Fig. 10.7, note the slight differences in the distribution of codes to categories between coder 302 and the average across all coders who coded item "Unit1QC." This cell is highlighted yellow to alert the coding team to an inconsistent Chi square distribution for coder 302 on this item.

10.6 Discussion

Ensuring the validity and reliability of human-coded data is a vital component for assuring the accuracy of the assessment and the comparability of test scores across countries/economies, languages, and assessment cycles. Computer-based systems to support coding teams and assessment administrators by monitoring coding and reporting coding statistics are vital to managing and protecting human-coded data quality in ILSAs. PISA's OERS and OECS provide national coding teams with formative statistics on between-coder agreement and overall coding reliability throughout the coding process. The OERS supports PBA coding, and codes for paper-based responses are manually input following coding on paper or in test booklets, and coding reliability statistics are reviewed at designated evaluation points. The OECS operates in conjunction with the computer-based mode of assessment, allowing coders to code all responses to one item at a time rather than a variety of items within a paper test booklet. Because all coding is conducted through an online system, the OECS can report real-time coder agreement and proportion agreement. Both systems produce standardized reports that monitor the progress of human coding throughout the coding process by flagging patterns of coder inconsistency and alerting coding managers to low levels of between- and within-coder agreement. This allows managers to quickly identify and react to problems with items, coding guides, and coders. Following coding, these systems expedite the review of coding reliability across countries.

In many respects, the OECS can be seen as an improvement over the OERS, as there are several more opportunities available when working with digital data than there are when working with paper-based data. For example: Real-time coding reliability updates in the OECS allow lead coders to monitor coding as it is ongoing and more quickly react to unacceptably low reliability estimates; individual coding checkpoints, which may be disruptive if all coding must cease before moving onto the next checkpoint, were eliminated because all responses (for anchor coding, multiple-coding, and single-coding) are distributed to coders automatically in the OECS; and having coders work from a computer terminal, rather than from a coding sheet passed around with the original test booklet (and possibly other coders' coding sheets), naturally enforces independent coding. Once improvements are made, however, it is still crucial to critically reflect on those changes and to listen to feedback from those working with the computer system on a regular basis.

As a general recommendation, both assessment designers and assessment support-system developers should collect feedback from coding teams using the computer-based coding systems throughout the coding process. Such feedback may identify system issues or recommend cycle-level improvements. In PISA, two examples that have been mentioned throughout this paper as sources of error flagged by the OECS are problems with items and problems of insufficient training. Poor items can be identified by low anchor agreement across countries and low between-coder agreement within and across countries, and item quality can often be verified in item analyses (e.g., by poor item parameter fit). If there are no item-related issues, there are several approaches to improving coder training to the benefit of increased coding and scoring reliabilities on the item. Feedback regarding the OECS is also collected. From coding team feedback, it was discovered that coding teams appreciated being able to review their anchor coding agreement early during coding, which was a feature of the OERS, but not the OECS until PISA 2022. While the OECS collected the coding data for anchor responses in PISA 2015 and 2018, the results were not visible to coding teams themselves. As a result, this functionality was built into the OECS for PISA 2022.

A caveat to this is that just because certain functionalities are requested, does not mean that they should be implemented. For example, the OECS was intentionally designed to have limited overviews of individual responses, e.g., limited ability to manipulate response codes by anyone but the original coder. Coding teams frequently request improvements in the flexibility and visibility of coding overviews, such as being able to search for an item and see all codes that have been assigned to a specific response or the coders who have coded specific responses. These requests are made for varying reasons, such as wanting to evaluate the workload of each coder or for better identifying the source of low coding agreement. However, the concern for this type of functionality is that allowing greater control over the specifics of data can be an additional source of data corruption. For example, PBA countries/economies using the OERS tend to report higher levels of coding agreement, sometimes perfect agreement, among their coders. Independent coders who adhere to and correctly apply the coding guidelines, generally, should be able to reach an item-level agreement of about 92–98%, and agreement much lower or much higher than this may be a sign of unreliable coding. In most cases where we see 99–100% coding agreement, it is highly unlikely that independent coding took place. In these cases, it is not unreasonable to assume that the high level of agreement was reached either through team coding (coding as a group) or through coding resolution, in which one or multiple codes are changed post-hoc to artificially inflate the agreement statistics. Such manipulation occurs more often when the system used to record coding relies on manual, rather than automated, recording.

The sheer volume of constructed response data collected in ILSAs, across countries/economies, languages, and cycles of assessment necessitates computer-based coding systems to facilitate independent coding and support coding teams. With efficient national coding designs and quality check procedures, coding systems, particularly those using digital (CBA) data, are able to minimize coding errors within and between coding teams. In longitudinal or long-term cohort surveys, it is also important to regularly identify areas for system and training improvements. As the shift toward computer-based and digital assessment continues, improvement in long-term assessments and their supporting systems is also vital for the assessment to remain accurate and relevant across cycles and through technological advancements. These systems can further be used to improve coder training, coding guides, and thereby, coding reliability. As such, the OECS and OERS are vital to PISA for ensuring the comparability of constructed-response scores across large, diverse samples.

References

Cohen, J. (1960). A coefficient of agreement for nominal scales. *Educational and Psychological Measurement, 20*, 37–46. https://doi.org/10.1177/001316446002000104

Mazzeo, J., & von Davier, M. (2014). Linking scales in international large-scale assessments. In L. Rutkowski, M. von Davier, & D. Rutkowski (Eds.), *Handbook of international large-scale assessment: Background, technical issues, and methods of data analysis* (pp. 229–258). CRC Press. https://doi.org/10.1201/b16061

Oliveri, M., & von Davier, M. (2011). Investigation of model fit and score scale comparability in international assessments. *Psychological Test and Assessment Modeling, 53*(3), 315–333. http://www.psychologie-aktuell.com/fileadmin/download/ptam/3-2011_20110927/04_Oliveri.pdf

Organisation for Economic Co-operation and Development (OECD). (2014). *PISA 2012 technical report.* https://www.oecd.org/pisa/pisaproducts/PISA-2012-technical-report-final.pdf

Organisation for Economic Co-operation and Development (OECD). (2017). *PISA 2015 technical report.* https://www.oecd.org/pisa/data/2015-technical-report/

Organisation for Economic Co-operation and Development (OECD). (2019). *PISA 2018 assessment and analytical framework.* https://www.oecd-ilibrary.org/education/pisa-2018-assessment-and-analytical-framework_b25efab8-en

Organisation for Economic Co-operation and Development (OECD). (2020). *PISA 2018 technical report.* https://www.oecd.org/pisa/data/pisa2018technicalreport/

Pearson, K. (1900). On the criterion that a given system of deviations from the probable in the case of a correlated system of variables is such that it can be reasonably supposed to have arisen from random sampling. *Philosophical Magazine Series 5, 50*(302), 157–175. https://doi.org/10.1080/14786440009463897

Shin, H. J., Yamamoto, K., He, Q., & von Davier, M. (2025). Automatic scoring of constructed responses in PISA using the machine-supported coding system. In L. Khorramdel, M. von Davier, & K. Yamamoto (Eds.), *Computer-based international large-scale assessments: Concepts, methodologies and quality assurance procedures.* Springer.

Chapter 11
Automated Scoring of Open-Ended Written Responses: Possibilities and Challenges

Michael Flor and Aoife Cahill

Abstract Automated scoring of open-ended constructed responses has become a mature technology that is widely used in English-language educational settings, including international assessments, such as TOEFL. Research on automated scoring in other languages is also rapidly developing. It might be tempting to consider the relevance of those developments for large-scale international assessments, including those conducted in multiple languages. Such considerations might benefit from being informed on the current state of scoring technologies and their important related aspects. In this chapter we describe the history and the present state of automated scoring technologies for written responses, with a distinction between essay scoring and content scoring. We outline the current technologies of such systems for scoring responses in English. We also present an overview of research on automated scoring for responses in other languages. Beyond the core technologies, we outline the advantages of automated scoring and the major lines of critique that had been expressed against it. We also describe the importance of additional aspects, such as validity, fairness, and dealing with atypical responses.

Technology for automated scoring of open-ended (constructed) responses has been evolving since the 1960s. Since its inception, there have been many developments in terms of understanding when to use automated scoring, how to evaluate it in the context of different applications, as well as the development of techniques to score responses other than essays (e.g. spoken responses, short responses, dialogues, etc.). The vast majority of research has focused on automated assessment of English responses. It remains an open question whether automated scoring can be applied

M. Flor (✉)
ETS, Princeton, NJ, USA
e-mail: mflor@ets.org

A. Cahill
Dataminr Inc, New York, NY, USA

L. Khorramdel et al. (eds.), *Innovative Digital-Based International Large-Scale Assessments*, Methodology of Educational Measurement and Assessment,
https://doi.org/10.1007/978-3-031-90951-1_11

to open-ended responses in the context of an international assessment administered in multiple languages.

Natural Language Processing (NLP) technology typically forms the foundation for most automated scoring capabilities. In the field of NLP, too, much focus has been on automated processing of English texts, though processing of multiple languages has been a prominent aspect of at least some major sub-fields of research (e.g. in machine translation). The difficulty in taking multiple languages into account in a particular domain is that techniques that work well for some languages do not work at all for others. Take, for example, the sub-task of syntactic analysis and identification of the main subject in a sentence. For some languages that are strongly constrained in their word order (like English, where the subject appears first, followed by the main verb, followed by optional arguments), the techniques to do this may rely on constituent order. However, for a morphologically-rich language with very flexible word order (e.g. Polish or Hungarian), constituent order is less likely to be useful, and a comprehensive morphological analysis of the words in the sentence would be required.[1]

In this chapter we focus on written responses and first give an overview of the history of automated scoring for both essays and content-specific responses. We then describe some of the more common techniques used in these applications in some detail, while also discussing how and whether they could be applied to assessments in languages other than English. Next, we briefly summarize some relevant work on automated scoring for other languages, and, finally, we conclude by presenting the main considerations and challenges associated with developing automated scoring techniques for large-scale assessments.

11.1 Historical Perspective

The earliest example of research on automated essay scoring tools is Project Essay Grade (PEG), developed by Ellis Page (1966, 1968). Page (1968) distinguished two types of variables that may be of interest in scoring an essay: what he termed 'proxes' and 'trins.' A *prox* is a variable that can be programmed to be recognized in a text (i.e., an indicator variable), while a *trin* (from 'intrinsic') is a kind of variable that an expert human rater would be concerned with (e.g., a latent variable). For example, the total number of words in an essay is a prox, which may be related to a theoretical variable "amount of elaboration." While the terms 'prox' and 'trin' are not used anymore, the distinction between indicator and latent/construct variables in writing assessment is still very much alive. That early project was not concerned

[1]Text-analysis techniques often rely on language-specific resources (e.g., dictionaries, morphology and grammar analysis tools, etc.). Although such tools have become more widely available for many languages, their performance is not equal for every language. Thus, for automated scoring, the equivalence of NLP processing across languages is an empirical research question for each given task.

with the latent variables that are commonly discussed in writing theory, such as organization or meaning. It was not concerned with whether the surface features that were chosen as indicators had any validity with respect to theories of writing. PEG was designed to predict an average holistic score given to an essay by human raters. Already in 1968, Page reported a correlation coefficient of .71 between machine scoring and expert raters, over a sample of 276 essays written by high school students.

By the 1990s, when computers became ubiquitous in education and more and more student-written responses became available in electronic form, the promise of automated scoring had motivated a flurry of research. Several commercial scoring engines became available by the turn of the twenty-first century, including e-rater™ from ETS (Burstein et al., 1998), Pearson Knowledge Technologies (Landauer et al., 2003), and IntelliMetric™ from Vantage Learning (Elliot, 2003). By the end of the first decade of the new millennium, automated scoring became widely used for formative assessment in K-12 education and in high-stakes assessments (Elliot & Williamson, 2013; Shermis et al., 2010). By that time, a new application of automated scoring also emerged—scoring of open-ended responses to content-oriented questions (Leacock & Chodorow, 2003; Pulman & Sukkarieh, 2005). For an extended historical perspective on utilization of NLP in automated scoring, see Chapelle and Chung (2010).[2]

11.1.1 Use Areas

At present, automated scoring of essays is widely used for high-stakes assessments, such as college and graduate school admission, for example GRE, TOEFL, GMAT, Pearson Test of English (PTE), postsecondary placement (ACCUPLACER), professional licensure and certification (Uniform Certified Public Accountants Examination). Notably, TOEFL and PTE measure proficiency in academic English and are used, among other purposes, to inform admission decisions for international students seeking to study at institutions of higher learning where English is the language of instruction. For many of those tests, assessment of essay writing forms a part of a larger testing construct, which also includes assessment of reading, listening, and speaking skills. Such large-scale assessments are usually delivered via the internet and are conducted at test centers around the world.

Automated scoring is also widely used for formative assessment, from the elementary school level to the college level. Many automated systems are used for instructional feedback in the classroom and online, such as Criterion (2008),

[2] Another influential tool that debuted at the same time was Coh-Metix, a tool that analyzes English texts on multiple measures of cohesion, complexity and readability (Graesser et al., 2004). Such automated measures can also be useful for estimating aspects of language proficiency (McNamara et al., 2015).

MyAccess (Vantage Learning, 2016), MyWritingLab (Pearson, 2016), Turnitin Revision Assistant (2019), etc. Automated scoring has also been evaluated for use in massive open online courses (MOOCs; Balfour, 2013; Reilly et al., 2014), but its adoption is not yet widespread in that area. Automated scoring also has a long history of use in intelligent tutoring systems, where it is used for assessment of knowledge of specific topics and for measuring student learning outcomes (Magliano & Graesser, 2012).

The main advantages of automated scoring over human scoring are lower cost, fast turnaround, and consistency of scores over time (Dikli, 2006; Weigle, 2013; Williamson et al., 2012; Xi, 2010a; Zhang 2013). Faster scoring can also facilitate faster score reporting. Those factors are particularly important for large-scale assessments, evaluating millions of written responses. Additional advantages are constant availability of scoring and a reduced need to coordinate large teams of raters. For large-scale assessments, large groups of human scorers typically need to be trained, and further, their scoring consistency needs to be monitored (Zhang, 2013). The use of an automated system as a second/control scorer in large scale assessment has been successfully demonstrated for such tests as TOEFL and GRE (Enright & Quinlan, 2010; Monaghan & Bridgeman, 2005). Another important advantage of automated scoring is the potential of real-time feedback to students, where feedback may include the scores, highlighting of errors, detailed comments about various aspects of the composition, and suggestions for improvement (Dikli & Bleyle, 2014; Stevenson, 2016; Weigle, 2013; Williamson et al., 2012).

In international large-scale assessments (ILSA), such as the Programme for International Student Assessment (PISA, n.d.), constructed responses are used to measure content knowledge, as well as different cognitive processes that are part of reading, math, and science skills and cannot be easily measured with multiple-choice item types. Moreover, constructed response items are supposed to measure facets of the skills of interest that are closer to real life situations. Presently, the measurement of writing skills seems to be missing in ILSAs. One reason is probably that they are not easy to measure, especially across multiple languages. A related concern is comparability across languages, which may also affect issues of fairness and validity. An additional concern is the reliability of the measurement, as only a limited number of essays could be included in the assessment. Moreover, rating constructed responses would increase the workload for human raters and, therewith, the costs for each participating country.

However, to participate successfully in society, in the labor market, or in school, reading, listening, and writing skills are needed. Consider writing. Even in the new 'digital' economy, people of various occupations and in various positions are still expected to write reports, summaries, overviews, proposals, claims, reviews, opinions, and commentaries. Thus, the measurement of writing skills is a very relevant issue. Automatic scoring might be useful in that respect, especially for large scale assessments. To include the measurement of writing skills in ILSAs, we would need automated scoring procedures that are efficient, reliable, valid (with regard to measuring the construct of interest), and comparable across languages. This does not necessarily mean that essay formats have to be included in future ILSAs; formats with shorter texts, such as tweets or emails, might be an option as well. The choice depends on the construct definition of the test and on how well-developed the automatic scoring technology is for the particular type of writing.[3]

11.2 Assessment of Writing vs. Assessment of Content

In educational practice, there is a fundamental distinction between assessing the writing skills of students and assessing content knowledge via writing (Weigle, 2013). For assessment of overall writing skills, students write essays on a given topic. On the other hand, short written answers are used for assessment of the students' knowledge and opinions on a particular piece of subject matter, a practice that is often seen as better and deeper than multiple choice questions (Ackerman & Smith, 1988; Bennett et al., 1990; Klein et al., 2002; for a historical review see Bejar, 2017). The past two decades have seen an increased interest in the use of automated systems for scoring and analysis of constructed responses for both kinds of assessment.

The term *Automated Essay Scoring* (AES) refers to the use of computer programs to score essays written by students (Elliot & Williamson, 2013). A synonymous term is *Automated Essay Grading*. A closely related term is *Automated Writing Evaluation* (AWE), which refers to automated analysis and evaluation of written texts for educational assessment (Vitartas et al., 2016), e.g. detection of writing problems for instructional feedback purposes (Stevenson, 2016). AES and AWE typically deal with student-written essays. A related field is known as *Automated Short Answer Grading* (ASAG), *Short Answer Scoring* (SAS), *content scoring*, and *content-focused assessment*,[4] This field focuses on assessing constructed responses to specific questions (Burrows et al., 2015).

Content scoring is more concerned with correctness and accuracy of a written response (relative to a given rubric), while AES/AWE is concerned with overall quality of writing performance (again, relative to a given rubric), i.e. how well can one express ideas in writing (e.g. in English). In the extreme case, content-scoring might care only whether a written answer is correct, disregarding style, grammar, and spelling, whereas AES might be deployed for writing tasks where there is no correct answer (e.g. "write about your recent summer vacation"). However, the distinction is not always clear cut. Some rubrics for general writing proficiency may be rather close to content assessment, for example, where students are expected to demonstrate the use of analytic and argumentative skills beyond basic writing

[3]Using more of the shorter texts, instead of fewer long essays, may also contribute to increasing the reliability of the construct measurement. However, that also depends on the construct that is being measured; e.g., does it ask for short or long responses?

[4]Zehner et al. (2016) suggested the term "Automatic (Short Answer) Coding," noting that grading (scoring) is a special case of coding, i.e. assigning some category X to a text. In scoring, the categories are ordered, with some presumption of an interval scale.

proficiency (e.g. when writing essays to such tests as GRE or TOEFL). On the other hand, content scoring is not only applied for assessing specific knowledge, but also for assessing the expression of opinions and reasoning about particular situations (e.g. explaining some aspects of a science experiment), cases that do not have a clear cut 'correct' answer. AES systems typically address a broad notion of writing quality that encompasses vocabulary, grammar, organization, and conventions, and some aspects of content. Writing tasks routinely scored by AES include expository, persuasive, and narrative essays, source-dependent reading comprehension essays, and English Language Arts (ELA) writing assignments. Successful use of AES has been demonstrated for tasks that require critical thinking in comprehending and analyzing source text materials (Deane et al., 2013; Steedle & Elliot, 2012). Content-scoring tasks tend to be from science, mathematics, history, and ELA reading comprehension (Burrows et al., 2015; Liu et al., 2016; Riordan et al., 2017). Responses in content-scoring tasks are typically shorter than in writing proficiency assessments. Essays typically average between about 100 and 600 words or longer (Shermis, 2014), while short answer scoring may work with constructed responses of just several words or a few sentences (Horbach & Zesch, 2019; Shermis, 2015). In practice, the technologies used for AES and content scoring systems are often closely related, especially in extracting features from texts and using machine learning approaches for scoring.

11.3 Methods for Automated Scoring

Automated scoring applications involve statistical models that are tuned to predict human-assigned scores. Traditional approaches to automated essay scoring utilize two fundamental steps: (a) extraction of linguistically-motivated features from text and (b) learning to combine features into statistical models that can be used to produce scores.

Feature extraction utilizes Natural Language Processing methods and tools. It is often driven by two considerations: what features are considered construct-relevant for a given assessment, and what features can be reliably and accurately computed from a text. For example, the proportion of spelling errors in a text can often be reliably computed. It may be a relevant measure (among other measures) for assessing the competence of a non-native writer in a target language, but it would be construct-irrelevant for assessing knowledge on a specific topic. For scoring models, features are also evaluated as to how much they contribute to the overall scoring model (i.e. predicting human scores). Irrespective of predictive utility, some features may be developed and retained for other purposes, such as feedback on writing quality or for detecting attempts to cheat and game the system.

For overall writing quality, AES systems typically utilize the following types of features: grammar errors, mechanics errors (e.g. spelling and punctuation), usage errors (for example, using wrong prepositions), use of stylistic devices, organization and development of the text, richness and sophistication of vocabulary, lexical complexity, cohesion, transitions and coherence of text, and use of idiomatic phraseology. The presence and proportion of use for specific words, expressions, and n-grams (sequences of *n* words) are also often utilized as features.

There is an important distinction between positive and negative linguistic features. Positive features address aspects of sophistication, such as vocabulary sophistication, syntactic complexity, use of discourse markers and connectors, etc. Typically, more proficient essays have higher values for those features. A simple example of such features is word frequency: An essay that uses a larger amount of less-common words is typically considered to show some vocabulary sophistication and might receive a higher grade. Negative linguistic features reflect errors: counts or proportions of various types of errors in text. Typically, more proficiently written essays have lower values for those features. Common error-oriented features include misspellings, errors in punctuation, fragmented and garbled sentences, run-on sentences, subject–verb agreement errors, pronoun and preposition errors, omitted words, and missing articles. Many linguistically-motivated features are described in the papers presenting various AES systems (Shermis & Burstein, 2013).

Some AES systems compute hundreds of (micro) features from text and then aggregate them into macro-features aligned with specific aspects of the writing construct. For example, the e-rater system incorporates macro-features that measure grammar, usage, mechanics, organization, development, and vocabulary (Burstein et al., 2013; Chen et al., 2017). The use of multiple micro-features to measure construct-defined macro-features can contribute to greater construct validity, as the macro-feature is measured from multiple aspects. It also contributes to better stability of the overall model, the values of individual micro-features may fluctuate greatly (or even be absent for any given text), but aggregation into macro-features prevents too-strong influences from such fluctuations on the whole scoring model.

The same types of features are often used for assessing writing quality for native speakers of a language and for language learners (Quinlan et al., 2009). Crossley et al. (2014) studied the use of linguistically-motivated features for predicting TOEFL scores. Crossley et al. (2016) investigated the use of cohesive discourse markers for scoring essays written by English learners. Vajjala (2018) presented a comprehensive study with a large variety of linguistic features used for scoring EFL/ESL[5] essays for writing quality. Overall, many linguistically-motivated features are useful, but their predictive value (and weights in AES models) differs in different data sets. Thus, AES models need to be developed and tuned depending on the nature of the writing task, the population of examinees who will be evaluated, and the expected range of writing proficiency levels.

Larkey (1998) and Attali (2011) have noted that paying attention to the probabilities of using specific words across essays of different score levels may contribute to better score prediction. However, this comes at the expense of scoring models

[5]EFL stands for English as Foreign Language, and ESL stands for English as Second Language, reflecting whether non-English speaking students study English in an English speaking country (ESL) or in a non-English speaking country (EFL).

being very topic-specific. Prompt-specificity is an important aspect for AES. Usually, scoring systems are trained and tuned separately for each writing prompt; given a target prompt, rated essays for this particular prompt are required for training (Attali & Burstein, 2006; Dikli, 2006; Foltz et al., 1999). Attali et al. (2010) describe an approach to automated essay scoring that avoids using prompt-specific information or vocabulary by shifting the focus 'towards form rather than content,' i.e. more emphasis on vocabulary richness, aspects of grammar, and presence of errors. In addition to prompts, one may consider the difference in writing tasks, such as theme-based vs. source-based writing.[6] Beigman Klebanov et al. (2016) have found that estimating topic-specificity of an essay can contribute to improving the overall automated score for a prompt-independent AES system. Beigman Klebanov et al. (2014) have also shown that prompt-specific information can be especially important for scoring essays in situations where students need to analyze prompt-specific materials (such as in the TOEFL Integrated writing task). Zesch et al. (2015) describe task-dependent and task-independent linguistic features for essay scoring. They found that using only task-independent features leads to worse AES performance than using all features. Thus, the trade-offs of using prompt-specific or task-specific information for automated scoring needs to be considered for any specific assessment.

11.3.1 Machine Learning in Traditional Methods of AES

Development of AES is usually approached as a supervised machine learning task for tuning feature weights and producing a single score for an essay (Valenti et al., 2003). The typical methodology is to divide a set of human-scored essays into three sets: training, tuning (development), and testing. NLP algorithms are used to extract the proposed linguistic features. Notably, the quality of feature extraction varies, depending on the feature and on the tools involved. Feature weighting and tuning of model parameters for score prediction is typically done over the development set. This may require anywhere from a few hundred to thousands of essays that have been prescored by human raters. A standard approach is to regard essay scoring as a regression problem (Attali & Burstein, 2006). Other approaches regard essay scoring as a classification task (Larkey, 1998; Rudner & Liang, 2002) or as hierarchical classification (McNamara et al., 2015). Classification approaches develop models for each level on a nominal grading scale; the text of a new response is then compared to the model of each grading level and is assigned the score of the model to which it is most similar. Yannakoudakis et al. (2011) approach essay

[6]Theme-based writing describes a situation in which students are tasked to write on a specific theme (or topic), usually the one given as a prompt. Source-based writing describes a situation in which students are given some source materials (some of which may be texts) and are required to base their writing on those materials (e.g. summarize, compare and contrast, etc.).

scoring as a ranking problem, i.e. considering that essay scores are on an ordinal scale, rather than an interval scale. In a rare example, Chen and He (2013) demonstrated that the agreement between human raters and the machine can be incorporated into the learning process to improve the scoring model.

A different method of automated scoring uses Latent Semantic Analysis (LSA; Foltz et al., 1999). In LSA-based approaches, a large matrix of word-document co-occurrences is collected, which is then algebraically abstracted while reducing the number of dimensions in the matrix. As a result, it learns to associate a word (type) with a vector of numbers that serves as its 'meaning representation' (Landauer et al., 1998). To build a scoring model, human-scored essays are collected. For each essay, the words are converted to vectors and aggregated (typically by averaging the vectors). A new essay about to be scored is similarly converted to a set of vectors and then aggregated. Then, the task is to find to which model (set of essays prescored by experts) the new essay is most similar—by comparing the aggregated vector components of the new essay and the vectors(s) of each score-level model (Kakkonen et al., 2005; Landauer et al., 2003; Wild et al., 2005). Alternatively, a score for a new essay may be the average of several most closely matched prescored essays (nearest neighbors). Due to abstraction beyond specific words, such models can be very powerful for capturing semantic similarity, but they do tend to be topic-specific.

For a detailed description of many current AES systems that use the traditional features and learning architectures, as well as LSA-based approaches, see Zupanc and Bosnić (2015) and Shermis and Burstein (2013). For an up-to-date review, see Ke and Ng (2019).

11.3.2 Content Scoring

Automated content scoring systems are developed to assign a score or a correctness label to student responses. They are typically used for grading specific content-related questions or to asses reading comprehension. Such systems need to deal with many sources of diversity in student responses (Horbach & Zesch, 2019). One factor is response length; longer responses allow for more diversity. Another factor is the target student population and their language proficiency. While more proficient students and native speakers of a language may be expected to have greater variety in their answers, considerable diversity may arise for other groups as well. Student responses may contain spelling and grammatical errors, which make it more difficult to process by automated systems. A very important factor is the nature of the question/topic itself; some questions call for very constrained answers, while other questions may allow wider ranges of acceptable answers.

There are two main techniques for evaluating whether the content of a student response is accurate (or partially accurate). The decision on which approach is more suitable often depends on the complexity of the item/prompt/scoring-rubric, as well as the amount of human-scored data available.

The first approach—sometimes called *reference-based*—compares new responses to a small number of "exemplar" or "reference" answers at each score point. This approach works well when an expert can easily enumerate all possible ways to respond to an item, or when the response requires a few simple concepts. An automated system measures the overlap between a student answer and the teacher/instructor reference material. Automated similarity measurement techniques (see Agirre et al., 2013 for an overview) are often used to compare responses to the exemplar content. Early systems that utilized this approach were described by Leacock and Chodorow (2003), Pulman and Sukkarieh (2005), and Sukkarieh and Blackmore (2009). Modern techniques usually rely on the notion of distributional similarity, which works well for detecting paraphrases. An advanced approach, using sentence alignment and semantic similarity measures was described by Mohler et al. (2011). One disadvantage of the reference-based approach is that enumerating all potentially correct responses might be relatively easy for some types of questions (typically for restricted factual questions), but quite difficult or even impossible for other types of questions. For example, when a question invites the writer to express opinions or evaluations (even for a given stimulus text), the response variability can be huge, but which (kinds of) responses are correct or 'adequate' is not often clear in advance. Another disadvantage of this approach is that students are generally creative, and preparing a system for diverse responses can be, in many cases, a very challenging and time-consuming task.

The second approach—sometimes called *response-based* or *instance-based*—takes advantage of a large enough set of human-scored responses to build an automated scoring model using machine learning techniques (Higgins et al., 2014; Horbach & Zesch, 2019; Madnani et al., 2017). The techniques used for this approach are usually either (1) traditional machine-learning models (such as support-vector machine models) that rely on vocabulary-based features in addition to deeper syntactic and semantic features (Ramachandran et al., 2015; Sakaguchi et al., 2015) or (2) neural network models that take vector-based representations of the responses as input (Riordan et al., 2017). The main advantage of the response-based approach is that it does not require explicit enumeration of all possible responses, but it allows the system to generalize reasonably well. Basically, a response that is more similar to the high-scored responses will be scored higher. The main disadvantage of this approach is that, depending on the complexity of the item, a large collection of human-scored data may be required in order for the model to automatically learn enough generalized patterns to be able to accurately score new responses.

A large corpus of questions and scored student responses was compiled in 2012 by Dzikovska et al. (2013), as part of the SemEval Shared Task. Another data set of scored, short-answer responses was released as part of the Hewlett Foundation sponsored ASAP (Automated Student Assessment Prize competition series; ASAP, 2012; Shermis & Hammer, 2013). These large datasets have been used to train and evaluate various automated scoring NLP systems, usually based on traditional machine learning or deep-learning approaches. For comprehensive reviews of methods and systems for automated, short-answer grading, see Burrows et al.

(2015), Galhardi and Brancher (2018), Roy et al. (2015). A comprehensive list of data sets used for evaluation of automated content scoring is presented by Horbach and Zesch (2019).

11.3.3 Deep Learning Approaches to Automated Scoring

Recent research has demonstrated the possibility of building scoring applications using neural networks and deep learning architectures (Alikaniotis et al., 2016; Dong & Zhang, 2016; Taghipour & Ng, 2016; Zhang & Litman, 2018; Zhao et al., 2017). In particular, words of an essay/response are recoded into distributed word representations (so called word-embeddings; Levy & Goldberg, 2014), and a neural network model is trained to combine the information, resulting in a single dense vector representing the whole essay. A score is given based on a non-linear neural layer over such representation. These models are trained using rather large sets of prescored essays. Riordan et al. (2017) investigated how several basic neural approaches suited for automated essay scoring (writing quality) perform on short answer scoring (content scoring). They found that neural architectures can outperform a strong non-neural baseline, but performance and optimal parameter settings vary across the prompts, due to wide diversity of content in the short-answer prompts.

Neural models were shown to achieve state-of the-art scoring results on some experimental datasets. For example, for the ASAP dataset, quadratic weighted kappa (QWK) agreement with human raters is at about 0.77 (Riordan et al., 2019). One of the often-touted advantages of neural architectures is that they eliminate the need for development of linguistic features (Taghipour & Ng, 2016). Deep learning systems excel in learning patterns in the data without much explicit guidance. Nevertheless, for specific tasks (like essay scoring), broad linguistic decisions are still required; for example, should the system involve syntax/grammar, or not, or discourse relations (Nadeem et al., 2019)?. With deep learning approaches, there is a shift in focus from engineering linguistic features towards neural-network construction and parameter tuning. For a review of recent neural network approaches in essay scoring, see Ke and Ng (2019).

The utility of neural models for large-scale assessments or even classroom assessments remains to be investigated. Since neural models are largely 'opaque,' it is very difficult to analyze what exactly they capture and what influences their performance. Unlike scoring systems that extract linguistic and content features, current neural scoring architectures cannot be used directly for feedback; they do not 'tell' you why the specific score was assigned and what might be wrong (or good) in your essay.

The opacity of AES systems is also related to their acceptability by educational experts and by the public in general. When a scoring engine relies on generally accepted types of feature/components (such as spelling, grammar, cohesion, etc.) and combines them in well-known ways (e.g. regression), it might be more 'explainable' and more acceptable. When the system architecture is more opaque, it becomes harder to explain exactly what the score is based on, and thus, people may become skeptical as they do not understand (even intuitively) what the automated score represents.

However, as this field of research is rapidly developing, both for general interpretability of neural models (Linzen et al., 2019), and for educational applications, such as Grammatical Error Correction (Bryant et al., 2019), the aspect of opacity may soon find some solutions.

Another important aspect is that current neural models can be quite brittle, in the sense that their performance can be greatly affected by the quality of training material and some aspects of the input data. This has been demonstrated for the task of machine translation (Belinkov & Bisk, 2018; Khayrallah & Koehn, 2018) and for the task of automatically detecting grammar errors (Htut & Tetreault, 2019). More research is required to understand how such systems can be made more robust and reliable to be used in operational assessments.

11.4 Automated Scoring Beyond English

The bulk of research in AES and ASAG has concentrated on writing in English, due to the widespread use and international importance of the English language, wide availability of NLP tools for English, and also due to availability of scored writing data sets in English. Research on AES and ASAG in other languages is emerging.

Lemaire and Dessus (2001) used LSA to build a prototype AES system for French. The IntelliMetric system from Vantage Learning was claimed to support AES in multiple languages, including Chinese (Schultz, 2013). Östling et al. (2013) developed an AES system for high school essays written by native speakers in Swedish; the system used both positive and negative linguistic features. Lilja (2018) presented a modern neural network architecture for scoring Swedish essays. Vajjala and Lõo (2013, 2014) developed an approach for automatically predicting writing proficiency in Estonian, with an emphasis on utilizing specific morphological aspects of that language. Ishioka and Kameda (2004, 2006) presented an early AES system that scored Japanese essays by comparing them to exemplar texts written by professional writers and assessing several sets of positive linguistic features. Yamamoto et al. (2018) described an AES system for scoring essays in Japanese, which uses positive linguistic features and error counts. Chang et al. (2006, 2007) presented an early attempt towards an AES system for Chinese, intended for scoring essays by native speakers. Peng et al. (2010) described Chinese AES systems that use vector-space models (including LSA). Ke et al. (2016) describe automatic scoring of university-level essays in Chinese using graph-theory approach. Yu et al. (2017) described an AES system for Chinese essays, that uses positive linguistic indicators (length, complexity, sophistication). Hsieh et al. (2017) describe a system designed to give feedback to writers who are learning Chinese.

Yang et al. (2020) describe a graph-based system for evaluating essays written by Chinese middle school students.

Wild et al. (2005) presented an essay scoring system for German, based on LSA. Hancke and Meurers (2013) described an approach to classifying the proficiency of German-language essays using a variety of positive linguistic features. Horbach et al. (2017) investigated automated scoring for a complex writing task with essays written by native German-speakers. Although they applied state-of-the-art machine learning methods, only moderate success of automated scoring was obtained (QWK of up to .49 with human rater scores), due to the reduced variety of language proficiency in that data set. Zehner et al. (2016) presented an approach to automated coding of short text responses in German, collected in PISA. They clustered similar responses using LSA before applying machine learning for scoring. Castro-Castro et al. (2008) described an essay scoring system for Spanish and Basque, based on extraction of both positive linguistic features and errors. Amorim and Veloso (2017) described an approach to automatic essay scoring for Brazilian Portuguese, using topical and positive linguistic features, but achieving only modest agreement with human scores. Ben-Simon and Cohen (2011) described an AES system for Hebrew; the same system was later used in a validation study by Cohen et al. (2018).

Berggren et al. (2019) investigated automated essay scoring for non-native learners of the Norwegian language, exploring traditional feature-based approaches and the newer deep-learning models. Westera et al. (2018) describe an automated scoring system for the Dutch language. Kakkonen and Sutinen (2004) presented an essay scoring system for Finnish that uses LSA to compare essays to subject-matter-specific text passages from the course materials. Kakkonen et al. (2005) extended that approach to compare new essays to essays scored by experts. There are several research publications about AES systems for Arabic. Al-Jouiea and Azmia (2017) presented a system that uses a combination of LSA and a rhetorical structure parser that estimates levels of cohesion in text. Alghamdi et al. (2014) described a system that uses LSA and error counts. A system presented by Mezher and Omar (2016) uses LSA but distinguishes words by their part-of-speech, so it is sensitive to some aspects of grammar. Loraksa and Peachavanish (2007) reported an experiment on building an AES system for the Thai language, using LSA and a neural network. Automated systems for scoring short-answer responses in Korean were presented by Cheon et al. (2015), Jang et al. (2014), and Kim et al. (2018).

Most of the non-English systems follow the general blueprint of traditional automated scoring systems for English—a combination of linguistic features (often language-specific) and a machine learning component for deriving a scoring model. For an extended discussion of automated scoring in languages other than English see Shermis (2020). As noted by Shermis, most of those systems are proofs of concept, not active operational systems.[7]

[7] We could not find any published reports on operational use of non-English AES/ASAG systems, although there are some signs that schools in China might be trying out such technology (e.g. Chen,

Some of the techniques used for automated scoring in English can be relatively straightforwardly applied to new languages, particularly those that rely on only the words that appear in the response. However, most techniques would require some adaptation to work with multiple languages. Some of the semantic similarity measurement tools rely on hand-crafted resources, like lists of synonyms or a thesaurus, which may not be available—or of sufficient size—in other languages. Techniques that rely on syntactic or semantic features would need access to parsers and other NLP tools in each language. Cohen et al. (2003) provide a rare discussion about unique features of the Hebrew language (grammar and lexicon) that need to be taken into account for an AES system. Zehner et al. (2016), working on short-answer scoring in German, note the need for umlaut-normalization (e.g. changing ä into a) for recognition of related words. Zesch et al. (2015) describe an experimental AES system that can be adjusted to work in either English or German.

There are increasing efforts to build large-scale training resources for grammar-analysis tools (e.g. the Universal Dependency Treebank initiative; Universal Dependencies, n.d.), though the quantity of available data still differs by language. Still other techniques rely on word-vector representations as their inputs, and though there are many pre-trained vector resources now available (e.g. Grave et al., 2018, describe word-embeddings for 157 languages), care would need to be taken that the multi-lingual resources perform similarly and equivalently across languages. Another trend that is gaining much traction in NLP research is cross-lingual transfer, in which tools are task-trained in one language and then trained to transfer the task-performing capability into another language (Chen et al., 2019; Lin et al., 2019). The promise of this approach for automated scoring is yet to be explored.

11.5 Considerations and Challenges for Operational Assessments

11.5.1 Atypical Responses

While most student responses typed on a computer can be processed and scored by an automated system, some responses may have atypical characteristics that make it difficult for an automated scoring system to assign a score (Lochbaum et al., 2013; Reinertsen, 2018; Yoon et al., 2018; Zhang et al., 2016). These are often called *non-scorable* or *aberrant* responses. Some of the categories of such responses include (a) response in non-target language, i.e., not in the targeted language of the examination; (b) off-topic responses, i.e., not on the assigned topic; (c) short generic responses, like "I don't know", etc.; (d) repetition of the prompt question; (e) reduplication of sentences or paragraphs; (f) canned responses, e.g., typically

2018), and AES is intended to be used on the MHK Chinese language proficiency test for non-native speakers (Zhang & Lin, 2017).

generic, memorized responses; (g) random characters, e.g., banging on the keyboard; and (h) no-response at all. In addition, even valid responses can have a certain amount of noise (e.g. emoticons, ASCII art, etc.), which can be problematic for NLP systems.

Unusual responses may be detected at several points in the processing stream, before scoring occurs, during scoring, or even after scoring. A pre-filtering stage may be utilized to flag out responses that do not have valid materials to be scored (e.g. categories *(a)*, *(d)*, *(g)*, and *(h)* above). For other types of aberrant responses, sophisticated processing may be required for detection.

To detect responses that are in a non-target language, Yoon and Higgins (2011) utilized language identification technology. "Banging on the keyboard" can be identified by analyzing character and part-of speech sequences and looking for ill-formed sequences (Higgins et al., 2006). If a system is equipped with a vocabulary database in the target language (a dictionary), a potential filter could just check what proportion of the input (words) is found in the dictionary.

Detecting other types of aberrant responses may be more akin to detecting of gaming and cheating inputs (Lochbaum et al., 2013). Increasing the length of essays by padding an essay with additional words is one gaming strategy aimed at achieving higher scores. If words are added more or less randomly, an entropy-based approach may be used for detection (since random sequences have low probability). Reduplication of many sentences and paragraphs requires a different approach, such as string repetition detection. Threshold values need to be considered beyond which a response would be flagged as aberrant.

Approaches to detecting off-topic essays have been developed based on prompt/question-specific content models, such as the classic vector space model (Higgins et al., 2006; Louis & Higgins, 2010). More recently, word-embeddings vector models were successfully used for flagging off-topic speech responses (Malinin et al., 2017; Yoon et al., 2017). Yamamoto et al. (2018) utilize vector-space cosine similarity for estimating topicality of written essays to assigned rubrics. Such approaches build generalized lexical or abstracted models of expected content. Responses that do not match such models beyond some predefined threshold levels are flagged as potentially off-topic.

Canned responses are often generic texts (or snippets) that are memorized by test-takers prior to the exam. Detection of canned responses can be partially achieved via off-topic essay detection, especially for writing tasks where content is expected to be very topical. Canned responses can also be considered under the rubric of plagiarism detection. Potthast et al. (2014) provide an overview of approaches to detecting plagiarism in written texts.

When a response is flagged as aberrant in a scoring situation that has a human in the loop, the standard approach is to route the response to a human rater for scoring. The score from the automated system may or may not be combined with human scores, depending on the use case and the kind of flag. If a response is flagged as aberrant when an AES system is a sole scorer, it might be handled in one of the following ways. Either a score is given, accompanied with a warning that it is unreliable, or no score is given, and a warning is registered. For a large-scale assessment, the amount of material flagged and rerouted for human decision makers might become a serious issue, with implications for cost and time that are needed to process the flagged data. On the other hand, failure to properly handle aberrant responses in a testing situation may negatively impact the validity and public perception of the test.

An automated scoring system needs to be robust towards aberrant responses. Improper handling of aberrant responses may reduce the accuracy, the success rate, and the perceived validity of a system. Moreover, automated scoring systems should cope with aberrant responses not only in the final deployment, but also during development and tuning (machine learning) for the task because aberrant responses may skew and distort the training process and, thus, contribute to decreased accuracy, validity, and reliability of an automated scoring system.

Reinertsen (2018) also mentions a small category of written responses that were rejected by an AES system, but were still scorable by human raters. For large-scale assessment, this might also become a problem; if a system flags too many responses as aberrant, it may place an increased burden on human raters/controllers to process all those responses manually.

11.5.2 Evaluation of Automated Scoring

A widely adopted, fundamental approach to evaluation of AES and ASAG systems is comparing the scores given by the computer to the scores assigned by human raters to the same texts (Bennett & Zhang, 2016; Burrows et al., 2015).

Technical aspects of evaluation may depend on the scale used for scoring. For many writing-proficiency tasks, the holistic scores of essay quality often lie on a scale of 1–5 points (for example, TOEFL) or 1–6 points (for example, GRE, and CEFR). In such cases, the standard measures are the Pearson correlation and the quadratic weighted kappa (Williamson et al., 2012). The latter is used to stress adjacent agreement and penalize gross differences in scores. For content scoring, Ziai et al. (2012) recommend the use of the Root Mean Squared Error (RMSE) and the Pearson correlation coefficient. Notably, content scoring tasks can often have a very limited scoring scale, such as, 0–2, or even simply correct/incorrect. Yannakoudakis and Cummins (2015) provide a critical discussion of the evaluation metrics and agreement measures for automated scoring systems.

Summarizing the accuracy of AES systems in the first decade of the twenty-first century, Vojak et al. (2011, p. 102) wrote: "AES systems have shown a high degree of accuracy—that is to say, the holistic scores assigned by computers have a high rate of exact or adjacent agreement (scoring within one point on either side of the exact score) with expert human raters. For example, e-rater scores within a single point of expert human scorers 97% of the time, which is similar to the discrepancy rate between two human scorers … . IntelliMetric scores within a single point of expert human raters 99% of the time …, and the Intelligent Essay Assessor (the scoring engine for WriteToLearn) scores a similar 98.1%". There were some exceptions. A study by Wang and Brown (2007), found that in grading college essays the IntelliMetric system gave significantly higher grades than two trained faculty members. Since then, extensive research has examined the validity and reliability of various AES systems. In 2012, the Hewlett Foundation sponsored a series of competitions called Automated Student Assessment Prize (ASAP, 2012). For an overview of that initiative, see Shermis (2014). In 2013, a special issue of the journal Assessing Writing was devoted to the issues of AES validity (Elliot & Williamson, 2013). The general finding has been that AES systems can show high agreement with human scores when grading essays written in good faith (Bennett & Zhang, 2016; Shermis, 2014; Stevenson, 2016). However, care must be taken as for how the systems are trained and what exactly they are deployed for. The volume edited by Shermis and Burstein (2013) provides some detailed results and in-depth discussion of the reliability and validity of several specific AES systems. Automatically assigned scores are considered reliable enough that they have replaced one out of two human annotators, even in high-stakes language proficiency tests such as TOEFL (Attali & Burstein, 2006; Enright & Quinlan, 2010). Indeed, the practice of utilizing automated scores together with human ratings can contribute to increased acceptability for deploying automated scoring (Attali, 2013; Bennett & Zhang, 2016). For large-scale assessments, this may require considerable organizational efforts.

A complementary perspective on validity is to take a good look at the human raters. Although human raters are a 'gold standard' for evaluating AES systems, characteristics of human raters that are used for evaluating an AES system do matter. Powers et al. (2015) compared e-rater scores for a set of persuasive essays written for the GRE exam to scores given to the same essays by different groups of human raters. They found that the AES matched the scores of trained expert raters much better than the scores assigned by naïve, untrained raters.

For a recent in-depth consideration of issues of validity in automated scoring, see Bennett and Zhang (2016). Issues of validity are still actively researched, as can be seen by a continuing stream of publications on this topic (see e.g. Cohen et al., 2018; Kieftenbeld & Boyer, 2017; Raczynski & Cohen, 2018). For automated short answer grading, Liu et al. (2016) state that the accuracy of automated scores depends on a number of factors, including the content domain, the complexity of the tasks, the levels of the scoring rubrics, and the number of responses available to build the automated scoring models.

Williamson et al. (2012) describe a detailed framework for evaluation and use of automated scoring and emphasize guidelines for aspects of implementation and maintenance of such technologies. They emphasize that the reliability and validity of a system may change when a system is updated, when it uses different settings or when it is used with a population of examinees for which it was not intended.

Issues of reliability and validity have been extensively explored for automated scoring in one language—English. For an international assessment, potentially administered in multiple languages, additional aspects may be relevant. Just as comparability of multilingual forms of a test needs to be considered (Ercikan et al., 2004), comparability of automated scoring in multiple languages would require scrutiny. Either if it is one engine applied for multiple languages, or a set of language-specific systems, comparability of their performance for a given assessment/test would need to be studied.

11.5.3 Fairness Considerations and Automated Scoring

The notion of fairness in the design and use of educational tests has a long history (Hutchinson & Mitchell, 2019; Poe & Elliot, 2019; Xi, 2010b; Zieky, 2016). For an extended discussion of the legal and social aspects of test fairness see Camilli (2005). In general terms, a test is considered to be fair if the differences in test scores between the test-takers are due only to differences in skills relevant to the task, i.e. skills that are part of the construct (Zieky, 2016). While fairness can be compromised by issues in test design and use, it may also be affected by the properties of the scoring procedure (Penfield, 2016). Bridgeman et al. (2012, p. 29) note: "Evidence of differential performance for subgroups has obvious fairness implications for individuals in high- to moderate-stakes assessments, but such evidence would also be of concern for … assessments … that attempt to track the progress of various subgroups. Such tracking could be compromised if automatic scoring were substituted for human scoring and the two scoring methods ranked subgroups differently."

Fairness of automatic essay scoring has been researched much less than validity issues. A common approach is to investigate mean score differences between relevant subgroups, then compare AES system scores with human-assigned scores (Ramineni et al., 2012a, 2012b; Williamson et al., 2012). A standard distinction is for subgroups by gender (Zhang et al., 2017). For tests administered in countries with considerable ethnic diversity, comparisons between ethnic groups are also included (Guo, 2009; Ramineni et al., 2012a). If language proficiency is not part of the construct, a comparison of fairness in scoring between native vs. non-native speakers of the tested language may be conducted (Guo, 2009). For international tests (such as TOEFL), a relevant comparison is between test-takers from different countries of origin, examinees with different L1 backgrounds, or by countries in which test centers are located (Bridgeman et al., 2012; Ramineni et al., 2012a, 2012b; Xi, 2010b). Zehner et al. (2018) used automatic NLP tools to analyze difference between genders in responses to the PISA reading assessment in German (this test was not scored by machine, but the use of automated technology for uncovering between-group differences is notable).

Recently, some additional aspects of fairness have been investigated, such as fairness to groups of people with disabilities. Buzick et al. (2016) examined the fairness of automatic essay scoring for examinees with or without learning disabilities and ADHD. They found that human raters and the AES system (e-rater) assigned highly similar essay scores for each of the subgroups. Loukina and Buzick (2017) investigated discrepancies in automated speech scoring for speakers with and without speech impairments.

Loukina et al. (2019) present a discussion of various forms of fairness that may be expected from AES systems that rely on machine learning, specifically due to the statistical nature of such systems. They distinguish three basic forms of fairness: (1) Are automated scores equally accurate for all groups of interest? (2) Are automated scores consistently higher or lower than human-assigned scores for members of different groups? (3) Are automated scores consistently higher or lower for members of different groups despite them having the same language proficiency? Their simulation study demonstrates that achieving total fairness in all perspectives may not be feasible.

In one particular sense, automated scoring is fairer than human scoring, in the sense that automated scoring is consistent: It applies exactly the same grading criteria to all submissions. This is not always the case with human raters, due to a variety of cognitive factors, such as leniency, fatigue, over-reliance on some restricted aspects of writing, etc. (Bejar, 2012; Zhang, 2013). Beyond that, the fairness of any particular automated scoring system needs to be established separately. Even the applicability of an established system to a new type of scoring task may raise concerns of fairness.

11.5.4 Critiques of Automated Scoring

The widening adoption of automatic essay scoring systems has raised a set of concerns about the adoption of such technologies (NCTE, 2013). Critiques of automated scoring can be roughly sorted into three categories. One concern is that automated scoring often focuses on objective 'surface' features of writing (such as spelling and grammar) and is less sensitive to ideas, argumentation, and creativity (Perelman, 2014; Rothermel, 2006). However, Deane (2013) argues that there is an important connection between surface features and deeper constructs of writing for automated scoring. Moreover, new methods for automatically measuring deeper aspects of writing have been developed, such as measuring sentiment (Farra et al., 2015), argument quality (Song et al. 2014; Stab & Gurevych, 2017), and narrative quality (Somasundaran et al., 2018). Integration of such novel aspects into operational essay scoring may be expected in the near future.

Another line of critique argues that, in writing to a machine, students lose the social purpose of writing. Alderson and Wall (1993) expressed concerns that awareness of scoring methods on the part of students and teachers will lead to negative consequences for instruction and study habits. This also applies to automated scoring. When students become aware that essays are scored by machine, they may attempt to write to the machine rather than to a human audience. They may be less motivated to convey meaning, bring interesting arguments, or engage the reader (Condon, 2013). Thus, awareness of automated scoring may lead students to suppress the creative, reader-oriented aspects of writing.

A third line of criticism argues that when writing for a machine, students might be aiming to "beat the machine" and score a higher grade (Herrington & Moran, 2001).

Depending on the algorithms that are used, computer-based scoring systems may be less sensitive than human scorers to some aspects of writing, for example, word order or vocabulary placement. Thus, the possibility of cheating and gaming the system to inflate scores becomes a widespread concern, especially when AES is used for high-stakes assessments. Lochbaum et al. (2013, p. 3) define gaming of AES as "deliberate injection of construct-irrelevant features in order to influence measures of performance".

Powers et al. (2001) conducted one of the earliest studies designed to test to what extent an AES system can be fooled by writing essays in bad faith. A team of writing experts were given detailed information on the linguistic features by the then-current version of the e-rater AES system. On the basis of this information, they composed two kinds of essays—one set designed to receive an artificially high score and another set designed to receive an artificially low score. Expert writers were successful in tricking the system into assigning overly high scores (26 of 30 essays) and somewhat successful in the other direction (10 of 24), as compared to ratings of the same essays by human raters. Notably, gaming a system by well-informed experts is not the same as gaming by test-takers. Nevertheless, test takers may resort to a variety of approaches (strategies) aimed at receiving undeservedly high scores. Some of the strategies can be focused at manipulating the syntax and vocabulary, features that are often used in AES. Bejar et al. (2014) studied to what extent e-rater scores might be inflated if test-takers memorize a list of rare words and sprinkle them in their essays. They found that only a small fraction of e-rater scores would increase when using this strategy. Higgins and Heilman (2014) presented a study on the susceptibility of automated scoring engines to gaming and cheating. They explored the following strategies for artificially improving the scores: (1) lengthening the text, (2) re-using words from the prompt, and (3) overusing general academic words. The study showed that some simple strategies can sway the scoring engine performance. They also noted that the estimation of the susceptibility cannot be made based on engine design and configuration alone.

Another approach to cheating is systematic plagiarism, where test-takers submit a high-quality, canned essay as if it were their own composition. In many testing situations, such an essay would be off-topic from an assigned prompt, thus an example of trying to beat the system with complex syntax and vocabulary but inaccurate content. Some operational AES systems already include algorithmic components for detecting and flagging off-topic essays (Higgins et al., 2006; Lochbaum et al., 2013), and new systems would be required to follow suit.

A different approach to gaming AES was demonstrated by the Babel system (BABEL Generator, 2014). Babel artificially generates essays based on several keywords, using sophisticated canned sentence structures and rare words. Content-wise, the resulting essays are nonsensical to the human reader. Such essays were shown to receive high scores from a number of scoring systems (Kolowich, 2014), which was considered to undermine the validity argument for AES. However, Cahill et al. (2018) have demonstrated algorithms that successfully distinguished between Babel-generated essays and a variety of essays written by test takers in good faith.

Their methods involved analysis of grammar, discourse coherence, and lexical cohesion in text.[8]

While Babel-generated essays are not a very practical cheating method, this case exemplifies a certain dynamic in the field of automated scoring. When specific weaknesses in automated scoring systems are discovered, developers try to build in filters or flags for detecting various kinds of cheating (Zhang et al., 2016).

When considering automated scoring in large-scale assessments, educators should be aware of the various cheating methods to which particular systems might be susceptible. While certain approaches to cheating, such as plagiarism, are generic, other approaches to gaming a system may be based on some understanding of how the scoring system works. When such knowledge becomes widespread (through public media or test-preparation coaching agents), the problem of cheating would require special attention.

11.6 Addendum

Since this chapter has been composed, the automatic scoring domain has seen a lot of new developments and publications. For a thorough introduction to automated essay scoring, we recommend the book by Beigman Klebanov and Madnani (2022). For a survey of deep learning approaches to essay scoring see Uto (2021). For some more advanced treatment of neural scoring, see Mayfield and Black (2020). Bai and Stede (2023) provide a good introduction to several kinds of scoring systems that use neural approaches. Litman et al. (2021) describe some approaches to fairness evaluation for automated scoring systems.

11.7 Conclusions

Inclusion of constructed writing responses in international large-scale assessments might be worthwhile, both because constructed writing responses can be useful for measuring complex cognitive processes and because writing skills are quite relevant in modern economy and education. Wider use of constructed responses in international testing might be facilitated by automated scoring technologies, if such technologies can provide cost-efficient solutions that are reliable, valid, and comparable across languages.

In this chapter we presented an overview of the current state of automated scoring technologies for written responses. We noted the distinction between essay scoring, which generally focuses on writing proficiency, and content scoring, which focuses

[8] Although those kinds of features are typically used in AES systems, the successful utilization of such features for detection of artificial essays was a new development.

on measuring specific aspects of knowledge and reasoning. For assessments in the English language, both fields are very advanced, in particular, showing good agreement with human-assigned scores. For essay scoring, several commercial engines are already well established, with extensive literature covering aspects of performance and validity, and with experience in high-stakes and international testing.

From a technical perspective, established approaches for automated scoring rely on NLP technologies for feature extraction from texts and on machine learning for developing scoring models from features. A new generation of scoring technologies is rapidly evolving, based on neural networks and deep-learning architectures. However, those newer methodologies are currently limited to research literature and have not been tested in operational environments. Both established and new approaches require some considerable amounts of data prescored by human raters. In general, automated scoring systems need to be tuned to the specific writing task they are set to score. In many cases, tuning must be specific to the topic and the writing prompt, especially for content scoring. For measuring general writing proficiency, non-topical approaches are also available.

Depending on the level of analysis required for a particular task, the adaptation of existing techniques available for one language may not be immediately transferable to other languages. This poses a considerable challenge to the task of developing automated scoring for large-scale international assessments administered in multiple languages. On the other hand, techniques that do not rely on sophisticated automated linguistic analysis might be easier to 'port' to other languages. For example, there have been considerable research efforts in the area of vector-based representations of words, which can be automatically trained from large amounts of raw data in a given language. In theory, if a large enough, relevant, digital corpus is available in a given language, such resources could be trained. Such resources then may be sufficient for certain techniques, including deep learning (neural networks) or vector-based distributional similarity. However, such techniques may not always be appropriate for a given task, and careful evaluations would be required to ensure adequate performance in the desired context. With both traditional and newer scoring technologies, it is very important to carefully consider the specific writing task, the target audience (s), and the expected proficiency levels.

Some additional dimensions should be taken into account for large-scale use of automated scoring. Just as with human-scored assessments, issues of validity and reliability of automated scoring need to be addressed. Those issues are not generic, they arise again and again for every specific scoring engine and for any particular use of such technology. Aspects of fairness to different groups of test-takers are also important, especially when unfairness might arise from the statistical nature of the scoring technology. Automated scoring systems also need to be robust in the face of various kinds of aberrant responses, especially for large-scale assessments where the amount of such responses might be non-negligible. An important aspect of validity is the potential susceptibility of an automated scoring system to cheating and gaming behavior. Sophisticated gaming by experts may contribute to negative public perception of automated scoring. On the other hand, a variety of practical cheating methods might be used by test-takers, especially on high-stakes assessments, which may lead to poor validity and reduced acceptability of test results. Methods of detecting cheating and gaming are evolving, and their inclusion in operational systems is highly recommended.

Overall, automated scoring of written responses is a mature area of research and a well-established area of practical applications for the English language. Research on automated scoring in other languages is progressing, while new developments in this field are posed to provide new opportunities and new challenges.

References

Ackerman, T. A., & Smith, P. L. (1988). A comparison of the information provided by essay, multiple-choice, and free-response writing tests. *Applied Psychological Measurement, 12*(2), 117–128. https://doi.org/10.1177/014662168801200202

Agirre, E., Cer, D., Diab, M., Gonzalez-Agirre, A., & Guo W. (2013). SEM 2013 shared task: Semantic textual similarity. In *Proceedings of the second joint conference on lexical and computational semantics (*SEM), Volume 1, Proceedings of the main conference and the shared task, Atlanta, GA, June 13–14, 2013* (pp. 32–43). Association for Computational Linguistics. https://www.aclweb.org/anthology/S13-1004

Alderson, J. C., & Wall, D. (1993). Does washback exist? *Applied Linguistics, 14*(2), 115–129. https://doi.org/10.1093/applin/14.2.115

Alghamdi, M., Alkanhal, M., Al-Badrashiny, M., Al-Qabbany, A., Areshey, A., & Alharbi, A. (2014). A hybrid automatic scoring system for Arabic essays. *AI Communications, 27*(2), 103–111. https://doi.org/10.3233/AIC-130586

Alikaniotis, D., Yannakoudakis, H., & Rei. M. (2016). Automatic text scoring using neural networks. In *Proceedings of the 54th annual meeting of the Association of Computational Linguistics*. https://doi.org/10.18653/v1/P16-1068

Al-Jouiea, M. F., & Azmia, A. M. (2017). Automated evaluation of school children essays in Arabic. *Procedia Computer Science, 117*, 19–22. https://doi.org/10.1016/j.procs.2017.10.089

Amorim, E. C. F., & Veloso, A. (2017). A multi-aspect analysis of automatic essay scoring for Brazilian Portuguese. In *Proceedings of the student research workshop at the 15th conference of the European chapter of the Association for Computational Linguistics, Valencia, Spain* (pp. 94–102). https://www.aclweb.org/anthology/E17-4010

ASAP. (2012). The William and Flora Hewlett Foundation Automated Student Assessment Prize (ASAP). https://www.kaggle.com/c/asap-aes

Attali, Y. (2011). *A differential word use measure for content analysis in automated essay scoring* (Research Report 11-36). Educational Testing Service. https://doi.org/10.1002/j.2333-8504.2011.tb02272.x

Attali, Y. (2013). Validity and reliability of automated essay scoring. In M. D. Shermis & J. Burstein (Eds.), *Handbook of automated essay evaluation* (pp. 181–198). Routledge.

Attali, Y., Bridgeman, B., & Trapani, C. (2010). Performance of a generic approach in automated essay scoring. *The Journal of Technology, Learning, and Assessment, 10*(3) https://ejournals.bc.edu/index.php/jtla/article/view/1603

Attali, Y., & Burstein, J. (2006). Automated essay scoring with e-rater V.2. *Journal of Technology, Learning, and Assessment, 4*(3) https://ejournals.bc.edu/index.php/jtla/article/view/1650

BABEL Generator. (2014). https://babel-generator.herokuapp.com

Bai, X., & Stede, M. (2023). A survey of current machine learning approaches to student free-text evaluation for intelligent tutoring. *International Journal of Artificial Intelligence in Education, 33*, 992–1030. https://doi.org/10.1007/s40593-022-00323-0

Balfour, S. P. (2013). Assessing writing in MOOCs: Automated essay scoring and calibrated peer review. *Research & Practice in Assessment, 8*, 40–48. https://www.rpajournal.com/assessing-writing-in-moocs-automated-essay-scoring-and-calibrated-peer-review/

Beigman Klebanov, B., Flor, M., & Gyawali, B. (2016). Topicality-based indices for essay scoring. In *Proceedings of the 11th workshop on innovative use of NLP for building educational applications (BEA11)* (pp. 63–72). https://doi.org/10.18653/v1/W16-0507

Beigman Klebanov, B., & Madnani, N. (2022). *Automated essay scoring* (Synthesis lectures on human language technologies, #52). Springer Nature.

Beigman Klebanov, B., Madnani, N., Burstein, J., & Somasundaran, S. (2014). Content importance models for scoring writing from sources. In *Proceedings of the 52nd annual meeting of the Association for Computational Linguistics (volume 2: Short papers), Baltimore, MD* (pp. 247–252). https://doi.org/10.3115/v1/P14-2041

Bejar, I. I. (2012). Rater cognition: implications for validity. *Educational Measurement: Issues and Practice, 31*(3), 2–9. https://doi.org/10.1111/j.1745-3992.2012.00238.x

Bejar, I. I. (2017). A historical survey of research regarding constructed-response formats. In R. E. Bennett & M. von Davier (Eds.), *Advancing human assessment, methodology of educational measurement and assessment* (pp. 565–633). Springer. https://doi.org/10.1007/978-3-319-58689-2_18

Bejar, I. I., Flor, M., Futagi, Y., & Ramineni, C. (2014). On the vulnerability of automated scoring to construct-irrelevant response strategies (CIRS): An illustration. *Assessing Writing, 22*, 48–59. https://doi.org/10.1016/j.asw.2014.06.001

Belinkov, Y., & Bisk, Y. (2018). Synthetic and natural noise both break neural machine translation. In *Proceedings of the sixth international conference on learning representations, ICLR 2018*. https://arxiv.org/pdf/1711.02173.pdf

Bennett, R. E., Ward, W. C., Rock, D. A., & LaHart, C. (1990). *Toward a framework for constructed-response items* (Research report). Educational Testing Service. https://doi.org/10.1002/j.2333-8504.1990.tb01348.x

Bennett, R. E., & Zhang, M. (2016). Validity and automated scoring. In F. Drasgow (Ed.), *Technology and testing: Improving educational and psychological measurement* (pp. 142–173). Routledge.

Ben-Simon, A., & Cohen, Y. (2011). The Hebrew language project: Automated essay scoring & readability analysis. In *Proceedings of the 37th annual IAEA conference, Manila*. https://iaea.info/documents/the-hebrew-language-project-automated-essay-scoring-readability-analysis/

Berggren, S. J., Rama, T., & Øvrelid, L. (2019). Regression or classification? Automated essay scoring for Norwegian. In *Proceedings of the fourteenth workshop on innovative use of NLP for building educational applications, Florence, Italy* (pp. 92–102). Association for Computational Linguistics. https://doi.org/10.18653/v1/W19-4409

Bridgeman, B., Trapani, C., & Attali, Y. (2012). Comparison of human and machine scoring of essays: Differences by gender, ethnicity, and country. *Applied Measurement in Education, 25*(1), 27–40. https://doi.org/10.1080/08957347.2012.635502

Bryant, C., Felice, M., Andersen, Ø. E., & Briscoe, T. (2019). The BEA-2019 shared task on grammatical error correction. In *Proceedings of the fourteenth workshop on innovative use of NLP for building educational applications, Florence, Italy* (pp. 52–75). Association for Computational Linguistics. https://doi.org/10.18653/v1/W19-4406

Burrows, S., Gurevych, I., & Stein, B. (2015). The eras and trends of automatic short answer grading. *International Journal of Artificial Intelligence in Education, 25*, 60–117. https://doi.org/10.1007/s40593-014-0026-8

Burstein, J., Kukich, K., Wolff, S., Lu, C., Chodorow, M., Braden-Harder, L., & Harris, M. D. (1998). Automated scoring using a hybrid feature identification technique. In *Proceedings of the annual meeting of the Association of Computational Linguistics, Montreal, Canada* (pp. 206–210). Association for Computational Linguistics. https://doi.org/10.3115/980845.980879

Burstein, J., Tetreault, J., & Madnani, N. (2013). The E-rater® automated essay scoring system. In M. D. Shermis & J. Burstein (Eds.), *Handbook of automated essay evaluation* (pp. 55–67). Routledge. https://doi.org/10.4324/9780203122761

Buzick, H., Oliveri, M. E., Attali, Y., & Flor, M. (2016). Comparing human and automated essay scoring for prospective graduate students with learning disabilities and/or ADHD. *Applied Measurement in Education, 29*(3), 161–172. https://doi.org/10.1080/08957347.2016.1171765

Cahill, A., Chodorow, M., & Flor, M. (2018). Developing an e-rater advisory to detect babel-generated essays. *Journal of Writing Analytics, 2*, 204–224. https://wac.colostate.edu/docs/jwa/vol2/cahill.pdf

Camilli, G. (2005). Test fairness. In R. L. Brennan (Ed.), *Educational measurement* (4th ed.). American Council on Education / Praeger. https://eduq.info/xmlui/handle/11515/34503

Castro-Castro, D., Lannes-Losada, R., Maritxalar, M., Niebla, I., Pérez-Marqués, C., Álamo-Suárez, N. C., & Pons-Porrata, A. (2008). A multilingual application for automated essay scoring. In *Advances in artificial intelligence* (Lecture notes in computer science, 5290, pp. 243–251). https://doi.org/10.1007/978-3-540-88309-8_25

Chang, T. -H., Lee, C. -H., & Chang, Y. -M. (2006). Enhancing automatic Chinese essay scoring system from figures-of-speech. In *Proceedings of the twentieth Pacific Asia conference on language, information, and computation* (pp. 28–34). http://ebooks.iospress.nl/volumearticle/3578

Chang, T. -H., Lee, C. H., & Tam, H. -P. (2007). On issues of feature extraction in Chinese automatic essay scoring system. In *Proceedings of the 2007 conference on artificial intelligence in education: Building technology rich learning contexts that work* (pp. 545–547). IOS Press.

Chapelle, C. A., & Chung, Y.-R. (2010). The promise of NLP and speech processing technologies in language assessment. *Language Testing, 27*(3), 301–315. https://doi.org/10.1177/0265532210364405

Chen, H., & He, B. (2013). Automated essay scoring by maximizing human-machine agreement. In *Proceedings of the 2013 conference on empirical methods in natural language processing, Seattle, WA* (pp. 1741–1752). https://www.aclweb.org/anthology/D13-1180

Chen, J., Zhang, M., & Bejar, I. I. (2017). *An investigation of the e-rater® automated scoring engine's grammar, usage, mechanics, and style microfeatures and their aggregation model* (Research Report ETS RR-17-04). Educational Testing Service. https://doi.org/10.1002/ets2.12131

Chen, S. (2018, May 27). China's schools are quietly using AI to mark students' essays ... but do the robots make the grade? *South China Morning Post*. https://www.scmp.com/news/china/society/article/2147833/chinas-schools-are-quietly-using-ai-mark-students-essays-do

Chen, X., Awadallah, A. H., Hassan, H., Wang, W., & Cardie, C. (2019). Multi-source cross-lingual model transfer: Learning what to share. In *Proceedings of the 57th annual meeting of the Association for Computational Linguistics* (pp. 3098–3112). https://doi.org/10.18653/v1/P19-1299

Cheon, M., Seo, H., Kim, H., Noh, E., Sung, K., & Lim, E. (2015). An automated scoring tool for Korean short-answer questions based on semi-supervised learning. In *Proceedings of the 2nd workshop on natural language processing techniques for educational applications, Beijing, China* (pp. 59–63). Association for Computational Linguistics and Asian Federation of Natural Language Processing. https://doi.org/10.18653/v1/W15-4409

Cohen, Y., Ben-Simon, A, & Hovav, M. (2003, October). *The effect of specific language features on the complexity of systems for automated essay scoring*. Paper presented at the International Association of Educational Administration. Manchester, England. http://files.eric.ed.gov/fulltext/ED482933.pdf

Cohen, Y., Levi, E., & Ben-Simon, A. (2018). Validating human and automated scoring of essays against "True" scores. *Applied Measurement in Education, 31*(3), 241–250. https://doi.org/10.1080/08957347.2018.1464450

Condon, W. (2013). Large-scale assessment, locally-developed measures, and automated scoring of essays: Fishing for red herrings? *Assessing Writing, 18*, 100–108. https://doi.org/10.1016/j.asw.2012.11.001

Criterion. (2008). Criterion. https://criterion.ets.org/criterion/default.aspx

Crossley, S. A., Kyle, K., Allen, L. K., Guo, L., & McNamara, D. S. (2014). Linguistic microfeatures to predict L2 writing proficiency: A case study in Automated Writing Evaluation. *The Journal of Writing Assessment, 7*(1) http://www.journalofwritingassessment.org/article.php?article=74

Crossley, S. A., Kyle, K., & McNamara, D. S. (2016). The development and use of cohesive devices in L2 writing and their relations to judgments of essay quality. *Journal of Second Language Writing, 32*, 1–16. https://doi.org/10.1016/j.jslw.2016.01.003

Deane, P. (2013). On the relation between automated essay scoring and modern views of the writing construct. *Assessing Writing, 18*, 7–24. https://doi.org/10.1016/j.asw.2012.10.002

Deane, P., Williams, F., Weng, V., & Trapani, C. S. (2013). Automated essay scoring in innovative assessments of writing from sources. *The Journal of Writing Assessment, 6*(1) http://www.journalofwritingassessment.org/article.php?article=65

Dikli, S. (2006). An overview of automated scoring of essays. *Journal of Technology, Learning, and Assessment, 5*(1), 4–3. https://ejournals.bc.edu/index.php/jtla/article/view/1640

Dikli, S., & Bleyle, S. (2014). Automated essay scoring feedback for second language writers: How does it compare to instructor feedback? *Assessing Writing, 22*, 1–17. https://doi.org/10.1016/j.asw.2014.03.006

Dong, F., & Zhang, Y. (2016). Automatic features for essay scoring—An empirical study. In *Proceedings of the 2016 conference on empirical methods in natural language processing*. https://doi.org/10.18653/v1/D16-1115

Dzikovska, M. O., Nielsen, R. D., Brew, C., Leacock, C., Giampiccolo, D., Bentivogli, L., Clark, P., Dagan, I., & Dang, H. T. (2013). SemEval-2013 Task 7: The joint student response analysis and 8th recognizing textual entailment challenge. In *Proceedings of second joint conference on lexical and computational semantics (*SEM), volume 2: Seventh international workshop on semantic evaluation (SemEval 2013), Atlanta, GA* (pp. 263–274). Association for Computational Linguistics. https://www.aclweb.org/anthology/S13-2045

Elliot, N., & Williamson, D. (2013). Assessing writing special issue: Assessing writing with automated scoring systems. *Assessing Writing, 18*, 1–6. https://doi.org/10.1016/j.asw.2012.11.002

Elliot, S. (2003). IntelliMetric: From here to validity. In M. D. Shermis & J. Burstein (Eds.), *Automated essay scoring: A cross disciplinary approach*. Lawrence Erlbaum. https://www.routledge.com/Automated-Essay-Scoring-A-Cross-disciplinary-Perspective/Shermis-Burstein/p/book/9781138964211

Enright, M. K., & Quinlan, T. (2010). Complementing human judgment of essays written by English language learners with e-rater® scoring. *Language Testing, 27*(3), 317–334. https://doi.org/10.1177/0265532210363144

Ercikan, K., Gierl, M. J., McCreith, T., Puhan, G., & Koh, K. (2004). Comparability of bilingual versions of assessments: Sources of incomparability of English and French versions of Canada's national achievement tests. *Applied Measurement in Education, 17*(3), 301–321. https://doi.org/10.1207/s15324818ame1703_4

Farra, N., Somasundaran, S., & Burstein, J. (2015). Scoring persuasive essays using opinions and their targets. In *Proceedings of the tenth workshop on innovative use of NLP for building educational applications, Denver, CO* (pp. 64–74). Association for Computational Linguistics. https://doi.org/10.3115/v1/W15-0608

Foltz, P. W., Laham, D., & Landauer, T. K. (1999). The intelligent essay assessor: Applications to educational technology. *Interactive Multimedia Electronic Journal of Computer-Enhanced Learning, 1*(2) http://www.imej.wfu.edu/articles/1999/2/04/index.asp

Galhardi, L., & Brancher, J. (2018, November 13–16). Machine learning approach for automatic short answer grading: A systematic review. In *Proceedings of the 16th Ibero-American conference on AI, Trujillo, Peru, IBERAMIA 2018, LNAI 11238* (pp. 380–391). Springer. https://doi.org/10.1007/978-3-030-03928-8_31

Graesser, A. C., McNamara, D. S., Louwerse, M. M., & Cai, Z. (2004). Coh-Metrix: Analysis of text on cohesion and language. *Behavior Research Methods, Instruments, and Computers, 36*(2), 193–202. https://doi.org/10.3758/BF03195564

Grave, E., Bojanowski, P., Gupta, P., Joulin, A., & Mikolov, T. (2018). Learning word vectors for 157 languages. In *Proceedings of the eleventh international conference on language resources and evaluation (LREC-2018), Miyazaki, Japan*. European Languages Resources Association. https://www.aclweb.org/anthology/L18-1550

Guo, F. (2009). *Fairness of automated essay scoring of GMAT® AWA* (RR-09-01). GMAC Research Reports. https://www.gmac.com/-/media/files/gmac/research/research-report-series/rr0901_awafairness.pdf

Hancke, J., & Meurers, D. (2013). Exploring CEFR classification for German based on rich linguistic modeling. In *Learner corpus research 2013, Book of abstracts. Bergen, Norway* (pp. 54–56). https://lcr2013.w.uib.no/files/2013/09/hanke-meurers-handout-2x2.pdf

Herrington, A., & Moran, C. (2001). What happens when machines read our students' writing? *College English, 63*(4), 480–499. https://doi.org/10.2307/378891

Higgins, D., Brew, C., Heilman, M., Ziai, R., Chen, L., Cahill, A., Flor, M., Madnani, N., Tetreault, J., Blanchard, D., Napolitano, D., Lee, C. M., & Blackmore, J. (2014, March). "Is getting the right answer just about choosing the right words?" The role of syntactically-informed features in short answer scoring. https://arxiv.org/abs/1403.0801v2

Higgins, D., Burstein, J., & Attali, Y. (2006). Identifying off-topic student essays without topic specific training data. *Natural Language Engineering, 12*(2), 145–159.

Higgins, D., & Heilman, M. (2014). Managing what we can measure: Quantifying the susceptibility of automated scoring systems to gaming behavior. *Educational Measurement: Issues and Practice, 33*(3), 36–46. https://doi.org/10.1017/S1351324906004189

Horbach, A., Scholten-Akoun, D., Ding, Y., & Zesch, T. (2017). Fine-grained essay scoring of a complex writing task for native speakers. In *Proceedings of the 12th workshop on innovative use of NLP for building educational applications* (pp. 357–366). https://doi.org/10.18653/v1/W17-5040

Horbach, A., & Zesch, T. (2019). The influence of variance in learner answers on automatic content scoring. *Frontiers in Education, 4*(28), 1–14. https://doi.org/10.3389/feduc.2019.00028

Hsieh, Y., Hiew, C. K., & Tay, Y. X. (2017). Computer-mediated corrective feedback in Chinese as a second language writing: Learners' perspectives. In D. Zhang & C. H. Lin (Eds.), *Chinese as a second language assessment* (pp. 225–246). Springer. https://doi.org/10.1007/978-981-10-4089-4

Htut, P. M., & Tetreault, J. (2019). The unbearable weight of generating artificial errors for grammatical error correction. In *Proceedings of the fourteenth workshop on innovative use of NLP for building educational applications, Florence, Italy* (pp. 478–483). Association for Computational Linguistics. https://doi.org/10.18653/v1/W19-4449

Hutchinson, B., & Mitchell, M. (2019). 50 years of test (un)fairness: Lessons for machine learning. In *Proceedings of FAT* '19: Conference on fairness, accountability, and transparency*. https://doi.org/10.1145/3287560.3287600

Ishioka, T., & Kameda, M. (2004). Automated Japanese essay scoring system: JESS. In *Proceedings of the 15th international workshop on database and expert systems applications, Tokyo, Japan* (pp. 4–8). https://doi.org/10.1109/DEXA.2004.1333440

Ishioka, T., & Kameda, M. (2006). Automated Japanese essay scoring system based on articles written by experts. In *Proceedings of the 21st international conference on computational linguistics and 44th annual meeting of the ACL, Sydney, Australia* (pp. 233–240). Association for Computational Linguistics. https://doi.org/10.3115/1220175.1220205

Jang, E., Kang, S., Noh, E., Kim, M., Sung, K., & Seong, T. (2014). KASS: Korean automatic scoring system for short-answer questions. In *Proceedings of the 6th international conference on computer supported education (CSEDU-2014)* (pp. 226–230). https://doi.org/10.5220/0004864302260230

Kakkonen, T., Myller, N., Timonen, J., & Sutinen, E. (2005). Automatic essay grading with probabilistic latent semantic analysis. In *Proceedings of the 2nd workshop on building educational applications using NLP, Ann Arbor, MI* (pp. 29–36). https://www.aclweb.org/anthology/W05-0206

Kakkonen, T., & Sutinen, E. (2004). Automatic assessment of the content of essays based on course materials. In *Proceedings of the 2nd IEEE international conference on information technology: Research and education, London, UK*. https://doi.org/10.1109/ITRE.2004.1393660

Ke, X., Zeng, Y., & Luo, H. (2016). Autoscoring essays based on complex networks. *Journal of Educational Measurement, 53*(4), 478–497. https://doi.org/10.1111/jedm.12127

Ke, Z., & Ng, V. (2019). Automated essay scoring: A survey of the state of the art. In *Proceedings of the twenty-eighth international joint conference on artificial intelligence (IJCAI-19), Macao* (pp. 6300–6308). https://doi.org/10.24963/ijcai.2019/879

Khayrallah, H., & Koehn, P. (2018). On the impact of various types of noise on neural machine translation. In *Proceedings of the 2nd workshop on neural machine translation and generation, Melbourne, Australia* (pp. 74–83). Association for Computational Linguistics. https://doi.org/10.18653/v1/W18-2709

Kieftenbeld, V., & Boyer, M. (2017). Statistically comparing the performance of multiple automated raters across multiple items. *Applied Measurement in Education, 30*(2), 117–128. https://doi.org/10.1080/08957347.2017.1283316

Kim, J., Park, K., Chae, J., Jang, H., Kim, B., & Jung, S. (2018). Automatic scoring system for short descriptive answer written in Korean using lexico-semantic pattern. *Soft Computing, 22*(13), 4241–4249. https://doi.org/10.1007/s00500-017-2772-7

Klein, D. C. D., Chung, G. K. W. K., Osmundson, E., & Herl, H.E. (2002). *Examining the validity of knowledge mapping as a measure of elementary students' scientific understanding* (CSE Report 557). University of California, National Center for Research on Evaluation, Standards, and Student Testing. https://cresst.org/wp-content/uploads/TR557.pdf

Kolowich, S. (2014, April 28). Writing instructor, skeptical of automated grading, pits machine vs. machine. *The Chronicle of Higher Education*. https://www.chronicle.com/article/writing-instructor-skeptical-of-automated-grading-pits-machine-vs-machine

Landauer, T. K., Foltz, P. W., & Laham, D. (1998). Introduction to latent semantic analysis. *Discourse Processes, 25*, 259–284. https://doi.org/10.1080/01638539809545028

Landauer, T. K., Laham, D., & Foltz, P. W. (2003). Automated scoring and annotation of essays with the intelligent essay assessor. In M. D. Shermis & J. Burstein (Eds.), *Automated essay scoring: A cross-disciplinary perspective* (pp. 87–112). Erlbaum. https://www.routledge.com/Automated-Essay-Scoring-A-Cross-disciplinary-Perspective/Shermis-Burstein/p/book/9781138964211

Larkey, L. (1998). Automatic essay grading using text categorization techniques. In *Proceedings of the 21st ACM-SIGIR conference on research and development in information retrieval, Melbourne, Australia* (pp. 90–95). https://doi.org/10.1145/290941.290965

Leacock, C., & Chodorow, M. (2003). C-rater: Automated scoring of short-answer questions. *Computers and the Humanities, 37*(4). https://doi.org/10.1023/A:1025779619903

Lemaire, B., & Dessus, P. (2001). A system to assess the semantic content of student essays. *Journal of Educational Computing Research, 24*(3), 305–320.

Levy, O., & Goldberg, Y. (2014). Linguistic regularities in sparse and explicit word representations. In *Proceedings of the eighteenth conference on computational natural language learning, Ann Arbor, MI* (pp. 171–180). Association for Computational Linguistics. https://doi.org/10.3115/v1/W14-1618

Lilja, M. (2018). *Automatic essay scoring of Swedish essays using neural networks*. Department of Statistics, Uppsala University. http://www.diva-portal.org/smash/record.jsf?pid=diva2%3A1213688&dswid=6452

Lin, H. -Y., Chen, C. Y., Lee, J., Li, Z., Zhang, Y., Xia, M., Rijhwani, Sh., He, J., Zhang, Zh., Ma, X., Anastasopoulos, A., Littell, P., & Neubig, G. (2019). Choosing transfer languages for cross-lingual learning. In *Proceedings of the 57th annual meeting of the Association for Computational Linguistics* (pp. 3125–3135). https://doi.org/10.18653/v1/P19-1301

Linzen, T., Chrupała, G., Belinkov, Y., & Hupkes, D. (2019). *Proceedings of the 2019 ACL workshop BlackboxNLP: Analyzing and interpreting neural networks for NLP, Florence, Italy*. Association for Computational Linguistics. https://www.aclweb.org/anthology/W19-4800.pdf

Litman, D., Zhang, H., Correnti, R., Matsumura, L. C., & Wang, E. (2021). A fairness evaluation of automated methods for scoring text evidence usage in writing. In I. Roll, D. McNamara, S. Sosnovsky, R. Luckin, & V. Dimitrova (Eds.), *Artificial intelligence in education. AIED 2021* (Lecture notes in computer science) (Vol. 12748). Springer. https://doi.org/10.1007/978-3-030-78292-4_21

Liu, O. L., Rios, J. A., Heilman, M., Gerard, L., & Linn, M. C. (2016). Validation of automated scoring of science assessments. *Journal of Research in Science Teaching, 53*(2), 215–233. https://doi.org/10.1002/tea.21299

Lochbaum, K. E., Rosenstein, M., Foltz, P. W., & Derr, M. A. (2013). *Detection of gaming in automated scoring of essays with the IEA*. Paper presented at the National Council on Measurement in Education Conference (NCME), San Francisco, CA. https://www.apparent-wind.com/mbr/papers/Gaming_Detection_NCME2013.pdf

Loraksa C., & Peachavanish, R. (2007). Automatic Thai-language essay scoring using neural network and latent semantic analysis. In *First Asia international conference on modelling & simulation (AMS '07), Phuket* (pp. 400–402). https://doi.org/10.1109/AMS.2007.19

Louis, A., & Higgins, D. (2010). Off-topic essay detection using short prompt texts. In *Proceedings of the 5th workshop on innovative use of NLP for building educational applications* (pp. 92–95). https://www.aclweb.org/anthology/W10-1013

Loukina, A., & Buzick, H. (2017). *Use of automated scoring in spoken language assessments for test takers with speech impairments* (ETS RR-17-42). Educational Testing Service. https://doi.org/10.1002/ets2.12170

Loukina, A., Madnani N., & Zechner, K. (2019). The many dimensions of algorithmic fairness in educational applications. In *Proceedings of the fourteenth workshop on innovative use of NLP for building educational applications, Florence, Italy* (pp. 1–10). Association for Computational Linguistics. https://doi.org/10.18653/v1/W19-4401

Madnani, N., Loukina, A., & Cahill, A. (2017). A large scale quantitative exploration of modeling strategies for content scoring. In *Proceedings of the 12th workshop on innovative use of NLP for building educational applications*. https://doi.org/10.18653/v1/W19-4401

Magliano, J. P., & Graesser, A. C. (2012). Computer-based assessment of student-constructed response. *Behavioral Research Methods, 44*, 608–621. https://doi.org/10.3758/s13428-012-0211-3

Malinin A., Knill, K., Ragni, A., Wang, Y., & Gales, M. (2017). An attention-based model for off-topic spontaneous spoken response detection: An initial study. In *Proceedings of the 7th workshop on speech and language technology in education* (pp. 144–149). https://doi.org/10.21437/SLaTE.2017-25

Mayfield, E., & Black, A.W. (2020). Should you fine-tune BERT for automated essay scoring? In *Proceedings of the fifteenth workshop on innovative use of NLP for building educational applications, Seattle, WA* (pp. 151–162). Online. Association for Computational Linguistics. https://aclanthology.org/2020.bea-1.15.pdf

McNamara, D. S., Crossley, S. A., Rosco, R. D., Allen, L. K., & Dai, J. (2015). A hierarchical classification approach to automated essay scoring. *Assessing Writing, 23*, 35–59. https://doi.org/10.1016/j.asw.2014.09.002

Mezher, R., & Omar, N. (2016). A hybrid method of syntactic feature and latent semantic analysis for automatic Arabic essay scoring. *Journal of Applied Sciences, 16*, 209–215. https://doi.org/10.3923/jas.2016.209.215

Mohler, M., Bunescu, R., & Mihalcea, R. (2011, June 19–24). Learning to grade short answer questions using semantic similarity measures and dependency graph alignments. In *Proceedings of the 49th annual meeting of the Association for Computational Linguistics, Portland, OR* (pp. 752–762). Association for Computational Linguistics. https://www.aclweb.org/anthology/P11-1076

Monaghan, W., & Bridgeman, B. (2005). E-Rater as a quality control on human scores. *R&D Connections*. Educational Testing Service. https://www.ets.org/research/policy_research_reports/publications/periodical/2005/cwyf

Nadeem, F., Nguyen, H., Liu, Y., & Ostendorf, M. (2019). Automated essay scoring with discourse-aware neural models. In *Proceedings of the fourteenth workshop on innovative use of NLP for building educational applications, Florence, Italy* (pp. 484–493). Association for Computational Linguistics. https://doi.org/10.18653/v1/W19-4450

NCTE. (2013). *NCTE position statement on machine scoring*. National Council of Teachers of English. http://www2.ncte.org/statement/machine_scoring

Östling, R., Smolentzov, A., Tyrefors Hinnerich, B., & Höglin, E. (2013). Automated essay scoring for Swedish. In *Proceedings of the 8th workshop on innovative use of NLP for building educational applications, Atlanta, GA* (pp. 42–47). Association for Computational Linguistics. https://www.aclweb.org/anthology/W13-1705

Page, E. B. (1966). The imminence of grading essays by computer. *Phi Delta Kappan*, 238–243. https://www.jstor.org/stable/20371545

Page, E. B. (1968). The use of the computer in analyzing student essays. *International Review of Education, 14*, 210–225. https://doi.org/10.1007/BF01419938

Pearson. (2016). MyWritingLab. https://www.pearsonmylabandmastering.com/northamerica/mywritinglab

Penfield, R. D. (2016). Fairness in test scoring. In N. J. Dorans & L. L. Cook (Eds.), *Fairness in educational assessment and measurement* (pp. 55–76). Routledge. https://www.routledge.com/Fairness-in-Educational-Assessment-and-Measurement/Dorans-Cook/p/book/9781138026193

Peng, X., Ke, D., Chen. Z., & Xu, B. (2010). Automated Chinese essay scoring using vector space models. In *Proceedings of the 4th international universal communication symposium, Beijing, China*. IEEE. https://doi.org/10.1109/IUCS.2010.5666229

Perelman, L. (2014). When "the state of the art" is counting words. *Assessing Writing, 21*, 104–111. https://doi.org/10.1016/j.asw.2014.05.001

PISA. (n.d.). https://www.oecd.org/pisa.

Poe, M., & Elliot, N. (2019). Evidence of fairness: Twenty-five years of research in assessing writing. *Assessing Writing*. https://doi.org/10.1016/j.asw.2019.100418

Potthast, M., Hagen, M., Beyer, A., Busse, M., Tippmann, M., Rosso, P., & Stein, B. (2014). Overview of the 6th international competition on plagiarism detection. In *Proceedings of the CLEF conference on multilingual and multimodal information access evaluation*. http://ceur-ws.org/Vol-1180/CLEF2014wn-Pan-PotthastEt2014.pdf

Powers, D. E., Burstein, J., Chodorow, M., Fowles, M. E., & Kukich, K. (2001). *Stumping E-rater: Challenging the validity of automated essay scoring* (ETS Research Report 01-03). Educational Testing Service. https://doi.org/10.1002/j.2333-8504.2001.tb01845.x

Powers, D. E., Escoffery, D. S., & Duchnowski, M. P. (2015). Validating automated essay scoring: A (modest) refinement of the "gold standard". *Applied Measurement in Education, 28*(2), 130–142. https://doi.org/10.1080/08957347.2014.1002920

Pulman, S. G., & Sukkarieh, J. Z. (2005). Automatic short answer marking. In *Proceedings of the second workshop on building educational applications using NLP, Ann Arbor, MI* (pp. 9–16). Association for Computational Linguistics. https://www.aclweb.org/anthology/W05-0202

Quinlan, T., Higgins, D., & Wolff, S. (2009). *Evaluating the construct-coverage of the e-rater® scoring engine* (Research Report RR-09-01). Educational Testing Service. https://doi.org/10.1002/j.2333-8504.2009.tb02158.x

Raczynski, K., & Cohen, A. (2018). Appraising the scoring performance of automated essay scoring systems—Some additional considerations: Which essays? Which human raters?

Which scores? *Applied Measurement in Education, 31*(3), 233–240. https://doi.org/10.1080/08957347.2018.1464449

Ramachandran, L., Cheng, J., & Foltz, P. (2015). Identifying patterns for short answer scoring using graph-based lexico-semantic text matching. In *Proceedings of the tenth workshop on innovative use of NLP for building educational applications, Denver, CO, June 4, 2015* (pp. 97–106). Association for Computational Linguistics. https://doi.org/10.3115/v1/W15-0612

Ramineni, C., Trapani, C. S., Williamson, D., Davey, T., & Bridgeman, B. (2012a). *Evaluation of the e-rater® scoring engine for the GRE® issue and argument prompts* (ETS RR-12-02). Educational Testing Service. https://doi.org/10.1002/j.2333-8504.2012.tb02284.x

Ramineni, C., Trapani, C. S., Williamson, D., Davey, T., & Bridgeman, B. (2012b). *Evaluation of the e-rater® scoring engine for the TOEFL® independent and integrated prompts* (ETS RR-12-06). Educational Testing Service. https://doi.org/10.1002/j.2333-8504.2012.tb02288.x

Reilly, E. D., Stafford, R. E., Williams, K. M., & Corliss, S. B. (2014). Evaluating the validity and applicability of automated essay scoring in two massive open online courses. *The International Review of Research in Open and Distributed Learning, 15*(5). https://doi.org/10.19173/irrodl.v15i5.1857

Reinertsen, N. (2018). Why can't it mark this one? A qualitative analysis of student writing rejected by an automated essay scoring system. *English in Australia, 53*(1), 52–60. https://search.informit.org/doi/10.3316/informit.702250811785762

Riordan, B., Flor, M., & Pugh, R. (2019). How to account for mispellings: Quantifying the benefit of character representations in neural content scoring models. In *Proceedings of the fourteenth workshop on innovative use of NLP for building educational applications, Florence, Italy* (pp. 116–126). Association for Computational Linguistics. https://doi.org/10.18653/v1/W19-4411

Riordan, B., Horbach, A., Cahill, A., Zesch, T., & Lee, C. -M. (2017). Investigating neural architectures for short answer scoring. In *Proceedings of the 12th workshop on innovative use of NLP for building educational applications* (pp. 159–168). https://doi.org/10.18653/v1/W17-5017

Rothermel, B. A. (2006). Automated writing instruction: Computer-assisted or computer-driven pedagogies? In P. Freitag Ericsson & R. H. Haswell (Eds.), *Machine scoring of student essays: Truth and consequences* (pp. 199–210). University Press of Colorado, Utah State University Press. https://digitalcommons.usu.edu/usupress_pubs/139/

Roy, S., Narahari, Y., & Deshmukh, O. D. (2015). A perspective on computer assisted assessment techniques for short free-text answers. In E. Ras & D. Joosten-ten Brinke (Eds.), *Computer assisted assessment. Research into E-assessment. TEA 2015. Communications in computer and information science* (Vol. 571). Springer. https://www.springer.com/gp/book/9783319086569

Rudner, L. M., & Liang, T. (2002). Automated essay scoring using Bayes' theorem. *The Journal of Technology, Learning and Assessment, 1*(2), 3–21. https://ejournals.bc.edu/index.php/jtla/article/view/1668

Sakaguchi, K., Heilman, M., & Madnani, N. (2015). Effective feature integration for automated short answer scoring. In *Proceedings of the 2015 annual conference of the North American chapter of the ACL, human language technologies, Denver, CO* (pp. 1049–1054). https://doi.org/10.3115/v1/N15-1111

Schultz, M. (2013). The IntelliMetric™ automated essay scoring engine—A review and an application to Chinese essay scoring. In M. D. Shermis & J. Burstein (Eds.), *Handbook of automated essay evaluation* (pp. 89–98). Routledge. https://www.routledge.com/Automated-Essay-Scoring-A-Cross-disciplinary-Perspective/Shermis-Burstein/p/book/9781138964211

Shermis, M. D. (2014). State-of-the-art automated essay scoring: Competition results and future directions from a United States demonstration. *Assessing Writing, 20*, 53–76. https://doi.org/10.1016/j.asw.2013.04.001

Shermis, M. D. (2015). Contrasting state-of-the-art in the machine scoring of short-form constructed responses. *Educational Assessment, 20*(1), 46–65. https://doi.org/10.1080/10627197.2015.997617

Shermis, M. D. (2020). International applications of machine scoring. In D. Yan, A. A. Rupp, & P. W. Folz (Eds.), *Handbook of automated scoring: Theory into practice* (pp. 113–132). Chapman & Hall/CRC. https://www.routledge.com/Handbook-of-Automated-Scoring-Theory-into-Practice/Yan-Rupp-Foltz/p/book/9781138578272

Shermis, M. D., & Burstein, J. (2013). *The handbook of automated essay evaluation: Current applications and new directions*. Routledge. https://www.routledge.com/Automated-Essay-Scoring-A-Cross-disciplinary-Perspective/Shermis-Burstein/p/book/9781138964211

Shermis, M. D., Burstein, J., Higgins, D., & Zechner, K. (2010). *Automated essay scoring: Writing assessment and instruction. International encyclopedia of education* (3rd ed., pp. 20–26). Elsevier. https://doi.org/10.1016/B978-0-08-044894-7.00233-5

Shermis, M. D., & Hamner, B. (2013). Contrasting state-of-the-art automated scoring of essays. In M. D. Shermis & J. Burstein (Eds.), *Handbook of automated essay evaluation*. Routledge. https://www.routledge.com/Automated-Essay-Scoring-A-Cross-disciplinary-Perspective/Shermis-Burstein/p/book/9781138964211

Somasundaran, S., Flor, M., Chodorow, M., Molloy, H., Gyawali, B., & McCulla, L. (2018). Towards evaluating narrative quality in student writing. *Transactions of the Association for Computational Linguistics, 6*, 91–106. https://doi.org/10.1162/tacl_a_00007

Song, Y., Heilman, M., Beigman Klebanov, B., & Deane, P. (2014). Applying argumentation schemes for essay scoring. In *Proceedings of the first workshop on argumentation mining, Baltimore, MD* (pp. 69–78). Association for Computational Linguistics. https://doi.org/10.3115/v1/W14-2110

Stab, C., & Gurevych, I. (2017). Parsing argumentation structures in persuasive essays. *Computational Linguistics, 43*(3), 619–659. https://doi.org/10.1162/COLI_a_00295

Steedle, J. T., & Elliot, S. (2012). *The efficacy of automated essay scoring for evaluating student responses to complex critical thinking performance tasks*. Council for Aid to Education.

Stevenson, M. (2016). A critical interpretative synthesis: The integration of automated writing evaluation into classroom writing instruction. *Computers and Composition, 42*, 1–16. https://doi.org/10.1016/j.compcom.2016.05.001

Sukkarieh, J., & Blackmore, J. (2009). c-rater: Automatic content scoring for short constructed responses. In *Proceedings of the twenty-second international FLAIRS conference* (pp. 290–295). https://www.aaai.org/ocs/index.php/FLAIRS/2009/paper/view/122

Taghipour, K., & Ng, H. T. (2016). *A neural approach to automated essay scoring*. In *Proceedings of the 2016 conference on empirical methods in natural language processing, Austin, TX*. https://doi.org/10.18653/v1/D16-1193

Turnitin Revision Assistant. (2019). https://www.turnitin.com/products/revision-assistant

Universal Dependencies. (n.d.). https://universaldependencies.org

Uto, M. (2021). A review of deep-neural automated essay scoring models. *Behaviormetrika, 48*, 459–484. https://doi.org/10.1007/s41237-021-00142-y

Vajjala, S. (2018). Automated assessment of non-native learner essays: Investigating the role of linguistic features. *International Journal of Artificial Intelligence in Education, 28*, 79–105. https://doi.org/10.1007/s40593-017-0142-3

Vajjala, S., & Lõo, K. (2013). Role of morpho-syntactic features in Estonian proficiency classification. In *Proceedings of the 8th workshop on innovative use of NLP for building educational applications, Atlanta, GA*. https://www.aclweb.org/anthology/W13-1708

Vajjala, S., & Lõo, K. (2014). Automatic CEFR level prediction for Estonian learner text. In *Proceedings of the third workshop on NLP for computer-assisted language learning. NEALT Proceedings series 22/Linköping electronic conference proceedings* (Vol. 107, pp. 113–127). https://www.aclweb.org/anthology/W14-3509

Valenti, S., Neri, F., & Cucchiarelli, A. (2003). An overview of current research on automated essay grading. *Journal of Information Technology Education, 2*. https://doi.org/10.28945/331

Vantage Learning. (2016). MyAccess. http://www.vantagelearning.com/products/my-access-school-edition

Vitartas, P., Heath, J., Midford, S., Ong, K., Alahakoon, D., & Sullivan-Mort, G. (2016). Applications of automatic writing evaluation to guide the understanding of learning and teaching. In S. Barker, S. Dawson, A. Pardo, & C. Colvin (Eds.), *Show me the learning. Proceedings of ASCILITE 2016, Adelaide* (pp. 592–601). https://2016conference.ascilite.org/wp-content/uploads/ascilite2016_vitartas_full.pdf

Vojak, C., Kline, S., Cope, B., McCarthey, S., & Kalantzis, M. (2011). New spaces and old places: An analysis of writing assessment software. *Computers and Composition, 28*, 97–111. https://doi.org/10.1016/j.compcom.2011.04.004

Wang, J., & Brown, M. S. (2007). Automated essay scoring versus human scoring: A comparative study. *Journal of Technology, Learning, and Assessment, 6*(2) https://ejournals.bc.edu/index.php/jtla/article/view/1632

Weigle, S. C. (2013). English language learners and automated scoring of essays: Critical considerations. *Assessing Writing, 18*, 85–99. https://doi.org/10.1016/j.asw.2012.10.006

Westera, W., Dascalu, M., Kurvers, H., Ruseti, S., & Trausan-Matub, S. (2018). Automated essay scoring in applied games: Reducing the teacher bandwidth problem in online training. *Computers & Education, 123*, 212–224. https://doi.org/10.1016/j.compedu.2018.05.010

Wild, F., Stahl, C., Stermsek, G., & Neumann, G. (2005). Parameters driving effectiveness of automated essay scoring with LSA. In *Proceedings of the 9th CAA conference*. Loughborough University. https://repository.lboro.ac.uk/articles/conference_contribution/Parameters_driving_effectiveness_of_automated_essay_scoring_with_LSA/9490403

Williamson, D. M., Xi, X., & Breyer, F. J. (2012). A framework for evaluation and use of automated scoring. *Educational Measurement: Issues and Practice, 31*(1), 2–13. https://doi.org/10.1111/j.1745-3992.2011.00223.x

Xi, X. (2010a). Automated scoring and feedback systems: Where are we and where are we heading? *Language Testing, 27*(3), 291–300. https://doi.org/10.1177/0265532210364643

Xi, X. (2010b). How do we go about investigating test fairness? *Language Testing, 27*(2), 147–170. https://doi.org/10.1177/0265532209349465

Yamamoto, M., Umemura, N., & Kawano, H. (2018). Automated essay scoring system based on rubric. In R. Lee (Ed.), *Applied computing & information technology. ACIT 2017* (Studies in computational intelligence) (Vol. 727). Springer. https://doi.org/10.1007/978-3-319-64051-8_11

Yang, L., Xin, T., & Cao, C. (2020). Predicting evaluations of essay by computational graph-based features. *Frontiers in Psychology, 11*. https://doi.org/10.3389/fpsyg.2020.531262

Yannakoudakis, H., Briscoe, T., & Medlock, B. (2011). A new dataset and method for automatically grading ESOL texts. In *Proceedings of the 49th annual meeting of the association for computational linguistics: Human language technologies* (pp. 180–189). https://www.aclweb.org/anthology/P11-1019

Yannakoudakis, H., & Cummins, R. (2015). Evaluating the performance of automated text scoring systems. In *Proceedings of the tenth workshop on innovative use of NLP for building educational applications, Denver, CO* (pp. 213–223). Association for Computational Linguistics. https://doi.org/10.3115/v1/W15-0625

Yoon, S. -Y., Cahill, A., Loukina, A., Zechner, K., Riordan, B., & Madnani, N. (2018). Atypical inputs in educational applications. In *Proceedings of NAACL-HLT 2018* (pp. 60–67). https://doi.org/10.18653/v1/N18-3008

Yoon, S. -Y., & Higgins, D. (2011). Non-English response detection method for automated proficiency scoring system. In *Proceedings of the 6th workshop on innovative use of NLP for building educational applications* (pp. 161–169). https://www.aclweb.org/anthology/W11-1420

Yoon, S. -Y., Lee, C. M., Choi, I., Wang, X., Mulholland, M., & Evanini, K. (2017). Off-topic spoken response detection with word embeddings. In *Proceedings of the 18th annual conference of the International Speech Communication Association* (pp. 2754–2758). https://doi.org/10.21437/INTERSPEECH.2017-388

Yu, J. -L., Kuo, B. -C., & Pai, K. C. (2017). Developing Chinese automated essay scoring model to assess college students' essay quality. In *Proceedings of the 10th international conference on educational data mining* (pp. 430–432). https://educationaldatamining.org/EDM2017/proc_files/papers/paper_151.pdf

Zehner, F., Goldhammer, F., & Sälzer, C. (2018). Automatically analyzing text responses for exploring gender-specific cognitions in PISA reading. *Large-Scale Assessment in Education, 6*(7). https://doi.org/10.1186/s40536-018-0060-3

Zehner, F., Sälzer, C., & Goldhammer, F. (2016). Automatic coding of short text responses via clustering in educational assessment. *Educational and Psychological Measurement, 76*(2), 280–303. https://doi.org/10.1177/0013164415590022

Zesch, T., Wojatzki, M., & Scholten-Akoun D. (2015). Task-independent features for automated essay grading. In *Proceedings of the tenth workshop on innovative use of NLP for building educational applications* (pp. 224–232). https://doi.org/10.3115/v1/W15-0626

Zhang, D., & Lin, C. H. (2017). *Chinese as a second language assessment*. Springer. https://doi.org/10.1007/978-981-10-4089-4

Zhang, H., & Litman, D. (2018). Co-attention based neural network for source-dependent essay scoring. In *Proceedings of the thirteenth workshop on innovative use of NLP for building educational applications* (pp. 399–409). https://doi.org/10.18653/v1/W18-0549

Zhang, M. (2013). Contrasting automated and human scoring of essays. *R&D Connections, 21*. Educational Testing Service. https://www.ets.org/Media/Research/pdf/RD_Connections_21.pdf

Zhang, M., Chen, J., & Ruan, C. (2016). *Evaluating the advisory flags and machine scoring difficulty in the e-rater® automated scoring engine* (ETS RR-16-30). Educational Testing Service. https://doi.org/10.1002/ets2.12116

Zhang, M., Dorans, N. J., Li, C., & Rupp, A. (2017). Differential feature functioning in automated essay scoring. In H. Jiao & R. W. Lissitz (Eds.), *Test fairness in the new generation of large-scale assessment* (pp. 186–208). Information Age. https://www.infoagepub.com/products/Test-Fairness-in-the-New-Generation-of-Large-Scale-Assessment

Zhao, S., Zhang, Y., Xiong, X., Botelho, A., & Heffernan, N. (2017). A memory augmented neural model for automated grading. In *Proceedings of the fourth ACM conference on learning @ scale* (pp. 189–192). https://doi.org/10.1145/3051457.3053982

Ziai, R., Ott, N., & Meurers, D. (2012, June). Short answer assessment: establishing links between research strands. In *Proceedings of the seventh workshop on building educational applications using NLP, Montréal, Canada*. https://www.aclweb.org/anthology/W12-2022

Zieky, M. J. (2016). Fairness in test design and development. In N. Dorans & L. Cook (Eds.), *Fairness in educational assessment and measurement* (pp. 9–32). Routledge. https://www.routledge.com/Fairness-in-Educational-Assessment-and-Measurement/Dorans-Cook/p/book/9781138026193

Zupanc, K., & Bosnić, Z. (2015). Advances in the field of automated essay evaluation. *Informatica, 39*, 383–395. http://www.informatica.si/index.php/informatica/article/view/81

Chapter 12
Automatic Scoring of Constructed Responses in PISA Using the Machine-Supported Coding System

Hyo Jeong Shin, Kentaro Yamamoto, Qiwei He, and Matthias von Davier

Abstract This chapter illustrates how a machine supported coding system (MSCS) was developed and implemented for automatic scoring of constructed responses in the Programme for International Student Assessment (PISA) and presents its performance. As an international large-scale survey, PISA 2018 collected data from students across more than 80 countries and in about 50 languages in the computer-based mode. Conceptually, the MSCS can be viewed as a supervised learning of natural language data that is generalizable across multiple languages to be used for PISA data. Without extracting language-dependent features, the MSCS utilizes the exact-matching method and automatically applies a verified label to a subset of incoming uncoded responses, which guarantees the highest accuracy in the prediction. In the PISA 2018 assessment cycle, the MSCS has shown to alleviate scoring burden as much as 25% on average across all participating countries and across domains. This chapter starts with the motivation for developing the MSCS to benefit a large-scale setting with multiple languages of assessment. We also present the findings from an analysis of PISA 2018 data and explore possible avenues for continuing to improve.

H. J. Shin (✉)
Sogang University, Seoul, South Korea
e-mail: hshinedu@sogang.ac.kr

K. Yamamoto
Independent Consultant for Psychometric Methodology and Large-Scale Assessment (formerly ETS), Princeton, NJ, USA

Q. He
Graduate School of Arts and Science, Georgetown University, Washington, DC, USA

M. von Davier
Lynch School of Education and Human Development, TIMSS & PIRLS International Study Center at Boston College, Chestnut Hill, MA, USA
e-mail: vondavim@bc.edu

L. Khorramdel et al. (eds.), *Innovative Digital-Based International Large-Scale Assessments*, Methodology of Educational Measurement and Assessment,
https://doi.org/10.1007/978-3-031-90951-1_12

12.1 Constructed Responses in PISA

Cognitive domains, including mathematics, reading, and science, in the Programme for International Student Assessment (PISA) are measured through both multiple-choice (MC) items and constructed-response (CR) items. The inclusion of CR items helps to vary the types of cognitive processes triggered by the assessment items and provides a more balanced and complete assessment of heterogenous participants' abilities (Organisation for Economic Co-operation and Development [OECD], 2019, p. 49). In each cycle of PISA, CR items that require scoring by humans make up nearly a third of items across the cognitive domains. Therefore, accurate and reliable scoring[1] of CR item responses is a key component to assure data quality, which leads to the reliability and validity of test results (ETS, 2021; Shin et al., 2019).

However, scoring CR item responses by humans is typically time consuming, expensive, and can be prone to error due to a lack of coding consistency between (and within) human raters. Particularly, in cross-country survey settings, such scoring or coding tasks can be even more complicated when considering the number of participating countries and the possibility of item adaptations (see Lennon & Kirsch, 2025 and Ou et al., 2025 of this volume) due to unique linguistic, cultural, or national interpretations of the item. For example, in PISA 2018, students from more than 80 countries participate in the assessment (over 6000 students per country) in over 50 different languages. Thereby, it is critical to ensure that the scoring process of CR item responses is efficient and accurate, not only within each country but also across countries.

In the past, it was almost impossible to monitor the scoring process of CR item responses at the international level due to the difficulty of establishing a database of raw responses, including students' original text responses written on paper. However, the shift from paper- to computer-based assessment (CBA) in PISA 2015 enabled fully computer-based data collection and management, which allowed the associated scoring process to be technology-driven. To facilitate the efficient and accurate coding in a CBA environment, a machine-supported coding system (MSCS) was developed for PISA to automatically rate text responses on CR items (Yamamoto et al., 2017, 2018). Conceptually, the MSCS can be viewed as a type of supervised learning of natural language data (e.g., students' raw text responses to the CR items) that assigns codes (labels) automatically to a subset of incoming unlabeled data. To be generalizable across multiple languages and assure the highest prediction of scoring process, simple but straightforward exact-matching methods are used. That is, the MSCS identifies text data that have been scored consistently and frequently by humans in the past PISA cycles, and extracts those labeled text responses with the associated labels (scores). Subsequently, rather than having

[1]Typically, PISA uses the term "coding" as a process to assign a code or score (full credit, partial credit if applicable, or no credit, or double-digit codes if applicable) by raters to the constructed response. In this chapter, we use "scoring" as a more general term, and sometimes "coding" and "rating" are used interchangeably.

human raters score the exact same unlabeled data multiple times, the coding process is automated so that the appropriate and verified labels (scores) are applied to unlabeled duplicate responses, avoiding unnecessary repetition in human coding leading to coding efficiency, and reducing the chance for human error.

While the concept for such a tool is quite limited in defining the similarities of responses through exact matching, the biggest benefit of this approach is precise implementation to ensure the continued accuracy of applied codes and verification that the tool does, in fact, reduce coding burden, across several unique countries and language groups. In the following sections, we describe the motivation for and development of the MSCS and present the results of its implementation in PISA 2018, specifically looking at gains in coding efficiency. Finally, we propose possible methods for improving upon the MSCS using more advanced natural language processing (NLP) techniques.

12.2 Motivation for Automatic Scoring of CR Items in PISA

The magnitude of scoring work by humans in PISA cannot be underestimated. In-person and virtual international coder training meetings are held for all participating countries to familiarize the lead and other bilingual coders from each country with the items to be coded. Coding materials, including scoring rubrics, are then translated from a common source language into the local language for national teams to reference, and expert coders are expected to maintain coding hotlines during scheduled national coding sessions to answer any difficult coding questions that arise in the field. Thousands of coders are hired each PISA cycle to manually score written text responses on the CR items. In the 2018 PISA main survey, for example, approximately 2000 coders across all CBA countries and economies were responsible for coding all CR item responses in the cognitive domains of mathematics, reading, and science. By design, each human rater is assumed to code roughly 1000 responses per day to ensure the completion of over the course of 1–2 weeks, within the PISA cycle timeframe. Given these figures, the motivation behind the development of the automatic scoring of text responses for PISA is clear: to reduce coding burden and to reduce the opportunity for human error (e.g. fatigue effect, lack of understanding in scoring rubric, etc.).

Automatic scoring of text responses is a popular task in the field of NLP. Undoubtedly, it is challenging because of the variance observed in the text responses provided freely by test-takers unlike the fixed selections as in MC items. In this regard, Padó (2016) showed that the more complex scoring model is required as the more variance is found in the text responses. For example, within a single language setting, expression variance (i.e. there can be many different expressions to deliver the same meaning) as well as orthographic variance (i.e., non-native speakers can make more spelling errors) can be important factors of the linguistic variance observed from the text responses (Bhagat & Hovy, 2013; Ellis & Barkhuizen, 2005). Recently, Horbach and Zesch (2019) outlined various influencing factors that determine the linguistic variance and conducted experiments using the type of prompt, language in the data, average length of answers, and number of available training instances. In particular, for PISA which is administered in multiple languages, language is an important factor that increases the variance in the text. For example, morphological richness and the word order vary greatly across different languages, and some basic techniques, even removing the capitalization or trimming white spaces may not completely work, or work at all, in some languages. In fact, removing the capitalization is irrelevant in several languages, including Arabic, Chinese, Japanese, and Korean, and trimming white spaces can potentially alter the meaning of the words in some language contexts more severely than in others.

While the language factor increases the variance, the fact that PISA CR items elicit only short-text responses (typically less than 100 words) limits the linguistic variance. How much linguistic variance in a response plays a large role in the type of automated scoring approach used. Two widely known automated scoring engines, *c-rater*™ (Leacock & Chodorow, 2003; Sukkarieh & Blackmore, 2009) and *e-rater*® (Attali & Burstein, 2006; Burstein, 2003), both developed at Educational Testing Service (ETS), are used differentially based on the expected length, and thereby variance, of answers, such as *c-rater*™ for short-text responses and *e-rater*® for essays. Although the response fields of PISA CR items are open-ended, and the respondent has the potential to give any response to the item, only short-text responses are elicited by items in PISA; however, the precise length of expected short-text answers vary across CR items. Yamamoto et al. (2017) examined the magnitude of the regularities in a pilot study using eleven Reading items from PISA 2015 data. It was found that the item with the most instances of repeated raw responses resulted in a 94–98% workload reduction across country-by-language groups, whereas the sample item with the fewest repeated responses reduced coding workload by only 5–29%.

12.2.1 Linguistic Variance in the Text Responses to CR Items in PISA 2015

Extending this effort, we analyzed all CR items in four domains (mathematics, reading, science, and financial literacy) in the PISA 2015 main survey, using two quantities, reduction rate and missing rate, as a limited but straightforward way to evaluate the linguistic variance of text responses across all country-by-language groups. Nearly 150 items (21 from mathematics, 58 from science, 51 from reading, and 16 from financial literacy) across 59 countries (90 country-by-language groups) were analyzed. For a given item in each country-by-language group, reduction rate and missing rate were calculated as follows.

$$Reduction\ rate = \frac{(Number\ of\ raw\ responses - Number\ of\ unique\ responses)}{Number\ of\ raw\ responses} \times 100$$

$$Missing\ rate = \frac{Number\ of\ blank\ responses}{Number\ of\ raw\ responses} \times 100$$

The *reduction rate* is an index of how much workload can be reduced given on the number of identical responses. The larger the regularity or homogeneity exists among text responses, the greater the reduction rate is. For example, if an item has 500 identical text responses (therefore, contributing only one unique response), the human coder only has to code the first, unique, response, rather than the whole set of 500, resulting in a 99.8% scoring reduction. In this formula, we defined "unique response" using the exact-matching method without any pre-processing of string variables (e.g., removal of double-spaces or white space). For instance, "11km" (no space) and "11 km" (a space in front of *km*) are viewed as two unique responses, not as identical responses. Also, note that the blank response ("") is also considered a type of unique response.

Next, the *missing rate* is an index of how often blank or *nonresponses* are observed in the data. While students' text responses must first be seen by human raters to be assigned the appropriate score, nonresponses do not have to be presented to human coders if a proper missing code can be assigned by the computer system alone. In the past, when PISA was administered as PBA, all text responses, even blank responses, had to be scored by humans. Thus, filtering out blank responses in the CBA already helps to reduce the scoring burden and improve the scoring efficiency. Furthermore, blank responses are generally supposed to be scored as "not-reached" but occasionally are scored incorrectly as "wrong" by human raters due to a lack of understanding of the coding guide. Coding "not-reached" or "wrong" has important implications in the item parameter estimation stage (OECD, 2020, Chapter 9), and thus, automatic scoring of missing responses correctly as "not-reached" further improves the quality of scores.

Table 12.1 reports the distribution of reduction rates and missing rates across participating country-by-language groups in each construct. To obtain more stable

Table 12.1 Distribution of reduction rate and missing rate in each domain from PISA 2015 main survey

	Domain	Number of items	Mean	Median	Min	Max	SD
Reduction rate (%)	Math	21	38.69	34.40	0.00	96.74	23.81
	Reading	51	26.66	17.52	0.00	98.03	24.61
	Science	58	21.26	17.96	0.00	87.45	14.74
	Financial	16	17.47	15.10	0.00	63.19	11.05
Missing rate (%)	Math	21	17.66	15.20	0.00	63.18	12.34
	Reading	51	11.34	9.00	0.00	59.70	9.15
	Science	58	18.21	15.67	0.00	77.21	12.27
	Financial	16	15.51	12.44	0.00	63.89	10.89

results when aggregated, only items with 30 or more text responses (in the target country-by-language group) were included. The results show that there were significant regularities in the observed text responses: coding workload could potentially be reduced by about 39% in mathematics, 27% in reading, 21% in science, and 17% in financial literacy when all identical instances were automatically coded. Note that the potential reduction values range from a minimum of zero to a maximum of 90, or nearly 90%, in the core domains of mathematics, reading, and science.

Such variation reiterates that the expected level of regularities of text responses is item-dependent and is dependent on languages and the population of students (Horbach & Zesch, 2019; Shin et al., 2024). For example, the regularity in students' text responses is likely to be greater if the item asks the student to locate the correct response in the passage because the answer may be directly copied from the provided passage. In contrast, if the item requires inference, for which the correct response must be extrapolated from given information, the regularity in students' answers is likely to be small. Further, the level of commonality is also dependent on country-by-language group, which indicates factors of languages, populations, and performance. For the same CR item, students' answers could be more similar to each other if their performance on the item or utilized cognitive processes are more homogenous. Also, for the same CR item, some languages could have more diverse ways of expressing the same meaning than other languages, leading to fewer regularities in the observed text responses. Accepting such variations across items and country-by-language groups, significant regularities are present in students' text responses, and the amount of regularity suggests that automatic scoring of students' text responses on CR items in PISA is beneficial.

12.3 Development of the Machine-Supported Coding System (MSCS)

According to Bennett (2011), there are two general approaches of automated scoring of text responses. The first approach mostly applies to CR items that can be scored using exact-matching techniques, often short CR items. For this case, the challenge is relatively trivial: correct/incorrect answer(s) are known or verified in advance and can be used to evaluate the student's response. The second approach applies to items for which the responses are too complex to be scored through an exact-matching approach. In such cases, automated scoring of complex responses is generally accomplished via a scoring "model." Quite often, such models indicate supervised machine learning that extracts feature from students' original text response and uses those features to predict a score, such as the *c-rater*™ (Leacock & Chodorow, 2003; Sukkarieh & Blackmore, 2009) and *e-rater*® (Attali & Burstein, 2006; Burstein, 2003) scoring engines.

One outstanding benefit of the first approach, based on exact matching, is perfect accuracy, in theory, if the associated label is verified to be the gold standard. However, the exact matching method can be of limited use due to the rigid definition of the similarity between text responses and, thus, can only be applicable to identical responses that were already observed in the data. On the contrary, the latter scoring engines, based on supervised machine learning, allow sufficiently similar responses to be labeled although text strings are not identical. The huge benefit of this second approach is that all new unlabeled data can be assigned scores with varying accuracy. However, the second approach can never reach perfect accuracy in the prediction, and the performance of the scoring engine needs to be evaluated (e.g., Loukina et al., 2020).

12.3.1 Considerations for PISA Automatic Scoring Model

In the context of PISA, in which students' answers to CR items are collected in multiple languages, the first approach based on the exact-matching method was preferred as the first automatic scoring model to be introduced for several reasons. First, scores (labels) assigned to the CR item responses are used for estimating item parameters in the following stage, which is the critical phase in establishing the comparability across participating countries and cycles (OECD, 2017, 2020). In this regard, when coding was solely done by human raters in PISA 2015, monitoring inter-rater reliability, within-country and across-countries, was considered an important component to assuring the quality of scores given to CR item responses. Likewise, accuracy and the quality of the scores on CR items were treated as the priority in introducing the automatic scoring for the first time in PISA.

Second, the automatic scoring model at the international level had to be generalizable enough for multiple languages spoken in participating countries, at least for the introduction of automated scoring to PISA. Deploying an automatic scoring engine based on supervised machine learning often requires tremendous effort, even for a single item in a single language. Such automated scoring models are typically developed based on one language (e.g., English) to derive accurate scoring at the satisfactory level in the specific language environment. Because of the language diversity in spelling, grammar, wording, and so on, it is very challenging, and highly inadvisable, to generalize one single language model to other languages. Therefore, to profit all participating countries in PISA, a simple but straightforward approach that is less language-dependent was preferred.

Lastly, scores assigned to the CR item responses are published in the public use file (PUF) for secondary analyses conducted by policy makers and researchers across the world. In the PUF for CR items, labels of correct, partially correct, or incorrect according to a coding rubric are provided without students' original text responses. Furthermore, the coding rubric for some items defines qualitatively different reasoning processes as different scoring categories for the partial credit or full credit. For example, according to the coding rubric, {11, 12, 13} can be assigned for each of *qualitatively* different reasoning for the same partial credit score of "1", and those so-called double-digit codes become publicly available as well. In such cases, data become sparse, which can be even more challenging to deploy any working automatic scoring engines. Therefore, for reporting accurate scores in the PUF without original text responses, laborious manual scoring by human experts was preferred, as it was assumed to have a higher accuracy than predicting scores through machine.

Given the reasons explained above, the automated scoring approach based on the exact-matching method was defensible for use by the MSCS in a multilingual context, offering potential coding workload reduction for every participating country and language group. Technically speaking, this approach skips extracting features from students' text responses, thus, not using normalizing or advanced machine learning techniques. Therefore, this automatic scoring based on the exact matching for CR item responses in PISA was named the machine-*supported* coding system to emphasize that the machine plays a supplementary role to *support* coding by human raters.

12.3.2 Workflow of the MSCS

Instead of extracting features and using the feature space to predict a label, as is usually done in NLP and supervised machine learning, the MSCS harvests *coded unique response* (CUR) data from the historical database for application in new, unlabeled data, as described by Yamamoto et al. (2018). As shown in Fig. 12.1, the workflow of the MSCS can be divided into two phases: (a) *creation of the CUR pool* that extracts text responses associated with verified scores from the past data and

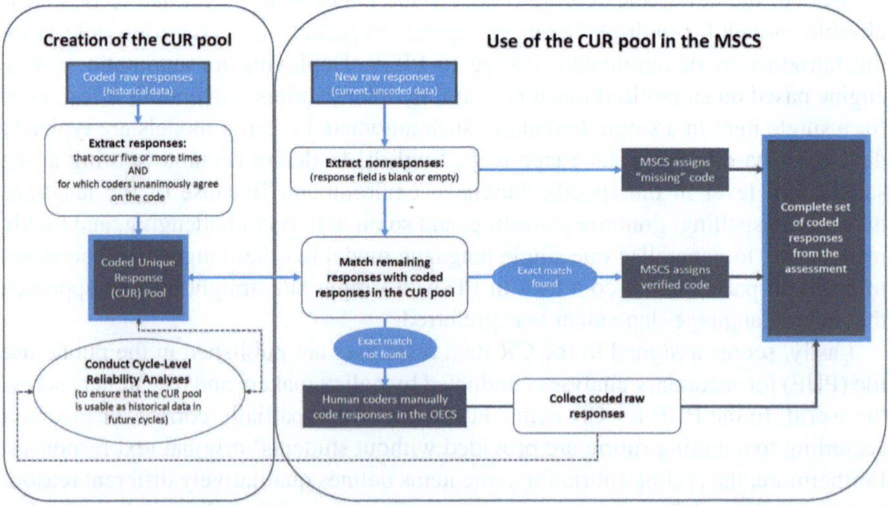

Fig. 12.1 Machine-supported coding system workflow for PISA constructed-response items

(b) *use of the CUR pool* that applies the verified scores to a new set of matched but unlabeled responses.

In the first phase, historical data are combined and analyzed together; for example, the labeled text responses from PISA 2015 and PISA 2018 were combined and analyzed together, and unique responses were extracted based on the exact matching method. Then, a simple machinery sorts each unique text response by score categories: "0" for incorrect, "1" for either fully correct for dichotomous items and for partially correct for polytomous items, "2" for fully correct for polytomous items, and "9" for missing. These labels for unique responses are viewed as "verified" only if the unique response is both frequently observed (i.e., at least five times) and the human-awarded code is exclusively observed (e.g., the response exists in only a single score category; for example, the response has only ever been coded as "correct" and never "incorrect"). These two major rules were used in determining label verification to ensure only accurate (e.g., unanimously agreed upon) codes for each unique response to be contained. Labeled responses with verified codes that have been extracted from the historical data are stored in the CUR pool. Each country-by-language group has its own CUR pool with a unique set of text responses and associated labels (codes or scores).

Additionally, there is a special scoring rule for nonresponses. Usually, blank responses are not of interest in automatic scoring, but assigning the correct codes for nonresponses in the PISA as "not-reached" is important for the subsequent item calibration stage. The code for a nonresponse is known in advance and, thus, can be added directly to a CUR pool as a verified label for all items across all domains (including new items or new domains) and for all country-by-language groups (including new participating countries or language groups) regardless of the availability of historical data. This approach ensures that a nonresponse is noted as a unique response for each CR item in the CUR pool, and as such, nonresponses can be directly filtered out of new unlabeled data and automatically scored by the MSCS rather than being assigned to human coders.

Finally, some special cases for two-part items (e.g., *radio button* items, which require an explanation following an MC selection) require a combination of response strings for accurate labeling; these cases are entered into the CUR pool alongside the full unique response string with a bar (|) notation indicating where the MC selection and the CR field were combined.

In the end, the integrated CR scoring system of PISA works as follows. Once new unlabeled data come in, the operational procedure in PISA is to place them into the Open-Ended Coding System (OECS; Kerzabi et al., 2025) for distribution to human raters. This happened to all text responses for CR items in PISA 2015 without exception. With the introduction of MSCS, prior to being sent to human coders, nonresponses are filtered out automatically, and the appropriate label (code or score) for a nonresponse is applied automatically. Removing nonresponses before human coding reduced the total raw response pool by as much as 12% on some items and in some domains, as reported in Table 12.1. The remaining text responses are matched to labeled text responses stored in the CUR pool from historical data, specific to the country and language. If an exactly matching pair, including spaces, spelling mistakes, and punctuation, is found in the CUR pool, the verified label is automatically applied to the unlabeled data. Any responses not filtered out are sent to human raters for manual scoring in the OECS.

Between PISA cycles, labeled data from the most recent cycle are merged with historical data from previous cycles and analyzed as part of quality assurance measures. These steps are indicated by dashed lines in Fig. 12.1. Frequencies and consistencies in the labels are checked to make sure that the unique text response has occurred at least five times with an exclusive, consistent code (i.e., human raters unanimously agree). In this process, all verified labels are also reviewed to make sure that they are aligned with the item's coding guide (e.g., verified labels are among the possible codes for that item in the coding guide), and that there have been no significant modifications to the coding guides (e.g., a response that was partial credit during the field trial is changed to full credit during the main survey). During this phase, the system can also identify any unique text responses with disagreements by human raters. This merging and review process not only verifies the quality of historical data, but it may also increase the volume of unique text responses in a country-by-language CUR pool. The count (number of occurrences) of an original text response that were insufficiently observed (fewer than five times) in a previous cycle are combined with the counts of that response in the current cycle, and if the newly combined set of responses meet the criteria of both frequency of observation and consistency of labels, the unique text response and its label can be included to the CUR pool for use in future cycles. For example, in preparation for PISA 2022, all available previous-cycle data, including the PISA 2015 main survey, the PISA 2018 field trial, and the 2018 main survey data were merged and processed. Because of the continued use of CBA in PISA, accumulated historical data can be used to further expand, re-verify, and prepare CUR pools for future cycles of assessment. Overall, the application of the MSCS has been effective in reducing the workload of human raters, and the automatic application of verified labels is expected to improve overall coding accuracy, if applicable, by reducing the opportunities for human coding error. In the next section, the gains in efficiency and accuracy of coding afforded by the implementation of the MSCS are illustrated.

12.4 Performance of the Machine-Supported Coding System in PISA 2018

For CBA countries participating in the PISA 2018 cycle, about 2000 coders across countries were responsible for coding all CR item responses in the core domains of mathematics, reading, and science. To facilitate the described scoring process, the MSCS was operationalized in 2018 cycle for the first time, and the MSCS was implemented in both the PISA 2018 field trial (Yamamoto et al., 2018) and the main survey. For the field trial, the CUR pools consisted only of responses collected during the 2015 main survey; for the main survey, the CUR pools consisted of

Table 12.2 Number of CR items and groups where MSCS was implemented in PISA 2018 field trial and main survey

	Category	2018 field trial	2018 main survey
Items	Mathematics	21	21
	Reading (Trend)	51	43
	Reading (New)	71	46
	Science	32	32
	Financial Literacy	18	13
Groups	Countries	50	71
	Languages	34	50
	Country-by-language groups	65	112

responses from both the 2015 main survey and the 2018 field trial. CUR pools are unique to each country-by-language group, with responses provided by that country's own students and with verified labels provided by that country's own coding team. Following the described quality assurance procedures, any inconsistent codes that were identified in the historical data necessarily excluded the unique response from being used in the CUR pool for the 2018 main survey. It should be noted that by screening students' text responses through the MSCS first, it was expected that the responses seen by human raters would be more complex responses that cannot be easily scored by machine. With the accompanying reduction in workload, raters can focus their efforts on these potentially confusing responses more so than in previous cycles.

Table 12.2 presents the number of CR items and participating countries and economies where MSCS was implemented in the field trial and main survey, respectively. New items were developed for the major domain, reading (71 items) and for financial literacy (2 items). Note that no historical data was available for these newly developed items, thus, coding efficiency is not expected except for filtering out blank responses and for scoring special known cases. For reading and financial literacy, a larger number of items were administered during the field trial than were ultimately selected for use in the main survey, but some of them were dropped between the field trial and main survey. For the domains of mathematics and science, only trend items were used (e.g., items from previous cycles), so there were no changes in the items administered between the field trial and the main survey for these domains.

For the operational use for PISA 2018, the CUR pool was established based on the 2015 main survey data for the field trial, while it was based on the 2015 main survey and 2018 field trial together for the main survey. Thus, new countries and new language groups participating in PISA 2018 cycle for the first time could have the advantage of MSCS during the main survey.

12.4.1 Efficiency in Coding

Through the application of the MSCS, all CBA domains in PISA saw a considerable reduction in human coding burden. The extent to which scoring by human raters was reduced we refer to as *efficiency*, and efficiency gains are summarized by domain in Table 12.3 for the field trial and in Table 12.4 for the main survey. The tables report the percentage of responses coded by the MSCS and by human raters across all items in the four domains and across country-by-language groups with both mean and median values. The proportion of responses that were coded by the MSCS varied by item, country, and language, such that the distribution of the percentages

Table 12.3 Percentage of responses coded by the MSCS and by human coders across countries in the 2018 field trial

		Human-coded	Machine-coded		
	Stats	Total (%)	Nonresponse (%)	Valid (%)	Total (%)
Mathematics	Mean	66.1	17.4	16.5	33.9
	Median	70.3	14.4	7.0	29.7
Reading (new)	Mean	84.3	11.1	4.6	15.7
	Median	86.7	8.3	0.0	13.3
Reading (trend)	Mean	79.3	10.0	10.7	20.7
	Median	87.2	7.6	0.0	12.8
Science	Mean	75.3	12.4	12.3	24.7
	Median	77.5	8.8	3.0	22.5
Financial Literacy	Mean	87.1	12.1	0.8	12.9
	Median	88.9	10.6	0.0	11.1

Note. Because median percentage values were pooled across items, countries, and language groups, the nonresponse and valid percentages coded do not necessarily sum to the total

Table 12.4 Percentage of responses coded by the MSCS and by human coders across countries in the 2018 main survey

		Human-coded	Machine-coded		
	Stats	Total (%)	Nonresponse (%)	Valid (%)	Total (%)
Mathematics	Mean	67.5	18.3	14.2	32.4
	Median	71.5	15.2	4.9	28.5
Reading (new)	Mean	76.7	14.7	8.5	23.3
	Median	81.7	10.5	0.0	18.2
Reading (trend)	Mean	71.2	18.4	10.3	28.7
	Median	78.9	15.4	0.0	20.9
Science	Mean	74.6	12.6	12.7	25.3
	Median	77.2	8.9	3.4	22.6
Financial Literacy	Mean	79.9	13.4	6.2	19.6
	Median	83.3	10.3	0.0	16.3

Note. Because median percentage values were pooled across items, countries, and language groups, the nonresponse and valid percentages coded do not necessarily sum to the total automatically scored for each item per group can be skewed; therefore, medians are reported in addition to the mean values.

The first two columns under the machine-coded header, *nonresponse* and *valid*, indicate the mean and median percentages of responses across CBA participants that were automatically scored by the MSCS as either a nonresponse or as a verified response (correct, with full or partial credit, and incorrect). The total of these values is also presented, which can be compared to the percentage of human-coded responses, reported in the first column. Note that prior to the MSCS, all responses to CR items had to be coded by humans without exception, including nonresponses. On average, the scoring burden for human raters was reduced for the 2018 field trial by a minimum of approximately 13% in financial literacy to nearly 34% in mathematics. Similarly, for the 2018 main survey, the reduction in scoring burden ranged from 19% in financial literacy to 33% in mathematics.

For both the field trial and main survey, an average of approximately 10–17% of the total responses across all domains were nonresponses, which is similar to what was observed in the PISA 2015 main survey, shown in Table 12.1. On new reading items, for which no verified labels were available from 2015, the extraction of nonresponses alone reduced the human coding burden by 11% in the field trial and by 15% on average in the main survey. In addition, labels for some special cases of two-part items (e.g., radio button items, which require an explanation following a selection on the MC part) were known. Therefore, pre-defined scoring rule (i.e., valid selection on the MC part, but empty response on the open-field was labeled as "incorrect") was applied and it showed the efficiency of 4.6% on reading responses for new items. When comparing Tables 12.3 and 12.4, the relative gains of adding the 2018 field trial data are noticeable for new reading items (4.6–8.5%) and financial literacy items (0.8–6.2%) in particular. While losses are not expected when new data are introduced, the percentage of responses coded by the MSCS did drop slightly (less than 1%) in mathematics between the field trial and main survey. This may be due to having removed some responses from the CUR pool due to observed inconsistencies in human coding when the 2015 main survey data and 2018 field trial data were combined ahead of the 2018 main survey (see the discussion and conclusion section for more on this issue). Another possibility is a cohort effect such as certain words being used more often or triggered by changes in the platform (e.g., different lengths or sizes of the text fields).

As a graphical comparison of the MSCS performance between field trial and main survey, Fig. 12.2 shows the percentage of responses coded during the PISA 2018 field trial (using only 2015 main survey data) and the percentage coded during the 2018 main survey (using both 2015 main survey data and 2018 field trial data). The black lines indicate the PISA 2018 field trial data, and the red lines indicate the 2018 main survey data, with the thick lines representing the total percentage of responses scored by the MSCS. In decomposing that figure, the thin lines represent the total percentage of valid responses (i.e., correct, partially correct, or incorrect) automatically scored by the MSCS, and the dashed lines represent nonresponses that were also automatically coded. The total percentage of responses automatically coded between the main survey and the field trial was quite similar in the domains of

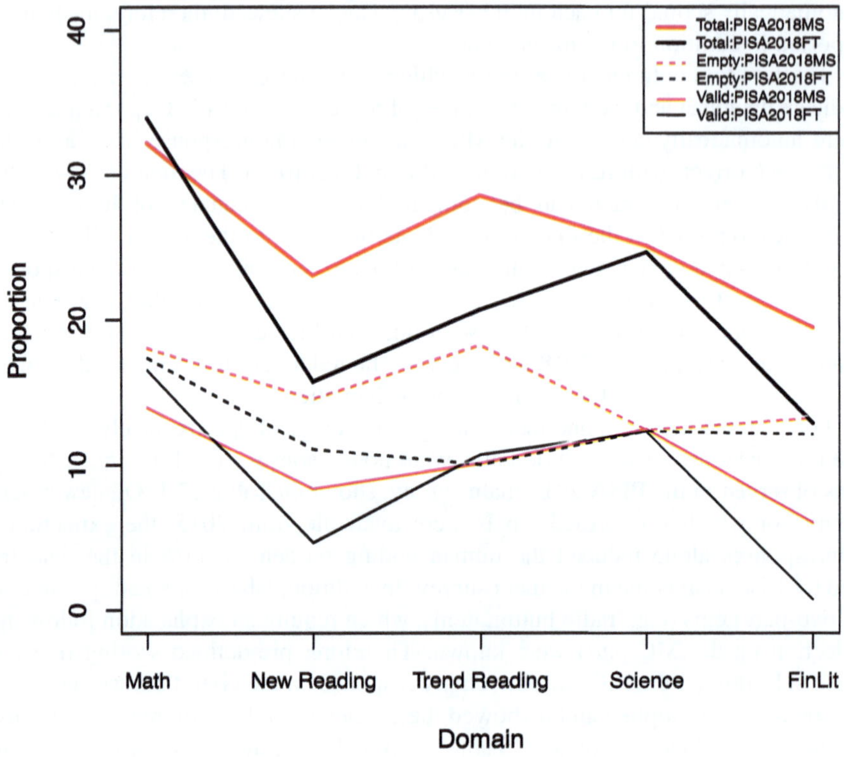

Fig. 12.2 Percentage of responses coded by the MSCS in the PISA 2018 field trial and main survey

mathematics (about 33–34%) and science (about 25%). New reading items in the field trial did not have any historical data in the CUR pools prior to assessment (new reading items did not appear in the 2015 main survey); however, about 18% of responses to new reading items were collected and added to the CUR pools during field trial coding, which contributed to an 8% increase in the total number of responses automatically scored during the main survey. Trend reading items also saw an 8% increase in the percent coded between cycles, and financial literacy, with a mix of both trend and new items, saw a 7% increase.

12.4.2 Variation in Efficiency Across Items and Participating Groups

The amount of workload reduction, or the number of unlabeled data that were scored by the MSCS, is item-dependent as well as country-by-language-dependent as explained in the Sect. 12.2.1. An item with high regularity of text responses, thus, less variance, is expected to be catalogued in the CUR pool and more automatically

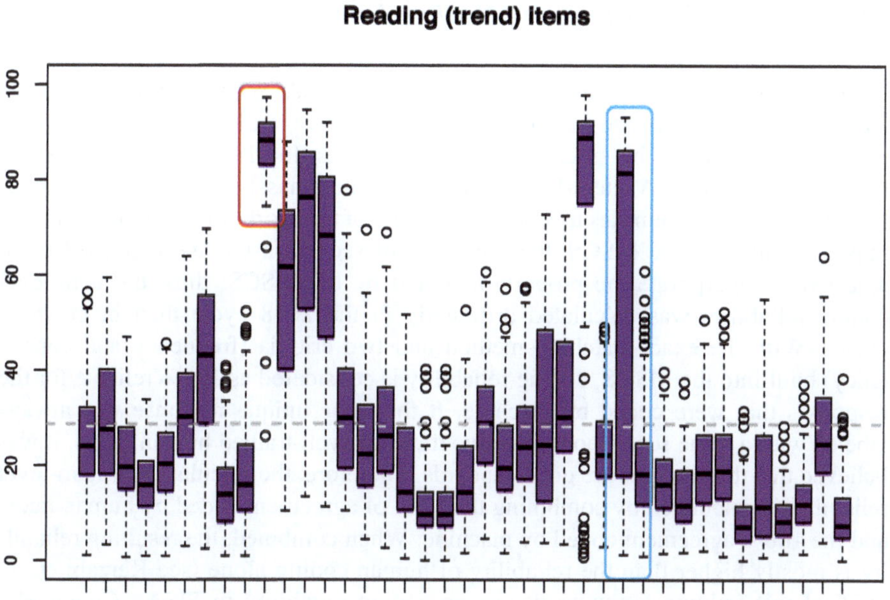

Fig. 12.3 Distribution of efficiency gains for each item across country-by-language groups

scored by the MSCS. Figure 12.3 presents the distribution of efficiency gains for trend reading items from the main survey; individual items are located on the x-axis and the efficiency gains are presented on the y-axis with the grey dashed horizontal line indicating the average efficiency across items. Each boxplot represents the distribution of efficiency of MSCS for the given item across country-by-language groups. As shown, the performance of MSCS in reducing the manual scoring by human varied across CR items in PISA, as well as across country-by-language groups. Two extreme items are noted with red and blue outlines—the item outlined in red showed high efficiency in most country-by-language groups, while the item outlined in blue showed a huge variation across country-by-language groups. Regarding the latter, given the same item, some country-by-language groups may have a relatively larger CUR pool due to larger proportions of identical responses (e.g., observed more than five times) or more consistent scores given by human raters to the same unique responses or a substantially large proportion of blank responses. On the other hand, some country-by-language groups may have a relatively smaller pool due to less uniformity in students' answers or inconsistent scores given to the same raw responses or a few blank responses. In particular, the inconsistency of the codes given by human raters can be monitored to investigate the accuracy of coding through the MSCS.

12.4.3 Potential to Improve Coding Accuracy

In international large-scale assessments, including PISA, scoring of text responses are expected to be accurate for all CR item responses regardless of assigned human raters or students' origins of countries and languages. In PISA, after switching the primary mode to CBA, item-level coding reliability has been monitored within-country and across countries to assure the quality of the scores of CR items. With the implementation of the MSCS, the reliability of scores on CR items needed to be adjusted to incorporate the proportion scored by the MSCS; thus, the item-level coding reliability was calculated differently in the 2018 cycle than in previous cycles. With the exact-matching method and two criteria (frequency and consistency) built into the MSCS, coding reliability is considered as 100% reliable for the responses that were coded by machine. If the same unique response was always labeled with a same score more than five times by well-trained human raters, it was believed that those scores are gold standards. Therefore, the calculation of item-level reliability is calculated by combining the level of agreement coded by human raters and the 100% agreement coded by machine. When combined, the resulting reliability is mostly higher than the reliability of human coding alone (see Kerzabi et al., 2025 of this volume for more details on coding reliability in PISA). Concretely, item-level coding reliability would continue to increase if a larger portion of responses were scored by the MSCS.

However, it is important to note that coding reliability does not always guarantee coding accuracy. The concept of coding reliability is similar to consistency because it quantifies the extent to which multiple raters provide the same judgement in labeling the text responses in a relative manner (Myford & Wolfe, 2003). In contrast, coding accuracy presumes that there is a gold standard and comparison is conducted for each rater's judgment to that key in an absolute manner. For example, there can be cases where the wrong score was labeled consistently (e.g., confusing description in the coding rubric) although it has been rare in the case of PISA.

In this regard, the MSCS has the potential to evaluate the coding accuracy and eventually to improve accuracy by utilizing the accumulated data from the past and supplying the verified labels. The current verification process is very conservative, as it requires that exclusively consistent codes should have been observed at least five times. The quality assurance procedure is built in to detect any inconsistencies in human coding by comparing the most recent cycle-level data to the CUR pool which points to a specific source of coding error: misapplication of the coding guide. By reviewing data collected by the MSCS and the latest cycle's data, additional improvements in coding accuracy can be conceived.

In particular, coding guides and coder training materials can elaborate and exemplify common misconceptions in the coding using the observed students' responses that were inconsistently coded. At the international level, some items that show highest disagreements can be identified and communicated with content experts and item developers to refine the coding guides and update coder training materials where applicable. Subsequently, a combination of real student responses and expected student responses can be used in the international coder training as examples for what coders are likely to observe. More importantly, with the harvested real responses from the MSCS, the coder training materials can be tailored by each country-by-language group to improve coder training within-country. The example responses in coding guides used by coders in the field, which are directly translated from the international coder training materials, may not be practical because they may not resemble the real responses that coders are likely to observe in their country from students speaking their language (Zehner et al., 2015). With real observed responses, national coding teams could train with responses that are actually expected from students, harvested from the historical data.

Furthermore, with the help from MSCS and human expert coders from the national coding teams, "confusing to code" cases can be included directly to the CUR pool so that verified label can be readily applied by machine. First example includes CR items that have double-digit codes in the coding rubric. It is more common to observe qualitative coding disagreements among coders: raters agree on whether the response is correct or partially correct but disagree on what type of correct or partially correct response is provided given the coding guides. Coding such cases are often quite complex, and the lack of understanding about the coding guides can easily result in human coding disagreement. If a verified code is submitted and included in the CUR pool ahead of coding, fewer coding errors are expected due to a lack of understanding. Another example is with radio button items which require a secondary explanation to an initially selected multiple-choice part. The coding guide in PISA defines "incorrect" if the student selected a multiple-choice option but failed to type any explanatory response text box (blank). Although infrequent, we have observed that human raters incorrectly score such responses as "missing," due to a nonresponse in the text box. In order to avoid such coding error, a pre-defined code appropriate for the radio button selection and blank text box combination was included as a part of the CUR pool beginning in the PISA 2018 main survey.

12.5 Concluding Remarks and Outlook

The MSCS was developed to improve the efficiency and, potentially, accuracy of the coding process for CR items in PISA, exploiting the new CBA environment. Conceptually, the MSCS can be viewed as a supervised learning of natural language data that is generalizable across multiple languages to be used for a subset of PISA data. The system capitalizes on the regularity of observed responses, specifically, the extent to which a small number of unique responses proportionately occur across all student responses on a particular item. Without extracting language-dependent features, the MSCS utilizes the exact-matching method and automatically applies a verified label to incoming unscored responses, which guarantees the highest accuracy in the prediction, the applicability to all different languages in PISA, and the accurate reporting of scores aligned with the coding rubric.

This chapter illustrated the motivation behind the development of this system, how it works in practice, and what lessons were learned from the analyses. When all CR items across four domains in the PISA 2015 main survey were analyzed, it was shown that the manual human coding could be potentially reduced by about 39% in mathematics, 27% in reading, 21% in science, and 17% in financial literacy when all identical instances were automatically coded. Interpreting these figures as the motivation, since PISA 2018 cycle, the workflow of MSCS was built and operationalized. On average, a low of approximately 13% in financial literacy to a high of 34% in mathematics of the total students' text responses were coded by MSCS. For the 2018 main survey, the coding burden was reduced by a low of approximately 19% in financial literacy to a high of 33%, in mathematics. By including a verified code for responses that were left blank, as well as special known type of codes (e.g., radio button items), the MSCS was applicable even to new items and newly participating country-by-language groups where no historical data were available. In addition to efficiency, the MSCS has shown to improve the item-level coding reliability and proven potential to improve the human coding accuracy by updating coder training materials and coding guides.

The MSCS has already shown to alleviate manual coding burden and has the potential to further improve coding accuracy across all participating countries and languages. The system is based on the simple but straightforward algorithm, which can be readily applicable for any international large-scale assessments involving CR items that require scoring by humans. With the experiences and insights obtained from PISA 2018 implementation, further improvements can be explored and made based on well-designed research studies. We want to conclude this chapter by providing two interesting future research directions extending the current MSCS.

12.5.1 Extension of the MSCS Using NLP Techniques

Extension of the MSCS to further improve the efficiency and accuracy can be pursued in two ways. The first approach is more readily applicable by modifying the current conservative operation practices, keeping the current exact-matching methods. Across PISA cycles, it was expected that the CUR pool would continue enlarging by adding new unique responses with more instances from the most recent cycle if the associated label is consistent with the historical data. However, the opposite could happen when a CUR pool is merged with data containing inconsistent codes from the most recent cycle of data collection. This was observed during the PISA 2018 main survey, although rare, in which the historical data logged consistent codes but fewer than five responses, and the new cycle data contained additional responses, but inconsistent coding by idiosyncratic coder's miscode. Applying the current conservative threshold, such cases were necessarily excluded in future cycle CUR pools due to that inconsistent coding. A repeated review and evaluation process at each cycle is beneficial to maintain the quality of CUR pools, but the process is quite sensitive to such idiosyncrasies in coding. Therefore, modifying the current conservative thresholds and continuously evaluating the quality of the CUR pools can be useful to exclude idiosyncratic codes, while maintaining perfect accuracy at the same time. The simplest and most straightforward operational way of enlarging the CUR pool is to decrease the threshold of frequency of response observation. Currently, the MSCS requires identical responses to be observed at least five times, but lowering the frequency to fewer than five times can be defensible if the associated labels are still robustly verified. Another readily operationalizable approach is to build the CUR pool at the language-group level, rather than country-by-language group. If the CUR pool could be established at the language level so that the same CUR pool could be applied to multiple countries that speak the same language, more instances of identical responses could be observed across different countries but from the same language group. Of course, we cautioned such an approach at the beginning of this chapter, as it would need to be approached very carefully and would require further verification that no item-level changes have been made by a particular country (i.e., if the French assessment in Canada uses "miles" while the French assessment in France uses "kilometers").

The second approach is to exploit more advanced NLP techniques that have been rapidly evolved in a last few decades. While any stochastic methods including machine learning techniques cannot achieve perfect accuracy as is done through exact-matching methods, the type and amount of error in the prediction through advanced NLP algorithms may still be acceptable and promising, considering that the coding inconsistencies are observed in the labels produced by human raters. For example, in the context of large-scale assessment, pattern clustering through the latent semantic analyses (e.g., Zehner et al., 2016), coding rules based on the longest common sequence (e.g., Sukkarieh et al., 2012), basic preprocessing such as removing white space, punctuation, and capitalization (e.g., Manning & Schütze, 1999; Zehner et al., 2016) can be systematically evaluated and explored for the operational uses; see also Zehner and Goldhammer (2025) of this volume for additional possibilities for coding text responses in PISA called ReCo (Automatic Text Response Coder). Most recently, a study by Zehner et al. (2021) revealed that automatic normalization steps implemented through the extension of MSCS with ReCo could improve coding efficiency by an additional 5.1% with a minor loss in accuracy of –0.5%. Further collaborative research is in progress to relax the definition of an exact-matching response through various machine learning techniques and to evaluate the applicability to multiple languages.

More broadly, Flor and Hao (2021) explained that an automated scoring engine usually has two major components: a feature representation that turns the raw texts into numerical forms, and the second is to map the feature representation to a numerical score or categories via statistical modeling or machine learning methods. In this regard, MSCS skips the first major component to extract features from the original text responses. This was strategically chosen for MSCS, as the first automatic scoring model in PISA, to be more robust in a multilingual setting, but more investigation about the feature representation would be valuable to extend the scope of automatic scoring of text responses in PISA. Such feature representations would include language-dependent and language-independent features as well as micro-features (specific to certain item) and macro-features (applicable to larger item sets).

12.5.2 Evaluation of the MSCS Performance Across Participating Groups

Although the MSCS has shown promise to alleviate the manual coding by human, the huge variation in the performance across CR items and participating country-by-language groups was noticeable (Fig. 12.3). The MSCS was designed to be a standardized procedure that are applicable for all countries, languages, and CR items. Thus, essentially, the variation in the performance of MSCS depends on various factors, including the quality of the CUR pool (e.g., how consistent human raters in the national coding team was, how homogeneous students' text responses were), and how similar the new incoming unlabeled data are to the established CUR pool from the past data. Given that the text responses are collected in multiple languages for the same set of CR items, evaluation of the MSCS performance across participating countries would be an ideal testbed to compare learner populations with different languages (Horbach & Zesch, 2019). Once features are extracted from the original text responses per item from each country-by-language group, the performance of MSCS can become even more widespread, depending on the availability of NLP resources for each language group. With careful experiments and evaluation across multiple languages and countries, further research and more principled operationalization can be pursued. At the same time, focusing on the variation across items, characterizing CR items in PISA using the quantities that are standardizable across multiple languages, (more strictly, country-by-language groups), such as type/token ratio, would be particularly useful to anticipate manual coding burden to facilitate the overall coding procedures supplemented by the MSCS.

References

Attali, Y., & Burstein, J. (2006). Automated essay scoring with e-rater® V.2. *The Journal of Technology, Learning, and Assessment, 4*(3) https://ejournals.bc.edu/index.php/jtla/article/view/1650

Bennett, R. E. (2011). *Automated scoring of constructed-response literacy and mathematics items.* Advancing Consortium Assessment Reform (ACAR). Arabella Philanthropic Advisors.

Bhagat, R., & Hovy, E. (2013). What is a paraphrase? *Computational Linguistics, 39*(3), 463–472.

Burstein, J. (2003). The E-rater scoring engine: Automated essay scoring with natural language processing. In M. D. Shermis & J. Burstein (Eds.), *Automated essay scoring: A cross-disciplinary perspective* (pp. 113–121). Lawrence Erlbaum.

Ellis, R., & Barkhuizen, G. P. (2005). *Analysing learner language.* Oxford University Press.

ETS. (2021). *Best practices for constructed-response scoring.* Educational Testing Service. https://www.ets.org/s/about/pdf/cr_best_practices.pdf

Flor, M., & Hao, J. (2021). Text mining and automated scoring. In A. A. von Davier, R. J. Mislevy, & J. Hao (Eds.), *Computational psychometrics: New methodologies for a new generation of digital learning and assessment. Methodology of educational measurement and assessment*. Springer. https://doi.org/10.1007/978-3-030-74394-9_14

Horbach, A., & Zesch, T. (2019, April). The influence of variance in learner answers on automatic content scoring. *Frontiers in Education, 4*, 28. Frontiers.

Kerzabi, E., Yamamoto, K., & Chen, C. (2025). Tools and platforms for the improvement of coder reliability: OERS and OECS. In L. Khorramdel, M. von Davier, & K. Yamamoto (Eds.), *Computer-based international large-scale assessments: Concepts, methodologies and quality assurance procedures*. Springer.

Leacock, C., & Chodorow, M. (2003). C-rater: Automated scoring of short-answer questions. *Computers and the Humanities, 37*, 389–405. https://doi.org/10.1023/A:1025779619903

Lennon, M., & Kirsch, I. (2025). Innovations in item development for computer-based international assessments. In L. Khorramdel, M. von Davier, & K. Yamamoto (Eds.), *Computer-based international large-scale assessments: Concepts, methodologies and quality assurance procedures*. Springer.

Loukina, A., Madnani, N., Cahill, A., Yao, L., Johnson, M. S., Riordan, B., & McCaffrey, D. F. (2020). Using PRMSE to evaluate automated scoring systems in the presence of label noise. In *Proceedings of the fifteenth workshop on innovative use of NLP for building educational applications*. https://www.aclweb.org/anthology/2020.bea-1.2/

Manning, C. D., & Schütze, H. (1999). *Foundations of statistical natural language processing*. MIT Press.

Myford, C. M., & Wolfe, E. W. (2003). Detecting and measuring rater effects using many-facet Rasch measurement: Part I. *Journal of Applied Measurement, 4*(4), 386–422.

Organisation for Economic Co-operation and Development (OECD). (2017). *PISA 2015 technical report*. OECD. https://www.oecd.org/pisa/data/2015-technical-report/

Organisation for Economic Co-operation and Development (OECD). (2019). *PISA 2018 assessment and analytical framework*. OECD. http://www.oecd.org/education/pisa-2018-assessment-and-analytical-framework-b25efab8-en.htm

Organisation for Economic Co-operation and Development (OECD). (2020). *PISA 2018 technical report*. OECD. https://www.oecd.org/pisa/data/pisa2018technicalreport/

Ou, L., Kerzabi, E., Rosen, Y., & von Davier, A. A. (2025). New constructs and item types in large-scale assessments. In L. Khorramdel, M. von Davier, & K. Yamamoto (Eds.), *Computer-based international large-scale assessments: Concepts, methodologies and quality assurance procedures*. Springer.

Padó, U. (2016, December). Get semantic with me! The usefulness of different feature types for short-answer grading. In *Proceedings of COLING 2016, the 26th international conference on computational linguistics: Technical papers* (pp. 2186–2195).

Shin, H. J., von Davier, M., & Yamamoto, K. (2019). Investigating rater effects in international large-scale assessments. In B. Veldkamp & C. Sluijter (Eds.), *Theoretical and practical advances in computer-based educational measurement. Methodology of educational measurement and assessment*. Springer.

Shin, H. J., Andersen, N., Horbach, A., Kim, E., Baik, J., & Zehner, F. (2024). Operational automatic scoring of text responses in 2016 ePIRLS: Performance and linguistic variance.

Sukkarieh, J. Z., & Blackmore, J. (2009). c-rater: Automatic content scoring for short constructed responses. In *Proceedings of the twenty second international FLAIRS conference* (pp. 290–295). http://www.aaai.org/ocs/index.php/FLAIRS/2009/paper/download/122/302

Sukkarieh, J. Z., von Davier, M., & Yamamoto, K. (2012). *From biology to education: Scoring and clustering multilingual text sequences and other sequential tasks* (Research Report No. RR-12-25). Educational Testing Service. https://doi.org/10.1002/j.2333-8504.2012.tb02307.x

Yamamoto, K., He, Q., Shin, H. J., & von Davier, M. (2017). *Developing a machine-supported coding system for constructed-response items in PISA* (Research Report No. RR-17-47). Educational Testing Service. https://doi.org/10.1002/ets2.12169

Yamamoto, K., He, Q., Shin, H. J., & von Davier, M. (2018). Development and implementation of a machine-supported coding system for constructed-response items in PISA. *Psychological Test and Assessment Modeling, 60*(2), 145–164. https://www.psychologie-aktuell.com/fileadmin/download/ptam/2-2018_20180627/02_PTAM-2-2018_Yamamoto_v2.pdf

Zehner, F., & Goldhammer, F. (2025). Automatically coding PISA text responses. In L. Khorramdel, M. von Davier, & K. Yamamoto (Eds.), *Computer-based international large-scale assessments: Concepts, methodologies and quality assurance procedures*. Springer.

Zehner, F., Goldhammer, F., & Sälzer, C. (2015). Using and improving coding guides for and by automatic coding of PISA short text responses. In *2015 IEEE international conference on data mining workshop (ICDMW), Atlantic City, NJ* (pp. 280–287). https://doi.org/10.1109/ICDMW.2015.189

Zehner, F., Sälzer, C., & Goldhammer, F. (2016). Automatic coding of short text responses via clustering in educational assessment. *Educational and Psychological Measurement, 76*(2), 280–303. https://doi.org/10.1177/0013164415590022

Zehner, F., Shin, H. J., Kerzabi, E., Andersen, N., Goldhammer, F., & Yamamoto, K. (2021, June). *Automatic normalization for advancing response coding consistency and effort in PISA*. Paper to be presented at the annual meeting of the National Council on Measurement in Education (NCME), Virtual Meeting.

Chapter 13
Automatically Coding PISA Text Responses

Fabian Zehner and Frank Goldhammer

Abstract Automatic coding (e.g., scoring) of short-text responses constitutes a promising method for tackling challenges associated with the constructed-response format (e.g., in adaptive testing). At the same time, it opens up plenty of new opportunities. This chapter lists such opportunities along with challenges and contextualizes the task of automatic coding. It elaborates on the functioning of ReCo, a software package for automatically coding PISA responses. An empirical demonstration shows that, once trained, the system can be reused for most items in repeated measurements without substantial effort. The chapter concludes with an outlook on how ReCo could be blended with PISA's Machine-Supported Coding System and how it could help to harmonize human codes across different test languages.

13.1 The New Role of Text Responses

Text responses are no longer the unwanted child of measurement. In the past, only brave researchers with access to abundant resources would set up instruments involving constructed responses. Masses of text responses would need to be evaluated by a horde of raters and would then linger in deep layers of dust in data archives.

While this intentionally exaggerated statement may be read with an impish grin, its core still hits an uncomfortable truth. Educational large-scale assessments, in particular, come with large volumes of data. Thus, manual work on single open-ended responses uses up substantial amounts of financial time and resources. Additionally, there is some skepticism as to the measurement's objectivity.

F. Zehner (✉) · F. Goldhammer
DIPF | Leibniz Institute for Research and Information in Education, Frankfurt am Main, Germany

Centre for International Student Assessment (ZIB), Frankfurt, Germany
e-mail: f.zehner@dipf.de

L. Khorramdel et al. (eds.), *Innovative Digital-Based International Large-Scale Assessments*, Methodology of Educational Measurement and Assessment,
https://doi.org/10.1007/978-3-031-90951-1_13

In this chapter, we describe one solution to the issue: an approach for automatically coding short text responses by computers (Zehner et al., 2016).[1] While we also contextualize and elaborate on the concept, we furthermore present performance results of an automatic coding system that has been trained on data from PISA 2012 (Zehner et al., 2016) and evaluate it on data with the same items from PISA 2015. This way, we demonstrate its generalizability beyond the actual training data and that, once trained, the system can be reused for repeated measurements in (quasi-) longitudinal assessments without substantial effort.

When constructed responses were becoming more popular in large-scale assessments in the 1990s (Bennett, 1993), several concerns and fears were discussed. The constructed responses' (supposedly) marginal gains in construct coverage were thought to be easily outweighed by their lack of objectivity, reliability, and efficiency (cf. Bejar, 2017; Bennett & Ward, 1993). The Programme for International Student Assessment (PISA), however, has included constructed responses since its inception (Adams & Wu, 2002). Proponents of the open-ended format mainly pointed out that measurement of higher-order cognitive skills is lacking in closed response formats (e.g. Guthrie, 1984). In recent times, an incremental value of open-ended response formats for construct validity—or construct coverage to put it more precisely—could indeed be shown for some constructs; for example, for computer and information literacy (Ihme et al., 2017), as well as for literacy in mathematics (e.g., Birenbaum & Tatsuoka, 1987; Bridgeman, 1991) and reading (e.g., Griffo, 2011; Lim, 2019; Millis et al., 2011; Rauch & Hartig, 2010; Rupp et al., 2006), whereas only sometimes specific psychometric properties have been found to differ (e.g., differing reliabilities for different test taker performance levels; Schult & Sparfeldt, 2018).

With the rise of natural language processing techniques over the last few decades, text responses no longer need to be an additional burden on resources—quite the opposite, as they can now constitute a new source of information. Today, large-scale assessments produce a rich set of supplementary information that can be used for answering methodological, as well as substantive, research questions. Researchers can dust off text responses from data archives and perform secondary data analyses. For example, one study using this approach found the reading gender gap mirrored the linguistic information in text responses in a secondary analysis and could derive more detailed inferences (Zehner et al., 2018), information that had been lost in scored data. Another study demonstrated how patients' narratives were indicative for the diagnosis of posttraumatic stress disorder (He, 2013).

[1] Please note that this chapter was written and accepted for publication in 2019; 6 years before its actual publication. In the meantime, the methodology behind ReCo is available in the R-package shinyReCoR with an interactive graphical user interface (https://www.reco.science, Andersen & Zehner, 2021). Similarly, the chapter was written before transformers and pre-trained large language models had their vast impact on the field of automatic scoring and are thus not included. For recent overviews on factors affecting whether automatic coding is effective and feasible, see Zesch et al. (2023), and for an outline of prevailing methodological paradigms, see Bexte et al. (2024).

Automatic coding of open-ended responses can be traced back to Page (1966, 1968) who attempted to automatically grade essays. At that time, associated research, as well as its application, mainly remained in the field of essay grading. Coinciding with the growing demand for the open-ended response formats described above, computational linguists and psychometricians found their interest in automatically scoring shorter responses only in the 1990s (e.g., Bejar, 1991; Kaplan, 1991). This interest already included a diverse set of measured constructs like programming skills (Braun et al., 1990), mathematical problem solving (Sebrechts et al., 1991), and constructs involving natural language (Burstein et al., 1996). Over the following two decades, several systems for automatically coding short text responses have emerged, also triggered by two data competitions (Hewlett Foundation, 2012; Dzikovska et al., 2013); for an overview, see Burrows et al. (2014) and Shermis (2020).

Over the last few years, a huge body of literature has been evolving around the automatic coding of short answers, mainly stimulated by computational linguists. Yet, educational, international large-scale assessments, which constitute a particularly attractive field of application, have hardly been involved in these developments. As one of two exceptions, Zehner et al. (2016) demonstrated how well a program, called ReCo (Automatic Text Response Coder), performs in coding PISA text responses. ReCo clusters responses into groups by their coarse semantics. Although it only incorporates baseline techniques that can be scaled to a vast number of languages, most investigated PISA items could be scored with a fair to good up to excellent agreement beyond chance compared to the human codes (evaluated only for German data so far). For three out of ten investigated items, the automatic coding was rather weak. A second development—also from the PISA context—comes from Yamamoto et al. (2018), who proposed an even simpler approach without any natural language processing called MSCS (PISA's Machine Supported Coding System; cf. Shin et al., 2024 of this volume). The MSCS automatically assigns codes to new responses if the exact same text response has been consistently coded by five human raters before. While the MSCS, theoretically, has perfect accuracy, its coverage in terms of the number of responses that can be coded automatically can be severely constrained, depending on the diversity of item responses. ReCo, on the other hand, has varying accuracy, but can be applied to all responses. Therefore, the two systems complement each other nicely. For potential operational usage in PISA, the MSCS and ReCo could work together in order to combine MSCS's optimal accuracy and ReCo's perfect coverage.

As mentioned before, short text responses have certain potential that could not have been made use of hitherto. In the following sections, we describe how ReCo uses semantic vectors and machine learning in order to automatically code PISA text responses (Sect. 13.2). Given this, we describe how using natural language processing techniques on text responses makes them more desirable than the unwanted child they had been before (Sect. 13.3). Besides illuminating the opportunities that come with automatic coding, we also shed light on the challenges it is facing, and we further contextualize ReCo's design decisions (Sects. 13.4 & 13.5). An empirical demonstration shows ReCo's robustness (Sect. 13.6). The final discussion concludes how ReCo could be blended with PISA's MSCS and also outlines how it might be utilized to overcome cross-lingual inconsistencies in human coding across countries (Sect. 13.7).

Finally, before descending down the rabbit hole, we need to emphasize that we deliberately do not use the term automatic grading or scoring, but automatic coding instead. That is because the first two are special cases of coding, and, from our perspective, the scope of natural language processing in assessment should not be limited to ordinal classifications of the response product.

13.2 ReCo: The Automatic Response Coder

ReCo is a free program that enables us to automatically code text responses (Zehner et al., 2016). Also, it is capable of extracting construct-relevant features in the context of reading literacy assessment (Zehner et al., 2018). In this chapter, we will focus on automatic coding.

It is worth noting that, unlike other comparable products (e.g., c-rater; Sukkarieh & Blackmore, 2009), ReCo exclusively addresses semantics and does not include additional features (such as response length). While some other products utilize far more powerful natural language processing techniques (e.g., coreference resolution) and partly more powerful machine learning methodology (thus yielding higher accuracy), ReCo sticks to baseline methods on purpose, achieving three main goals. (i) It scales easily to other languages, (ii) it is robust towards improper language usage like that typically found in low-stakes assessment, and (iii) it exclusively applies transparent methods and, thus, (a) can be used even without annotated training data for specific purposes due to its unsupervised training part and (b) allows substantive conclusions to be drawn if set up accordingly. For automatically assigning codes, ReCo continues the analysis with a supervised learning and evaluation step.

13.2.1 Preprocessing: Normalizing Diversity

How does ReCo work and make sense of a response? Let us consider a sample response: "A girl falling into and wandering through a fentasy world." First of all, several preprocessing steps are applied for normalizing purposes (i.e., making responses more comparable). Among others, punctuation is removed (full stop), upper case is transformed to lower case (a), misspellings are corrected (fantasy), stop words are removed (so-called synsemantic words without semantic significance of their own; e.g., and), and stemming is applied, which cuts off affixes (falling). Further techniques (e.g., compound splitting) can be considered depending on the language, measurement goal, and need for portability to other languages. For our sample response, the preprocessing phase ends up with five tokens: [girl], [fall], [wander], [fantasy], and [world].

13.2.2 Semantics: Getting a Grip on Words' Meaning

In the next step, ReCo needs a way of distinguishing the semantics of such tokens. For this, it utilizes a *vector space model*, a type of compiled dictionary with vectors as numerical representation of their semantics. For computing the vector space model, ReCo first collects a text corpus from Wikipedia by crawling articles that are related to an item's topics (e.g., starting at the page *Alice in Wonderland* and subsequently adding pages that are connected by in- and out-going links). Next, this text corpus is input for Latent Semantic Analysis (LSA; Deerwester et al., 1990). A semantically representative, tailored corpus is important for LSA (Quesada, 2011). LSA takes (non-)co-occurrences of words into account for estimating semantic similarity. The basic idea is that semantically similar words occur in semantically similar contexts (i.e., documents or articles), which need not be the same ones. At the technical level, LSA builds a co-occurrence matrix A, representing the number of occurrences of each term in each document. Applying a singular value decomposition to this co-occurrence matrix, LSA then linearly decomposes A into a product of two orthogonal matrices U and V^T, representing term and document loadings respectively, as well as a diagonal matrix Σ holding the singular values (cf. Martin & Berry, 2011).

$$A' = U\Sigma V^T \tag{13.1}$$

Before the decomposition, A is weighted by log-Entropy (Dumais, 1991). The idea of this is to give more impact to words that are specific to particular contexts over words that are scattered diffusely across contexts (in the sense of relevance feedback; Salton & Buckley, 1990). For cell a_{ij} in matrix A,

$$a'_{ij} = \log_2(tf_{ij} + 1) * \sum_j \frac{p_{ij} \log_2 p_{ij}}{\log_2 n_j} \tag{13.2}$$

with

$$p_{ij} = \frac{tf_{ij}}{gf_i} \tag{13.3}$$

and tf_{ij} referring to the frequency of term i in document j. gf_i represents the frequency of term i across all n_j documents. After having decomposed the weighted co-occurrence matrix, LSA reduces the dimensionality to typically 300 dimensions, with the idea that the corpus documents' topics are correlated themselves (Martin &

Berry, 2011). Using the cropped and decomposed matrix, LSA ends up with the vector space model. It holds a 300-dimensional vector for each term in the corpus (given by the first 300 columns in U). Due to similar loadings on the singular values, words' vectors that are semantically similar (such as wander and walk) would turn out to point into similar directions whereas unrelated words' vectors would point into different directions (e.g., wander and tax). Here, similarity of two terms is defined as cosine, meaning the ratio of the two vectors' dot product (Martin & Berry, 2011) and the product of the vector lengths, which means applying an L2-norm (Gabrilovich & Markovitch, 2009).

13.2.3 Machine Learning: Relating Vectors to Codes

Making use of this numerical representation of semantics, ReCo next computes the centroid vector for each response's tokens. This step, again, ignores word order (so-called, *bag of words*). All responses are then grouped by an unsupervised cluster analysis into response types, meaning only semantic vectors go into the analysis, but the cluster solution is not optimized towards separating instances with different codes. In the cluster analysis, cosine's inverse (arccos) is used as the distance metric. If categorizing the responses is the overall goal (e.g., for data sifting), the analysis stops here. Otherwise, if the identified categories, or response types, are associated with given codes, ReCo then enters the supervised stage. For this, a naïve Bayes classifier decides which response type is typically associated with which code (e.g., correct / incorrect).

$$P(c_k|g_l)_{l,r} = \frac{P(g_l|c_r) * P(c_r)}{P(g_l)} \tag{13.4}$$

$P(c_k| g_l)$ resembles the probability that response r, belonging to cluster g_l, has been assigned to code c_k by the external criterion (e.g., human coder). A response type's highest probability $P(c_k| g_l)_{l,\ r}$ determines its code.

If ReCo needs to code a new response, the response's centroid vector is compared to the classifier's cluster centroids. The most similar response type then determines which code is assigned.

13.3 Opportunities: Why Bother with Automatic Processing of Text Responses?

As stated in the introduction, compared to closed response formats with a limited number of predefined response options, open-ended formats can turn out to be beneficial. The main reason for assessment designers to opt for closed formats, though, are the practical implications of human coding when using open-ended formats; particularly if natural language is involved like it is in text responses. Furthermore, even if efforts are made to recruit raters, train them rigorously (Bejar, 2012), maybe even statistically calibrate their characteristics (Myford et al., 1995), and evaluate the inter-rater agreement, the measurement will still be less objective in terms of the response evaluation. Would another rater have assigned the same code? Would the same rater, but in a different state (e.g., earlier in the process), have assigned the same code? Would another test designer have decided for the same rubrics that raters use for determining the code?

However, just because closed formats provide the code or score naturally does not mean that the measured information is of any practical use. Thus, decisions for open-ended or closed formats should be based on, beside empirical matters, construct considerations: Which cognitive processes are intended to be evoked by the question and, thus, define what is being measured via the observation of the responses?

While automatic coding does not change the nature of open-ended response formats, it allows for the reduction of several of the discussed disadvantages. Does the project aim at an adaptive test, for which constructed-response formats that requires human scoring are hardly feasible? Are cross-country comparisons the very essence of the large-scale assessment, making constructed responses difficult to include because of comparability issues in coding across cultures and test languages? If so, automatic coding might be the solution for unlocking the door to those endeavors. Even if the automatic coding is not meant to be a substitute for human codes, it can help to support humans to code more consistently and, with this, increase data quality. Also, it allows new procedures to be implemented during the assessment. While this taps into the areas of assessment for learning and intelligent tutoring systems, where interesting innovations also are being proposed (e.g., Bywater et al., 2019; Fang et al., 2019; Nye et al., 2014; Paiva et al., 2014), we focus here on assessment of learning. In the following, we list a few selected examples of potential innovations (adding to Zehner et al., 2016) that may—or already do—enhance educational assessment by automatically coding constructed responses.

- **Preparing data for raters.** Before any text data is actually coded by humans, it can be categorized by computer. One usage of this is to streamline the human coding process by grouping the responses and supporting human raters with annotations by raising awareness to particularly relevant response characteristics. This might reduce the raters' cognitive load or simply reduce the required time (e.g., Wolska et al., 2014), thus, improving data quality. A second use is to create coding guides based on the categorized empirical data (e.g., Marvaniya et al., 2018). This way, it is very easy to create rubrics that sufficiently mirror the empirical diversity of responses, including prototypical reference texts for each rubric. Zehner et al. (2015) showed that the reference texts in PISA coding guides, by far, do not comprehensively mirror the empirical diversity raters face. In both scenarios, text data is sufficient for most methodologies; that is, no manually scored data is required for training purposes.

- **Accompanying the rater's code with a machine code.** When trained, human raters are quite good in evaluating single responses. They are not very good in scaling this effort up to large-scale dimensions like in PISA, where hundreds of thousands of responses have to be coded. Fortunately, this is a task at which computers are brilliant. Thus, humans and computers can make a good team for coding tasks, successfully having been put into practice in automatic essay grading (Enright & Quinlan, 2010). Horbach and Pinkal (2018), for example, let raters code prototypical responses before and during clustering for enhancing the result of automatic coding. Mieskes and Pado (2018) elaborate on several characteristics of coding procedures that determine how automatic coding and human raters can work together. Basically, the most prominent approach is to reduce efforts of manual coding and inconsistencies of coding by propagating single human codes to responses that are found to be similar according to the machine. Other options are to simply flag responses for additional manual review or, with confidence in the accuracy of the automatic system, one can also incorporate both codes, the machine's and the human's, into the final code if human bias is sought to be reduced. In some cases, the strengths of machine and human coding can be combined this way.
- **Comparing responses across test languages.** International large-scale assessments, such as PISA, mainly focus on international comparisons. Naturally but also problematically, constructed responses are always coded by different national teams who are native speakers of the test language (Organization for Economic Co-operation and Development [OECD], 2017). This is unavoidable, but it yields uncertainty with respect to the comparability of cross-country and -language coding. The PISA coding teams within the countries strive towards building a consensus for borderline responses, and they are trained jointly in initial international meetings, but as soon as the teams are working on their actual country-specific data, they are hardly connected across countries and do not really exchange their decisions based on the empirical data. While, to date, automatic coding systems are typically designed for a single specific language, there are recent developments for bridging the gap. For example, Horbach et al. (2018) mapped responses from German to English by means of machine translation. The coding task was then carried out jointly with the translated English responses, as well as for the original English responses. While this does not work too well so far, there are other promising methodological developments that allow languages to be directly compared (for a comprehensive overview see Ruder et al., 2019). In an ongoing project, the ReCo software applies those methods for making responses from different languages in PISA directly comparable, not only allowing the consistency of coding to be checked across test languages, but also enabling substantive research questions to be incorporated at the country level (e.g., which response type / line of reasoning is more prevalent in a given test language or country).
- **Coding On-the-Fly** (e.g., for Computerized Adaptive Testing; CAT). Because PISA has been moving toward adaptive testing for the respective major domains since 2018 (OECD, 2019), it now uses item scoring information during test

administration. This way, proficiency estimation is sought to be more efficient by tailoring the items' difficulties to the test taker's ability. Items with a constructed response format, however, are not included in this on-the-fly estimation of ability because their scoring is not available during test administration. Given PISA's relatively high number of such responses, automatically coded text responses would substantially increase the test information that is available for adaptive routing. An open research question is, however, how accurate does the automatic coding have to be in order not to undermine the benefits of adaptive testing? Or to put it differently, how can adaptive testing account for uncertainty in scoring? A second scenario in which on-the-fly coding (not scoring) might be beneficial is when test takers give a borderline response (cf. idea in Zehner et al., 2016). In an individual assessment, the test administrator would typically dig deeper and probe the test taker with the relevant bit of information that was missing or too vague. By means of automatic coding, the computerized test environment might similarly be able to identify such borderline responses. Even in group assessments, it could then ask the test taker to provide the clarifying bit of information. The effect of selectively intervening during the assessment would have to be investigated carefully.

13.4 Challenges: Different Settings, Different Responses

While the previous section sheds light on the benefits of automatic coding, it, of course, comes with its own costs and limitations. When a system for the automatic coding of text responses is designed, the developers have to make several types of decisions. These decisions determine, among others, what is being measured, whether the captured information is sufficiently indicative for this, and to which kinds of data the system is applicable.

13.4.1 The Nature of Different Kinds of Text Data

Depending on several characteristics of the assessment, the nature of the written language in responses will differ notably. By nature, we refer to a complex set of response characteristics that are inherently induced by characteristics of the assessment. Based on our experiences over the past years, we here discuss a selection of characteristics that most prominently and most potentially affect text response qualities and are relevant to the PISA context: language, respondents' age, low- vs. high-stakes, input modality, and individual vs. group testing.

- **Language.** Obviously, the test language has a high impact on how text responses look in the end. Design decisions need to be adjusted accordingly. That is, is the system designed for one particular language? Which components are required for this particular language (e.g., frequency of compounds, are word boundaries present)? Are the employed components well-scalable to a large number of languages?

Less obvious and less-often considered, but really crucial, is whether the assessment's target group comprises (large numbers of) non-native speakers. Less conventional written language poses challenges for the normalizing steps and increases their importance.

- **Age.** Text responses from primary school students entail problems similar to those written by non-native speakers. Furthermore, responses by 15-year-olds also differ from those by adults, not so much with respect to standardized language, but more aberrant responses will be found in the former than the latter group. Aberrant (openly disengaged) responses pose a very special challenge, as they can be so off-topic that often little is known about them from the training data. Teen slang is not only to be found in aberrant responses, but also otherwise suitable responses. Moreover, 15-year-olds are still exposed to a standardized formal education and are, thus, very much used to being tested. This can lead to partly stereotypical response styles, like always repeating the question in the first half of the response. Adults, on the other hand, are exposed to less formalized and standardized structures in their every-day life and, hence, can be expected to write in more heterogeneous ways.
- **Low vs. high stakes.** Assessments like PISA are low stakes for the individual test taker, which leads to lower engagement (Schlosser et al., 2019). Lower engagement not only entails more empty and more aberrant responses, but it also results in less thorough writing. All facets of writing can be affected by this; that is, orthography, proper use of syntax, thoroughness in writing complete sentences, as well as response length. All this has a huge impact on the methods that are applicable to the data because the partly fragmentary and non-conventional responses in low-stakes assessments preclude the use of the most powerful natural language processing techniques (e.g., textual entailment) that require very accurate annotations (like part-of-speech tagging) and complex language models. If ill-formed language is entered into these powerful models, they will simply fail (Dzikovska et al., 2016; Higgins et al., 2014).

  On the other hand, responses from high-stakes assessments typically feature more proper language, allowing for more powerful and accurate natural language processing techniques, but they also come with the risk of cheating intentions. Automatic systems can be prone to cheating if their mechanisms are known and if they rely on superficial features (e.g., response length). Bejar et al. (2014) showed how changing single words with synonyms raised the automatic coding's estimation moderately.
- **Input Modality.** Another influential assessment characteristic is the input modality. In paper-based assessments, the input mode is the writing with a pen and pencil; in computer-based ones, it is keyboarding or even software-based keyboarding for touch devices. It is known that responses from computer-based assessments can be substantially longer (Horkay et al., 2006; White et al., 2015; Zehner et al., 2019; Zehner et al., 2020). At the same time, keyboarding

introduces entirely different kinds of misspellings (e.g., through typos), and test takers tend to write a bit more carelessly (i.e., not so much caring about capital letters and typos), maybe going back to their every-day habits of writing on digital devices.

13.4.2 *What to Measure, Which Information to Capture?*

Most crucial for determining different design decisions is the intention of what should be measured, but surprisingly, most publications in the area tend not to really reflect on how the selected methods relate to the measurement goal. For example, reading literacy is more easily captured in basic indicators than is the construct of scientific literacy. It lies in the nature of each construct which information in a written response tends to indicate correct and incorrect responses if, for example, scoring is the task at hand.

When measuring reading literacy, test takers typically read a stimulus text and answer questions about the text or associated topics based on the text. With reading being a very generic and rather domain-independent skill, the stimuli contain most of the required domain-knowledge and sometimes require some crude declarative world knowledge. Thus, the majority of reading items ask for naming some semantic concepts. Fine-grained, in-depth relationships of these semantic concepts are less important because the instrument is intended to assess reading skills and not the understanding of some mechanics in the field or logic skills. Of course, there are items that do not match this coarse statement, but for the majority of typical items, it will hold true.

For example, an item might ask why the protagonist cried out loud, requiring the test taker to understand (and respond) that Sam was surprised to find his uncle ringing at the door, though Sam thought the uncle was dead. Such an item would require the test taker to mentally integrate different pieces of the story as well as the world knowledge that dead people tend to not ring doorbells and that death is associated with strong emotions. For a correct response, concatenating the central semantic concepts, surprise, uncle, doorbell, and death (in some combination), would do. This is in line with the QUEST model that postulates that responses are produced by concatenating selected propositions that are filtered according to the question focus and category by applying some linguistic template (Graesser & Clark, 1985; also cf. Zehner et al., 2018, for a more detailed description of the assumed cognitive processes).

And now, here comes the crux. The nature of this prevalent reading testing paradigm implies that test takers would typically name semantic concepts or not, and this determines the response correctness. Even in low-stakes assessments like PISA, responses to these types of items, such as, "Sam was not surprised," are highly unlikely. The test taker understood that Sam's scream was associated with the state of being surprised, and, hence, they would not negate this proposition. Neither are responses such as, "Sam was surprised to find the doorbell lying dead on the doormat," observed. These two examples would be responses that name relevant concepts but confuse their relationships. That is why testing for the bare presence of semantic concepts suffices to assess response correctness for typical reading literacy items.

Now let us switch to the second example construct, scientific literacy. Testing for this, instruments require test takers to already come prepared with some strategic knowledge about how to go about scientific reasoning. The stimuli would typically elaborate on specific domain-knowledge and explain mechanisms in detail. Analogous to the reading example, a science item might ask why engineers chose to reverse the flow of the Chicago River, which requires students to understand the semantic concepts water flow, hygiene, Lake Michigan, and Chicago River, among others. However, and most notably, the simple concatenation of these semantic concepts will not do for responding correctly, but the concepts have to be put into correct spatial and temporal order. In text responses, these relationships would be indicated by words without strong semantic specificity, such as [flowing] into, originally, and then. Test takers' confusion about which direction water flows and into what body is likely to be seen in responses. Thus, we can see how different constructs lead to the different pieces of information we actually need to capture when coding responses.

In technical terms, the described exemplary difference would show up if developers decide to use the so-called bag of words, which neglects word order and regards words as isolated units. For the described reading literacy item, this would be perfectly fine, as it is very unlikely that test takers confuse the relationship of concepts in their responses in a way that would indicate they did not properly comprehend the text; for example, if they correctly figured out that the scream was associated with surprise, they would typically not respond that Sam's uncle was astonished instead of Sam himself. Thus, the bag of words with all its shortcomings would probably still work well for this item. The same would be true for stop word removal, as synsemantic words would not be indicative of response correctness here. On the other hand, our scientific literacy item would probably turn out to be poorly coded by the machine if the bag of words–paradigm was applied. Here, it simply makes a difference whether Lake Michigan is allowed to flow into the Chicago River or the other way around, a difference that cannot be distinguished in the bag of words. Furthermore, it is very likely that both response types would be encountered in empirical data.

13.5 Ockham's Razor Revisited: Advocacy for Coarse Modeling

Natural language is often vague, but there are ways to have natural language be more specific. For example, scientific writing attempts to be precise and to avoid ambiguity through rigid structures, technical language, and register markers. While this is a special characteristic of the scientific genre, this applies neither to every-day usage of natural language nor to responses in low-stakes assessments. Rather, in the latter, the context becomes more dominant, the verbal utterance has to express less and can imply more; this is a linguistic phenomenon dealt with in pragmatics (Bakker, 2010, p. 151):

> In its most principled (some would say, radical) form, pragmatics sees "linguistic meaning" not as something that inheres in the words and sentences themselves of the language, but in the strategies by which speakers convey through language what they mean or intend to achieve. Words don't mean, speakers do; and a pragmatic account of a language is necessarily cognitive, in taking account not only the contextual reality within which the speech action takes place, but also speakers' perception and assessment of that reality.

This means that utterances transport more than their word basis because the context (in the form of the senders' perceived reality during the action of utterance) determines implications. In a trivial example, a test taker could respond with a single word: "scared." Given the context (stimulus text), it might be clear that the adjective can only refer to the story's protagonist, as he is the only agent with feelings in the story. Thus, the response implies a full proposition, specified by a subject and a predicate: "the protagonist is [scared]." Since low-stakes assessments evoke more informal language than high-stakes assessments, a single-word response like this will be far more common in the former, whereas the latter would more likely evoke the utterance of the full-fledged proposition ("The protagonist is scared."), probably additionally elaborating on the statement. Less formal writing, or oral speech, is known to be more fragmented and less integrative (Chafe, 1982). That is, responses bound to less-formal contexts are prone to constituting a chain of fragmented ideas opposed to linguistically clever integrations of several ideas into a single phrase in formal written language.

Put differently, this effect of less-engaged responding on the prevalence of pragmatics comes with further typical characteristics. For example, informal or less engaged responding is more prone to misspelling, poor grammar, incomplete statements, and the like.

To wrap up the point, it is clear that the observed language in (low-stakes) assessments does not always comply with linguistic conventions (deviating from the norm) and will often leave room for different interpretations of the response's intention. While powerful computational language models are available these days that provide a better understanding of utterances, they are strictly bound to proper language use and the availability of such a computational language model. Moreover, there are promising techniques like textual entailment (cf. Androutsopoulos & Malakasiotis, 2010) that allows one statement to transform into another. However, such techniques are highly dependent on proper language, too. In an international assessment, such as PISA, it is impossible to create such models for all test languages. And even imagining a utopian world in which perfect normalizing techniques can be used for treating poor language, and computational language models are available for all test languages, the scoring systems could still not deal with the issue raised by pragmatics. For this, some computational model of the test taker's perceived context (built situation model; Kintsch, 1998) is needed to constrain the set of plausibly implied propositions to a finite number. If the latter were possible, we would have a far better measurement of reading than posing questions. The ReCo software, on the other hand, takes the approach of coarse techniques that do not need to resolve which proposition might be implied and to which subject a correlate (e.g., pronoun) it is referring. Thus, ReCo does not attempt to model responses' (intended) propositions, but only addresses the responses' rough overall semantics.

Therefore, for international, low-stakes assessments, we advocate for coarse techniques like those employed in ReCo (e.g., bag of words and stemming). Obviously, language does not work the way ReCo utilizes it for analysis. If somebody exclaimed, "Less Bread, More Taxes!" it would make quite a difference to demand more bread and less taxes. However, the amodal, purely symbolic and non-grounded language understanding of ReCo makes use of several phenomena around language in assessment. Responses (specifically correct ones) are tightly bound to the semantic universe defined by the stimulus (text and question). Test takers typically do not identify the correct solution and write it down in its negated form. Plausible exceptions to this occur; for example, if an item is prone to such differences (e.g., if it made a difference stating that a person had witnessed a scene vs. the person had not) or if the construct is tuned to fine-grained differences (e.g., at the logical level; cf. Sect. 4.2). In these cases, ReCo's methodology is not appropriate. The big advantage of the named baseline techniques (stop word removal, stemming, bag of words, LSA) is that they are readily available off-the-shelf for large numbers of languages (if dependent on a specific language at all) and are, thus, easily scaled up if the according resources are available (text corpus, [e.g., stemming] algorithms, data).

Another rather coarse-grained element of ReCo is the first, unsupervised learning step for identifying response types. While a supervised learning would attain higher accuracy in the first place, the coarse, unsupervised approach comes with several advantages: transparency, flexibility, and robustness. For going one step further and assigning codes to the identified response types, ReCo requires manually coded data for carrying out supervised learning. However, the supervised classifier is still based on the imprecise unsupervised step. The empirical part of this chapter demonstrates how ReCo's coarse-grained methodology yields more robust results for applications in new settings.

13.6 Empirical Demonstration: Classifiers for Repeated Measurements

We have demonstrated how ReCo performed on PISA items with German data from PISA 2012 (Zehner et al., 2016). The main advantage of automatic coding, however, only holds if the resulting classifiers can be used over and over again in repeated measurements without much additional effort. This means that the classifiers should be appropriately accurate across time and robust towards changes in the assessment design. As for the latter, PISA 2015 introduced computer-based assessment, meaning that text responses are prone to changing in some ways (Zehner et al., 2019, 2020), partly due to the new input modality.

13.6.1 Procedure

In the previous study (cf. Zehner et al., 2016), responses to ten items from the paper-based PISA 2012 were transcribed for analysis. For each item, a classifier for identifying correct and incorrect responses was built. These classifiers consist of a uniquely-built vector space model on the semantics of terms associated to the item topics. Additionally, the classifier provides the partitioning and respective codes of the vector space (i.e., which areas in the vector space models are more associated with correct or incorrect responses).

Eight items stemmed from the reading test and one each from mathematics and science. So-called *Access & Retrieve* items, which address one facet of reading literacy, typically evoke a strictly limited diversity of responses (and are, thus, easy to code) were, therefore, intentionally under-represented in the study design (only one item). On average, about $n = 4200$ responses were available for each item. For more details on the items and the 2012 data, see Zehner et al. (2016).

In the present empirical demonstration, the German PISA 2015 data are evaluated using the classifiers built on the data from PISA 2012 with the same ten items. This way, the classifiers are cross-validated using independent data, and the methodology is checked for its robustness towards design changes like the shift from paper- to computer-based assessment. In the 2015 data, $n = 1220$ responses were available for each item, on average.

13.6.2 Results

Table 13.1 provides the agreement between human and computer (ReCo) as percentages. The baseline here constitutes the 2012 data, on which the classifiers were trained. The main interest of this analysis lies in the change in agreement from this baseline when the classifiers are applied to the 2015 data, on which they were not trained. Since percentage of agreement is a mediocre coefficient at best, Table 13.2 provides agreement in terms of Cohen's Kappa.

Overall, the changes from 2012 to 2015 look rather stable in terms of percentage of agreement. While a decline in performance on the independent, 2015 data set could have been expected, six items range in the lower magnitude of decline [−3.2%, −0.6%], whereas three items actually show a performance increase of similar magnitude of [+1.0%, +3.6%], but the math item shows a huge drop in performance, by −10.0%. Taking the marginal code distribution into account, the kappa values

Table 13.1 Agreement (%): classifiers trained on PISA 2012 data

Item	$\%_{h:c_{2012}}$	$\%_{h:c_{2015}}$	Δ
1·Explain protagonist's feeling	91.2 [±0.2]	90.6	−0.6
2·Evaluate statement	86.5 [±0.3]	87.5	+1.0
3·Interpret the author's intention	90.4 [±0.2]	88.2	−2.2
4·List recall	98.4 [±0.1]	95.9	−2.5
5·Evaluate stylistic element	76.2 [±0.4]	79.8	+3.6
6·Verbal production	92.7 [±0.2]	89.5	−3.2
7·Select and judge	86.6 [±0.3]	88.9	+2.3
8·Explain story element	86.7 [±0.3]	86.0	−0.7
9·Math	85.0 [±0.3]	75.0	−10.0
10·Science	88.4 [±0.3]	86.2	−2.2

Note. 95% confidence intervals given in brackets

Table 13.2 Agreement (κ): classifiers trained on PISA 2012 data

Item	$\%_{h:c_{2012}}$	$\%_{h:c_{2015}}$	Δ
1·Explain protagonist's feeling	.533 [.519; .547]	.614	+.081
2·Evaluate statement	.729 [.723; .735]	.750	+.021
3·Interpret the author's intention	.458 [.444; .472]	.357	−.101
4·List recall	.955 [.952; .959]	.905	−.050
5·Evaluate stylistic element	.503 [.495; .511]	.521	+.018
6·Verbal production	.704 [.695; .713]	.726	+.022
7·Select and judge	.679 [.672; .686]	.739	+.060
8·Explain story element	.676 [.668; .683]	.612	−.064
9·Math	.669 [.661; .676]	.481	−.188
10·Science	.761 [.754; .767]	.725	−.036

Note. 95% confidence intervals given in brackets

show five items with a decline in agreement and five items with an increase in agreement.

13.6.3 Discussion

The comparison of automatic coding performance on 2012 and 2015 data, using classifiers trained on the 2012 data, shows no overall decline when excluding the math item, for which a substantial performance drop is obvious. Apart from this, the finding that there are even some items for which the classifiers perform better on a dataset with the same items but from an independent cohort and a different study design (e.g., differing input modalities) than which they were trained is surprising. This underlines the robustness of ReCo's design due to its conservative learning approach. Part of the reason for the performance increase observed in the 2015 data is due to human coding mistakes in the 2012 data, which resulted in lower accuracy in training and testing. This shows the importance of the conservative modeling approach in order to not incorporate noise from the training data into the classifier. Furthermore, this illustrates how automatic coding could accompany, instead of replace, human coding in order to alleviate the coding process by flagging inconsistencies post-hoc or grouping responses a-priori.

The large drop in performance for the math item is another important finding. Potential reasons for the performance drop are manifold. The most crucial one is a decrease in empty responses by 4%. Empty responses are naturally coded accurately as incorrect (or missing) by both human and computer. Moreover, the median length of responses changed substantially from 91 to 74 characters between the paper- and computer-based assessment modes, respectively, meaning that there might be new (shorter) response types not taken into account by the classifier. Another indicator for a change in the underlying data is an increase in correct responses by 5.5%. All of these findings show that coding performance, automatic or manual, is always item-specific due to certain mechanics at the item level. Therefore, coding procedures have to be monitored closely.

Thus, we can conclude that using classifiers from previous rounds of measurement for automatic coding of short text responses is feasible without much effort. Given that PISA administers a large number of trend items (the same items administered over several cycles of assessment), automatic coding could save substantial amounts of resources and improve data quality. It will be a matter for future developments and research to investigate the feasibility of ongoing quality checks. That is, human coding would partly need to be kept in place for checking whether classifiers continue to work as they have in previous cycles of assessment. The automatic system would need to provide a level of certainty for the codes so that it would become apparent if new data for trend items came with a large share of new response types that had not been seen before and, thus, needed to be manually coded. The components necessary for such a certainty coefficient would address cluster purity and heterogeneity, as well as the distance of the new responses to their assigned cluster centroids. This way, some kind of temporally accumulated bias of old classifiers could be monitored. The system could select those responses that need special manual treatment.

13.7 Concluding Discussion and Outlook

In this chapter, we described the ReCo software, which is designed for automatically coding text responses in assessments and contextualized its baseline methodological approach. In an empirical demonstration, we showed that the employment of automatic coding can build on classifiers trained on previous data for most items, whereas there might be single items that need further precautions. This way, substantial amounts of resources and effort could be saved, to the extent that automatic coding could largely improve the quality of human coding when used for indicating coding inconsistencies. In this final section, we conclude our discussion on automatic coding and provide a brief glimpse into two promising ongoing developments in the realm that could further innovate assessment.

If the system is used for automatic coding (opposed to mere text categorization for data sifting), it has to be trained in a supervised way. Currently, this means responses have to be manually coded by humans, initially. The fact that automatic coding is optimized and evaluated with respect to humans as the gold standard bears risks. Both model training (Loukina et al., 2018; Reidsma & Carletta, 2008) and evaluation (Loukina et al., 2020) are hampered if the manually coded data are prone to containing noise due to inconsistencies, stemming from differing subjective perspectives of coders, annotation fatigue, varying coding experience, bias, and more (Bauer & Zapata-Rivera, 2020; Bejar, 2012). At the same time, the machine is susceptible to other sources of error (e.g., negation or improper language use). Thus, the field needs to identify ways to replace humans as the gold standard in coding (e.g., using psychometric quality criteria for model optimization, such as the prediction of an external criterion). That said, we do not argue for removing humans entirely from the coding process. Rather, they should be involved where tasks play to their strengths, and the machine should support human coding in improving data quality.

Automatic coding is not constrained to binary classification. For example, automatic essay grading typically yields interval-scaled evaluations of texts (Attali & Burstein, 2006), but the described approach by ReCo for short answers is also capable of handling polytomous coding without adjustments. Naturally, with more classes, the chance of disagreement between human and machine increases, and accuracy can be expected to drop.

Finally, another interesting notion refers to the nature of different codes. To put it differently, the data generating mechanism for responses of different codes differ by definition. Obviously, in tests, proficiency determines a response's code, but so does test engagement. ReCo treats all codes and their responses equally. In future developments, ReCo will handle this issue by an innovative machine learning approach that treats one code (i.e., incorrect) as a special default class. Furthermore, it will include a specialized classifier for identifying aberrant responses.

In the following two subsections, we conclude the chapter by sketching two future directions to be taken in the context of ReCo.[2]

[2] As stated in the previous footnote, this chapter was submitted to the present book 6 years prior to its publication. Thus, the directions outlined here have already been implemented in the meanwhile. Related publications by the authors are available.

13.7.1 Direction I: ReCo Teams Up with PISA's Coding System

Yamamoto et al. (2018) proposed PISA's Machine-Supported Coding System (MSCS) that builds on the very simple and very effective idea that responses that have already been seen before can be automatically coded with the same code. For being ultimately language-independent, the system builds on exact matching, which means that all characters must be identical between two responses in order to be assumed to have been seen before.

This rigid definition of similarity yields theoretically perfect accuracy, but for most items, the coverage of responses that can be automatically coded is severely limited. Therefore, we are currently investigating how far the MSCS's definition of similarity can be relaxed towards ReCo's fuzzy definition to still retain very high accuracy but gain more coverage.

In PISA's operational procedure, a fallback system would be plausible with MSCS in the beginning. If it cannot code a response, ReCo could do it and provide a level of certainty. In cases in which a specified threshold of certainty is not passed, the response could be handed on to the human coders.

13.7.2 Direction II: Cross-Language Comparisons

The described evaluations in this chapter were based on analyses with German data. In the future, it will not only be necessary to extend ReCo's scope to additional languages, but other developments will further unlock new possibilities. Instead of only comparing responses from the same test language, a model is being developed that allows direct comparisons of responses across different languages. This means it would be possible to analyze whether a given type of response is more prevalent in, for example, English than in Spanish data. Also, it would allow human coding inconsistencies across test languages to be revealed and corrected. Moreover, from a technical perspective, the automatic coding classifiers would be enhanced by having training data available from several different languages for the same model. Furthermore, automatic coding could be applied to languages for which no training data is available.

For such a system, ReCo will be extended. One of the analytical approaches would be to build separate semantic vector space models for the required additional test language (in line with the procedure described in Sect. 2.2). Next, specific words relating to language-independent concepts are used to align the vector space models (cf. Beinborn & Choenni, 2019). Based on this, a transformation hyperplane is learned in order to generalize the transformation from one into the other language.

13.7.3 Conclusion

In summary, automatically coding short-text responses is a promising method for addressing the challenges of efficiently and accurately coding constructed responses in international large-scale assessments. Among others, such automation could be fully utilized in adaptive test designs, in which on-the-fly scoring could inform routing decisions. An additional challenge of international assessments is coding data across multiple languages and countries, which we will be able to tackle in the future by simultaneous automatic coding across languages. Beside these future directions, ReCo has shown to have accuracy similar to human coding. The empirical demonstration in this chapter shows that once trained, the system can be reused for most items in repeated measurements without substantial effort. The current possibility of blending ReCo with PISA's MSCS for automatically coding text responses could help to improve coding efficiency and accuracy in PISA, make response codes available for use in adaptive testing, and harmonize human codes across different testing languages.

Acknowledgements Special thanks go to our colleague Helge Einspanier who provides us with impressive computational capabilities, not just by buying fancy hardware, but also by combining what is available and making it most powerful. Another big thanks go to Emily Kerzabi, who provided most thorough, competent, and smooth copyediting.

References

Adams, R. J., & Wu, M. (Eds.). (2002). *PISA 2000 technical report*. OECD Publishing. https://doi.org/10.1787/9789264199521-en

Andersen, N., & Zehner, F. (2021). shinyReCoR: A shiny application for automatically coding text responses using R. *Psych, 3*(3), 422–446. https://doi.org/10.3390/psych3030030

Androutsopoulos, I., & Malakasiotis, P. (2010). A survey of paraphrasing and textual entailment methods. *Journal of Artificial Intelligence Research, 38*, 135–187. https://doi.org/10.1613/jair.2985

Attali, Y., & Burstein, J. (2006). Automated essay scoring with e-rater® V.2. *The Journal of Technology, Learning, and Assessment, 4*(3) https://ejournals.bc.edu/index.php/jtla/article/view/1650

Bakker, E. J. (2010). Pragmatics: Speech and text. In E. J. Bakker (Ed.), *A companion to the ancient Greek language* (pp. 151–178). https://doi.org/10.1002/9781444317398.ch11

Bauer, M. I., & Zapata-Rivera, D. (2020). Cognitive foundations of automated scoring. In D. Yan, A. A. Rupp, & P. W. Foltz (Eds.), *Handbook of automated scoring* (pp. 13–28). CRC Press, Taylor & Francis Group. https://doi.org/10.1201/9781351264808-2

Beinborn, L., & Choenni, R. (2019). Semantic drift in multilingual representations (Research Report, CoRR). https://arxiv.org/abs/1904.10820

Bejar, I. I. (1991). A methodology for scoring open-ended architectural design problems. *Journal of Applied Psychology, 76*(4), 522–532. https://doi.org/10.1037/0021-9010.76.4.522

Bejar, I. I. (2012). Rater cognition: Implications for validity. *Educational Measurement: Issues and Practice, 31*(3), 2–9. https://doi.org/10.1111/j.1745-3992.2012.00238.x

Bejar, I. I. (2017). A historical survey of research regarding constructed-response formats. In R. E. Bennett & M. von Davier (Eds.), *Advancing human assessment* (pp. 565–633). Springer. https://doi.org/10.1007/978-3-319-58689-2_18

Bejar, I. I., Flor, M., Futagi, Y., & Ramineni, C. (2014). On the vulnerability of automated scoring to construct-irrelevant response strategies (CIRS): An illustration. *Assessing Writing, 22*, 48–59. https://doi.org/10.1016/j.asw.2014.06.001

Bennett, R. E. (1993). On the meanings of constructed response. In R. E. Bennett & W. C. Ward (Eds.), *Construction versus choice in cognitive measurement* (pp. 1–27). Lawrence Erlbaum Associates. https://doi.org/10.4324/9780203052518-4

Bennett, R. E., & Ward, W. C. (Eds.). (1993). *Construction versus choice in cognitive measurement: Issues in constructed response, performance testing, and portfolio assessment*. Lawrence Erlbaum Associates. https://doi.org/10.4324/9780203052518-17

Bexte, M., Horbach, A., & Zesch, T. (2024). Strengths and weaknesses of automated scoring of free-text student answers. *Informatik Spektrum, 47*, 78–86. https://doi.org/10.1007/s00287-024-01573-z

Birenbaum, M., & Tatsuoka, K. K. (1987). Open-ended versus multiple-choice response formats—It does make a difference for diagnostic purposes. *Applied Psychological Measurement, 11*(4), 385–395. https://doi.org/10.1177/014662168701100404

Braun, H. I., Bennett, R. E., Frye, D., & Soloway, E. (1990). Scoring constructed responses using expert systems. *Journal of Educational Measurement, 27*(2), 93–108.

Bridgeman, B. (1991). A comparison of open-ended and multiple-choice question formats for the quantitative section of the graduate record examinations general test. *ETS Research Report Series, 1991*(2), i–25. https://doi.org/10.1002/j.2333-8504.1991.tb01402.x

Burrows, S., Gurevych, I., & Stein, B. (2014). The eras and trends of automatic short answer grading. *International Journal of Artificial Intelligence in Education, 25*, 1–58. https://doi.org/10.1007/s40593-014-0026-8

Burstein, J., Kaplan, R. M., Wolff, S., & Lu, C. (1996). Using lexical semantic techniques to classify free-responses. In E. Viegas (Ed.), *Proceedings of the ACL SIGLEX workshop on breadth and depth of semantic lexicons* (pp. 20–29). Association for Computational Linguistics. https://www.aclweb.org/anthology/W96-0304.pdf

Bywater, J. P., Chiu, J. L., Hong, J., & Sankaranarayanan, V. (2019). The teacher responding tool: Scaffolding the teacher practice of responding to student ideas in mathematics classrooms. *Computers & Education, 139*, 16–30. https://doi.org/10.1016/j.compedu.2019.05.004

Chafe, W. L. (1982). Integration and involvement in speaking, writing, and oral literature. In D. Tannen (Ed.), *Spoken and written language: Exploring orality and literacy* (pp. 35–53).

Deerwester, S., Dumais, S. T., Furnas, G. W., Landauer, T. K., & Harshman, R. (1990). Indexing by latent semantic analysis. *Journal of the American Society for Information Science, 41*(6), 391–407. https://doi.org/10.1002/(SICI)1097-4571(199009)41:6<391::AID-ASI1>3.0.CO;2-9

Dumais, S. T. (1991). Improving the retrieval of information from external sources. *Behavior Research Methods, Instruments, & Computers, 23*(2), 229–236. https://doi.org/10.3758/BF03203370

Dzikovska, M. O., Nielsen, R. D., Brew, C., Leacock, C., Giampiccolo, D., Bentivogli, L., et al. (2013). Semeval-2013 task 7: The joint student response analysis and 8th recognizing textual entailment challenge. In M. Diab, T. Baldwin, & M. Baroni (Eds.), *Proceedings of the second joint conference on lexical and computational semantics* (pp. 263–274) https://www.aclweb.org/anthology/S13-2045.pdf

Dzikovska, M. O., Nielsen, R. D., & Leacock, C. (2016). The joint student response analysis and recognizing textual entailment challenge: Making sense of student responses in educational applications. *Language Resources and Evaluation, 50*(1), 67–93. https://doi.org/10.1007/s10579-015-9313-8

Enright, M. K., & Quinlan, T. (2010). Complementing human judgment of essays written by English language learners with e-rater® scoring. *Language Testing, 27*(3), 317–334. https://doi.org/10.1177/0265532210363144

Fang, Y., Lippert, A., Cai, Z., Hu, X., & Graesser, A. C. (2019). A conversation based intelligent tutoring system benefits adult readers with low literacy skills. In R. A. Sottilare & J. Schwarz (Eds.), *Adaptive instructional systems* (pp. 604–614). Springer International Publishing. https://doi.org/10.1007/978-3-030-22341-0_47

Gabrilovich, E., & Markovitch, S. (2009). Wikipedia-based semantic interpretation for natural language processing. *Journal of Artificial Intelligence Research, 34*, 443–498. https://doi.org/10.1613/jair.2669

Graesser, A. C., & Clark, L. F. (1985). *Structures and procedures of implicit knowledge* (Vol. 17). Ablex.

Griffo, V. (2011). Examining NAEP: The effect of item format on struggling 4th graders' reading comprehension (Doctoral dissertation, UC Berkeley). https://escholarship.org/uc/item/8273224n

Guthrie, J. T. (1984). Testing higher level skills. *Journal of Reading, 28*(2), 188–190.

He, Q. (2013). Text mining and IRT for psychiatric and psychological assessment (Dissertation, University of Twente, Twente). https://doi.org/10.3990/1.9789036500562.

Hewlett Foundation. (2012). Automated student assessment prize: Phase two—Short answer scoring. https://www.kaggle.com/c/asap-sas/

Higgins, D., Brew, C., Heilman, M., Ziai, R., Chen, L., Cahill, A., et al. (2014). Is getting the right answer just about choosing the right words? The role of syntactically-informed features in short answer scoring (Research Report, CoRR). https://arxiv.org/abs/1403.0801

Horbach, A., & Pinkal, M. (2018). *Semi-supervised clustering for short answer scoring*. In *Proceedings of the Eleventh International Conference on Language Resources and Evaluation (LREC-2018)* (pp. 4065–4071) https://www.aclweb.org/anthology/L18-1641/

Horbach, A., Stennmanns, S., & Zesch, T. (2018). Cross-lingual content scoring. In *Proceedings of the Thirteenth Workshop on Innovative Use of NLP for Building Educational Applications* (pp. 410–419) https://www.aclweb.org/anthology/volumes/W18-05/

Horkay, N., Bennett, R. E., Allen, N., Kaplan, B., & Yan, F. (2006). Does it matter if I take my writing test on computer? An empirical study of mode effects in NAEP. *Journal of Technology, Learning, and Assessment, 5*(2), n2. https://www.researchgate.net/publication/28798515_Does_it_Matter_if_I_take_My_Writing_Test_on_Computer_An_Empirical_Study_of_Mode_Effects_in_NAEP

Ihme, J. M., Senkbeil, M., Goldhammer, F., & Gerick, J. (2017). Assessment of computer and information literacy in ICILS 2013: Do different item types measure the same construct? *European Educational Research Journal, 16*(6), 716–732. https://doi.org/10.1177/1474904117696095

Kaplan, R. M. (1991). *Using a trainable pattern-directed computer program to score natural language item responses (ETS Research Report: RR-91-31)*. Educational Testing Service. https://doi.org/10.1002/j.2333-8504.1994.tb01581.x

Kintsch, W. (1998). *Comprehension: A paradigm for cognition*. Cambridge University Press.

Lim, H. (2019). Test format effects: A componential approach to second language reading. *Language Testing in Asia, 9*(6), 1–22. https://doi.org/10.1186/s40468-019-0082-y

Loukina, A., Madnani, N., Cahill, A., Yao, L., Johnson, M. S., Riordan, B., & McCaffrey, D. F. (2020). Using PRMSE to evaluate automated scoring systems in the presence of label noise. In *Proceedings of the fifteenth workshop on innovative use of nlp for building educational applications*. https://www.aclweb.org/anthology/2020.bea-1.2/

Loukina, A., Zechner, K., Bruno, J., & Beigman Klebanov, B. (2018). Using exemplar responses for training and evaluating automated speech scoring systems. In *Proceedings of the Thirteenth Workshop on Innovative Use of NLP for Building Educational Applications* (pp. 1–12). https://doi.org/10.18653/v1/W18-0501

Martin, D. I., & Berry, M. W. (2011). Mathematical foundations behind latent semantic analysis. In T. K. Landauer, D. S. McNamara, S. Dennis, & W. Kintsch (Eds.), *Handbook of latent semantic analysis* (pp. 35–55). Erlbaum. https://citeseerx.ist.psu.edu/viewdoc/download?doi=10.1.1.538.3692&rep=rep1&type=pdf

Marvaniya, S., Saha, S., Dhamecha, T. I., Foltz, P., Sindhgatta, R., & Sengupta, B. (2018). Creating scoring rubric from representative student answers for improved short answer grading. In A. Cuzzocrea et al. (Eds.), *Proceedings of the 27th ACM International Conference on Information and Knowledge Management – CIKM '18* (pp. 993–1002). ACM Press. https://doi.org/10.1145/3269206.3271755

Mieskes, M., & Pado, U. (2018). Work smart – Reducing effort in short-answer grading. In *Proceedings of the 7th Workshop on NLP for Computer Assisted Language Learning (NLP4CALL 2018) at SLTC, Stockholm, 7th November 2018* (pp. 57–68). Linköping University Electronic Press, Linköpings universitet. https://www.aclweb.org/anthology/W18-7107

Millis, K., Magliano, J., Wiemer-Hastings, K., Todaro, S., & McNamara, D. S. (2011). Assessing and improving comprehension with latent semantic analysis. In T. K. Landauer, D. S. McNamara, S. Dennis, & W. Kintsch (Eds.), *Handbook of latent semantic analysis* (pp. 207–225). Erlbaum. https://doi.org/10.4324/9780203936399.ch11

Myford, C. M., Marr, D. B., & Linacre, J. M. (1995). *Reader calibration and its potential role in equating for the Test of Written English (ETS Research Report: RR-95-40)*. Educational Testing Service. https://doi.org/10.1002/j.2333-8504.1995.tb01674.x

Nye, B. D., Graesser, A. C., & Hu, X. (2014). AutoTutor and family: A review of 17 years of natural language tutoring. *International Journal of Artificial Intelligence in Education, 24*(4), 427–469. https://doi.org/10.1007/s40593-014-0029-5

Organization for Economic Co-operation and Development. (2017). PISA 2015 technical report. https://www.oecd.org/pisa/data/2015-technical-report/PISA2015_TechRep_Final.pdf

Organization for Economic Co-operation and Development. (2019). *PISA 2018 assessment and analytical framework*. https://doi.org/10.1787/19963777.

Page, E. B. (1966). The imminence of grading essays by computer. *Phi Delta Kappan, 48*, 238–243. https://www.jstor.org/stable/20371545?seq=1#metadata_info_tab_contents

Page, E. B. (1968). The use of the computer in analyzing student essays. *International Review of Education, 14*(2), 210–225. https://doi.org/10.1007/BF01419938

Paiva, F., Glenn, J., Mazidi, K., Talbot, R., Wylie, R., Chi, M. T. H., et al. (2014). Comprehension SEEDING: Comprehension through self explanation, enhanced discussion, and INquiry generation. In *International Conference on Intelligent Tutoring Systems* (pp. 283–293). https://doi.org/10.1007/978-3-319-07221-0_35

Quesada, J. (2011). Creating your own lsa spaces. In T. K. Landauer, D. S. McNamara, S. Dennis, & W. Kintsch (Eds.), *Handbook of latent semantic analysis* (pp. 71–85). Erlbaum. https://doi.org/10.4324/9780203936399.ch4

Rauch, D. P., & Hartig, J. (2010). Multiple-choice versus open-ended response formats of reading test items: A two-dimensional IRT analysis. *Psychological Test and Assessment Modeling, 52*(4), 354–379. https://www.researchgate.net/publication/49593722_Multiple-choice_versus_open-ended_response_formats_of_reading_test_items_A_two-dimensional_IRT_analysis

Reidsma, D., & Carletta, J. (2008). Reliability measurement without limits. *Computational Linguistics, 34*(3), 319–326. https://doi.org/10.1162/coli.2008.34.3.319

Ruder, S., Vulić, I., & Søgaard, A. (2019). A survey of cross-lingual word embedding models. *Journal of Artificial Intelligence Research, 65*, 569–631. https://doi.org/10.1613/jair.1.11640

Rupp, A. A., Ferne, T., & Choi, H. (2006). How assessing reading comprehension with multiple-choice questions shapes the construct: A cognitive processing perspective. *Language Testing, 23*(4), 441–474. https://doi.org/10.1191/0265532206lt337oa

Salton, G., & Buckley, C. (1990). Improving retrieval performance by relevance feedback. *Journal of the American Society for Information Science, 41*(4), 288–297. http://doi.wiley.com/10.1002/(SICI)10974571(199006)41:4<288::AID-ASI8>3.0.CO;2-H

Schlosser, A., Neeman, Z., & Attali, Y. (2019). *Differential performance in high vs. low stakes tests: Evidence from the GRE test. CESifo* (Working Paper No. 7590). https://ssrn.com/abstract=3374582

Schult, J., & Sparfeldt, J. R. (2018). Reliability and validity of PIRLS and TIMSS. *European Journal of Psychological Assessment, 34*(4), 258–269. https://doi.org/10.1027/1015-5759/a000338

Sebrechts, M. M., Bennett, R. E., & Rock, D. A. (1991). *Agreement between expert system and human raters' scores on complex constructed-response quantitative items (ETS Research Report: RR-91-11)*. Educational Testing Service. https://doi.org/10.1002/j.2333-8504.1991.tb01377.x

Shermis, M. D. (2020). International applications auf automated scoring. In D. Yan, A. A. Rupp, & P. W. Foltz (Eds.), *Handbook of automated scoring* (pp. 113–131). CRC Press, Taylor & Francis Group. https://doi.org/10.1201/9781351264808-7

Shin, H. J., Yamamoto, K., He, Q., & von Davier, M. (2024). Automatic scoring of constructed responses in PISA using the machine-supported coding system. In L. Khorramdel, M. von Davier, & K. Yamamoto (Eds.), *Innovative digital-based international large-scale assessments—Foundations, methodologies and quality assurance*. Springer.

Sukkarieh, J. Z., & Blackmore, J. (2009). C-rater: Automatic content scoring for short constructed responses. In *Proceedings of the twenty second international FLAIRS conference* (pp. 290–295) http://www.aaai.org/ocs/index.php/FLAIRS/2009/paper/download/122/302

White, S., Kim, Y. Y., Chen, J., & Liu, F. (2015*). Performance of fourth-grade students in the 2012 NAEP computer-based writing pilot assessment: Scores, text length, and use of editing tools* (Working Paper Series, NCES 2015–119). https://files.eric.ed.gov/fulltext/ED562627.pdf

Wolska, M., Horbach, A., & Palmer, A. (2014). Computer-assisted scoring of short responses: The efficiency of a clustering-based approach in a real-life task. In A. Przepiórkowski & M. Ogrodniczuk (Eds.), *Advances in natural language processing* (pp. 298–310). Springer International Publishing. https://doi.org/10.1007/978-3-319-10888-9_31

Yamamoto, K., He, Q., Shin, H. J., & von Davier, M. (2018). Development and implementation of a machine-supported coding system for constructed-response items in PISA. *Psychological Test and Assessment Modeling, 60*(2), 145–164. https://www.psychologie-aktuell.com/fileadmin/download/ptam/2-2018_20180627/02_PTAM-2-2018_Yamamoto_v2.pdf

Zehner, F., Goldhammer, F., Lubaway, E., & Sälzer, C. (2019). Unattended consequences: How text responses alter alongside PISA'S mode change from 2012 to 2015. *Education Inquiry, 10*(1), 34–55. https://doi.org/10.1080/20004508.2018.1518080

Zehner, F., Goldhammer, F., & Sälzer, C. (2015). *Using and improving coding guides for and by automatic coding of PISA short text responses.* In Proceedings of the IEEE ICDM Workshop on Data Mining for Educational Assessment and Feedback (ASSESS 2015). http://www.aspiringminds.com/pages/assess/2015/papers/PID3896317.pdf

Zehner, F., Goldhammer, F., & Sälzer, C. (2018). Automatically analyzing text responses for exploring gender-specific cognitions in PISA reading. *Large-Scale Assessments in Education, 6*(7), 1–26. https://doi.org/10.1186/s40536-018-0060-3

Zehner, F., Kroehne, U., Hahnel, C., & Goldhammer, F. (2020). PISA reading: Mode effects unveiled in text responses. *Psychological Test and Assessment Modeling, 62*, 55–75. https://www.researchgate.net/publication/339849105_PISA_reading_Mode_effects_unveiled_in_short_text_responses

Zehner, F., Sälzer, C., & Goldhammer, F. (2016). Automatic coding of short text responses via clustering in educational assessment. *Educational and Psychological Measurement, 76*(2), 280–303. https://doi.org/10.1177/0013164415590022

Zesch, T., Horbach, A., & Zehner, F. (2023). To score or not to score: Factors influencing performance and feasibility of automatic content scoring of text responses. *Educational Measurement: Issues and Practice, 42*, 44–58. https://doi.org/10.1111/emip.12544

Part IV
New and Advanced Methodologies for Analyzing Response and Process Data

Chapter 14
Linking Large-scale Assessments Over Time and Administration Modes: Scaling, Linking, and Mode Effects in PISA 2015

Matthias von Davier, Lale Khorramdel, and Kentaro Yamamoto

Abstract This chapter outlines the steps necessary for balancing innovation with maintaining reliable and valid trend measures in international large-scale assessments (ILSAs), using PISA 2015 as an example. New scaling methods are introduced that involve concurrent calibration, accounting for item-by-country, item-by-language, and item-by-cycle interactions, and more general IRT models. The goal is to establish approaches that allow for more comprehensive measurement invariance testing, the exploration of the impact of different assessment modes when moving from a paper-based to a computer-based administration, and to provide a better model-data fit and more robust item parameter estimates while maintaining trend measure integrity. The presented methods can be adapted for other large-scale assessments and can also be used for further research on measurement invariance.

Scale comparability as the basis for cross-country comparisons and the reporting of trends in domains, such as reading, mathematics and science, is an important feature of international large-scale assessments (ILSAs), like the Programme for International Student Assessment (PISA), and of high interest for countries and economies. While comparisons across countries within a given assessment cycle allow to contrast different education systems and their impact on what students know and can do, trend measures provide stake holders with information about changes in students' proficiencies and their learning environments over the course of multiple assessment cycles, as well as the impact of different variables, such as changes in

---

M. von Davier (✉) · L. Khorramdel
Lynch School of Education and Human Development, TIMSS & PIRLS International Study Center at Boston College, Chestnut Hill, MA, USA
e-mail: vondavim@bc.edu

K. Yamamoto
Independent Consultant for Psychometric Methodology and Large-Scale Assessment (formerly ETS), Princeton, NJ, USA

L. Khorramdel et al. (eds.), *Innovative Digital-Based International Large-Scale Assessments*, Methodology of Educational Measurement and Assessment,
https://doi.org/10.1007/978-3-031-90951-1_14

school systems, curricula, teacher trainings, or phenomena like the Covid-19 pandemic. To enable both types of comparisons, scales must be developed that allow for comparable, valid, and reliable test scores.

ILSAs also aim at providing meaningful data to policy makers and educational researchers that do not just emphasize comparability to the past but are also relevant to the future. Therefore, ILSAs undergo updates and innovations regarding their frameworks, instrument developments, and administration modes to stay up to date, leading to the development and piloting of new items. To ensure that trend items from previous assessment cycles and newly developed items can be placed on the same reporting scale, both need to be administered together through a linked booklet design, allowing different sets of items to be administered in each assessment cycle. In addition, ILSAs must guarantee the coverage of comprehensive frameworks, leading to very large item pools overall. Consequently, individual students receive different subsets of the item pool to avoid overburdening them, while the entire item pool is administered at the country level (note that large-scale assessments provide group-level test scores, not individual test scores). To link across students and assessment cycles, ILSAs use balanced incomplete block designs that utilize multiple linked booklets with overlapping items (Mazzeo & von Davier, 2008; Rutkowski et al., 2010; Weeks et al., 2013).

To analyze the data resulting from such complex designs, ILSAs rely on the use of item response theory (IRT; Lord & Novick, 1968) for scaling and linking. IRT methods allow the linking of different sets of items on a common scale and for evaluating the comparability of items across countries and assessment cycles based on concepts of measurement invariance (Mazzeo & von Davier, 2013). IRT models can be used to examine and address differential item functioning (DIF) due to differences in school curricula and instructional practices, language or cultural differences, translation errors, changes in the mode of administration, different scoring practices, and item drift across assessments over.

In addition to new item development and technological innovations, changes and updates to the scaling methodologies are sometimes necessary, as well, to facilitate content innovations and to minimize measurement and linking errors (von Davier et al., 2019b). However, any changes to the assessment or the analysis methods can challenge the comparability of the data and compete with the goal of measuring trends. Therefore, their implementation should be strategic and well thought through and should be accompanied by research studies to evaluate any impact on the trend measures (von Davier et al., 2019a).

PISA 2015, for example, moved to a fully computer-based administration mode for most participating countries, introduced the measurement of collaborative problem solving as a new innovative domain, and included additional countries, leading to the participation of a more diverse group of economies with larger variability in average student achievement scores. To ensure scale comparability across a large group of countries and languages, PISA introduced a new, more general IRT modeling approach to improve the data analysis in two ways. First, the new modeling approach improved the stability of item parameter estimates by utilizing the entire PISA database from all available assessment cycles in a concurrent calibration. Second, moving to a more general IRT model with relaxed model constraints increased the model-data fit and allowed measurement scientists to examine the assumption of measurement invariance at the item level more thoroughly and to improve the treatment of DIF. However, before implementing this new scaling and linking approach operationally in PISA, an extensive linking study was conducted that compared the new methodology to the approach historically taken in PISA, to examine whether the measurement of trend is impacted (von Davier et al., 2019b). Moreover, a comprehensive mode effect study was conducted to also examine the impact of the change in administration mode on the item parameter estimation of trend items (von Davier et al., 2019a). Information regarding both studies is also available in the PISA 2015 Technical Report (Chaps. 9 and 12; OECD, 2017).

This chapter illustrates both the linking study (Study 1) and the mode effect study (Study 2) that were conducted in the PISA 2015 cycle to showcase these specific examples how to implement content innovations and analytical changes in ILSAs in general without losing trend, while strengthening scale linking and improving measurement precision.

14.1 Study 1: Linking PISA from 2012 to 2015

PISA was launched by the Organisation for Economic Co-operation and Development (OECD) in 2000 and aims to measure the proficiency of 15-year-old students in the domains of reading, mathematics, and science, as well as a new innovative domain in each assessment cycle (e.g., collaborative problem solving in 2015, global competence in 2018, creative thinking in 2022, etc.). The study was conducted in six assessment cycles between 2000 and 2015 and, since the first cycle, the number of participating countries has increased from 43 to 72 countries in 2015. In 2015, PISA moved from a paper-based to a computer-based assessment administration to allow the use of new interactive item types and the measurement of innovative domains. With this increasing diversity of languages, cultures, and average proficiencies, along with the change in administration mode, PISA needed to revisit its scaling and linking methodologies.

In prior PISA cycles (2000–2012), the Rasch model (Rasch, 1960) for dichotomously scored items and the partial credit model (PCM; Masters, 1982) for polytomous items was used to estimate national (country-level) and international item parameters (across all countries). The national item parameters were estimated separately for each country while the international item parameters were estimated based on a reduced sample of 500 randomly drawn students from each of the OECD participating countries (OECD calibration sample). The national and international item parameter estimates were compared via graphical model checks to evaluate their consistency across countries. If the relative national difficulty of an item was significantly higher or lower than international calibration results, that item was flagged for further examination and potential deletion from the analysis. To ensure comparability across assessment cycles, trend items were used, which are items from previous cycles that are re-administered based on a complex linking design (Yamamoto & Mazzeo, 1992; Mazzeo & von Davier, 2008, 2013). Trend item parameters were estimated separately for each cycle, and these estimations were adjusted in subsequent PISA cycles to match the average difficulty of trend items from previous cycles. This linear transformation, using mean/mean equating, ensured that subsequent calibrations were on the same scale, based on the assumption that the item parameters of trend items remained invariant across cycles.

However, two concerns were raised regarding this approach. First, the Rasch model was considered insufficient to adequately address the complexity of the PISA data (Oliveri & von Davier, 2011; Kreiner & Christensen, 2014, among others) and other large-scale assessments utilize more general IRT models (Mazzeo & von Davier, 2013; von Davier & Sinharay, 2013) like the two-parameter-logistic model (2PLM; Birnbaum, 1968) for dichotomously scored responses and the generalized partial credit model (GPCM; Muraki, 1992) for items with more than two ordered response categories. Second, the linear transformation approach based on estimates from separate IRT calibrations does not provide an adequate tool for examining measurement invariance underlying the common scale to achieve comparability, as it does not account for item-by-country and item-by-cycle interactions (von Davier et al., 2019b).

To address these concerns, PISA 2015 implemented two significant changes. It used a more flexible IRT modelling approach that relaxes constraints and invariance assumptions for establishing a more integrated scale and a more rigorous link across multiple countries and assessment cycles (e.g., von Davier & von Davier, 2007; von Davier et al., 2011; Yamamoto & Mazzeo, 1992). Moreover, it introduced a linking approach similar to the one used in the National Assessment of Educational Progress (NAEP; Yamamoto & Mazzeo, 1992) and the Programme for the International Assessment of Adult Competencies (PIAAC; Yamamoto et al., 2019) for investigating measurement invariance by utilizing a concurrent calibration that accounts for item-by-country, item-by-cycle, and item-by-mode effects. This alternative linking approach allows scientists to gain a better understanding of the invariance properties of items and scales over time.

Both changes to the scaling methodology needed to be compared to the approaches used historically in PISA to examine whether there was any impact on the trend measures. This was done in a comprehensive linking study by von Davier et al. (2019b). We illustrate the approaches and the main results of the linking study in the following sections.

14.1.1 Implementing a Concurrent Calibration in PISA

The linking study (von Davier et al., 2019b) examined whether a concurrent calibration via multiple-group IRT model that uses the data from all available PISA cycles can replace the historic analytic methods, which used separate calibrations based on data from single assessment cycles and a linear linking method.

In this concurrent calibration approach, item parameters are estimated concurrently across country-by-language and country-by-cycle groups (e.g., Canada-French-2003, Canada-French-2006, Canada-French-2009, etc., and Canada-

English-2003, Canada-English-2006, Canada-English-2009, etc.), together with group-specific parameters (means and variances) of ability distribution. This directly implements an assumption of measurement invariance across groups. The resulting common set of item parameters (e.g., Glas & Verhelst, 1995; Oliveri & von Davier, 2011; von Davier & von Davier, 2007; Yamamoto, 1998) are identical across groups and cycles and referred to as international item parameters. The measurement invariance assumption is evaluated by examining DIF based on item fit statistics like the root mean square deviation (RMSD) and mean deviation (MD). Both fit statistics quantify the magnitude of deviations in the observed data from the estimated item characteristic curves (ICC) for each item-by-group combination. Invariance of all item parameters across all groups is referred to as strong invariance or scalar invariance (e.g., Byrne et al., 1989; Steenkamp & Baumgartner, 1998). If certain items exhibit misfit or DIF for particular groups, measurement invariance cannot be upheld for those cases. Instead, partial invariance models are considered by incorporating item-by-group interactions. This is achieved by allowing unique group-specific parameter estimates for items that violate scalar invariance in a parsimonious way, retaining a very large portion of the same (international) item parameter estimates across different groups for a comparable scale. This approach was described by, for example, Glas and Jehangir (2013), Glas and Verhelst (1995), Oliveri and von Davier (2011), and Yamamoto (1998), and allows for the retention of more information compared to an approach where items exhibiting DIF are dropped from the analysis. While the items with DIF treated in this way no longer contribute to the international set of comparable responses, they continue to contribute to the reduction of measurement uncertainty for the specific group(s). If multiple groups show item misfit in the same direction, the same unique parameter can be estimated for these groups; this unique parameter would be different from the common or international parameters across all groups but would be common across a subset of groups still maintaining a certain degree of comparability.

The concurrent calibration in the PISA linking study was applied to a comprehensive database that combines all available data from participating countries over the five assessment cycles conducted between 2000 and 2012. Item parameters were estimated concurrently across assessment cycles and countries within assessment cycles using a multiple-group approach (Bock & Zimowski, 1997; von Davier & Rost, 2016) and constrained to be equal across countries and cycles, while each sample was allowed to have a different latent ability distribution. To compare the concurrent calibration approach with the historic approach, the Rasch model and PCM were used. Analyses were done separately for each content domain (math, reading, and science) and results were compared to the published PISA results by calculating Spearman's rank correlations between item parameter estimates for the historic approach and the new approach, country means were also correlated across approaches (see Table 14.1).

Table 14.1 shows that the correlations per cycle as well as the correlations across all cycles are high for both the country means and the item parameter estimates (> 0.9 for all domains) indicating that the concurrent calibration approach does not compromise the previously published results when used in place of the historic

Table 14.1 Correlations between the concurrent calibration (Rasch model and PCM) and published results

	Country means			Item parameter estimates		
PISA cycle	Math	Reading	Science	Math	Reading	Science
2000	0.997	0.996	0.992	0.986	0.979	0.987
2003	0.996	0.988	0.981	0.991	0.982	0.982
2006	0.998	0.992	0.998	0.991	0.980	0.987
2009	0.999	0.997	0.997	0.994	0.994	0.991
2012	0.999	0.991	0.997	0.999	0.995	0.996
Overall	0.994	0.991	0.995	0.995	0.989	0.985

method. Differences in results from both approaches can be thought of as a combination of differences in the analysis sample (historic separate calibrations versus concurrent calibration) and measures used in the computation (published group means are based on plausible values, group means from the concurrent calibration are direct model-based estimates).

While the country means and item parameter estimates based on both linking approaches seem very similar, the concurrent calibration method has important advantages over the linear transformation linking method. Because the linear linking approach is based on separate calibrations for each assessment cycle, it produces multiple parameter estimates and leads to questions about which set of item parameter estimates should be used and the extent to which differences between national and international calibration results are acceptable. It involves various aggregations and mean–mean adjustments that finally may lead to an amplification of linking errors and greater statistical uncertainty. Oliveri and von Davier (2011) showed that a model using separate calibrations fits less well than a model that uses concurrent calibration when looking at customary information criteria.

On the contrary, the concurrent calibration allows for the direct estimation of country means, more stable item parameter estimates and a reduction in linking error because it utilizes data from multiple assessment cycles, and it starts from a constrained joint model that estimates a set of common item parameters across all groups and then allows a small number of group-specific parameters for a minimum number of groups based on the item fit statistics. Moreover, the concurrent calibration allows for the examination of item-by-country, item-by-cycle, and item-by-mode effects, as previously mentioned, and therefore is better suited to examine violations to measurement invariance caused by issues such as incorrect translations, mode effects or design flaws.

14.1.2 Implementing the 2PLM and GPCM in PISA

After establishing that the concurrent calibration was well suited for the PISA data without impacting the reporting of trends, the linking study (von Davier et al.,

2019b) also explored whether a more general IRT modeling approach with relaxed model constraints and invariance assumptions provides a better model–data fit to the PISA data than the Rasch model. For this purpose, the following IRT models were compared to each other using the large comprehensive data set consisting of the five PISA cycles from 2000–2012:

1) The Rasch Model and PCM (Rasch, 1960; Masters, 1982),
2) the 2PLM and GPCM (Birnbaum, 1968; Muraki, 1992),
3) a Hybrid Model that is a combination of the Rasch model/PCM and the 2PLM/GPCM; and
4) a generalized Hybrid Model, referred to as Hybrid IBGI[1]

The Hybrid model provides a combination of the Rasch model/PCM and the 2PLM/GPCM. It starts with estimating the Rasch model/PCM for all items and relaxes the Rasch model assumption of equal slopes in cases where the item discrimination is markedly different based on model-data fit indices. This allows for the estimation of slope parameters for those items that are found to discriminate more (or less) well than the items that follow the Rasch model.

The generalized Hybrid model is an extended version of the Hybrid model and relaxes model assumptions further by incorporating item-by-group interactions (IBGI). Given that international assessments are translated into multiple target languages, item-by-country-by-language interactions are a potential threat to validity. As such, some items in some countries may function differently from how the item generally functions in the majority of countries. The same holds for shifts in item characteristics across assessment years. Therefore, this model was estimated as a multiple group model using groups divided by countries, languages, and cycles, resulting in over 300 distinctive groups for the PISA 2000–2012 data. It relaxes the measurement invariance assumption across groups to some degree based on item fit statistics and allows for the inclusion of a small number of group-specific item parameters, if needed. This approach addresses threats to scale invariance due to partial item invariance across groups in a parsimonious way, maintaining a large number of item parameters as common across groups, for comparability (Glas & Jehangir, 2013; Glas & Verhelst, 1995; Oliveri & von Davier, 2011, 2014; Yamamoto, 1998).

The different IRT models were estimated separately for each content domain (mathematics, reading, science) and the Bayesian Information Criterion (BIC; Schwarz, 1978) was used for overall model fit evaluation to determine which model fits relatively best to the 2000–2012 database. Table 14.2 presents the BIC for each model and content domain.

Results show that the Hybrid IBGI model, the most general model, fits the data relatively best in every content domain, indicating that partial invariance holds for some items. However, the number of items with partial invariance for which unique parameters were estimated is quite small and most items exhibit scalar invariance

[1] Note that this model is called "Hybrid IBCI" in the paper by von Davier et al. (2019b).

Table 14.2 Mode-Data Fit Statistics (BIC) for different IRT models for the PISA 2000–2012 data

Model	Mathematics	Reading	Science
Rasch and PCM	26,400,730	30,968,125	29,908,518
2PLM and GPCM	26,118,134	30,675,531	29,585,732
Hybrid	26,175,012	30,691,983	29,591,677
Hybrid IBGI	25,946,516	30,472,304	29,302,806

Table 14.3 Percentage of item parameters holding scalar invariance (i.e., show high comparability across groups)

Domain	Rasch model/PCM	2PLM/GPCM	Hybrid	Hybrid IBGI
Math	96.52%	97.49%	97.37%	99.97%
Reading	93.52%	94.84%	94.83%	98.66%
Science	94.32%	95.58%	95.64%	99.95%

Table 14.4 Correlations between published results (Rasch model) and concurrent calibration results (Hybrid IBGI model)

PISA cycle	Math	Reading	Science
2000	0.996	0.993	0.991
2003	0.991	0.983	0.977
2006	0.994	0.985	0.993
2009	0.998	0.994	0.994
2012	0.998	0.990	0.994
Overall	0.993	0.989	0.991

and retained international parameters for high comparability across groups in every domain (see Table 14.3). More detailed comparisons of the different IRT models can be found in von Davier et al. (2019b).

To examine whether the concurrent calibration based the Hybrid IBGI model provides country means comparable with the published ones, Spearman's rank correlations of country means were computed for each domain in each cycle. The results provided in Table 14.4, show high correlations (>0.97) in all domains and assessment years, indicating highly consistent country means across methods.

After examining the fit of the new modeling approach developed for PISA 2015 to trend items and data from past PISA cycles (2000–2012), the fit of the Rasch model/PCM versus the 2PLM/GPCM was tested for items newly developed for PISA 2015 (i.e., items for the domains Science and Collaborative Problem Solving). This comparison was carried out based on the PISA 2015 field test data in preparation of the main survey. Item parameters for trend items were fixed to the estimates obtained from the linking study, and for new items, parameters were estimated using a multigroup Rasch/PCM model and a multigroup 2PLM/GPCM, carried out as a concurrent calibration across country-by-language groups. Both model approaches were compared (see Table 14.5) and it was found that the 2PLM/GPCM showed better overall model fit than the Rasch/PCM model.

In summary, the Hybrid IBGI model showed to be the comparably best fitting model for the PISA 2000–2012 trend data, and starting with the PISA 2015 cycle,

Table 14.5 Mode-Data Fit Statistics (BIC) for the Rasch model/PCM and the 2PLM/GPCM for new items in the PISA 2015 Field Test

Domain	Rasch model/PCM	2PLM/GPCM
Science	4438373.06	4395491.75
Collaborative problem solving	1974832.24	1948035.21

the 2PLM/GPCM was implemented for new items, with both, the trend and new item parameters being estimated on a common and comparable scale. This new scaling and linking approach (i.e., the concurrent calibration based on more general IRT models), provides highly comparable item parameter estimates across countries, languages, and assessment cycles. When applied to the trend data, it provides country means very similar to the published ones (based on the Rasch model) and does not jeopardize the reporting of trends while at the same time improving the overall model-data fit and allowing for a more in-depth evaluation of measurement invariance assumptions at the item level.

14.1.3 Reducing Linking Error in PISA

The new scaling and linking approach also allowed for revisiting and reducing the *linking error* in PISA 2015. The linking error estimation is used to quantify the uncertainty that arises from differences and changes in the assessment design and scaling models used in different assessment cycles and is a key tool in accounting for cycle-to-cycle differences.

The linking error estimated in past PISA cycles (2000–2012) accounted only for differences across trend items observed for the re-estimated difficulty parameter from one cycle to the next (based on separate IRT calibrations). Hence, it quantified the item-by-cycle interactions and assumes that all country measures are affected in the same way by linking errors, which results in scale-level linking error.

In contrast, the new concurrent calibration approach in PISA based on a more general model and a larger data set accounts for item-by-country-by-cycle interactions and takes the differences due to model error into account, for example, DIF across countries that is not fully accounted for in the historic modeling approach. For more information on the new and improved linking error approach, see the PISA 2015 Technical Report (Chaps. 9 and 12 in OECD, 2017).

14.2 Study 2: Examining Mode Effects in PISA 2015

To further enhance item parameter estimation and ensure a stable trend measure, a well-designed mode effect study was implemented in the PISA 2015 field test (von Davier et al., 2019a; OECD, 2017). This study examined whether transitioning trend items from a paper-based to a computer-based administration mode altered the psychometric qualities of the items.

Transitioning ILSAs to a computer-based administration mode offers several advantages. It improves the standardization of test administration and data collection, increases test-taker engagement through innovative item types, for example, and allows for the inclusion of constructs that measure twenty-first-century skills, such as collaborative problem solving and creative thinking. Additionally, a computer-based administration enables the collection of log file and process data, providing insights into how test-takers navigate and respond to tasks and items.

However, items administered on a computer in one assessment cycle may not retain the same characteristics or measure the same skills as those administered on paper in a previous cycle. Therefore, to maintain comparability across administrations, it is crucial to evaluate whether features of the paper-based format change when items are presented on the computer in ways that could jeopardize comparability.

When PISA 2015 transitioned to a computer-based administration, a mode effect study was designed for the field test prior to the main survey, linking the paper-based items to their computer-based counterparts. Trend items were administered in both formats to each participating country, and the resulting data were analyzed for potential mode effects using different IRT mode effect models (von Davier et al., 2019a). Students in each country were assigned randomly to one of the two administration modes independently of proficiency, prior experience, or other student variables. This equivalent groups design allowed to test the null hypothesis of "no mode effect."

The following sections describe these models and their implementation, illustrating how the results of the mode effect study were used to address differential item functioning (DIF) due to mode differences and improve the scaling and linking of the PISA 2015 data.

14.2.1 A First Look: Comparing Classical Item Statistics and IRT Item Parameter Estimates

In a first step, classical item statistics were computed for the paper-based (PBA) and computer-based (CBA) administration modes for a total of 318 items to compare nonresponse (omissions) rates and item cluster position effects between these modes (a very small number of items that exhibited technical issues were not included in this analysis).

14.2.1.1 Omission Rates and Omission Effects

Two types of nonresponse effects were examined: (1) omission rates and (2) omission effects. Omission rates were defined as the number of omitted responses divided by the total number of test takers who received the item in each country, and omission rates larger than 10% were considered as high. Omission effects were investigated by calculating the percent of correct responses (P+) by including or not including omitted response information. More precisely, the following treatments of omitted responses were considered: omitted responses treated as "wrong" response (W rule) and omitted responses being excluded from the P+ calculation (X rule). A difference of more than 5% between P+ calculated under the W rule and the X rule was considered as an omission effect being present.

The omission rate in all content domains (reading, mathematics, science) was higher for the PBA than the CBA, except for financial literacy, which had a 1.1% higher omission rate in the CBA than the PBA. Considering items overall domains, high omission rates were found for 47 items in the CBA and 54 items in the PBA (when counting items with omission rates higher than 10% across at least 20 countries). Constructed response items accounted for the largest proportion of items with high omission rates (98% in the CBA and 100% in the PBA) and impacted the *P+* value calculation the most.

High omission effects were detected for 13 countries in the PBA and only 3 countries in the CBA. Further, 23 countries were found to have more than 50 items with high-omission-effects in the PBA, while similar conditions were found for only 9 countries in the CBA. On average across countries, high omission effects were found for 32.5 items in the CBA and 44.4 items in the PBA.

Overall, findings showed a reduction in omission rates and omission effects in the CBA over the PBA, with correlations of omission rates and omission effects between the CBA and PBA of 0.88 and 0.80, respectively.

14.2.1.2 Item Cluster Position Effects

Item position effects are a common issue in large assessment programs because substantial position effects can increase measurement error and introduce bias. The PISA 2015 field test design balanced the position of item clusters (note each cluster consists of multiple items) across test booklets to monitor its impact on the average proportions of correct responses (P+).

To examine position effects in the PBA versus the CBA, data from the PISA 2015 field test, which included both PBA and CBA, were compared to the paper-based PISA 2009 data as a reference point. The comparison is illustrated in Table 14.6, showing a decrease in average P+ values (across item clusters and countries) with increasing item position (i.e., from position 1 to 4) in all content domains for the PISA 2009 data. In contrast, the PISA 2015 field test data display only a slight tendency of decreasing P+ values, remaining relatively stable across positions. These

Table 14.6 Average P+ values across item clusters and countries for the PISA 2009 main study and the PISA 2015 field test by item cluster position

Cluster position:	1	2	3	4	4–1
PISA 2009 (PBA)					
Mathematics	0.411	0.401	0.384	0.371	−0.040
Reading	0.581	0.557	0.532	0.499	−0.083
Science	0.490	0.478	0.457	0.435	−0.055
PISA 2015 (CBA/PBA)					
Mathematics	0.439	0.454	0.432	0.427	−0.012
Reading	0.575	0.580	0.566	0.544	−0.031
Science	0.409	0.418	0.401	0.385	−0.024

results indicate that position effects are less pronounced in the CBA compared to the PBA.

In conclusion, results of the PISA 2015 field test data provide evidence that the CBA provides better overall data quality with respect to improved omission rates and reduced position effects compared to the PBA.

14.2.1.3 IRT Item Parameter Estimates: Graphical Model Checks and Correlations

In a next step, item parameters estimated for the paper-based (PBA) and computer-based (CBA) administration modes were compared using correlations and graphical model checks (Rasch, 1960). Item parameters for the CBA mode were estimated based on the PISA 2015 field test data and item parameters for the PBA mode were estimated based on the PISA 2015 field test data combined with the PISA 2000–2012 data set from the linking study (for more stable estimates). Parameter estimates are based on data from 68 field test countries that submitted their data through November 2014 ($n > 150{,}000$).

The PBA-based item parameters were fixed to the estimates obtained from the linking study, while the item parameters for the CBA items were freely estimated simultaneously but constrained to be equal across countries (i.e., international item parameters were estimated).

Due to the random assignment of students to administration modes, the underlying ability distributions of the PBA and CBA field test samples are assumed to be identical. As such, the computer-based parameters were not expected to differ significantly, or systematically, from the item parameters from the linking study and were verified using the PBA Field Trial sample.

Figure 14.1 shows item parameter comparisons between the PBA and CBA modes. The distinction among the different content domains (Reading, Mathematics and Science, Financial Literacy) was ignored because the parameters obtained across modes appeared to vary consistently across all domains. These graphical model checks provide evidence of overall general agreement between the parameters based on the PBA and CBA modes. While there are differences, it appears that the level of difficulty of an item remains largely the same between PBA parameters.

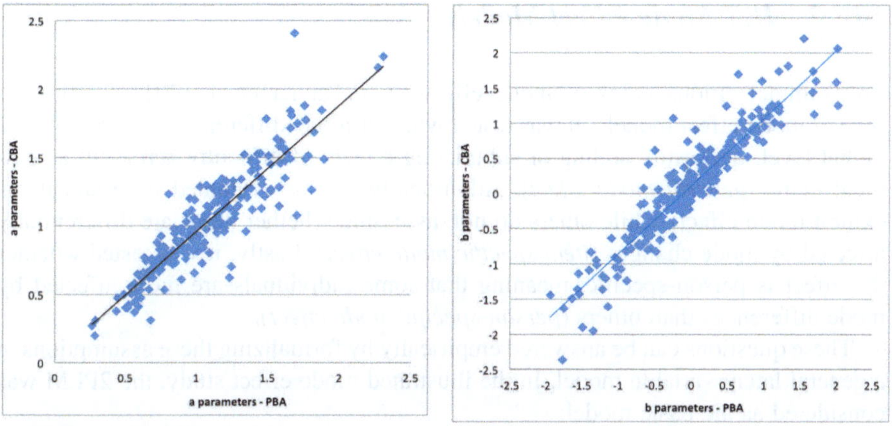

Fig. 14.1 Comparison of parameter estimates across PBA and CBA modes for all content domains (reading, mathematics, science, financial literacy). Item Slope Parameter Estimate (**a**) are on the left side, Item Difficulty Parameter Estimate (**b**) are on the right side

Table 14.7 Correlations of item slope (a) and difficulty (b) parameters between PBA and CBA trend items within and across content domains

Domain	a	b
Mathematics	0.91	0.95
Reading	0.90	0.95
Science	0.94	0.92
Financial literacy	0.92	0.94
All domains	0.91	0.94

Moreover, correlations (see Table 14.7) between the item parameter estimates for PBA and CBA trend items are high within each domain, ranging from 0.90 to 0.94 for slope parameters[2] and from 0.92 to 0.95 for difficulty parameters as well as across domains (0.91 for slope parameters, 0.94 for difficulty parameters). These high correlations as well as the graphical model tests suggest that the same constructs are being measured under both modes and that PBA and CBA results can be reported on the same scales.

After it was confirmed that PBA and CBA items can generally be linked and estimated on a common scale, IRT mode effect models were used to examine the type and level of mode effects to establish the best DIF treatment in the IRT scaling for increased comparability and measurement precision.

[2]Note that only the 2PLM-based slope parameters were used to calculate correlations.

14.2.2 *IRT Mode Effect Models*

To examine various types of mode effects, von Davier et al. (2019a) estimated several mode effect models. It was tested whether mode differences are present on a global level, uniformly adding or subtracting a level of difficulty across all assessment items (*general mode effect*). Additionally, it was evaluated if certain items exhibit mode effects while others do not, assessing whether items are differentially affected by mode changes (*item-specific mode effect*). Lastly, it was tested whether the effect is person-specific, meaning that some individuals are more affected by mode differences than others (*person-specific mode effect*).

These questions can be answered empirically by formalizing these assumptions in a general latent variable model. In the illustrated mode effect study, the 2PLM was considered as the basic model:

$$P(x=1|\theta,\alpha_i,\beta_i)=\frac{\exp(\alpha_i\theta+\beta_i)}{1+\exp(\alpha_i\theta+\beta_i)} \tag{14.1}$$

denotes the probability of a correct response by a respondent with proficiency θ for an item with parameters α_i, β_i. The notation in (14.1) can be transformed to the customary notation by letting $a = \alpha/1.7$ and $b = -\beta/\alpha$. Additional model parameters can be introduced to formalize different assumptions of how mode effects may impact item functioning. This was done to examine different item-level and person-level effects.

14.2.2.1 General and Item-specific Mode Effects

A common (but maybe overly simple) mode effect assumption is that all items are "shifted" by a certain amount with respect to their difficulty. The reason could be that reading or processing the item stem is generally harder or easier on the computer (by the same amount for all items), or entering a response is more complex or simpler in the digital version than on paper. A mode effect that homogeneously applies to all items can be modeled by adding the same constant to all difficulty parameters. A general mode effect parameter δ_m represents how much more difficult (or easy) solving any item is when presented in a different mode relative to the reference model (14.1). For items, we assume that

$$P(X=1|\theta,\alpha_i,\beta_i,\delta_m)=\frac{\exp\left(\alpha_i\theta+\beta_i-1_{\{I+1,\ldots,2I\}}(i)\delta_m\right)}{1+\exp\left(\alpha_i\theta+\beta_i-1_{\{I+1,\ldots,2I\}}(i)\delta_m\right)}. \tag{14.2}$$

This can be thought of as a model for twice the number of items with each item being administered twice, once on paper and once on computer. The indicator function $1_{\{I+1,\ldots,2I\}}(i)$ equals 1 if the item index is in the second half, that is, the range $I+1$, …, $2I$. The first 1, …, I items are the PBA items without mode effect, and the CBA items are indexed by $I + 1, \ldots, 2I$. In this notation it is assumed that item i and item $i + I$ are the same but administered in different modes. This leads to a model with 2I items (instead of I items for each delivery mode separately) in which the difficulty parameters for items presented in one mode (PBA) are assumed to be β_i for $i = 1, \ldots, I$ and the item parameters for the other mode (CBA) are appended as parameters β_j for $j = I + 1, \ldots, 2I$ and arranged in the same order and constrained to follow $\beta_{j+1} = \beta_j - \delta_m$. In the equivalent groups design, each respondent receives half of the items, either paper-based items, indexed by $i = 1, \ldots, I$, or the computer-based items indexed by $i = I + 1, \ldots, 2I$.

In contrast to the assumptions of a general mode effect parameter, one could argue that not all items change difficulty when moving from a PBA to a CBA. Some could be more difficult, some could be at the same difficulty level, and some could get easier. This leads to a model that adds an item-specific effect δ_{mi} to the difficulty parameter. This can be written as a DIF parameter, quantifying item-specific changes:

$$P(X = 1|\theta, \alpha_i, \beta_i, \delta_{mi}) = \frac{\exp\left(\alpha_i\theta + \beta_i - 1_{\{I+1,\ldots,2I\}}(i)\delta_{mi}\right)}{1 + \exp\left(\alpha_i\theta + \beta_i - 1_{\{I+1,\ldots,2I\}}(i)\delta_{mi}\right)}. \tag{14.3}$$

The difference in comparison to the model of metric (or "weak") factorial invariance (Meredith, 1993) is that the CBA difficulties that are defined in reference to the PBA mode are decomposed into two components, that is, $\beta_{i+I} = \beta_i - \delta_{mi}$, while it is still assumed that $\alpha_{i+1} = \alpha_i$ for the slope parameters. This decomposition indicates that the difficulties are shifted by some item or item feature-dependent amount. The shift is being applied to one mode on an item-by-item basis in comparison to the reference mode with no shift. For items that are not significantly affected by the administration mode a model constraint may be imposed assuming that $\delta_{mi} = 0$.

The model in Eq. (14.3) with constraints across both modes on slope parameters, as well as potential constraints on the DIF parameters, establishes a measurement invariance (e.g., Meredith, 1993) IRT model. This model can be viewed as representing weak factorial invariance. The larger the number of constraints of the type $\delta_{mi} = 0$ can be assumed, the more a model with strong factorial invariance is approached. Note that the equality of means and variances of the latent variable is already assumed in both modes because it is expected that respondents receiving the test in computer or paper mode are randomly selected from a single population.

The link between modes in (14.3) can be viewed as increasingly more invariant the more that slope parameters can be assumed to be $\alpha_{mi} = 0$ for items in the new mode. Each constraint $\alpha_{mi} = 0$ makes the respective item response functions for items i and $i + I$ identical across modes.

14.2.2.2 Person-specific Mode Effects

If it **cannot** be assumed that the mode effect is a constant (even if item dependent) shift in difficulty for all respondents, one may assume that an additional proficiency ϑ is required to accurately model response probabilities for the new (computer) mode. This leads to a multidimensional model with a second latent variable that is added to the item function for items administered in the computer mode:

$$P(X=1|\theta,\alpha_i,\alpha_{mi},\beta_i,\vartheta)=\frac{\exp(\alpha_i\theta+\beta_i-\alpha_{mi}\vartheta)}{1+\exp(\alpha_i\theta+\beta_i-\alpha_{mi}\vartheta)}. \tag{14.4}$$

The expression $\alpha_{mi}\vartheta$ in (14.4) indicates that there is a second slope parameter α_{mi} for items ($i=I+1, \ldots, 2I$) administered in the computer mode and that the effect of the mode is person dependent and quantified through a second latent variable ϑ. Note that slope parameters α_i and item difficulties β_i are, as before in models (14.2) and (14.3), equal across modes. However, an additional "mode-slope" parameter α_{mi}, for $i=I+1, \ldots, 2I$, needs to be estimated, with constant $\alpha_{mi}=0$ for $i \leq I$. For the joint distribution $f(\theta, \vartheta)$ uncorrelated latent variables are assumed, $cov(\theta, \vartheta)=0$, to ensure identifiability.

In Eq. (14.4), it is assumed that the effect of the person "mode" variable varies across items. Less plausible but feasible is a model with item-invariant effects $\alpha_m\vartheta$ (a Rasch variant of a random mode effect).

14.2.3 Results of the Mode Effect Study

All models presented above were vetted by the PISA 2015 Technical Advisory Group for operational use after thorough review and applied to the PISA 2015 field test data as multigroup models in a concurrent calibration. Table 14.8 shows the results of the model comparison separately by content domain (reading, mathematics, science, financial literacy) based on overall model-fit statistics. Lower values indicate better fit, but when the magnitude of the statistics is similar, the more parsimonious model should be preferred.

An evaluation of the log-penalty shows that the simple general mode effect model does not fit as well as the item-specific model (14.3) and the person-specific model (14.4). Models (14.3) and (14.4) appear to fit the data almost equally well, both accounting for item-specific effects in slightly different ways. Clearly, the mode effect is not a homogenous shift of difficulties, but rather one that affects some items more than others.

Moreover, the median reliability for scores in all content domains for both models (14.3 and 14.4) was quite similar across groups, with median values ranging from 0.8 to 0.85. Additionally, the correlation between estimated scores for both models in each content domain was $r=0.999$, which suggests that there was little added

Table 14.8 Model-Fit statistics for the different mode effect models by content domain

Content domain	Model	Penalty AIC	AIC	Penalty BIC	BIC	Penalty CAIC	CAIC	Log penalty	Akaike
Read	(14.2)	818	1,770,885	4877	1,774,944	5286	1,775,353	0.534144	0.534391
Read	(14.3)	990	1,760,709	5903	1,765,622	6398	1,766,117	0.531022	0.531320
Read	(14.4)	1104	1,758,594	6583	1,764,073	7135	1,764,625	0.530349	0.530682
Math	(14.2)	620	1,416,987	3697	1,420,064	4007	1,420,374	0.526304	0.526534
Math	(14.3)	674	1,409,948	4019	1,413,293	4356	1,413,630	0.523668	0.523919
Math	(14.4)	714	1,409,235	4257	1,412,778	4614	1,413,135	0.523388	0.523654
Science	(14.2)	1694	5,378,045	10,100	5,386,451	10,947	5,387,298	0.586249	0.586433
Science	(14.3)	1984	5,361,306	11,830	5,371,152	12,822	5,372,144	0.584392	0.584608
Science	(14.4)	2180	5,356,556	12,998	5,367,374	14,088	5,368,464	0.583852	0.584090
FinLit	(14.2)	192	253,996	1003	254,807	1099	254,903	0.564498	0.564925
FinLit	(14.3)	236	251,899	1233	252,896	1351	253,013	0.559736	0.560260
FinLit	(14.4)	248	251,744	1295	252,792	1419	252,916	0.559365	0.559917

utility in using Model (14.4). Based on these findings it was concluded that model (14.3) provides appropriate fit to the data and that item-specific, but not person- (or country[3]-) specific, mode effect parameters are needed.

Further, the results obtained with model (14.3) showed that while most mode effects on individual items were positive, some were negative. This finding indicates that a common linear adjustment-based equating method would not be appropriate. This also opens opportunities to optimize the linking between PBA and CBA: items that are impacted by mode effects can receive different parameter estimates between PBA and CBA, while equality constraints can be used for those items that are least affected to estimate common item parameters across modes.

Based on the distribution of the mode effect sizes in model (14.3) it was found that for most items strong measurement invariance holds, and no significant mode effects could be detected. These items formed the basis for linking the CBA assessment to past paper-based PISA cycles through the estimation of common item parameters. Weak factorial invariance could be assumed for the remaining trend items administered in the CBA.

Based on the results from model (14.3), which assumes the same parameters for the PBA as for the CBA and adjusted the PBA item difficulty parameters by a DIF parameter for a subset of items, the concurrent calibration used in the PISA 2015 scaling was adjusted. Items that were identified to have the same characteristics across modes received a common item parameter while a small subset of items that were identified to differ between modes received unique mode-specific item parameters. This modeling approach allowed the linking of the PBA and CBA trend items on a common scale and to establish strong measurement invariance and produced a stable trend measure.

14.2.4 The Impact of Mode Effects on Country Means

Some countries and stake holders were concerned that changing the mode of administration might impact the overall achievement results by making the PISA instruments more difficult. To address this concern, the impact of mode effects relative to other variables of interest was examined using the PISA 2015 field test data.

Country means based on the domain-specific skills obtained from a simplified version of model (14.4) were split by three variables and compared to one another: gender, mode, and a random split of schools within each country. Model (14.4) was simplified for this analysis so that it incorporates scalar invariance for those items that showed little or no mode difficulty differences and assumes metric invariance

[3] Note that the random effect in model 14.4 could be adjusted for each country separately, so this model picks up country differences as well. The similarity between the fit of models 14.4 and 14.3 shows, that no country specific constraints are needed.

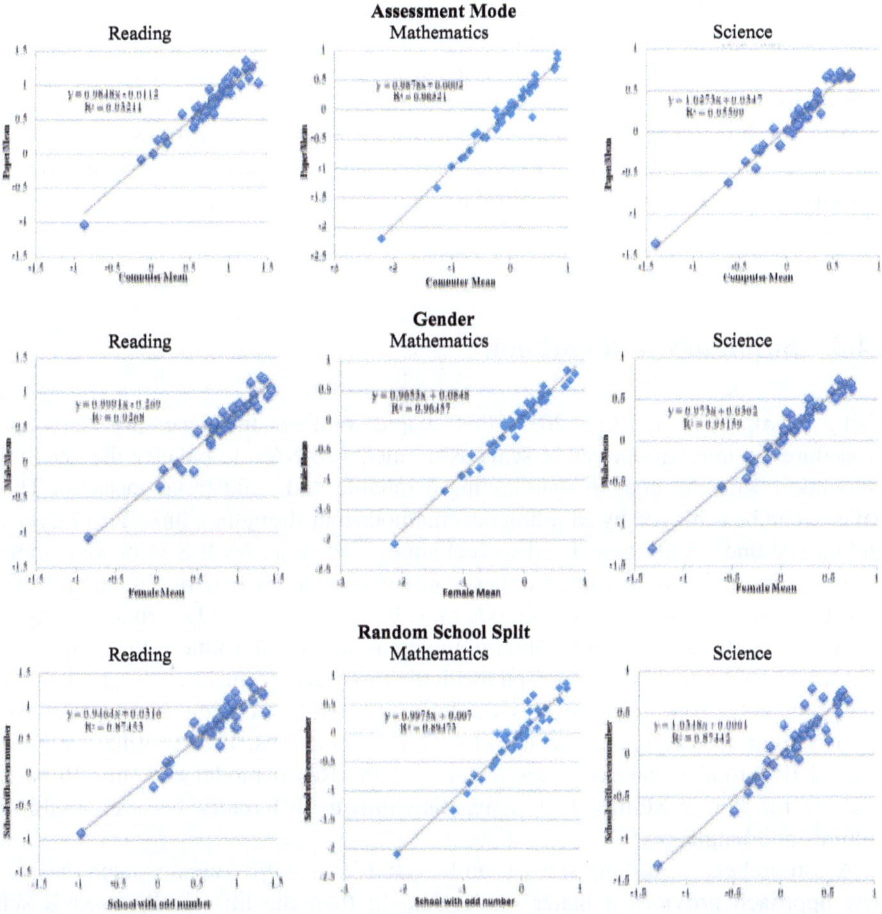

Fig. 14.2 Comparison of country means by assessment mode, gender, and random school split for reading, mathematics, and science

for the remaining items. There were no country-specific mode effects needed or applied in these analyses. This ensures comparability across countries while accounting for item-specific difficulty differences for a subset of items only, with these differences applied across all countries in the same way. This approach ensured that comparability is maximized, while mode effects that affected different items in different directions were accounted for so that potential effects on scale comparisons were minimized.

The comparisons are illustrated in Fig. 14.2 separately for content domains (reading, mathematics, science) and show that good agreement between country means by assessment mode could be achieved. The differences and variability between gender groups, and also the two groups formed by randomly splitting schools, were at the same level or larger than the differences obtained by splitting the sample by administration mode. The largest differences between means were observed based on a random school split, not based on mode. Thus, differences between countries might be due more to differences between students and schools than to differences based on the assessment mode.

Additionally, given the smaller sample size of the field test, differences that one may be tempted to attribute to mode differences are at the same order of magnitude as what could be observed if we split the Field Trial sample randomly by some other criterion.

14.3 Summary and Outlook

Using examples from PISA 2015, this chapter outlines the necessary steps and procedures in international large-scale assessments (ILSAs) to balance the need for innovation with the goal of maintaining a reliable and valid trend measure. This balance can be achieved by adopting new methods that strengthen links across cycles and reduce uncertainty associated with changes over time. As ILSAs become more inclusive, involving a growing number of countries, and moving into the digital era, their procedures, features, and methods must be revisited and adjusted accordingly.

In PISA 2015, a new scaling and linking approach was introduced that employs a concurrent calibration which uses all available PISA data from all cycles, rather than analyzing data from single cycles as done previously, accommodating item-by-country, item-by-language, and item-by-cycle interactions. It also utilizes a more general IRT model that combines features of the Rasch model and the 2PLM to account for the possibility that items discriminate differently between multiple cultures and languages.

A comprehensive linking study (von Davier et al., 2019b) demonstrated that this new approach provides a better model-data fit than the historically used Rasch model, offers more robust item parameter estimates, and allows for the investigation of measurement invariance across groups (split by cycle, country, and language). Moreover, this improved approach maintains the integrity of the trend measure, a crucial tool for stakeholders and policymakers, and reduces the linking error related to changes in assessment design and methodology across cycles by more comprehensively accounting for DIF across groups.

Additionally, PISA transitioned from a paper-based assessment (PBA) to a computer-based assessment (CBA) in 2015. To address this change, a well-designed mode effect study was implemented in the PISA 2015 field test, where participating countries administered the same instruments in both modes. Using graphical model checks and various IRT mode effect models (von Davier et al., 2019a), it was found that most items were invariant across modes, meaning they exhibited the same psychometric characteristics in both formats and could be used to link PBA and CBA through the estimation of common item parameters.

Only a subset of items was affected by the administration mode, either being more or less difficult on the computer compared to the paper version. For these items, unique mode-specific parameters were estimated in the concurrent calibration to account for mode effects. The identified mode effect was item-specific, with no general effect impacting all items uniformly. Because only a small subset of items was affected, a robust link and stable trend measure could be established between PISA 2015 and past assessment cycles.

Moreover, it was demonstrated that the magnitude of the mode effects did not significantly impact the estimation of country means more than other variables, such as gender or a random split of schools. In fact, a random split of schools showed a larger impact on country means than mode effects.

The presented PISA examples and lessons learned can be adopted and adjusted for other large-scale assessments. For example, a similar mode effect study with an improved design was implemented for TIMSS 2019 when it moved to a fully computer-based administration (von Davier, 2020). Moreover, the different modeling approaches can be used to conduct further research and additional in-depth analyses of measurement invariance related to different variables such as content and cognitive domains, item type, or respondent characteristics like computer familiarity or gender (e.g., Khorramdel et al., 2020).

It is important to note that there is 'no one size fits all' approach as different assessments and tests have different goals, serve different target populations and are measuring different constructs routed in different frameworks and based on different test designs. Hence, linking approaches and mode effect comparisons must be adjusted to the specific needs and requirements of an assessment. Some studies might find significant mode effects and others barely any (Lowrie & Logan, 2015), and the ones that do find effects might encounter different types (e.g., item-level versus overall mode effects) and extent of mode effects that require different treatments in the linking and scaling approaches. Different test designs, for example, computer-adaptive tests versus linear tests or linking across groups within a test administration versus linking across multiple testing cycles over time, might require (slightly) different considerations in the linking approach. The beauty of the IRT models and approaches proposed in this chapter is their flexibility and adaptability to different situations and needs, and the possibility of adding further extensions.

References

Birnbaum, A. (1968). Some latent trait models and their use in inferring a student's ability. In F. M. Lord & M. R. Novick (Eds.), *Statistical theories of mental test scores*. Addison-Wesley.

Bock, R. D., & Zimowski, M. F. (1997). Multiple group IRT. In W. J. van der Linden & R. K. Hambleton (Eds.), *Handbook of modern item response theory* (pp. 433–448). Springer. https://doi.org/10.1007/978-1-4757-2691-6_25

Byrne, B. M., Shavelson, R. J., & Muthén, B. O. (1989). Testing for equivalence of factor covariance and mean structures: The issue of partial measurement invariance. *Psychological Bulletin, 105*, 456–466. https://doi.org/10.1037/0033-2909.105.3.456

Glas, C. A. W., & Jehangir, K. (2013). Modeling country specific differential item functioning. In L. Rutkowski, M. von Davier, & D. Rutkowski (Eds.), *Handbook of international large-scale assessment*. CRC Press. https://doi.org/10.1201/b16061

Glas, C. A. W., & Verhelst, N. D. (1995). Testing the Rasch model. In G. H. Fischer & I. W. Molenaar (Eds.), *Rasch models: Foundations, recent developments, and applications* (pp. 69–95). Springer. https://doi.org/10.1007/978-1-4612-4230-7_5

Khorramdel, L., Pokropek, A., Joo, S.-H., Kirsch, I., & Halderman, L. (2020). Examining gender DIF and gender differences in the PISA 2018 Reading literacy scale: A partial invariance approach. *Psychological Test and Assessment Modeling, 62*, 179–231. https://www.psychologie-aktuell.com/fileadmin/Redaktion/Journale/ptam-2020-2/02_Khorramdel.pdf

Kreiner, S., & Christensen, K. B. (2014). Analyses of model fit and robustness. A new look at the PISA scaling model underlying ranking of countries according to reading literacy. *Psychometrika, 79*(2), 210–231. https://doi.org/10.1007/s11336-013-9347-z

Lord, F. M., & Novick, M. R. (1968). *Statistical theories of mental test scores*. Addison-Wesley.

Lowrie, T., & Logan, T. (2015). The role of test-mode effect: Implications for assessment practices and item design. In *In pursuit of quality mathematics for all. Proceedings of the 7th ICMI-East Asia Regional Conference on Mathematics Education* (pp. 649–656). http://mathted.weebly.com/uploads/7/8/5/0/7850000/pp_649_tom_lowrie_final_paper_edited.pdf

Masters, G. N. (1982). A Rasch model for partial credit scoring. *Psychometrika, 47*, 149–174.

Mazzeo, J., & von Davier, M. (2008). Review of the Programme for International Student Assessment (PISA) test design: Recommendations for fostering stability in assessment results, Doc. ref. EDU/PISA/GB(2008)28. Retrieved from http://www.oecd.org/dataoecd/44/49/41731967.pdf

Mazzeo, J., & von Davier, M. (2013). Linking scales in international large-scale assessments. In L. Rutkowski, M. von Davier, & D. Rutkowski (Eds.), *Handbook of international large-scale assessment: Background, technical issues, and methods of data analysis*. CRC Press. https://doi.org/10.1201/b16061

Meredith, W. (1993). Measurement invariance, factor analysis and factorial invariance. *Psychometrika, 58*, 525–543. https://doi.org/10.1007/BF02294825

Muraki, E. (1992). A generalized partial credit model: Application of an EM algorithm. *Applied Psychological Measurement, 16*(2), 159–177. https://doi.org/10.1002/j.2333-8504.1992.tb01436.x

OECD. (2017). *PISA 2015 technical report*. OECD Publishing. https://www.oecd.org/content/dam/oecd/en/about/programmes/edu/pisa/publications/technical-report/PISA2015_TechRep_Final.pdf

Oliveri, M. E., & von Davier, M. (2011). Investigation of model fit and score scale comparability in international assessments. *Psychological Test and Assessment Modeling, 53*(3), 315–333. Retrieved from http://www.psychologie-aktuell.com/fileadmin/download/ptam/3-2011_20110927/04_Oliveri.pdf

Oliveri, M. E., & von Davier, M. (2014). Toward increasing fairness in score scale calibrations employed in international large-scale assessments. *International Journal of Testing, 14*(1), 1–21. https://doi.org/10.1080/15305058.2013.825265

Rasch, G. (1960). *Probabilistic models for some intelligence and attainment tests*. Nielsen & Lydiche. (Expanded edition, Chicago, University of Chicago Press, 1980).

Rutkowski, L., Gonzalez, E., Joncas, M., & von Davier, M. (2010). International large-scale assessment data: Issues in secondary analysis and reporting. *Educational Researcher, 39*(2), 142–151. https://doi.org/10.3102/0013189X10363170

Schwarz, G. (1978). Estimating the dimension of a model. *Annals of Statistics, 6*, 461–464. Retrieved from https://www.jstor.org/stable/2958889?&seq=1#page_scan_tab_contents

Steenkamp, J. M., & Baumgartner, H. (1998). Assessing measurement invariance in crossnational consumer research. *Journal of Consumer Research, 25*, 78–90. https://doi.org/10.1086/209528

von Davier, M. (2020). TIMSS 2019 scaling methodology: Item response theory, population models, and linking across modes. In M. O. Martin, M. von Davier, & I. V. S. Mullis (Eds.),

Methods and procedures: TIMSS 2019 technical report (pp. 11.1–11.25) https://timssandpirls.bc.edu/timss2019/methods/chapter-11.html

von Davier, M., Khorramdel, L., He, Q., Shin, H. J., & Chen, H. (2019a). Developments in psychometric population models for technology-based large-scale assessments: An overview of challenges and opportunities. *Journal of Educational and Behavioral Statistics, 44*(6), 671–705. https://doi.org/10.3102/1076998619881789

von Davier, M., & Rost, J. (2016). Logistic mixture-distribution response models. In W. van der Linden (Ed.), *Handbook of item response theory* (Vol. 1, 2nd ed., pp. 393–406). CRC Press. https://doi.org/10.1201/9781315119144

von Davier, M., & Sinharay, S. (2013). Analytics in international large-scale assessments: Item response theory and population models. In L. Rutkowski, M. von Davier, & D. Rutkowski (Eds.), *Handbook of international large-scale assessment: Background, technical issues, and methods of data analysis* (pp. 155–174). CRC Press. https://doi.org/10.1201/b16061

von Davier, M., & von Davier, A. (2007). A unified approach to IRT scale linking and scale transformations. *Methodology: European Journal of Research Methods for the Behavioral and Social Sciences, 3*(3), 115–124. https://doi.org/10.1027/1614-2241.3.3.115

von Davier, M., Xu, X., & Carstensen, C. H. (2011). Measuring growth in a longitudinal large-scale assessment with a general latent variable model. *Psychometrika, 76*(2), 318–336. https://doi.org/10.1007/s11336-011-9202-z

von Davier, M., Yamamoto, K., Shin, H. J., Chen, H., Khorramdel, L., Weeks, J., Davis, S., Kong, N., & Kandathil, M. (2019b). Evaluating item response theory linking and model fit for data from PISA 2000–2012. *Assessment in Education: Principles, Policy & Practice, 26*(4), 466–488. https://doi.org/10.1080/0969594X.2019.1586642

Weeks, J., Yamamoto, K., & von Davier, M. (2013). Design considerations for the program for international student assessment. In L. Rutkowski, M. von Davier, & D. Rutkowski (Eds.), *Handbook of international large-scale assessment: Background, technical issues, and methods of data analysis*. CRC Press. https://doi.org/10.1201/b16061

Yamamoto, K. (1998). Scaling and scale linking. In T. S. Murray, I. S. Kirsch, & L. B. Jenkins (Eds.), *Adult literacy in OECD countries: Technical report on the first international adult literacy survey* (pp. 161–178). National Center for Education Statistics.

Yamamoto, K., Khorramdel, L., von Davier, M., Ali, U. S., & Robin, F. (2019). *Scaling PIAAC cognitive data: Technical report of the survey of adult skills (PIAAC), 3rd edition, chapter 17*. PIAAC, OECD Publishing. Retrieved from https://www.oecd.org/skills/piaac/publications/PIAAC_Technical_Report_2019.pdf

Yamamoto, K., & Mazzeo, J. (1992). Item response theory scale linking in NAEP. *Journal of Educational Statistics, 17*(2), 155–174.

Chapter 15
Effects of Mode Transition on Instruments and Subpopulation Performance in NAEP

Paul A. Jewsbury, Yue Jia, and Nuo Xi

Abstract In 2017, the National Assessment of Educational Progress (NAEP) transitioned the reading and mathematics assessments at grades 4 and 8 to a digitally-based platform. As part of the transition, a comprehensive mode evaluation study was conducted to provide evidence on whether and how to maintain trend reporting across the paper-based and digitally based administrations. The study involved administering the NAEP instruments in paper and digital formats to randomly equivalent groups. Although items retained many of their unique properties, such as approximate rank-order difficulty, items were generally more difficult when administered in digital format. Apart from constructed response items in grade 4 showing a larger increase in difficulty when adapted to digital format, the increased difficulty was fairly uniform across all items within a subject and grade, and almost all item mean scores were significantly different between modes. To account for differences in the measurement instruments, common population linking was applied. After linking, few significant differences in mean estimates for state/jurisdictions, districts, and subpopulations were found between the digital and paper results. Observed significant differences were associated with small effect sizes, and the effect sizes were inconsistent across grades and across subjects.

As the largest nationally representative and continuing assessment of what U.-S. students know and can do, the National Assessment of Educational Progress (NAEP) measures academic progress over time, or *trends*. As a measure of change, NAEP has a constant need to balance two competing pressures in its assessment design (Beaton, 1990a). First, to minimize changes in the instrument and administration so that estimated changes in proficiency may be interpreted as changes in a single latent construct (Beaton, 1990b). Second, to adapt to changes in the

P. A. Jewsbury (✉) · Y. Jia · N. Xi
ETS, Princeton, NJ, USA
e-mail: pjewsbury@ets.org

L. Khorramdel et al. (eds.), *Innovative Digital-Based International Large-Scale Assessments*, Methodology of Educational Measurement and Assessment,
https://doi.org/10.1007/978-3-031-90951-1_15

educational and policy environment to ensure NAEP measures skills that are relevant and valued (Zwick, 1992).

In 2017, the NAEP reading and mathematics assessments at grades 4 and 8 were administered digitally for the first time, following prior digital NAEP pilot assessments. To support the transition from a traditional paper-and-pencil-based assessment to a digitally-based assessment (DBA), a comprehensive mode study of the impact of digital administration on the NAEP instrument and subpopulation proficiency estimates was conducted.

The mode study was designed to inform decisions about how and whether to maintain score comparability between paper and digital administrations. To evaluate *how* to maintain score comparability, the psychometric properties of corresponding items in the digital and paper NAEP instruments were compared. To evaluate *whether* to maintain score comparability, the equivalence of scores for major reporting outcomes—means of subpopulations, such as gender and race/ethnicity, as well as means of state/jurisdictions and districts—were compared between the paper and digital administrations, after the digital results were linked to the paper metric.

A recent major shift in the educational and policy environment is the growing importance of computers in education, such as the emphasis on digital learning in federal education legislation (SRI International, 2018), the increasing use of digital technologies in the classroom (Pearson, 2015), and the digitalization of state assessments (National Center for Education Statistics [NCES], 2018a). For example, more than 95% of American students, regardless of location, report access to a computer in their classroom (NCES, 2018b). To accommodate this deeply digitalized educational environment, NAEP has embarked on studying a transition to digitally-based assessment since the early 2000s.

Early NAEP studies on the impact of digital administration on mathematics, writing, and problem solving in technology-rich environments, showed a consistent pattern, as we will discuss (Bennett et al., 2008; Bennett et al., 2007; Sandene et al., 2005). Most relevant to this chapter is the 2001 mathematics online study (Bennett et al., 2008; Sandene et al., 2005), involving approximately 1000 students. All students received an item block of 20 questions in paper format, and then received another item block of 26 questions in either a digital or a paper format.

In the 2001 mathematics online study with eighth graders, Bennett et al. (2008) found that items operated similarly when administered in either paper or digital format, with item response theory (IRT) item parameter estimates between the two formats correlated at $r = .86$ and $r = .96$ for discrimination and difficulty parameters, respectively. However, the difficulty parameter estimates were slightly higher for items in digital format relative to the same items administered in paper format. The greater difficulty parameter estimates reflected a reduction of 5% in the mean item score in the digital administration (Sandene et al., 2005), which translates to a difference of about four NAEP scale points or .14 of the population standard deviation (Bennett et al., 2008).

While the digital format of the mathematics item block was more difficult than the paper format, this increase in difficulty was uniform across subgroups based on gender, race/ethnicity, school urban locale (city, urban, rural), region of the country (northeast, southeast, central, west), and school type (public, private). However, the increase in difficulty was slightly larger for students reporting that at least one parent graduated from college (Bennett et al., 2008).

In 2015, NAEP conducted an extensive digital field trial involving the complete digital administration of the reading, mathematics, and science assessments to nationally representative samples at grades 4, 8, and 12 (Oranje & Moran, 2017; Xu & Jewsbury, 2017). For the more directly relevant reading and mathematics assessments at grades 4 and 8, the sample sizes were between 12,000 and 15,000. These digital field trials were administered in the same window of time as the 2015 operational NAEP assessments, administered in paper format to samples of approximately 150,000 each. The random equivalence of the samples facilitated a comparison of the differences in paper and digital instrument operation.

The far more comprehensive 2015 digital field trials replicated the results from the 2001 mathematics online study (Bennett et al., 2008; Sandene et al., 2005), despite the digital assessments in 2015 and 2001 differing in time and digital delivery systems. The differences in mean item score[1] between the digital administration (in the field trial) and the paper administration (in the operational assessment) were larger at grade 4 (6.7% and 5.2% for reading and mathematics, respectively) than at grade 8 (2.0% and 2.5%, respectively). Apart from a relatively uniform increase in difficulty, the items otherwise operated very similarly between the two modes; for example, the IRT item parameter estimates between the two modes were highly correlated. However, constructed response items in reading grade 4 had a larger increase in difficulty transitioning from paper to digital mode than the selected response items in the same assessment.

Aside from the 2015 study enabling a mode comparison of psychometric properties of the NAEP instrument, the large sample sizes enabled more power in testing if the increased difficulty of the digital administration was uniform across major subgroups than the prior NAEP studies (Bennett et al., 2007, 2008; Sandene et al., 2005). The 2015 study used a common population linking procedure that allows for uniform differences in the difficulty of the digital and paper instruments across subgroups (also used in the present study; see the Methods section). After applying the common population linking procedure, no statistically significant differences in proficiency estimates between the two modes were found across gender, race/ethnicity, English language learner status, disability status, or free or reduced-price school lunch status.

The present study, concerning the effect of mode transition of the 2017 NAEP assessments, was designed based on prior findings in NAEP reviewed above (e.g., Bennett et al., 2007, 2008; Oranje & Moran, 2017; Sandene et al., 2005; Xu & Jewsbury, 2017), including the substantial but relatively uniform impact of digital administration on the difficulty of the instrument. The goal of the 2017 mode study was to provide evidence on whether NAEP should maintain trend reporting across

[1] In percent correct units.

the 2015 paper-based administration and the 2017 digitally based administration of reading and mathematics at grades 4 and 8. Furthermore, unlike the 2015 field trial, the 2017 mode study involved sampling representative state/jurisdiction and district samples for both modes. Analyses in the 2017 mode study were largely designed to answer the following three questions:

1. What was the impact of digital administration on the psychometric properties of the NAEP instrument?
2. What was the mode impact on state/jurisdiction and district proficiency estimates?
3. What was the impact of the mode transition on subpopulation (e.g., gender, race/ethnicity) proficiency estimates?

15.1 Methods

15.1.1 Analysis Model of NAEP

The analysis of NAEP data is centered on an IRT-latent regression model to accommodate two primary design features of NAEP (von Davier et al., 2006). First, each student receives only a randomly assigned pre-assembled subset of all the items in the assessment. This is done to balance the competing desires of limiting testing time and maintaining a comprehensive item pool from a broad framework. Second, NAEP is required to report results only for aggregated groups of students, such as race/ethnicity, gender, states/jurisdictions, and districts, and NAEP is not allowed to report results for individual students (Johnson, 1992).

Among other benefits, the use of IRT methods enables NAEP to model the responses of students who received different, but overlapping, sets of items. In the IRT model, the response to item j for student i is dependent on the p latent proficiencies of the student, $\boldsymbol{\theta}_i = (\theta_1, \ldots, \theta_p)$, where $p = 2$ for reading and $p = 5$ for mathematics,[2] and the nature of this dependency is governed by the properties of the item, $\boldsymbol{\beta}_j$. The elements of $\boldsymbol{\beta}_j$ depend on the specific IRT model fit to item j. For example, when the three-parameter logistic model (3PL) is used, $\boldsymbol{\beta}_j = (a_j, b_j, c_j)$ where a_j, b_j, and c_j are the discrimination, difficulty, and pseudo-guessing parameters of the 3PL model. The responses to individual items are assumed to be conditionally independent such that,

[2] In the 2017 reading assessments, the latent proficiencies were Informational and Literary. In the 2017 mathematics assessments, the latent proficiencies were Number Properties and Operations; Measurement; Geometry; Data Analysis, Statistics, and Probability; and Algebra.

$$\Pr(\mathbf{Y}_i = \mathbf{y}_i|\boldsymbol{\theta}_i, \boldsymbol{\beta}) = \prod_{j=1}^{J} \Pr(Y_{ij} = y_{ij}|\boldsymbol{\theta}_i, \boldsymbol{\beta}_j),$$

where $\mathbf{Y}_i = (Y_{i1},\ldots,Y_{iJ})$ is a vector comprising the response variables for all J items. The dependency of the responses on the latent proficiencies and the item response functions are modelled with either the two-parameter logistic model (2PL), three-parameter logistic model (3PL), or the general partial credit model (GPCM; Muraki, 1992), depending on the type of item.

To estimate results for different student groups, a regression model is employed in which the NAEP proficiencies are the dependent variables and group indicators are the independent variables. Because the NAEP proficiencies are latent variables, the model is a latent regression (Mislevy, 1984, 1985). Conditional on $\mathbf{x}_i$, the vector of contextual variables and group indicators for student i, $\boldsymbol{\theta}_i$ is assumed to follow a multivariate normal distribution. That is, $\boldsymbol{\theta}_i|\boldsymbol{x}_i \sim N(\boldsymbol{\Gamma}^T\boldsymbol{x}_i,\boldsymbol{\Sigma})$, where $\boldsymbol{\Gamma}$ is the vector of regression parameters, and $\boldsymbol{\Sigma}$ is a variance-covariance matrix.

The complete NAEP model, combining IRT and the latent regression model, is obtained by assuming $\mathbf{Y}_i$ and $\mathbf{x}_i$ are independent conditional on $\boldsymbol{\theta}_i$, leading to the marginal likelihood,

$$\prod_{i=1}^{N} \int \Pr(\mathbf{Y}_i = \mathbf{y}_i|\boldsymbol{\theta}_i, \boldsymbol{\beta})\phi(\boldsymbol{\theta}_i; \boldsymbol{\Gamma}^T\mathbf{x}_i, \Sigma)d\boldsymbol{\theta}_i,$$

where N is the total number of students, and $\phi(\cdot| \cdot, \cdot)$ is the multivariate normal density function. The complete NAEP model provides an approximation to the conditional posterior distribution of latent proficiency for each test taker, $\boldsymbol{\theta}_i|\boldsymbol{x}_i,\boldsymbol{y}_i$, enabling the generation of plausible values (Mislevy, 1991).

15.1.2 Study Design

A central principle in the 2017 mode study design was that paper and digital instruments were administered to randomly equivalent and representative samples, to enable attributing systematic differences in performance to administration mode. The default method of linking scores from different assessments in NAEP is to assume the item parameters are invariant across the two assessments and to conduct a multigroup IRT or concurrent calibration (Bock & Zimowski, 1997; Mislevy et al., 1992). However, prior work suggests that item parameters are not invariant across modes (e.g., Bennett et al., 2008; Sandene et al., 2005; Xu & Jewsbury, 2017). In such a scenario, *common population linking* was considered as an alternative. Common population linking is often used for linking assessments with different modes (Eignor, 2007), and has been routinely used in NAEP when faced with a change in instrument or administration (e.g., Johnson & Zwick, 1990; NCES, 2009; Perie et al., 2005; Yamamoto & Mazzeo, 1992).

Both the paper and digital assessments followed a careful sampling design involving student-level weights to ensure both samples were representative and randomly equivalent, in accordance with standard NAEP procedures (Johnson & Rust, 1992; see also Jewsbury et al., 2020 for detail specific to the 2017 sampling and weighting design). For each assessment (e.g., reading grade 4), approximately 150,000 students participated in the digital assessment, and approximately 40,000 students participated in the paper assessment. Most sampled schools had both paper and digital sessions (for details, see Jewsbury et al., 2020). This is a more powerful and controlled sampling design, and the positive dependency between the paper and digital samples due to shared schools further increased power for cross-mode comparisons. All analyses conducted as part of the mode study accounted for the positive sample dependency, resulting in smaller standard errors on comparisons between the two samples than if the two samples were independent.

15.1.3 Linking

In accordance with the study design and previous research (Bennett et al., 2008; Sandene et al., 2005; Xu & Jewsbury, 2017), scores were linked between the paper and digital assessments following the common population design to allow for expected differences in item operation (specifically, non-invariance of item parameters across mode; Yamamoto & Mazzeo, 1992). More detail on the samples assigned to paper and digital assessments to satisfy the assumptions of the common population design is available in Jewsbury et al. (2020).

Prior to linking, the paper and digital data were analyzed separately. The paper data were analyzed by standard NAEP procedures that produced IRT-based scores on the existing NAEP *reporting metric* via concurrent calibration with data from the prior NAEP administration (see Mislevy et al., 1992; Yamamoto & Mazzeo, 1992). The digital data were analyzed with a single-group IRT model, producing IRT-based scores in an arbitrary metric referred to as a *theta metric*. The digital scores on the arbitrary theta metric were transformed onto the reporting metric with the linear transformation,

$$\widehat{Y} = \widehat{A}\widehat{\theta} + \widehat{B},$$

where,

$$\widehat{A} = \frac{\widehat{\sigma}_X}{\widehat{\sigma}_\theta}; \widehat{B} = \widehat{\mu}_X - \widehat{A}\widehat{\mu}_\theta,$$

$\widehat{\mu}_X$ and $\widehat{\sigma}_X$ are the estimated national mean and standard deviation in the reporting metric based on the paper results, and $\widehat{\mu}_\theta$ and $\widehat{\sigma}_\theta$ are the estimated national mean and standard deviation in the theta metric based on the digital results. For NAEP reading and mathematics, linking was applied at the subscale level (see the National Assessment Governing Board [NAGB] 2017a and 2017b for descriptions of the reading and mathematics subscales). The subscale scores were later linearly combined into a composite scale score, based on the weights defined in the corresponding subject framework (NAGB, 2017a, 2017b), to describe the overall attainment of students in the specific subject area.

15.1.4 Standard Errors

As NAEP produces multiple imputed scores (i.e., plausible values), the standard NAEP standard error calculation involves the sum of two variance components: the within-imputation variance component estimated with a jackknife balanced replication method-(for details, see Johnson & Rust, 1992), and the between-imputation variance component estimated via Monte Carlo. An additional component was added to the scores associated with uncertainty due to the estimation of $\widehat{A}$ and $\widehat{B}$ (for details on the 2017 procedure, see Mazzeo et al. (2025) of this volume; for a comprehensive theoretical account of errors, see Jewsbury, 2019; for more recently developed methods, see Jewsbury, 2024).

15.1.5 Instruments

The 2017 reading and mathematics paper-based assessments were a complete re-administration of the 2015 reading and mathematics operational assessments, following the standard NAEP procedures for paper-based assessments. To ensure a consistent testing experience across schools, the 2017 digitally based assessment was administered using NAEP-owned Microsoft Surface Pro tablets, running a web application called eNAEP to administer the assessment. The Surface Pro tablets were equipped with external keyboards, styluses, and a pair of ear buds. Trained NAEP administrators oversaw the testing process and were responsible for the tablet operations. Students watched a tutorial on the use of the device prior to the assessment. More information is available from NAEP's digitally based assessment webpage (NCES, n.d.).

The digital instruments were designed to measure the same skills with the same content coverage as the original paper-based instruments, while being appropriate for digital administration. Items in the digital instruments were either *transadapted items*, which were developed to resemble a specific item from the paper assessment, or *new digital items*, which were developed specifically for digital administration. In other words, transadapted items had corresponding paper-based versions or "parents," while new digital items did not. Approximately 91% of reading items and 67%

of mathematics items in the digital assessment were transadapted from paper, with the percentages approximately equal between grades.

Of the transadapted items, items were either directly transferred from paper to digital format or transferred to digital format with new tools. The former included most of the items and were as equivalent to the paper format item as reasonably possible. The items transferred to digital with new tools had the same measurement target as the original paper format versions but were modified to facilitate the digital administration environment. For example, some mathematics selected response items were enhanced with drag and drop functionality, only possible in a digital environment. Similarly, some reading items were enhanced with a look-back button that directed students to the relevant section in the passage.

As previously noted, NAEP employs matrix sampling designs, in which items are assembled into mutually exclusive *item blocks*, and students do not receive all item blocks in the assessment (see Johnson, 1992). In the 2017 reading and mathematics assessments, each student received two item blocks, each separately timed. For the reading assessment, 9 of 10 and 12 of 13 of the blocks in the digital assessment were transadapted from paper blocks, for grades 4 and 8, respectively. An additional block at each grade was developed comprising only new digital items. For the mathematics assessment, each grade had ten item blocks, six comprising transadapted items and four comprising new digital items. While the item blocks comprising transadapted items were designed to resemble paper-format item blocks as closely as reasonable, because some paper items could not be reasonably presented in digital format, the six item blocks comprising transadapted items do not have exactly the same item composition. For more details on the NAEP instruments in 2017, see Jewsbury et al. (2020).

15.2 Results

15.2.1 Impact of Digital Administration on the Psychometric Properties of the NAEP Instrument

The impact of the mode transition on item operation was examined by comparing the item properties of digital and paper versions of the same item. Because the digital and paper versions of the assessment were provided to randomly equivalent samples, the mean item score and other item properties could be directly compared between the two modes.

As described in the Instruments section, a majority of the digital assessment items were transadapted from paper, and their original paper versions were administered as part of the paper assessment in 2017. However, the mathematics blocks comprising transadapted items did not exactly correspond to the parent paper blocks. Some transadapted items were in different positions or different contexts (defined by which items were presented prior to the focal item), which is known to cause differences in item operation, known as context and position effects (e.g., Zwick, 1991). Therefore, to avoid potential confounding, only transadapted mathematics items that were in the same position or shifted by only one position from the corresponding paper-format item positions, were included in the item-level comparisons here.

For reading, all 88 and 120 transadapted items at grades 4 and 8, respectively, were included in the present analysis. For mathematics, 78 and 77 items were included at grades 4 and 8, respectively. Figure 15.1 compares the paper mean score to the digital mean score of each item included in the analysis. The mean item score was highly correlated between the two modes ($r = .97$ for reading, $r = .99$ for mathematics, at both grades), suggesting that items retain much of their relative difficulties within the item pool and, after accounting for a mostly uniform increase in difficulty, their unique properties. However, items were generally harder in the digital format, as indicated by most points being lower than the 45-degree diagonal line of equality (or $y = x$ line). With the exception of reading grade 8, almost all items were significantly harder in digital format.

Table 15.1 shows the average mean item scores for reading and mathematics. The *Overall* rows show the average across all items included in the analysis for each subject and grade combination. The overall difference (in the *DBA-PBA (SE)* column) is significantly different from zero at the $\alpha = .05$ level in each case and is notably larger for grade 4 than grade 8 at each subject. Table 15.1 also shows the average mean item score broken down by item type. At grade 4, there is a relatively large difference between selected response and constructed response items (3.0% and 4.5% for reading and mathematics, respectively). At grade 8, the difference is relatively smaller (.4% and 1.0%, respectively).

Aside from the overall scale, both reading and mathematics have subscales defined by NAEP frameworks (NAGB, 2017a, 2017b). Table 15.2 shows that the mode differences vary to a degree across subscales, although less distinct than the selected response versus constructed response difference at grade 4 (Table 15.1). All subscales show mode differences significantly different from zero at the $\alpha = .05$ level.

Comparisons of item parameter estimates showed a similar pattern as the mean item score comparisons. Generally, item parameter estimates were very highly correlated between modes, and most digital items were estimated as more difficult than the paper counterparts ($ps < .05$). As the pattern of results from the item parameter comparison can be anticipated based on the results provided above, the item parameter comparison is not shown here (but see Jewsbury et al., 2020 for full details).

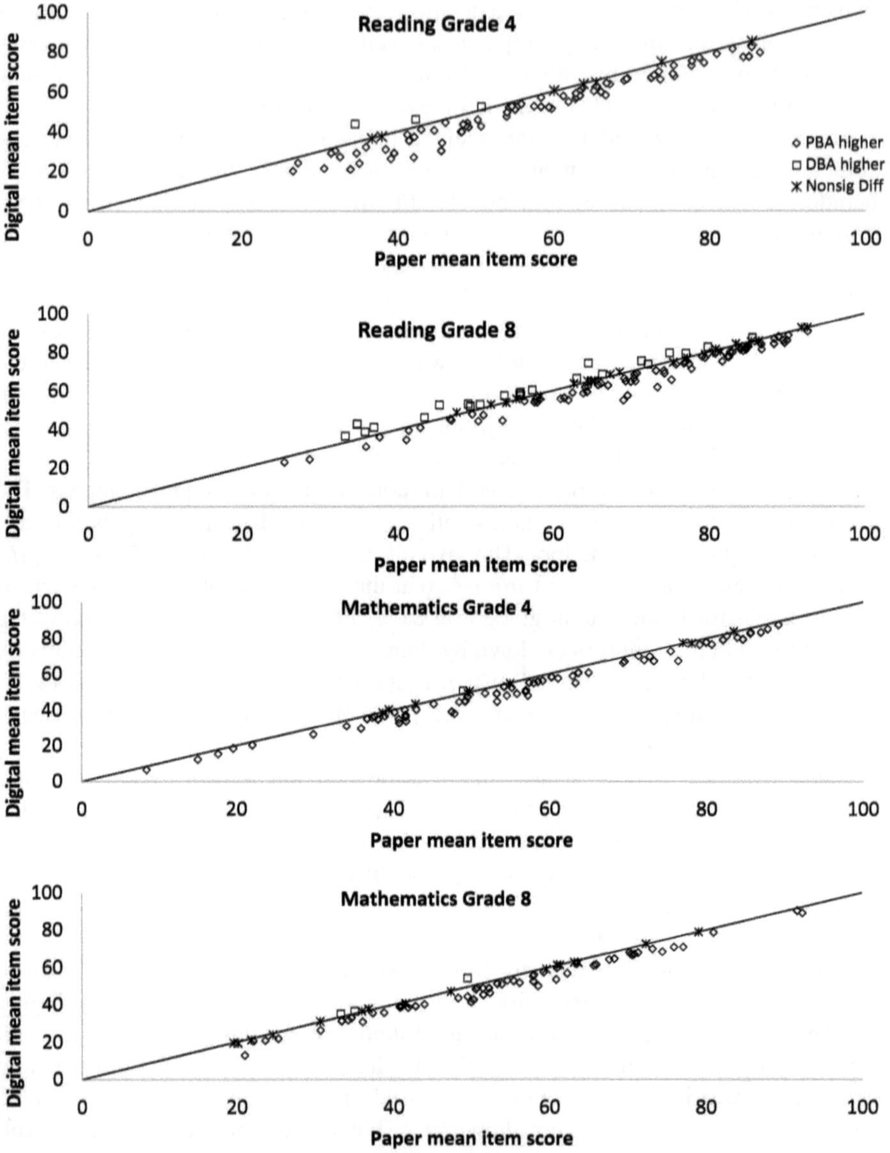

Fig. 15.1 Paper versus digital mean item scores on comparable items for each subject and grade combination

Table 15.1 Mean item score by item type for the paper and digital instruments on a comparable set of transadapted items

	Grade 4			Grade 8		
Subject	DBA	PBA	DBA-PBA (SE)	DBA	PBA	DBA-PBA (SE)
Reading						
Overall	47%	53%	−5.3% (.18)	61%	63%	−1.9% (.20)
Selected response	60%	64%	−3.8% (.22)	74%	76%	−1.6% (.19)
Constructed response	35%	42%	−6.8% (.18)	53%	55%	−2.0% (.24)
Mathematics						
Overall	52%	55%	−3.7% (.23)	49%	52%	−2.9% (.25)
Selected response	54%	56%	−2.4% (.24)	51%	53%	−2.5% (.26)
Constructed response	46%	52%	−6.9% (.31)	44%	48%	−3.5% (.30)

Note: DBA = digitally based assessment. PBA = paper-based assessment

Table 15.2 Mean item score by subscale for the paper and digital instruments on a comparable set of transadapted items: 2017

	Grade 4			Grade 8		
Subject	DBA	PBA	DBA-PBA (SE)	DBA	PBA	DBA-PBA (SE)
Reading						
Literary	48%	54%	−5.6% (.24)	62%	64%	−2.4% (.26)
Information	46%	51%	−5.0% (.23)	61%	62%	−1.3% (.25)
Mathematics						
Numbers & operations	52%	55%	−2.5% (.27)	56%	58%	−2.6% (.31)
Measurement	54%	60%	−5.9% (.29)	53%	55%	−2.1% (.39)
Geometry	52%	56%	−3.6% (.35)	41%	44%	−2.8% (.39)
Data analysis	39%	43%	−4.3% (.36)	35%	39%	−3.8% (.37)
Algebra	55%	60%	−4.6% (.30)	51%	54%	−2.8% (.26)

Note: DBA = digitally based assessment. PBA = paper-based assessment

15.2.2 Mode Impact on State/Jurisdiction and District Proficiency Estimates

Although common population linking ensures that the estimated national subscale means and standard deviations are equal between the two modes of assessment, other estimates are not guaranteed to be equal. We define the term *mode residual* as the difference between an estimate based on the digital assessment results and an estimate of the same parameter based on the paper assessment results. The use of the term mode residual was meant to clarify that residuals are differences calculated *after* applying a linking procedure that accounts for the overall greater difficulty of the digital assessment. Ideally, the greater difficulty of the digital assessment is

Table 15.3 Number of states/jurisdictions with significant mode residuals

Grade and subject	DBA > PBA	DBA < PBA	Nonsignificant
Grade 4 reading	3	1	48
Grade 8 reading	4	4	44
Grade 4 mathematics	3	2	47
Grade 8 mathematics	3	1	48

Note: Significance was defined as $\alpha < .05$ with no multiple comparison adjustment. DBA = scale score estimate based on the digitally based assessment. PBA = scale score estimate based on the paper-based assessment

Table 15.4 Number of districts with significant mode residuals

Grade and subject	DBA > PBA	DBA < PBA	Nonsignificant
Grade 4 reading	0	0	27
Grade 8 reading	0	3	24
Grade 4 mathematics	1	2	24
Grade 8 mathematics	0	0	27

Note: Significance was defined as $\alpha < .05$ with no multiple comparison adjustment. DBA = scale score estimate based on the digitally based assessment. PBA = scale score estimate based on the paper-based assessment

uniform for all groups of interest, resulting in few mode residuals significantly different from zero.

The large number of state/jurisdictions and districts imply an elevated false positive rate when testing each mode residual separately without correcting for multiple comparisons. The Benjamini-Hochberg procedure that controls the false discovery rate (FDR; Benjamini & Hochberg, 1995) was used, following standard NAEP procedures. The results with the FDR correction are more suited for evaluating whether any state/jurisdiction or district has significant mode residuals, while the results without the FDR correction are more suited to questions related to only one state/jurisdiction or district. The family size was all states/jurisdictions or all districts within the subject and grade combination (i.e., 52 for state/jurisdictions and 27 for districts).

Table 15.3 lists the number of state/jurisdictions with mode residuals significantly different from zero without correction for multiple comparisons and $\alpha = .05$. Without correcting for multiple comparisons, 8–15% of the state/jurisdiction mode residuals were significant within each subject and grade combination, for a total of 21 significant residuals. The results were mostly inconsistent across grade and subjects, with only two states/jurisdictions showing positive significant mode residuals in more than one grade and subject combination. With the FDR correction applied, only two states/jurisdictions for grade 8 reading had significant mode residuals (one positive and one negative).

Table 15.4 lists the number of districts with mode residuals significantly different from zero without correction for multiple comparisons. No district had a significant mode residual after applying the FDR correction. Without correcting for multiple

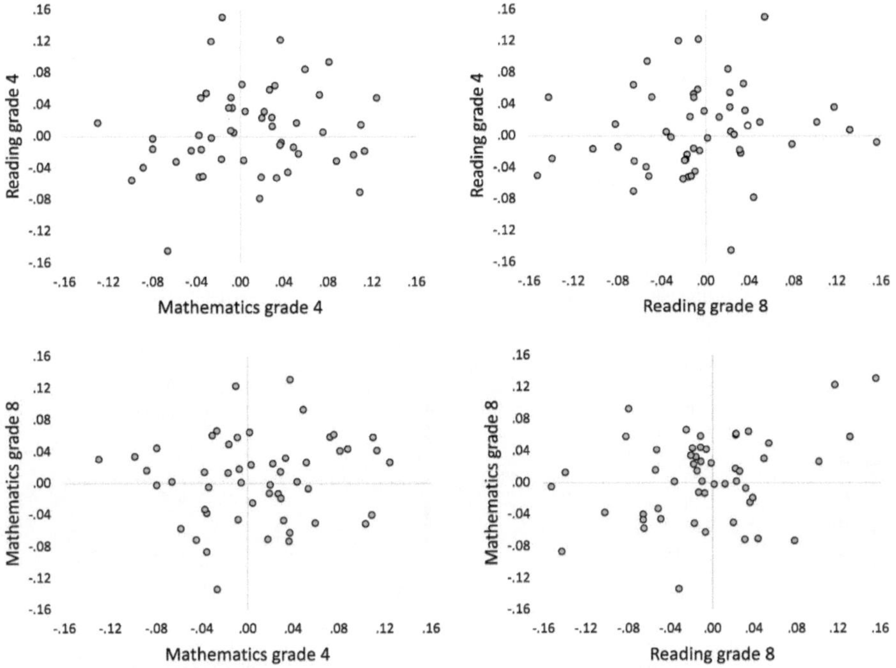

Fig. 15.2 Mode residual effect sizes for state/jurisdictions, across subject and across grade

comparisons, 0–11% of the district mode residuals were significant within each subject and grade combination, for a total of six significant residuals.

To examine the consistency of the state/jurisdiction and district mode residuals across subject and grade combinations, and to examine whether each particular state/jurisdiction or district was systematically advantaged or disadvantaged across subject and grades, scatterplots comparing the mode residuals across subject and grade combinations were generated. To standardize the analysis, mode residual effect sizes were calculated as the mode residual divided by the estimated standard deviation for the relevant state/jurisdiction or district. The estimated standard deviation was based on the digital-based assessment results because these were the reported results.

Figure 15.2 shows the scatterplots of mode residual effect sizes for state/jurisdictions. The correlations between mode residuals between subjects were .15 and .32 for grades 4 and 8, respectively. The correlations between mode residuals between grades were .12 and .10 for reading and mathematics, respectively. Only the correlation between grade 8 reading and mathematics was significantly different from zero ($p < .05$). Note that some artificial correlation is expected due to most schools in the sample participating in both reading and mathematics, resulting in spurious correlations.

Figure 15.3 shows the scatterplots of mode residual effect sizes for districts. The correlations between mode residuals between subjects were .47 and .05 for grades

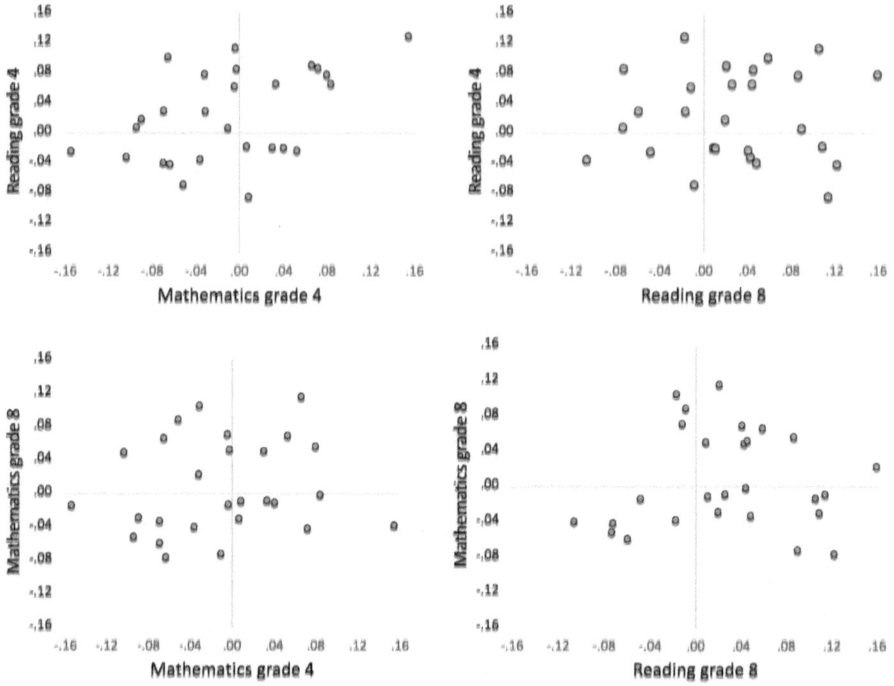

Fig. 15.3 Mode residual effect sizes for districts, across subject and across grade

4 and 8, respectively. The correlations between mode residuals between grades were −.02 and .13 for reading and mathematics, respectively. Only the correlation between grade 4 reading and mathematics was significantly different from zero ($p < .05$). The magnitude of the correlation is partially driven by an extreme case in the top right of the plot. Excluding the extreme case resulted in a correlation of .36.

15.2.3 Mode Impact on Subpopulation Proficiency Estimates

Like state/jurisdiction and district results, subgroup results are not guaranteed to be equivalent across modes of assessment by the common population linking procedure. Mode residuals for subpopulations were also examined to evaluate how well the scores from the two modes aligned.

One major concern was whether students were advantaged or disadvantaged depending on their ability, such as whether low performing students were disadvantaged by digital administration. Figure 15.4 shows Q-Q (quantile-quantile) plots of every percentile for the digital and paper scores. The results suggest that the scores between the two assessments were highly aligned after the common population linking.

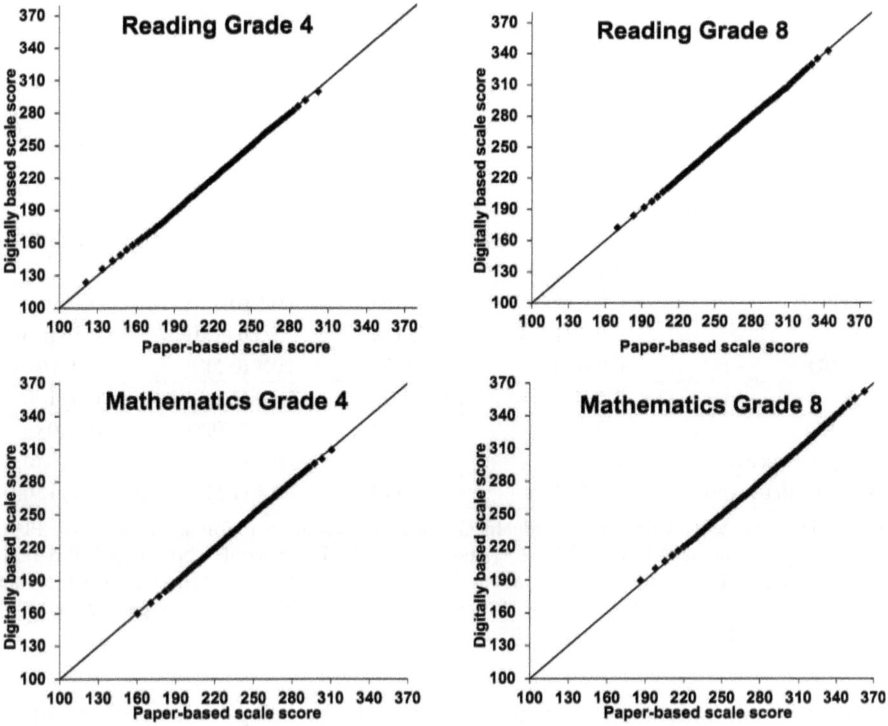

Fig. 15.4 Q-Q (quantile-quantile) plots of every percentile for the digital and paper scores

Like the state/jurisdiction and district analysis, above, the mode residuals of major reporting student groups were tested for significant differences from zero. Table 15.5 shows the comparison results of the mode residuals for major reporting groups and a computer access variable. The major reporting groups are defined by the five main contextual variables NAEP is federally mandated to measure, specifically, race/ethnicity, gender, socioeconomic status (National School Lunch Program eligibility, NSLP), disability status (SD), and English language learning status (ELL; No Child Left Behind Act of 2001, 2002). The computer access variable is a student self-reported variable on access to a computer at home. Although not a major reporting group, computer access was included as especially relevant to the mode study comparison.

Following standard NAEP procedures, the Benjamini-Hochberg False Discovery Rate (FDR) procedure was used, in which all categories within a variable form a family, as described in NAEP Technical Documentation on the Web (NCES, 2000–2018). Only two groups had significant results. Students with an Individualized Education Program or on a Section 504 Plan (SD) had a significant mode residual favoring higher performance on the grade 4 mathematics paper assessment by 3.6 points. Students reporting computer access at home had a significant mode

Table 15.5 Major reporting student groups and computer access mode residuals: 2017

Group	Reading		Mathematics	
	Grade 4	Grade 8	Grade 4	Grade 8
White	−0.9 (0.6)	−1.4 (0.6)	−0.1 (0.6)	−0.1 (0.6)
Black	0.2 (0.9)	1.5 (1.0)	−0.5 (0.9)	0.6 (1.0)
Hispanic	0.9 (0.9)	1.7 (0.9)	0.0 (0.8)	0.0 (1.1)
Male	0.5 (0.7)	0.7 (0.6)	−0.2 (0.6)	0.5 (0.7)
Female	−0.5 (0.6)	−0.7 (0.6)	0.2 (0.5)	−0.5 (0.7)
NSLP	0.1 (0.6)	1.1 (0.6)	0.0 (0.5)	0.0 (0.8)
No NSLP	0.2 (0.6)	−0.5 (0.6)	0.6 (0.6)	0.1 (0.7)
SD	−0.5 (1.5)	−0.5 (1.2)	−3.6 (1.0)*	1.8 (1.2)
Non-SD	0.0 (0.5)	0.1 (0.5)	0.4 (0.5)	−0.2 (0.6)
ELL	2.5 (1.5)	3.6 (1.8)	0.0 (1.0)	0.6 (1.8)
Non-ELL	−0.3 (0.5)	−0.3 (0.5)	−0.1 (0.5)	0.1 (0.6)
Computer access	1.1 (0.5)*	0.2 (0.5)	0.9 (0.5)	0.3 (0.6)
No computer access	−1.3 (2.1)	−0.1 (1.9)	0.4 (1.5)	2.5 (1.9)

Note: SD = students with an Individualized Education Program or on a Section 504 Plan. ELL = English language learner. NSLP = students eligible for National School Lunch Program. Students with no information available about their status in the National School Lunch Program were not included in either the NSLP or No NSLP categories. Standard errors in parentheses
*Significantly different from zero ($p < .05$)

residual favoring higher performance on the grade 4 reading digital assessment by 1.1 points.

15.3 Discussion

15.3.1 Impact on the Measurement Instrument

The results comparing item-level performance suggest three general conclusions. First, the digital versions of items were more difficult than the original paper versions, consistent with previous research (Bennett et al., 2008; Sandene et al., 2005; Xu & Jewsbury, 2017). This result supported common population linking, which allows for differences in item operation and difficulty between the two assessments, rather than assuming invariant item parameters (Yamamoto & Mazzeo, 1992).

Second, apart from being more difficult in digital format, items retained much of their original relative properties, such as the approximate rank-order of the item difficulties. In combination with other results in the mode comparison, the highly correlated item properties across modes provides evidence for construct similarity between the paper and digital assessments.

Third, there was no theoretically demarcated grouping of items without mode differences. Instead, almost all items had statistically significant mode differences.

Apart from item type at grade 4, there were no clear or predictable pattern suggesting that certain items were more or less affected by the mode transition. This informed the decision for NAEP not to assume that a subset of items are equivalent across modes. Instead, NAEP used common population linking without assuming measurement equivalence between modes on the item-level.

Note that by not modelling mode differences at the item-level, the linking procedure employed by NAEP ignores information about the pairings between the paper items and their corresponding transadapted digital versions. Thus, the approach may be less efficient than an approach that models the paper-digital item pairs. However, error variance introduced into NAEP group score estimators due to the common population linking procedure was a small contribution to the error variance of most estimators (Jewsbury, 2019; Mazzeo et al., 2025). Ultimately, the potential to introduce bias from inaccurately modelling the paper-digital item pairs was judged to overweigh the potential reduction of error variance that may result from modelling the item pairs, and the common population linking procedure was used.

15.3.2 Impact on State/Jurisdiction and District-Level Results

With a multiple comparison adjustment (FDR; Benjamini & Hochberg, 1995), only one of the four subject and grade combinations had significant mode residuals for state/jurisdictions (two, in opposite directions) and none had any significant mode residuals for districts. Specifically, one state/jurisdiction had a significant mode residual favoring digital performance and one state/jurisdiction had a significant mode residual favoring paper performance for reading grade 8.

When evaluating each jurisdiction individually, without multiple comparison adjustment, 8% of the 316 statistical tests were significant at the $\alpha = .05$ level. Of these, only two cases were significant in more than one subject and grade combination, with better digital scores at grade 8 for both subjects. Furthermore, the mode residual effect sizes were largely inconsistent across grade and across subject.

15.3.3 Impact on Group-Level Results

After common population linking, the digital and paper-based scores were highly aligned across the proficiency range, suggesting that low proficiency students were not disadvantaged differentially by the digital administration. Of the NAEP major reporting groups, only students with an Individualized Education Program or on a Section 504 Plan (SD) had a mode residual significantly different from zero and only for one subject and grade combination (mathematics grade 4). For students reporting access or no access to a computer at home, only students reporting access had a mode residual significantly different from zero, and only for one subject and grade combination (reading grade 4).

15.3.4 Summary of the Results

The mean item scores on transadapted digital items was significantly lower than their original paper administration counterparts, despite that both assessments were administered to randomly equivalent samples. This result suggested that the digitally administered NAEP instruments were more difficult than the original paper versions, replicating results of previous studies (Bennett et al., 2008; Sandene et al., 2005; Xu & Jewsbury, 2017). Although the reduced mean item score was largely consistent across items (see Fig. 15.1), there was still evidence that some items were more affected by the mode transition than other items (see Table 15.1). These results reinforce the importance of using a linking procedure that allows for differences in difficulties of the instruments when comparing results across assessments delivered in different modes.

After the common population linking procedure was applied, scores between the digital and paper assessments were very highly aligned. As the common population linking procedure does not force the ability distribution to be the same across digital and paper for all groups, expecting there to be no instances of groups more or less affected by the mode transition may be unreasonable. Nevertheless, few significant differences between the linked digital and paper scores were detected, mode residuals were largely inconsistent between subjects and between grades, and significant mode residuals were associated with small effect sizes.

The results had and will have relevance to other assessments transitioning to digitally based assessment which may have similar paper-digital differences (e.g., as found by the Trends in International Mathematics and Science Study; TIMSS; Fishbein et al., 2018). The observed differences in psychometric properties between the digital and paper instruments that were found across multiple devices and across several years in carefully designed mode studies for NAEP (Bennett et al., 2007, 2008; Jewsbury et al., 2020; Oranje & Moran, 2017; Sandene et al., 2005; Xu & Jewsbury, 2017) guided the design of NAEP digital transitions and the selection of the method to link assessments administered in different modes (for more discussion, see Jewsbury et al., 2020). In particular, the large uniformity of the mode differences enabled a psychometrically valid approach to linking the paper and digital assessments. However, even within NAEP, the mode differences were not consistent across different grades and different subjects, underscoring the importance of carefully designed bridge studies specific for each assessment transitioning to digital administration.

References

Beaton, A. E. (1990a). Introduction. In E. G. Johnson & R. Zwick (Eds.), *Focusing the new design: The NAEP 1988 technical report* (Vol. Report No. 19-TR-20, pp. 3–9). Educational Testing Service.

Beaton, A. E. (1990b). Introduction. In A. E. Beaton & R. Zwick (Eds.), *The effect of changes in the national assessment: Disentangling the NAEP 1985–86 reading anomaly* (Vol. Report No. 17-TR-21, pp. 1–13). Educational Testing Service.

Benjamini, Y., & Hochberg, Y. (1995). Controlling the false discovery rate: A practical and powerful approach to multiple testing. *Journal of the Royal Statistical Society, Series B, 57*, 289–300.

Bennett, R. E., Braswell, J., Oranje, A., Sandene, B., Kaplan, B., & Yan, F. (2008). Does it matter if I take my mathematics test on computer? A second empirical study of mode effects in NAEP. *Journal of Technology, Learning, and Assessment, 6*, 1–38.

Bennett, R. E., Persky, H., Weiss, A. R., & Jenkins, F. (2007). *Problem solving in technology-rich environments: A report from the NAEP technology-based assessment project (NCES 2007–466)*. National Center for Education Statistics, US Department of Education. Retrieved from http://nces.ed.gov/pubsearch/pubsinfo.asp?pubid=2007466

Bock, R. D., & Zimowski, M. F. (1997). Multiple group IRT. In W. J. van der Linden & R. K. Hambleton (Eds.), *Handbook of modern item response theory* (pp. 433–448). Springer.

Eignor, D. R. (2007). Linking scores derived under different modes of test administration. In N. J. Dorans, M. Pommerich, & P. W. Holland (Eds.), *Linking and aligning scores and scales* (pp. 135–159). Springer.

Fishbein, B., Martin, M. O., Mullis, I. V., & Foy, P. (2018). The TIMSS 2019 item equivalence study: Examining mode effects for computer-based assessment and implications for measuring trends. *Large-scale Assessments in Education, 6*(11), 1–23.

Jewsbury, P. A. (2019). *Error variance in common population linking bridge studies (Research Report No. RR-19-42)*. Educational Testing Service.

Jewsbury, P. A., Finnegan, R., Xi, N., Jia, Y., Rust, K. F., & Burg, S. (2020). *2017 NAEP transition to digitally-based assessment: Mode evaluation study (white paper)*. National Center for Education Statistics. https://nces.ed.gov/nationsreportcard/subject/publications/main2020/pdf/transitional_whitepaper.pdf

Jewsbury, P. A. (2024). Generally applicable variance estimation methods for common-population linking. *Journal of Educational and Behavioral Statistics*. OnlineFirst. https://doi.org/10.3102/10769986241263976

Johnson, E. G. (1992). The design of the National Assessment of educational Progress. *Journal of Educational Measurement, 29*, 95–110.

Johnson, E. G., & Rust, K. F. (1992). Population inference and variance estimation for NAEP. *Journal of Educational Statistics, 17*, 175–190.

Johnson, E. G., & Zwick, R. (1990). *Focusing the new design: The NAEP 1988 technical report* (Vol. Report No. 19-TR-20, pp. 3–9). Educational Testing Service.

Mazzeo, J., Liu, B., Donoghue, J., & Xu, X. (2025). Approximate standard errors for NAEP results that incorporate linking error under the random group design. In L. Khorramdel, M. von Davier, & K. Yamamoto (Eds.), *Innovative digital-based international large-scale assessments—Foundations, methodologies and quality assurance procedures*. Springer.

Mislevy, R. J. (1984). Estimating latent distributions. *Psychometrika, 49*, 359–381.

Mislevy, R. J. (1985). Estimation of latent group effects. *Journal of the American Statistical Association, 80*, 993–997.

Mislevy, R. J. (1991). Randomization-based inference about latent variables from complex samples. *Psychometrika, 56*(2), 177–196.

Mislevy, R. J., Johnson, E. G., & Muraki, E. (1992). Chapter 3: Scaling procedures in NAEP. *Journal of Educational Statistics, 17*, 131–154.

Muraki, E. (1992). A generalized partial credit model: Application of an EM algorithm. *Applied Psychological Measurement, 16*, 159–176.

National Assessment Governing Board. (2017a). *Mathematics framework for the 2017 National Assessment of Educational Progress*. https://www.nagb.gov/content/nagb/assets/documents/publications/frameworks/mathematics/2017-math-framework.pdf

National Assessment Governing Board. (2017b). *Reading framework for the 2017 National Assessment of Educational Progress*. https://www.nagb.gov/content/nagb/assets/documents/publications/frameworks/reading/2017-reading-framework.pdf

National Center for Education Statistics. (2000–2018). *NAEP technical documentation on the web*. https://nces.ed.gov/nationsreportcard/tdw/

National Center for Education Statistics. (2009). *The nation's report card: Reading 2009* (NCES 2010–458). Institute of Education Sciences, U.S. Department of Education, Washington, DC. https://nces.ed.gov/nationsreportcard/pdf/main2009/2010458.pdf.

National Center for Education Statistics. (2018a). *Administration of computer-based state assessments in English language arts (ELA) and mathematics in grades 4 and 8*. National Center for Education Statistics. https://nces.ed.gov/programs/statereform/tab2_22.asp

National Center for Education Statistics. (2018b). *Students' computer access and use*. National Center for Education Statistics. https://www.nationsreportcard.gov/sq_computer/

National Center for Education Statistics. (n.d.). NAEP: Digitally based assessments. https://nces.ed.gov/nationsreportcard/dba/

No Child Left Behind Act of 2001. (2002). Pub. L. No. 107–110, 115 Stat. 1425.

Oranje A., & Moran, R. (2017, April). Design principles and analysis methodology for mode transitions of group score assessments. Paper presented at the meeting of the National Council on Measurement in Education, San Antonio, TX.

Pearson. (2015). Student mobile device survey 2015, National report: Students in grades 4–12. https://www.pearsoned.com/wp-content/uploads/2015-Pearson-Student-Mobile-Device-Survey-Grades-4-12.pdf

Perie, M., Moran, R., & Lutkus, A. D. (2005). *NAEP 2004 trends in academic progress: Three decades of student performance in reading and mathematics (NCES 2005–464)*. Government Printing Office.

Sandene, B., Horkay, N., Bennett, R. E., Allen, N., Braswell, J., Kaplan, B., & Oranje, A. (2005). *Online assessment in mathematics and writing: Reports from the NAEP technology-based assessment project (NCES 2005–457)*. National Center for Education Statistics, U.-S. Department of Education. http://nces.ed.gov/pubsearch/pubsinfo.asp?pubid=2005457

SRI International. (2018). An overview of current digital curricula and instructional tools in mathematics, reading, and writing (Unpublished commissioned paper). Prepared by Everson, H., Kim, H. J., Yarnall, L., & Ziker, C. Under contract ED-IES-13-C-0017. SRI International.

von Davier, M., Sinharay, S., Oranje, A., & Beaton, A. (2006). The statistical procedures used in National Assessment of Educational Progress: Recent developments and future directions. In C. R. Rao & S. Sinharay (Eds.), *Handbook of statistics* (*Psychometrics*) (Vol. 26, pp. 1039–1055). Elsevier.

Xu, X., & Jewsbury, P. A. (2017, April). Comparison of trend calibration approaches across modes with (un)common items and random samples. Paper presented at the meeting of the National Council on Measurement in Education, San Antonio, TX.

Yamamoto, K., & Mazzeo, J. (1992). Item response theory scale linking in NAEP. *Journal of Educational Statistics, 17*, 155–173.

Zwick, R. (1991). Effects of item order and context on estimation of NAEP reading proficiency. *Educational Measurement: Issues and Practice, 10*(3), 10–16.

Zwick, R. (1992). Chapter 7: Statistical and psychometric issues in the measurement of educational achievement trends: Examples from the National Assessment of Educational Progress. *Journal of Educational Statistics, 17*, 205–218.

Chapter 16
Approximate Standard Errors for NAEP Results That Incorporate Linking Error Under the Random Group Design

John Mazzeo, Bingchen Liu, John Donoghue, and Xueli Xu

Abstract An important methodological challenge for survey assessments is how to manage the requirements to evolve and innovate with respect to assessment content, test delivery, and analysis methods while maintaining the core purpose of providing valid trend results. In NAEP, one way this challenge has been addressed over the years is through the conducting of special linking studies that use a random groups design (Kolen, 2007, Data collection designs and linking procedures. In N. J. Dorans, M. Pommerich, & P. W. Holland (Eds.), *Linking and aligning scores and scales* (pp. 31–55). Springer). In this chapter, we describe an operationally feasible approximation of NAEP standard errors that incorporates an estimate of the linking error. We present the method developed and used in the reporting of the 2017 NAEP results. We provide an illustration of the method using the 2015 and 2017 NAEP results, including a comparison to an exact formula-based method appropriate for group means. Lastly, we describe ongoing theoretical and practical work that builds on and improves the methods used in 2017.

An important methodological challenge for survey assessments—like the Programme for International Student Assessment (PISA) and the United States National Assessment of Educational Progress (NAEP)—is how to manage the requirements to evolve and innovate with respect to assessment content, test delivery, and analysis methods while maintaining the core purpose of providing valid trend results. In NAEP, one way this challenge has been addressed over the years is through the conducting of special studies designed to support the linking of

J. Mazzeo (✉) · B. Liu · J. Donoghue · X. Xu
Educational Testing Service, Princeton, NJ, USA
e-mail: mazzeo123@comcast.net; bingchen_liu@outlook.com; jdonoghue@ets.com; xxu@ets.com

L. Khorramdel et al. (eds.), *Innovative Digital-Based International Large-Scale Assessments*, Methodology of Educational Measurement and Assessment,
https://doi.org/10.1007/978-3-031-90951-1_16

results from more current revised versions of the assessment to the existing trend lines.[1,2]

Most NAEP linking studies have been carried out using a random group design (Kolen, 2007) in which distinct randomly equivalent samples of test takers from a common population are administered one of the two assessments to be linked. One sample of test takers is administered the assessment under the original administration design, while a second sample is administered the revised version of the assessment. Data from the two samples are analyzed separately, and the score distributions from the two samples—expressed in terms of an item response theory (IRT; Bock & Moustaki, 2007) latent variable scale—are aligned. The alignment is typically achieved by calculating a linear linking function that equates the means and standard deviations from the two samples. The *linking function* is then used to place results obtained under the revised version of the assessment on scales comparable to those used to report results for the original administration. We refer to the situation in which the results being reported come from a sample that was not involved in the estimation of the linking function as the *external case*. We refer to the situation in which the sample whose results are being reported are used to estimate the linking function as the *internal case*.

Over the last four decades, the NAEP program has made extensive use of random group designs for linking studies. Such studies were first used in NAEP to study the so-called 1985–1986 NAEP reading anomaly (Beaton & Zwick, 1992). Analogous approaches were used to link results between NAEP assessments that employed different approaches to including special-needs test takers and English language learners (Lutkus & Mazzeo, 2003; Mazzeo et al., 2000). Random group designs were used from 1990 through 2003 to link the results from separately conducted national and state-by-state assessments (Yamamoto & Mazzeo, 1992) and to link NAEP results to international assessments like TIMSS (Jia et al., 2014; Johnson et al., 2005; Phillips, 2007). In 2017, the NAEP program again used a random group study design to transition two of its most important assessments—reading and mathematics—from paper-and-pencil assessments (PBA) to administration on tablets (digitally based assessments, or DBA). The design of the 2017 assessments and the results of the linking study are described in Jewsbury et al. (2024) of this volume.

[1]This chapter assumes a basic familiarity with the design and analysis methodologies of group-score assessments like NAEP, PISA, and Trends in International Mathematics and Science Study (TIMSS). See Beaton and Johnson (1992), Beaton and Zwick (1992), Mazzeo (2018), and von Davier et al. (2007) for overviews.

[2]In typical testing situations, where individual scores for test takers are generated and reported, linking of scores occurs between individual forms of each assessment. For survey assessments like NAEP and PISA, each assessment consists of a collection of test forms that are typically randomly assigned to test takers, and group-level results, rather than test-taker scores, are reported. Despite these differences, many of the concepts and methods associated with test-score linking can be readily adapted to the task of linking results between different versions of an assessment by considering the collection of forms that make up each version as an individual form.

NAEP, like other large-scale survey assessments, accompanies the reporting of results (e.g., means, standard deviations, percentages above achievement levels, and percentiles) with measures of uncertainty—standard errors—that reflect its complex survey design. The estimation of the linking functions from random group designs is also based on sample data, and its use changes the nature and magnitude of uncertainty associated with the reported results. Ideally this should be accounted for when comparisons are made from results obtained under revised procedures to results obtained under the original assessment format. When reporting results from NAEP-TIMSS linking studies, efforts have been made to incorporate various kinds of linking errors (Jia et al., 2014; Johnson, 1998; Johnson et al., 2005; Phillips, 2007). However, until recently, NAEP had not developed practical approaches to quantifying and reflecting this additional source of uncertainty when reporting its standard operational results.

In this chapter, we describe an operationally feasible approximation of NAEP standard errors that incorporates an estimate of the linking error associated with *the random group linking design*. The approach described here was used in reporting the 2017 mathematics and reading results (Jewsbury et al., 2024). As will become evident, the approximation is a resampling-based approach, a variant of the jackknife (Quenouille, 1956; Tukey, 1958; Miller, 1964), and is built around extending current operational procedures. Earlier related efforts to use resampling-based methods to estimate and incorporate linking error for *common-item linking designs* include work by Cohen et al. (2001), Haberman et al. (2009), Xu and von Davier (2010), and Wang et al. (2018). Work by Hsieh et al. (2009), which looked at extending the current NAEP jackknife approach to more comprehensively account for additional sources of variance beyond those captured in the current NAEP approach, also made use or resampling-based methods.

The chapter is organized as follows. We first review some basic information about the calculation of standard errors in NAEP. Similar approaches are used in other large-scale survey assessments, such as PISA. We then introduce additional terminology and notation to discuss the linking studies and the uncertainty that use of their results adds. Next, we present the method developed. We provide an illustration of the method using the 2015 and 2017 NAEP reading results, including a comparison to an exact formula-based method appropriate for group means. Lastly, we describe ongoing theoretical and practical work that builds on and improves the methods used in 2017.

16.1 The Basic Standard Error Estimation (BSEE) Approach in Survey Assessments

Estimation of the error variance for a typical survey assessment statistic (e.g., a NAEP scale score mean, standard deviation, percentile, or achievement-level percentage for national reporting subgroups, states, and districts) must account for the facts that: (a) the test-taker samples are multistage probability samples of schools and students, and (b) the direct estimation methods, based on IRT (Bock & Moustaki, 2007), are derived directly from test-taker item responses through the use of multivariate latent variable regression models (Mislevy, 1991). Thus, for any scale score statistic, estimated error variances are generated for two distinct components: (a) school/student sampling design and (b) latency of the dependent variable (i.e., scale scores).

Because NAEP makes use of complex multistage sampling designs and employs various adjustments for school- and student-level nonresponse and group-level statistics, the results of interest are calculated using case weights that appropriately reflect the designs and nonresponse adjustments. We will denote the set of weights used to calculate results for the full sample as $\boldsymbol{W_F}$. To provide estimates of the sampling error of these results, NAEP employs a jackknife-based approach to estimate a component of variance due to the complex sampling design and nonresponse adjustments (Johnson & Rust, 1992). For each sample, a set of N_r replicate weights are produced. Each set of replicate weights involves deleting a portion of the data set and reweighting the remaining data so that the full replicate data set remains representative of the total jurisdiction to which it refers. This is accomplished in practice by setting the case weights for the units within a primary sampling unit[3] (PSU) equal to 0 and upweighting the data in a matched demographically similar PSU. Denote the multiple sets of replicate weights as $\boldsymbol{W} = (\boldsymbol{W_1}, \ldots, \boldsymbol{W_r}, \ldots, \boldsymbol{W_{N_r}})$, $r = 1, 2, \ldots N_r$.

Estimates of group-level scale-score results are approximated through the use of plausible values (PVs; Mislevy & Sheehan, 1987). PVs are random draws from the distribution implied by a latent variable model used in NAEP.[4] More specifically, the set of PVs[5] that is generated for each test taker is a random draw from a (possibly multivariate) normal distribution[6] conditional on estimated IRT item parameters, estimates of linear latent regression model parameters, test takers' responses to assessment questions, and a full array of demographic and contextual variables (von Davier et al., 2007; Mazzeo, 2018) used as predictors in the linear latent regression. For each test taker, a set of N_p PVs are drawn. We will denote these sets of PVs for the sample as $\tilde{\boldsymbol{\theta}} = \left(\tilde{\boldsymbol{\theta}}_1, \tilde{\boldsymbol{\theta}}_2, \ldots, \tilde{\boldsymbol{\theta}}_p, \ldots, \tilde{\boldsymbol{\theta}}_{N_p}\right)$. Note that pairing the sets of weights (full sample and replicate) with the PVs effectively induces a series of $(N_r + 1) \times N_p$ replicate data sets. Let $t_p \equiv t\left(\tilde{\boldsymbol{\theta}}_p, \boldsymbol{W_F}\right)$ denote a statistic of interest (e.g., mean, standard deviation) calculated using the pth set of PVs and a set of *full sample weights*. The full sample estimate of the statistic of interest is $t_F = \frac{\sum_{p=1}^{N_p} t_p}{N_p}$.

[3]PSUs in NAEP are typically individual schools or, in some cases, geographical areas.

[4]See, for example, von Davier et al. (2007) or Mazzeo (2018) for a description of the full NAEP latent variable model.

[5]PVs are described further in ensuing sections.

[6]For ease of explication, the discussion in this section assumes a univariate latent variable, but the basic approach generalizes in straightforward fashion to the multivariate case.

Let $t_{rp} = t(\tilde{\boldsymbol{\theta}}_p, \boldsymbol{W}_r)$ refer to that same statistic of interest but calculated with the rth set of replicate weights, $\boldsymbol{W}_r$, and $\tilde{\boldsymbol{\theta}}_p$. Error variance due to sampling is estimated with

$$\widehat{V}_{smp}(t_F) = \sum\nolimits_{r=1}^{N_r} \frac{1}{N_p} \sum\nolimits_{p=1}^{N_p} \left(t_{rp} - t_p\right)^2. \tag{16.1}$$

To keep matters computationally feasible, NAEP estimates the sampling component based only on the first PV. So, in practice

$$\widehat{V}_{smp}(t_F) \approx \sum\nolimits_{r=1}^{N_r} \left(t_{r1} - t_1\right)^2. \tag{16.2}$$

Adapting the work of Rubin (1987) to the context of latent variable models, error variance due to latency is

$$\widehat{V}_{lat}(t_F) = \frac{\sum\nolimits_{p=1}^{N_p} \left(t_p - t_F\right)^2}{N_p - 1}, \tag{16.3}$$

and the basic standard error estimation (BSEE) estimate of the total error variance for statistic t_F becomes

$$\widehat{V}(t_F) = \widehat{V}_{smp}(t_F) + \left(\frac{N_p + 1}{N_p}\right) \widehat{V}_{lat}(t_F). \tag{16.4}$$

The approach generalizes in a straightforward fashion to the covariance between two different statistics— t and t' calculated on this same sample. The sampling component of the covariance can be estimated with

$$\widehat{C}_{smp}\left(t_F, t'_F\right) = \sum\nolimits_{r=1}^{N_r} \frac{1}{N_p} \sum\nolimits_{p=1}^{N_p} \left(t_{rp} - t_p\right)\left(t'_{rp} - t'_p\right). \tag{16.5}$$

Similarly, the latency component can be estimated with

$$\widehat{C}_{lat}\left(t_F, t'_F\right) = \frac{\sum\limits_{p=1}^{N_p} \left(t_p - t_F\right)\left(t'_p - t'_F\right)}{N_p - 1}, \tag{16.6}$$

and the total estimate of the covariance is estimated with

$$\widehat{C}\left(t_F, t'_F\right) = \widehat{C}_{smp}\left(t_F, t'_F\right) + \left(\frac{N_p + 1}{N_p}\right) \widehat{C}_{lat}\left(t_F, t'_F\right). \tag{16.7}$$

16.2 Expanding the BSEE Approach to Incorporate Random Group Linking Error

When linking studies are carried out under the random group design, the estimation of error becomes more complicated. To make matters more concrete, let $\overline{\theta}_{Rev}$ and s^2_{Rev} denote the mean and variance of the PVs for the linking study sample taking the revised assessment, calculated using the appropriate full sample weights as described in the previous section. Similarly, let $\overline{Y}_{Orig}$ and S^2_{Orig} denote the mean and variance of the PVs of the linking study sample taking the original version of the assessment. Let $\overline{Y}_{Rev}$ refer to mean of the sample taking the revised versions of the assessment, after transformation to the scale on which results of the original assessment are reported, and let $S^2_{\mathrm{Re}\,v}$ refer to corresponding variance after transformation. Transformation constants are obtained by constraining the means and variances of the linking study samples to be equal after transformation, i.e., $\overline{Y}_{Rev} = \overline{Y}_{Orig}$ and $S^2_{Rev} = S^2_{Orig}$. The definitions of the transformation constants implied by these equality constraints are $\widehat{a} = \frac{S_{Orig}}{s_{Rev}}$ and $\widehat{b} = \overline{Y}_{Orig} - \frac{S_{Orig}}{s_{Rev}}\overline{\theta}_{Rev}$.

The estimated linking coefficients $(\widehat{a}, \widehat{b})$ are used to transform the data from a sample that has been administered the revised assessment to a metric that is comparable to that of the original assessments. In the situation where the transformation coefficients are arbitrarily chosen constants, they can simply be applied to the sample data, $\tilde{\boldsymbol{\theta}}$, and the procedures of the previous section apply directly. But in the situation at hand, the transformation constants are realized values of the estimators $\left(\widehat{a}, \widehat{b}\right)$. Hence, procedures for estimating standard errors should be modified as the statistics of interest now become functions not only of the sample data and weights, but of the coefficient estimators:

$$t_F \equiv t\left(\tilde{\boldsymbol{\theta}}, \mathbf{W}_F, \widehat{a}, \widehat{b}\right). \tag{16.8}$$

For the 2017 assessment, we developed an approximation of these standard errors built on the external case assumptions. The external case (i.e., the case where the sample for which results are being reported was not involved in the linking study) is relevant for at least two reasons. First, the external case describes the situation when linking coefficients are part of the process for expressing the results of future assessments on existing scales. For example, in the context of NAEP, the estimated linking function in the DBA-to-PBA mode linking work described by Jewsbury et al. (2024) will be used for future DBA assessments (2019 and beyond) because these assessments, in turn, will be linked to the 2017 DBA results through the usual common item linking carried out by NAEP. Second, the external case can be used as a conservative approximation to the approach required for the internal case in the sense that it will often provide overestimates of the linking error. Hence statistical tests of small differences between DBA and PBA results that might be due to the change in assessment methodology will be less likely to be declared significant, albeit with a corresponding loss in power to detect some "true" construct-relevant changes.

Our approach was built on the simplifying *assumption* that the correlation between the linking error and other sources of error (sampling and latency) would be sufficiently small to allow for a practical and reasonably accurate approximation of the exact standard error. The motivation for this assumption is that the principle source of linking errors are the errors in estimating the transformation constants, which, in the external case, does not involve data from the samples to which it is being applied. In ensuing sections, results from the approximation are also compared to those obtained with an exact formula that does not make that assumption. The discussion section of the paper contains additional material on the implications of the assumption for the development of improved methods.

With the assumption of independent linking error, an approximation could be obtained by adding an additional component of variance to the usual BSEE estimate of the *transformed* statistic (i.e., the statistic that results from treating the linking constant as fixed, rather than random). The additional component is intended to reflect the uncertainty associated with estimating the linking coefficients, which we denote as $\widehat{V}_{link}(t_F)$. Thus,

$$\widehat{V}(t_F) \approx \left[\widehat{V}_{smp}(t_F) + \frac{(N_p+1)}{N_p}\widehat{V}_{lat}(t_F)\right] + \widehat{V}_{link}(t_F). \tag{16.9}$$

16.2.1 A Monte Carlo Estimator of Linking Variance

As described by Jewsbury et al. (2024), the NAEP program typically reports results within each subject area on several distinct IRT-based subscales, as well for an overall composite scale that is defined as a weighted average of the subscale results. For example, NAEP reading results are reported on two subscales—literary reading and informational reading, as well as an overall composite reading scale. For each of the subscales and the composite scale, group-level results and associated standard errors are reported for average scores, percentiles (10th, 25th, 50th, 75th, and 90th percentile), and standard deviations. Moreover, NAEP also reports on the percentages of test takers at or exceeding three predetermined cut scores on the composite scale. These percentages are referred to as *achievement level percentages*.

The previous work by Johnson (1998) developed formulas based on Taylor series approximations for a linking error component for group means. The approach we used to approximate the linking error component was developed to be more generally applicable to the full range of statistics that the program reports and to be applicable for results at both the subscale and composite scale levels. To simplify the presentation, we first describe the approach as it would be applied at the NAEP subscale level for any statistic of interest. We subsequently discuss how the procedure was then generalized to be appropriate to composite scale results.

16.2.1.1 Estimating Linking Error for NAEP Subscale Results

Let $\boldsymbol{\theta}_g$ denote the set of PVs available from the analysis for group g in the original DBA IRT metric. Let

$$\tilde{Y}_g = t\left(\boldsymbol{\theta}_g, \widehat{a}, \widehat{b}\right), \tag{16.10}$$

be any statistic of interest calculated on the PVs after applying the estimated linear transformation. The idea behind the Monte Carlo (MC) approach is to attempt to estimate the linking error variance component by repeatedly replacing the point estimates of the linking coefficients with series of N_v draws from their estimated sampling distribution, recalculating the statistic of interest with each draw, and then calculating the variance among this set of statistics. More specifically, the method proceeds as follows.

Let $\widehat{\boldsymbol{\Sigma}} = \begin{pmatrix} \widehat{\sigma}^2_{\widehat{a}} & \widehat{\sigma}_{\widehat{a},\widehat{b}} \\ \widehat{\sigma}_{\widehat{a},\widehat{b}} & \widehat{\sigma}^2_{\widehat{b}} \end{pmatrix}$ and $\widehat{\boldsymbol{a}} = \left(\widehat{a}, \widehat{b}\right)$ and let $\tilde{\mathbf{a}}$ and $\tilde{\boldsymbol{\Sigma}}$ denote realized sample values of these estimators. We make the simplifying assumption that the distribution of $\widehat{\boldsymbol{a}}$ is bivariate normal (BVN).[7] Using the realized sample estimates, we take N_v random draws from a $BVN\left(\tilde{\mathbf{a}}, \tilde{\boldsymbol{\Sigma}}\right)$ distribution, which we denote as $\widehat{\boldsymbol{a}}_v = \left(\widehat{a}_v, \widehat{b}_v\right)$. These random draws are used in turn to create N_v replicates of the statistic of interest, which we denote as $\tilde{Y}_{gv} = t\left(\boldsymbol{\theta}_g, \widehat{a}_v, \widehat{b}_v\right)$. We estimate the linking variance as

$$\widehat{V}_{link,mc}\left(\tilde{Y}_g\right) = \frac{1}{N_v - 1} \sum_{v=1}^{N_v} \left(\tilde{Y}_{gv} - \frac{1}{N_v} \sum_{v=1}^{N_v} \tilde{Y}_{gv}\right)^2, \tag{16.11}$$

and obtain an estimate of total variance as

[7] The rationale for the normal distribution assumption was motivated by the literature on linear regression (see, e.g., Searle, 1971, p. 99) with normally distributed errors. As shown in Sect. 16.2.2, the results of the Monte Carlo estimator for means agree quite well with those from formulas derived without reference to normality assumptions (see Eq. 16.25). However, the robustness of the agreement between the Monte Carlo and formula-based results to violations of the normality assumption is an area that would benefit from additional study.

$$\widehat{V}(\tilde{Y}_g) = \widehat{V}_{smp}(\tilde{Y}_g) + \frac{(N_p + 1)}{N_p}\widehat{V}_{lat}(\tilde{Y}_g) + \widehat{V}_{link}(\tilde{Y}_g). \tag{16.12}$$

Similarly, the proficiency estimation phase of the NAEP analysis involves the production of multivariate PVs (20 sets of two-dimensional PVs for reading and five-dimensional PVs for mathematics).

16.2.1.2 Estimating Linking Error for NAEP Composite Scale Results

Although NAEP analysis and linking is rooted in subscales, the most visible NAEP results are the composite scale results and achievement level percentages that are typically reported. So, it was imperative that we generalize the linking error for composite scale results and achievement level percentages, as well as subscale results.

NAEP composite scale PVs are weighted averages of the subscale PVs (National Center for Education Statistics, 2017). Therefore, there are 20 sets of composite PVs associated with each of the sets of subscale results. The achievement levels are the percentages of test takers with composite scale PVs at or above particular values. Therefore, there is a set of 20 estimates of the achievement level percentages—one for each set of composite scale values. Because of this structure, it was quite straightforward to generalize the Monte Carlo procedure to the composite score scale.

The need to calculate the composite linking error required that we adjust the estimation of error variance for the linking coefficients to include additional covariance terms, all of which, as described in the next section, are obtainable through a generalized version of the BSEE procedure. Let $\left(\widehat{a}_s, \widehat{b}_s\right)$ refer to the linking coefficients for subscale s, and N_s refer to the number of subscales (two in the case of reading, five in the case of mathematics). To prepare for the composite case, we need to redefine the estimated error covariance matrix to include covariance terms like $\widehat{\sigma}_{\widehat{a_s a_{s'}}}$, $\widehat{\sigma}_{\widehat{b_s b_{s'}}}$, and $\widehat{\sigma}_{\widehat{a_s b_{s'}}}$. The linking coefficient vector expands to length $2N_s$ and becomes $\widehat{\boldsymbol{a}}^* = \left\{\left(\widehat{a}_1, \widehat{b}_2\right) \ldots \left(\widehat{a}_{N_s}, \widehat{b}_{N_s}\right)\right\}$. In analogous fashion, the estimated error covariance matrix expands from a 2-by-2 matrix to a $2N_s$-by-$2N_s$ matrix and becomes

$$\widehat{\Sigma}^* \equiv \begin{pmatrix} \hat{\sigma}^2_{\hat{a}_1} & \cdots & \hat{\sigma}_{\hat{a}_1 \hat{b}_{N_S}} \\ \vdots & \ddots & \vdots \\ \hat{\sigma}_{\hat{b}_{N_S} \hat{a}_1} & \cdots & \hat{\sigma}_{\hat{b}_{N_s}} \end{pmatrix}. \tag{16.13}$$

The generalization of the Monte Carlo procedure is then straightforward. When the statistics relate to the composite scale, it involves the DBA-to-PBA linking coefficients for all the scales. When calculating the N_v replicates, $\tilde{Y}_{gv}$ it is necessary to draw samples of $\widehat{\boldsymbol{a}}_v = \left\{ \left(\widehat{a}_{1v}, \widehat{b}_{2v}\right) \ldots \left(\widehat{a}_{N_s v}, \widehat{b}_{N_s v}\right) \right\}$ from a multivariate normal distribution with mean $\widetilde{\mathbf{a}}_{\mathbf{c}}$ and variance $\widetilde{\boldsymbol{\Sigma}}^*$. The remainder of the procedure operates as previously defined.

16.2.1.3 Applying the BSEE Procedure to Estimate Error of the Linking Coefficients

The procedure described in the previous section requires that we have estimates of the error variances and covariances of the linking coefficients. In NAEP, such estimates are typically obtained in bridge studies that employ the same kinds of complex matrix sample designs used to for main reporting purposes. Because they are functions of sample means and standard deviations, the variance of the linking coefficient estimates, as well as the covariance between them, can be obtained using a variation of the same general BSEE approach employed by the program (Johnson & Rust, 1992) to estimate variance errors for means and standard deviations (i.e., the jackknife approach for estimating the impact of sampling error plus an additional between-imputations component due to the latent variable nature of the NAEP reporting scales).

In the typical NAEP situation, the BSEE is applied to the results from a single sample.[8] However, the estimators of the linking coefficients, $\widehat{a}$ and $\widehat{b}$, are functions of the sample means and variances of both samples in the linking study, each of which has its own set of replicate data sets and sets of PVs. How should the BSEE procedure be adapted to this situation? The principal issue to be addressed is how to pair the replicate data sets and multiple sets of PVs from the two samples to obtain the appropriate jackknife-based estimates. Given these, Eqs. 16.4 and 16.7 can be used to obtain the estimated error variances for each of the various linear transformation coefficients and their covariances.

First, consider the situation in which the linking study samples have been selected independently of one another. The NAEP 2015 assessment design—which was carried out as a trial in preparation for the operational linking conducted in 2017—is an example of this first situation. For both grades 4 and 8, the NAEP program conducted both a full PBA assessment in one sample of schools and a smaller DBA assessment in a different, independently selected, set of schools. At each grade, the two samples were both representative of probability samples from a common population. However, the sampling was such that the original (PBA) and revised (DBA) assessments were conducted in mutually exclusive sets of schools.

Beginning first with the sampling component for $\widehat{a}$ or $\widehat{b}$—because the school samples are completely distinct—any pairing of replicate data sets should be acceptable. Let $\boldsymbol{W}$ denote the N_r sets of replicate weights for the PBA sample and $\boldsymbol{W}'$ denote the N'_r sets of replicate weights for the DBA sample. Denote the number of these

[8] For comparisons, such as t-tests, the means and variances arise from a single sample.

pairings as N_z. When the number of sets of replicate weights are not equal, and $N_r \geq N'_r$, then there are $N_z = \binom{N_r}{N'_r} N'_r!$ possible pairings, each of which contains two sets of replicate weights—one to be used for calculating the DBA statistics of interest, the other to be used for calculating the PBA statistic. We denote the potential pair replicates as $\mathbf{Z}_z$, where z = 1, 2...N_z. Any pair should be acceptable to use in obtaining an estimate of the sampling component of variance. Using the a coefficient as an example, we denote these acceptable estimators as

$$\widehat{V}_{smp}(\widehat{a}|Z_z) = \sum_{r=1}^{N_r} \frac{1}{N_p} \sum_{p=1}^{N_p} (\widehat{a}_{rp} - \widehat{a}_p)^2. \tag{16.14}$$

Ideally, one would like to take

$$\widehat{V}_{smp}(\widehat{a}) = \frac{1}{N_z} \sum_{z=1}^{N_z} \widehat{V}_{smp}(\widehat{a}|Z_z). \tag{16.15}$$

In practice, of course, N_z will often be prohibitively large. For example, typically in NAEP (and explicitly in the 2015 linking study) $N_r = N'_r = 62$, therefore $N_z = 62!$ It may not be practical to implement Eq. 16.15. A pragmatic approach would be to approximate Eq. 16.15 by averaging over a small manageable set of $N_z^* < N_z$ of random pairings. Thus,

$$\widehat{V}_{smp}(\widehat{a}) \approx \frac{1}{N_z^*} \sum_{z=1}^{N_z^*} \widehat{V}_{smp}(\widehat{a}|Z_z). \tag{16.16}$$

The analogous approach could also be applied to the covariance term by defining

$$\widehat{C}_{smp}\left(\widehat{a}, \widehat{b}|Z_z\right) = \sum_{r=1}^{N_r} \frac{1}{N_p} \sum_{p=1}^{N_p} (\widehat{a}_{rp} - \widehat{a}_p)\left(\widehat{b}_{rp} - \widehat{b}_p\right) | Z_z, \tag{16.17}$$

and proceeding as in Eq. 16.16 to average over a set of random pairings.

Next, consider the situation in which the test takers in the linking study samples come largely from the same set of schools. The NAEP 2017 assessment design, described by Jewsbury et al. (2024), is an example. The random assignment of test takers to DBA or PBA conditions was carried out—to the extent that school size permitted—at the individual school level. Because the sets of replicate weights are generated by pairing schools and adjusting the weight, the only pairing of weights that will produce an appropriate estimate is the one in which the same school has its weights set to 0 for both the PBA and DBA sample. In other words, there is only a single value of $\boldsymbol{Z}^* \equiv \boldsymbol{Z}_{z^*} = \left[(\boldsymbol{W}_1, \boldsymbol{W}_1'), \ldots, (\boldsymbol{W}_r, \boldsymbol{W}_r'), \ldots, \left(\boldsymbol{W}_{N_r}, \boldsymbol{W}_{N_r}'\right)\right]$ that will produce an appropriate estimate. Therefore, in this case,

$$\widehat{V}_{smp}(\widehat{a}) = \widehat{V}_{smp}(\widehat{a}|\boldsymbol{Z}^*), \tag{16.18}$$

and

$$\widehat{Cov}_{smp}\left(\widehat{a}, \widehat{b}\right) = \widehat{Cov}_{smp}\left(\widehat{a}, \widehat{b}|\boldsymbol{Z}^*\right). \tag{16.19}$$

The situation is more straightforward with respect to the latency component. In a typical random group linking design, as applied in the NAEP context, the linking study samples are separately analyzed, and the multiple sets of PVs are generated from separate independent random processes. Thus, the situation with the latency component should mirror the situation described above for independently selected samples. There are N_p sets of PVs for the sample taking the original assessment and N_p' sets for the sample taking the revised assessment and, in NAEP, $N_p = N_p'$. Therefore, we have $N_q = N_p!$ possible pairings, $\boldsymbol{Q}_q$, $q = 1, 2, \ldots N_q$ and we can estimate the latency component as.

$$\widehat{V}_{lat}(t) = \frac{1}{N_q}\sum\nolimits_{q=1}^{N_q} \widehat{V}_{lat}(t|\boldsymbol{Q}_q), \tag{16.20}$$

and any required covariance terms as

$$\widehat{C}_{lat}\left(\widehat{a}, \widehat{b}\right) = \frac{1}{N_q}\sum\nolimits_{q=1}^{N_q} \widehat{C}_{lat}\left(\widehat{a}, \widehat{b}|\boldsymbol{Q}_q\right). \tag{16.21}$$

In practice, a practical alternative is to approximate Eq. 16.21 by averaging over a more manageable set of N_q^* random pairings.

Following standard practice in NAEP, the BSEE estimates of the linking coefficient errors are

$$\widehat{V}(\widehat{a}) = \widehat{V}_{smp}(\widehat{a}) + \frac{N_p + 1}{N_p}\widehat{V}_{lat}(\widehat{a}), \tag{16.22}$$

$$\widehat{V}\left(\widehat{b}\right) = \widehat{V}_{smp}\left(\widehat{b}\right) + \frac{N_p + 1}{N_p}\widehat{V}_{lat}\left(\widehat{b}\right), \tag{16.23}$$

$$\widehat{C}\left(\widehat{a}, \widehat{b}\right) = \widehat{C}_{smp}\left(\widehat{a}, \widehat{b}\right) + \frac{N_p + 1}{N_p}\widehat{C}_{lat}\left(\widehat{a}, \widehat{b}\right). \tag{16.24}$$

It is worth noting that the procedure described above for obtaining variances and covariances is generally applicable and could be applied to other statistics that are functions of two samples.

16.2.2 *Empirical Example: 2015 and 2017 NAEP Grade 8 Reading Results*

To illustrate the approach, we use the data from the 2015 and 2017 reading assessments for grade 8 to estimate linking coefficients and their associated error variances. The 2017 assessment along with the basic analysis plan for both data sets were described in Jewsbury et al. (2024). The data collection design for the 2015 assessments differed slightly from that of 2017. For the 2015 assessment, representative school samples were selected in each of 52 jurisdictions (states, the District of Columbia, and the Department of Defense schools), along with an additional nationally representative sample of private schools. Random samples of test takers within each of these schools were selected in turn and administered the PBA assessment. With appropriate weights, the data from these jurisdiction-level samples were aggregated to form a large nationally representative sample of test takers for the PBA results and were analyzed with the usual NAEP approaches (see, e.g., Mazzeo, 2018). The analysis mirrored that described by Jewsbury et al. (2024) and results in a set of PVs for each test taker that, when aggregated, provides national results on the NAEP reporting scales.

For DBA, a separate nationally representative sample of public and private schools was also selected. Again, random samples of test takers within each of these schools in turn were selected and administered the DBA assessment. The DBA data were separately analyzed with the usual NAEP approach. The analysis resulted in a set of PVs for each test taker initially on interim scales (in a separate IRT θ-metric with a mean of 0 and standard deviation of 1). When aggregated, the PVs provide national results on separate DBA interim scales.

Table 16.1 shows the school and test-taker sample sizes for the grade 8 PBA and DBA samples in each of the two assessment years. For the 2015 assessment, where the samples were independent, the total number of schools, 7395, is the sum of the number in each assessment condition. For the 2017 assessment, random assignment to assessment condition was carried out within-school whenever the number of grade-eligible students made it feasible. As a result, 5913 schools provided students for both the PBA and DBA assessment conditions.

Table 16.1 School (to the nearest 10) and test-taker sample sizes (to the nearest 100) for 2015 and 2017 grade 8 reading assessments

	2015			2017		
Sample	PBA	DBA	**Total**	PBA	DBA	**Total**
School	1250	6150	**7400**	6050	6500	**6640**
Test taker	136,500	16,400	**152,900**	38,400	141,800	**180,200**

Table 16.2 Estimates of 2015 and 2017 reading subscale linking coefficients and their standard errors

	2015				2017			
Reading subscale:	*a*	SE(*a*)	*b*	SE(*b*)	*a*	SE(*a*)	*b*	SE(*b*)
Literary	37.5	0.42	264.5	0.63	38.9	0.35	262.9	0.39
Informational	34.8	0.37	269.4	0.57	36.9	0.29	268.3	0.36

16.2.2.1 Example: Estimating Linking Coefficient Error

For both assessment years, we estimated linking constants by equating the means and standard deviations of assessment results for the aggregate (i.e., national-level) DBA and PBA assessments. To mimic the typical situation in NAEP linking studies and to simplify discussion in this paper, we treated the larger sample (PBA in 2015 and DBA in 2017) as the one taking the revised assessment and used the smaller sample as the one taking the original assessment (i.e., the sample for which results in the reporting metric were available).[9]

For the 2015 data, where the samples used to estimate the link came from different school samples, we used 50 different pairings of replicate data sets and averaged these to obtain our estimate of sampling variance. Exploratory analyses suggested that averaging over 50 pairings produced stable estimates while remaining computationally feasible. For the 2017 data, where the samples mostly came from the same schools, as noted earlier, only a single pairing makes sense. For estimating the latency component, we again used 50 pairings of PVs from the DBA and PBA samples for both the 2015 and 2017 results.

Table 16.2 shows the point estimates for the linking coefficients and their standard errors for both reading subscales. The standard errors of the linking coefficients were notably small relative to the size of the linking coefficients themselves—which is not a surprising result given the large sample sizes used to obtain the estimates. They are also quite small relative to the units of the scale, as indicated by the standard deviation. For example, the standard deviations for the national sample were 38 points in both assessment years for the literary reading subscale and 35 (2015) and 36 (2018) for the informational reading subscale.

Table 16.3 shows the error variances and covariances for each subscale, along with their constituent estimates of the sampling component and latency component, as well as total (co)variance estimates.[10] As is typically the case for NAEP statistics, the principal source of uncertainty in estimating the linking coefficients comes from

[9]For illustrative purposes, we artificially placed the 2015 DBA results on a NAEP-like reporting metric by using a transformation similar to the one derived operationally to convert the 2015 PBA IRT-metric result on the NAEP-like reporting scale. We then subsequently linked the 2015 PBA IRT-metric results to the same NAEP-like scale by linking to these artificial DBA scale score results.

[10]It should be noted that the square roots of the variances of *a* and *b* in the Total row of Table 16.3 are the same as the corresponding standard errors shown in Table 16.2, within rounding error.

Table 16.3 Estimated sampling, latency, and total error variances for 2015 and 2017 reading subscale linking coefficients

		2015			2017		
Variance:	Reading subscale:	*Var(a)*	*Var(b)*	*Cov(a,b)*	*Var(a)*	*Var(b)*	*Cov(ab)*
Sampling	Literary	0.126	0.324	−0.039	0.081	0.105	−0.015
	Informational	0.110	0.291	−0.027	0.060	0.091	−0.005
Latency	Literary	0.051	0.071	−0.003	0.042	0.047	−0.013
	Informational	0.029	0.040	−0.001	0.024	0.040	−0.006
Total	Literary	**0.177**	**0.395**	**−0.042**	**0.123**	**0.152**	**−0.028**
	Informational	**0.139**	**0.330**	**−0.029**	**0.085**	**0.131**	**−0.011**

the sampling of schools and students—accounting for 80 to 90% of the total error variance for 2015 data, in which the PBA and DBA samples were independent, and close to 70% for the 2017 data, in which dependent samples were used to estimate the linking coefficients. For both data sets, the estimated cross-scale covariances between the *a* coefficients are positive, as are the cross-scale covariances for the *b* coefficients. Both within and across scales, the covariances between the *a* and *b* coefficients were close to 0 and slightly negative, and the sampling and latency components were more comparable in magnitude. The effect of the sampling design is also evident. The sampling components of the error variances for the 2017 data are considerably smaller than those estimated with the 2015 data, while the latency components, which do not depend on the sample design, are more similar across years. There are likely two design features that contribute to the smaller sampling components in 2017. First, the number of schools and test takers in the target sample in 2017 is larger than it was in 2015. Second, as noted earlier, in 2017, 5913 of the 6641 schools contributed test takers to both conditions, whereas, in 2015, the different modes of assessment were carried out in distinct samples of schools.

16.2.2.2 Example: Estimating Error for Subscale Means

Using the linking coefficient estimates (and covariances) from the previous section, we produced MC approximations for the standard errors for NAEP subscale means.[11] As part of our development of the MC approach, we conducted a study to investigate the number of draws required in the MC procedure to achieve stable estimation of the linking error variance components using the 2017 data sets. We compared the stability of estimates with N_ν ranging from 50 to 2000. With N_ν = 500, the range of estimates on the linking error were quite small relative to the size of the linking error estimate itself, and even smaller relative to magnitude of the statistics they were associated with. Based on these analyses, we chose 500 draws

[11] Jewsbury (2019) showed that these same formulas can be applied to a class of NAEP subscale statistics that he calls AB estimators, which include NAEP scale score means and percentiles.

as a reasonable trade-off between computational burden and stability of the MC estimation of the linking error.

16.2.2.2.1 Method-of-Moments Estimators for Group Means

In order to provide some evaluation of the proposed approximation, the results produced by the MC method were compared to two method-of-moments estimators (Mood et al., 1973) derived from exact analytic expressions for the standard errors for the subscale means. Estimators were derived for both the external case and the internal case.

Let $\overline{\theta}_g$ represent the sample group mean prior to transformation and $\overline{Y}_g$ represent the mean after transformation. Let $\overline{\theta}_g$ and $\overline{Y}_g$ represent the corresponding population quantities. Similarly, let a and b refer to the population linear transformation constants. The exact formulas were derived for the case in which both the mean and transformation coefficients are considered random variables. For the external case, the exact expression for the standard error of the transformed mean is

$$V_{EX-E}\left(\overline{Y}_g\right) = a^2 V\left(\overline{\theta}_g\right) + \left\{V\left(\overline{\theta}_g\right) + \overline{\Theta}_g^2\right\} V(\widehat{a}) + V\left(\widehat{b}\right) + 2\Theta_g C\left(\widehat{a}, \widehat{b}\right). \quad (16.25)$$

Note that the first term to the right of the equality is just the usual expression for the variance of a linear transformed variable when a fixed arbitrary transformation coefficient is used. This term can be estimated by carrying out the NAEP BSEE on the linearly transformed scale scores. The sum of the remaining terms on the right then represents the adjustment to the BSEE that must be estimated due to random nature of the linking components, as follows:

$$Adj_{EX-E}\left(\overline{Y}_g\right) \equiv V_{EX-E}\left(\overline{Y}_g\right) - a^2 V\left(\overline{\theta}_g\right) = \left\{V\left(\overline{\theta}_g\right) + \overline{\Theta}_g^2\right\} V(\widehat{a}) + V\left(\widehat{b}\right) + 2\overline{\Theta}_g C\left(\widehat{a}, \widehat{b}\right). \quad (16.26)$$

Analogous expressions can be derived for the internal case.

$$V_{EX-I}\left(\overline{Y}_g\right) = V_{EX-E}\left(\overline{Y}_g\right) + C\left(\widehat{a}^2, \overline{\theta}_g^2\right) + C^2\left(\widehat{a}, \overline{\theta}_g\right) + 2\left\{C\left(\widehat{ab}, \overline{\theta}_g\right) - \overline{Y}_g C\left(\widehat{a}, \overline{\theta}_g\right)\right\}. \quad (16.27)$$

This expression requires additional covariance terms due to the dependence between the source sample statistics and the linking constants that is present for the internal case.

By substituting sample estimates of the needed population quantities, along with estimates of the error variances and covariances of the various estimators reflected in Eqs. 16.25 and 16.27—which can be obtained with the NAEP two-sample jackknife

Table 16.4 Sample sizes (*N* to the nearest 100), group means, and estimated standard errors with and without linking error: 2015 reading informational subscale

Group	*N*	Mean	SE	SE(MC)	SE(MM-E)	SE(MM-I)
All students	136,500	268.7	0.22	0.63	0.62	0.53
Male	69,500	264.9	0.25	0.64	0.63	0.53
Female	67,000	272.6	0.29	0.65	0.64	0.58
White	72,100	277.7	0.21	0.62	0.61	0.55
Black	23,500	250.9	0.50	0.81	0.81	0.74
Hispanic	26,600	256.8	0.44	0.76	0.75	0.71
A/PI	7100	283.3	1.27	1.40	1.40	1.22
AI/AN	3600	255.4	1.55	1.67	1.66	1.72
SwD	4400	237.0	1.23	1.42	1.42	1.47
LEP	1700	229.6	1.85	2.00	2.00	2.03
State 1	6000	262.5	1.19	1.33	1.32	1.20
State2	5700	263.9	1.06	1.21	1.21	1.11
State 3	5300	267.5	1.06	1.21	1.21	1.09
State 4	3500	265.0	1.35	1.48	1.47	1.49
State 5	3300	270.1	1.05	1.20	1.20	1.15
State 6	2000	271.2	0.81	0.99	0.99	1.06
State 7	1900	264.2	1.22	1.36	1.35	1.35
State 8	1800	277.2	0.85	1.02	1.02	1.04
State 9	1700	250.6	1.06	1.24	1.23	1.13
State 10	1300	280.3	0.83	1.01	1.00	1.03

Note. SE = BSEE, SE(MC) = the Monte Carlo estimator, SE(MM-E) = the method of moments formula for the external case, SE(MM-I) = the method of moments formula for the internal case, A/PI = Asian/Pacific Islander, AI/AN = American Indian/Alaskan Native, SwD = students with disabilities, LEP = limited English proficiency

approach described earlier—method-of-moments estimates of the total variance that incorporate the linking error variance can be obtained for both the internal and external case.

16.2.2.2.2 Comparing Results for MC and Method-of-Moments Estimators

Tables 16.4 (2015) and 16.5 (2017) present estimates of the mean scores and standard errors with and without the linking error component. Results are presented for the informational subscale, but findings were comparable for the literary subscale. Results are presented for the overall national mean, by gender, race/ethnicity groups, groups qualifying for accommodations, and 10 states (the five largest and five smallest sample sizes). For each reporting group, the table contains the test-taker sample size, the estimate of the mean and the NAEP estimated standard error without linking error, and three versions of the standard error that includes linking error, one based on the MC method and two method-of-moments estimators.

Table 16.5 Sample sizes (*N* to the nearest 100), group means, and estimated standard errors with and without linking error: 2017 reading informational subscale

Group	*N*	Mean	SE	SE(MC)	SE(MM-E)	SE(MM-I)
All students	141,800	269.0	0.27	0.46	0.45	0.38
Male	72,300	265.2	0.30	0.48	0.47	0.42
Female	69,500	272.9	0.35	0.51	0.50	0.44
White	73,000	277.5	0.29	0.47	0.46	0.39
Black	24,000	250.7	0.47	0.62	0.62	0.59
Hispanic	29,900	257.6	0.49	0.62	0.62	0.58
AS/PI	7800	285.0	1.04	1.11	1.11	1.11
AI/AN	2600	254.9	1.29	1.34	1.34	1.33
SwD	17,500	235.0	0.54	0.71	0.72	0.66
LEP	8700	228.9	0.92	1.04	1.04	1.05
State 1	7100	263.3	1.28	1.33	1.33	1.28
State 2	5900	264.9	1.06	1.12	1.12	1.04
State 3	5600	269.4	1.05	1.10	1.11	1.06
State 4	4100	265.6	1.07	1.13	1.13	1.11
State 5	3400	268.5	1.13	1.18	1.19	1.18
State 6	2100	276.3	1.16	1.21	1.21	1.27
State 7	2000	268.5	1.31	1.35	1.36	1.40
State 8	2000	258.2	1.14	1.20	1.20	1.20
State 9	1600	283.5	1.63	1.67	1.67	1.68
State 10	1400	249.4	0.99	1.06	1.07	1.04

Note: SE = BSEE, SE(MC) = the Monte Carlo estimator, SE(MM-E) = the expectation-based formula for the external case, SE(MM-I) = the expectation-based formula for the internal case, A/PI = Asian/Pacific Islander, AI/AN = American Indians/Alaskan Native, SwD = students with disabilities, LEP = limited English proficiency

There are several patterns or results worth noting. First, for the total (all students) group and largest subgroups, the size of the standard errors that incorporate linking error remains quite small considering that the standard deviation for the nation on this scale is 39 points. Second, the incorporation of linking error had a proportionally larger effect on the 2015 results than it did on the 2017 results. For example, the 2015 external case standard errors for the total group are close to three times larger than the normal NAEP standard errors, whereas in 2017 the increase is less than two times larger. This result is predictable from the previous section, which showed the estimates of the linking coefficient errors to be larger in 2015. Third, in percentage terms, the increases in standard errors are less noticeable for the states. This reflects the fact that linking error, which depends on the large linking samples, is modest in size relative to the uncertainty that arises due to school and student sampling in the state results.

For all groups and states, the estimates produced by the MC method closely match the method-of-moments estimates for the external case. This is no doubt partly due to the fact that both estimators depend on the same estimates of the error (co)variances of the linking coefficients. For the total group result, the two internal

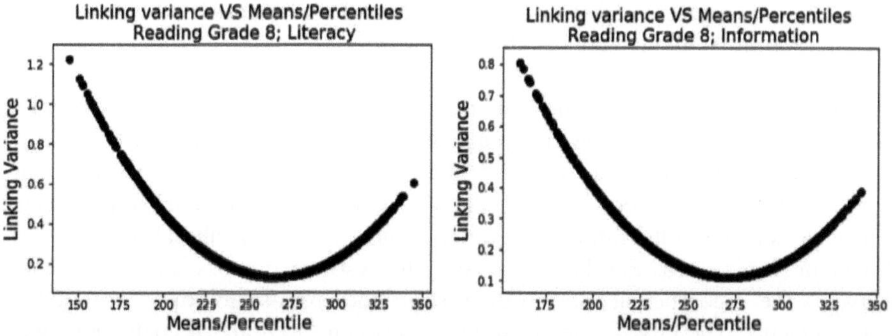

Fig. 16.1 Linking error variance plotted against grade 8 reading means/percentiles: Literary and informational subscales

case formulas are also virtually identical and, predicted by the Taylor series formula and Jewsbury's (2019) result, the target sample standard error calculated with the normal BSEE matches these numbers. The two internal-case formulas also agree closely with each other for the other subgroups and states, though not quite as closely as in the external case.

It is of further interest to note for groups with sample sizes below 10,000 and for the states, differences between the internal and external are either negligible or, in some cases, slightly smaller for the internal case. There are several possible reasons why this is occurring. First, Eqs. 16.30 and 16.33, 16.34, 16.35 show that the error variance for the mean depends on covariance terms, which can be positive or negative, and have coefficients that are sometimes positive and sometimes negative. Thus, it is certainly possible that, for any particular group, the true standard error is, in fact, smaller in the internal case than it is in the external case. This is consistent with Jewsbury's (2019) recent theoretical work for the internal case. Moreover, the results in Tables 16.4 and 16.5 are estimates. For smaller groups, where one might expect little difference between the internal and external case standard errors, the BSEE-based estimates of the covariance terms may not be sufficiently accurate to detect small differences between the external and internal cases.

In the interest of further exploring operational simplification, we decided to investigate the behavior of the MC estimates of linking error. To this end, we defined a collection of 1060 subgroups based on the cross-classification of several background variables. Using the 2017 grade 8 reading data, we used the MC method to calculate the NAEP statistics of interest (means and percentiles) and their associated linking error estimates.

Figure 16.1 shows a plot of the MC linking error estimate against the group means and percentiles (10th, 25th, 50th, 75th, and 90th) for each of the two reading subscales. Thus, in each plot, there are 6360 points plotted (i.e., 6 * 1060). The plots at the top of the figure show that MC linking errors display a quadratic relationship to the reference statistic. We obtained least-squares estimates of the quadratic coefficients based on the MC linking errors and, as discussed below, these regression estimates were used operationally in 2017 as part of the NAEP Data Explorer.

16.2.2.3 Example: Estimating Error for Composite Means and Achievement Levels

As noted earlier, one of the motivations for developing the MC approach was its potential applicability to the full range of statistics on which NAEP reports—specifically, composite scale statistics like means and achievement level percentages. As discussed in Sect. 16.2.1.2, the MC procedure can be readily generalized given an estimate of the full error covariance matrix for both subscales. These were given in Table 16.6. For both data sets, the estimated cross-scale covariances between the *a* coefficients are positive as are the cross-scale covariances for the *b* coefficients. The cross-scale covariances between the *a* and *b* coefficients were close to 0 and slightly negative.

Table 16.7 shows the 2015 and 2017 composite means and associated standard errors, with and without the linking error component, for the same set of national subgroups and states used in discussing the subscale results. The results follow the same general pattern as was observed for the subscales. The magnitude of the standard errors that include linking error remains modest relative to the reporting scale standard deviation (36 points in 2017). The relative increase in standard errors due to the incorporation of linking error is greater for 2015 than for 2017, due to the larger estimated error variance in the linking coefficients for that year. The relative increase in standard errors is greater for the larger sized groups and states, due to the smaller sampling error component associated with these groups.

Table 16.8 presents results for one of the NAEP program's key benchmark points, the percentage of students at or above the proficient achievement level, along with associated standard errors with and without the linking error component. The pattern of results largely mirrors those found for composite means. The impact of including linking error was larger for 2015 than 2017 and more pronounced, in percentage terms, for larger groups where sampling errors were small.

Using the same 1060 groups from earlier analyses, we again examined the relationship between composite scale means, percentiles, and achievement level percentages and the MC estimates of their linking error. Figure 16.2 shows a plot of linking error against its point estimate for composite scale means/percentiles and

Table 16.6 Estimates of error covariance matrices of estimated linking coefficients: 2015 and 2017 NAEP reading

		Full covariance matrix—2015				Full covariance matrix—2017			
		Literary		Informational		Literary		Informational	
Reading subscale:	*Coefficient*	*a*	*b*	*a*	*b*	*a*	*b*	*a*	*b*
Literary	*a*	0.177				0.123			
	b	−0.042	0.395			−0.028	0.152		
Informational	*a*	0.098	−0.026	0.139		0.062	−0.017	0.085	
	b	−0.024	0.262	−0.029	0.330	−0.013	0.110	−0.011	0.131

Table 16.7 Sample sizes (*N* to nearest 100), composite means, and their standard errors with and without linking error: 2015 and 2017 NAEP reading

	2015				2017			
Group	*N*	Mean	SE	SE(MC)	*N*	Mean	SE	SE(MC)
All students	13,650	266.7	0.21	0.59	141,800	266.6	0.26	0.45
Male	69,500	262.0	0.24	0.61	72,300	261.6	0.28	0.47
Female	67,000	271.5	0.28	0.62	69,500	271.7	0.35	0.50
White	72,100	275.5	0.21	0.59	73,100	274.9	0.29	0.46
Black	23,500	249.5	0.47	0.77	24,000	249.1	0.45	0.61
Hispanic	26,600	254.9	0.41	0.71	29,900	255.1	0.49	0.63
AS/PI	7100	280.8	1.25	1.37	7800	282.1	0.99	1.05
AI/NA	3600	253.7	1.68	1.78	2600	253.0	1.23	1.29
SwD	4400	233.8	1.14	1.33	17,500	231.9	0.54	0.72
LEP	1700	226.6	1.61	1.77	8695	225.4	0.88	1.02
State 1	6000	260.3	1.11	1.24	7100	260.2	0.95	1.02
State 2	5700	261.9	1.08	1.22	5900	262.4	0.99	1.06
State 3	5300	264.6	1.14	1.26	5600	266.6	0.90	0.97
State 4	3500	263.5	1.23	1.35	4100	262.9	0.80	0.88
State 5	3300	268.0	1.17	1.30	3400	267.1	1.01	1.07
State 6	2000	270.0	1.14	1.26	2100	273.4	1.21	1.26
State 7	1900	261.6	0.89	1.05	2000	266.3	0.87	0.94
State 8	1800	275.0	1.13	1.25	2000	256.5	0.88	0.94
State 9	1700	249.1	0.73	0.91	1600	280.8	0.82	0.89
State 10	1300	278.2	0.71	0.90	1400	246.7	0.84	0.91

Note: SE = BSEE, SE(MC) = the Monte Carlo estimator, A/PI = Asian/Pacific Islander, AI/AN = American Indian/Alaskan Native, SwD = students with disabilities, LEP = limited English proficiency

the percentages of at/above the *basic* and *proficient* achievement levels cut scores. For the composite means and percentiles, the MC linking errors behave similarly to the corresponding statistics on the subscales.

The situation is less tidy for achievement level percentages. The MC estimates of linking error for the percentages of students at or above each achievement level, along with their respective complements (i.e., the percentage below each level) plotted against the linking errors for those achievement level percentages. Though the general shape of the relationship again appears to be quadratic, the relationship is far less deterministic. The MC estimates of linking error clearly depend on factors in addition to the actual percentages.

16.2.2.4 Operational Implementation in NAEP

The NAEP Data Explorer (NDE) is a central component of NAEP's system of reporting results. The NDE is an Internet application hosted on NCES's website, which, among other things, allows the user to define groups of interest on the basis of

Table 16.8 Sample sizes (*N* to the nearest 100), percentages of test takers at or above proficient, and standard errors with and without linking error: 2015 and 2017 NAEP reading

	2015				2017			
Group	*N*	> = PRFNT	SE	SE (MC)	*N*	> = PRFNT	SE	SE (MC)
All students	136,500	35.5	0.32	0.72	141,800	36.0	0.36	0.53
Male	69,500	30.5	0.38	0.73	72,300	31.1	0.41	0.55
Female	67,000	40.6	0.48	0.83	69,500	41.1	0.47	0.62
White	72,100	45.3	0.39	0.84	73,100	45.0	0.47	0.64
Black	23,500	16.4	0.55	0.71	24,000	17.5	0.56	0.63
Hispanic	26,600	21.8	0.54	0.78	29,900	23.1	0.58	0.68
AS/PI	7000	53.1	1.78	1.90	7800	54.6	1.33	1.38
AI/NA	3600	22.9	1.77	1.85	2600	21.5	1.86	1.88
SwD	4400	10.0	0.87	0.90	17,500	9.2	0.53	0.55
LEP	1700	4.3	0.92	0.93	141,800	36.0	0.36	0.53
State 1	6000	29.5	1.32	1.45	7100	25.7	1.07	1.09
State 2	5700	29.2	1.59	1.71	5900	27.5	1.33	1.37
State 3	5300	31.5	1.76	1.86	5590	31.7	1.40	1.44
State 4	3500	31.3	1.62	1.73	4100	28.5	1.17	1.20
State 5	3300	36.3	1.59	1.69	3400	31.6	1.21	1.23
State 6	2000	37.3	1.57	1.69	2100	37.9	1.57	1.60
State 7	1900	32.5	1.34	1.48	2000	31.3	1.27	1.31
State 8	1800	45.1	1.67	1.80	2000	22.9	1.51	1.53
State 9	1700	19.3	1.21	1.42	1600	44.2	1.25	1.29
State 10	1300	48.3	1.64	1.84	1400	15.9	1.47	1.50

Note: PRFNT = proficient, SE = BSEE, SE(MC) = the Monte Carlo estimator, A/PI = Asian/Pacific Islander, AI/AN = American Indian/Alaskan Native, SwD = students with disabilities, LEP = limited English proficiency

background variables and request any of several statistics (e.g., means, standard deviations, percentiles, percentages at or above achievement levels) for subscales or the composite scale. NDE calculates the requested results, including associated standard errors, on the fly in real time. Development work determined that carrying out the MC calculations as part of NDE would not be feasible without unduly degrading the response time required for the system to provide a user with the requested statistic. An approach was needed that was computationally tractable and that provided results with good fidelity to the results produced by the MC method.

As is evident from Figs. 16.1 and 16.2, for subscale and composite means and percentiles, the MC estimated linking error is a quadratic function of the reported statistic. These results suggested a solution to NDE operational implementation for subscale and composite statistics is to estimate the coefficients of the quadratic formula for each grade/subject/statistic combination to compute the estimate of linking error. This was done by using the least squares fit to the MC linking error

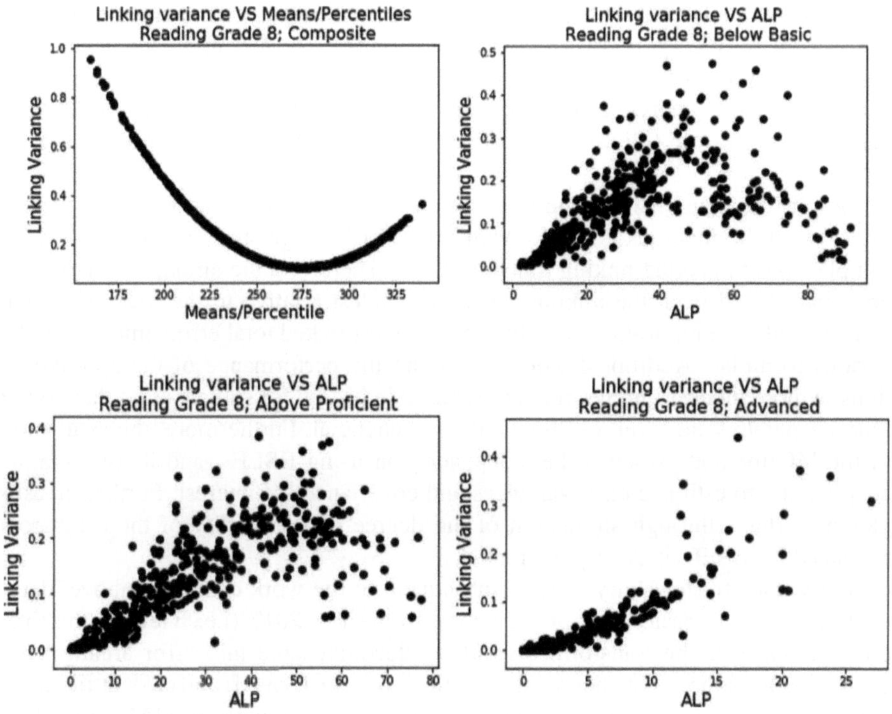

Fig. 16.2 Linking error variance plotted against the grade 8 reading composite scale means and achievement level percentages (ALP)

variances described earlier. While such an approach works quite well for means, percentiles, and standard deviations, it functions less well for achievement level percentages. Despite the somewhat weaker relationship between the linking error estimates and the achievement level, the program decided to proceed operationally with the same quadratic approximation approach for the proficient achievement-level percentage linking errors. Further work on operationally feasible approaches to estimating linking error for achievement level approaches remains an important priority for the NAEP program.

16.3 Discussion

The incorporation of linking error in NAEP standard errors represents a step forward for the program. In the past, when linking studies were conducted, operationally tractable procedures for approximating linking error had yet to be developed. In 2017, NAEP used a practical approximation that is adaptable to other linking situations where the program uses randomly equivalent group designs. The results from the MC approach and the quadratic approximation derived from it that was used operationally and in NDE match results obtained with available analytic expressions for incorporating this source of variance. The MC approach generalizes easily to support not just comparisons of subscale means and percentiles, but also composite scale means, percentiles, standard deviations,[12] and achievement level percentages (i.e., the full range of statistics on which the NAEP program typically reports).

Despite this step forward, additional research and development work is needed to improve on the methods going forward. The 2017 bridge study made use of large samples to estimate the linking transformation. Therefore, the amount of additional error introduced from the linking was quite modest relative to the scale on which NAEP results are reported, as are differences in estimated total errors implied by the various formulas. Additional work examining the performance of the approximations in more modest samples (e.g., those used in NAEP for subject areas that do not report separate state-level results) would be beneficial. Furthermore, the evaluation of the MC method presented here depended on using BSEE—and its two-sample adaptation—to estimate error variances and covariances of interest. Further investigation, perhaps through simulation of the degree of the quality of the estimators produced by BSEE, is clearly warranted.

Work done by Jewsbury (2019), subsequent to the work described above, demonstrates that the basic logic of the approach used in 2017 (i.e., viewing the total error variance of the transformed result as decomposable into error arising from estimating the source statistic, or the statistic to be transformed) and error attributable to estimating the linking coefficients does not generally apply to NAEP reporting statistics, including subscale means and percentiles. That is to say, approaches like our MC approach based on or consistent with this decomposition logic are inherently approximations to the exact standard errors. In contrast, the method of moments formulas for group means based on expectations are exact. The fact that the results for the MC method differ little from those obtained with the method of moment estimates suggests that, at least under the assessment designs used in 2017, the approximation based on variance decomposition logic performs acceptably for subscale group means and percentiles under external case assumptions.

The degree to which similar conclusions can be asserted for composite means/percentiles and achievement levels is less clear. On the one hand, the results for composite means and percentiles seemed to follow a similar pattern to those obtained with subscales. So perhaps there is reason to be optimistic that the approximations used are reasonable as well for these types of composite statistics, at least under 2017 design conditions. The situation for achievement level percentages is far less clear, so more work on developing a practical approach that improves upon the external case approximation presented here for the full range of statistics and for designs that differ from that used in 2017 remains a priority.

[12]Though results were not shown here to keep the presentation concise, results for estimating standard error for standard deviations were similar to those obtained for means and percentiles.

The NAEP program decided to begin its incorporation of linking error with approximate methods based on the external case for the practical reasons discussed above. Moreover, as demonstrated in earlier sections, the standard errors for subscale means were consistently larger than those based on formulas for the internal case. Assuming that such a finding would generalize to other statistics, the program felt it would be better to err on the side of overestimating the standard errors—thereby declaring comparisons to PBA results from previous assessments nonsignificant—because there are other sources of variance we are not currently capturing (e.g., trend linking error due to item sampling, differential DBA to PBA linking relationships for states and/or subgroups), so incorporating a slightly larger linking error seemed defensible.

Pragmatic considerations notwithstanding, having a practical approach appropriate to the internal case would clearly be of benefit, either for use in bridge study years or as a reference point for better understanding the degree of overestimation provided by the external case. Jewsbury (2024) presented exact formulas, appropriate to the internal case, for a variety of NAEP subscale statistics. He noted, however, that the practicality of using these and extending the developments to cover the full range of NAEP statistics is unclear. He suggested an approach based on resampling methods like those used in NAEP BSEE might provide a more tractable solution. To that end, Liu and Mazzeo (2019) have developed an alternative modification to the BSEE approach that is again applicable to the full range of NAEP statistics and incorporates linking error into both the sampling and latency component estimates in a way that avoids the need to rely on variance decomposition approximations. This newer approach was used for the reporting of the 2018 NAEP assessment results for U.S. history, geography, and civics—all of which were transitioned to DBA assessments in 2018.

Another area in which additional development may be warranted is that concerning subgroup invariance of the linking coefficients, or lack thereof, and how and whether to incorporate lack of invariance as a source of error. The linking NAEP used in 2017 was estimated at the national level. The data collection designs, with separate samples of PBA and DBA test takers within each reporting jurisdiction, also supported an evaluation of the residual differences between DBA and PBA results, after linking, for national subgroups, states, and jurisdictions. As part of the linking error development work, ETS did undertake several exploratory investigations of whether error components reflecting lack of invariance could be estimated and, if so, what would be the appropriate way to incorporate such components into the standard errors. The work conducted did not yield satisfactory approaches or sufficiently consistent evidence of lack of invariance to warrant operational implementation. Moreover, even if sound methods were available, it is unclear whether a lack of invariance should be considered an additional source of error variance or as potential systematic bias.

Lastly, the general question of estimating linking error for the typical NAEP common-item linking used to maintain scales for trend comparisons has yet to be adequately addressed. While at least one proposal for a method that might, in principle, accomplish this has been made (Hsieh et al., 2009), there is not yet an operationally tractable approach in the context of NAEP. This general issue remains an important area for future study.

Acknowledgements The authors wish to thank Dan McCaffrey, Helena Jia, Paul Jewsbury, and Mei-Jang Lin at Educational Testing Service for contributions to this work. We would also like to thank Keith Rust at Westat and Sophia Rabe Hesketh at the University of California for useful discussions.

Funding Sources The research reported here was supported under the National Assessment of Educational Progress (ED-IES-13-C-0017) to Educational Testing Service (ETS) as administered by the Institute of Education Sciences, U.S. Department of Education. The opinions expressed are those of the authors and do not represent views of the Institute or the U.S. Department of Education.

References

Beaton, A. E., & Johnson, E. G. (1992). Overview of scaling methodology used in the national assessment. *Journal of Educational Measurement, 29*, 163–175. https://doi.org/10.1111/j.1745-3984.1992.tb00372.x

Beaton, A. E., & Zwick, R. (1992). Overview of the National Assessment of Educational Progress. *Journal of Educational Statistics, 17*, 95–109. https://doi.org/10.3102/10769986017002095

Bock, R. D., & Moustaki, I. (2007). Item response theory in a general framework. In C. R. Rao & S. Sinharay (Eds.), *Handbook of statistics* (Psychometrics) (Vol. 26, pp. 469–514). Elsevier.

Cohen, J., Johnson, E., & Angeles, J. (2001). *Estimates of the precision of estimates from NAEP using a two-dimensional jackknife procedure* [Paper presentation]. National Council of Measurement in Education, Seattle, WA.

Haberman, S. H., Lee, Y. H., & Qian, J. (2009). *Jackknife techniques for the evaluation of equating accuracy (RR-09-39)*. Educational Testing Service. https://doi.org/10.1002/j.2333-8504.2009.tb02196.x

Hsieh, C., Xu, X., & von Davier, M. (2009). Variance estimation for NAEP data using a resampling-based approach: An application of cognitive diagnostic models. *IERI Monograph Series: Issues and Methodologies in Large Scale Assessments, 2*, 161–174. https://doi.org/10.1002/j.2333-8504.2010.tb02233.x

Jewsbury, P. A. (2019). *Error variance in common population bridge studies (RR-19-42)*. Educational Testing Service. https://doi.org/10.1002/ets2.12279

Jewsbury, P. A. (2024). Generally applicable variance estimation methods for common-population linking. Journal of Educational and Behavioral Statistics, Online First. https://doi.org/10.3102/10769986241263976

Jewsbury, P. A., Jia, Y., & Xi, N. (2024). Effects of mode transition on instruments and subpopulation performance in NAEP. In L. Khorramdel, M. von Davier, & K. Yamamoto (Eds.), *Innovative digital-based international large-scale assessments—Foundations, methodologies and quality assurance procedures*. Springer.

Jia, Y., Phillips, G., Wise, L. L., Rahman, T., Xu, X., Wiley, C., & Diaz, T. E. (2014). *2011 NAEP-TIMSS linking study: Technical report on the linking methodologies and their evaluations (NCES 2014–461)*. National Center for Education Statistics. https://nces.ed.gov/nationsreportcard/subject/publications/studies/pdf/2014461.pdf

Johnson, E. G. (1998). *Linking the National Assessment of Educational Progress (NAEP) and the Third International Mathematics and Science Study (TIMSS): A technical report (NCES 98–499)*. Government Printing Office. https://nces.ed.gov/pubs98/98499.pdf

Johnson, E. G., Cohen, J., Chen, W.-H., Jiang, T., & Zhang, Y. (2005). *2000 NAEP–1999 TIMSS linking report (NCES 2005–01)*. National Center for Education Statistics. https://nces.ed.gov/pubs2005/200501.pdf

Johnson, E. G., & Rust, K. F. (1992). Population inference and variance estimation for NAEP. *Journal of Educational Statistics, 17*, 175–190. https://doi.org/10.2307/1165168

Kolen, M. J. (2007). Data collection designs and linking procedures. In N. J. Dorans, M. Pommerich, & P. W. Holland (Eds.), *Linking and aligning scores and scales* (pp. 31–55). Springer. https://doi.org/10.1007/978-0-387-49771-6

Liu, B., & Mazzeo, J. (2019, April). *New jackknife variance estimators for NAEP during transitions from paper-based to digital-based assessments* [Paper presentation]. National Council on Measurement in Education, Toronto: Ontario.

Lutkus, A. D., & Mazzeo, J. (2003). *Including special-needs students in the NAEP 1998 reading assessment, Part I: Comparison of overall results with and without accommodations (NCES 2003–467)*. National Center for Education Statistics. https://nces.ed.gov/nationsreportcard//pdf/main1998/2003467.pdf

Mazzeo, J. (2018). Large-scale group-score assessments. In W. J. van der Linden (Ed.), *Handbook of item response theory* (pp. 297–312). CRC Press.

Mazzeo, J., Carlson, J. E., Voelkl, K. E., & Lutkus, A. D. (2000). *Increasing the participation of special needs students in NAEP: A report on 1996 NAEP research activities (NCES 2000–473)*. National Center for Education Statistics. https://nces.ed.gov/nationsreportcard//pdf/main1996/2000473.pdf

Miller, R. G. (1964). A trustworthy jackknife. *The Annals of Mathematical Statistics, 35*, 1594–1605.

Mislevy, R. J. (1991). Randomization-based inference about latent variables from complex samples. *Psychometrika, 56*, 177–196. https://doi.org/10.1007/BF02294457

Mislevy, R. J., & Sheehan, K. J. (1987). Marginal estimation procedures. In A. E. Beaton (Ed.), *Implementing the new design: The NAEP 1983–84 technical report (15-TR-20)* (pp. 293–360). Educational Testing Service.

Mood, A. M., Graybill, F. A., & Boes, D. C. (1973). *Introduction to the theory of statistics* (3rd ed.). McGraw-Hill.

National Center for Education Statistics. (2017). *National Assessment of Educational Progress (NAEP): NAEP Technical Documentation: Definition of composite scales*. https://nces.ed.gov/nationsreportcard/tdw/analysis/scaling_determination_composite.asp

Phillips, G. W. (2007). *Expressing international educational achievement in terms of U.-S. performance standards: Linking NAEP achievement levels to TIMSS*. American Institutes for Research. https://www.air.org/sites/default/files/downloads/report/naep-timss_0.pdf

Quenouille, M. H. (1956). Notes on bias in estimation. *Biometrika, 43*, 353–360. https://doi.org/10.1093/biomet/43.3-4.353

Rubin, D. B. (1987). *Multiple imputation for nonresponse in surveys*. Wiley.

Searle, S. R. (1971). *Linear models*. John Wiley & Sons.

Tukey, J. W. (1958). Bias and confidence in not-quite large samples [abstract]. *The Annals of Mathematical Statistics, 29*, 561.

von Davier, M., Sinharay, S., Oranje, A., & Beaton, A. (2007). The statistical procedures used in National Assessment of Educational Progress: Recent developments and future directions. In C. R. Rao & S. Sinharay (Eds.), *Handbook of statistics* (*Psychometrics*) (Vol. 26, pp. 1039–1055). Elsevier.

Wang, L., Qian, J., & Lee, Y. H. (2018). *Grouping effects on jackknifed variance estimation for item response theory scaling and equating with cluster-based assessment data (RR-18-16)*. Educational Testing Service. https://doi.org/10.1002/ets2.12204

Xu, X., & von Davier, M. (2010). *Linking errors in trend estimation in large scale surveys (RR-10-10)*. Educational Testing Service. https://doi.org/10.1002/j.2333-8504.2010.tb02217.x

Yamamoto, K., & Mazzeo, J. (1992). Item response theory scale linking in NAEP. *Journal of Educational Statistics, 17*, 155–173. https://doi.org/10.2307/1165167

Chapter 17
Scoring and Scaling Items from Innovative Domains

Matthias von Davier, Hyo Jeong Shin, and Lale Khorramdel

Abstract In recent years, there is an increasing interest in measuring more complex and innovative constructs through interactive and gamified tasks. To guide scoring, scaling, and reporting aspects of such innovative constructs in international large-scale assessments, this chapter starts with illustrating the data collected from innovative tasks as human-computer interactions. We then explain the general approach to understand and analyze data from innovative domains and illustrate two prominent examples: automated scoring of text sequence responses and feature generation of PISA 2015 collaborative problem solving (CPS) items. We conclude the chapter with the cautionary note on analyzing and interpreting the results of innovative domains with respect to the reliability, validity, and comparability of the assessment.

In any scientific field, data does not emerge independent of human observation and interpretation. Similarly, item scores, which are the arguably most common data type in educational measurement, are the result of rules that help humans or computers categorize student behaviors shown in reaction to tasks. Person attributes, such as scale scores, are either functions or aggregates of item scores or estimates of a latent variable given these item scores. These function values or estimates are often the main target of statistical inference and the main tool to compare test takers or describe and compare populations of test takers.

M. von Davier (✉) · L. Khorramdel
Lynch School of Education and Human Development, TIMSS & PIRLS International Study Center at Boston College, Chestnut Hill, MA, USA
e-mail: vondavim@bc.edu

H. J. Shin
Sogang University, Seoul, South Korea

L. Khorramdel et al. (eds.), *Innovative Digital-Based International Large-Scale Assessments*, Methodology of Educational Measurement and Assessment,
https://doi.org/10.1007/978-3-031-90951-1_17

Since the emergence of computer-based assessments (CBA), data has been collected using computers, and there are increasing interests in measuring more complex and innovative cognitive skills that were not afforded in the paper-based assessments. In the context of international large-scale assessments (ILSAs) where participating countries and economies are ranked based on their performance, producing scale scores in innovative domains that are meaningful, reliable, and valid at the group-level is a challenging task. Moreover, the requirement of comparability is also of paramount importance. Item developers and domain experts provide information about how responses to innovative items are expected to relate to the latent trait attributes being inferred from measurement. These experts have a good understanding of the domain, but as the saying goes: “nobody is perfect.” As it is the case in traditional scale development: Some responses to some tasks may be less related to the attributes measured than is desirable, compared to responses on other tasks or to other attributes measured.

There are recent developments that attempt to take advantage of techniques developed in computer science, machine learning, and modern applied statistics, such as educational data mining (Baker & Inventado, 2014), computational psychometrics (von Davier et al., 2021; Cipresso et al., 2019), computational psychology (Sun, 2008), gamification, and the use of stealth assessments (e.g. Shute, 2011), as well as attempts to use artificial intelligence in psychometrics and assessment development (e.g., von Davier, 2018, 2019). However, the main aim of this chapter is to be agnostic about the methods used as long as they can be replicated and satisfy professional standards. That is, the focus of this chapter is to address the issue: “Can the data we analyze be assumed immutable and to be accepted as they are?” or “are data indeed products of human (or machine) interpretation and in that sense, functions of sequences of observations that are transformed into the input of psychometric modeling and educational measurement?” (which is what we propose as the more reasonable approach, given that scoring is known to be imperfect).

The guiding assumptions of this chapter can be described as follows:

1. The data points, scores, category choices, that are often understood as direct observations are a product of more atomic behaviors or artifacts of human behavior, such as writing in and erasing, marking options, or on the computer, keystrokes, mouse-moves, or eye movements (when using an eye tracker, e.g., Eraslan et al., 2019).
2. The scoring rules provided by content experts can be viewed as human generated feature extraction rule from these atomic behaviors. These feature extraction rules can be either taken for granted (classical approach) or can be scrutinized, improved, and emulated (automated scoring of writing tasks, e.g., Attali & Burstein, 2006). Feature extraction and scoring can be considered essentially the same, and educational measurement and psychometrics can learn from natural language processing (NLP) and machine learning (ML) techniques to improve these processes (e.g., Han et al., 2019).
3. The construction of specific feature extraction rules are often implicitly guided by the type of latent construct that one aims to measure. While both unsupervised and supervised extraction rules are applicable, guiding the construction by the

type of differences the measurement construct is intended to assess is key to success. When aiming at a continuous latent trait, considerations of monotonicity of derived scores need to be incorporated, while for categorical latent variables and latent attribute profiles, homogeneity considerations may be more relevant (e.g., Tatsuoka, 1983; Gitomer & Yamamoto, 1991).

17.1 Views on the Data and Innovative Domains

17.1.1 Distilling the Data: Categorical vs. Continuous Latent Traits

Scholars and researchers in different disciplines have developed methods to derive indicators based on human behavior with a variety of data collection and analytic methods. In the field of educational measurement, test-taker responses to different types of questions in surveys or achievement tests (measuring cognitive skills or knowledge) are collected and scaled using state of the art statistical models to produce quantitative scores. The primary goal is to 'line up on a string', or sort, or order, or quantify in the sense of a numerical assessment of 'how much more or less', or to measure some kind of understanding. Another distinctive line is based on the desire for comparisons of the type 'A is different from B is different from C' without an implied order. The former represents *quantitative* test theory, while the latter can be described as *qualitative* test theory (Rost, 1988).

Regarding *qualitative test theory*, there is a long tradition of categorical or qualitative person variables going back to ancient times, including the description of four humors by Hippocrates (460–370 BC). Some recent examples include the Myers-Briggs Type Indicator (MBTI; Myers, 2016), and learner types and cognitive styles (Riding & Rayner, 2013), all of which are subject to serious and in some cases well-founded criticism. However, there is still solid interest in typological theories in education and psychology, that have some good empirical as well as theoretical support. Theories of coping behavior (Krohne, 1989), cognitive styles (Sternberg & Zhang, 2001), developmental stages (Piaget, 2000), even self-monitoring (Snyder, 1974) have been argued to be typological rather than dimensional constructs (Gangestad & Snyder, 1985; Rost & von Davier, 1995; von Davier & Rost, 1995). A more general discussion of the seemingly discrepant views of continuous versus categorical person attributes is given by DeBoeck et al. (2005), as well as more recently, von Davier et al. (2012). In terms of methods for analyzing response data while assuming categorical person attributes, there are cluster analytic methods (e.g., Punj & Stewart, 1983), taxometric analyses (e.g., Meehl, 1992; Meehl & Yonce, 1994, 1996; Ruscio et al., 2006.), as well as more generally, the well-developed statistical theory of discrete mixture distribution models (e.g., Peel & McLachlan, 2000) have been adapted to be included in educational and psychological research (Yamamoto, 1989; Rost, 1990; Rost & von Davier, 1995; von Davier & Rost, 1995; Wilson & Draney, 1997; for an overview: von Davier & Carstensen, 2007).

While philosophical differences exist between the two favoring, quantitative or qualitative person traits, the methodological issues surrounding the distinction of these two types of conceptualizations can be understood as the duality of the underlying assumptions of continuous versus discrete latent traits (DeBoeck et al., 2005; von Davier et al., 2012). Haberman et al. (2008) show how modeling approaches that assume discrete classes rather than a continuous trait can be statistically indistinguishable when only examining response data. External evidence and validity arguments are needed to support the decision in favor of a latent continuum versus a latent classification variable.

Related examples come from so-called mastery testing and learning where it is assumed that certain skills are either present or absent, and can be acquired, but not—or rarely—unlearned. One early attempt to develop a formal model that implements this assumption was presented by Macready and Dayton (1977). Tatsuoka (1983) with her rule-space approach had a strong impact on the field and in some ways revolutionized the thinking about how assessment development and modeling must go hand in hand if indeed a hypothesized set of skills or attributes are to be identified. Another relevant approach in this domain is the conceptualization of attributes as knowledge spaces (e.g., Doignon & Falmagne, 2012). These early conceptualizations have in common that several skills or knowledge states are looked at simultaneously, and that each respondent is characterized by their vector or responses only, which are the basis for inferring the not directly observed (latent) state of skills or knowledge. These states were subsequently referred to as attribute vectors or attribute patterns (e.g., Leighton et al., 2004), and models have been developed that allow a probabilistic approach to making inferences about these states. Mathematically, these models are multidimensional discrete latent variable models, as the states (attributes) typically only have two possible categories (mastery, non-mastery) while there may be multiple latent variables. The models became to be known as diagnostic models, or diagnostic classification models, and there is an abundance of these (e.g., Stout et al., 1996; von Davier, 2005a; De La Torre & Douglas, 2004; Hanson et al., 2000). An edited volume by von Davier and Lee (2019) collected contributions of many of the original authors of diagnostic classification models and provides an in-depth overview of the field of diagnostic modeling.

17.1.2 Data Is a Product of Decisions and Aggregation

Back in the 1990s, it was not unusual to see psychometricians, even very well-regarded senior applied psychometricians, argue that psychometrics is agnostic and does not care where the responses, or the sum score that is derived from these often dichotomous responses is coming from and how it came about. Essentially it was argued that "the score is the score is the score." The sum score, being a high-level data summary should be taken for granted, and not questioned. You may hear from proponents of this approach "it is what it is," so we do study the properties of the total score as it was given to us and do not question the rules that produced the score. This is of course, to state it very clearly, wrong, from the ground up. First, even classically trained psychometricians do item analysis and may suggest an item score to be eliminated if the item is too hard or too easy or does not correlate with the total, or better, the residual score. Second, items are found to be deficient, even in a high-profile case. For example, it was recently found that an SAT item had a scoring error.[1] Whether it is the machinery, moist paper when scanning SAT answers[2] or something else, data is not free from issues that can be traced either to human or machine issues, and only some errors can be declared processing or procedural errors.

In contrast to the sum score approach, item-level analytic approaches, including item factor analysis, item response theory (IRT), structural equation modeling (SEM), or latent class or latent profile models, have the benefit to enable scrutinizing response variables. However, in the age of CBA, to gain an understanding of student's behavior, it is not sufficient to look at item level scores alone. It is imperative to utilize the additional data collected by technology based test delivery and to examine what types of different behavioral patterns may lead to the same aggregates (von Davier, 2009, 2010). That is, the item-score or even more so the sum-score one may be given as an aggregate of correct or incorrect choices of what students do is only high-level summary of how humans interact with computers. Profoundly different action patterns and test-taking strategies (e.g., Cohen, 2007) can be behind the same overall sum score of correct or incorrect responses:

As an example: One student may try to answer an open-ended question, write an initial incorrect response, go to the next item, get an insight based on the information given, return to the open-ended item and replace the initial response with the correct one after thinking for two minutes. Another student reads the item, immediately recalls exactly this item he saw on a test prep website where students share memorized content and enters the correct response in less than 10 seconds.

The issue is that early educational measurement methods were largely developed by scholars and practitioners who either chose to use or were given only the total score. Also, separation of tasks in large testing organizations may have exacerbated the issue: Around that time, tests were paper based, the results were tallied by others, and all the analyst saw was the total score. This separation of tasks leads to the naïve approach that ignores that responses are given in a context, at a certain time, after having seen other tasks that came before or were presented after the current one (if students are allowed to revisit). However, CBA allows us to record a detailed trace of behaviors in the form of process data (such as response time, revisits, and action sequences), so there is more information about how students responded, and not only what they typed.

Experimental psychology has recognized early on that observations are obtained in a context, and that performance even on simple tasks suffers under fatigue and other factors that may impede the ability to perform optimally on the task at hand. In the field of educational measurement, this was the rationale behind standardized testing, ideally, so that all test takers would be administered the test under exactly the

[1] https://www.insidehighered.com/admissions/article/2019/06/03/college-board-admits-error-mathematics-scoring-may-sat.

[2] https://www.nytimes.com/2007/08/25/us/25sat.html.

same conditions. However, fatigue, anxiety, test-preparation, or lack thereof, and other intervening variables cannot be fully controlled. In addition, there are well known systematic differences in the statistical quality of different types of items that largely depend on their response format. Bejar (2017) provides an overview of the history of using constructed response items and describes how scoring reliability can be considered a major challenge of the use of data coming from human judgement of open responses.

One of the central phenomena we want to focus on is known as speed-accuracy tradeoff (Luce, 1986; Heitz, 2014) studied in the field of experimental psychology. A task attempted faster or even under time pressure is in many cases less likely to be performed correctly. Given more time, it is a well-known phenomenon that respondents have a higher likelihood to find the correct answer on a wide range of tasks. This implies that one should not only look at whether a student gets an item right, but also at the speed of completion, relative to similar tasks taken by the same or other respondents. However, unfortunately it was found that there are inter-individual differences with respect to this dependency. While one respondent may gain a great deal from having another minute to answer, another respondent may still not see improved chances to solve the same task. Instead of looking any further for fictitious examples, empirical evidence can be called upon. The analysis of data from the USMLE (United States Medical Licensing Examination) showed that native speakers, on average, complete all items in the test faster than second-language speakers, and that the latter typically perform not quite at the level of native speakers (Harik et al., 2020). Does that mean that non-natives will necessarily be slower, less proficient physicians? Physicians do not have to interact with real patients through multiple choice items, in the same format they were presented when taking the USMLE. In real world interactions, they may still take a little more time understanding the patient, which is why this component may be a valid part of the validity evidence. However, there may be medical disciplines where language processing speed is less relevant compared to other disciplines, so that a lower reading speed and maybe a different form of speed accuracy tradeoff should not be as relevant in some domains.

Examples or questions with respect to "what constitutes the correct response" are abundant from many real assessments: Do we take into account only the final answer in a mathematics item, or do we only count a response as correct if a student showed the calculations and how the solution was obtained? In a CBA reading task that presents text and answer fields on multiple pages, do we trust an answer to be correct if the data shows that the respondent spent less than 1/10 of a second reading the 500-word text? In a reading item presented on a computer, do we count a response collected by marking/highlighting words with a mouse as correct if the correct sentence was highlighted, but omitting to highlight the last two characters in the next sentence (Sukkarieh et al. 2012)? In a complex multiple-choice item, do we count an answer as correct if 3 out of 10 choices were correct, but a respondent selected all correct response options, as well as additional incorrect response options?

One can argue in all these cases that we will be agnostic and rely on how experts and item developers have designed the item and how scoring rules were specified.

This can be described as the 'contest' view of educational testing. In contrast, the 'measurement' perspective of testing would argue that scores (or even the scoring of items) should be optimized so that the rules are being adjusted to provide an aggregate statistic with highest possible psychometric qualities. Among these qualities that measurement typically aims at are reliability, validity, and comparability, all of which can be expressed as objective functions based on probabilistic tools, such as prediction and classification (Haberman, 2004).

In light of these considerations, measuring 'new constructs' or 'innovative domains' may appear even more challenging. If measuring mathematics skills has to consider testing situations, and process data, and whether human scoring went well, etc., how can innovative domains such as 'global competency' or 'creativity' be measured successfully? The next sections make an attempt to provide a few examples of how this can be accomplished.

17.1.3 Twenty-First Century Skills and Innovative Domains

The measurement of so-called twenty-First century skills and innovative domains—which are considered crucial for success in today's world characterized by fast emerging technological and digital innovations and a competitive global market (e.g., Kennedy & Sundberg, 2020)—may be particularly challenging, as these are constructs representing a broad set of knowledge, abilities, work habits, and non-cognitive constructs that are often intertwined. Besides the issue of coming to an agreement on how these constructs should be defined (e.g., Chen, 2023), there is also the question how they can be measured (e.g., Geisinger, 2016). While this is beyond the scope of this article, one major issue is that these labels refer to rather broad domains of human behavior and interaction. Examples in PISA are collaborative problem solving in the 2015 cycle, global competence in the 2018 cycle, and creative thinking in the 2022 cycle. Another issue is that the definition of measurement frameworks often depends on the evaluation from others, be it scholars, experts in the field or lay-people. Taking creative thinking as an example, it appears that a data point representing a creativity assessment would be more complex than a data point on a math quiz that assesses whether someone masters multi-digit multiplication. More so than traditional skills, one may expect that there is more disagreement whether a response was creative or not compared to whether an answer to a multiplication problem was considered correct or not. Another issue for a domain, such as global competence, is whether it is a skill that can be acquired and trained, a behavioral disposition that depends on personality variables, such as resilience and subjective well-being (OECD, 2019a), or a combination of both. The same could also be asked about creativity, which could be considered a combination of artistic talent, intelligence, openness and extraversion (OECD, 2019b). In contrast to that, a personality component contributing to reading literacy or mathematics competency seems to be less plausible.

However, assume we have an assessment framework that lays out (using evidence-centered design or other forms of rationales for evidence gathering) a set of observables expected to be indicative of interindividual differences in an innovative domain. How do we derive variables from these observables and construct or engineer scores (Luecht, 2013) that can be used to measure this *new* construct? In simple words, how do we attach numbers to observables ('score' the items) to get item responses that can be used to measure what we aim to measure? In most cases, experts or the item developers will be able to contribute to developing a scoring guide. Ideally, this would be sufficient, but most practitioners will experience that the expected behavior of test takers is not necessary the observed behavior. Commonly, some scores or some of the observed variables (item responses) do not behave as expected, at least for some items, and there seems to be no obvious statistical association with answers to other items, or the association is weak, or even reversed. In the extreme, there may be no answer key at all, and the responses that were observed require a complete sorting out, which means that a completely free problem space arises. Batchelder and Romney (1988) described a method of 'scoring a test without an answer key' for cases where no answer key is known, and the correct choice for each of the items on a test must be inferred. A couple of creative thinking items in PISA 2022 cycle follow this approach because no scoring rule is pre-defined but determined based on a data-driven way to indicate the novelty or unusualness (OECD, 2019b).

The case we illustrate below is less extreme and falls more into the category of optimizing the scoring of observables for which some hypothesis exists, and for which we have some evidence collected on a subset of observables that reliably allows us to assign test takers to lower or higher performers or different types of responders. This general approach will be outlined below and illustrated with a few examples from the PISA 2015 collaborative problem solving (CPS) items.

17.2 General Approach and Illustrations

17.2.1 A General Approach to Scoring Innovative Items

In this section, we assume that there is a vector of J of observables $X_1, \ldots, X_J$ that was constructed to measure a latent variable τ which is not directly observed. Each observable can be a random variable of different types, X_1 can be an audio clip, X_2 can be a category selection on a multiple-choice item, X_3 can be a string typed as a response to an 'explain why' item, X_4 can be a graph, and so on. Let $n = 1, \ldots, N$ represent the index set of the respondents and let τ_n denote the value the variable takes on for each respondent, and x_{nj} represents the responses of respondent n. Note that there can be multiple respondents where $n, m \in \{1, \ldots, n\}$ with $\tau_n = \tau_m$ so that there are equivalence classes of respondents who share the same 'level' with respect to the latent variable τ.

A scoring or feature function is a function that takes a subset of the observed variables and maps it onto the score or feature space we want to use for the measurement model to obtain an estimate. As a concrete and frequently used example, the feature function of a multiple-choice item with four response categories

A, B, C, D maps all responses that match a certain category called the key to 1 whereas all other responses are mapped to zero. More formally, if the key is B for item j, we have

$$Y_m = F_{MC}(X_j|B) = 1_{\{B\}}(X_j)$$

for $X_j \in \{A, B, C, D\}$. There are K different inputs for a simple MC item with K choices, but only two outputs of the function, namely correct (1) and incorrect (0). For a simple MC test with J items, there will be $X_1, \ldots, X_J \in \{A, B, C, D\}$ unscored observables representing the different category choices and $Y_1, \ldots, Y_J \in \{0, 1\}$ scored responses that are interpreted as representing the correct versus incorrect choices made by the respondents.

A complex multiple-choice item with more than on correct response can also be easily represented as a scoring function of the type above. Now X_j represents the set of selected responses among the total set, $X_j \subseteq \{A, B, C, D\}$ can be a subset rather than only a single selected element. As an example, if either one of the two response A, B is considered correct, we can write

$$Y_m = F_{MC}(X_j|A, B) = 1_{\{\{A\},\{B\},\{A,B\}\}}(X_j)$$

where the set-theoretic notation $\{\{A\}, \{B\}, \{A, B\}\}$ indicates that either A or B or selecting A, B is considered as a correct response, and all other responses are considered as incorrect.

Moreover, the scoring can also involve multiple observables, say X_i, X_j. One rather common example are item sums for partial credit scoring, which are often used to combine multiple closely related binary selections (select all that apply) or to reduce local dependencies (e.g., Verhelst, 1997). For the two items above, assuming their keys are D and C respectively, one can write

$$F_{PC}(X_i, X_j|D, C) = 1_{\{D\}}(X_i) + 1_{\{C\}}(X_j).$$

Obviously more complex feature functions are needed for other item types. Simple string comparisons are easily implemented in this framework, as well as comparisons of numerical input against a correct solution, or against an acceptable interval of solutions.

Also, while the index function produces only values in $\{0, 1\}$, a multiplication by a constant and adding simple scoring functions can produce more complex mappings into any set of real numbers. If for example a scoring rule requires answer A to be worth 1.781 points and answer B to be worth 0.91 points, one can write

$$F_{1.781,0.91}(X_i|A, B) = 1.781 * 1_{\{A\}}(X_i) + 0.91 * 1_{\{B\}}(X_j).$$

The output of scoring or feature functions are essentially what psychometricians use in their analyses. IRT uses scored item functions of the type shown for MC and partial credit (PC) items, while classical test theory (CTT) uses sums over MC and/or

PC functions and restricts analyses on the statistical properties of totals, largely ignoring item level information.

For simplicity, we focus on examples of a single quantitative latent variable, often called as unidimensional or univariate model, as it is the most common case to be confronted with in educational applications. Alternatively, variables that represent different types of respondents, such as the ones underlying latent structure and latent class models (von Davier, 2009) are also easily implemented by replacing the order relations below by inequalities. For a quantitative latent variable, there are important assumptions of commonly used factor analytic methods or IRT models that can be utilized to derive appropriate feature extraction rules or scoring functions. Recall that the exercise here is to develop a scale based on expert guidance or information about how items should be scored received from item developers, and that the feature functions are developed in support of this expert generated hypothesis of a quantitative ordering of respondents.

One common footing on which most quantitative latent trait models rest is the assumption of strict monotonicity (e.g., Junker, 1993; Mokken, 1971; Rasch, 1960; Stout, 1990; von Davier et al., 2007; Xu, 2007). For any two respondents with $\tau_n < \tau_m$ we postulate

$$E\left[F\left(X_{nj}\right)\right] < E\left[F\left(X_{mj}\right)\right]$$

for all items $j = 1, \ldots, J$. That is, if one person has a lower level on the latent variable than another person, $\tau_n < \tau_m$, the expected value on any scored item is also lower for this person is also lower than the expected score of the other. This is an assumption of latent variable models that can be used to derive the construction principle for scoring functions. It implies a positive correlation between expected item score and underlying score as a necessary, but not sufficient condition to be met by all scoring functions.

In addition to the strict monotonicity, in the subsequent sections, we use a generalized version of the customary assumption of local independence made in latent variable models, namely that the scoring/feature of observables given the latent variable are locally independent. This will be discussed in more detail in the sections below.

17.2.2 Illustrations

We assume that there is a *seed* of reliable information about differences between respondents. This seed can take on multiple forms. One simple case is that we can operate on the basis of known group differences. There may be experts and novices in the database, say the first 1,…,n/2 of the sample are novices and the second n/2 + 1,…,n of the sample are experts. More generally, it is assumed that there is a grouping, a random variable that assigns every respondent to exactly one of multiple equivalency classes. More formally, for any respondent n, there is a group membership function, i.e.,

$$g_n = G(x_{n1}, \ldots, x_{nJ}, z_{n1}, \ldots, z_{nK})$$

where $x_{n1}, \ldots, x_{nJ}$ denote the observables, and $z_{n1}, \ldots, z_{nK}$ denote other respondent covariates that are not part of the scoring or feature generation functions. These can be demographic variables, responses to attitude questions, self-reports, or observations and judgements form parents, teachers, supervisors, peers, etc.

Most commonly, when using the grouping $g = G()$ in the process of feature generation, it will be a function of existing scoring rules that were provided by developers and content experts. For the standard example of a quantitative scale, the grouping often has the form of a partial sum score

$$g_n = G(x_{n1}, \ldots, x_{nJ}) = \sum_{j_1, \ldots, j_l} F_{MC}\left(x_{nj_i} | K_{j_i}\right)$$

The item scores $F_{MC}\left(x_{nj_i} | K_{j_i}\right)$ that are part of this sum come from a set $j_1, \ldots, j_l$ of *gold-standard* items that are considered well-designed, without scoring rule flaws (e.g., the key K_{j_i} is assumed to be unique and correct), and with a strict monotonic association representing the ordering of respondents relative to the construct that is the target of inferences. This allows an evaluation of the other items, those that are not part of the gold standard, in reference to this ordinal grouping (Armitage, 1955; Cochran, 1954; Mantel & Haenszel, 1959).

Note that, in this example, the grouping variable is a partial score, based on *trusted* item functions. This means that

$$g_n \in \{0, \ldots, l\}$$

orders the sample based on their partial scores obtained on the l item score functions Y_{j_i}:

$$\sum_{j_1, \ldots, j_l} F_{MC}\left(x_{nj_i} | K_{j_i}\right) = \sum_{j_1, \ldots, j_l} Y_{j_i} = S_{j_1, \ldots, j_l}$$

Instead of partial scores, grouping variables may be based on pre-existing categorizations, such as a split of the sample into novices and experts, different leagues (local, regional, national, international), that can be assumed to be ordered with respect to the construct (e.g., Armitage, 1955).

Given a grouping variable $g_n = G()$ as defined above, additional scores (features) can be created

$$Y_k = F(X_1, \ldots, X_J | G)$$

from the observables $X_1, \ldots, X_J$. As before, consider a simple example of a MC item for which the key is either undefined, or initial data showed that the key provided by the developers does not work as expected. Examples are that the scored response variable based on this improper key is either uncorrelated or even negatively correlated with the partial sum $S_{j_1,\ldots,j_l}$ or the expected response is larger for a group that is expected to perform poorly than for a group that is expected to be proficient. These kinds of observations are made in exploratory analyses, as well as in distracter analysis.

Instead of discarding the item during the analyses, one option would be to define a new key as

$$K_j^* = \arg\max\left\{cor\left(S_{j_1,\ldots,j_l}, F_{MC}\left(X_j|k\right)\right) : k \in \{A, B, C, D\}\right\}$$

which is that choice of a key $k \in \{A, B, C, D\}$ that maximizes the correlation of the scored response on item j and the partial score based on the gold-standard items j_1, …, j_l. Then, the new MC variable would be defined as

$$Y_j^* = F_{MC}\left(X_j|k\right)$$

which guaranteed that the correlation to the partial score $S_{j_1,\ldots,j_l}$ is maximized. Content experts would immediately argue that this is an atheoretical choice, and that it is a data-driven approach without theoretical justification. However, the approach can be viewed as an exploratory that provides an empirical remedy in that experts can review the new variable and decide whether to discard it or retain it if there is a rationale for the 'new' key to be considered as a correct choice.

Accordingly, any other imaginable combination of observables and mappings into the score space can be considered. This opens the possibility to utilize machine learning techniques more actively, or any other statistical model. The basic idea is still unaltered: the objective function is either the correlation of the engineered item score with the *gold-standard* partial score, or a modified criterion that is also applicable in the case of qualitative/categorical or multivariate latent variables. This can be expressed as

$$\varphi^* = \arg\max\left\{1 - \frac{{\sigma_{e|\varphi}}^2}{{\sigma_{e|\varphi}}^2 + {\sigma_{b|\varphi}}^2} : \varphi \in \Omega\right\}$$

where φ is some parameter and ${\sigma_{e\,|\,\varphi}}^2 = E[V(Y_\varphi|\,S)]$ and ${\sigma_{b\,|\,\varphi}}^2 = V[E(Y_\varphi|\,S)]$ are the expectation of the conditional variance of the derived score/feature $Y_\varphi = F(X_1, \ldots, X_J \mid \varphi)$ given the parameter φ and the variance of the conditional expectations, respectively. The closer the maximized expression is to 1, the more homogeneous are the respondents in partial score groups with respect to the derived score.

17.2.2.1 Example 1: Adjusting Automated Scoring of Text Sequence Responses

An example of how scoring can be adjusted based evidence on how test takers responded was presented by Sukkarieh et al. (2012). The aim of their study was to improve the scoring of constructed response (CR) items that were automatically scored based on model responses. The example concerns highlighting items, in which a respondent is asked to mark a piece of text with the mouse during a reading task on a computer. One part of the reading passage contained the information that was considered the correct choice, and a machine scoring algorithm was devised to automatically assign correct and incorrect scores based on whether the highlighted sequence agrees with one out of several expert-provided model responses. The study ended up adjusting the scoring based on an algorithm that quantifies the agreement between a given sequence (the expert generated correct responses) and the observed sequence (what a respondent highlighted with the mouse). The algorithm used for this purpose is known as the longest common subsequence (LCS; Hirschberg, 1977) which has been mainly used in gene sequence analysis for decades.

Figure 17.1 shows an example of such a comparison of two short DNA strings. The LCS algorithm has been implemented in many programming languages and can be used as a standard function call to compare sequences in any (finite) alphabet. The LCS algorithm was used to calculate the overlap between highlighted string and model response, while choosing the maximum overlap across multiple model responses for every observed response. A normalized measure was derived as

$$d = \frac{2LCS(response, standard)}{len(response) + len(standard)}$$

and can be shown to vary between 0.0 and 1.0 for any observed *response* and any expert generated *standard*. If the normalized distance is 0, there is no overlap between standard (expert generated) response and the observed answer, if $d = 1.0$, the response is identical to the standard.

The original scoring process assigns credit only if an observed response agrees completely with one of the expert-provided standards. A modified feature extraction was devised that assigns credit also for responses with a constant somewhat lower distance threshold $C < d < 1.0$. As a function in the framework above, each engineered feature or combined item score can be viewed as

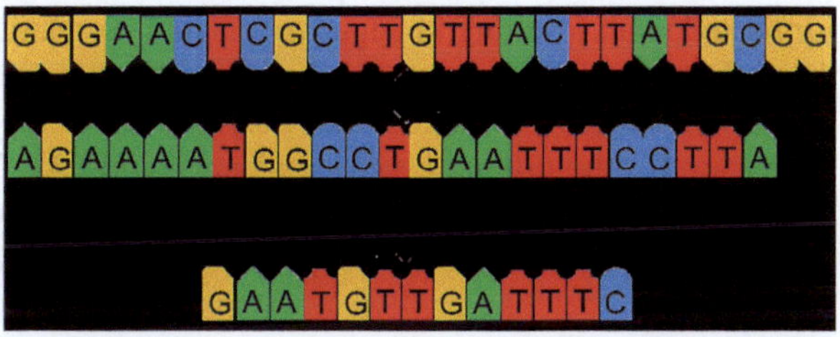

Fig. 17.1 Example of longest common subsequence (LCS)

$$Y = 1_{[C,1.0]}\left\{\frac{2LCS(response, standard)}{len(response) + len(standard)}\right\}$$

where $\frac{2LCS(response, standard)}{len(response)+len(standard)}$ can be viewed as a normalized sum of binary items that state whether a target character was highlighted or not. This is similar to scoring each character that should be highlighted individually, and finally defining a derived score if more than C% was scored as being highlighted.

Similar to the above feature sub-score correlation, the threshold C internationally in the PIAAC analysis, was chosen for each highlight item depending on whether the reliability of the overall score improved (OECD, 2013). The results presented in Sukkarieh et al. (2012) show that the adjusted scoring using the LCS approach with empirically determined threshold was able to improve the reliability of the engineered highlighting scores.

17.2.2.2 Example 2: Feature Engineering of Collaborative Problem-Solving Items

Another variant of feature generation was used in the analysis of the collaborative problem solving (CPS) data of the PISA 2015 assessment. In this assessment, students were asked to navigate a simulated online chat, interacting with simulated conversation partners that were providing pre-specified responses as reaction to student selected answers. This is, for all practical purposes, nothing but a series of MC items for which experts provided a response option for each stage that was considered the 'collaborative' choice. To make this more seem like a real interaction, students were able to correct their choices in subsequent stages of the scenario.

The PISA 2015 CPS scale used in the Field Trial and Main Survey will be presented here. At the Field Trial phase, this scale consisted of seven units that comprised 188 items. The units are based on simulated conversations with one or more computer-based agents that were designed to provide a virtual collaborative conversation. Students select an optimal sentence from a list of answers to navigate the conversation with agents or choose one or more actions programmed in the unit.

For two of the seven units (CPS01 and CPS05, shown in Table 17.1), changes to the scoring of responses were necessary before the data could be used for unidimensional IRT scaling. Using path analyses it was found that—due to the nature of the CPS items—data from these two units showed item dependencies in the responses. This was because of different paths that could be taken by students through the simulated chat, resulting in negative residual correlations. Since such dependencies have the potential to introduce bias into the results, the CPS chat items exhibiting dependencies were combined into polytomous "composite items" by summing the responses for the different paths students could take. Table 17.1 provides an overview of the combination rules used for these composite items. The use of these new scores reduced the number of dependent scores, but also led to combinations that

Table 17.1 Combination of items of Units CPS01 and CPS05 to achieve locally independent scoring in the PISA 2015 analyses. The variable names reflect internal naming conventions

New item ID for composite items	Combinations of CPS items
CPS01:201C	CPS01:201 + CPS01:202
CPS01:203C	CPS01:203 + CPS01:204 + CPS01:205
CPS01:206C	CPS01:206 + CPS01:207
CPS01:301C	CPS01:301 + CPS01:302 + CPS01:303
CPS01:304C	CPS01:304 + CPS01:305
CPS01:307C	CPS01:307 + CPS01:308 + CPS01:309A + CPS01:309B+ CPS01:310 + CPS01:311+ CPS01:312A
CPS01:312 BC	CPS01:312B + CPS01:313
CPS01:317C	CPS01:317 + CPS01:318 + CPS01:319
CPS05:103C	CPS05:103 + CPS05:104
CPS05:105C	CPS05:105 + CPS05:106 + CPS05:107
CPS05:201C	CPS05:201 + CPS05:202
CPS05:208C	CPS05:208 + CPS05:209 + CPS05:210
CPS05:212C	CPS05:212 + CPS05:213
CPS05:304C	CPS05:304 + CPS05:305

consequently reduced the number of item scores available for the IRT scaling was 164.

Note that this again fits into the framework presented above, namely that the new derived variables are function of individual scores obtained from other scored items. That is, for example,

$$Y_{\mathrm{CPS01:201C}} = Y_{\mathrm{CPS01:201}} + Y_{\mathrm{CPS01:202}}$$

a partial credit item was engineered by combining two binary items. A derived item score is a combination of categories considered correct based on expert input. The dependencies discovered based on the path design in the CPS chat items led to combination of scores.

One of the reasons why this was necessary is hinted at above. The structure of a chat item aims at simulating a natural conversation around a topic. Conversations, unlike knowledge questions, are somewhat redundant and may address the same issue multiple times over the course of the chat. Hence, there may be dependencies between responses even after the overall CPS skill is accounted for. This made it necessary to combine multiple chat responses that were tapping into the same issue.

For the PISA 2015 Main Survey, 134 items were selected out of the 164 (partly combined) items for the CPS domain (unit 101 was not selected). After fitting the IRT-based measurement models, response residuals were computed using the *mdltm* software (von Davier, 2005b). Item-level response residuals were calculated for each item by respondent interaction for all observed responses, and pairwise correlations among these residuals were computed for the different country samples. Several pairs of items were identified with highly correlated residuals; the pattern was quite

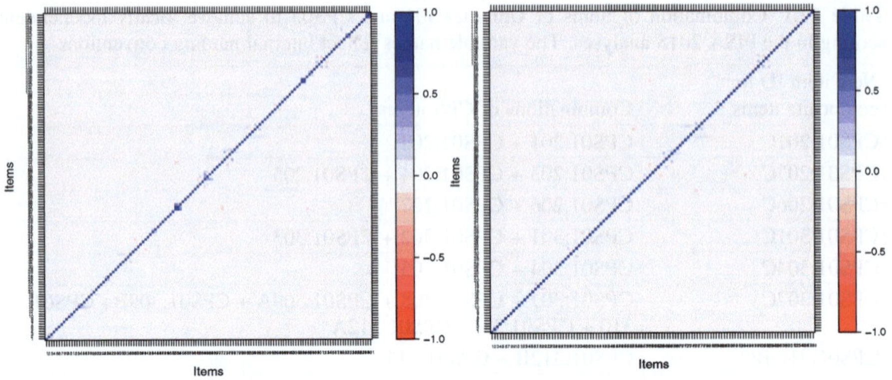

Fig. 17.2 Residual correlation plot among CPS items averaged across countries before further combining item scores based on locally dependent observables (**left**) and after treating the dependencies (**right**)

Table 17.2 List of composite items based on residual analysis

New item ID for composite items	Combinations of CPS items
CPS04:301C	CPS04:301 + CPS04:302 + CPS04:304
CPS06:107C	CPS06:107 + CPS06:108
CPS02:102C	CPS02:102 + CPS02:103
CPS02:209C	CPS02:209 + CPS02:210 + CPS02:211
CPS03:108C	CPS03:108 + CPS03:109 + CPS03:110 + CPS03:111
CPS05:108C	CPS05:108 + CPS05:109
CPS05:203C	CPS05:203 + CPS05:204
CPS05:308C	CPS05:308 + CPS05:309
CPS05:408C	CPS05:408 + CPS05:409

consistent across countries. Figure 17.2 (left panel) shows the correlations among CPS items averaged across countries. Relatively highly correlated item pairs are indicated by blue diamonds and were mainly found near the diagonal line. This indicates that the dependencies (high item pair correlations of response residuals) were mainly localized and taking place within a few selections. Rather than accounting for these in generalized latent traits measured through all responses in a unit, these localized dependencies were treated by item combinations as described above.

Based on the findings from the residual analyses, additional items were combined into composite items to remove the remaining local dependencies. Table 17.2 shows the combination of these items into composite items. Details about the items included in this rescoring can be found in the databases containing country-specific data, as well as variable and value labels.

After the combination into additional composite items, the number of CPS items was reduced to 121 (from the initial set or 134 items) for inclusion in the IRT scaling. To evaluate the performance of the composite items, residual analyses were repeated using the same set of countries. The resulting residual correlation plot in Fig. 17.2

(right panel) shows no remaining 'blue diamonds' indicating a successful elimination of the remaining residual dependencies in the new set of scores. This set was used in the analyses and operational reporting of the CPS scale for PISA 2015.

17.2.2.3 Issues Related with Bias and Equity of Engineered Scores

The above methods aim at providing expert guided, but empirically informed improvements to the scoring of innovative domains. The main tool that was described in some detail is the generation of item scores or features using unscored nominal item categories, or even response sequences, as input. These inputs can come from one observable or multiple observables that are grouped based on content, context, or some other criteria.

This content or contextual grouping is crucial, as it provides the rationale for why a derived score based on these grouped observables may have any validity. Combining randomly observed variables into some conglomerate of score summaries may seem an attractive option and is what an exploratory approach, such as machine learning, may do, but it comes at a cost. Combining items is already a threat to score interpretation. Combining context-wise unrelated items is most certainly not conducive to producing anything human experts can make sense of or would likely accept as an improvement.

However, there is an additional concern that needs to be addressed when scoring items, whether it is by feature generation and combination of multiple observables or not. This concern is related to the question whether the derived features are fair and do not function differently for certain subpopulations defined by language, citizenship, gender, ethnicity, and other policy relevant variables. While item developers and international assessment experts implement a range of measures to ensure items are fair, translatable, and unbiased, additional statistical checks are required to ensure that the engineered features as functions of observables remain unbiased.

Methods to assure this have been described in the context of differential item functioning (DIF) from the classical test theory perspective (Dorans, 1989; Holland & Thayer, 1986), as well as from the IRT perspective in a form of item fit and measurement invariance assessment (e.g., von Davier & Rost, 1995; von Davier et al., 2019; Oliveri & von Davier, 2014). These approaches can be applied in straightforward ways to derived item scores that were engineered as features derived from multiple observables.

17.3 Next Steps and Outlook

Tasks used in international large-scale assessments evolve and the desire to assess innovative domains is getting promoted with the technical delivery platforms available to these programs. Innovation takes place by improving these digital-based tasks to make them visually more appealing, more interactive, and more informative regarding how test-takers approach problems in both more traditional domains as well as more innovative domains.

As a by-product, more interactive tasks require students to show how they arrived at a response. This increases the amount of information collected on each respondent beyond a simple multiple-choice selection or the entering of a response using the keyboard. Often, several steps need to be taken before a response is recorded. While some of these steps may be more of a side-product and do not uniquely relate to whether a respondent mastered the task or understood what was asked, these steps or response sequences can be crucial parts of the task in many other instances.

One example is the problem-solving task used in PISA 2012 described by Greiff et al. (2013) and reanalyzed by Han et al. (2019). This task required students to conduct simulated experiments before drawing a graph as the response. It is conceivable (and was observed in the data) that there may be students who decided not to run any simulated experiment but rather directly move to answering the item. As the number of choices in graphing the response is finite, it is equally conceivable that the selections made for the graph can agree with the scoring rule even if a student did not do any experiment. In this case, either the student was clairvoyant, or did find the correct solution by a lucky guess or knew the response from another student who took the same item a day before. To score a response as correct even if the student did not follow the instructions and, therefore, must have guessed the answer is less than ideal. The methods described in this chapter can easily be adjusted to derive a scoring (feature) function that includes the requirement to have conducted at least one experiment.

For the combination of multiple observables that were initially scored individually and binary, the reader is referred to Verhelst (1997) for historical context and theoretical justification. Verhelst (1997) has shown that sums of binary item scores that follow the Rasch model are consistent with the Rasch partial credit model. For more complex combinations, model-based approaches that operate on subsets of observables can be envisioned. These may combine unsupervised learning for variable reduction, for example, based on latent class analysis (e.g., Wetzel et al., 2015), and models estimated with regularization methods (Chen et al., 2017) to ensure convergence to well defined and interpretable combinations of observable categories, as well as methods for optimal scaling. In all these approaches, however, it is deemed crucial to combine only those groups of observables that are considered to share a common context as identified by item development experts. This ensures that algorithms for the improvement of scoring are utilized on interpretable sets of variables.

In summary, the realization that data is never emerging without active observation and interpretation of these observations can break new ground and help teams of content developers and quantitative researchers to fine-tune and craft more reliable, valid, and fair assessments. The features or derived scores need to be scrutinized and carefully tested for invariance across subgroups of test takers and vetted against validity and comparability arguments so that inadmissible variants are discarded and only those engineered scores are retained that do not modify the construct of interest.

References

Armitage, P. (1955). Tests for linear trends in proportions and frequencies. *Biometrics (International Biometric Society), 11*(3), 375–386. https://doi.org/10.2307/3001775.JSTOR3001775

Attali, Y., & Burstein, J. (2006). Automated essay scoring with e-rater® V. 2. *The Journal of Technology, Learning and Assessment, 4*(3).

Baker, R. S., & Inventado, P. S. (2014). Educational data mining and learning analytics. In *Learning analytics* (pp. 61–75). Springer.

Batchelder, W. H., & Romney, A. K. (1988). Test theory without an answer key. *Psychometrika, 53*(1), 71–92.

Bejar, I. I. (2017). A historical survey of research regarding constructed-response formats. In R. Bennett & M. von Davier (Eds.), *Advancing human assessment. Methodology of educational measurement and assessment*. Springer. https://doi.org/10.1007/978-3-319-58689-2_18

Chen, D. (2023). Toward an understanding of 21st-century skills: From a systematic review. *International Journal for Educational and Vocational Guidance, 23*, 275–294. https://doi.org/10.1007/s10775-021-09511-1

Chen, Y., Li, X., & Liu, J. (2017). Regularized latent class analysis with application in cognitive diagnosis. *Psychometrika, 82*, 660–692. https://doi.org/10.1007/s11336-016-9545-6s11336-016-9545-6

Cipresso, P., Colombo, D., & Riva, G. (2019). Computational psychometrics using psychophysiological measures for the assessment of acute mental stress. *Sensors (Basel), 19*(4), 781. https://doi.org/10.3390/s19040781. PMID: 30769812; PMCID: PMC6412878.

Cochran, W. G. (1954). Some methods for strengthening the common $\chi 2$ tests. *Biometrics, 10*, 417–451.

Cohen, A. D. (2007). The coming of age for research on test-taking strategies. In J. Fox, M. Wesche, D. Bayliss, L. Cheng, C. E. Turner, & C. Doe (Eds.), *Language testing reconsidered* (pp. 88–111). University of Ottawa Press. https://books.openedition.org/uop/1568

De Boeck, P., Wilson, M., & Acton, G. S. (2005). A conceptual and psychometric framework for distinguishing categories and dimensions. *Psychological Review, 112*(1), 129–158. https://doi.org/10.1037/0033-295X.112.1.129

De La Torre, J., & Douglas, J. A. (2004). Higher-order latent trait models for cognitive diagnosis. *Psychometrika, 69*(3), 333–353.

Doignon, J. P., & Falmagne, J. C. (2012). *Knowledge spaces*. Springer Science & Business Media.

Dorans, N. J. (1989). Two new approaches to assessing differential item functioning: Standardization and the Mantel–Haenszel method. *Applied Measurement in Education, 2*(3), 217–233.

Eraslan, S., Yaneva, V., Yesilada, Y., & Harper, S. (2019). Web users with autism: Eye tracking evidence for differences. *Behaviour & Information Technology, 38*(7), 678–700.

Gangestad, S., & Snyder, M. (1985). To carve nature at its joints: On the existence of discrete classes in personality. *Psychological Review, 92*(3), 317–349.

Geisinger, K. F. (2016). 21st century skills: What are they and how do we assess them? *Applied Measurement in Education, 29*(4), 245–249. https://doi.org/10.1080/08957347.2016.1209207

Gitomer, D. H., & Yamamoto, K. (1991). Performance modeling that integrates latent trait and class theory. *Journal of Educational Measurement, 28*(2), 173–189. ETS Research Report Series. https://doi.org/10.1002/j.2333-8504.1991.tb01367.x

Greiff, S., Holt, D. V., & Funke, J. (2013). Perspectives on problem solving in educational assessment: Analytical, interactive, and collaborative problem solving. *Journal of Problem Solving, 5*(2), 71–91.

Haberman, S. J. (2004). Probability prediction and classification. *ETS Research Report Series, 2004*(1), i–23. https://doi.org/10.1002/j.2333-8504.2004.tb01946.x

Haberman, S. J., von Davier, M., & Lee, Y. H. (2008). Comparison of multidimensional item response models: Multivariate normal ability distributions versus multivariate polytomous ability distributions. *ETS Research Report Series, 2008*(2), i–25. https://doi.org/10.1002/j.2333-8504.2008.tb02131.x

Han, Z., He, Q., & von Davier, M. (2019). Predictive feature generation and selection using process data from PISA simulation-based environment: An application of tree-based ensemble methods. *Frontiers in Psychology, 10*, 1421. https://doi.org/10.3389/fpsyg.2019.02461

Hanson, T. E., Johnson, W. O., & Gardner, I. A. (2000). Log-linear and logistic modeling of dependence among diagnostic tests. *Preventive Veterinary Medicine, 45*(1–2), 123–137. https://doi.org/10.1016/S0167-5877(00)00120-3

Harik, P., Feinberg, R. A., & Clauser, B. E. (2020). How examinees use time: Examples from a medical licensing examination. In *Integrating timing considerations to improve testing practices* (pp. 73–89). Routledge.

Heitz, R. P. (2014). The speed-accuracy tradeoff: History, physiology, methodology, and behavior. *Frontiers in Neuroscience, 8*, 150.

Hirschberg, D. S. (1977). Algorithms for the longest common subsequence problem. *Journal of the ACM (JACM), 24*(4), 664–675.

Holland, P. W., & Thayer, D. T. (1986). Differential item functioning and the Mantel-Haenszel procedure. *ETS Research Report Series, 2*, i–24.

Junker, B. W. (1993). Conditional association, essential independence and monotone unidimensional item response models. *The Annals of Statistics, 21*(3), 1359–1378. http://www.jstor.org/stable/2242199

Kennedy, T. J., & Sundberg, C. W. (2020). 21st century skills. In B. Akpan & T. J. Kennedy (Eds.), *Science education in theory and practice. Springer texts in education.* Springer. https://doi.org/10.1007/978-3-030-43620-9_32

Krohne, H. W. (1989). The concept of coping modes: Relating cognitive person variables to actual coping behavior. *Advances in Behaviour Research and Therapy, 11*(4), 235–248. https://doi.org/10.1016/0146-6402(89)90027-1

Leighton, J. P., Gierl, M. J., & Hunka, S. M. (2004). The attribute hierarchy model for cognitive assessment: A variation on Tatsuoka's rule-space approach. *Journal of Educational Measurement, 41*, 205–237. https://doi.org/10.1111/j.1745-3984.2004.tb01163.x

Luce, R. D. (1986). *Response times: Their role in inferring elementary mental organization (No. 8).* Oxford University Press on Demand.

Luecht, R. M. (2013). Assessment engineering task model maps, task models and templates as a new way to develop and implement test specifications. *Journal of Applied Testing Technology, 14*(1), 1–38.

Macready, G. B., & Dayton, C. M. (1977). The use of probabilistic models in the assessment of mastery. *Journal of Educational Statistics, 2*(2), 99–120. https://doi.org/10.3102/10769986002002099

Mantel, N., & Haenszel, W. (1959). Statistical aspects of the analysis of data from retrospective studies of disease. *Journal of the National Cancer Institute, 22*(4), 719–748.

Meehl, P. E. (1992). Factors and taxa, traits and types, differences of degree and differences in kind. *Journal of Personality, 60*(1), 117–174.

Meehl, P. E., & Yonce, L. J. (1994). Taxometric analysis: I. Detecting taxonicity with two quantitative indicators using means above and below a sliding cut (MAMBAC procedure). *Psychological reports.*

Meehl, P. E., & Yonce, L. J. (1996). Taxometric analysis: II. Detecting taxonicity using covariance of two quantitative indicators in successive intervals of a third indicator (MAXCOV procedure). *Psychological Reports, 78*(3_suppl), 1091–1227.

Mokken, R. J. (1971). *A theory and procedure of scale analysis with applications in political research.* de Gruyter (Mouton). https://doi.org/10.1515/9783110813203

Myers, S. (2016). Myers-Briggs typology and Jungian individuation. *Journal of Analytical Psychology, 61*(3), 289–308.

OECD. (2013). Technical report of the survey of adult skills (PIAAC). https://www.oecd.org/skills/piaac/_Technical%20Report_17OCT13.pdf

OECD. (2019a). PISA 2018 assessment and analytical framework. https://doi.org/10.1787/19963777.

OECD. (2019b). PISA 2021 creative thinking framework. https://www.oecd.org/pisa/publications/PISA-2021-creative-thinking-framework.pdf

Oliveri, M. E., & von Davier, M. (2014). Toward increasing fairness in score scale calibrations employed in international large-scale assessments. *International Journal of Testing, 14*(1), 1–21. https://doi.org/10.1080/15305058.2013.825265

Peel, D., & McLachlan, G. J. (2000). Robust mixture modelling using the t distribution. *Statistics and Computing, 10*, 339–348. https://doi.org/10.1023/A:1008981510081

Piaget, J. (2000). Piaget's theory of cognitive development. *Childhood Cognitive Development: The Essential Readings, 2*, 33–47.

Punj, G., & Stewart, D. W. (1983). Cluster analysis in marketing research: Review and suggestions for application. *Journal of Marketing Research, 20*(2), 134–148.

Rasch, G. (1960). *Probabilistic models for some intelligence and attainment tests*. Nielsen & Lydiche. (Expanded edition, Chicago, University of Chicago Press, 1980).

Riding, R., & Rayner, S. (2013). *Cognitive styles and learning strategies: Understanding style differences in learning and behavior*. David Fulton Publishers.

Rost, J. (1988). Test theory with qualitative and quantitative latent variables. In R. Langeheine & J. Rost (Eds.), *Latent trait and latent class models* (pp. 147–171). Plenum.

Rost, J. (1990). Rasch models in latent classes: An integration of two approaches to item analysis. *Applied Psychological Measurement, 14*(3), 271–282. https://doi.org/10.1177/014662169001400305

Rost, J., & von Davier, M. (1995). Mixture distribution Rasch models. In G. H. Fischer & I. W. Molenaar (Eds.), *Rasch Models: Foundations, recent developments, and applications*. Springer. https://doi.org/10.1007/978-1-4612-4230-7_14

Ruscio, J., Haslam, N., & Ruscio, A. M. (2006). *Introduction to the taxometric method: A practical guide*. Lawrence Erlbaum.

Shute, V. J. (2011). Stealth assessment in computer-based games to support learning. *Computer Games and Instruction, 55*(2), 503–524.

Snyder, M. (1974). Self-monitoring of expressive behavior. *Journal of Personality and Social Psychology, 30*(4), 526–537. https://doi.org/10.1037/h0037039

Sternberg, R. J., & Zhang, L. F. (2001). *Perspectives on thinking, learning, and cognitive styles (Edited)*. Lawrence Erlbaum.

Stout, W. F. (1990). A new item response theory modeling approach with applications to unidimensionality assessment and ability estimation. *Psychometrika, 55*, 293–325. https://doi.org/10.1007/BF02295289

Stout, W., Nandakumar, R., & Habing, B. (1996). Analysis of latent dimensionality of dichotomously and polytomously scored test data. *Behaviormetrika, 23*, 37–65. https://doi.org/10.2333/bhmk.23.37

Sukkarieh, J., von Davier, M., & Yamamoto, K. (2012). From biology to education: Scoring and clustering multilingual text sequences and other sequential tasks. *ETS Research Report Series. ETS RR-12-25, 2012*(2), i–43.

Sun, R. (2008). *The Cambridge handbook of computational psychology*. Cambridge University Press.

Tatsuoka, K. K. (1983). Rule space: An approach for dealing with misconceptions based on item response theory. *Journal of Educational Measurement, 20*(4), 345–354.

Verhelst, N. (1997). *Modeling sums of binary responses by the partial credit model*, Cito (Measurement and Research Department Report No. 97–7). Arnhem, The Netherlands.

von Davier, M. (2005a). A general diagnostic model applied to language testing data. *ETS Research Report Series, 2005*(2), i–35. https://doi.org/10.1002/j.2333-8504.2005.tb01993.x

von Davier, M. (2005b). *mdltm: Software for the general diagnostic model and for estimating mixtures of multidimensional discrete latent traits models [Computer software]*. ETS.

von Davier, M. (2009). Some notes on the reinvention of latent structure models as diagnostic classification models. *Measurement: Interdisciplinary Research and Perspectives, 7*(1). https://doi.org/10.1080/15366360902799851

von Davier, M. (2010). Why sum scores may not tell us all about test takers. *Wang, L. (Ed.): Special issue on Quantitative Research Methodology. Newborn and Infant Nursing Reviews, 10*(1), 27–36.

von Davier, M. (2018). Automated item generation with recurrent neural networks. *Psychometrika, 83*, 847–857. https://doi.org/10.1007/s11336-018-9608-y

von Davier, M. (2019). Training Optimus Prime, M.D.: Generating medical certification items by fine-tuning OpenAI's gpt2 transformer model. *Computation and Language*. https://doi.org/10.48550/arXiv.1908.08594

von Davier, M., & Carstensen, C. H. (2007). *Multivariate and mixture distribution Rasch models: Extensions and applications*. Springer. https://doi.org/10.1007/978-0-387-49839-3

von Davier, M., Khorramdel, L., He, Q., Shin, H. J., & Chen, H. (2019). Developments in psychometric population models for technology-based large-scale assessments: An overview of challenges and opportunities. *Journal of Educational and Behavioral Statistics, 44*(6), 671–705. https://doi.org/10.3102/1076998619881789

von Davier, M., & Lee, Y.-S. (2019). *Handbook of diagnostic classification models: Models and model extensions, applications, software packages*. Springer Nature. https://doi.org/10.1007/978-3-030-05584-4

von Davier, A. A., Mislevy, R. J., & Hao, J. (2021). Introduction to computational psychometrics: Towards a principled integration of data science and machine learning techniques into psychometrics. In A. A. von Davier, R. J. Mislevy, & J. Hao (Eds.), *Computational psychometrics: New methodologies for a new generation of digital learning and assessment. Methodology of educational measurement and assessment*. Springer. https://doi.org/10.1007/978-3-030-74394-9_1

von Davier, M., Naemi, B., & Roberts, R. D. (2012). Factorial versus typological models: A comparison of methods for personality data. *Measurement: Interdisciplinary Research and Perspectives, 10*(4), 185–208.

von Davier, M., & Rost, J. (1995). Polytomous mixed Rasch models. In G. H. Fischer & I. W. Molenaar (Eds.), *Rasch models: Foundations, recent developments, and applications*. Springer. https://doi.org/10.1007/978-1-4612-4230-7_20

von Davier, M., Rost, J., & Carstensen, C. H. (2007). Introduction: Extending the Rasch model. In *Multivariate and mixture distribution Rasch models. Statistics for social and behavioral sciences*. Springer. https://doi.org/10.1007/978-0-387-49839-3_1

Wetzel, E., Xu, X., & von Davier, M. (2015). An alternative way to model population ability distributions in large-scale educational surveys. *Educational and Psychological Measurement, 75*(5), 739–763. https://doi.org/10.1177/0013164414558843

Wilson, M., & Draney, K. (1997). Partial credit in a developmental context: The case for adopting a mixture model approach. *Objective Measurement: Theory into Practice, 4*, 333–350.

Xu, X. (2007). *Monotone properties of a general diagnostic model (ETS Research Report RR-07-25)*. Educational Testing Service. https://doi.org/10.1002/j.2333-8504.2007.tb02067.x

Yamamoto, K. (1989). *HYBRID model of IRT and latent class model (ETS Research Report RR-89-41)*. Educational Testing Service.

Chapter 18
Preparing and Analyzing Log and Process Data in Large-Scale Assessments

Saskia Keskpaik, Reinaldo Dos Santos, and Franck Salles

Abstract Over the last decade, France, like many other countries, has been undergoing a transition from paper-based to digital assessments in order to measure student performance in education. There is a rising interest in France in technology-enhanced items, which offer new ways to assess traditional competencies, as well as address higher-order skills and the so-called twenty-first century skills. The rich data captured by these items, often referred to as process data, allow insight into how students tackle the item and their response strategies. "Big data" solutions were set up in France in order to handle the large volume and complexity of these process data from technology-based assessments. Data-driven approaches, stemming from the domain of educational data mining, as well as theory-driven approaches, drawing on expert knowledge of the field, were applied to the process data. Our experience has taught us that making process data meaningful necessitates integrating three paramount aspects of our work: didactics, technology, and analytics. Coordinating findings, constraints, and needs of these three aspects is essential to devise process indicators, draw hypotheses, and allow for valid interpretation. Valid process data interpretation enables findings to be actionable in the field, from an "assessment for learning" perspective.

18.1 Transition to Computer-Based Assessment

The Directorate for Evaluation, Forecast, and Performance (DEPP) of the French Ministry of Education has been implementing standardized student assessments for more than 50 years (Trosseille & Rocher, 2015). The DEPP is also responsible for managing various international studies in France: PISA, PIAAC, PIRLS, TIMSS,

S. Keskpaik · R. D. Santos · F. Salles (✉)
Directorate for Evaluation, Forecast and Performance (DEPP), Ministry of Education, Paris, France
e-mail: franck.salles@education.gouv.fr

L. Khorramdel et al. (eds.), *Innovative Digital-Based International Large-Scale Assessments*, Methodology of Educational Measurement and Assessment,
https://doi.org/10.1007/978-3-031-90951-1_18

and ICILS, among others. Since the beginning of the 2000s, due to a growing demand in France for indicators on student performance on three different levels (national, local/regional and classroom level), the number of large-scale standardized assessments conducted by the DEPP has been increasing significantly.

During the last few years, France, like many other countries, has been undergoing a transition from paper-based to digital assessments in order to measure student performance in education. This transition has been favored by the national context, improvement of information technology infrastructure in schools, growing public demand for local student performance indicators, as well as by global efforts towards the digitalization of international studies (PISA in 2015, TIMSS in 2019, PIRLS in 2021) and the development of assessment-related educational technology.

The transition from paper-based to digital assessments has brought along new opportunities: cost reduction, use of multimedia and simulations in tasks, adaptive testing, automated scoring, etc. However, the digitalization of our assessments has also given rise to new challenges: comparability with paper-based tests for ensuring trends, usability, security, equipment, digital divide, etc. The two major aspects of this transition have been scalability and innovation. The first very large-scale digital assessment took place in France in 2015 when 160,000 sixth-grade students were assessed on desktop computers in French, mathematics, and science (Andreu et al., 2016). Today, the DEPP is conducting census-based assessments at all grade-levels from grade 1 to grade 10 (around 800,000 students per grade level), including computer-based assessments from grade 6 to grade 10.

The transition to the digital mode has paved the way for the use of innovative content and functionality in student assessments. There is a rising interest in France in technology-enhanced items (TEI), which offer new ways to assess traditional competencies, as well as to address higher order skills (Drijvers, 2019) and the so-called twenty-first century skills. Moreover, these items have the potential to link assessment feedback to learning. The TEIs can be extremely valuable when measuring problem solving skills. Compared to traditional assessments, they not only provide scoring information, whether the response is correct or not, but allow rich data to be collected that enable the method by which students arrived at their answers to be determined (Greiff et al., 2015).

In 2016, collaboration was initiated between the DEPP and the Ministry of Education of Luxembourg in order to conceptualize and design innovative interactive items in mathematics. Nearly 30 interactive items were field-tested in France in May 2017; around 15,000 students in grade 9 were sampled to participate in this pilot test (3000 students per item). Students' digital traces were recorded in log data files.

The presentation of materials in this chapter follows the article Salles et al. (2020) but puts some more emphasis on the data processing infrastructure and workflow, as well as on the technical and methodological choices involved in the collection, handling, and the analysis of the log data.

18.1.1 *Towards New Types of Data…*

The rich data produced by digital assessments are often referred to as process data. "In cognitive ability testing, process data can be defined as empirical information about the cognitive (as well as meta-cognitive, motivational, and affective) states and related behavior that mediate the effect of the measured construct(s) on the task product (i.e., item score). Thus, operationally, process data can be regarded as the empirical data reflecting the course of working on a test item," (Goldhammer & Zehner, 2017, p. 128).

A lot of effort was put into conceptualizing and designing the assessment framework for the first TEIs. Nevertheless, the involvement of the domain experts in the technical development of these items, particularly in the decisions regarding the data collection, was rather marginal in the early days of the transition from paper-based to digital assessments in France. The development process also lacked the supervision from the data analyst's point of view. Regarding data collection, the emphasis was given to the ability to reconstruct (replay) the student response.[1] Hence, the whole process of how students tackle the item was captured in the log files: clicks, mouse movements, drawing and deleting lines, dragging and dropping on a scale or an object, keystrokes, etc. The actual technical decisions and solutions of how to construct the logs and record the students' interactions remained mainly the responsibility of the software developers.

The minimal involvement of domain experts and data analysts in the phase of items' technical development had its implications at a later data processing stage: the decisions taken by the software developers did not always prove to be optimal from the data analysis perspective.[2] These innovative items brought along increased complexity in terms of data processing. The data collected with these items came in new shapes and structures; students' responses were captured in a semi-structured JSON format, relatively unfamiliar to the data analysts working at the DEPP. Moreover, the data from the interactive items included large amounts of low level process data—all mouse movements, clicks, keystrokes being recorded—that needed to be sorted out in order to make sense of the data.

Reflections and frameworks on how to conceptualize, represent and analyze log data collected in digital assessments have started to emerge (Goldhammer & Zehner, 2017; Kroehne & Goldhammer, 2018). Instead of collecting the complete log data, the focus is shifting towards the importance of collecting relevant log events in order to address a certain research question (Kroehne & Goldhammer, 2018). The

[1]Kroehne and Goldhammer (2018) define this as the replay-completeness of the log data, which "will be fulfilled, if all information is contained in the log data required to recreate a hypothetical 'screen cast' (i.e., representing the progression of visual information presented on screen during the whole session)," (Kroehne & Goldhammer, 2018, p. 556).

[2]For example, a slider value or a position on a scale was recorded in pixels instead of the actual value on the slider/scale. The pixel values needed to be recoded into actual response values for the subsequent data analysis, adding an extra step in the data pre-processing phase.

emphasis in these frameworks is given to theory-driven or top-down approach according to which the use of log data for assessment purposes is already considered in the design phase of the measurement to ensure and improve the interpretation of derived process indicators (Goldhammer et al., 2021). It is important to note though that a balance needs to be found between these two approaches: "logging everything," which brings along the risk of collecting very noisy data sets, and "logging only according to theoretical assumptions," thereby limiting what can be analyzed later in secondary use of the process data.

Since the early days of the transition to digital assessment, the DEPP has considerably strengthened the collaboration between domain experts, data analysts, and software developers in order to identify which interactions need to be captured regarding the assessment framework and the research questions. Tightening the collaboration in the phase of items' technical development has also helped to ensure that the technical implementations under consideration are optimal from the data processing point of view.

18.1.2 …and "Big Data" Solutions

Interactive items generate very large data sets if process data are collected at a fine-grained level. Our first large-scale field-trial of interactive items in 2017 produced nine gigabytes of data. The data processing infrastructure used at the DEPP back then—consisting mostly of local machines, limited in their capacity of in-memory processing—was not fully adapted to deal with this kind of volume. Therefore, a need to look into *big data* solutions was rapidly identified.

The concept of big data is often explained in terms of three Vs: volume, variety and velocity.[3] Out of the three Vs, the first two ones, volume and variety, are particularly relevant in the context of the digital transition of the DEPP's assessments.[4] Big data are often measured in terabytes (1000 gigabytes) or petabytes (1,000,000 gigabytes). The DEPP still faces relatively small big data, expressed mainly in gigabytes, but the volume is growing fast. This is due to the increasing number of census-based assessments conducted by the DEPP, as well as the growing number of interactive items used in these assessments. Also, the number of test-takers involved in DEPP's assessments has been rising rapidly. DEPP's national census-based assessments produce around 30 gigabytes of data per assessment, and some of their projects are expected to collect terabytes of data. Such large volumes cannot be processed on a single local machine.

---

[3] Sometimes five or more Vs are being brought out, we will concentrate on the three most commonly mentioned ones here.

[4] Assessments conducted by the DEPP do not yet involve real-time data processing, although this might become a requirement in the near future due to the growing demand for reducing the delay between the test administration and the production of reports given back to the participating schools.

Big data are also characterized by their variety. The data arrive in different types, forms, and from different sources. The data can be unstructured (such as text, images and sounds) or partially structured (recorded in JSON or XML format). The process data collected in the very first field-test of interactive items are semi-structured, recorded in JSON format. The DEPP's assessment projects today additionally involve collecting and processing students' responses in the form of text, image and audio files.

Big data can, thus, be defined as the collection and storage of data so large and complex that it exceeds the processing capabilities of existing techniques and systems. These challenges in data storage and data processing have made us look into distributed infrastructure solutions which enable allocating the data and data processing on several interconnected machines, or a "cluster." A cluster is composed of the so-called worker nodes whose work is organized and monitored by a "master" node.

One of the very first infrastructure solutions that was used for processing the data from the field-test of interactive items was an experimental "sandbox" platform set up at the National Institute of Statistics and Economic Studies (INSEE). Among other technologies, this big data platform makes use of the document-oriented database program MongoDB. Early solutions also included cloud computing services, such as Microsoft Azure: a data science virtual machine, comprised of 16 cores and 64 gigabytes of random-access memory (RAM), was set up for the DEPP by the consulting company CapGemini as part of a collaboration undertaken in 2017/2018. Several in-house experimental clusters have been set up at the DEPP since then. Amongst others, the DEPP has been using Cloudera Hadoop distribution (version 5.13.3); this big data infrastructure is comprised of two master servers, four workers, and an edge node for interfacing with clients. Each machine has 48 central processing units (CPUs), except for the edge node, which has 16 CPUs.

Hadoop is a free and open source framework, designed to deal with large volumes of data in a distributed environment. It is a whole ecosystem comprised of many different components and services. These include the Hadoop Distributed File System (HDFS), a distributed, scalable and portable file system, written in java for the Hadoop framework. Optimized for large data files, the HDFS allows the segmentation of data into blocks (the block size is set to 128 megabytes by default but can be configured).[5] These blocks are then replicated (by default three times) and distributed on several cluster nodes. This highly fault-tolerant design ensures the integrity of the data in case a cluster node fails.

The DEPP's big data infrastructure also includes the Apache Spark™, an open source distributed computing framework. Spark does not have its own file management system; it makes use of the HDFS. Spark's fundamental data structure is the resilient distributed dataset (RDD): an immutable distributed collection of objects. Each dataset in an RDD is divided into logical partitions, which may be computed on

[5]For instance, with a block-size of 128 megabytes and a data file of 10 gigabytes, the data will be segmented in $10 \times 1024 \div 128 = 80$ blocks.

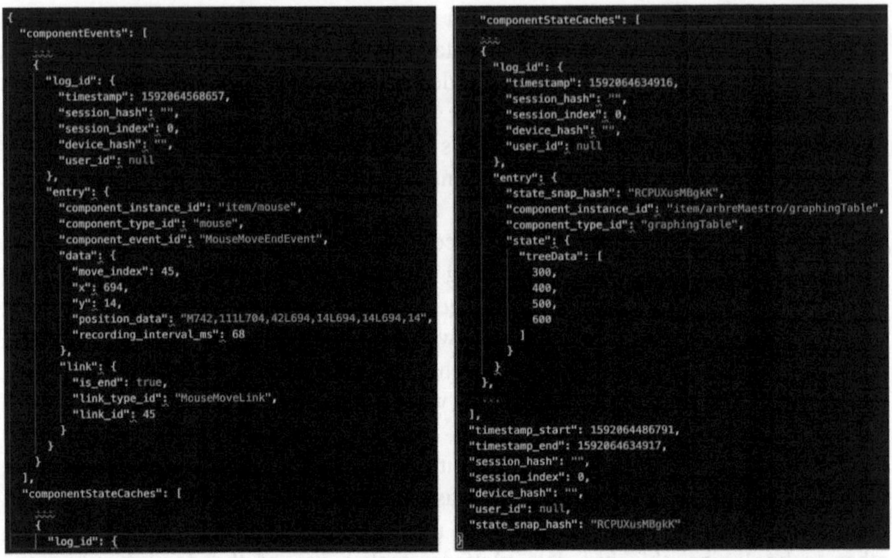

Fig. 18.1 Extract of a log file

different nodes of the cluster. Computer calculations are carried out within the RAM, making the data processing faster and more efficient. Spark has APIs in Python and in R languages; it has several ready-to-use data processing libraries and makes use of the existing Python and R libraries (Apache Spark, n.d.).

18.2 From Pre-Processing to Machine Learning: General *Vs.* Item-Specific Workflow

When describing the process data workflow, it is important to distinguish between general and item-specific data processing. Some of the data treatments can be applied to all of the items, whereas other phases of data handling are specific to each item. The raw data file received by the DEPP from its contractor responsible for the platform hosting is double-encoded (in base64 and lzstring; GitHub, n.d.). The decoding of the data is common to all items and is achieved by using Spark and parallel processing. After decoding, the data is stored on HDFS.

Many interactive items share a common toolbox, containing tools such as a pencil, an eraser, a ruler, a compass, and a calculator. However, each interactive item also has a specific set of tools, unique to the item. Depending on the item, this could be a graphing or a calculation tool, a slider component, a scale… The logged components, therefore, vary according to the item. Each recorded log file has an *events* and a *state* component (Fig. 18.1): the events include all student-by-item interactions, and the state represents the final state or the screen shot of the item the moment it was submitted by the student. For each event, a timestamp is collected, which allows for reconstructing student response processes in time. Each log file also contains a start and an end timestamp, expressing the total time spent on the item.

Scripts were developed (in the form of Jupyter notebooks) for each phase of the data processing, from data preparation and through pre-analysis to the application of machine learning algorithms. Before being able to run any analysis, we need to transform the data from the tree-like "noisy" semi-structure (JSON) to a tabular (columns and rows) structure. During this pre-processing phase we use Spark, in particular the Spark SQL module for structured data processing, in order to run the transformations and the computations efficiently in a distributed manner.

The next data processing steps include gradual unwrapping and creating of variables of interest. First, all events and the associated timestamps and values are being extracted from the events component of the logs. Each data row in this intermediary data table represents a single interaction, with columns expressing the properties of the interactions, such as the event's type (ex: mouse) and name (mouse move start event), value (x and y coordinates), and timestamp.

In order to facilitate the exploration and the interpretation of our analysis, we make use of feature engineering to create new meaningful variables. These are constructed mostly following the top-down scheme that generates features based on hypotheses developed by content experts (Han et al., 2019). The final output of these data pre-processing steps is a reduced data frame that can be analyzed on a single machine, using data analysis tools such as python pandas (pandas development team, 2021), numpy (NumPy, 2020), and scikit-learn (n.d.). Each data row in the final transformed data frame represents a single student, and columns contain variables that have either been extracted from the log files or created through the feature engineering.

Once the data were transformed into a usable format, we first implemented a data-driven or a bottom-up approach, applying various methods and algorithms to the dataset that were at the time being explored in the domain of educational data mining.

18.2.1 *Data-Driven Approach to Process Data: A Toolbox*

Data-driven analyses are made quite easy by the wide range of methods and algorithms available. These methods mobilize quite diverse areas of mathematics, computer science, or data science. In the course of our various studies, we have thus mobilized graph theory on an item. We have also tried to describe task sequences using different methods derived from natural language processing. Finally, and most importantly, we made extensive use of different machine learning methods, to carry out both supervised and unsupervised learning.

18.2.2 Graph Theory

Graph theory is the mathematical and computer science discipline that studies graphs, which are abstract models of patterns of networks connecting objects (Barnes & Harary, 1983). These models are made up of vertices (also called nodes or points) and edges (called links or lines) between these vertices. The edges may be directed or undirected, meaning that vertices describe an equal or unequal relation. In network theory, all agents are equal and, therefore, describe an undirected graph. In flow theory, the edges have a start node and an end node and, therefore, represent a directed graph. Traditionally, the edges of a directed graph are called arrows (Riaz & Ali, 2011).

A cyclic graph or circular graph is a graph that consists of a single cycle, or in other words, a number of vertices (at least 3) connected in a closed chain. The cyclic graph with n vertices is called Cn. The number of vertices in Cn is equal to the number of edges, and each vertex has a degree 2, i.e. each vertex has exactly two edges incident to it.

In one of the interactive items that we studied, called "The Lake," students were asked to approximate the perimeter of a lake on a map, using a digital ruler, available to them within the item environment. One of the research objectives was to distinguish students who used mathematical properties of the rectangular form of the lake and, therefore, measured the sides to add and multiply, from the students who used the polygonal approximation of a complex form by circling the lakes with the ruler before summing the distances.

The use of graph theory consisted of replacing each stroke made with the digital ruler by a graph node and connecting two strokes at their edge if their extremities were in each other's neighborhood. Through this adaptation, the question "are students making use of the polygonal approximation of a complex form?" (or, in practical terms, "are the ruler strokes roughly forming a circuit?") could become "is the graph generated by the strokes an undirected cyclic graph, depending on the neighborhood cut?" which is a mathematically answerable question.

18.2.3 Natural Language Processing (NLP)

In order to determine the strategy that students used to solve items, we thought about using NLP methods, assuming that each digital tool used by the student would form a character into a string. Once these strings were generated, patterns could be found using different methods. NLP is an extremely popular range of methods nowadays. It can be implemented in a wide range of situations, from sentiment analysis to pattern recognition (Sukkarieh et al., 2012; Hao et al., 2015; Yamamoto et al., 2016; He et al., 2019a, 2019b).

In computer science, the longest common substring problem is to find the longest string that is a substring of two or more strings. Its solution is to build the generalized suffix tree of the strings and to find the longest common subtree (Weiner, 1973). This method is frequently used in plagiarism detection or pattern recognition. However, its strictness has the drawback of offering less resistance to noise than other methods as n-grams or Levenshtein distance matrices.

An n-gram is a sub-sequence of n elements constructed from a given sequence. From a given word sequence, it is possible to obtain the likelihood function of the appearance of the next word. From a learning corpus, it is possible to construct a probability distribution for the next word with a history of size n (Sidorov et al., 2014). The n-gram models are a possible way to categorize a sequence of operations performed by the student, in order to extract patterns in terms of sequences of tasks (He & von Davier, 2015).

The Levenshtein distance is a measure (in the mathematical sense of the term) of the difference between two strings of characters. It is also known as editing distance or time dynamic distortion. This distance is greater as the number of differences between the two strings is greater (Ristad & Yianilos, 1998).

The Levenshtein distance between two words A and B is the minimum cost to transform A into B by performing the following elementary operations only: (i) substitution of a character of A by a character different from B; (ii) insertion (or addition) in A of a character of B; (iii) deletion (or erasure) of a character of A. Each of these operations is associated with a cost. Generally, the cost is equal to one for all three operations.

18.2.4 Supervised Learning

In our situation, the use of supervised learning was guided by one main idea: to find the features that could predict success or failure on the item. As the variable of interest was categorical (pass or fail), we chose to focus on classification rather than regression. We then explored two different classifiers: a binary regressor with logistic regression and an ensemble model with the random forests.

Logistic regression is a form of binomial regression (Cramer, 2002). Mathematically, a binary logistic model has a dependent variable with two possible values, represented by an indicator variable, where the two values are labelled "0" and "1". In the logistic model, the log-odds (the logarithm of the odds) for the value labelled "1" is a linear combination of one or more independent variables. The variables can each be a binary variable (two classes, coded by an indicator variable) or a continuous variable (any real value).

As a regression, the model describes the probability of output (a binary variable) in terms of input (features derived from the click and keystroke stream) and does not perform statistical classification (it is not a classifier). However, it can be used to produce a classifier, for instance, by choosing a cut-off value and classifying inputs with probability greater than the cut-off as one class and below the cut-off as the other.

Random forests (Breiman, 2001) are an ensemble learning method that works by building a multitude of decision trees during learning, before returning the class mode (classification), or the average forecast (regression) of individual trees (Han et al., 2019). The purpose of a decision tree is to create a model that predicts the value of a target variable from several input variables. A decision tree or classification tree is a tree in which each internal node (non-leaf) is marked with an input characteristic. Arcs from a node labelled with an input characteristic are labelled with each of the possible values of the target or output characteristic, or the arc leads to a subordinate decision node on a different input characteristic. Each leaf of the tree is labelled with a class or probability distribution on the classes, which means the dataset has been classified by the tree either in a specific class or in a particular probability distribution.

Decision trees are nevertheless known for their many disadvantages. The first is the tendency to overfit the training set. The second is its non-deterministic aspect; the order of use of the functionalities generates a completely different tree structure.

The training algorithm for random forests applies the general technique of bootstrap aggregating, or bagging (Breiman, 1996), to tree learners. Given a training set $X = x_1, \ldots, x_n$ with responses $Y = y_1, \ldots, y_n$, bagging repeatedly (B-times) selects a random sample with replacement of the training set and fits trees to these samples. After training, predictions for unseen samples x' can be made by averaging the predictions from all the individual regression trees on x', or by taking the majority in the case of classification trees.

The number of samples or trees, B, is a free parameter. An optimal number of trees B can be found using cross-validation. However, the question of the order of the features remains unsolved. That is why random forests differ slightly from the general bagging; random forests use a modified tree-learning algorithm that selects a random subset of the features at each candidate split in the learning process. This process is sometimes called *feature bagging* or *random subspace method* (Ho, 1998).

18.2.5 Unsupervised Learning

Unsupervised learning is a machine learning technique used to detect patterns in a data set with no prior information about this data. It is mainly used as a clustering technique, to group or segment the data, in order to identify communalities. Clustering algorithms can be understood as a family of different approaches. The main ones are the density-based or centroid-based algorithms. When analyzing the process data from the first interactive items, we chose to use the most widely used algorithm of each of those families (Wierzchoń, & Kłopotek, 2018).

DBSCAN (density-based spatial clustering of applications with noise) (Ester et al., 1996) is a density-based data partitioning algorithm in that it relies on the estimated density of clusters to perform partitioning. The DBSCAN algorithm uses two parameters: the distance ε and the minimum number of points, MinPts, which must be within a radius ε for these points to be considered as a cluster. The input parameters are, therefore, an estimate of the point density of the clusters. The basic idea of the algorithm is then, for a given point, to retrieve its ε-neighborhood and check that it contains MinPts (or more). This point is then considered as part of a cluster. We go then through the ε-neighborhood systematically to find all the points in the cluster.

Advantages of the DBSCAN algorithms are numerous. DBSCAN does not require one to specify the number of clusters in the data *a priori*, as opposed to k-means. DBSCAN can find arbitrarily shaped clusters. Due to the MinPts parameter, the single-link effect (different clusters being connected by a thin line of points) is reduced. DBSCAN has a notion of noise and is robust to outliers. Parameters, MinPts, and, ε can be set by a domain expert, if the data are well understood.

Drawbacks, however, can also be listed: DBSCAN is not entirely deterministic, in that border points that are reachable from more than one cluster can be part of either cluster, depending on the order the data are processed. The approach also does not work well with data sets with large differences in densities because the MinPts-ε combination cannot then be chosen appropriately for all clusters. Finally, if the data and scale are not well understood, choosing a meaningful distance threshold ε can be difficult.

Partitioning into k-means is a data partitioning method and a combinatorial optimization problem (MacQueen, 1967). Given points and an integer k, the problem is to divide the points into k groups, often called clusters, in order to minimize a certain function. We consider the distance from a point to the average of the points of its cluster (called the centroid); the function to minimize is the sum of the squares of these distances.[6]

The initialization is a determining factor in the quality of the results (local minimum). There are two common initialization methods: Forgy's algorithm (Lloyd, 1982) and random partitioning. Forgy's algorithm assigns the k points of the initial averages to k randomly selected input data. Random partitioning randomly assigns a cluster to each data point and then (pre-first) calculates the initial mean points.

The k-means method is a fast and simple method for clustering. It is systematically convergent and offers an easy visualization. On the other hand, the number of clusters k is an input parameter, and an inappropriate choice of k may yield poor results. That is why, when performing k-means, it is important to determine the number of clusters in the data set through cross-validation. The Silhouette coefficient (Rousseeuw, 1987) is generally considered as a good measure of consistency in data partitioning.

Moreover, convergence to a local minimum may produce counterintuitive results. The k-means++ algorithm (Arthur & Vassilvitskii, 2006) tackles this by specifying a

[6]The k-means algorithm is quite popular for its simplicity and effectiveness; however, there are other methods with equal or better power, in particular the Latent Class Analysis (Magidson & Vermunt, 2002).

procedure to initialize the cluster centers before proceeding with the standard k-means optimization iterations.

These different methods of data analysis have proven successful in the past (Jain et al., 1999). They constitute a powerful methodological toolkit for extracting and interpreting complex data with little or no structure. However, as we have witnessed, they are not sufficient on their own. In many cases, data cannot be simply separated from noise without prior knowledge of the studied field. For this reason, we have turned our efforts towards a hybrid, more theory-oriented approach, drawing on expert knowledge of the field to derive an analytical approach to the data.

18.3 Towards a Theory-Driven Approach

Given the DEPP's missions, detailed in the first part of this chapter, ensuring validity of content in assessment requires close collaboration with institutional stakeholders, teachers, and researchers with regards to national curriculum. It enables DEPP's assessment constructs, instruments, publications, and reports to fit several levels of the educational system, from policy makers to the classroom. In this context, purely data-driven analysis is not suitable for ministerial communication, as validating the interpretation of process data in the light of these constructs is paramount.

In 2017, most DEPP's TEI development concerned France's national assessment CEDRE. CEDRE (Subject-related sample-based assessment cycle) is a sample-based large-scale assessment aiming to measure students' abilities in several school subjects at the end of grade 9 every 5 years in France. CEDRE 2019 framework is based on the national French curriculum in mathematics. Administered in 2008 and 2014, CEDRE in mathematics was administered again in May 2019. This new cycle is computer-based for the first time. As trends must be secured so that the comparability with the previous cycles is guaranteed, a large part of the test instruments are similar to formerly paper-based items. However, the DEPP developed TEIs, very different from more classical item formats in order to profit fully from the potentialities of assessing mathematics with digital tools (Stacey & Wiliam, 2013). Offering students the possibility to use digital tools during the assessment allows for outsourcing basic procedural work to such tools (Drijvers, 2019); therefore, opportunities are given to students to engage more fully in higher order skills, such as problem solving, devising a strategy, and carrying out mathematical thinking, which the CEDRE aims to capture (MEN DEPP, 2017).

In support of the work of defining a construct, as well as developing test instruments for CEDRE, we convened several structuring concepts of research in mathematical didactics and more generally in educational research (notion of "conceptions," task analysis, content analysis, semiotic analysis, assessment, etc.). Our theoretical background was specifically issued from educational research aiming at tackling and describing student cognitive models and assessment task analysis in general. This theoretical framework fed DEPP's log data reasoning in order to propose a valid interpretation of patterns and regularities of features derived from data-driven methods (both unsupervised and supervised) in terms of the same theoretical background that was operated in the first steps of the test development process.

18.3.1 A Theoretical Framework in Mathematics Assessment

Research in mathematical didactics has developed powerful theoretical frames and notions accounting for students' mathematical conceptions (Sfard, 1991; Robert, 1998; Douady, 1991; Brousseau, 2006). These frames can be operated to develop student models according to the context of the national curriculum (El-Kechaï et al., 2011). Additionally, and in order to take into consideration the ways in which students can express their mathematical knowledge in DEPP's computer-based test environment and TEIs, these notions have been completed by others (Rabardel, 2002; Drijvers et al., 2016; Trouche, 2003) aiming at accounting for this modality in our construct. Task didactical *a priori* analysis was then undertaken during item development according to this theoretical background. Measuring how students can express their qualitative understanding of mathematical concepts (Kilpatrick et al., 2001) in CEDRE's TEIs is of utmost interest for the DEPP. For this purpose, process data became an important outcome (Bergner & von Davier, 2018) along with the scored responses on specific TEIs.

18.3.2 A Priori Task Analysis

An example of such TEI, "Tree growth" is given below (Fig. 18.2), chosen among CEDRE's new interactive items (Salles et al., 2020). In this task, two nonlinear functions model two different tree growths. Both are given in linked numerical and graphical representations (Stacey & Wiliam, 2013). Students act on the task by entering the age of the trees in months, presenting a numerical representation (a table of values). A calculation tool returns the corresponding tree heights and a graphing tool spots the points in the graph. Both actions are realized when one presses the "Calculate and graph" button. By default, values for 300, 500, and 600 months are given. Additional tools can be used: a pencil, an eraser (not allowing to erase given information), a ruler, a compass, and a calculator. The question can be translated as: "At what age (other than 0 months) do both trees have the same height?" The correct response is 390 months.

From a grade 9 student's point of view, this task requires conceptual understanding of functions and their representations (table of values, graph). Function conception has two different characters as Sfard (1991) and Doorman et al. (2012) showed: "In lower secondary grades, functions mainly have an operational character and are seen as an input–output 'machine' that process input values into output values. In higher grades, functions have a more structural character with various properties

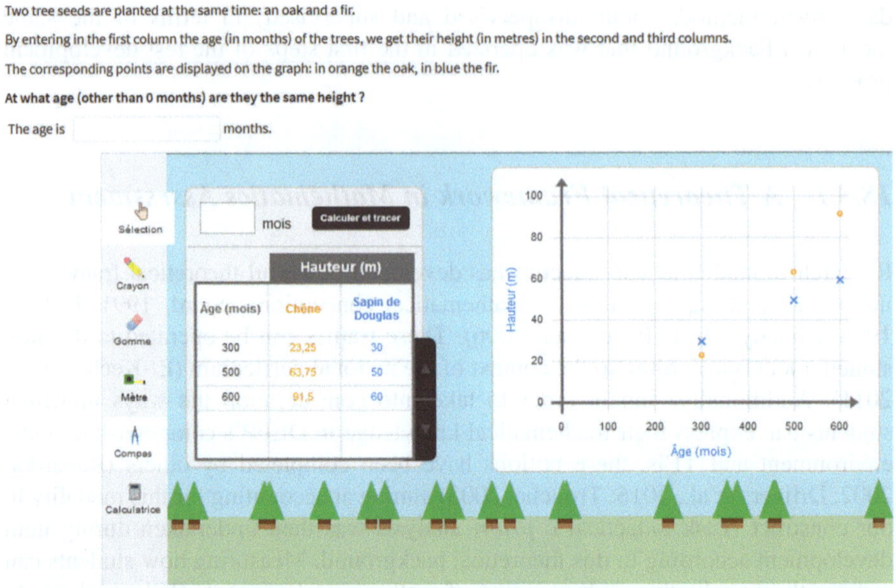

Fig. 18.2 Interactive item "Tree growth". Note. Partly translated in English for the purpose of this publication. Hauteur = Height; Âge (en mois) = Age (in months); Chêne = Oak; Sapin de Douglas = Douglas fir; Calculer et tracer = Calculate and graph)

(Sfard, 1991). They become mathematical objects that are represented in different ways, are ordered into different types according to their properties, and are submitted to higher-order processes such as differentiation and integration. We argue that the transition from functions as calculation operations to functions as objects is fundamental for conceptual understanding in this domain," (Doorman et al., 2012, p. 1243). The distinction of operational and structural characters of conceptions in algebra is also present in institutional French curriculum resources in mathematics (MEN DGESCO, 2016).

Students can adapt the way they work with these conceptions in the problem-solving process by adding intermediate information using the calculation and graphing tool (Roditi & Salles, 2015). On one hand, they can then opt for a *trial and error* method. This would consist of entering a number of months, comparing the results returned either in the numerical or graphical representation, and deciding to enter another number of months until the solution (390) is found. Alternating attempts around the target value or aiming at it from below or above could improve the trial and error process. These students show essentially good understanding of the concept of functions in their operational characteristics. This method could imply a relatively large number of tries. On the other hand, students who understand that both functions are increasing, notably from studying them in the graphical representation, can quickly aim at the target number of months. The pencil, for example, can be used to draw lines and introduce a continuous representation of the functions. The inversion of tree heights between 300 and 500 months can also be noticed.

Table 18.1 Core elements of instrumented action schemes in relation to the operational character of functions

1	Knowing the difference between input and output in a contextual use of a function as a model.
2	Understanding that the tool returns unique heights (outputs) for both trees (numerical and graphical representations), when choosing a number of months (inputs).
3	Entering a number of months within the domain [0; 600].
4	Comparing outputs either in the numerical or graphical representation. Validating by linking to the real life situation.
5	Deciding on the next number of months to enter considering the comparison to the previous one.
6	Iterating the process.

These students understand functions as objects with properties in their structural character.

In addition to the computer's keyboard and mouse, the following digital tools are embedded within the item environment: a calculation and graph tool, specific to the item; a pencil; an eraser; a compass; and a calculator.

The calculation and graph tool does not require complex usage schemes (Rabardel, 2002). Two usage schemes are identified for this tool: enter a number of months within the domain [0; 600] via an input box and a popup number pad and understand that the tool returns unique heights (outputs) for both trees (numerical and graphical representations) when the button "calculate and graph" is clicked. It is to be noted that no tutorial or tool training was provided to students, as the assumption is made that usage schemes are close to usual tools, such as a currency converter or regular graphing tools. Nevertheless, we can imagine some students feeling the need to check out the tool by first using it for testing purposes, for example, entering extreme values or values not directly connected to the primary task. In reference to the instrumental approach, we describe next how the tool can be utilized in this situation in order to solve the task, once assumed that students will build an instrument (Trouche, 2003) from the "calculation and graph" tool within the item environment.

Goal-oriented action schemes are organized around the core elements shown in Table 18.1. This type of instrumentation is linked to an operational approach of the concept of functions. As mentioned earlier, the pencil can also be used in order to get a continuous model on the domain or part of it. Students can link points together using the pencil. This is an intermediate step towards the primary task. The following core elements are part of action sequences using the pencil, as well as the calculate and graph tool, and are characteristic of a structural understanding of the concept of a function (Table 18.2).

Of course, students may operate composite goal-oriented action schemes, mixing the use of both principal tools. For example, one can use the pencil to sketch continuous graphs (potentially after trying 100 or 200 months) to get a more complete view of the graph shapes or rely on points' colors and choose 400 months in the first tries and then use a trial and error strategy to aim precisely at the target with the calculate and graph tool.

Table 18.2 Core elements of instrumented action schemes in relation to the structural character of functions

1	Understanding that the growth phenomenon is a continuous one. Hence, the functions modelling it are continuous.
2	Assuming both functions will be strictly increasing.
3	Using the pencil to link consecutive points together.
4	Decide on the next number of months to enter considering line intersection.
5	Going back to numerical values to aim at accuracy.

18.3.3 Theory-Driven Process Data Analysis

CEDRE's interactive items, the "Tree growth" item among them, were included in the May 2017 pilot test with a sample of 3000 grade 9 students per item. Process indicators (Goldhammer et al., 2021) were either defined or constructed from the log data via feature engineering, as a result of the *a priori* didactical analysis. They could potentially give evidence of success or failure in the task considering either the operational or structural character of functions used by students. Problem solving strategies based on an operational understanding of functions imply that the calculation and graph tool was used more often, potentially through a dichotomous strategy, hence testing many numbers of months and spending more time on the item. Features such as the month list length, the number of alternating attempts within this list, and the time spent on the item could then shed light on such strategies in the data. Symmetrically, a strategy based on a structural conception could lead to optimizing both the number of tries and the time spent, aiming at the target interval from the very first tests, perhaps using the pencil tool. Features such as the standard deviation of the values in the month list and the distance between first inputs and the target are of specific interest. Accordingly, the main features used in the analytical models and their nature are displayed in Table 18.3.

Classical machine learning analysis was then applied to the data. Supervised learning was implemented to determine the predictive power of the model on student achievement and to estimate the weight of the variables in the prediction. Unsupervised learning aimed at clustering the sample. Mean values of the most important variables were then calculated for each of the clusters derived from unsupervised learning.

The analysis of the variables of importance in this model allows us to select the parameters that are indicative of the student's success in completing the item (Fig. 18.3).

The first important characteristic is the number of values tested in the tree data table. The second is the variance of the values tested, i.e. the extent to which these values are concentrated around the target value. In the same spirit, we also find the difference between the first input value and the target value as well as the difference between the second input value and the target value (to avoid the noise of the first value). In addition, we also find the number of alternations around the target value, a characteristic that expresses the choice of a dichotomous search procedure. Finally,

Table 18.3 Main features used in the analytical models

Feature	Nature	Design
Month list length	Integer variable	Present in the log data
First input between 200 and 600	Boolean	Constructed through feature engineering
Number of alternating attempts within the month list	Integer variable	Constructed through feature engineering
Time spent on the item, in seconds	Continuous variable	Present in the log data
Distance between first input and target value	Integer variable	Constructed through feature engineering
Distance between second input and target value	Integer variable	Constructed through feature engineering
Distance between the last input and the target	Integer variable	Constructed through feature engineering
Standard deviation of the values in the month list	Continuous variable	Constructed through feature engineering
Target value is in the month list	Boolean	Present in the log data
Pencil use	Boolean	Present in the log data

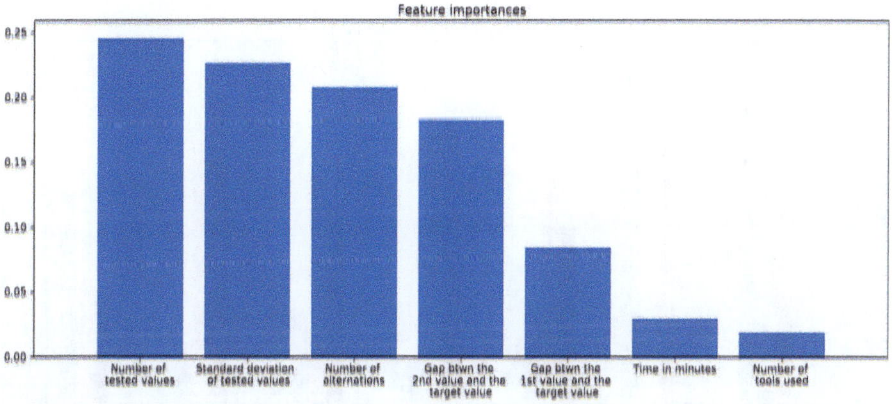

Fig. 18.3 Feature importance

but to a lesser extent, we find the time spent resolving the exercise and the number of tools used.

Clustering analysis distinguishes 4 clusters corresponding to 4 different students' profiles (Figs. 18.4 and 18.5). In the case of this specific item, each cluster's size represents 25% of the responding students. Two profiles (green and orange on the graph) achieved the task. The other two (blue and red) correspond to students who failed. Apart from the obvious variable "value 390 tested," no other variable allows for discrimination between success and failure. This result is disappointing in the sense that the model cannot explain students' achievement on the item. However, variables of interest in the *a priori* analysis allow profiles along a dimension other

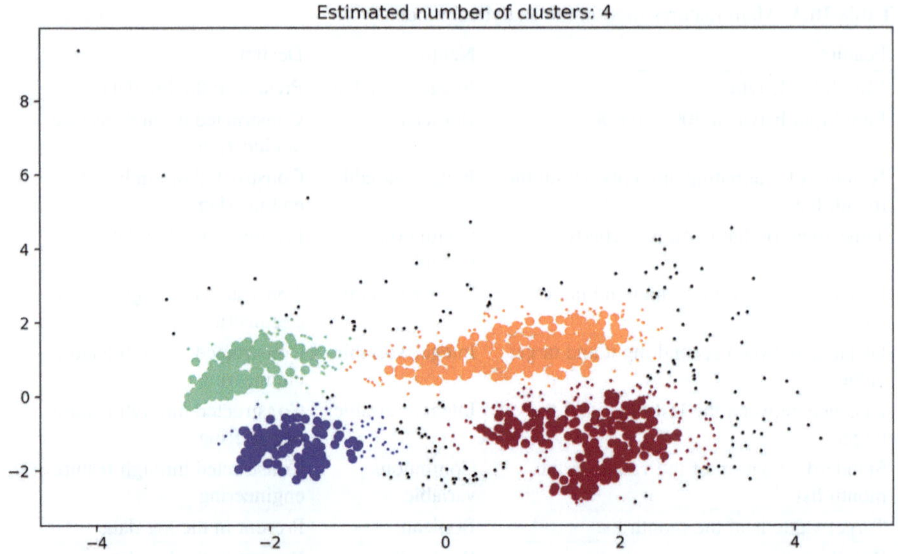

Fig. 18.4 Result of DBSCAN

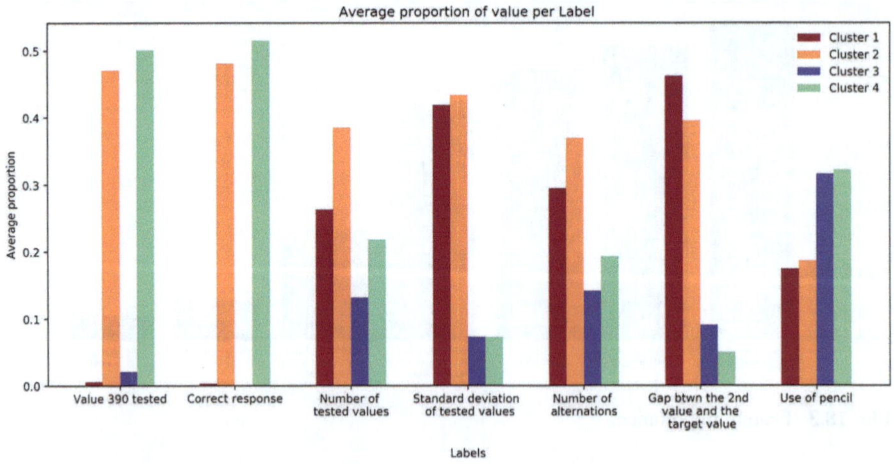

Fig. 18.5 Average proportion of value per Label

than achievement to be described. Two clusters (orange and red in Fig. 18.4) show a large number of inputs, a large number of alternating inputs, a first input far from the target, and a large distribution of inputs. These characteristics can be interpreted as students from these groups preferring a trial and error problem-solving strategy, approaching the underlying concept of function in its operational aspect. Half of them achieved the task successfully; the other half did not. The other two groups share a different and opposite profile description according to the same variables: a small number of inputs, a small number of alternating inputs, a first input close to the target, and a narrow distribution of inputs. Moreover, this second category of students used the pencil more often, altogether identifying strategies related to the structural aspect of functions. Like the first two groups, the structural approach led to failing the task for half of the students who favored it.

18.4 Discussion

While the didactical and analytical models used in this example could not predict grade 9 students' achievement on this item, nevertheless, they could help us identify two solution strategies well known in the didactics literature. From a curriculum designer perspective, this kind of result on a national level could be very valuable. It could help decision making based on the evidence of the manifestation of both structural and operational understanding among grade 9 students. Of course, the analysis should be replicated and extended to a full set of items potentially aiming at discriminating between the two conceptions within the variation and relationships domain or even in other domains. Furthermore, as it is strongly based on research findings in didactics, such a result can be fruitfully disseminated towards subject matter experts, as well as policy makers and stakeholders in charge of teacher training, and contribute to better addressing the use of large-scale assessment findings at the classroom level. And last but not least, this type of qualitative result, along with corresponding released items, could be of paramount use within the classroom in service of improving teaching practices, such as assessment for learning (Grapin & Sayac, 2017).

One of the lessons drawn from DEPP's experience in dealing with process data considers the working methodology employed when it comes to data mining in education. Making log data meaningful necessitates integrating three paramount aspects of our work: didactics, technology, and analytics. Coordinating findings, constraints, and needs of these three aspects is essential to devise process indicators, draw hypotheses, and allow for valid interpretation. As Goldhammer et al. (2021) state: "The valid interpretation of process indicators depends on a careful and clear definition of how the targeted attribute, empirical evidence (behavioral low-level features and derived process indicators), and the activity space that can elicit the desired behavior (actions) are linked to each other."

In our case, linking construct definitions, TEIs development, and feature selection or engineering has to be thought through at every step of the process. The definition of cognitive models and targeted conceptions feed the test environment and item development. In return, interactions and tool use are carefully accounted for in *a priori* task analysis based on the theoretical frame. At this stage, important modifications in the item development occur. Simultaneously, software developers, along with data scientists and educators, need to design what low-level features need to be logged in order to give evidence of targeted conceptions or cognitive processes. This very sensitive step consists of finding a delicate balance between "log everything," risking collecting very noisy data sets, and "log only according to theoretical assumptions," risking limited serendipitous investigations *a posteriori*. Besides, the defined log data should allow for feature engineering and construction of process indicators.

One of the biggest challenges of process data handling has been the multitude of log structures. On one hand, this is due to the lack of cooperation between the data analysts and the software developers during the early days of the transition from paper-based to digital assessments. Different software developers have taken different roads when making decisions about the technical implementation of log files. On the other hand, standardization is a complicated task due to the specificity of the items. The DEPP is looking into international educational technology standards in order to find optimal solutions for capturing logs in a more consistent way across items and assessments.

The handling of the process-data at the DEPP has been a relatively manual process so far. Aiming for the standardization of log files would allow the data processing workflow to be automated in the future. As a member of the FLIP+ e-assessment community (2021), the DEPP is searching for international partners and collaborations aspiring to develop standards and best practices in the domain of process data analysis. For instance, the DEPP has an ongoing partnership with CITO (n.d.) from the Netherlands to explore ways to automate or industrialize data processing workflows. In the context of this collaboration, large-scale sample-based data collection was organized in France during fall 2020. New TEIs in mathematics were included in this assessment.

Today, the TEIs are mainly included in the DEPP's sample-based assessments, and the data are still small enough such that secondary analyses can be conducted on a single machine. As more and more of these interactive items are being used in census-based assessments, it may be prudent to look into distributed machine learning modules in the near future. In the long run, the DEPP is aspiring to set up a fully integrated system with an automated workflow that would allow actionable reports on student performance to be produced as quickly as possible after the assessment.

One of the many possible uses of process data in an educational context is fundamentally related to large-scale assessments. Psychometrics, and in particular item response theory, are commonly used methods for determining the ability of individuals to respond to a latent trait measured by the test. Nevertheless, process data are likely to be good assets for improving the quality of the model.

The first idea that comes to mind is to compare the speed of students with their performance. Various studies show that "the faster, the better" is not an entirely accurate hypothesis (Weeks et al., 2016; Naumann, 2019; Ulitzsch et al., 2019), and this is certainly a promising area of research. On the other hand, the idea that beyond performance, students would employ different strategies to respond to items is a shared observation within the teaching community. Among the DEPP's future research objectives is investigating if there is a way to connect student ability to the cognitive approach used, that is, if the students are more likely to use a particular strategy based on their overall ability.

When it comes to accounting for a student's problem-solving strategy in large-scale assessments, process data is a definitely a better indicator than students' self-reported responses to the "show your work" instruction in paper-and-pencil assessments. However, one has to keep in mind that log data can hardly give a complete and accurate picture of student thinking and activity. A lot is happening outside of the computer-based test environment. As part of the validation strategy, the approach must be weighed in the light of what can be observed in classrooms or cognitive laboratories via fine-grained qualitative data collection (eye tracking, think aloud sessions, etc.). This validation process, even if it cannot be administered in large scale, is of high empirical value for several purposes. It strengthens the interpretation of process data and is important in the course of the didactical *a priori* task analysis, capturing students' attitudes and the use of tools external to the technology-based test environment. It helps to identify observable evidence within features that are suitable to target students' conceptions. Moreover, it is a way to inform on the potential relationship of our approach with pedagogy. Confronting this approach with what teachers can observe from students' work processes or behaviors and what is related to successful learning in the classroom can only improve the validity of our interpretations.

After all, students' learning is what matters. Valid process data interpretation enables findings to be actionable in the field, from an "assessment for learning" perspective. The formative assessment platform mathemaTIC (Vretta Inc., 2025), from the Ministry of Education in Luxembourg, involves numerous TEIs, developed in cooperation with the DEPP. The extra value of implementing automated process data analysis based on the framework and methodology presented in this chapter on such a platform is undoubted. Where information from traditional paper-pencil assessments would limit the understanding of students' reasoning in solving complex tasks, digital assessment opens up avenues for teachers to shift their pedagogical practices towards understanding and creating new ways of learning.

References

Andreu, S., Ben Ali, L., & Rocher, T. (2016). Evaluation numérique des compétences du socle en début de sixième: des niveaux de performance contrastés selon les académies. Note d'information, *n° 16.18*, MENESR-DEPP. https://archives-statistiques-depp.education.gouv.fr/Default/doc/SYRACUSE/10147/evaluation-numerique-des-competences-du-socle-en-debut-de-sixieme-des-niveaux-de-performance-contras?_lg=fr-FR

Apache Spark. (n.d.). Spark overview. http://spark.apache.org/docs/latest/

Arthur, D., & Vassilvitskii, S. (2006). K-means++: The advantages of careful seeding. http://ilpubs.stanford.edu:8090/778/1/2006-13.pdf

Barnes, J. A., & Harary, F. (1983). Graph theory in network analysis. *Social Networks, 5*(2), 235–244. https://doi.org/10.1016/0378-8733(83)90026-6

Bergner, Y., & von Davier, A. (2018). Process data in NAEP: Past, present, and future. *Journal of Educational and Behavioral Statistics, 44*(6), 706–732. https://doi.org/10.3102/1076998618784700

Breiman, L. (1996). Bagging predictors. *Machine Learning, 24*(2), 123–140. https://doi.org/10.1023/A:1018054314350

Breiman, L. (2001). Random forests. *Machine Learning, 45*(1), 5–32. https://doi.org/10.1023/A:1010933404324

Brousseau, G. (2006). *Theory of didactical situations in mathematics: Didactique des Mathématiques, 1970–1990*. Springer Science & Business Media.

CITO. (n.d.). We are CITO. https://www.cito.com/

Cramer, J. S. (2002). The origins of logistic regression. Tinbergen Institute Discussion Paper, 119/4. https://papers.tinbergen.nl/02119.pdf

Doorman, M., Drijvers, P., Gravemeijer, K., Boon, P., & Reed, H. (2012). Tool use and the development of the function concept: From repeated calculations to functional thinking. *International Journal of Science and Mathematics Education, 10*(6), 1243–1267. https://doi.org/10.1007/s10763-012-9329-0

Douady, R. (1991). Tool, object, setting, window: Elements for analysing and constructing didactical situations in mathematics. In A. J. Bishop, S. Mellin-Olsen, & J. Van Dormolen (Eds.), *Mathematical knowledge: Its growth through teaching* (pp. 107–130). Springer. https://doi.org/10.1007/978-94-017-2195-0_6

Drijvers, P. (2019). Digital assessment of mathematics: Opportunities, issues and criteria. *Mesure et évaluation en éducation, 41*(1), 41–66. https://doi.org/10.7202/1055896ar

Drijvers, P., Ball, L., Barzel, B., Heid, M. K., Cao, Y., & Maschietto, M. (2016). *Uses of digital technology in lower secondary mathematics education*. Springer. https://link.springer.com/book/10.1007/978-3-319-33666-4

El-Kechaï, N., Delozanne, É., Prévit, D., Grugeon, B., & Chenevotot, F. (2011). Evaluating the performance of a diagnosis system in school algebra. In H. Leung, E. Popescu, Y. Cao, R. W. H. Lau, & W. Nejdl (Eds.), *Advances in web-based learning—ICWL 2011* (Vol. 7048, pp. 263–272). Springer. https://doi.org/10.1007/978-3-642-25813-8_28

Ester, M., Kriegel, H.-P., Sander, J., & Xu, X. (1996). A density-based algorithm for discovering clusters in large spatial databases with noise. In *Kdd* (Vol. 96, pp. 226–231). https://doi.org/10.5555/3001460.3001507

FLIP+ e-assessment community. (2021). FLIP+ Home. https://flip-plus.org/

GitHub. (n.d.). lz-string. https://github.com/pieroxy/lz-string

Goldhammer, F., Hahnel, C., Kroehne, U., & Zehner, F. (2021). From byproduct to design factor: On validating the interpretation of process indicators based on log data. *Large-scale Assessments in Education, 9*, 20. https://doi.org/10.1186/s40536-021-00113-5

Goldhammer, F., & Zehner, F. (2017). What to make of and how to interpret process data. *Measurement: Interdisciplinary Research and Perspectives, 15*(3–4), 128–132. https://doi.org/10.1080/15366367.2017.1411651

Grapin, N., & Sayac, N. (2017). Using external assessments for improving assessment practice of primary school teachers: A first study and some methodological questions. CERME10. Dublin, Ireland. hal-01949277. https://hal.archives-ouvertes.fr/hal-01949277/document

Greiff, S., Wüstenberg, S., & Avvisati, F. (2015). Computer-generated log-file analyses as a window into students' minds? A showcase study based on the PISA 2012 assessment of problem solving. *Computers & Education, 91*, 92–105. https://doi.org/10.1016/j.compedu.2015.10.018

Han, Z., He, Q., & von Davier, M. (2019). Predictive feature generation and selection using process data from PISA interactive problem-solving items: An application of random forests. *Frontiers in Psychology, 10*, 2461. https://doi.org/10.3389/fpsyg.2019.02461

Hao, J., Shu, Z., & von Davier, A. (2015). Analyzing process data from game/scenario-based tasks: An edit distance approach. *Journal of Educational Data Mining, 7*(1), 33–50. https://doi.org/10.5281/ZENODO.3554705

He, Q., Borgonovi, F., & Paccagnella, M. (2019a). Using process data to understand adults' problem-solving behaviour in the Programme for the international assessment of adult competencies (PIAAC): Identifying generalised patterns across multiple tasks with sequence mining. OECD Education Working Papers, 205. https://doi.org/10.1787/650918f2-en.

He, Q., Liao, D., & Jiao, H. (2019b). Clustering behavioral patterns using process data in PIAAC problem-solving items. In *Theoretical and practical advances in computer-based educational measurement. Methodology of educational measurement and assessment* (pp. 189–212). Springer. https://doi.org/10.1007/978-3-030-18480-3_10

He, Q., & von Davier, M. (2015). Identifying feature sequences from process data in problem-solving items with *N*-grams. In *Quantitative psychology research* (pp. 173–190). Springer. https://doi.org/10.1007/978-3-319-19977-1_13

Ho, T. K. (1998). The random subspace method for constructing decision forests. *IEEE Transactions on Pattern Analysis and Machine Intelligence, 20*(8), 832–844. https://doi.org/10.1109/34.709601

Jain, A. K., Murty, M. N., & Flynn, P. J. (1999). Data clustering: A review. *ACM computing surveys (CSUR), 31*(3), 264–323. https://doi.org/10.1145/331499.331504

Kilpatrick, J., Swafford, J., Findell, B., & National Research Council (U.S.), & Mathematics Learning Study Committee. (2001). *Adding it up: Helping children learn mathematics*. National Academy Press. http://site.ebrary.com/id/10038695

Kroehne, U., & Goldhammer, F. (2018). How to conceptualize, represent, and analyze log data from technology-based assessments? A generic framework and an application to questionnaire items. *Behaviormetrika, 45*, 527–563. https://doi.org/10.1007/s41237-018-0063-y

Lloyd, S. (1982). Least squares quantization in PCM. *IEEE Transactions on Information Theory, 28*(2), 129–137. https://ieeexplore.ieee.org/abstract/document/1056489

MacQueen, J. (1967). Some methods for classification and analysis of multivariate observations. *Proceedings of the Fifth Berkeley Symposium on Mathematical Statistics and Probability, 1*, 281–297. Oakland. https://digitalassets.lib.berkeley.edu/math/ucb/text/math_s5_v1_article-17.pdf

Magidson, J., & Vermunt, J. (2002). Latent class models for clustering: A comparison with K-means. *Canadian Journal of Marketing Research, 20*(1), 36–43. https://jeroenvermunt.nl/cjmr2002.pdf

MEN DEPP. (2017). Cedre 2014; Mathématiques en fin de collège (No. 209). https://www.education.gouv.fr/cid122693/cedre-2014-mathematiques-en-fin-de-college.html

MEN DGESCO. (2016). Du numérique au littéral. *Ressources d'accompagnement des anciens programmes*. http://media.eduscol.education.fr/file/Programmes/17/3/du_numerique_au_litteral_109173.pdf

Naumann, J. (2019). The skilled, the knowledgeable, and the motivated: Investigating the strategic allocation of time on task in a computer-based assessment. *Frontiers in Psychology, 10*, 1429. https://doi.org/10.3389/fpsyg.2019.01429

NumPy. (2020). NumPy documentation. https://numpy.org/doc/

Pandas Development Team. (2021). Pandas documentation. https://pandas.pydata.org/docs/

Rabardel, P. (2002). People and technology: A cognitive approach to contemporary instruments. *Université Paris, 8*, 188. https://hal.archives-ouvertes.fr/file/index/docid/1020705/filename/people_and_technology.pdf

Riaz, F., & Ali, K. M. (2011). Applications of graph theory in computer science. In *Third International Conference on Computational Intelligence, Communication Systems and Networks* (pp. 142–145). IEEE. https://doi.org/10.1109/CICSyN.2011.40

Ristad, E. S., & Yianilos, P. N. (1998). Learning string-edit distance. *IEEE Transactions on Pattern Analysis and Machine Intelligence, 20*(5), 522–532. https://ieeexplore.ieee.org/abstract/document/682181

Robert, A. (1998). Outils d'analyses des contenus mathématiques à enseigner au lycée et à l'université. *Recherches en didactique des mathématiques, 18*(2), 139–190. https://hal.archives-ouvertes.fr/hal-01809685/

Roditi, E., & Salles, F. (2015). Nouvelles analyses de l'enquête PISA 2012 en mathématiques, un autre regard sur les résultats. *Éducation & formations, 86–87*, 235–257. https://hal.archives-ouvertes.fr/hal-01157937/

Rousseeuw, P. J. (1987). Silhouettes: A graphical aid to the interpretation and validation of cluster analysis. *Journal of Computational and Applied Mathematics, 20*, 53–65. https://doi.org/10.1016/0377-0427(87)90125-7

Salles, F., Dos Santos, R., & Keskpaik, S. (2020). When didactics meet data science: Process data analysis in large-scale mathematics assessment in France. *Large-scale Assessments in Education, 8*(1), 7. https://doi.org/10.1186/s40536-020-00085-y

scikit-learn. (n.d.). Machine Learning in Python. https://scikit-learn.org/stable/

Sfard, A. (1991). On the dual nature of mathematical conceptions: Reflections on processes and objects as different sides of the same coin. *Educational Studies in Mathematics, 22*(1), 1–36. https://doi.org/10.1007/BF00302715

Sidorov, G., Velasquez, F., Stamatatos, E., Gelbukh, A., & Chanona-Hernández, L. (2014). Syntactic N-grams as machine learning features for natural language processing. *Expert Systems with Applications, 41*(3), 853–860. https://doi.org/10.1016/j.eswa.2013.08.015

Stacey, K., & Wiliam, D. (2013). Technology and assessment in mathematics. In M. A. Ken, A. J. B. Clements, C. Keitel, J. Kilpatrick, & F. K. S. Leung (Eds.), *Third international handbook of mathematics education* (pp. 721–751). https://doi.org/10.1007/978-1-4614-4684-2_23

Sukkarieh, J., von Davier, M., & Yamamoto, K. (2012). From biology to education: Scoring and clustering multilingual text sequences and other sequential tasks. *ETS Research Report Series, 2012*(2), i–43. https://doi.org/10.1002/j.2333-8504.2012.tb02307.x

Trosseille, B., & Rocher, T. (2015). Les évaluations standardisées des élèves. Perspective historique. *Éducation & Formations, 86–87*, 15–35. https://www.education.gouv.fr/evaluation-des-acquis-principes-methodologie-resultats-11573

Trouche, L. (2003). From artifact to instrument: Mathematics teaching mediated by symbolic calculators. *Interacting with Computers, 15*(6), 783–800. https://doi.org/10.1016/j.intcom.2003.09.004

Ulitzsch, E., von Davier, M., & Pohl, S. (2019). A hierarchical latent response model for inferences about examinee engagement in terms of guessing and item-level non-response. *British Journal of Mathematical and Statistical Psychology, 73*, 83–112. https://doi.org/10.1111/bmsp.12188

Vretta Inc. (2025). MathemaTIC: Personalized learning in mathematics. https://www.mathematic.lu/

Weeks, J. P., von Davier, M., & Yamamoto, K. (2016). Using response time data to inform the coding of omitted responses. *Psychological Test and Assessment Modeling, 58*, 671–701. https://www.psychologie-aktuell.com/fileadmin/download/ptam/4-2016_20161219/06_Weeks.pdf

Weiner, P. (1973). Linear pattern matching algorithms. In *14th Annual Symposium on Switching and Automata Theory (swat 1973)* (pp. 1–11). IEEE. https://ieeexplore.ieee.org/document/4569722

Wierzchoń, S. T., & Kłopotek, M. (2018). *Modern algorithms of cluster analysis*. Springer International Publishing. https://doi.org/10.1007/978-3-319-69308-8_2

Yamamoto, K., von Davier, M. & Sukkarieh, J. (2016). Systems and methods for evaluating multilingual text sequences. Patent No. US 9,471,667 B2. Date of Patent: Oct. 18, 2016.

Chapter 19
Using Response Times for Improving the Understanding and Handling of Missing Responses in Low-Stakes Assessments

Esther Ulitzsch, Tobias Alfers, Jing Lu, Chun Wang, and Steffi Pohl

Abstract In large-scale assessments, examinees commonly do not provide answers to all items administered. Such (unplanned) missing responses can occur either due to item omissions or due to not reaching the end of the assessment. If not appropriately accounted for, item and person parameter estimates as well as group statistics can be biased and, for example, lead to distorted performance rankings. Considering additional information on test-taking behavior, such as response times, may critically improve the understanding of the occurrence of missing responses and assist a more nuanced handling. After a short review of approaches for handling missing responses that employ information solely from item responses, we review four current model-based approaches that incorporate response time information. Each approach is focused on different missingness mechanisms. In an empirical example, using data from PISA 2018, the approaches are illustrated and compared in terms of the insight on test-taking behavior they provide, as well as their assumptions on the occurrence of missing responses.

E. Ulitzsch (✉)
IPN–Leibniz Institute for Science and Mathematics Education, Kiel, Germany
e-mail: ulitzsch@leibniz-ipn.de

T. Alfers · S. Pohl
Freie Universität Berlin, Berlin, Germany

J. Lu
Northeast Normal University, Changchun, China

C. Wang
University of Washington, Seattle, WA, USA

L. Khorramdel et al. (eds.), *Innovative Digital-Based International Large-Scale Assessments*, Methodology of Educational Measurement and Assessment,
https://doi.org/10.1007/978-3-031-90951-1_19

In large-scale assessments (LSAs), it is rather common that examinees do not provide answers to all items administered, either due to omitting items or due to not reaching the end of the assessment. Items are said to be omitted when the examinee has seen an item but, for whatever reason, has decided not to respond (Mislevy & Wu, 1996). In the case that an examinee failed to attempt a sequence of items presented at the end of a test, the resulting missing responses are referred to as not-reached items (NRIs; Mislevy & Wu, 1996). There is commonly strong variation in the extent of missing responses. In the Programme for International Student Assessment (PISA) 2018 (OECD, 2020), for instance, the occurrence of NRIs and omissions ranged from 0.55% (China) to 13.80% (Peru) and from 0.62% (Malaysia) to 14.94% (Bosnia and Herzegovina), respectively.

Having a sound understanding of how these missing responses occurred is important for at least two reasons. First, missing responses force researchers to explicate their beliefs on the nature of the underlying missingness mechanisms, which becomes evident in their decisions on how missing data are treated in the analysis. These decisions may heavily impact country rankings (Köhler et al., 2017; Robitzsch, 2014; Rose et al., 2010), as well the conclusions drawn from trend analyses (Sachse et al., 2019) and, therefore, need to be taken carefully. For an adequate decision, a comprehensive understanding of missingness mechanisms is paramount. Second, variation in the occurrence of missing responses indicates differences in how examinees approached the assessment. Understanding missingness mechanisms thus supports a better understanding of differences in test-taking behavior in general and allows considering these differences when drawing inferences on examinee competencies.

Understanding the behavior underlying missing responses is not a trivial endeavor since, by definition, no information on a key outcome of test-taking behavior—the item responses themselves—is available. Against this background, the rise of computer-based assessment and the opportunity to log additional information on test-taking behavior may considerably assist understanding and handling missing responses.

In what follows, we first shortly review approaches for dealing with and investigating missing responses, employing information from item responses and information on which items are missing. We then discuss limitations of these approaches. Subsequently, we delineate the potential of leveraging response times for a more nuanced handling and understanding of the occurrence of missing responses. We review four current model-based approaches for missing responses that incorporate information from response times and illustrate the insights that can be gained with these approaches in an empirical example using data from PISA 2018. Finally, we discuss future directions that may aid their applicability.

19.1 Limitations of Currently Applied Approaches for Dealing with Missing Responses in Large-Scale Assessments

Operationally in LSAs, it is common to deal with missing values due to item omissions and NRIs by either ignoring missing responses, by scoring them as (partially) incorrect, or by using model-based approaches that draw on information from item responses and information on which items are missing, e.g., regress ability on the number or proportion of NRIs or employ missingness indicators as indicators of an additional latent variable within multidimensional IRT (MIRT) models (see Debeer et al., 2017; Glas & Pimentel, 2008; Holman & Glas, 2005; Mislevy & Wu, 1996; Pohl et al., 2014; Rose et al., 2010, for overviews and examples). All three approaches have been used for dealing with NRIs in educational LSAs (Rose et al., 2017, 2010). While these approaches provide a way to deal with missing values and have many advantages, they also have some relevant limitations (Pohl & von Davier, 2018). We introduce these approaches and delineate their limitations exemplarily focusing on approaches for dealing with NRIs due to lack of speed. In discussing these approaches, we assume that a) examinees may differ in the level of speed with which they approach the assessment, as well as b) that there is an intraindividual trade-off between the effective speed level examinees choose to operate on and the level of accuracy (or effective ability) examinees can show under the chosen speed level (speed-accuracy trade-off; see van der Linden, 2007, for an introduction). This implies that if examinees would want to put a stronger focus on accuracy, they could do so at the expense of working at a slower pace, resulting in higher effective ability and lower effective speed (van der Linden, 2009). The impact of this phenomenon on assessment results under different missing value approaches has been discussed in Pohl and von Davier (2018) and is reviewed in the following.

Ignoring not-reached items: When ignoring missing responses, missing values are assumed to be missing at random (MAR; see Mislevy & Wu, 1996; Rubin, 1976) and the missingness mechanism is assumed to be unrelated to ability. When ignoring NRIs, ability is estimated solely based on responses given to reached items. When NRIs occur due to lack of speed, this practice disadvantages examinees who, on the expense of lower accuracy, worked with a speed level sufficient to reach the end of the test. Conversely, it advantages examinees who, by working with a slower pace, did not reach the end of the test and had more time available for the items they succeeded to attempt in the given time.

Incorrect scoring of not-reached items: Scoring missing values as incorrect conveys to the idea that items left unanswered are not correctly solved. When NRIs occur due to lack of speed, scoring NRIs as incorrect penalizes examinees who did not allocate their time such that they could reach the end of the test. However, this confounds differences in test-taking behavior and effective ability since, with NRIs scored as incorrect, low test scores might stem from rather different scenarios. For instance, examinees might either have shown low ability but worked sufficiently fast to finish the test on time, or they might not have reached the end of the test due to working too slowly but displayed higher levels of ability on approached items.

Model-based approach for not-reached items: Common model-based approaches for missing responses (e.g., Glas & Pimentel, 2008; Holman & Glas, 2005; O'Muircheartaigh & Moustaki, 1999; Rose et al., 2010) perceive missing data patterns to be a result of a person-specific tendency to (not) respond. In the approach for NRIs by Rose et al. (2010), for instance, this tendency is operationalized as the number (or proportion) of NRIs and included as an explanatory variable in the model of ability. While this approach supports investigating how the tendency to not reach the end of the test is related to competency, it is agnostic to why examinees did not reach the end of the assessment. When NRIs occurred due to lack of speed, the number of NRIs can be seen as a rough proxy for speed—however, only for examinees that did not reach the end of the test (Pohl et al., 2019). Information on differences in speed is not preserved in the number of NRIs for examinees that reached the end of the test. Further, when multiple mechanisms underlie the occurrence of NRIs, say lack of speed and quitting, these different issues are confounded in the mere number of NRIs (Ulitzsch et al., 2019b). Speed issues and quitting are, however, two different mechanisms that may need to be treated differently.

All of the discussed approaches rely on information retrieved from item responses and information on which items are missing. These sources of information can, however, only provide some information on missingness mechanisms. For a more nuanced understanding of examinees' test-taking behavior and missingness mechanisms, further information is needed. Process data, and specifically response times may provide such information.

19.2 Current Missing Data Approaches that Incorporate Response Times

In recent years, multiple approaches have been developed that incorporate response time information for handling missing responses and investigating their occurrence in LSA data. These approaches may be classified into approaches that aim at identifying response time thresholds to score omitted items as either missing or incorrect (Frey et al., 2018; Weeks et al., 2016; Yamamoto et al., 2013) and approaches that incorporate response times directly into item response theory (IRT) models (Lu & Wang, 2020; Pohl et al., 2019; Ulitzsch et al., 2019a, 2019b, 2019c). We here focus on model-based approaches. Existing models differ in which kind of missing values they consider, as well as in their assumptions on the mechanisms underlying their occurrence. In the following, we will review and illustrate four recent approaches. Approach 1 (Pohl et al., 2019) and 2 (Ulitzsch et al., 2019b) employ response times for modeling and accounting for mechanisms underlying NRIs. The former models and accounts for NRIs due to lack of speed, while the latter allows for disentangling, modeling, and accounting for NRIs due to lack of speed and due to quitting. Approach 3 (Ulitzsch et al., 2019c) is focused on modeling and accounting for non-ignorable item omissions. Approach 4 (Lu & Wang, 2020) uses response times for the joint modeling of NRIs due to lack of speed and item omissions.

19.2.1 Approach 1: Modeling Not-Reached Items Due to Lack of Speed

Pohl et al. (2019) delineated that the information contained in response times on examinees' time allocation strategies can be leveraged for modeling and accounting for the mechanism underlying NRIs due to lack of speed. They showed that the hierarchical speed-accuracy (SA) model proposed by van der Linden (2007) provides a well-suited means to that end. In the SA model, responses and response times comprise indicators for first-level measurement models of ability and speed, respectively. On the second level, a joint distribution is assumed for ability and speed.[1] The model is depicted in Fig. 19.1a).

For item responses u_{ij}, containing examinee i 's, $i = 1, \ldots, N$, response to item $j, j = 1, \ldots, K$, customary IRT models can be employed. When employing a Rasch model, the probability of observing a correct response is modeled as

$$p(u_{ij}=1) = \frac{\exp(\theta_i - b_j)}{1 + \exp(\theta_i - b_j)}, \tag{19.1}$$

with θ_i denoting examinee i 's ability and b_j denoting item j 's difficulty. For response times t_{ij}, giving the time examinee i required to generate an answer to item j, van der Linden (2007) suggested a lognormal model, governed by the examinee's speed τ_i and the item's time intensity β_j

$$\ln(t_{ij}) \sim \mathcal{N}\left(\beta_j - \tau_i, \alpha_j^{-2}\right). \tag{19.2}$$

Here, α_j represents the inverse of the logarithmized response times' residual standard deviation and is commonly interpreted as a time discrimination parameter. The joint distribution of person parameters is modeled as multivariate normal with mean vector and covariance matrix

[1] Note that item parameters can be modeled either as random effects (van der Linden, 2007), assuming a joint distribution, or as fixed effects (Molenaar et al., 2015; see De Boeck, 2008, for a discussion of conceptual differences). Here, we model item parameters as fixed effects, as done in major LSAs (Foy, 2017, 2018; OECD, 2013, 2020).

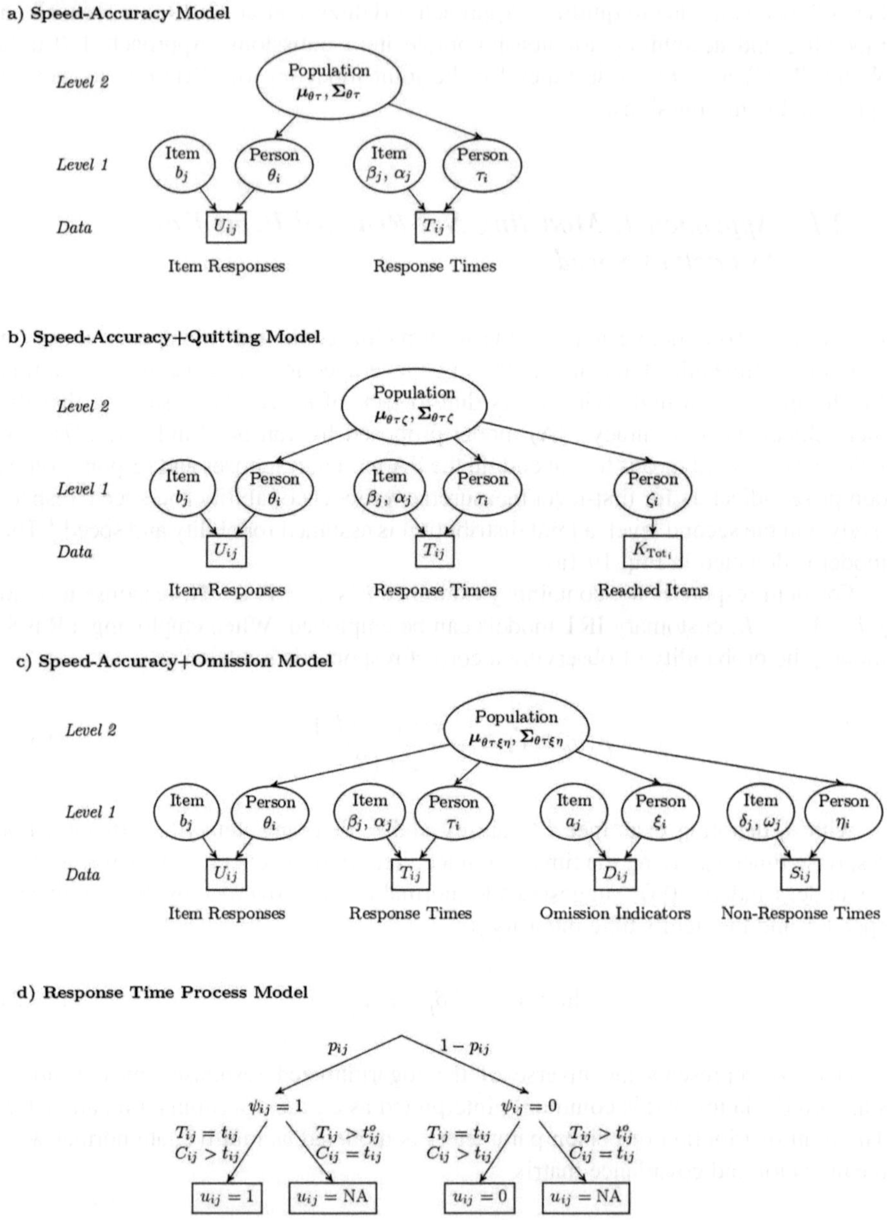

Fig. 19.1 Model-based approaches utilizing response times for modeling missing responses

$$\boldsymbol{\mu}_{\mathcal{P}} = (\mu_\theta, \mu_\tau) \qquad \text{and} \qquad \boldsymbol{\Sigma}_{\mathcal{P}} = \begin{pmatrix} \sigma_\theta^2 & \sigma_{\theta\tau} \\ \sigma_{\theta\tau} & \sigma_\tau^2 \end{pmatrix}. \tag{19.3}$$

For model identification, person parameter means are set to zero. Investigating the joint distribution of person parameters supports a better understanding of how the mechanism resulting in NRIs due to lack of speed is related to ability. For instance, in the case of a negative correlation, examinees displaying higher ability levels are less likely to complete the test within the allocated time.

Pohl et al. (2019) delineated that for NRIs due to lack of speed, the SA model provides a closer description of the missingness mechanism than the model-based approach for modeling NRIs by Rose et al. (2010). First, by accounting for examinee speed, the SA model includes a direct measure of the mechanism underlying NRIs due to lack of it, as compared to the number of NRIs as a rough proxy measure. Second, the SA model considers and accounts for differences in working speed also for examinees who reached the end of the test. The number of NRIs, in contrast, is zero and, as such, does not preserve information on differences in working speed for all examinees reaching the end of the test. As such, the approach by Rose et al. (2010) accounts for these differences only for examinees with NRIs.

19.2.2 Approach 2: Disentangling and Modeling Not-Reached Items Due to Lack of Speed and Due to Quitting

For an even richer description of the occurrence of NRIs, response times provide the opportunity to distinguish NRIs occurring due to lack of speed from those occurring due to quitting, that is, stopping to work before the time limit is reached while there are still unanswered items. To that end, Ulitzsch et al. (2019b) suggested to construct an indicator c_i of observed quitting behavior as

$$c_i = \begin{cases} 1 & \text{if } k_{tot_i} < K \text{ and } t_{tot_i} < T_{lim} \\ 0 & \text{otherwise,} \end{cases} \tag{19.4}$$

where K and T_{lim} give the number of administered items and time allocated for their completion, and k_{tot_i} and t_{tot_i} give the number of items reached by examinee i as well as the time required for doing so. The indicator c_i takes the value 1 for examinees who quit the assessment and 0 for examinees who reached the time limit or finished the test before quitting.

For modeling and accounting for the quitting process, Ulitzsch et al. (2019b) suggested an extension of van der Linden's SA model by an additional component. In the speed-accuracy+quitting (SA + Q) model, the number of reached items before quitting is taken as an indicator for quitting behavior. If quitting behavior is observed ($c_i = 1$), a Poisson lognormal model is employed for k_{tot_i}. This component model assumes that the probability that examinee i quits the assessment after attempting k_{tot_i} items, i.e., that his or her quitting position k_{Q_i} corresponds to k_{tot_i}, can be described by a Poisson process as a function of the examinee's test endurance ζ_i. That is,

$$p\left(k_{Q_i} = k_{tot_i}\right) = \frac{\exp(-\exp(\zeta_i))\exp\left(\zeta_i\right)^{k_{tot_i}}}{k_{tot_i}!}. \tag{19.5}$$

When no quitting behavior has been observed ($c_i = 0$), quitting behavior is exposed to right censoring and k_{tot_i} marks the censoring item position. For modeling the censoring, a survival function $S(k_{tot_i})$ is employed, giving the probability that examinee i will continue the assessment beyond k_{tot_i} items, given his or her test endurance ζ_i. Hence,

$$S(k_{tot_i}) = p\left(k_{Q_i} > k_{tot_i}\right) = 1 - \sum_{l=0}^{k_{tot_i}} \frac{\exp(-\exp(\zeta_i))\exp\left(\zeta_i\right)^{l}}{l!}. \tag{19.6}$$

As in the SA model, the SA + Q model assumes a joint multivariate distribution of ability, speed, and test endurance, allowing to account for test speed and test endurance in the estimation of ability, as well as to investigate how the different processes underlying NRIs are related to ability as well as to each other. For instance, a negative correlation between speed and test endurance would indicate that examinees who complete more items before quitting are more likely to run out of time. A positive correlation between ability and test endurance would indicate that examinees displaying higher ability levels complete more items before quitting. The model is depicted in Fig. 19.1b).

19.2.3 Approach 3: Modeling Item Omissions

Different than for NRIs, for item omissions timing information may be recorded. The speed-accuracy+omission (SA + O) model (Ulitzsch et al., 2019c) leverages this information for jointly assessing and modeling omission behavior and response behavior. The SA + O model extends the SA model by including a process model for omission behavior that considers information on which items have been omitted as well as response times associated with item omissions. It allows for different data-generating processes underlying response times associated with response and omission behavior. For modeling response behavior, the SA model is employed. Omission behavior is modeled in terms of two latent variables: a) an omission speed variable ξ_i describing the speed with which items are omitted and b) a latent omission propensity η_i which describes the tendency to omit items and accounts for item omissions violating the distinctness (or latent ignorability) assumption (i.e., omissions going back to a mechanism potentially related to ability; see Holman & Glas, 2005; O'Muircheartaigh & Moustaki, 1999; Pohl & Becker, 2019). The model is depicted in Fig. 19.1c).

Omission indicators d_{ij} constitute the measurement model for examinees' latent omissions propensity, taking the value 1 if examinee i omitted item j and 0 if a response was observed. The probability that examinee i omits item j is assumed to be a function of the examinee's omission propensity ξ_i and the item's omission difficulty a_j

$$p(d_{ij}=1)=\frac{\exp(\xi_i-a_j)}{1-\exp(\xi_i-a_j)}. \tag{19.7}$$

Different data-generating processes are allowed to underlie response times associated with observed responses and item omissions. Here, response times associated with observed responses are modeled according to Eq. 19.2. Non-response times s_{ij}, i.e., response times associated with item omissions, are modeled as

$$\ln\left(s_{ij}\right)\sim\mathcal{N}\left(\delta_j-\eta_i,\omega_j^{-2}\right), \tag{19.8}$$

where δ_j and ω_j^{-2} denote item j 's omission time intensity and omission time discrimination, and η_i gives examinee i 's omission speed. As such, the SA + O model allows for examinees to operate on different speed levels when generating responses and omitting items. Again, the model assumes a joint multivariate distribution of person parameters, allowing to investigate how the different components describing response and omission behavior are related to each other. For instance, researchers can investigate whether examinees employ different pacing strategies when generating responses or omitting items or how the tendency to omit items is related to ability and speed. Likewise, researchers can investigate item characteristics, such as format, position, or difficulty, potentially associated with increased omission behavior (i.e., lower omission difficulty parameters). Further, the SA + O model accounts for differences in omission propensity as well as for pacing strategies when providing responses and omitting items in the estimation of ability. Although not explicitly stated or shown, as speed is included in the model, the SA + O model not only accounts for item omissions, but also for items not reached due to time limits.

19.2.4 Approach 4: Jointly Modeling Not-Reached and Omitted Items

Approaches 1 to 3 are focused on either NRIs or omitted items. The response time process (RTP) model by Lu and Wang (2020), in contrast, jointly models mechanisms underlying NRIs and item omissions. As in Approach 1, the model assumes that NRIs occurred due to lack of speed. NRIs due to quitting (as in Approach 2) are not considered in the model. Regarding item omissions, Approaches 3 and 4 differ in how item omissions and working speed are conceptualized. Approach 4 assumes that examinees differ in the amount of time, or effort, they are willing to invest in a given item. Different than in Approach 3, item omissions are assumed to occur whenever the time an examinee requires for generating a response exceeds the time he or she is willing to invest (Lee & Ying, 2015; Wolf et al., 1995).

Figure 19.1d) shows the tree plot for the RTP model. Here, $\psi_{ij} = 1$ denotes that examinee i knows the correct answer to item j and $p(\psi_{ij} = 1)$ is modeled according to Eq. 19.1. The time examinee i is willing to invest on item j is denoted with the censoring variable C_{ij}, and the solution-based response time, i.e., the time the examinee requires for generating a response to that item is denoted with T_{ij}. When $T_{ij} < C_{ij}$, a response is provided, and T_{ij} corresponds to the observed response time t_{ij}. An item omission occurs when $T_{ij} > C_{ij}$, in which case the examinee stops working on the item before he or she gives an answer, leading to a right-censored event. In this case, T_{ij} corresponds to the omission response time t^o_{ij}. Thus, the omission indicators can be expressed as $d_{ij} = I(T_{ij} > C_{ij})$. The solution-based response time T_{ij} is assumed to follow a lognormal distribution as specified in Eq. 19.2. Omission response times are modeled according to the survival function

$$S\left(t^o_{ij}\right) = p\left(T_{ij} > t^o_{ij}\right) = \phi\left(\alpha_j(\beta_j - \tau_i) - \ln\left(t^o_{ij}\right)\right), \tag{19.9}$$

where $\phi(\cdot)$ denotes the cumulative normal distribution function. Again, the ability and speed parameters are assumed to follow a bivariate normal distribution. Due to the censoring process, the observed response time t_{ij} corresponds to the solution-based response time only under $d_i = 0$ and marks the censoring time c_{ij}, otherwise.

In comparison to Approaches 2 and 3, no further latent variables are introduced in addition to the ability and speed parameters, which may ease the estimation procedure. In contrast to Approach 3, in which different response time processes are modeled for item responses and omissions, in this approach only the solution-based response time is modeled, whereas the omission response time is not explicitly modeled. This is because the omission response time is considered a censoring variable (i.e., C_{ij}) whose distribution is independent of T_{ij}. That is, the censoring mechanism (i.e., the time examinees are willing to invest) is assumed to be unrelated to ability and speed, i.e., is assumed to be ignorable. The different distributional assumptions for timing data also mirror the different considerations on how examinees interact with assessments inherent to Approaches 3 and 4.

19.3 Illustration of Missing Data Approaches that Incorporate Response Times

To illustrate the application of the missing data approaches and their potential to gain a better understanding of different missingness mechanisms, we applied all models to one data set. We made use of responses and response time data from a single mathematics item block from PISA Peru 2018, comprising $K = 12$ items. To also investigate NRIs, we focused only on a subsample of $N = 272$ examinees that were administered the item block at the end of the 120 min assessment. In the item block, 8.20% of the items were not-reached and 14.46% were omitted, going back to 25.74% of examinees not reaching the end of the item block and 68.01% of examinees showing omission behavior, i.e., omitting at least one item. Item-level rates of omissions and NRIs are given in Table 19.1. For reasons of simplicity, partial credit items were dichotomized, scoring partially correct as correct.

For the application of Approach 2, for classifying NRIs as occurring due to lack of speed or due to quitting, we drew on information contained in cumulative response times. Note that in PISA, response times are adjusted for loading times, such that cumulative response times are commonly smaller than the total time spent on the test. To accommodate this issue, we employed a heuristic approach and classified NRIs as due to quitting when examinees did not reach the end of the item block and their total time spent on the assessment fell below the time limit by less than 10% of the allocated testing time, that is, below $120 - 12 = 108$ minutes. This led to classifying 59 out of 70 examinees with NRIs as quitters.

As a baseline for comparison, we analyzed the data in line with the missing data handling currently implemented in PISA (OECD, 2020). Here, omissions are treated as wrong in the scaling procedure. In the population model, ability is regressed on the number of NRIs (Rose et al., 2010). To illustrate how different handling of missing data may impact results, we compared the distribution of person parameters (i.e., ability and, if available, speed), as well as item parameters (i.e., item difficulties and, if available, time intensity parameters) retrieved from all approaches against the baseline analysis and against each other.

R and Stan code to reproduce the empirical illustration are provided in the OSF repository accompanying this chapter.

19.3.1 Descriptive Analyses of Response Times and Missing Responses

The vast potential of response times for better understanding the occurrence of missing responses already becomes evident in merely descriptive analyses. Figure 19.2 displays the time spent on the cognitive assessment as a function of the number of NRIs on the last item block. The time limit of 120 min is marked with a dashed horizontal line. Multiple insights on the missingness mechanisms can

Table 19.1 Item-level rates of omissions and not-reached items and difficulty and time intensity parameters for all approaches

			Baseline	Approach 1		Approach 2		Approach 3		Approach 4	
Item	% O	% NR	*b*	*b*	β	*b*	β	*b*	β	*b*	β
1	0.01	0	−1.10	−1.11	3.93	−1.11	3.93	−1.12	3.93	−1.11	3.94
			[−1.47, −0.76]	[−1.48, −0.78]	[3.87, 4.00]	[−1.46, −0.77]	[3.86, 4.00]	[−1.47, −0.78]	[3.85, 4.01]	[−1.47, −0.77]	[3.87, 4.01]
2	0.03	0	0.42	0.34	4.10	0.34	4.10	0.36	4.10	0.34	4.11
			[0.08, 0.76]	[0.02, 0.68]	[4.03, 4.17]	[0.02, 0.67]	[4.02, 4.18]	[0.04, 0.69]	[4.02, 4.17]	[0.01, 0.66]	[4.04, 4.19]
3	0.34	0.01	0.85	−0.03	5.18	−0.03	5.19	0.05	5.19	−0.05	5.26
			[0.50, 1.22]	[−0.42, 0.36]	[5.10, 5.27]	[−0.42, 0.35]	[5.10, 5.27]	[−0.32, 0.44]	[5.09, 5.28]	[−0.45, 0.33]	[5.18, 5.35]
4	0.16	0.02	0.62	0.32	4.45	0.32	4.46	0.35	4.45	0.30	4.49
			[0.28, 0.98]	[−0.03, 0.69]	[4.38, 4.53]	[−0.05, 0.68]	[4.38, 4.54]	[0.01, 0.71]	[4.37, 4.53]	[−0.05, 0.67]	[4.41, 4.57]
5	0.44	0.03	3.38	2.57	4.77	2.57	4.78	2.68	4.78	2.54	4.88
			[2.83, 3.95]	[1.98, 3.18]	[4.66, 4.88]	[1.96, 3.15]	[4.66, 4.89]	[2.12, 3.26]	[4.67, 4.88]	[1.96, 3.14]	[4.78, 4.99]
6	0.03	0.04	1.34	1.25	4.27	1.26	4.28	1.27	4.27	1.25	4.28
			[0.96, 1.73]	[0.90, 1.62]	[4.20, 4.35]	[0.90, 1.61]	[4.20, 4.36]	[0.93, 1.64]	[4.19, 4.35]	[0.88, 1.63]	[4.20, 4.35]
7	0.11	0.06	2.50	2.30	5.01	2.30	5.02	2.33	5.01	2.29	5.04
			[2.02, 2.98]	[1.86, 2.78]	[4.92, 5.09]	[1.81, 2.79]	[4.94, 5.10]	[1.84, 2.81]	[4.93, 5.09]	[1.85, 2.75]	[4.96, 5.13]
8	0.03	0.07	1.60	1.50	3.81	1.50	3.82	1.52	3.81	1.49	3.82
			[1.22, 2.00]	[1.13, 1.87]	[3.73, 3.89]	[1.12, 1.89]	[3.74, 3.91]	[1.14, 1.89]	[3.73, 3.89]	[1.10, 1.89]	[3.74, 3.91]

9	0.07	0.12	3.93	3.75	4.54	3.76	4.56	3.77	4.55	3.74	4.58
			[3.26, 4.64]	[3.10, 4.48]	[4.48, 4.61]	[3.13, 4.49]	[4.49, 4.63]	[3.10, 4.49]	[4.46, 4.63]	[3.09, 4.47]	[4.51, 4.66]
10	0.15	0.17	2.86	2.53	5.04	2.54	5.05	2.57	5.04	2.52	5.07
			[2.34, 3.46]	[2.05, 3.09]	[4.95, 5.12]	[2.00, 3.11]	[4.96, 5.14]	[2.10, 3.08]	[4.96, 5.13]	[2.02, 3.05]	[4.98, 5.15]
11	0.31	0.2	3.33	2.66	4.62	2.66	4.64	2.75	4.63	2.63	4.81
			[2.76, 3.93]	[2.03, 3.32]	[4.50, 4.74]	[2.04, 3.32]	[4.52, 4.75]	[2.12, 3.44]	[4.52, 4.73]	[2.04, 3.27]	[4.69, 4.94]
12	0.04	0.26	1.97	1.80	4.74	1.81	4.77	1.81	4.75	1.79	4.77
			[1.53, 2.45]	[1.37, 2.26]	[4.63, 4.85]	[1.37, 2.27]	[4.66, 4.88]	[1.37, 2.27]	[4.67, 4.83]	[1.32, 2.26]	[4.65, 4.88]

Note: Bayesian credibility intervals are given in squared brackets. % O and % NR give item-level rates of omissions and not-reached items. b: difficulty; β: time intensity. Baseline: omissions scored as wrong and ability regressed on number of not-reached items; Approach 1: speed-accuracy model; Approach 2: speed-accuracy +quitting model; Approach 3: speed-accuracy+omission model; Approach 4: response time process Rasch model

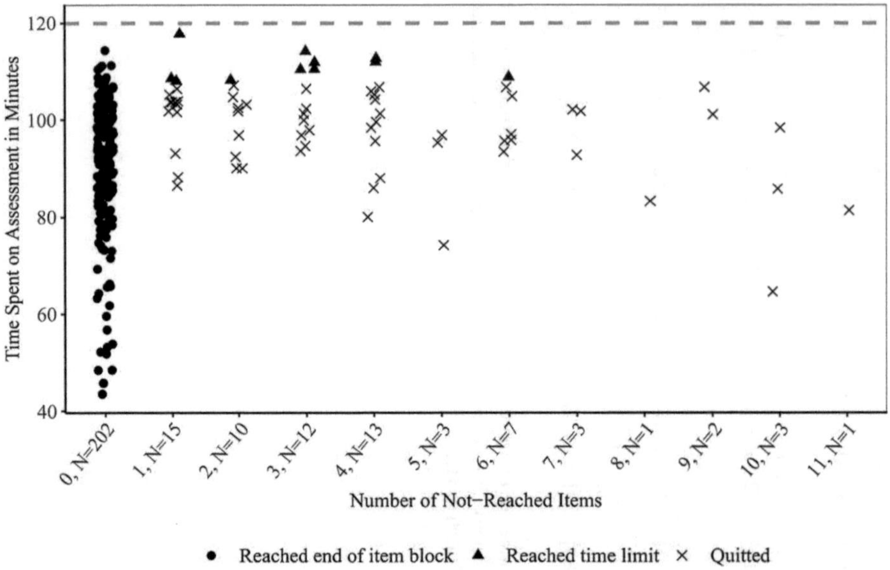

Fig. 19.2 Time spent on the assessment plotted against the number of not-reached items on the Mathematics Item Block 1 administered at the end of the PISA assessment session. The time limit is marked with a dashed horizontal line

be gained from this figure. First, and most importantly, it suggests the conclusion that while some examinees reached the time limit before reaching the end of the test, displaying NRIs due to lack of speed, others have quit the assessment before reaching the end of the test or the time limit. Hence, jointly investigating time spent on the assessment and the occurrence of NRIs allows disentangling these different mechanisms underlying NRIs. Second, time spent on the assessment also considerably varies across examinees completing the assessment, suggesting that there are vast differences in pacing. Such differences remain undetected when the mere information on the number of NRIs is employed as an indicator of how examinees allocated their time.

Following Weeks et al. (2016), we also investigated distributions of response times associated with correct, incorrect, and omitted responses. Figure 19.3 exemplarily depicts these analyses for a single item. Response times associated with item omissions displayed response time distributions considerably different from those associated with observed responses, being generally shorter and less variable. These different distributions may be indicative of different underlying processes, requiring different amounts of time. We observed such patterns across almost all items. This is in line with the results in Weeks et al. (2016), who reported such patterns to emerge across different domains and countries.

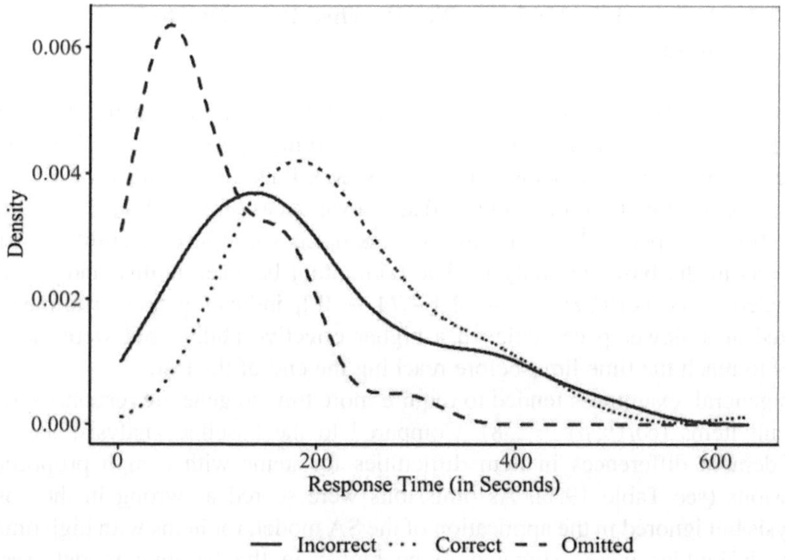

Fig. 19.3 Item response time distributions by response category for the third item of Mathematics Item Block 1

19.3.2 Estimation Procedure

All models were estimated using Bayesian estimation using Stan version 2.19 (Carpenter et al., 2017) employing the rstan package version 2.19.3 (Guo et al., 2018). In applications of Approaches 1 to 4, when not considered in the model, NRIs and item omissions were ignored. For all models, we ran three MCMC chains with 3000 iterations each, with the first half being employed as warm-up. The sampling procedure was assessed on the basis of trace plots and potential scale reduction factor (PSRF) values (Gelman & Rubin, 1992; Gelman & Shirley, 2011). All models showed good mixing of the chains and no PSRF values above 1.05.

19.3.3 Results

19.3.3.1 Baseline: PISA Procedure for Missing Responses

In the baseline analysis, scoring omissions as wrong and regressing ability on the number of NRIs, we estimated the variance of ability to be $var(\theta) = 2.11$, with a 95% Bayesian credibility interval of [1.56; 2.78]. We retrieved a standardized regression coefficient of 0.02 [−0.09; 0.13], indicating no relationship between ability and the tendency to not reach the end of the test, as operationalized by the number of NRIs. Item difficulty estimates are given in Table 19.1.

19.3.3.2 Approach 1: Modeling Not-Reached Items Due to Lack of Speed

Commonly, NRIs are the result of examinees running out of time and, as such, can be understood by studying the examinees' working speed. Van der Linden's SA model simultaneously models ability and speed, thus accounting for NRIs due to lack of speed. We observed ability and speed variances of 2.05 [1.47; 2.78] and 0.11 [0.09; 0.14], respectively. Note that the estimated variability of ability is slightly lower as in the baseline analysis. The correlation between ability and speed was estimated to be $\text{cor}(\theta, \tau) = -.62$ [−.74; −.48], indicating that examinees who worked at a slower pace achieved a higher effective ability, but were also more likely to reach the time limit before reaching the end of the test.

In general, examinees tended to require more time to generate responses to more difficult items ($cor(b, \beta) = .38$). Compared to the baseline analysis, there were considerable differences in item difficulties for items with a high proportion of omissions (see Table 19.1). As omissions were scored as wrong in the baseline analysis but ignored in the application of the SA model, for items with high omission rates, difficulties were estimated to be higher in the baseline model (see also Culbertson, April 2011; de Ayala et al., 2001; Finch, 2008; Köhler et al., 2015; Pohl et al., 2014; Rose et al., 2010, 2013).

19.3.3.3 Approach 2: Disentangling and Modeling Not-Reached Items Due to Lack of Speed and Due to Quitting

Applying the SA + Q model, we aimed at accounting for two different processes underlying NRIs, namely speed and quitting.

On average, we estimated that 24.82 [19.78; 33.39], of the items would be reached before examinees decide to quit.[2] Note that this is far beyond the censoring item position of $K = 12$. By and large, item parameters were similar to those retrieved from the SA model (see Table 19.1).

With $var(\theta) = 1.99$ [1.45; 2.71], $var(\tau) = 0.12$ [0.09; 0.14], and $cor(\theta, \tau) = -.62$ [−.74; −.48], we obtained similar variances and correlation for ability and speed as in Approach 1 (with $var(\theta)$ being slightly lower). Furthermore, we also found a negative correlation of ability θ with test endurance ζ of −.24 [−.45; −.01], indicating that examinees displaying a higher effective ability not only tended to display lower general working speed but also completed a smaller number of items before quitting. This suggests the conclusion that examinees displaying higher effective ability were more likely to display NRIs due to both reaching the time limit and quitting. The positive correlation between speed τ and test endurance ζ of .70 [.48; .87] indicates that examinees who worked faster had the tendency to complete more items before quitting the assessment. Note that in contrast to the baseline

[2] This information can be derived from mean test endurance as $\exp(\mu_\zeta)$.

analysis, both mechanisms presumably underlying NRIs were substantially related to ability. This may be due to the truncation of differences in test-taking behavior inherent to the number of NRIs, considering differences in test-taking behavior only for examinees who did not reach the end of the test, and/or the fact that the number of NRIs confounds NRIs occurring due to speed with those occurring due to quitting.

19.3.3.4 Approach 3: Modeling Item Omissions

The SA + O model is not only able to account for non-ignorable item omissions, but also provides valuable insights into the interplay of omission and pacing behavior, as well as how these behaviors are related to displayed ability. Again, results for ability and speed variances and correlations were similar to those of the previous two approaches, with $var(\theta) = 2.04$ [1.48; 2.74], $var(\tau) = 0.12$ [0.09; 0.15], and $cor(-\theta, \tau) = -.62$ [−.74; −.48]. With $var(\eta) = 0.48$ [0.32; 0.72], omission speed was more variable than general working speed. Likewise, with $var(\xi) = 2.71$ [1.93; 3.79], omission propensity showed considerable variation. Examinees displaying lower effective ability tended to omit more items and to do so faster, as indicated by the negative correlations of ability θ with both omission propensity ξ (−.39 [−.54; −.22]) and omission speed η (−.59 [−.80; −.35]). Further, the positive correlation between omission propensity ξ and omission speed η (.57 [.34; .74]) indicates that examinees who tended to omit more items also decided faster to do so. Speed τ, in contrast, was not related to omission propensity ξ (.02 [−.15; .18]). The non-perfect correlation between speed τ and omission speed η (.63 [.45; .78]) further indicates that different processes underlie pacing behavior for responses and omissions.

Item time intensities were comparable to those retrieved from Approaches 1 and 2. Item difficulties, however, were estimated to be somewhat higher in Approach 3. Since examinees displaying lower effective ability tended to omit more items, responses are more likely to be observed for examinees displaying higher effective ability. This is accounted for in Approach 3. When item omissions are ignored, however, as it is the case in Approaches 1 and 2, this results in difficulties to be underestimated (see Rose, 2013; Ulitzsch et al., 2019a).

Investigating relationships between item parameters provides additional insight into characteristics of items evoking omission behavior. Both difficult and time intense items tended to be omitted more often ($cor(b, a) = -.40$ and $cor(\beta, a) = -.72$). At the same time, examinees tended to require more time to omit items with a higher time intensity ($cor(\beta, \delta) = .43$).

19.3.3.5 Approach 4: Jointly Modeling Not-Reached and Omitted Items

The RTP model is able to jointly consider mechanisms underlying NRIs due to lack of speed and item omissions.

The estimated correlation between ability and speed was cor(θ, τ) = −.57 [−.70; −.43]. Ability and speed displayed variances of *var*(θ) = 1.99 [1.47; 2.69] and *var*(τ) = 0.12 [0.10; 0.15], respectively. While variances of ability and speed were similar to those retrieved from previous approaches, the correlation of ability and speed was markedly lower. This is due to the fact that in this approach a) speed is defined in terms of both, timing data for responses and omissions (whereas it is only defined based on timing data for observed responses in Approaches 1 to 3), and b) a different missingness mechanism underlying item omissions is assumed. Since item omissions are assumed to be ignorable in Approach 4, item difficulty estimates are comparable to those retrieved from Approaches 1 and 2. In addition, due to differences in whether response time data associated with omissions is considered in the measurement model for speed, Approach 4 yielded different estimates of time intensities for items with larger proportions of item omissions.

19.4 Future Directions

In LSA data, missing data mechanisms are highly plausible to be complex. Approaches for modeling missing data mechanisms inevitably need to focus on some aspects—ideally those deemed plausible and pertinent—and may neglect other missingness mechanisms. The SA + Q model, for instance, already accounts for NRIs due to speed and due to quitting simultaneously. It does, however, not incorporate item omissions. The RTP model considers NRIs due to lack of speed and item omissions; it does not, however, consider quitting. All three types of missing data—NRIs due to speed and quitting, as well as item omissions—occur in LSAs and need to be handled simultaneously. Considering NRIs due to lack of speed and quitting, as well as omission behavior simultaneously, could be achieved by combining the SA + O or the RTP model with the SA + Q model. When doing so, decisions on model choice should be guided by considerations on the mechanisms one assumes to underlie item omissions. While the RTP model conceptualizes omission behavior to occur when the time required to solve an item exceeds the implicit time an examinee allocates to the item, the SA + O model assumes omissions to go back to examinees' omission tendency. Further, both models make different assumptions on timing underlying responses and omissions. In either case, model combinations require to derive in which way different missingness mechanisms impact each other and to translate these considerations into an appropriate model. This will possibly result in rather complex, high-dimensional models. Theoretically deriving appropriate models for simultaneously considering multiple missingness mechanisms underlying NRIs and omissions and assessing conditions under which such models perform well is still open for investigation.

The discussed approaches support gaining a fine-grained understanding of different behavioral aspects associated with the occurrence of omissions and NRIs and allow to account for a broad array of complex missingness mechanisms. Nevertheless, each approach rests on strict assumptions on examinee behavior in general and the mechanisms underlying omissions and NRIs in particular. Violations of these assumptions pose a threat to both the validity of conclusions on the missingness mechanisms and trustworthiness of parameter estimates, as missing responses may not be accounted for adequately. For instance, current model-based approaches for NRIs due to lack of speed rest on the assumption that speed is stationary. Especially in the context of strict time limits, this assumption is plausible to be violated, as examinees may speed up toward the end of the test. In this case, assuming a constant level of speed misspecifies the missingness mechanism, potentially distorting conclusions on both examinees' adjusted ability estimates and the missingness mechanism itself. Model-based approaches incorporating a latent omission propensity assume, among others, that item omissions do not depend on the corresponding unobserved responses, i.e., that item omissions satisfy MAR given the latent omission propensity (Pohl & Becker, 2019; Robitzsch, 2020). This assumption is violated whenever item omissions occurred because examinees could not solve the items they omitted. Again, ignoring this aspect of the omission mechanism may potentially distort conclusions on both examinees' adjusted ability estimates and the mechanism underlying item omissions. It is, therefore, strongly warranted that researchers carefully evaluate the plausibility of the assumptions of the model employed and conduct sensitivity analyses on how violations of these assumptions may impact results (Robitzsch, 2021).

To support an even more nuanced understanding of complex missingness mechanisms and take them into account, future model developments may opt for further refinements. For instance, currently available model-based approaches for item omissions have in common that they assume a single, homogeneous mechanism to underlie all item omissions. This must not necessarily be the case, as examinees may change their omission behavior across the test or different subgroups of examinees may skip items for qualitatively different reasons. Indeed, when asked for explaining their motives for omitting items, examinees oftentimes name different reasons for item omissions, such as lack of confidence in the correct answer, fatigue, or lack of motivation (Jakwerth et al., 2003; OECD, 2013; van Barneveld et al., 2013). Thus, it is possible that in the course of the assessment, the same examinee might omit items for different reasons. Mixture extensions of the RTP model or the SA + O model that allow for different types of omissions, e.g., omissions due to lack of motivation and due to lack of confidence in the right answer, might support detecting and modeling such processes. It is reasonable to assume that different types of item omissions differ in the time required for their execution and are, as such, accompanied by different data-generating processes underlying the associated response times. This vein of thought is implemented in approaches leveraging response times for distinguishing observed responses stemming from good faith attempts to solve an item and disengaged rapid guessing behavior (Ulitzsch et al., 2019a; Wang & Xu, 2015; Wise & Gao, 2017). This rationale may also aid disentangling and modeling different omission mechanisms.

Research leveraging process data for understanding missingness mechanisms has primarily focused on response times. These allow researchers to get a better understanding of test-taking behavior by providing information on *how long* examinees interacted with an item until generating a response or deciding to omit it. In addition, computer-based LSAs also provide process data containing information on *how* examinees interacted with items. For instance, PISA provides information on the number of actions (that is, clicks, double clicks, key presses and drag/drop events) for each item-by-examinee interaction in its public database. Furthermore, in addition to the total response time per item, the first time to action, defined as the time between the first showing of the item and the first action recorded for the item, is provided (OECD, 2020). Oftentimes, for more complex tasks, e.g. simulations as implemented in the Programme for the International Assessment of Adult Competencies (PIAAC OECD, 2013) or tasks on collaborative problem solving as implemented in PISA, time-stamped action sequences are available. Analyzing such data in addition to observed final responses may facilitate understanding how examinees arrive at a certain response by understanding "how individuals plan, evaluate, and select operations" (He & von Davier, 2016, p. 72) and may critically improve our understanding of missing data mechanisms. For instance, omission processes may be further disentangled by identifying different patterns of examinee-by-item interactions. While omissions with short response times and no interactions might point towards disengaged omission behavior, omissions with longer response times and interaction patterns resembling those of observed responses might indicate that examinees tried to solve the item but due to, e.g., lack of confidence in the correct answer decided to omit it. A similar stream of thought is implemented in current approaches that augment response time information with information on the number of performed actions for classifying responses as stemming from solution or disengaged test-taking behavior (Sahin & Colvin, 2020). Likewise, for disentangling different mechanisms underlying quitting behavior, subgroups of examinees could be identified that differ in how they interact with the items they respond to and, as such, might also differ in whether, when, and for what reason they quit the assessment.

It should be noted that the discussed approaches have been developed for investigating missingness mechanisms in low-stakes assessments. While they may be applicable for investigating missing responses in high-stakes assessment data, they may describe markedly different aspects of test-taking behavior. For instance, quitting behavior in high-stakes assessments as compared to low-stakes assessments may occur because examinees feel overtaxed rather than due to fatigue effects of motivational issues. The same holds true for item omissions, with, e.g., strategic omission behavior for increasing test scores being considerably more plausible in high-stakes as compared to low-stakes assessment. Model-based approaches developed for low-stakes assessment data may not adequately depict these different missingness mechanisms, making model adaptations necessary. Most importantly, model-based approaches for missing values may not be suitable for dealing with missing values in high stakes assessments, as—when the employed approaches become known—this may impact test-taking behavior, undermining fair scoring.

19.5 Concluding Remarks

Examinees differ in how they approach assessments. In low-stakes LSAs, missing responses pose prominent and evident differences. Since the decision on how missing responses are treated in analyses can heavily impact results, understanding the underlying mechanisms is paramount for making appropriate decisions on how to deal with missing responses and drawing valid inferences on examinee competencies. Leveraging the additional information on test-taking behavior contained in response times allows for a nuanced modeling and understanding of test-taking behavior associated with the occurrence of missing responses in LSAs. Currently available approaches pose examples for how this can be done. As such, these approaches also initiate discussions on which missingness mechanisms can plausibly be assumed—and must then be modeled. Further discussions and theoretical developments concerning plausible missingness mechanisms, as well as approaches translating these considerations into statistical models, are encouraged and needed.

19.6 Author's Note

Supplemental online materials can be found in the OSF and are available via the following link: https://osf.io/5gbru/

References

Carpenter, B., Gelman, A., Hoffman, M. D., Lee, D., Goodrich, B., Betancourt, M., Brubaker, M., Guo, J., Li, P., & Riddell, A. (2017). Stan: A probabilistic programming language. *Journal of Statistical Software, 76*(1). https://doi.org/10.18637/jss.v076.i01

Culbertson, M. J. (2011). *Is it wrong? Handling missing responses in IRT* (tech. rep.) [Paper presented at the Annual Meeting of the National Council on Measurement in Education]. Paper presented at the Annual Meeting of the National Council on Measurement in Education.

De Boeck, P. (2008). Random item IRT models. *Psychometrika, 73*(4), 533–559. https://doi.org/10.1007/s11336-008-9092-x

de Ayala, R., Plake, B. S., & Impara, J. C. (2001). The impact of omitted responses on the accuracy of ability estimation in item response theory. *Journal of Educational Measurement, 38*(3), 213–234. https://doi.org/10.1111/j.1745-3984.2001.tb01124.x

Debeer, D., Janssen, R., & Boeck, P. (2017). Modeling skipped and not-reached items using IRTrees. *Journal of Educational Measurement, 54*(3), 333–363. https://doi.org/10.1111/jedm.12147

Finch, H. (2008). Estimation of item response theory parameters in the presence of missing data. *Journal of Educational Measurement, 45*(3), 225–245. https://doi.org/10.1111/j.1745-3984.2008.00062.x

Foy, P. (2017). *TIMSS 2015 user guide for the international database*. TIMSS & PIRLS International Study Center, Lynch School of Education, Boston College and International Association for the Evaluation of Educational Achievement (IEA). https://timssandpirls.bc.edu/timss2015/international-database/downloads/T15_UserGuide.pdf

Foy, P. (2018). *PIRLS 2016 user guide for the international database*. TIMSS & PIRLS International Study Center, Lynch School of Education, Boston College and International Association for the Evaluation of Educational Achievement (IEA). https://timssandpirls.bc.edu/pirls2016/international-database/downloads/P16_UserGuide.pdf

Frey, A., Spoden, C., Goldhammer, F., & Wenzel, S. F. C. (2018). Response time-based treatment of omitted responses in computer-based testing. *Behaviormetrika, 45*(2), 505–526. https://doi.org/10.1007/s41237-018-0073-9

Gelman, A., & Rubin, D. B. (1992). Inference from iterative simulation using multiple sequences. *Statistical Science, 7*(4), 457–472. https://doi.org/10.1214/ss/1177011136

Gelman, A., & Shirley, K. (2011). Inference from simulations and monitoring convergence. In S. Brooks, A. Gelman, G. Jones, & X.-L. Meng (Eds.), *Handbook of Markov Chain Monte Carlo*. Boca Raton, FL.

Glas, C. A. W., & Pimentel, J. L. (2008). Modeling nonignorable missing data in speeded tests. *Educational and Psychological Measurement, 68*(6), 907–922. https://doi.org/10.1177/0013164408315262

Guo, J., Gabry, J., & Goodrich, B. (2018). *Rjags: R interface to Stan* [R package version 2.18.2]. R package version 2.18.2. https://CRAN.R-project.org/package=rstan

He, Q., & von Davier, M. (2016). Analyzing process data from problem-solving items with n-grams: Insights from a computer-based large-scale assessment. In A. L. van der Ark, D. M. Bolt, W.-C. Wang, J. A. Douglas, & S.-M. Chow (Eds.), *Quantitative psychology research. The 79th annual meeting of the psychometric society, Madison, Wisconsin, 2014* (pp. 750–777). NY, Springer.

Holman, R., & Glas, C. A. W. (2005). Modelling non-ignorable missing-data mechanisms with item response theory models. *British Journal of Mathematical and Statistical Psychology, 58*(1), 1–17. https://doi.org/10.1111/j.2044-8317.2005.tb00312.x

Jakwerth, P. M., Stancavage, F. B., & Reed, E. D. (2003). An investigation of why students do not respond to questions. In *NAEP validity studies*. Department of Education, National Center for Education Statistics.

Köhler, C., Pohl, S., & Carstensen, C. H. (2015). Taking the missing propensity into account when estimating competence scores: Evaluation of item response theory models for nonignorable omissions. *Educational and Psychological Measurement, 75*(5), 850–874. https://doi.org/10.1177/0013164414561785

Köhler, C., Pohl, S., & Carstensen, C. H. (2017). Dealing with item nonresponse in large-scale cognitive assessments: The impact of missing data methods on estimated explanatory relationships. *Journal of Educational Measurement, 54*(4), 397–419. https://doi.org/10.1111/jedm.12154

Lee, Y.-H., & Ying, Z. (2015). A mixture cure-rate model for responses and response times in time-limit tests. *Psychometrika, 80*(3), 748–775. https://doi.org/10.1007/s11336-014-9419-8

Lu, J., & Wang, C. (2020). A response time process model for not-reached and omitted items. *Journal of Educational Measurement*. https://doi.org/10.1111/jedm.12270

Mislevy, R. J., & Wu, P.-K. (1996). *Missing responses and IRT ability estimation: Omits, choice, time limits, and adaptive testing (ETS Research Report RR-96-30-ONR)*. Educational Testing Service. https://doi.org/10.1002/j.2333-8504.1996.tb01708.x

Molenaar, D., Tuerlinckx, F., & van der Maas, H. L. (2015). A generalized linear factor model approach to the hierarchical framework for responses and response times. *British Journal of Mathematical and Statistical Psychology, 68*(2), 197–219. https://doi.org/10.1111/bmsp.12042

OECD. (2013). *Technical report of the survey of adult skills (PIAAC)*. OECD Publishing. https://www.oecd.org/skills/piaac/_TechnicalReport_17OCT13.pdf

OECD. (2020). *PISA 2018 technical report*. OECD Publishing. http://www.oecd.org/pisa/data/pisa2018technicalreport/

O'Muircheartaigh, C., & Moustaki, I. (1999). Symmetric pattern models: A latent variable approach to item non-response in attitude scales. *Journal of the Royal Statistical Society: Series A (Statistics in Society), 162*(2), 177–194. https://doi.org/10.1111/1467-985X.00129

Pohl, S., & Becker, B. (2019). Performance of missing data approaches under nonignorable missing data conditions. *Methodology*. https://doi.org/10.5964/meth.2805

Pohl, S., Gräfe, L., & Rose, N. (2014). Dealing with omitted and not-reached items in competence tests: Evaluating approaches accounting for missing responses in item response theory models. *Educational and Psychological Measurement, 74*(3), 423–452. https://doi.org/10.1177/0013164413504926

Pohl, S., Ulitzsch, E., & von Davier, M. (2019). Using response time models to account for not-reached items. *Psychometrika, 84*(3), 892–920. https://doi.org/10.1007/s11336-019-09669-2

Pohl, S., & von Davier, M. (2018). Commentary: "On the importance of the speed-ability trade-off when dealing with not reached items" by Jesper Tijmstra and Maria Bolsinova. *Frontiers in Psychology, 9*(1988). https://doi.org/10.3389/fpsyg.2018.01988

Robitzsch, A. (2014). Zu nichtignorierbaren Konsequenzen des (partiellen) Ignorierens fehlender Item Responses im Large-Scale Assessment. In B. Suchan, C. Wallner-Paschon, & C. Schreiner (Eds.), *PIRLS & TIMSS 2011–die Kompetenzen in Lesen, Mathematik und Naturwissenschaft am Ende der Volksschule: Österreichischer Expertenbericht*. Graz.

Robitzsch, A. (2020). *About still nonignorable consequences of (partially) ignoring missing item responses in large-scale assessment*. https://osf.io/hmy45/download

Robitzsch, A. (2021). On the treatment of missing item responses in educational large-scale assessment data: An illustrative simulation study and a case study using PISA 2018 mathematics data. *European Journal of Investigation in Health, Psychology and Education, 11*(4), 1653–1687. https://doi.org/10.3390/ejihpe11040117

Rose, N. (2013). *Item nonresponses in educational and psychological measurement (Doctoral dissertation)*. Friedrich-Schiller-Universität Jena. https://d-nb.info/1036873145/34

Rose, N., von Davier, M., & Nagengast, B. (2017). Modeling omitted and not-reached items in IRT models. *Psychometrika, 82*(3), 795–819. https://doi.org/10.1007/s11336-016-9544-7

Rose, N., von Davier, M., & Xu, X. (2010). *Modeling nonignorable missing data with item response theory (IRT) (ETS Research Report RR-10 11)*. Educational Testing Service. https://doi.org/10.1002/j.2333-8504.2010.tb02218.x

Rubin, D. B. (1976). Inference and missing data. *Biometrika, 63*(3), 581–592. https://doi.org/10.1093/biomet/63.3.581

Sachse, K., Mahler, N., & Pohl, S. (2019). Effects of changing nonresponse mechanisms on trends and group comparisons in international large-scale assessments. *Educational and Psychological Measurement, 79*(4), 699–726. https://doi.org/10.1177/0013164419829196

Sahin, F., & Colvin, K. (2020). Enhancing response time thresholds with response behaviors for detecting disengaged examinees. *Large-Scale Assessment in Education, 8*(5). https://doi.org/10.1186/s40536-020-00082-1

Ulitzsch, E., von Davier, M., & Pohl, S. (2019a). A hierarchical latent response model for inferences about examinee engagement in terms of guessing and item-level nonresponse. *British Journal of Mathematical and Statistical Psychology*. https://doi.org/10.1111/bmsp.12188

Ulitzsch, E., von Davier, M., & Pohl, S. (2019b). A multi-process item response model for not-reached items due to time limits and quitting. *Educational and Psychological Measurement*. https://doi.org/10.1177/0013164419878241

Ulitzsch, E., von Davier, M., & Pohl, S. (2019c). Using response times for joint modeling of response and omission behavior. *Multivariate Behavioral Research*. https://doi.org/10.1080/00273171.2019.1643699

van der Linden, W. J. (2007). A hierarchical framework for modeling speed and accuracy on test items. *Psychometrika, 72*(3), 287–308. https://doi.org/10.1007/s11336-006-1478-z

van der Linden, W. J. (2009). Conceptual issues in response-time modeling. *Journal of Educational Measurement, 46*(3), 247–272. https://doi.org/10.1111/j.1745-3984.2009.00080.x

van Barneveld, C., Pharand, S.-L., Ruberto, L., & Haggarty, D. (2013). Student motivation in large-scale assessments. In M. Simon & M. Rousseau (Eds.), *Improving large-scale assessment in education: Theory, issues and practice*. Routledge.

Wang, C., & Xu, G. (2015). A mixture hierarchical model for response times and response accuracy. *British Journal of Mathematical and Statistical Psychology, 68*(3), 456–477. https://doi.org/10.1111/bmsp.12054

Weeks, J. P., von Davier, M., & Yamamoto, K. (2016). Using response time data to inform the coding of omitted responses. *Psychological Test and Assessment Modeling, 58*(4), 671–701.

Wise, S. L., & Gao, L. (2017). A general approach to measuring test-taking effort on computer-based tests. *Applied Measurement in Education, 30*(4), 343–354. https://doi.org/10.1080/08957347.2017.1353992

Wolf, L. F., Smith, J. K., & Birnbaum, M. E. (1995). Consequence of performance, test, motivation, and mentally taxing items. *Applied Measurement in Education, 8*(4), 341–351. https://doi.org/10.1207/s15324818ame0804_4

Yamamoto, K., Khorramdel, L., & von Davier, M. (2013). Scaling PIAAC cognitive data. In *Technical report of the survey of adult skills (PIAAC)*. http://www.oecd.org/skills/piaac/Technical_Report_2nd_Edition_Chapters_17-23.pdf

Chapter 20
Incorporating Timing Data into PISA Population Modeling

Hyo Jeong Shin, Matthias von Davier, and Kentaro Yamamoto

Abstract The 2015 cycle of PISA was among the first attempts to collect process data, including response time (RT), in an international large-scale assessment. In particular, RT data have the potential to be incorporated, along with many other non-cognitive variables, at the population modeling stage, contributing to the improved accuracy of plausible values and less biased secondary analyses statistics. In this chapter, we provide an overview of the RT variables provided to the public in PISA, and we present the pilot study addressing two research questions: (1) how to handle the item-level RT data to incorporate them appropriately into the population modeling, and (2) whether or not the inclusion RT information contributes to improving the reporting of proficiency scores. For the pilot study, we describe the procedures and present the results using the data collected in selected countries during the PISA 2015 cycle, which eventually guided the operational decisions for the PISA 2018 population modeling. For the first research question, we present one practical and straightforward method (among many different approaches) to derive person-level characteristics out of the item-by-person RT variables recorded in the log files. For the second research question, we compare the empirical analyses results, depending on the inclusion or exclusion of derived RT variables in the generation of plausible values. We conclude the chapter with discussion about limitations, implications, and suggestions for future studies.

H. J. Shin (✉)
Sogang University, Seoul, South Korea
e-mail: hshinedu@sogang.ac.kr

M. von Davier
Lynch School of Education and Human Development, TIMSS & PIRLS International Study Center at Boston College, Chestnut Hill, MA, USA
e-mail: vondavim@bc.edu

K. Yamamoto
Independent Consultant for Psychometric Methodology and Large-Scale Assessment (formerly ETS), Princeton, NJ, USA

L. Khorramdel et al. (eds.), *Innovative Digital-Based International Large-Scale Assessments*, Methodology of Educational Measurement and Assessment,
https://doi.org/10.1007/978-3-031-90951-1_20

International large-scale assessments (ILSAs) fit statistical models that integrate various pieces of information about the examinees, including responses to cognitive assessment items and background questionnaire (BQ) information, such as gender and parental education, and so on. With the introduction of computer-based ILSAs, richer information, including timing and process data, is now available and is increasingly used by researchers. The Programme for International Assessment of Adult Competencies (PIAAC) from 2012 and the Programme for International Student Assessment (PISA) from 2015 are among the first attempts to collect timing and process data in ILSAs. The main goal of this chapter is to investigate how to incorporate novel timing and process data in the analytic procedures commonly used in ILSAs. To that end, we start with a brief introduction about the conventional analytic procedures in most ILSAs.

The statistical procedure used to integrate various sources of information collected in ILSAs involves an item calibration stage and a conditioning modeling stage (Organisation for Economic Co-operation and Development [OECD], 2017, 2020a; von Davier & Sinharay, 2014; von Davier et al., 2006). In the item calibration stage, an item response theory (IRT) model is fitted for a mixed format of items in each cognitive domain (e.g., mathematics, reading, science), and item parameters are estimated. Then, in the conditioning modeling stage, a latent regression model (LRM) is fitted to the data fixing the item parameters obtained in the first (item calibration) stage and specifying the principal components based on BQ variables as explanatory variables. Hence, when the LRM is fitted, the group-level ability distribution is assumed to depend not only on the item responses on cognitive items but also on a large number of BQ predictors. Based on the latent regression estimates (variance-covariance matrix and regression parameters), multidimensional plausible values (PVs; Mislevy, 1991; Mislevy et al., 1992; von Davier et al., 2009) are generated as a set of proficiency scores (e.g., 10 sets of PVs in PISA) and then transformed on the reporting scale that is linked to previous assessment cycles and across countries. This whole analytic procedure is often referred to in PISA as *population modeling*, or IRT-LRMs,[1] as a more general statistical term combining IRT and LRM.

Process data in ILSAs are relatively new information that became available with the initiation of computer-based assessment and can be useful for improving data quality, test security, reliability, and validity (Bergner & von Davier, 2019; Ercikan & Pellegrino, 2017; von Davier et al., 2019). More specifically, there are three major reasons for utilizing response time (RT) data, which is the focus of this chapter, and process data more broadly for ILSAs. First, if the RT variables are available in the public database but not used to generate the plausible values, secondary analyses may result in biased estimates of the relationship between proficiency and the process data variables, should the secondary data user choose to include RT

[1]Details about analytic procedures used in LSA, such as NAEP and PISA, are well-described in Mislevy et al. (1992), von Davier and Sinharay (2014), von Davier et al. (2006), von Davier et al. (2009).

variables in their analyses (Meng, 1994; Mislevy, 1991). Second, RT information may uniquely contribute to the prediction of proficiency and, therefore, improve the quality of the PVs and increase measurement precision when properly handled (Mislevy, 1991; von Davier et al., 2019). In particular, a recent study (Shin et al., 2022) revealed that incorporating item-level RT variables into the limited IRT-LRM, specifying only a subset of BQ variables, is promising because the inclusion of item-level RT variables resulted in a substantial increase in proficiency measurement precision (i.e., greater explained variance), about 16% in PISA 2015 data, with negligible differences in regression coefficient estimates specified in the LRMs. Thirdly, the RT and process data are informative for understanding how test-takers manage their time and solve tasks. For example, RT and process data can be used to evaluate the validity of survey operation and cognitive responses (Yamamoto et al., of this volume; Yamamoto & Lennon, 2018). In addition, process data can be used to provide insights on the mechanisms of non-responses in the low-stake assessments and how the missing data should be treated (e.g., Lu & Wang, 2020; Pohl et al., 2019; Ulitzsch et al., 2019a; Weeks et al., 2016). In particular, non-response behavior unveiled by process data can be useful for investigating student effort and engagement (Anaya & Zamarro, 2020; Michaelides et al., 2020; Ulitzsch et al., 2019b). Furthermore, classifying respondents into groups that may relate to process data contributes to deepening our understanding about test-taking strategies and student inquiry (e.g., Greiff et al., 2015; He et al., 2018; Teig et al., 2020). Therefore, RT and process data can be considered important covariates of proficiency and performance that can potentially contribute to better describing group differences (Ercikan et al., 2020).

In this chapter, we focus on the first two reasons for incorporating RT into the population model: (1) to provide unbiased estimates between RT and PVs in the secondary analyses and (2) to improve the measurement precision of PVs. As mentioned earlier, a large number of BQ variables (which become even larger after being contrast-coded) are incorporated into the IRT-LRMs. Therefore, incorporating RT in the IRT-LRMs, which is already quite complex, is not straightforward and can be challenging. In the following sections, we briefly introduce the RT variables published in the PISA database with the example of the 2015 cycle and beyond. We then explain the IRT-LRMs, specifically addressing the major challenges of incorporating RT variables into the models. Following this, we describe one practical and promising approach that was attempted with the empirical analyses of PISA 2015 data. Finally, we conclude the chapter with limitations and suggestions for the further studies.

20.1 Response Time Variables Published in the PISA Database

Since the switch to computer-based assessments (CBA) in the 2015 cycle, PISA has provided process data variables, in addition to traditional item responses, in the public user file (PUF; OECD, 2020b). Several steps are taken to publish process data variables in the PUF. First, how individual students interact with items is recorded as a log file for individual student sessions in the student delivery system (SDS). Among many different types of logged events, a subset of meaningful events are extracted, and pre-defined process data variables are computed at the item-level, which are then published in the official PUF. Typically, an RT variable refers to the time a respondent spends on each item (or certain aspects within the time interval) in an assessment. Specifically, in PISA, RT is viewed as time spent looking at a particular screen before clicking the "next" button, assuming that only a single item is presented on the screen at a time, as it is with most PISA items. However, there are several ways of defining, precisely, how much time a student has spent looking at the screen. One simple option is to take the difference in time stamps between when the item begins appearing on the screen (beginning) and when the item no longer appears on the screen (ending) from the log file captured in the SDS. However, there may be a time delay between the first appearance of the item screen and when the item has fully loaded, so another option is to exclude the item loading time so that the RT variable purely reflects the period when the item is fully present on the screen, and the student can see the complete item. Such a difference may not sound like a major issue, but timestamps in PISA are recorded in milliseconds, and whether or not loading time is taken into account can lead to different values and interpretations of the RT variables, especially when loading time can be differently affected by the computer system specifications from different countries. Clarifications and updates of the timing data, including more general details about the logged events in the PISA SDS can be found in Annex K associated with the PISA 2018 technical report (OECD, 2020a), which is also applicable to PISA 2015.

Officially, PISA databases provide RT variables at the item-level for cognitive items and at the scale-level for the BQ. This is mainly because the cognitive domain tends to show a single item per screen, while the BQ is administered with multiple items on one screen (usually all items for a BQ scale appear together on screen at a time). Focusing on the cognitive domain, Table 20.1 presents a list of variables that are available for individual items for each of the PISA 2015 and 2018 cycles. In this table, process data variables are shaded, and the timing variable, which is of interest in this chapter, is bolded. Note that two sets of RT variables are now publicly available in each of the 2015 and 2018 databases: RT variables ending with "T" are the length of time the student spent on the item during their *last visit*, while the updated RT variables ending with "TT" are the total time spent on the item over *multiple item visits* (e.g., if the student went back to the previous screen within a unit and visited the item multiple times). Based on Annex K of the PISA 2018 technical report (OECD, 2020a), total time variables ("TT" summed over multiple visits)

Table 20.1 List of item-level variables provided in the PUF for PISA 2015 and 2018 cycles

Item-level variables		PISA 2015	PISA 2018
Raw response (ending with "R")[a]	Option chosen (e.g., A, B, C, D)	√	√
Coded response (ending with "C") or Scored response (ending with "S")	0 for no credit 1 for full credit (dichotomous) 1 for partial credit (polytomous) 2 for full credit (polytomous)[c]	√	√
Timing (ending with "T")	**Time spent for that item on the last visit, recorded in milliseconds**	√	√
Number of actions (ending with "A")	Counts of actions, including – mouse click – mouse double-click – key press – drag and drop	√	√
Time to first action (ending with "F")	Time spent before the first action happens, recorded in milliseconds		√
Updated timing (ending with "TT")[b]	Total time spent for that item across visits, recorded in milliseconds	√	√
Number of visits (ending with "V")[b]	Number of visits for that item	√	√

[a]Raw response is the option chosen by the students for multiple-choice item types. Raw responses for open-ended, constructed-response item types are not provided in the PUF
[b]These updated timing variables ("TT" and "V" variables), cognitive items total time/visits data file, are provided in the additional PUF, which was uploaded on the PISA website in September 2020 for PISA 2018 (OECD, 2020c) and in November 2020 for PISA 2015 (OECD, 2020b)
[c]Some other scoring variations, often referred to as double-digit scoring (e.g., 11 and 12 for the partial credit), exist in PISA according to the scoring rubrics

differ from last visit time variables ("T" captured on the last visit) in 7% of the data, with varying degrees across countries. At the time of analyses, the "T" variable was the only RT variable available, so the empirical analysis in the following section utilizes the RT variables ending with "T" from PISA 2015.[2] Additionally, there are two other RT variables that are relatively new. The *time to first action*, ending with "F," was not provided in the 2015 cycle, but it became available in the 2018 cycle. Further, the *number of visits*, ending with "V," also became available alongside the "TT" variable for secondary researchers to easily identify multiple visits, if needed.

To illustrate how RT ("T") variables generally function in PISA, Fig. 20.1 presents the mean RT in seconds (along the y-axis) for each item (along the x-axis) in PISA 2015 science. Each line represents a country/economy that participated in the CBA mode of PISA 2015. The average time taken for most items ranges from about

[2]The pilot study presented in this chapter did not examine the impact of using "TT" variables instead of "T" variables. Annex K (Uses and reporting of process data) associated with PISA 2018 technical report (https://www.oecd.org/pisa/data/pisa2018technicalreport/) illustrates the impact of "TT" variables instead of "T" variables in generating reported PVs for the 2018 cycle.

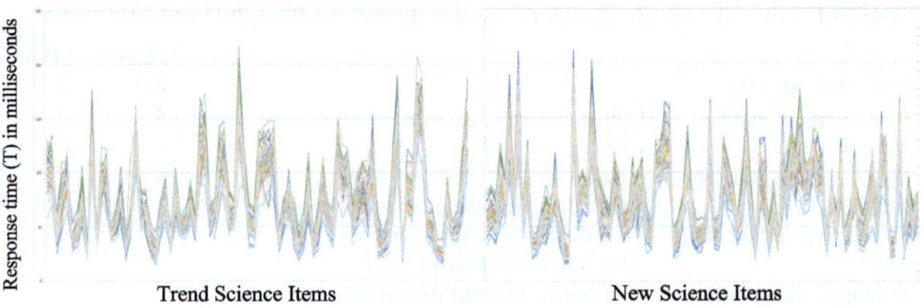

Fig. 20.1 Mean response time for each item (each line represents one country/economy) for science trend (left) and new (right) items in PISA 2015

30 seconds to more than 3 min, but more importantly, it is apparent that distributions of the mean item-level RT show considerably similar patterns across countries/economies, regardless of the country/economy's performance or languages. That is, an item that requires more time in one country/economy is likely to require more time in other countries/economies as well. This is supported by a recent study that found that there were no significant item-by-country interactions when identifying the minimum RT for valid item responses in PISA 2015 (Robin et al., 2018). Therefore, we assume that the variability in RT data is sufficiently accounted for by the respondents and items; this means that RT variables can be characterized either as features of the item or person characteristics, and this approach is applicable to all countries/economies. On one hand, when the RT variables are viewed as item features, RT item parameters can be estimated to measure the underlying latent RT scales, and indirect tests of measurement invariance of those item-level RT parameters (Shin et al., 2020). On the other hand, when RT variables are viewed as person characteristics, a better description of group differences can be derived, if the data are properly handled. More details about this latter approach and the proposed procedure is described in the following section.

20.2 Modeling Response Time in IRT-LRMs

There are various ways to incorporate RT variables in the IRT-LRMs, such as including RT data as additional covariates in the LRM, or RT variables can be incorporated at the IRT scaling stage with the item responses. With regard to the latter, Molenaar and van Rijn (2024) explore the more general and flexible statistical approach of incorporating RT variables at the item calibration stage. In the same line, previous studies (e.g., Bolsinova & Tijmstra, 2018; van der Linden et al., 2010) suggest that the RT as collateral information for modeling individual differences in ability can potentially improve the reliability of the assessment. This chapter, however, pertains to the first approach: the inclusion of RT variables as additional covariates in the LRMs. Similar to the benefits of the second approach, this approach can potentially improve the measurement precision of the reported PVs and is expected to reduce the bias in the estimates involving RT and PVs. Furthermore, the first approach is more aligned with the definition of framework where RT is not reflected in defining the construct (OECD, 2020a).

In the process of generating PVs in ILSAs, using all available information has been considered a good strategy (Little & Rubin, 2014; Mislevy, 1991; von Davier et al., 2009). Theoretically, under the data missing completely at random (MCAR) and data missing at random (MAR) assumptions, the rationale is to include all covariates that are available, which may be of analytic interest in order to minimize bias in the estimates of statistical quantities involving these imputations. As argued by Collins et al. (2001) regarding the use of auxiliary variables, the so-called *inclusive strategy* (an imputation model that contains as many as covariates as possible and available) is greatly preferred compared to a *restrictive strategy* (an imputation model with a limited number of variables) in terms of the increased effective sample size, reduced standard errors, and minimized biases.

Operationally, since the 2015 cycle of PISA, there have been three subordinate routines for accommodating a large number of BQ variables in the IRT-LRMs (OECD, 2017, 2020a). First, individual BQ variables are contrast-coded. Contrast coding allows for the inclusion of missing responses and avoids the necessity of assuming a linear relationship between the responses to any question and the outcome variable. Second, a principal components analysis (PCA) is conducted using the contrast-coded BQ variables to (1) remove collinearity among variables when present and (2) reduce the large number of variables into a smaller number of principal components that are sufficient to account for a large proportion of the variation in the BQ variables without over-parameterization. This process is conducted country-by-country to accommodate common BQ variables collected across all countries/economies, to accommodate optional BQ variables that are specific to a country/economy, and to allow for the estimation of country/economy-specific relationships between the BQ data and the proficiency. Third, the number of principal components (PCs) to be included in the PISA operational procedure is determined by the number of PCs that explain 80% of the total BQ variance for the country/economy or 5% of the raw student sample size per country/economy, whichever is less. These criteria ensure that the estimated model captures a large amount of information in the BQ variables and, at the same time, avoids overparameterization of the model, which could lead to unstable outcomes and erroneous inferences (OECD, 2017). Waylon et al. (2015) also concluded that the PCA strategy to reduce the number of auxiliary variables is an effective and practical way to reap the benefits of the inclusive strategy in the presence of possible auxiliary variables. With these steps in place, a large but limited number of extracted PCs are specified as covariates in the multivariate latent regressions.

In the PISA operational procedure, the country/economy-specific LRM can be expressed as follows for student proficiency distributions on multidimensional scales, conditional on covariates (**y**; selected principal components extracted from the contrast-coded BQ variables) and item responses (**x**) for multiple domains

(OECD, 2020a). Based on Bayes' theorem, the posterior distribution of skills given the observed item responses and covariates is constructed as follows:

$$P(\boldsymbol{\theta}_v|\boldsymbol{x}_v,\boldsymbol{y}_v,\Gamma,\Sigma) \propto P(\boldsymbol{x}_v|\boldsymbol{\theta}_v,\boldsymbol{y}_v,\Gamma,\Sigma)P(\boldsymbol{\theta}_v|\boldsymbol{y}_v,\Gamma,\Sigma)$$
$$= P(\boldsymbol{x}_v|\boldsymbol{\theta}_v)P(\boldsymbol{\theta}_v|\boldsymbol{y}_v,\Gamma,\Sigma), \quad (20.1)$$

where $\boldsymbol{\theta}_v$ is a vector of length d with scale values (these values correspond to performance on each of the skills) for student v. As shown, the posterior distribution of proficiency is proportional to the likelihoods of the item response data, $P(\boldsymbol{x}_v|\boldsymbol{\theta}_v)$, and prior distributions, $P(\boldsymbol{\theta}_v|\boldsymbol{y}_v,\Gamma,\Sigma)$. Given the conditional independence assumption, $P(\boldsymbol{x}_v|\boldsymbol{\theta}_v)$ is the product of independent likelihoods for the observed response to each cognitive item (estimated by IRT models) within each scale. Next, $P(\boldsymbol{\theta}_v|\boldsymbol{y}_v,\Gamma,\Sigma)$, which is a prior distribution, is the multivariate joint density of proficiencies of the scales, conditional on the extracted principal components $\boldsymbol{y}_v$ derived from BQ variables and regression parameters Γ and Σ. The item parameters are fixed and regarded as population values in the computation. This two-step approach is generally used to reduce computational burden. Here, latent proficiencies for each student v are assumed to follow multivariate normal distributions with mean $\Gamma'\boldsymbol{y}_v$ and covariance matrix Σ, $\boldsymbol{\theta}_v \sim N(\Gamma'\boldsymbol{y}_v,\Sigma)$, where Γ is the $K \times d$ matrix of regression coefficients, K is the number of principal components, and Σ is the $d \times d$ residual variance-covariance matrix.

Given that our interest lies in $\boldsymbol{y}_v$, where RT variables can be included in addition to the typical BQ variables, several methods can be considered as alternatives among various variable reduction methods or regularized regression methods, as summarized in von Davier et al. (2019). Further investigations are needed with regard to whether RTs on individual items can be directly included in the set of predictors for a PCA, or if items should be preprocessed separately from other BQs. However, not all methods are directly applicable or useful in the case of LRMs and the PISA design. More specifically, note that PISA aims at group-level score reporting and that the balanced incomplete block (BIB) designs in PISA 2015 and, more recently, multistage adaptive testing design from 2018 have been used to cover the broad framework without overburdening students' testing workload. As such, individual students only respond to a subset of items (e.g., two clusters out of twelve clusters in total in the science domain in PISA 2015). As a consequence, cognitive indicators, as well as process data variables, are available only for the items in the assigned form, while the observations on these variables are missing for all other items according to the BIB design. Taking this huge amount of missingness into account, a mixture of planned missingness (missing completely at random) and unplanned missingness (both missing at random and missing not at random), together with the sampling weights, performing some traditional and advanced statistical analyses are not fully afforded by many standard software programs and packages. Such complexities could be still applicable in the PISA 2018 cycle, with the missingness coming from the multistage-adaptive testing design. In the following, we introduce one practical and straightforward approach that incorporates RT information in the IRT-LRM, which is readily feasible in the current PISA operational procedure.

20.3 Empirical Analyses of the PISA 2015 IRT-LRMs Including RT Information

In our approach, RT variables can be included as a part of students' covariates along with other BQ variables. Necessarily, these population models are country/economy-specific multivariate LRMs, given that countries may vary in their use of time, have different pacing patterns and, hence, differ in the distribution of RTs and in how RTs may relate to performance. If there were a substantial relationship between RTs and proficiencies, the posterior standard deviations would be reduced, as would the standard errors associated with overall and subgroup estimates. The empirical analyses of PISA 2015 data are presented below to address two research questions: (1) how to appropriately handle item-level RT data to derive person characteristics and incorporate them in the LRMs, and (2) whether or not the inclusion of RT information contributes to improving the quality of PVs. We used the PISA 2015 data from nine countries with various performance levels and cultural backgrounds. Also, note again that, at the time of analyses, only the "T" variable was available, so the empirical analyses in the following section utilized the RT variables ending with "T" in the PISA 2015 PUF.

20.3.1 RQ1: Deriving Person Characteristics out of the Item-Level RT

RT data in PISA are recorded at the item-level for individual students, except for a few cases where two items are presented on the same screen. The data format itself resembles the item responses in the sense that each data point is captured as an item-by-person interaction. As a practical and straightforward solution, we can consider RT aggregation or summary statistics, such as means or medians of the RTs at some levels or a typology of students based on their RTs (e.g., students who used the maximum time allowed, students who quit the test quickly, or students' latent class membership based on their time use patterns). One of the important remaining concerns is a violation of the conditional independence assumption in the IRT-LRMs because RT data are recorded at the item-by-person level; incorporating both the item-level RT variable and item response for the same item, if correlated, could violate the conditional independence assumption underlying the IRT-LRM. More precisely, item response and RT from the same item may reflect item-specific performance, which can lead to conditional dependence. In this regard, a recent empirical study conducted by Shin et al. (2022) revealed that the differences in the regression coefficient estimates when RT was either included or excluded are negligible, and the benefits of incorporating RT outweigh the potential bias from violating the conditional independence assumption.

In this study, several layers of standardization and categorization were performed to derive an RT-related person covariate, or *proxy of working speed*, out of item-by-person interactions. Through several steps of aggregation and standardization, the relationship between RT and individual item responses becomes almost nonexistent, which makes the chance of violating conditional independence remote. More specifically, the following steps of categorization and standardization were conducted within each country.

1. In the first step, standardization was performed at the item level: Item-level continuous RT variables were discretized into decile rank values, ranging from 1 to 10. Note that in this process, RT variables associated with not-reached responses were excluded, while incorrect, correct, partial credit (if applicable), and omitted (skipped) responses were included, in the calculation of item-level decile ranks.
2. The next level of aggregation was at the cluster level. For the given cluster, for each person, means and standard deviations (SD) of item-level decile rank values were computed.
3. Third, the average and SD values were discretized into decile ranks again for each person for the given cluster. Note that in PISA 2015, all students took two clusters of science items, and they took different sets of items in each cluster position according to the BIB design. Eventually, four RT variables were produced at the cluster level: deciles of means across items administered in the (1) first cluster and (2) second cluster; and deciles of SDs across items administered in the (3) first cluster and (4) second cluster.

Table 20.2 presents several example cases of the derived person covariates out of item-level RT variables. The “mean” columns are decile ranks of the mean of item-level RT deciles at the cluster-level, and the “SD” columns are decile ranks of the *SD* of item-level RT deciles at the cluster-level. Viewing these four sets of variables as

Table 20.2 Example of derived person covariates out of item-level RT variables

		Cluster 1			Cluster 2		
Student ID	Cluster Position	Cluster	Mean	SD	Cluster	Mean	SD
02	1	S02	6	6	S04	1	5
03	2	S12	4	3	S07	4	3
04	1	S11	9	3	S02	0	0
05	2	S11	0	1	S02	0	0
07	2	S03	2	4	S09	0	3
08	1	S12	6	4	S09	5	5
09	2	S03	8	2	S12	7	0
10	1	S12	9	8	S09	6	2
11	2	S05	8	3	S11	NA	NA

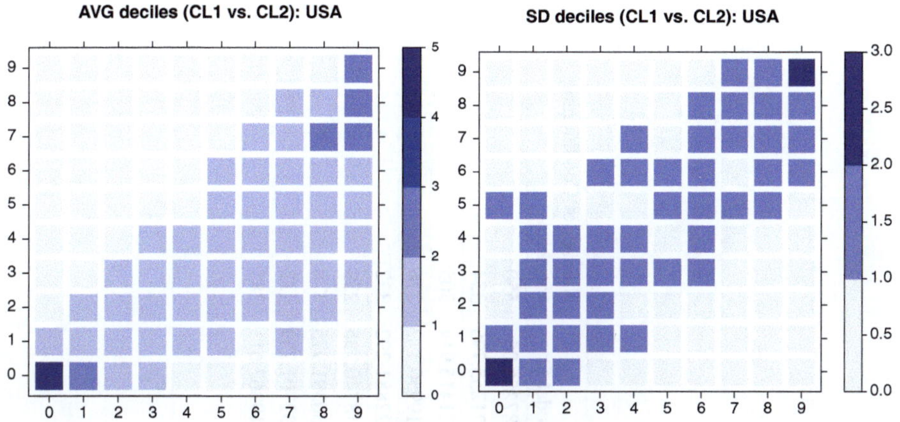

Fig. 20.2 Relationship among *AVG (average)* deciles and *SD* (standard deviation) deciles

proxies of working speed, low values in the mean column indicate that the student within the country tended to respond more quickly to items within the cluster, on average relative to others, while higher values indicate slower responding to items within the cluster. Similarly, lower values in the *SD* columns indicate that the student within the country maintained a consistent response speed within the cluster relative to others, while higher *SD* values indicate more variable response speed. Figure 20.2 illustrates the relationship among those four derived person-level RT covariates for the given cluster in one example country. Most of the students are located around the diagonal line, implying that there is a tendency for a student's RT data to be consistent between two cluster positions, although not exactly the same, which delivers unique information about how individual students proceed through the test.

20.3.2 RQ 2: Impacts on the Quality of PVs

In the next phase of analysis, four sets of RT covariates were contrast coded and included in the PCA, as with other BQ variables. In the current PISA operational procedure, the number of PCs to be included in the LRMs is determined by the number of PCs that explain 80% of the total BQ variance for the country/economy and do not to exceed 5% of the raw student sample size per country/economy to avoid over-parameterization (OECD, 2017). Table 20.3 compares the number of PCs retained for each model, and the corresponding fit statistics are the measure of data-model fit for each country/economy-level latent regression.

The PCA results show that the number of conditioning variables for all nine countries, except for one country (CNT2), is the same between the PISA 2015 operational analyses and the added RT-variable analyses (shaded columns). Given that added RT-variable analyses had greater total BQ variances with additional RT

Table 20.3 Comparison of the number of conditioning variables and model fit

		PISA 2015 MS			RT variables-added		
	N	Number of Cycles	Model fit	Number of Conditioning variables	Number of Cycles	Model fit	Number of Conditioning variables
CNT1	23,141	179	−85923.554	550	190	−87256.05	550
CNT2	6388	107	−21047.173	326	108	−21255.795	334
CNT3	6108	113	−21230.381	305	113	−21415.914	305
CNT4	5532	102	−19751.61	276	105	−19950.112	276
CNT5	6647	150	−22688.299	332	150	−22881.292	332
CNT6	5581	148	−18551.605	279	143	−18740.735	279
CNT7	7568	172	−30282.928	378	176	−30391.975	378
CNT8	8249	191	−32301.798	412	203	−32701.759	412
CNT9	5677	126	−19147.226	285	123	−19367.611	285

variables, this suggests that 5% of the raw student sample size per country/economy was used as the criteria to determine the number of PCs. That is, less than 80% of the total BQ variance was explained with this retained number of conditioning variables in each country for the sake of stable estimation. It is important to note that, although the same number of PCs were retained across two models for each dataset, the composition of PCs (loading patterns of the variables) changed. Therefore, the key question here becomes which set of PCs provides the relatively better predictive power.

Table 20.3 shows that the RT variables-added model always resulted in better model-data fit compared to the PISA 2015 operational model, in which RT variables were not included, and a similar number of iterations were needed to reach the convergence criteria. Lower values indicate better fit for the same item response data used and exactly the same number of PCs. Note that these fit statistics are used internally for the software called DGROUP (Educational Testing Service, 2012), which is used for the PISA operational IRT-LRMs. Improved fit statistics suggest that a set of PCs, when RT variables were included, provide better predictive power.

We further compared the performance of two models with respect to model parameters Γ and Σ. The differences in the standardized regression coefficients estimates (Γ) were negligible between the PISA 2015 operational model and RT variables-added model, ranging from approximately -0.06 to 0.05. Furthermore, Table 20.4 provides the differences in the estimated residual variance-covariance matrix (Σ) elements. When RT variables were added, the estimated residual variance decreased (i.e., measurement precision of the latent ability increased) in all countries for all domains, although the differences were very small. This pattern stood out in the science domain, where we processed the RT variables and incorporated them into the IRT-LRMs. Estimated correlations between domains were affected to varying degrees across countries, and the pattern of the change in the estimated correlations was also inconsistent across countries. In summary, the improved model fit and the decreased residual variances suggest that the predictive power increased in the LRM with PCs extracted under the RT variables-added model.

Although the inclusion of RT variables may look promising, the implications of doing so in terms of *trend* needs to be studied. Reporting of PISA scores aims at group-level, not at individual-level, comparisons. Therefore, stable group-level scores when the RT variables-included model is applied across domains and countries would be an indicator of stable trend measurement. If any significant differences result by having additional RT information in the IRT-LRMs, additional thorough studies would be desirable before updating the PISA operational procedures. Therefore, we compared the group-level statistics at the country/economy level with respect to the posterior means and posterior *SD*s. Table 20.5 presents the differences in the posterior means and posterior *SD*s in each group. This table shows that the inclusion of RT variables resulted in almost identical results as the PISA 2015 operational model. Thus, including RT information based on this practical and straightforward approach does not appear to impact trends, but further studies will be beneficial.

Table 20.4 Differences in estimated residual variance-covariance matrix (PISA 2015—RT added)

	Residuals				Correlations					
					Science			Reading		Math
	Science	Reading	Math	CPS	Math	Reading	CPS	Math	CPS	CPS
CNT1	0.009	0.002	0.003	0.001	−0.006	−0.005	−0.002	−0.008	−0.001	−0.004
CNT2	0.004	0.002	0.001	0	−0.001	0.004	0.005	0	0.004	−0.009
CNT3	0.004	0.002	0.001	0	−0.004	0	0.007	−0.001	−0.009	0.003
CNT4	0.004	0.001	0.002	0.001	0.007	0	0.006	0.003	0.009	0.002
CNT5	0.005	0.003	0.002	0.001	0.006	0.012	0.012	0.016	0.012	0.014
CNT6	0.001	0.001	0.001	0	−0.004	−0.001	−0.006	0	−0.002	−0.014
CNT7	0.005	0.002	0.004	0	−0.003	−0.001	−0.018	−0.004	−0.004	−0.001
CNT8	0.006	0.006	0.002	0.003	0.005	0.011	0.004	0.008	0.009	0.01
CNT9	0.008	0.003	0.001	0.002	0.003	0.01	0.003	0.008	0.013	−0.002

Table 20.5 Differences in posterior means and posterior SDs (PISA 2015—RT added)

	Math		Reading		Science		CPS	
	Post. Mean	Post. SD	Post. Mean	Post. SD	Post. Mean	Post. SD	Post. Mean	Post. SD
CNT1	0.000	0.000	0.001	0.001	−0.002	0.001	0.000	0.000
CNT2	0.002	−0.001	−0.001	0.002	0.001	−0.001	−0.001	0.001
CNT3	0.001	−0.001	−0.001	0.000	0.000	0.000	0.000	0.002
CNT4	0.002	−0.001	−0.001	0.002	0.001	0.001	0.001	−0.002
CNT5	0.002	−0.002	−0.001	0.000	0.000	0.000	−0.002	0.001
CNT6	−0.001	0.001	0.002	−0.001	−0.001	0.000	−0.002	−0.002
CNT7	−0.001	0.003	−0.001	0.000	0.000	0.000	0.002	−0.001
CNT8	0.001	−0.001	0.001	0.000	0.001	0.000	0.000	−0.002
CNT9	0.000	0.002	0.001	0.000	0.001	0.001	0.000	0.003

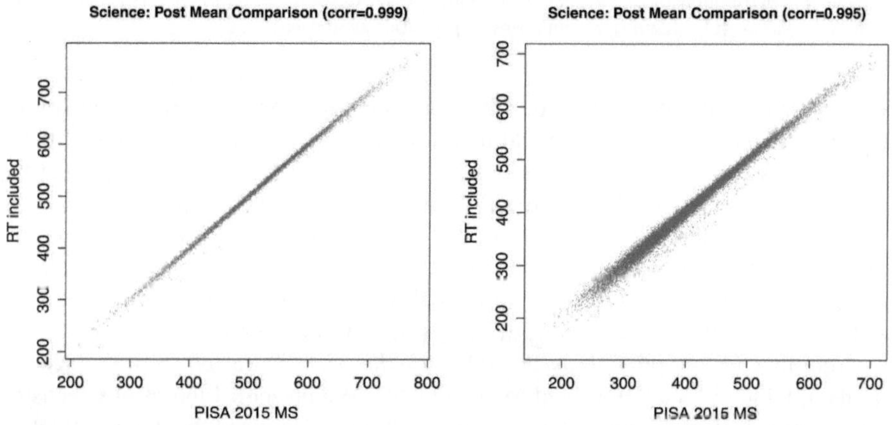

Fig. 20.3 Comparison of the individual-level posterior means (left: one of the nine countries with the highest correlation; right: one of the nine countries with the lowest correlation)

Given that group-statistics at the country/economy-level are stable regardless of the inclusion of RT variables, a further comparison at the individual level was conducted. As an example, one country's results in terms of posterior means and posterior SDs are presented in Fig. 20.3. This figure indicates that the PISA 2015 operational model and the RT variables-added model generated very similar individual-level posterior means. Most of the students that were administered the science domain received almost identical posterior means from both models, but with varying degrees across countries. In particular, lower performing students showed a relatively larger range of differences in the country on the right panel. Next, we compared the individual-level posterior *SD*s by calculating the differences between the PISA 2015 operational model and the RT variables-added model (Fig. 20.4). Note that most of the dots, representing individuals, are located above zero, which suggests that posterior *SD* at the individual level decreased when RT variables were added. Although the inclusion of RT variables did not show

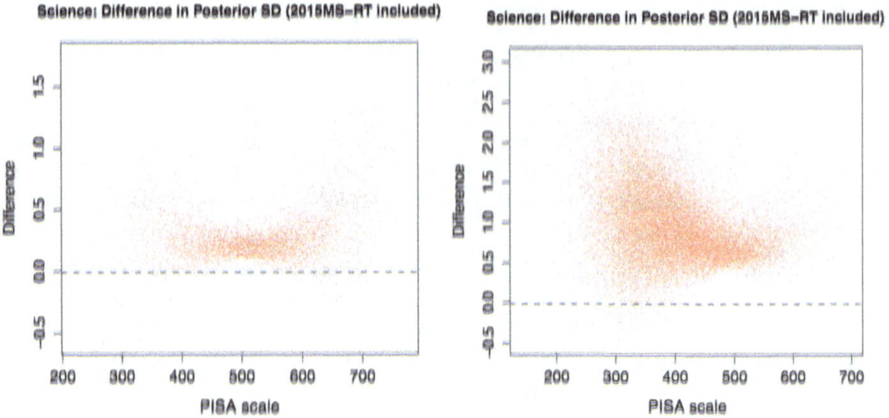

Fig. 20.4 Distribution of differences in individual-level posterior *SD*s (PISA 2015 main survey reported scores—RT added; same countries as Fig. 20.3 were used)

noticeable differences in posterior *SD* at the country/economy-level, it seems obvious that RT variables contributed to measurement precision at the individual level. Furthermore, the amount of posterior *SD* that showed a decrease was higher at the extreme proficiency levels. Because RT variables were added as person covariates, which construct the prior distribution, the impact of including RT variables seems more prominent for the performance range that is more distant from the test targets (around 500). Such a decrease in individual-level posterior *SD* was observed consistently across the nine countries of interest to varying degrees.

Note that in this pilot study the RT variables were extracted from the science domain, which was administered to all students. We presented the results focusing on the science domain due to the page limits, but we examined the impacts in other core domains (math and reading) as well. Briefly, posterior means were almost identical in other domains in which RT variables were not extracted. However, a decrease in posterior *SD* was not clearly observed in other domains. Most of the differences in posterior *SD*s were clustered around zero with random noise, rather than showing any systematic or consistent patterns.

Overall, this pilot study using the PISA 2015 data can be summarized as follows. First, country- and individual-level posterior means were almost identical. This implies that group-level reporting would not be impacted, and trend stability would not be threatened if the RT information was extracted based on the method described above and included as person covariates with other BQ variables in the IRT-LRM. Certainly, more investigation at the sub-group level would be desirable to pursue more comprehensive understanding and insights for reporting. Second, slightly smaller residual variance estimates, improved model fits, and decreased individual-level posterior *SD*s in all nine countries suggest that including RT information can improve the predictive power of the IRT-LRMs. Such measurement precision gains were mainly observed in the science domain from which RT variables were processed and extracted. As a follow-up, in the PISA 2018 operational procedures, RT variables from all domains were extracted and included in the IRT-LRMs, resulting in a decrease in individual-level posterior *SD*s across all domains (OECD, 2020a; Annex H from the PISA 2018 Technical Report).

20.4 Concluding Remarks

This chapter presents the pilot study that incorporated RT variables into the IRT-LRM of PISA. To that end, we began with an overview of the RT variables that are available in the PISA PUF and the primer for the IRT-LRM in the PISA operational procedures. Focusing on the PISA 2015 data, we introduced one practical and straightforward approach to incorporate RT variables into the IRT-LRM and evaluated the impacts at the country/economy-level and individual-level. Briefly, item-level RT variables were processed through several layers of aggregation and standardization to extract the person-level covariates, which can be viewed as proxies of working speed. This approach has two benefits: (1) the relationship between RT and individual item responses becomes almost nonexistent, which reduces the chance of violating conditional independence, and (2) practical enough to be readily operationalized in the current PISA procedure. Along with other BQ variables, these person-level RT-related covariates were entered through PCA, so that RT information could be a part of the composition of PCs, which are specified as explanatory variables in the multivariate LRMs. When RT information was added into the LRM, model fits were improved, residual variance estimates decreased slightly, and individual-level posterior *SD*s decreased slightly in all nine countries analyzed in the pilot study. This suggests that the PCs with the embedded RT information provided better predictive power, resulting in increased measurement precision at the country-level and individual-level. At the same time, almost identical posterior means at the country-level and individual-level was a reassurance that including RT information would not threaten trend reporting.

Consequentially, RT information was incorporated into the IRT-LRM during the PISA 2018 cycle (Annex H and Annex K of PISA 2018 Technical Report; OECD, 2020a). The fundamental approach has remained the same, except for further consideration of the item types. One recent study revealed that the average of latent correlations between the RT scales of constructed-response items and multiple-choice items were estimated to be about 0.5, and the highest correlation was about 0.7 (Shin et al., 2020). This implies that RT scales measured by different item types are distinct, and each RT scale provides somewhat unique information that is not captured by the RT scale of another item format. Therefore, in PISA 2018, RT variables were prepared similarly as described in this chapter, but item types were additionally considered. Thus, four RT variables for multiple-choice items and four RT variables for constructed-response items were extracted as person covariates and entered through the PCA.

Although the described approach in this chapter was promising and operationalized in generating PVs during PISA 2018, further studies will be beneficial. Most importantly, more thorough examinations on the conditional dependence between the RT variables and response for the given item would be desirable. Testing of observed conditional dependence in the real PISA data and evaluating the extent to which conditional dependencies are diminished by the described approach would be of particular interest. Furthermore, this chapter is limited in evaluating whether or not the inclusion of RT information into the IRT-LRM reduces the bias in the estimates involving RT and PVs. Well-designed simulation studies can corroborate empirical studies by evaluating the reduction in the bias in the estimates involving RT and PVs. Such simulation studies will also shed light on understanding the mechanism of the improved model fit. As illustrated in the empirical pilot study, improved fit of RT variables-added model seemed to result in only minimal changes in the associated quantities such as residual variance or almost no differences in the group-level posterior SD. The mechanism behind the improved fit can be revealed through manipulating relevant factors in the simulation studies. Furthermore, beyond the described practical and straightforward approach, advanced statistical approaches for variable reduction methods (e.g., machine-learning techniques) and regularized regressions as summarized in von Davier et al. (2019) can be considered alternative approaches to incorporate RT variables into the PISA population modeling.

Acknowledgement This chapter was based on the conference presentation by Shin et al. (2019). The authors would like to thank Harrison Gamble, Wei Zhao, Mat Kandathil, and Lokesh Kapur for their help in running the DGROUPs for the current study, Frederic Robin and Lale Khorramdel for their support, Dan McCaffrey and Jon Donoghue for their initial input and construct review of the preliminary findings, and Emily Kerzabi for her editorial help.

References

Anaya, L., & Zamarro, G. (2020). The role of student effort on performance in PISA: Revisiting the gender gap in achievement. *Education Reform Faculty and Graduate Students Publications.*. https://scholarworks.uark.edu/edrepub/116

Bergner, Y., & von Davier, A. A. (2019). Process data in NAEP: Past, present, and future. *Journal of Educational and Behavioral Statistics, 44*(6), 706–732. https://doi.org/10.3102/1076998618784700

Bolsinova, M., & Tijmstra, J. (2018). Improving precision of ability estimation: Getting more from response times. *British Journal of Mathematical and Statistical Psychology, 71*, 13–38. https://doi.org/10.1111/bmsp.12104

Collins, L. M., Schafer, J. L., & Kam, C.-M. (2001). A comparison of inclusive and restrictive strategies in modern missing data procedures. *Psychological Methods, 6*(4), 330–351. https://doi.org/10.1037/1082-989x.6.4.330

Educational Testing Service. (2012). *DGROUP* [Computer software].

Ercikan, K., Guo, H., & He, Q. (2020). Use of response process data to inform group comparisons and fairness research. *Educational Assessment., 25*, 179–197. https://doi.org/10.1080/10627197.2020.1804353

Ercikan, K., & Pellegrino, J. W. (Eds.). (2017). *Validation of score meaning for the next generation of assessments: The use of response processes*. Taylor & Francis.

Greiff, S., Wüstenberg, S., & Avvisati, F. (2015). Computer-generated log-file analyses as a window into students' minds? A showcase study based on the PISA 2012 assessment of problem solving. *Computers & Education, 91*, 92–105. https://doi.org/10.1016/j.compedu.2015.10.018

He, Q., von Davier, M., & Han, Z. (2018). Exploring process data in problem-solving items in computer-based large-scale assessments. In H. Jiao, R. W. Lissitz, & A. Van Wie (Eds.), *Data analytics and psychometrics: Informing assessment practices* (pp. 53–76). Information Age Publishing.

Little, R. J., & Rubin, D. B. (2014). *Statistical analysis with missing data*. John Wiley & Sons.

Lu, J., & Wang, C. (2020). A response time process model for not-reached and omitted items. *Journal of Educational Measurement., 57*, 584–620. https://doi.org/10.1111/jedm.12270

Meng, X.-L. (1994). Multiple-imputation inferences with uncongenial sources of input. *Statistical Science, 9*, 538–558. https://doi.org/10.1214/ss/1177010269

Michaelides, M. P., Ivanova, M., & Nicolaou, C. (2020). The relationship between response-time effort and accuracy in PISA science multiple choice items. *International Journal of Testing, 20*(3), 1–19. https://doi.org/10.1080/15305058.2019.1706529

Mislevy, R. J. (1991). Randomization-based inference about latent variables from complex samples. *Psychometrika, 56*(2), 177–196. https://doi.org/10.1007/BF02294457

Mislevy, R. J., Johnson, E. J., & Murakai, E. (1992). Scaling procedures in NAEP. *Journal of Educational Statistics, 17*(2), 131–154. https://doi.org/10.3102/10769986017002131

Molenaar, D., & van Rijn, P. (2024). Modeling item responses with response times as collateral information in large-scale educational assessments. In L. Khorramdel, M. von Davier, & K. Yamamoto (Eds.), *Innovative digital-based international large-scale assessments–foundations*, Methodologies and Quality Assurance Procedures. Springer.

Organisation for Economic Co-operation and Development. (2017). *PISA 2015 technical report*. OECD Publishing. http://www.oecd.org/pisa/data/2015-technical-report/

Organisation for Economic Co-operation and Development. (2020a). *PISA 2018 assessment and analytical framework*. OECD Publishing. http://www.oecd.org/education/pisa-2018-assessment-and-analytical-framework-b25efab8-en.htm

Organisation for Economic Co-operation and Development. (2020b). *PISA 2015*. PISA 2015 Database. http://www.oecd.org/pisa/data/2015database/

Organisation for Economic Co-operation and Development. (2020c). *PISA 2018*. PISA 2018 Database. http://www.oecd.org/pisa/data/2018database/

Pohl, S., Ulitzsch, E., & von Davier, M. (2019). Using response time models to account for not-reached items. *Psychometrika, 84*(3), 892–920. https://doi.org/10.1007/s11336-019-09669-2

Robin, F., Shin, H., Khorramdel, L., Yamamoto, K., & Pohl, S. (2018, July). Examining item specific response time thresholds for the coding of omitted responses in PISA. In *Paper presented at the 11th International Test Commission Conference (ITC)*.

Shin, H. J., Kerzabi, E., Joo, S. H., Robin, F., & Yamamoto, K. (2020). Comparability of response time scales in PISA. *Psychological Test and Assessment Modeling, 62*(1), 107–135.

Shin, H. J., Jewsbury, P., & van Rijn, P. W. (2022). Generating group-level scores under response accuracy-time conditional dependence. *Large-scale Assessments in Education, 10*(1), 1–19.

Shin, H. J., Yamamoto, K., Khorramdel, L., Robin, F., von Davier, M., Gamble, H., & Zhao, W. (2019, April). Incorporating response time into the PISA population model. In *Paper presented at the annual meeting of the National Council on measurement in education (NCME)*.

Teig, N., Scherer, R., & Kjærnsli, M. (2020). Identifying patterns of students' performance on simulated inquiry tasks using PISA 2015 log-file data. *Journal of Research in Science Teaching., 57*, 1400–1429. https://doi.org/10.1002/tea.21657

Ulitzsch, E., von Davier, M., & Pohl, S. (2019a). Using response times for joint modeling of response and omission behavior. *Multivariate Behavioral Research, 55*, 425–453. https://doi.org/10.1080/00273171.2019.1643699

Ulitzsch, E., von Davier, M., & Pohl, S. (2019b). A hierarchical latent response model for inferences about examinee engagement in terms of guessing and item-level non-response.

British Journal of Mathematical and Statistical Psychology., 73 Suppl 1, 83–112. https://doi.org/10.1111/bmsp.12188

van der Linden, W. J., Klein Entink, R. H., & Fox, J.-P. (2010). IRT parameter estimation with response times as collateral information. *Applied Psychological Measurement., 34*(5), 327–347. https://doi.org/10.1177/0146621609349800

von Davier, M., Gonzalez, E., & Mislevy, R. (2009). What are plausible values and why are they useful? *IERI Monograph Series, 2*, 9–36. https://www.ierinstitute.org/fileadmin/Documents/IERI_Monograph/IERI_Monograph_Volume_02.pdf

von Davier, M., & Sinharay, S. (2014). Analytics in international large-scale assessments: Item response theory and population models. In L. Rutkowski, M. von Davier, & D. Rutkowski (Eds.), *Handbook of international large-scale assessment: Background, technical issues, and methods of data analysis (pp. 155–174).*

von Davier, M., Sinharay, S., Oranje, A., & Beaton, A. (2006). The statistical procedures used in national assessment of educational progress: Recent developments and future directions. In C. R. Rao & S. Sinharay (Eds.), *Handbook of statistics: Psychometrics* (Vol. 26, pp. 1039–1055). Elsevier.

von Davier, M., Khorramdel, L., He, Q., Shin, H. J., & Chen, H. (2019). Developments in psychometric population models for technology-based large-scale assessments: An overview of challenges and opportunities. *Journal of Educational and Behavioral Statistics, 44*(6), 671–705. https://doi.org/10.3102/1076998619881789

Waylon, J. H., Rhemtulla, M., & Little, T. D. (2015). Using principal components as auxiliary variables in missing data estimation. *Multivariate Behavioral Research, 50*(3), 285–299. https://doi.org/10.1080/00273171.2014.999267

Weeks, J., von Davier, M., & Yamamoto, K. (2016). Using response time data to inform the coding of omitted responses. *Psychological Test and Assessment Modeling, 58*(4), 671–701.

Yamamoto, K., Ali, U. S., Robin, F., Kapur, L., & Kandathil, M. (2025). Procedures to evaluate and improve data quality using timing and process data. In L. Khorramdel, M. von Davier, & K. Yamamoto (Eds.), *Innovative digital-based international large-scale assessments–foundations, methodologies and quality assurance procedures*. Springer.

Yamamoto, K., & Lennon, M. L. (2018). Understanding and detecting data fabrication in large-scale assessments. *Quality Assurance in Education, 26*(2), 196–212. https://doi.org/10.1108/QAE-07-2017-0038

Chapter 21
Modeling Item Responses with Response Times as Collateral Information in Large-Scale Educational Assessments

Dylan Molenaar and Peter van Rijn

Abstract In large-scale educational assessments, it is becoming common practice to collect the item response times next to the actual item responses. These response times have many uses in large scale assessments, including data quality checks, checks on technical problems in data recording, and checks on aberrant response processes like guessing, careless responding, and responding with item preknowledge. In the present chapter, we focus on response times as collateral information concerning individual differences in the ability underlying the educational domains of an assessment. The main aim is to improve the reliability of large-scale educational assessments. We discuss various psychometric models that are available to simultaneously analyze responses and response times, we illustrate the added value of the response times to the reliability of the 2015 PISA cycle, and we discuss the current challenges and future directions of response time research in large-scale educational assessments.

International large-scale educational assessments provide valuable insights into the performance of educational systems across different countries over time. First, they can be used to improve educational systems by providing data on the link between student performance and system features in both low- and high-performing countries. In addition, since these assessment studies typically are conducted with some regularity (e.g., every 3 or 4 years), educational developments can be tracked and

---

D. Molenaar (✉)
University of Amsterdam, Amsterdam, The Netherlands
e-mail: d.molenaar@uva.nl

P. van Rijn
ETS Global, Amsterdam, The Netherlands

L. Khorramdel et al. (eds.), *Innovative Digital-Based International Large-Scale Assessments*, Methodology of Educational Measurement and Assessment,
https://doi.org/10.1007/978-3-031-90951-1_21

steered if necessary. Common large-scale educational assessments include PISA (OECD, 2015), PIRLS (Martin et al., 2017), and TIMSS (Mullis & Martin, 2017), which cover educational domains such as reading, mathematics, and science.

The rise of computerized testing has facilitated international educational assessments in different ways. Practical advantages are, for instance, that responses are automatically recorded such that item scores do not have to go through an error prone manual coding procedure (Allard et al., 1995). In addition, computerized testing opens the door for efficient testing procedures including multi-stage testing and adaptive testing to shorten test length while retaining estimation precision (van der Linden & Glas, 2010a). Besides these practical advantages, an important psychometric advantage of computerized testing is the possibility to gather the response times in addition to the item responses. The response times serve various goals in large-scale educational assessment. The first use of response times is as a check of the quality of the data (Oranje et al., 2017). Large-scale educational assessments are commonly low stakes for test takers but can be high stakes for educators and policy makers. As a result, data quality can differ considerably across different levels of the data collection (e.g., students, schools, regions, and countries). So, for example, response times may be used to establish whether the outlined testing procedures have been followed within each school, region, and country to enable meaningful inferences (see, e.g., OECD, 2015, Chap. 14). In addition, response time patterns may indicate other technical problems in data recording (Oranje et al., 2017), different processes behind omitted and not-reached items (Frey et al., 2018; Pohl et al., 2019; Weeks et al., 2016), disengaged students (Akyol et al., 2018; Soland et al., 2019), and, in the most extreme cases, test fraud (Cizek & Wollack, 2017) and data fabrication (Yamamoto & Lennon, 2018).

A second use of response times in large-scale educational assessments involves the evaluation of the test design. Although the items are carefully designed, some items may not function as intended (e.g., students take either less or more time to respond than expected). Many large-scale educational assessments make use of so-called matrix sampling designs in which subsets from the complete set of items are distributed across a set of test forms. Even though the difficulty level is typically designed to be similar across forms, the timing can turn out to be different, and speededness issues may occur (Lu & Sireci, 2007). In addition, if adaptive-testing designs are used, it may be investigated if students who follow the more difficult pathways take more time to respond compared to students who follow the easier pathways (e.g., Bridgeman & Cline, 2004). Based on the response time patterns, such issues may be investigated in more depth to evaluate whether this should be accounted for in the final analysis or not.

This chapter focuses on a third important application of response times in large-scale educational assessments: using response times as collateral information concerning individual differences in the ability underlying the educational domains (e.g., Bolsinova & Tijmstra, 2018; van der Linden et al., 2010; Molenaar et al., 2015). That is, if the responses and response times are correlated, the response times contain information about the ability of the respondents, which can potentially improve the reliability of an educational assessment. In this chapter, we first discuss the psychometric models that are available to simultaneously analyze responses and response times. Next, we illustrate the added value of the response times to the reliability of the 2015 PISA cycle. We end with a discussion concerning the current challenges and future directions of response time research in large-scale educational assessments.

21.1 Models

Three classes of response and response time models can be distinguished that differ in the intended aim of the model (see Molenaar et al., 2015; see De Boeck & Jeon, 2019, for an alternative categorization based on the statistical properties of the models). In the first class, the response times are treated as collateral information to the responses. That is, the response times are understood as covariates that can be used as an additional source of information concerning the ability underlying the responses. Examples include the models by Roskam (1987), van der Linden (2007), and Thissen (1983). As the main focus of this chapter is on this class of models, we will elaborate on these models later.

In the second class of models, the response times are used in a scoring rule to score the responses to the items of the test. For instance, responses can be scored by giving less credit to slower responses. That is, credit for a correct response can for instance decrease from 10 points in the first second to only 1 point after 30 seconds. If the researcher additionally wants to discourage fast guessing, negative credit can be given to fast incorrect responses. See Maris and van der Maas (2012) and van Rijn and Ali (2017, 2018) for possible statistical approaches.

In the third class, models for responses and response times are based on assumptions derived from theories about the psychological process that elicited the responses. For instance, in the diffusion model (Ratcliff, 1978), it is assumed that a decision process is driven by noisy accumulation of information that stops if evidence for one of the response alternatives exceeds a threshold. From this modeling perspective, the response times are understood as the most important source of information about this process. Psychometric variations of the diffusion model have been proposed by, for instance, Tuerlinckx and De Boeck (2005) and van der Maas et al. (2011). Other psychometric models based on theoretical processes include the race model (Ranger et al., 2015), the log-normal race model (Rouder et al., 2015), and the proportional-hazards model (Ranger & Kuhn, 2014).

In the context of modeling data from large-scale educational assessments where the focus lies on group-based inferences, response times can be incorporated either in the measurement model or in the population model. The three classes of modeling discussed above concern incorporating response time into the measurement model. Shin et al. (2019) discuss approaches for incorporating response times into the population model in the context of PISA. They included the contrast-coded decile ranks of the mean and standard deviation of the response times for items in a given cluster position as conditioning variables. These and many other conditioning variables were subsequently subjected to a principal component analysis (PCA), and the resulting principal components were entered into the population model.[1] They found that this approach of adding the response times to the model led to an increase in the measurement precision for ability. However, issues with this approach are that it neglects measurement error in the response times, possible group-related differences with respect to speed, and potential conditional dependencies between responses and response times (e.g., when the mean response time for a correct response is different from the mean response time for an incorrect response).

21.1.1 Response Times as Collateral Information

As discussed above, in the present chapter, we focus on increasing measurement precision in large-scale educational assessments. Specifically, in contrast to entering response time as a predictor in the population model, we will use an explicit measurement model for the response times to account for measurement error. Therefore, we consider the first class of models in which response times are a source of collateral information. An early contribution within this approach is by Roskam (1987); see Wang & Hanson, 2005, for a related approach). In this approach, a one-parameter logistic model is used for the responses with the response times residualized out, that is, if X_i denotes the random item response variable of item i and T_i denotes the corresponding response time, then

$$\text{logit}[P(X_i = 1|\theta)] = \theta - \beta_i + \ln T_i, \tag{21.1}$$

where θ is the latent ability variable, β_i is the item difficulty, and $\ln T_i$ is the log-transformed response time. The key feature of this model is that it includes the speed-accuracy tradeoff (Luce, 1986), where higher speed (smaller T_i) implies smaller accuracy [smaller $P(X_i|\theta)$].

The model above does not include a measurement model for the response times yet. That is, it treats the response times as observed covariates while it can be argued that they should be treated as random variables with an underlying latent speed variable similar to the treatment of the responses (see van der Linden, 2009). To this end, van der Linden (2007) developed a hierarchical model that includes measurement models for both the responses and the response times. The measurement models are connected by correlating the random item parameters and the random person parameters across the two measurement models. In the full model, it is assumed that the random person parameters are normally distributed and that, conditional on these random person parameters, all observed variables in the model are independent (see van der Linden & Glas, 2010b). Although van der

[1] See Rutkowski et al. (2013), for detailed descriptions of all typical steps in the modeling of large-scale educational assessment data.

Linden used a three-parameter model for the responses and an essentially tau-equivalent factor model (i.e., a linear factor model with equal factor loadings; Lord & Novick, 1968) for the log-response times, many variations on this general approach have been proposed.[2] That is, models based on the hierarchical model above have been proposed with alternative parametric distributions for the response times (e.g., Klein Entink et al., 2009; Molenaar & Bolsinova, 2017) and semi-parametric distributions (e.g., Loeys et al., 2014; Wang et al., 2013), with (non-) linear cross-loadings from the response times on the latent ability variable (Bolsinova & Tijmstra, 2018; Ferrando & Lorenzo-Seva, 2007; Molenaar et al., 2015; Thissen, 1983), with mixtures distributions for the responses and response times (e.g., Wang & Xu, 2015; Molenaar et al., 2016), and with residual correlations between the responses and response times (e.g., Ranger & Ortner, 2012; Bolsinova et al., 2017).

In the present chapter, we focus on (generalizations of) the hierarchical model of van der Linden (2007). We focus on this general modeling framework as it is arguably one of the most popular approaches to model response times to improve measurement precision in psychometric measurement models (as also argued by e.g., Ranger, 2013). Most other approaches (of which some examples are discussed above) serve different (key) modeling purposes. We focus on the following specification of the hierarchical model: If $f(X_i, \ln T_i | \theta, \tau)$ denotes the joint condition density of the responses and log-response times, then, by assuming

1. independence of X_i and $\ln T_i$ conditional on θ and τ,
2. independence of X_i and τ conditional on θ,
3. independence of $\ln T_i$ and θ conditional on τ,

this density is given by:

$$f(X_i, \ln T_i|\theta,\tau) = g(X_i|\theta)h(\ln T_i|\tau), \tag{21.2}$$

where $g(.)$ and $h(.)$ are the densities under the measurement models for the responses and response times respectively. For $g(.)$ we use a two-parameter logistic model (Birnbaum, 1968)

$$g(X_i|\theta) = \frac{\exp[x_i(\alpha_i\theta + \beta_i)]}{1 + \exp(\alpha_i\theta + \beta_i)}, \tag{21.3}$$

where α_i is a discrimination parameter and β_i is an easiness parameter. For $h(.)$ we use

[2]Note that van der Linden parameterizes the expected response times in the hierarchical model as E$(\ln T_i | \tau) = \nu_i - \tau$. As this parameterization implicitly fixes all factor loadings, λ_i, to -1, $\sigma_\tau^2 = \mathrm{Var}(\tau)$ is a free model parameter. This model can be rewritten as E$(\ln T_i | \tau) = \nu_i - \lambda\tau$ with $\sigma_\tau^2 = 1$ which explicates the form of an essentially tau-equivalent factor model.

$$h(\ln T_i|\tau) = \frac{1}{\sigma_{\varepsilon_i}} \phi\left(\frac{\ln T_i - (\nu_i + \lambda_i \tau)}{\sigma_{\varepsilon_i}}\right), \tag{21.4}$$

where ϕ (.) denotes the standard normal density function, $\sigma_{\varepsilon_i}^2$ is the residual response time variance, ν_i is a time intensity parameter, λ_i is a factor loading, and τ is a latent speed variable. For the distribution of θ and τ, $k(\theta, \tau)$, we use a bivariate normal distribution with correlation $\rho_{\theta\tau}$, in which the means of θ and τ are set to zero and their variances to one for identification purposes, which follows

$$k(\theta, \tau) = \frac{1}{2\pi\sqrt{1-\rho_{\theta\tau}^2}} \exp\left(\frac{\theta^2 - 2\rho_{\theta\tau} + \tau^2}{2(1-\rho_{\theta\tau}^2)}\right). \tag{21.5}$$

Our specification differs from the original specification by van der Linden (2007) in a couple of ways. First, we use a two-parameter model for the responses instead of a three-parameter model, and we include item-specific factor loadings for the response times instead of assuming the factor loadings to be equal across items. In addition, we parameterize the latent variable as $+\lambda_i\tau$ such that higher values on τ indicate slower responses (i.e., van der Linden uses $-\tau$ such that higher values on τ indicate faster responses). However, we will still refer to τ as a latent speed variable. Finally, we treat the item effects (captured by parameters α_i, β_i, ν_i, λ_i, and $\sigma_{\varepsilon_i}^2$) as fixed effects instead of correlated random effects, although we note that there could be substantive reasons to include random item effects (see, e.g., De Boeck, 2008). These adaptions above are intended to make the modeling framework fit (better) to the broader class of generalized linear random effect models.

In the modeling framework, the relation between speed and ability is strictly linear. The main aim of the log-transformation in Eq. 21.4 is to (1) linearize the relations between the transformed response times, the underlying latent speed variable, and the latent ability variable; and (2) to make the assumption of homoscedastic residual response time variance more plausible. However, as demonstrated by Chen et al. (2018) and Bolsinova and Molenaar (2019), the relation between the residual log-response times and the responses can demonstrate non-linearity. That is, in both studies, it was found that response accuracy first increases steeply across the log-response times of the faster responses, while it later decreases less steeply across the log-response times of the slower responses. Although this phenomenon is interesting from both a substantive and a measurement perspective, here, we focus on the generalized linear modeling framework above. That is, we assume that the linear relations between the log-transformed response times and the latent speed and ability variable capture the most important patterns in the data to demonstrate the potential use of response times in large-scale educational tests using the popular generalized linear random effects models discussed above. However, we note that testing the linearity assumption and accounting for possible violations is possible in practice (e.g., using the methods outlined in Bolsinova & Molenaar, 2019).

21.1.2 Improving Measurement Precision

As the latent ability variable underlying the responses is allowed to correlate with the latent speed variable underlying the response times, estimation of ability can draw from information about the ability in the response times. For instance, using the hierarchical model, van der Linden et al. (2010) showed for 17 items of a figural matrix test that the posterior standard deviations of the ability estimates are systematically lowered if response times are included in the analysis. In addition, Molenaar et al. (2015) showed in the hierarchical model that the correlation between the ability estimates on a chess test and a gold standard for chess ability increased if response times were added to the analysis.

The gain in estimation precision in the hierarchical model, thus, runs through the correlation between τ and θ, $\rho_{\theta\tau}$. A different way to extract information about the latent ability from the response times is to use cross-loadings. For instance, Thissen (1983) proposed the following model for the log-transformed response times:

$$h(\ln T_i|\theta, \tau^*) = \frac{1}{\sigma_{\varepsilon_i}} \phi\left(\frac{\ln T_i - (\nu_i + \lambda_i \tau^*) - \rho(\alpha_i \theta + \beta_i)}{\sigma_{\varepsilon_i}}\right), \tag{21.6}$$

where ρ is a cross-loading from the response times on the ability, and $(\alpha_i\theta + \beta_i)$ are the parameters from the two-parameter model in Eq. 21.3. In this model, we use τ^* to denote that this latent speed variable is residualized with respect to θ i.e.,

$$\tau = \rho_{\theta\tau}\theta + \sqrt{1 - \rho_{\theta\tau}^2}\,\tau^* \tag{21.7}$$

4. where we assume $VAR(\theta) = 1$ and $VAR(\tau^*) = 1$ for identification reasons so that $\rho_{\theta\tau}$ retains the interpretation of $COR(\tau, \theta) = \rho_{\theta\tau}$ and so that $COR(\tau, \tau^*) = \sqrt{1 - \rho_{\theta\tau}^2}$. Note that in this model, independence assumption 3 above (of $\ln T_i$ and θ conditional on τ) is relaxed. That is, Eq. 21.6 can be expressed in terms of τ (by rewriting τ^* as a function of τ in the transformation in Eq. 21.7 and inserting the result in Eq. 21.6). If the resulting expression is conditioned on τ, it will still depend on θ which violates the third independence assumption above.

Despite this difference, the conceptual idea of the model by Thissen is the same as that of the hierarchical model above in that information about θ in the log-response response times is modelled through a single parameter (i.e., ρ in Eq. 21.6 and $\rho_{\theta\tau}$ in Eq. 21.4). The difference between the models is that the amount of information in the log-response times is constrained differently. That is, as both models are linear random effects models, the decomposition of the log-response time variance can be derived using standard (co)variance algebra.[3] If we formulate the hierarchical model in terms of τ^* to facilitate comparison to the Thissen model, it follows in the hierarchical model that:

$$\text{Var}(\ln T_i) = \lambda_i^2 \sigma_{\tau^*}^2 + \rho_{\theta\tau}^2 \lambda_i^2 \sigma_\theta^2 + \sigma_{\varepsilon_i}^2. \tag{21.8}$$

Note that if $\sigma_\tau^2 = \sigma_\theta^2 = 1$ for identification, $\sigma_{\tau^*}^2 = 1 - \rho_{\theta\tau}^2$. Next, for the Thissen model, it follows that:

$$\text{Var}(\ln T_i) = \lambda_i^2 \sigma_{\tau^*}^2 + \rho^2 \alpha_i^2 \sigma_\theta^2 + \sigma_{\varepsilon_i}^2. \tag{21.9}$$

As can be seen by comparing Eqs. 21.8 and 21.9, the amount of θ-variance in the log-response times is constrained to be proportional to λ_i^2 in the hierarchical model and to α_i^2 in the Thissen model. This proportionality constraint can be relaxed by freely estimating item specific cross-loadings from the log-response times on the latent ability θ (Bolsinova & Tijmstra, 2018), i.e.,

$$h(\ln T_i | \theta, \tau^*) = \frac{1}{\sigma_{\varepsilon_i}} \phi\left(\frac{\ln T_i - (\upsilon_i + \lambda_i \tau^* - \rho_i \theta)}{\sigma_{\varepsilon_i}} \right). \tag{21.10}$$

As shown by Bolsinova and Tijmstra, doing so may increase the precision of the ability estimates. Note that in this model, β_i does not occur in the model for the log-response times, contrary to the model by Thissen (1983) above. However, as pointed out by Molenaar et al. (2015), this term is absorbed in υ_i and, therefore, does not cause any misfit.

21.1.3 Adding Predictors

As in large-scale educational measurement practice, predictors are commonly included in the population model to improve estimation precision. Hence, we allow for predictors in the population model of the approach outlined above. In this chapter, we are mainly interested in studying how much the response times add to the estimation precision of θ as compared to predictors that are common in practice (e.g., country and gender). Thus, to allow for predictors in the population model, we denote both latent variables by $\boldsymbol{\eta} = (\theta, \tau)'$ in the hierarchical model and by

[3] In fact, here we only need the rule that if $Y = a + \sum_{i=0}^{n} b_i X_i$ where Y and $X_1, X_2, \ldots, X_n$ are random variables, then $(Y) = \sum_{i=0}^{n} \sum_{j=0}^{n} b_i b_j COV(X_i)$.

$\eta = (\theta, \tau^*)'$ in the extended hierarchical model. Then, the latent variables can be linearly regressed on the predictor variables by:

$$\eta = B'z + \varepsilon, \tag{21.11}$$

where $\boldsymbol{\varepsilon}$ is a normally distributed 2-dimensional vector of residuals, z is a k-dimensional vector of predictors, and $\boldsymbol{B}$ is a $k \times 2$-matrix of regression slopes. Note that for identification purposes, the regression does not include intercepts, and the diagonal elements of $COV(\boldsymbol{\varepsilon})$ are fixed to 1.

21.1.4 Estimation

The models discussed above can be fit using various estimation techniques. For instance, Molenaar et al. (2015) used weighted least squares estimation (see Muthén, 1984) to fit the hierarchical model. This approach is based on minimizing the weighted squared difference between the polychoric correlation matrix and thresholds/means of the observed data and the model-implied polychoric covariance matrix and threshold/means. The weight matrix can be an identity matrix (unweighted least squares), or the (diagonal of the) asymptotic covariance matrix of the polychoric covariances (see e.g., Li, 2016). WLS is available in various software packages including Mplus (Muthén & Muthén, 2017), AMOS (Arbuckle, 2014) and R-packages Lavaan (Rosseel, 2012) and sem (Fox, 2006). In addition, Bayesian sampling schemes have been proposed to draw samples from the posterior parameter distribution (e.g., Fox et al., 2007). The sampling schemes differ in whether they use data augmentation to linearize the model (e.g., Fox et al., 2007) or whether they use explicit logistic parameterizations for the response part of the model (e.g., Bolsinova et al., 2017). Software packages that enable Bayesian estimation of the models above include Mplus (Muthén & Muthén, 2017) and R package cirt (Fox et al., 2007). Finally, marginal maximum likelihood (MML) procedures have been adopted to estimate the parameters (see e.g., Molenaar & Bolsinova, 2017). In MML, the latent ability and speed variables are integrated out of the likelihood function. The resulting marginal likelihood is then maximized with respect to the free model parameters. That is, if we have n items, the likelihood of the response vector $\boldsymbol{w}$ = X_1, …X_n, ln T_1, …, ln T_n) is given by (assuming local independence as discussed above):

$$L(\boldsymbol{w}) = \iint_{-\infty}^{\infty} \prod_{i=1}^{n} g(X_i|\theta)h(\ln T_i|\tau)k(\theta, \tau)d\theta d\tau, \tag{21.12}$$

Table 21.1 Frequency table of country-by-gender for the data used in the illustration

	Female	Male	Total
UK	6667	6839	13,506
USA	2809	2785	5594
Total	**9476**	**9624**	**19,100**

where n is the number of items. MML is available in Mplus (Muthén & Muthén, 2017).[4] All approaches above are full information approaches (in the case of WLS, estimation of the polychoric correlations and thresholds/means is full information). Therefore, the presence of missing data (which is common in large-scale educational assessments) poses no problem as long as the missing data can be considered missing at random (Little & Rubin, 2002).

21.2 Illustration

In the present section, we illustrate the modeling ideas presented above. That is, we will analyze responses and response times from the PISA 2015 cycle using the presented models to investigate the added value of including response times to a large-scale assessment. Specifically, we study how the reliability of the θ estimates is improved if response times are added to the more conventional analyses using the responses and predictors only.

21.2.1 Data

We make use of the PISA 2015 public data from the United States (USA) and the United Kingdom (UK). Here, we analyzed the science domain, which was the major domain in PISA 2015. The science domain consists of 183 items. To ease the analysis and reduce the number of parameters, we used dichotomized item scores. As predictors, we used 'country' (USA and UK), 'gender' (female and male), and 'ESCS', the PISA index of economic, social, and cultural status. The reported science mean for the UK is 509 (*SD*: 100) with females scoring 509 (*SD*: 98) and males 510 (*SD*: 101). For USA, the mean is 496 (*SD*: 99) with females scoring 493 (*SD*: 98) and males 499 (*SD*: 102). The original PISA 2015 data includes 19,834 students from the USA and the UK, and final student weights were used. We removed 734 cases due to missingness on either all item responses, all responses

[4]The marginal likelihood in Equation 21.5 corresponds to the way Molenaar and Bolsinova et al. (2017) implemented a non-linear version of the hierarchical model in Mplus. By default (i.e., the linear case), Mplus will only use a single dimension of integration of the likelihood in Equation 21.5, as there is a closed form solution for the integral over τ.

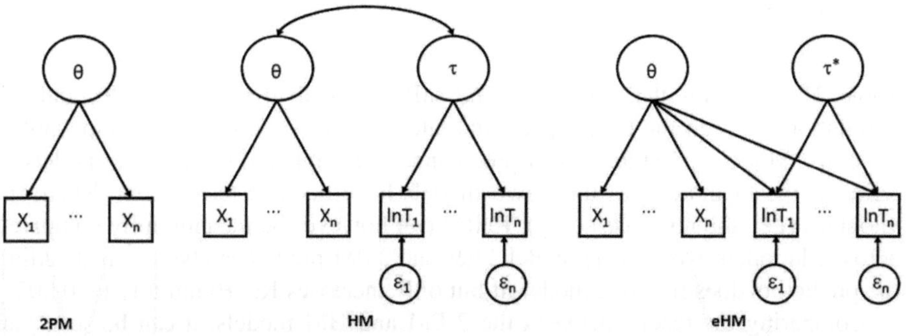

Fig. 21.1 Graphical representations of the models for responses only (*2 PM*), the hierarchical model with a correlation between the latent variables (*HM*), and the extended hierarchical model with cross-loadings from the response times to the latent ability variable (*eHM*)

times, or the predictors, which resulted in a sample of 19,100 respondents. Table 21.1 contains the frequency table of the resulting data for country-by-gender.

21.2.2 Models

We consider the following models:

1. 2 PM: A conventional two-parameter model with the responses only given by Eq. 21.3.
2. HM: A hierarchical model with correlated θ and τ given by Eqs. 21.3 and 21.4.
3. eHM: An extended hierarchical model with item specific cross-loadings given by Eqs. 21.3 and 21.6.

See Fig. 21.1 for a graphical representation of these three models. For each of these three models, we do and do not include predictors of the latent variables θ and τ, resulting in a total of six models (i.e., 2 PM, 2 PM-pred, HM, HM-pred, eHM, and eHM-pred). Each model is fit to the data using MML estimation in Mplus (Muthén & Muthén, 2017). Next, the reliability of the expected a posteriori (EAP) estimates for θ is evaluated, which can be calculated using (Kim, 2012):

$$\text{Rel}(\theta) = \frac{\text{Var}[\text{EAP}(\theta)]}{\text{Var}[\text{EAP}(\theta)] + \text{Mean}[\text{VAP}(\theta)]}, \tag{21.13}$$

where VAP(θ) denotes the variance a posteriori associated with EAP(θ). In addition, we establish the relative fit of the different models using the Akaike Information Criterion (AIC; Akaike, 1974), the Bayesian Information Criterion (BIC; Schwarz, 1978), and the sample size adjusted BIC (saBIC; Sclove, 1987). For these indices it holds that smaller values indicate better model fit.

21.2.3 Results

Table 21.2 contains the AIC, BIC, and saBIC model fit results, Rel(θ), and the correlation between the latent speed and latent ability variables $\rho_{\theta\tau}$ (if applicable). Note that Models 2 PM and 2 PM-pred cannot be compared to the other models in terms of the fit indices, as these two models have been fit to different dependent variables (i.e., the responses only). Rel(θ) can however be meaningfully compared across all models. Comparing model 2 PM and 2 PM-pred, it can be seen that adding the predictors does improve model fit but only increases Rel(θ) minorly by 0.002.

Comparing the results between the 2 PM and HM models, it can be seen that adding the response times in the HM model increases Rel(θ) by only 0.005. Subsequently adding the predictors in the HM-pred model improves model fit and increases Rel(θ) by 0.003. The largest increase in reliability is realized if item specific cross-loadings are allowed from the response times on θ. That is, from the table, it can be seen that Rel(θ) increases from 0.869 without response times to 0.896 with response times (a 0.027 increase). Subsequently adding the predictors (i.e., eHM-pred model) does not affect the reliability any further, although this model does fit the best in terms of the AIC, BIC, and saBIC.

What is interesting in Table 21.2 is that the models with cross-loadings (eHM and eHM-pred) outperform the models with a correlation between the latent variables (HM and HM-pred) in terms of model fit. From the parameter estimates (not depicted), we observed that the cross-loadings in the eHM and eHM-pred models are highly different from item to item, with some loadings significantly positive, some loadings significantly negative, and some loadings insignificant. Such a pattern of cross-loadings cannot be accounted for using the single latent correlation parameter, $\rho_{\theta\tau}$, in the HM and HM-pred models, as this restricts the relation between θ and the response times to be proportional to $\rho_{\theta\tau}$ for the various items. To illustrate this effect further, Fig. 21.2 shows scatter plots of the estimated discrimination parameters under different models, where the left panel shows the 2 PM-pred vs. the HM-pred, and the right panel shows the 2 PM-pred vs. the eHM-pred. For the 2 PM and the HM, the discrimination parameters align very well. However, for the 2 PM and the eHM, they do not. This is not surprising, due to the cross-loadings in the eHM. Interestingly, the differences seem to be larger for higher values (> 1.25).

Table 21.2 Model fit results for the six models considered in the illustration

Model	LL	#par	AIC	BIC	saBIC	Rel(θ)	$\rho_{\theta\tau}$ (*SE*)
2PL	−306407.4	366	613,547	616,423	615,260	.869	–
2PL-pred	−305331.1	372	611,406	614,329	613,147	.871	–
HM	−813510.9	916	1,628,854	1,636,051	1,633,140	.875	.414 (.014)
HM-pred	−811854.4	928	1,625,565	1,632,856	1,629,907	.878	.436 (.013)
eHM	−802388.0	1098	1,606,972	1,615,599	1,612,110	.907	–
eHM-pred	−801121.2	1110	1,604,462	1,613,184	1,609,657	.908	–

Note. LL: log-likelihood; #par: number of estimated parameters in the model

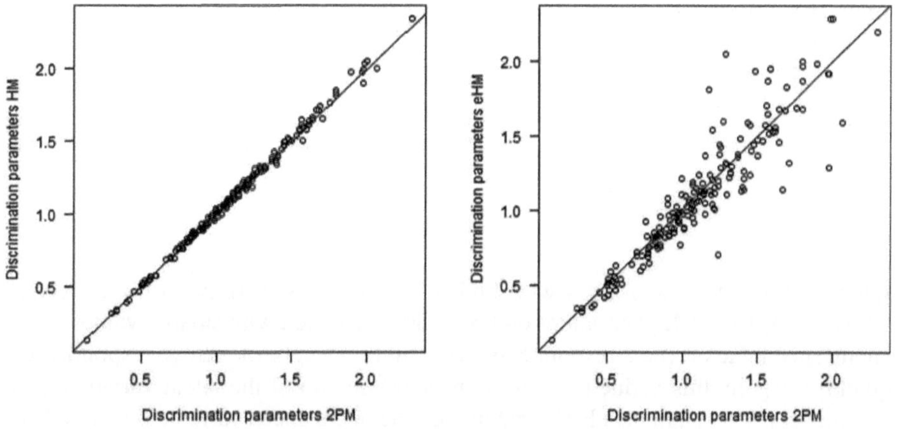

Fig. 21.2 Scatter plots of estimated discrimination parameters under different models (left: 2 PM-pred vs. HM-pred, right: 2 PM-pred vs. eHM-pred)

Table 21.3 Estimated slopes in the regression of $\boldsymbol{\theta}$ on the predictors (SE) for the different models considered in the illustration

Model	Country	Gender	ESCS
2 PM-pred	−.083 (.011)	.033 (.011)	.381 (.012)
HM-pred	−.079 (.011)	.029 (.011)	.381 (.012)
eHM-pred	−.042 (.012)	.001 (.011)	.335 (.013)

Note. For the Country and Gender predictors: UK and females are coded as '−1', and USA and males are coded as '1', respectively

Table 21.3 displays the results concerning the regression of θ on the predictors in the different models. As can be seen for the HM-pred and 2PL-pred models, the pattern of results is the same in that the UK has on average a higher level of θ compared to the USA, males have on average a higher level of θ compared to females, and higher levels of ESCS are associated with higher levels of θ. That is, adding the response times using a correlation between latent speed and ability does not seem to affect the regression importantly. However, adding the response times using the cross-loadings in eHM-pred does affect the regression slopes importantly compared to the HM-pred and 2PL-pred models. Most notably, the gender difference is now insignificant, and the estimated regression slopes for country and ESCS are lower. This is due to the different decomposition of the variance in the data in the eHM model (and eHM-pred) and the HM model (and HM-pred). That is, as noted in Eq. 21.8, in the HM model, the variance of θ in the response times is restricted to be proportional to λ_i^2, a restriction that is relaxed in the eHM model.

A similar (and more substantial) change occurs in the regression of τ on the predictors; see Table 21.4. It can be seen that the regression slopes for a country increases if the response times are modeled using the cross-loadings model (eHM-pred). However, the relation with ESCS is completely reversed. That is, in HM-pred, higher values on ESCS are associated with higher values on the latent

Table 21.4 Estimated slopes in the regression of $\boldsymbol{\tau}$ on the predictors (SE) for the different models considered in the illustration

Model	Country	Gender	ESCS
HM-pred	.191 (.013)	−.103 (.011)	.053 (.011)
eHM-pred	.213 (.014)	−.104 (.012)	−.118 (.014)

Note. For the Country and Gender predictors: UK and females are coded as '−1', and USA and males are coded as '1' respectively

speed factor (i.e., respondents with higher levels of ESCS respond more slowly), while in eHM-pred, higher values on ESCS are associated with smaller values on the latent speed factor (i.e., respondents with higher levels of ESCS respond more quickly). Again, this is due to a different interpretation of the latent variable in the two models. That is, in the eHM-pred model, the latent speed variable is residualized with respect to θ; see Eq. 21.6. Thus, the results of the HM-pred and eHM-pred models in Table 21.4 indicate that there is an unconditional positive effect of ESCS on speed, but that after accounting for individual differences in θ, ESCS has a negative effect on speed.

21.3 Discussion

In this chapter, we discussed a modeling approach to analyze responses and response times in large-scale assessments. Although there are different reasons to add response times to the conventional analysis of the responses, here we focused on the objective to increase the measurement precision of the latent ability underlying the responses. In our application to PISA 2015 data, we found that by adding the response times by introducing a correlation between latent speed and latent ability, reliability of the ability estimates increased by 0.007. By allowing item specific relations between ability and response times, reliability of the ability estimates increased by 0.037.

Ranger (2013) already pointed out that the information about ability in the response times is limited in models that only use a correlation between speed and ability. By adding item specific cross-loadings to the model, Bolsinova and Tijmstra (2018) showed that more information can be extracted from the response times. The present application also illustrates this point. That is, our results show that the relation between response times and ability are not uniform but depend highly on the item (e.g., due to item content, item type, item position, etc.). Therefore, the model with item specific cross-loadings added the most to the estimation precision of the latent ability. However, the question arises whether an increase of 0.037 in reliability is worth the effort. In smaller scale (scientific) applications, it may not be, as the increase will not practically change any conclusion substantially. However, in international large-scale educational assessments, an increase of this size may be valuable, given that many predictors commonly used sometimes add considerably less than 0.037 in terms of reliability.

Another aspect of the different models used in the present chapter is the interpretation of the latent variables across the different models. In the original hierarchical model by van der Linden (2007), both the latent ability and the latent speed variables have a substantive psychometric interpretation. That is, the latent ability variable represents individual differences in the overall proficiency of the respondents in the educational domain covered by the test, while the latent speed variable represents individual differences in overall speed of the respondents in responding to the test. Adding the unconstrained cross-loadings in the extended hierarchical model of Bolsinova and Tijmstra (2018) changes the statistical decomposition of the observed variance into that of the latent variables. As a result, the latent speed factor is a residual, which is not correlated to latent ability anymore and can better be interpreted as a methods factor from a multi-trait multi-method perspective (Eid, 2000; Eid et al., 2003). Note that in the extended hierarchical model, the interpretation of the latent ability factor remains unchanged, as compared to the original hierarchical model (see also Eid et al., 2003). However, due to the substantial difference in the interpretation of the speed factor between the models, it is advisable to use the original hierarchical model if the aim is to operationalize a substantively interpretable speed factor and to use the extended hierarchical model if the aim is to improve measurement accuracy of latent ability.

We end this chapter by discussing a number of limitations of our study, which, at the same time, can serve as directions for future research. First, we assumed that all parameters were invariant across groups (country and gender). Addressing measurement invariance becomes more involved as models become more complex and the invariance of an increasing number of parameters is to be determined. Second, in the applications, we have assumed homoscedasticity in the regression of the latent variables on the predictors. Although violations of this assumption may not have an important impact on the structure of the main effects as presented here, it may be valuable to consider heteroscedastic latent regression models to see whether reliability can be improved more. Third, we did not evaluate the models in terms of absolute fit. For example, in large-scale educational assessments, items typically occur in different positions across booklets, and this may have an impact on fit (Debeer & Janssen, 2013). Since most students do not pace their time evenly throughout the test, the fit of the response time part of the model may be affected most (e.g., Wise, 2015, Fig. 21.1). Fourth, we did not extensively study the patterns in the cross-loadings, but there may be interesting relationships with item type and/or content (see, e.g., Shin et al., 2020). Fifth, most large-scale educational assessments employ multistage sampling, and although we used sampling weights in our analyses, we did not account fully for the two-stage stratified sampling design that is used in PISA. While we doubt that this has an impact on the outcomes of our latent-structure analyses, depending on the size of the design effect, it almost surely will affect some of our results with respect to uncertainty (e.g., the standard errors of the latent regression parameters). Finally, we made some pragmatic decisions in our analysis: two English speaking countries (USA and UK) only, three predictors (country, gender, and ESCS), only one substantive domain (science), and only one PISA cycle (2015). These decisions were intended to keep the analysis focused on our main goal of demonstrating the potential use of response times in large-scale educational assessment. Taking into account multiple predictors, countries, domains, and data cycles will come with their own modeling challenges (e.g., the issue of measurement invariance, see above) that are important in practice but beyond the scope of the present study.

In closing, we want to stress that when including process data, such as response times, leads to models with more complicated latent structures, it is necessary to re-evaluate reporting options. That is, it may be necessary to look into alternatives, such as market-basket reporting (Mislevy, 1998), where, instead of reporting on latent variables, proficiency is expressed as an expected score on a set of representative items. This opens up the possibility of dealing with much more complex models, while keeping the reporting comprehensive (Zwitser et al., 2017).

Author's Note The research by Dylan Molenaar was made possible by a grant from the Netherlands Organization for Scientific Research (NWO VENI- 451-15-008).

References

Akaike, H. (1974). A new look at the statistical model identification. *Automatic Control, IEEE Transactions on, 19(6)*, 716–723.

Akyol, Ş. P., Krishna, K., & Wang, J. (2018). *Taking PISA seriously: How accurate are low stakes exams?* (No. w24930). National Bureau of Economic Research. https://www.nber.org/papers/w24930

Allard, G., Butler, J., Faust, D., & Shea, M. T. (1995). Errors in hand scoring objective personality tests: The case of the personality diagnostic questionnaire-revised (PDQ-R). *Professional Psychology: Research and Practice, 26*, 304–308. https://doi.org/10.1037/0735-7028.26.3.304

Arbuckle, J. L. (2014). *Amos 23.0 User's Guide*. IBM SPSS. ftp://public.dhe.ibm.com/software/analytics/spss/documentation/amos/23.0/en/Manuals/IBM_SPSS_Amos_User_Guide.pdf

Birnbaum, A. (1968). Some latent trait models and their use in inferring an examinee's ability. In E. M. Lord & M. R. Novick (Eds.), *Statistical theories of mental test scores* (pp. 17–20). Addison Wesley.

Bolsinova, M., De Boeck, P., & Tijmstra, J. (2017). Modelling conditional dependence between response time and accuracy. *Psychometrika, 82*(4), 1126–1148. https://doi.org/10.1007/s11336-016-9537-6

Bolsinova, M., & Molenaar, D. (2019). Nonlinear indicator-level moderation in latent variable models. *Multivariate Behavioral Research, 54*(1), 62–84. https://doi.org/10.1080/00273171.2018.1486174

Bolsinova, M., & Tijmstra, J. (2018). Improving precision of ability estimation: Getting more from response times. *British Journal of Mathematical and Statistical Psychology, 71*(1), 13–38. https://doi.org/10.1111/bmsp.12104

Bridgeman, B., & Cline, F. (2004). Effects of differentially time-consuming tests on computer-adaptive test scores. *Journal of Educational Measurement, 41*, 137–148. https://www.learntechlib.org/p/94205/

Chen, H., De Boeck, P., Grady, M., Yang, C. L., & Waldschmidt, D. (2018). Curvilinear dependency of response accuracy on response time in cognitive tests. *Intelligence, 69*, 16–23. https://doi.org/10.1016/j.intell.2018.04.001

Cizek, G. J., & Wollack, J. A. (2017). *Handbook of quantitative methods for detecting cheating on tests*. Routledge.

Debeer, D., & Janssen, R. (2013). Modeling item-position effects within an IRT framework. *Journal of Educational Measurement, 50*(2), 164–185. https://doi.org/10.1111/jedm.12009

De Boeck, P. (2008). Random item IRT models. *Psychometrika, 73*(4), 533. https://doi.org/10.1007/s11336-008-9092-x

De Boeck, P., & Jeon, M. (2019). An overview of models for response times and processes in cognitive tests. *Frontiers in psychology, 10*, 102. https://doi.org/10.3389/fpsyg.2019.00102

Eid, M. (2000). A multitrait-multimethod model with minimal asumptions. *Psychometrika, 65*(2), 241–261. https://link.springer.com/content/pdf/10.1007/BF02294377.pdf

Eid, M., Lischetzke, T., Nussbeck, F. W., & Trierweiler, L. I. (2003). Separating trait effects from trait-specific method effects in multitrait-multimethod models: A multiple-indicator CT-C (M-1) model. *Psychological methods, 8*(1), 38. https://doi.org/10.1037/1082-989x.8.1.38

Ferrando, P. J., & Lorenzo-Seva, U. (2007). An item response theory model for incorporating response time data in binary personality items. *Applied Psychological Measurement, 31*, 525–543. https://doi.org/10.1177/0146621606295197

Fox, J. (2006). Teacher's corner: Structural equation modeling with the sem package in R. *Structural Equation Modeling, 13*(3), 465–486. https://doi.org/10.1207/s15328007sem1303_7

Fox, J. P., Klein Entink, R. H., & van der Linden, W. J. (2007). Modeling of responses and response times with the package cirt. *Journal of Statistical Software, 20*(7), 1–14. https://EconPapers.repec.org/RePEc:jss:jstsof:v:020:i07

Frey, A., Spoden, C., Goldhammer, F., & Wenzel, S. F. C. (2018). Response time-based treatment of omitted responses in computer-based testing. *Behaviormetrika, 45*, 505–526. https://doi.org/10.1007/s41237-018-0073-9

Kim, S. (2012). A note on the reliability coefficients for item response model-based ability estimates. *Psychometrika, 77(1)*, 153–162. https://doi.org/10.1007/s11336-011-9238-0

Klein Entink, R., van der Linden, W. J., & Fox, J. P. (2009). A Box–Cox normal model for response times. *British Journal of Mathematical and Statistical Psychology, 62*(3), 621–640. https://doi.org/10.1348/000711008X374126

Li, C. H. (2016). Confirmatory factor analysis with ordinal data: Comparing robust maximum likelihood and diagonally weighted least squares. *Behavior research methods, 48*(3), 936–949. https://doi.org/10.3758/s13428-015-0619-7

Little, R. J. A., & Rubin, D. B. (2002). *Statistical analysis with missing data* (2nd ed.). Wiley.

Loeys, T., Legrand, C., Schettino, A., & Pourtois, G. (2014). Semi-parametric proportional hazards models with crossed random effects for psychometric response times. *British Journal of Mathematical and Statistical Psychology, 67*, 304–327. https://doi.org/10.1111/bmsp.12020

Lord, F. M., & Novick, M. R. (1968). *Statistical theories of mental test scores*. Addison-Wesley.

Luce, R. D. (1986). *Response times*. Oxford University Press.

Lu, Y., & Sireci, S. (2007). Validity issues in test speededness. *Educational Measurement: Issues and Practice, 26*, 29–37. https://doi.org/10.1111/j.1745-3992.2007.00106.x

Martin, M. O., Mullis, I. V., & Hooper, M. (2017). *Methods and procedures in PIRLS 2016. International Association for the Evaluation of Educational Achievement.*. https://timssandpirls.bc.edu/publications/pirls/2016-methods.html

Maris, G., & van der Maas, H. (2012). Speed-accuracy response models: Scoring rules based on response time and accuracy. *Psychometrika, 77*(4), 615–633. https://doi.org/10.1007/s11336-012-9288-y

Mislevy, R. J. (1998). Implications of market-basket reporting for achievement-level setting. *Applied Psychological Measurement, 11*(1), 49–63. https://doi.org/10.1207/s15324818ame1101_3

Molenaar, D., & Bolsinova, M. (2017). A heteroscedastic generalized linear model with a non-normal speed factor for responses and response times. *British Journal of Mathematical and Statistical Psychology, 70*(2), 297–316. https://doi.org/10.1111/bmsp.12087
Molenaar, D., Oberski, D., Vermunt, J. K., & De Boeck, P. (2016). Hidden Markov IRT models for responses and response times. *Multivariate Behavioral Research, 51*, 606–626. https://doi.org/10.1080/00273171.2016.1192983
Molenaar, D., Tuerlinckx, F., & van der Maas, H. L. (2015). A bivariate generalized linear item response theory modeling framework to the analysis of responses and response times. *Multivariate Behavioral Research, 50*(1), 56–74. https://doi.org/10.1080/00273171.2014.962684
Mullis, I. V. S., & Martin, M. O. (Eds.). (2017). *TIMSS 2019 assessment frameworks*. TIMSS & PIRLS International Study Center. http://timssandpirls.bc.edu/timss2019/frameworks/
Muthén, B. (1984). A general structural equation model with dichotomous, ordered categorical, and continuous latent variable indicators. *Psychometrika, 49*(1), 115–132. https://doi.org/10.1007/BF02294210
Muthén, L. K., & Muthén, B. O. (2017). *Mplus user's guide. Eighth Edition*. Muthén & Muthén. https://www.statmodel.com/download/usersguide/MplusUserGuideVer_8.pdf
OECD (2015). *PISA 2015 technical report*. https://www.oecd.org/pisa/data/2015-technical-report/
Oranje, A., Gorin, J., Jia, Y., & Kerr, D. (2017). Collecting, analyzing, and interpreting response time, eye tracking and log data. In K. Ercikan & J. W. Pellegrino (Eds.), *Validation of Score Meaning for the Next Generation of Assessments* (pp. 39–51). Routledge.
Pohl, S., Ulitzsch, E., & von Davier, M. (2019). Using response times to model not-reached items due to time limits. *Psychometrika. Online First.* https://doi.org/10.1007/s11336-019-09669-2
Ranger, J. (2013). A note on the hierarchical model for responses and response times in tests of van der Linden (2007). *Psychometrika, 78*(3), 538–544. https://doi.org/10.1007/S11336-013-9324-6
Ranger, J., & Kuhn, J. T. (2014). An accumulator model for responses and response times in tests based on the proportional hazards model. *British Journal of Mathematical and Statistical Psychology, 67*(3), 388–407. https://doi.org/10.1111/bmsp.12025
Ranger, J., Kuhn, J. T., & Gaviria, J. L. (2015). A race model for responses and response times in tests. *Psychometrika, 80*(3), 791–810. https://doi.org/10.1007/s11336-014-9427-8
Ranger, J., & Ortner, T. (2012). The case of dependency of responses and response times: A modeling approach based on standard latent trait models. *Psychological Test and Assessment Modeling, 54*(2), 128. https://www.psychologie-aktuell.com/fileadmin/download/ptam/2-2012_20120628/02_ranger.pdf
Ratcliff, R. (1978). A theory of memory retrieval. *Psychological Review, 85*(2), 59. https://doi.org/10.1037/0033-295X.85.2.59
Roskam, E. E. (1987). Toward a psychometric theory of intelligence. In E. E. Roskam & R. Suck (Eds.), *Progress in Mathematical Psychology* (pp. 151–171). North-Holland.
Rosseel, Y. (2012). Lavaan: An R package for structural equation modeling and more. Version 0.5–12 (BETA). *Journal of Statistical Software, 48*(2), 1–36. https://www.jstatsoft.org/article/view/v048i02
Rouder, J. N., Province, J. M., Morey, R. D., Gomez, P., & Heathcote, A. (2015). The lognormal race: A cognitive-process model of choice and latency with desirable psychometric properties. *Psychometrika, 80*(2), 491–513. https://doi.org/10.1007/s11336-013-9396-3
Rutkowski, L., von Davier, M., & Rutkowski, D. (Eds.). (2013). *Handbook of international large-scale assessment: Background, technical issues, and methods of data analysis*. CRC Press.
Schwarz, G. (1978). Estimating the dimension of a model. *The Annals of Statistics, 6*, 461–464. https://doi.org/10.1214/aos/1176344136
Sclove, L. (1987). Application of model-selection criteria to some problems in multivariate analysis. *Psychometrika, 52*, 333–343. https://doi.org/10.1007/BF02294360
Shin, H.-J., Kerzabi, E., Joo, S.-H., Robin, F., & Yamamoto, K. (2020). Comparability of response time scales in PISA. *Psychological Test and Assessment Modeling.*. https://www.psychologie-aktuell.com/fileadmin/Redaktion/Journale/ptam-2020-1/06_Shin.pdf

Shin, H.-J, Yamamoto, K., Khorramdel, L., Robin, F., von Davier, M., Gamble, H., Zhao, W. (2019, April). *Incorporating response times into population modeling in large-scale assessments*. Paper presented at the annual meeting of NCME.

Soland, J., Wise, S. L., & Gao, L. (2019). Identifying disengaged survey responses: New evidence using response time metadata. *Applied Measurement in Education, 32*, 151–165. https://doi.org/10.1080/08957347.2019.1577244

Thissen, D. (1983). Timed testing: An approach using item response testing. In D. J. Weiss (Ed.), *New horizons in testing: Latent trait theory and computerized adaptive testing* (pp. 179–203). Academic Press.

Tuerlinckx, F., & De Boeck, P. (2005). Two interpretations of the discrimination parameter. *Psychometrika, 70*(4), 629–650. https://doi.org/10.1007/s11336-000-0810-3

van der Linden, W. J., & Glas, C. A. W. (2010a). *Elements of adaptive testing*. Springer.

van der Linden, W. J., & Glas, C. A. W. (2010b). Statistical tests of conditional independence between responses and/or response times on test items. *Psychometrika, 75*(1), 120–139. https://doi.org/10.1007/s11336-009-9129-9

van der Linden, W. J. (2007). A hierarchical framework for modeling speed and accuracy on test items. *Psychometrika, 72*(3), 287. https://doi.org/10.1007/s11336-006-1478-z

van der Linden, W. J., Klein Entink, R. H., & Fox, J. P. (2010). IRT parameter estimation with response times as collateral information. *Applied Psychological Measurement, 34*(*5*), 327–347. https://doi.org/10.1177/01466216093498

van der Linden, W. J. (2009). Conceptual issues in response-time modeling. *Journal of Educational Measurement, 46*(3), 247–272. https://doi.org/10.1111/j.1745-3984.2009.00080.x

van der Maas, H. L. J., Molenaar, D., Maris, G., Kievit, R. A., & Borsboom, D. (2011). Cognitive psychology meets psychometric theory: On the relation between process models for decision making and latent variable models for individual differences. *Psychological Review, 118*(*2*), 339. https://doi.org/10.1037/a0022749

van Rijn, P. W., & Ali, U. S. (2017). A comparison of item response models for accuracy and speed of item responses with applications to adaptive testing. *British Journal of Mathematical and Statistical Psychology, 70*(2), 317–345. https://doi.org/10.1111/bmsp.12101

van Rijn, P. W., & Ali, U. S. (2018). A generalized speed–accuracy response model for dichotomous items. *Psychometrika, 83*(1), 109–131. https://doi.org/10.1007/s11336-017-9590-9

Wang, C., Chang, H. H., & Douglas, J. A. (2013). The linear transformation model with frailties for the analysis of item response times. *British Journal of Mathematical and Statistical Psychology, 66*, 144–168. https://doi.org/10.1111/j.2044-8317.2012.02045.x

Wang, T., & Hanson, B. A. (2005). Development and calibration of an item response model that incorporates response time. *Applied Psychological Measurement, 29*, 323–339. https://doi.org/10.1177/0146621605275984

Wang, C., & Xu, G. (2015). A mixture hierarchical model for response times and response accuracy. *British Journal of Mathematical and Statistical Psychology, 68*, 456–477. https://doi.org/10.1111/bmsp.12054

Weeks, J. P., von Davier, M., & Yamamoto, K. (2016). Using response time data to inform the coding of omitted responses. *Psychological Test and Assessment Modeling, 58*(4), 671. https://www.psychologie-aktuell.com/fileadmin/download/ptam/4-2016_20161219/06_Weeks.pdf

Wise, S. L. (2015). Response time as an indicator of test taker speed: Assumptions meet reality. *Measurement: Interdisciplinary Research and Perspectives, 13*(*3-4*), 186–188. https://doi.org/10.1080/15366367.2015.1105062

Yamamoto, K., & Lennon, M. L. (2018). Understanding and detecting data fabrication in large-scale assessments. *Quality Assurance in Education, 26*, 196–212. https://doi.org/10.1108/QAE-07-2017-0038

Zwitser, R. J., Glaser, S. S. F., & Maris, G. (2017). Monitoring countries in a changing world: A new look at DIF in international surveys. *Psychometrika, 82*(1), 210–232. https://doi.org/10.1007/s11336-016-9543-8

Chapter 22
Extreme Response Style and Rapid Responding: Two Sides of the Same Coin? A Joint Modeling Approach Using IRTree and Mixture IRT Models

Ulf Kroehne, Lale Khorramdel, Frank Goldhammer, and Matthias von Davier

Abstract This study examines the relationship between extreme response style (ERS) and rapid responding (RR) behavior in non-cognitive measures. Both phenomena can decrease the comparability, validity, and reliability of large-scale assessments (LSA), and different methods have been established and investigated to reduce the bias introduced by them. For example, RR can be identified using item-level response time thresholds, and ERS can be identified using binary pseudo items modeled in IRTree models. Previous studies could show that both response styles (acquiescence and ERS) and RR behavior can be successfully related to shorter response times and lower response time effort (RTE), respectively. Therefore, the question arises whether these two phenomena are similar types of behavior that potentially could be identified more effectively with a joint modeling approach, or whether they are behaviors with a different quality that must be treated separately. Analyzing the PISA 2015 data from Switzerland, we differentiate between groups based on their RR behavior shown as RTE and test them for ERS, and vice versa. We examine how the person variables for ERS and RTE are related and if the two phenomena can be empirically identified simultaneously. Results are compared regarding the validity of the interpretation of ERS and RTE.

U. Kroehne (✉) · F. Goldhammer
DIPF | Leibniz Institute for Research and Information in Education, Frankfurt am Main, Germany
e-mail: u.kroehne@dipf.de

L. Khorramdel · M. von Davier
Lynch School of Education and Human Development, TIMSS & PIRLS International Study Center at Boston College, Chestnut Hill, MA, USA
e-mail: vondavim@bc.edu

L. Khorramdel et al. (eds.), *Innovative Digital-Based International Large-Scale Assessments*, Methodology of Educational Measurement and Assessment,
https://doi.org/10.1007/978-3-031-90951-1_22

22.1 Introduction to Response Styles and Rapid Responding

Survey research uses standardized questionnaires for measuring noncognitive constructs. They usually contain questions with predefined sets of response options, such as rating or Likert-type scales where statements are judged based on categories reflecting the agreement or disagreement intensity towards the statement. Rating or Likert-type scales are believed to provide standardized, comparable information across subjects but do not come without problems. Respondents may not use and interpret the given response categories in the same way or give biased responses. Moreover, the used rating scale may not present an appropriate range to measure the actual degree of the construct of interest.

Furthermore, *response styles (RS)* may occur, reducing the validity and comparability of the measurement. RS bias is a well-known phenomenon in survey research (cf. Van Vaerenbergh & Thomas, 2013) and is often defined as construct-irrelevant responses that are not directly related to the construct of interest (Paulhus, 1991; Rost, 2004). RS comprise phenomena, such as the tendency to agree (acquiescence, ARS), the tendency to choose the midpoint of the rating scale (midpoint response style, MRS), and the tendency to mainly select the extreme response categories (extreme response style, ERS), are assumed to be largely stable individual characteristics within (Javaras & Ripley, 2007; Nunnally, 1967) and even across survey administrations and over time (Weijters et al., 2010; Wetzel et al., 2013). RS can harm the validity (Baumgartner & Steenkamp, 2001; De Jong et al., 2008; Dolnicar & Grun, 2009; Weijters et al., 2008) and dimensionality of the measurement (Rost, 2004) by contributing to systematic errors (Van Vaerenbergh & Thomas, 2013). They can especially be a problem in cross-cultural studies (Chun et al., 1974; Morren et al., 2012) and large-scale assessments (Bolt & Newton, 2011; Buckley, 2009), as they can lead to biased group comparisons and artifacts when examining the relation between the measured constructs.

Particularly when the assessment is low stakes, the respondents may not be motivated to fully engage with the survey questions (Khorramdel et al., 2017). This is related to another phenomenon that can occur in survey data and that is *rapid responding (RR)* behavior comparable to the well-known concept of rapid guessing (RG) in cognitive tests (Goldhammer et al., 2016; Lee & Jia, 2014; Ulitzsch et al., 2020). For cognitive tests, response time distributions often show two modes, indicating a mixture of two distinct processes (e.g., Schnipke & Scrams, 1997) labeled as rapid guessing and solution behavior. RG is so quick that actual solution behavior (SB, i.e., providing a thoughtful answer to the presented task or questions) is impossible even for fast test-takers. The differentiation of RG and SB is mostly investigated in cognitive assessments (e.g., Lee & Jia, 2014; Goldhammer et al., 2016) and is also known to affect the validity of LSA data (e.g., Wise, 2017). When similar behavior is examined for survey data, it can be referred to as RR (Ulitzsch et al., 2021, 2023), careless responding (Ulitzsch et al., 2021, 2023), or generally as careless or insufficient effort (e.g., Curran, 2016). Comparable to cognitive tests, qualitative differences are assumed for very fast answers (RR) and for responses associated with a longer response time, allowing for solution behavior. RR can be identified similar to RG using item-level response time thresholds, but RR has not yet been investigated routinely. The reason is that response times are not well defined if multiple questions are administered on one single computer screen. Typically, only the screen-level total time is provided for assessments with multiple items per screen, such as item batteries with Likert-type items. However, as shown by Kroehne et al. (2019), item-level response times can be extracted from log data files and used to identifying RR in questionnaires based on item-level response time thresholds. Moreover, the same study found that RR in questionnaires can explain a noticeable amount of variance in cognitive measures. Other researchers could also show that response time measures can be used for identifying careless responding, not only in cognitive assessment (Ulitzsch et al., 2021, 2023) but also in noncognitive assessment data (Ulitzsch et al., 2021, 2023); in other cases, it is assumed that careless responding does not reflect the trait of interest and, on average, requires less time than attentive responding. Regarding RS, it was found that ARS and ERS in questionnaires can be related to shorter response times obtained from the same questionnaire (Henninger & Plieninger, 2020; Mayerl, 2013) and even to shorter response times obtained from cognitive measures administered to the same examinees (Khorramdel et al., 2019). The latter finding indicates that careless responding might overlap across different parts of the assessment and that RG in cognitive tests could be related to RR and RS in prior or subsequent questionnaires.

Because both RS and RR behavior can be related to shorter response times and lower response time effort (RTE), respectively, the question arises whether these two phenomena are similar types of behavior that potentially could be identified more effectively with a joint modeling approach, or whether they are behaviors with a different quality that must be treated separately. Unfortunately, little is known so far about how RS and RR compare to each other. The presented study aims to fill this gap by investigating the relation between these two phenomena. IRTree and mixture IRTree models were used to test for RS behavior in selected scales of the PISA 2015 student questionnaire and a response time-based approach was utilized to identify RR in the same scales similar like identifying RG in cognitive data. Moreover, a novel IRTree modeling approach for the joint modeling of RS and RR will be introduced. Empirical results that shed light on the interplay of RS and RR are presented, and potential consequences for future research are derived and discussed.

22.1.1 Modeling Response Styles

Two very promising approaches to test and account for RS are mixture distribution item response theory (IRT) models (Rost et al., 1999; Austin et al., 2006; Eid & Rauber, 2000; Gollwitzer et al., 2005; Meiser & Machunsky, 2008; Wetzel et al., 2016; Wetzel et al., 2013) and multiprocess item response tree (IRTree) models (Böckenholt, 2012; Böckenholt & Meiser, 2017; DeBoeck & Partchev, 2012; Jeon

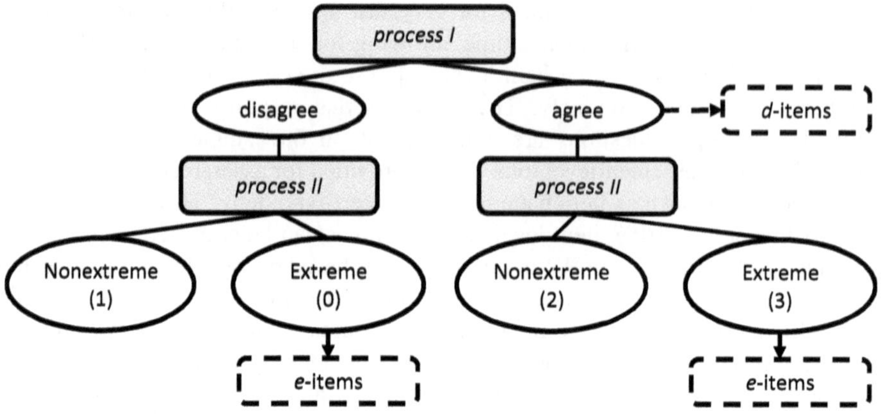

Fig. 22.1 Assumed processes in IRTree model for four-categorical items

& De Boeck, 2016; Khorramdel & von Davier, 2014; von Davier & Khorramdel, 2013). Mixture distribution approaches try to identify latent groups of respondents who show responses driven by RS while IRTree approaches try to disentangle RS from trait-related responses by identifying response subprocesses and continuous latent variables related to those processes. Both modeling approaches were successfully applied to LSA data to increase the validity of the data and to improve the comparability of scores (e.g., Plieninger & Meiser, 2014; Khorramdel et al., 2017; Khorramdel et al., 2019; Dïbek, 2019; Primi et al., 2019). RS models were also used to examine and better understand RS behavior, for example by relating RS variables to external variables or specific features of the instrument (e.g., Böckenholt, 2019; Lechner et al., 2019; Khorramdel et al., 2019).

Recently, mixture IRTree models were introduced (Khorramdel et al., 2019; Kim & Bolt, 2020) which combine those two approaches utilizing the strengths of both; that is, they can differentiate between different types of RS as well as between different latent classes of respondents who show different response behaviors. Khorramdel et al. (2019) introduced a multidimensional version of this combined modeling approach, but it can easily be applied to single questionnaire scales as well which the study in this chapter illustrates.

22.1.2 IRTree Models for ERS

IRTree models aim to separate inter-individual differences in RS from a robust trait estimate that is less biased by RS. The rationale of IRTree models is the assumption of distinct decisions underlying the answering process of Likert-type items. These response subprocesses account for the direction of the response (agreement or disagreement) and the quality of the response (extreme or moderate) and they can be used to separate RS from construct-related responses. Figure 22.1 illustrates how

Table 22.1 Coding of binary pseudo items (BPI) for four-categorical items

Original Response	Pseudo Item e (ERS)	Pseudo Item d (DI)	RR Indicators (r Items)	
			RT > Threshold	RT <= Threshold
0	1	0	0	1
1	0	0	0	1
2	0	1	0	1
3	1	1	0	1

Note. Rapid Responding (RR) indicators used for research question 3 to identify a latent Rapid Responding Propensity (RRP)

such response subprocesses can be used to create a tree structure for a four categorical item resulting in two subprocesses also called decision nodes. Each decision node can be coded as a binary pseudo item (BPI) like illustrated in Table 22.1 and the rating data are decomposed into these BPIs. RS and trait related responses are represented by different BPIs and can be modeled using IRT models.

The current study focusses on examining ERS in questionnaire scales where items were administered with a four-point Likert-type scale (i.e. comprise four response options). Responses to these scales were decomposed assuming two sequential processes according to the decision tree in Fig. 22.1. The first response process constitutes the decision to agree or disagree with a statement. The second response process accounts for the intensity of the decision made in the first response process (extreme versus nonextreme response option).

Thus, every questionnaire item was recoded into two different BPIs (see Table 22.1). One BPI comprises responses in extreme response categories only (e-items); responses to extreme categories are scored as 1 and the remainder as 0 (1-0-0-1). The second BPI comprises directional or construct-related responses controlled for a potential ERS (d-items); to account for ERS, the coding for d-items does not differentiate between moderate and extreme responses towards the trait and both are coded as 1 (0-0-1-1). Note that negatively scored items are rescored before being scored for BPIs.

Both pseudo item types (e-items, d-items) are modeled using unidimensional and multidimensional IRT (MIRT) models based on the 2PLM (Birnbaum, 1968) to test for the presence of ERS in the data. Moreover, mixture distribution IRT models, again based on the 2PLM, are applied to the BPI data to differentiate between latent classes of test-takers who show different response behaviors.

In MIRT models, the 2PLM can be specified for multiple scales. It is assumed that the 2PLM holds, with the qualifying condition that it holds with a different person parameter for each of a set of distinguishable subsets (scales) of items (Reckase, 2009; von Davier et al., 2007). For the case of a multidimensional 2PLM with between-item multidimensionality (each item loads on only one scale), the probability of response ($x_{iv} = 1$) to item i in scale k by respondent v can be defined as:

$$P(x_{iv}=1|\boldsymbol{\theta}_v,\beta_i,\boldsymbol{\alpha}_i)=\frac{\exp\left[\sum_{k=1}^{K}\boldsymbol{\alpha}_{ik}(x_{iv}\boldsymbol{\theta}_{vk}-\beta_i)\right]}{1+\exp\left[\sum_{k=1}^{K}\boldsymbol{\alpha}_{ik}(x_{iv}\boldsymbol{\theta}_{vk}-\beta_i)\right]}, \tag{22.1}$$

where $\boldsymbol{\theta}_{\mathrm{v}}$ is a vector of latent variables and $\boldsymbol{\alpha}_{ik}$ is the item loading for item i on scale k with the restriction that each item loads on only one scale. Unidimensional IRT models used in our analysis might be treated as special case of MIRT where $\boldsymbol{\theta}_{\mathrm{v}}=\theta_{\mathrm{v}}$ that is one latent dimension is assumed ($K = 1$).

For each scale, a one-dimensional IRT model (1D model) is compared to a two-dimensional IRT model (2D model), testing whether e-items and d-items can best be described by the latent construct of interest in the 1D model or by assuming an additional latent ERS variable in the 2D model. If the 1D model shows a better overall model-data fit than the 2D model, it would be assumed that there is no ERS and that e-items are measures of the construct of interest rather than a response style. If the 2D model shows a better overall model-data fit than the 1D model, ERS would be assumed in the data.

The mixture 2PLM is applied as mixture MIRT model with two separate factors for e-items and d-items to each selected questionnaire scale. It is assumed that the multidimensional 2PLM holds for multiple populations or latent classes simultaneously. That is, the 2PLM holds with a different set of parameters (β_{ic}, $\boldsymbol{\alpha}_{ic}$) in different populations, $c = 1,..,C$ classes with mixing proportions π_c and

$$P(x_{iv}=1|\boldsymbol{\theta}_v,\beta_{ic},\boldsymbol{\alpha}_{ic})=\sum_{c=1}^{C}\pi_{cv}\frac{\exp\left[\sum_{k=1}^{K}\boldsymbol{\alpha}_{ikc}(x_{iv}\boldsymbol{\theta}_{vk}-\beta_{ic})\right]}{1+\exp\left[\sum_{k=1}^{K}\boldsymbol{\alpha}_{ikc}(x_{iv}\boldsymbol{\theta}_{vk}-\beta_{ic})\right]}. \tag{22.2}$$

This approach aims to test whether there are different latent classes that differ with respect to their response behavior in the BPI data.

In the following section, the chapter introduces how RR can be added to the described IRTree model for a joint modeling of RR and ERS. Before illustrating this joint IRTree model, the chapter describes how item-level response times can be extracted from log data and used to derive indicators for flagging responses as RR in questionnaire scales.

22.1.3 Extracting Response Time from Log Data

Two things are necessary to apply the concept of RG to RR in questionnaires. The first requirement is to have item-level response times. For that purpose, Kroehne et al. (2019) applied methods developed for the analysis of log data from computer-based assessments, namely a framework that rests on the definition of meaningful states and the use of finite-state machines to process raw log events while

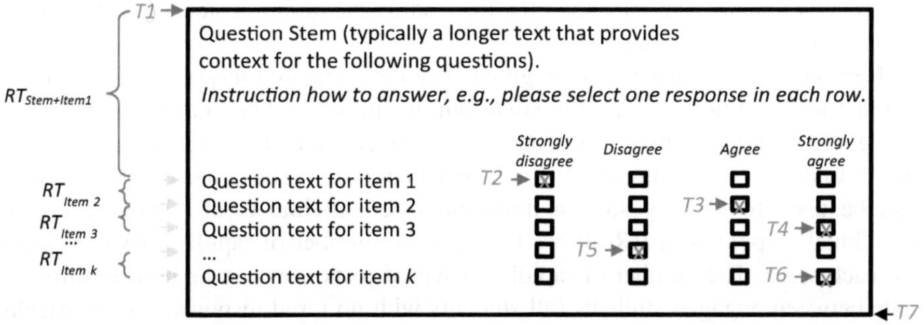

Fig. 22.2 Extraction of response times for item-batteries

extracting indicators based on states (Kroehne & Goldhammer, 2018). The application of this framework for extracting item-level response times from Likert-type item batteries uses the time difference between selected event types, typically collected in computer-based questionnaires as shown in the simple example in Fig. 22.2.

A questionnaire screen in a computer-based survey typically contains a question stem, and a couple of items. The time between the timestamp of a first log event (that indicates that the screen was loaded) and the timestamp for the answer-change event for question one gives the item-level response time for the first item (see *T1* and *T2* in Fig. 22.2, combined to the time difference $\text{RT}_{\text{Stem + Item 1}}$). This response time also includes the time for reading the question stem. The time difference to the next answer-change event allows computing an item-level time measure for the second item ($\text{RT}_{\text{Item 2}}$). The same idea is applied as an algorithm for all subsequent questions until the last item is answered (T6). Note that the time between the last answer-change event and the log event that indicates that the screen is unloaded (T7) is ignored. This procedure describes the basic idea for a prototypical and complete linear response pattern, which was for practical applications extended so that incomplete sequences, skipped responses, and changes to already provided answers are also taken into account (see Kroehne & Goldhammer, 2018).

22.1.4 Flagging Rapid Responses in Questionnaire Scales

The second crucial component necessary to investigate RR in questionnaire scales is a qualitative difference between rapid responses and regular responses at different, self-paced speed levels. In previous analyses using data from the PISA 2015 ICT familiarity questionnaire, Kroehne et al. (2019) observed that individual-level response times in non-cognitive measures show a bimodal distribution, similar to the bimodal response time distributions known from RG in cognitive tests. Such a bimodal distribution was even found for the average answering time (AAT), defined as the sum of all item-level response time components (ignoring the first time-component that also contains the time for reading the question stem), divided by the number of responses minus one.

Responses in the questionnaire given very fast, meaning responses belonging to the first mode of the response time distribution, are labeled and flagged as RR. This procedure requires, similar to the basic methods developed for RG in cognitive test items, to derive a time threshold. Visual inspection of the response time distribution can be used to identify such a threshold (see Kroehne et al., 2019). Flagging individual responses as RR allows to count the number of rapidly answered items for each scale. The number of rapidly answered responses can be used to differentiate between engaged students (all students with no rapid responses in a particular scale) and disengaged students with one ore multiple rapid responses. The classification in regular responses and RR can also be applied to item batteries using the average answering time (AAT). All responses to an item battery with an AAT that belongs to a first mode of the bimodal average answering time distribution are classified as RR.

Since questionnaire items cannot be answered incorrectly in contrast to cognitive items, but fast and with insufficient effort, identifying time thresholds for RR lacks the information provided by the approximated probability for a correct response conditional on the response time (known as conditional P+, see, e.g., Soland et al., 2019) compared to the concept of RG. Nevertheless, to keep the terminology in-line with the well-established methods for RG in cognitive assessments, we also derive a person-level variable that corresponds to the response time effort (RTE). An RTE measure can be computed as the total number of item-level responses not flagged as RR (for questionnaire items) or rapid guessing (for cognitive items), divided by the total number of responses. This variable is 1.0 for test-takers that show no RR or guessing behavior, and an RTE measures smaller than one informs about careless/insufficient effort (RR) for questionnaires similar to disengagement (RG) for cognitive tests.

The computation of RTE is conducted using flagged responses at the item-level. A similar measure can also be obtained for item batteries or scales when response times are aggregated within item sets. This can be done by using the AAT to classify RR versus regular responding for all scales that show a bimodal average answering time distribution using a threshold for the averaged answering time measure. Then, using RR for each scale, an RTE measure can be constructed at the scale level. Aggregating item-level time measures to the scale-level can lead to a more robust measure by reducing the impact of single item-level time measures. However, this comes at the price that disengaged RR for individual questions cannot be analyzed and, hence, changes in the engagement while answering an item battery remain undetected when using the AAT. A potential benefit of using scale-level classification becomes obvious by comparing AAT and the total time per screen. The AAT is the total time per screen minus the time for reading the item stimulus minus the time after the last response. Hence, the AAT can be considered as a measure of the time required for responding to survey questions and is less affected by reading speed. Note that the AAT can be approximated using the total time when the first response time (FRT) and the number of answered questions of an item battery or scale are known. Moreover, scale-level response times may be more frequently available in public use files of large-scale assessments than the item-level response times. The flagged rapid responses for each questionnaire scale can be used to derive person variables by aggregating the scale-level RTE measures to the person-level. More sophisticated modeling of item-level RR that takes measurement error of the indicators into account is described in the following.

22.1.5 *Joint IRTree Model for ERS and Rapid Responding Propensity*

Beyond the number of questions with rapid responses and the derived person-level RTE measure, indicator variables can be used to identify a latent *rapid responding propensity* (RRP) variable. The RR indicator variables at the item-level are observed, similar to the e-items and the d-items as binary variables for each question of an item battery. Indicators for RR are created by comparing the item-level response times with the item-specific time thresholds. The structure of the RR indicators is shown in Table 22.1 in addition to the ERS and direction variables (e-items and d-items). In the IRTree model, the item-level RR indicators are added as indicators for an additional latent variable. Similar to an approach developed in the context of missing values (e.g., Köhler et al., 2015; Zehner et al., 2021) a latent RRP is created using RR variables as indicator for a latent RRP that is expected to measure the tendency of test-takers to provide rapid responses when answering questions of an item-battery. Indicators for RR are assumed to be measures of RRP with measurement error.

Adding the latent RRP to the IRTree model in the single construct approach results in a three-dimensional (3D) model that can be used to estimate the relationship between ERS and RRP as well as between the trait and RRP (see Fig. 22.3). The latent correlation of RRP and ERS is considered as disattenuated measure of the relationship between a response style measure and RR.

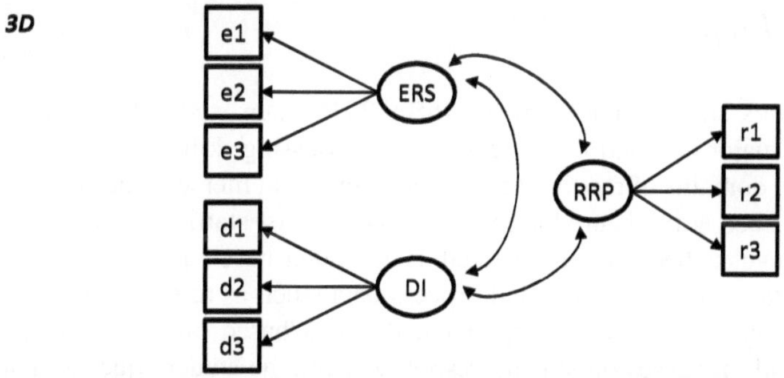

Fig. 22.3 3D IRTree model for jointly modeling the Trait (DI), extreme response style (ERS) and rapid responding propensity (RRP)

22.2 Research Questions

The overall research question in this study is to investigate, if RR can be differentiated from ERS or whether these two phenomena are similar types of behavior that potentially could be identified more effectively with a joint modeling approach. This question is approached from three different directions: First, the study sample was divided ad-hoc into two different engagement groups (engaged versus rapid responding or disengaged) based on the RTE measure, and IRTree models were used to test for ERS in both groups and to differentiate between ERS and the trait (DI). Then, it was investigated whether the correlation between ERS and DI is different for engaged versus disengaged students:

RQ 1: IRTree models: Is the correlation between DI and ERS different for engaged vs. disengaged students?

Second, mixture IRTree models were applied to differentiate between latent classes of respondents who show different response behaviors (i.e., classes with different item parameters in the BPIs). Then, the amount of RR based on the item-level thresholds and the AAT was examined in the different latent classes. In particular, we expect that respondents with different ERS behavior also show a different amount of RR:

RQ 2: Mixture IRTree models: Is the amount of RR different for students who show more ERS compared to students who show less ERS?

Third, a new IRTree model is proposed which allows the joint modeling of ERS and RRP and to examine the dimensionality and relationship between these variables:

RQ 3: Joint IRTree model: Is the latent RRP correlated with ERS in a joint modeling approach?

22.3 Method

22.3.1 Data

Data for this study come from the PISA 2015 main study student questionnaire. PISA is a major international student survey assessing skills of 15-year-old students in the core domains of reading literacy, mathematical literacy, and scientific literacy. PISA has been administered in cycles every 3 years, starting in 2000. Starting with the 2015 cycle, PISA is administered as computer-based assessment in most countries providing additional log and process data (such as response times and number of actions per screen) that supplement the data based on student responses. The current study focusses on student responses from the student questionnaire and the related response time measures. In specific, all scales with a 4-point Likert-type scale (see Table 22.2) were selected and the Swiss sample from the PISA 2015 main study

Table 22.2 Investigated scales/screens of the PISA 2015 student questionnaire with four-category items

Screen	Scale(s)	Items	Position
ST123	Parents emotional support (EMOSUPS)	4	14
ST118	Test anxiety (ANXTEST)	5	24
ST119	Achieving motivation (MOTIVAT)	5	25
ST082	Enjoy cooperation (COOPERATE) / value cooperation (CPSVALUE)	8	27
ST034	Sense of belonging to school (BELONG)	6	28
ST039	Teacher fairness (unfair teacher)	6	29
ST097	Disciplinary climate in science classes (DISCLISCI)	5	40
ST098	Inquiry-based science teaching and learning practices (IBTEACH)	9	41
ST100	Teacher support in a science classes (TEACHSUP)	5	43
ST103	Teacher-directed science instruction (TDTEACH)	4	44
ST104	Perceived feedback (PERFEED)	5	45
ST107	Adaption of instruction (ADINST)	3	46
ST092	Environmental awareness (ENVAWARE)	7	47
ST094	Enjoyment of science (JOYSCIE)	5	49
ST095	Interest in broad science topics (INTBRSCI)	5	50
ST113	Instrumental motivation (INSTSCIE)	3	51
ST129	Science self-efficacy (SCIEEFF)	8	52
ST131	Epistemological beliefs (EPIST)	6	53
ST146	Science activities (SCIEACT)	9	54
ST012	Cultural possessions at home (CULTPOSS), family WEALTH (WEALTH), ICT resources (ICTRES) and home possessions (HOMEPOS)	8	10
ST121	(anchoring vignette: Motivation)	3	26

was analyzed. For our analyses we removed all students without response time information and restricted the data set to students with complete response data for all investigated scales ($N = 3849$). Because the public use data file from PISA 2015 only provides the screen-level response times (where multiple items are administered on one screen), item-level response times had to be extracted from the Swiss log file data available to the authors.

22.3.2 *Software*

The package *LogFSM* was used for the extraction of item-level response times from log file data using finite-state machines (Kroehne & Goldhammer, 2018).[1] The IRT

[1]The particular LogFSM syntax for the decomposition of the response process can be requested from the authors.

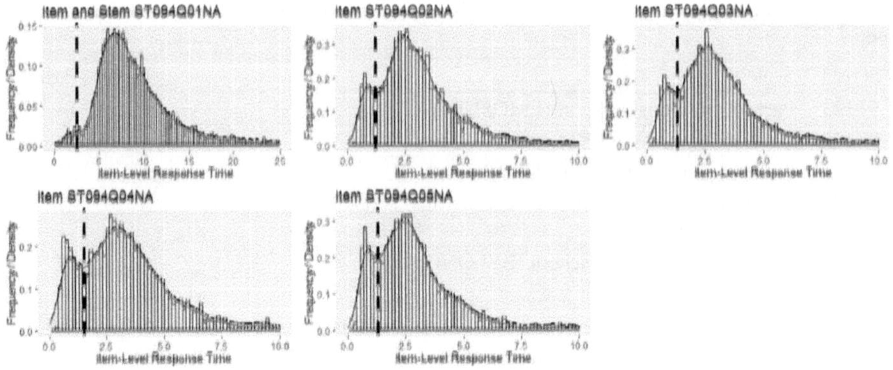

Fig. 22.4 Bimodal Item-Level Response Time Distribution Illustrated Using Scale ST094

models applied in this study were estimated by applying the mixture general diagnostic modeling framework (MGDM; von Davier, 2008, 2010), which allows specification of a discrete mixture model with a hierarchical component (von Davier, 2010) using the software *mdltm* (von Davier, 2005) for multidimensional discrete latent trait models. The software provides marginal maximum likelihood (MML) estimates obtained using the customary expectation-maximization (EM) algorithm, with optional acceleration.

22.3.3 *Visual Inspection*

A crucial prerequisite for analyzing RR using response times, as employed in this chapter, is a bimodal response time distribution, which enables the derivation of time thresholds through visual inspection. Bimodal distributions were found for selected items of all scales. Figure 22.4 provides an example where the bimodal distribution was present in all items, including the first item for which the item-level time measure also includes the time for reading the item stem. As illustrated with the dashed lines in Fig. 22.4, visual inspection of the response time distribution allows to identify response time thresholds at the item level that can be used to separate engaged responses and rapid responses.

As expected from previous work using the data from the ICT familiarity questionnaire (Kroehne et al., 2019), bimodal average answering time (AAT) distribution were found for all scales of the PISA student questionnaire investigated in the empirical application (see Fig. 22.5 for the selected scales “ST094”, “ST094” and “ST097”). As illustrated with the dashed lines in Fig. 22.5, visual inspection of the AAT distributions also allows to identify response time thresholds at the scale level that can be used to separate engaged responses to all items of an item-battery and item-batteries that were answered with RR.

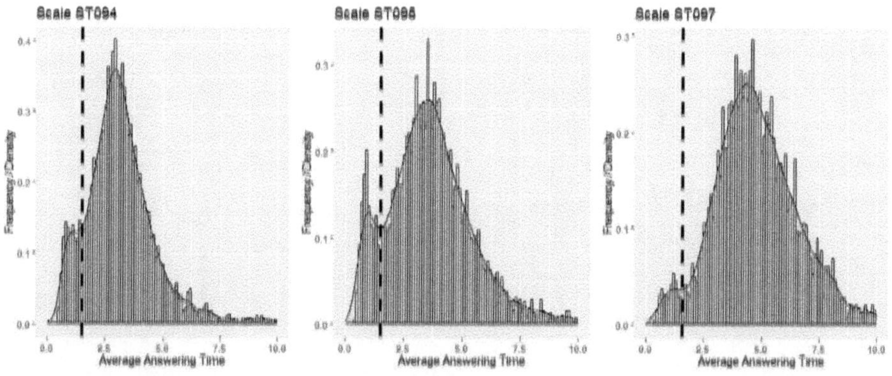

Fig. 22.5 Bimodal average answering time Distribution Illustrated Using Scales ST094, ST094 and ST097

22.3.4 Response Time Effort

By aggregating the RR indicators within persons, we create measures similar to the response time effort (RTE) known from research on RG for cognitive tests (Wise & Kong, 2005). RTE is a person-level variable computed by subtracting the number of classified rapid responses divided by the total number of responses from one. An RTE of 1.0 is interpreted as perfect engaged and a smaller RTE indicates lower engagement. The classification of responses is done using the item-level response time thresholds, compared to the item-level response times.

In previous research using RTE measures from cognitive tests, systematic group differences in RTE were found. Replicating these group differences but using an RTE measure based on questionnaire data can be understood as validity evidence regarding the interpretation of the RR behavior in questionnaires as low test-taking engagement. For cognitive tests it is known that male students show a lower RTE than female students (e.g., DeMars et al., 2013). As shown in the upper left part of Fig. 22.6, we found this gender effect also for the non-cognitive RTE measure used in this study. Replicating previous findings by Kroehne et al. (2019) for the PISA ICT Familiarity questionnaire, language differences in the RTE (using the test language which was either French, German or Italian in the Swiss data) can be identified as well (see upper right part of Fig. 22.6).

To provide further evidence that the RTE measure extracted from the response times in the student questionnaire shows patterns that can be expected from an effort measure, the lower part of Fig. 22.6 also shows RTE differences across different groups created by quartiles of the economic, social and cultural status (on the left) and the highest parental occupational status (on the right). As expected (e.g., Penk & Richter, 2017), the response time effort is lower for students from a lower socio-economical background and for students with parents having a lower occupational status.

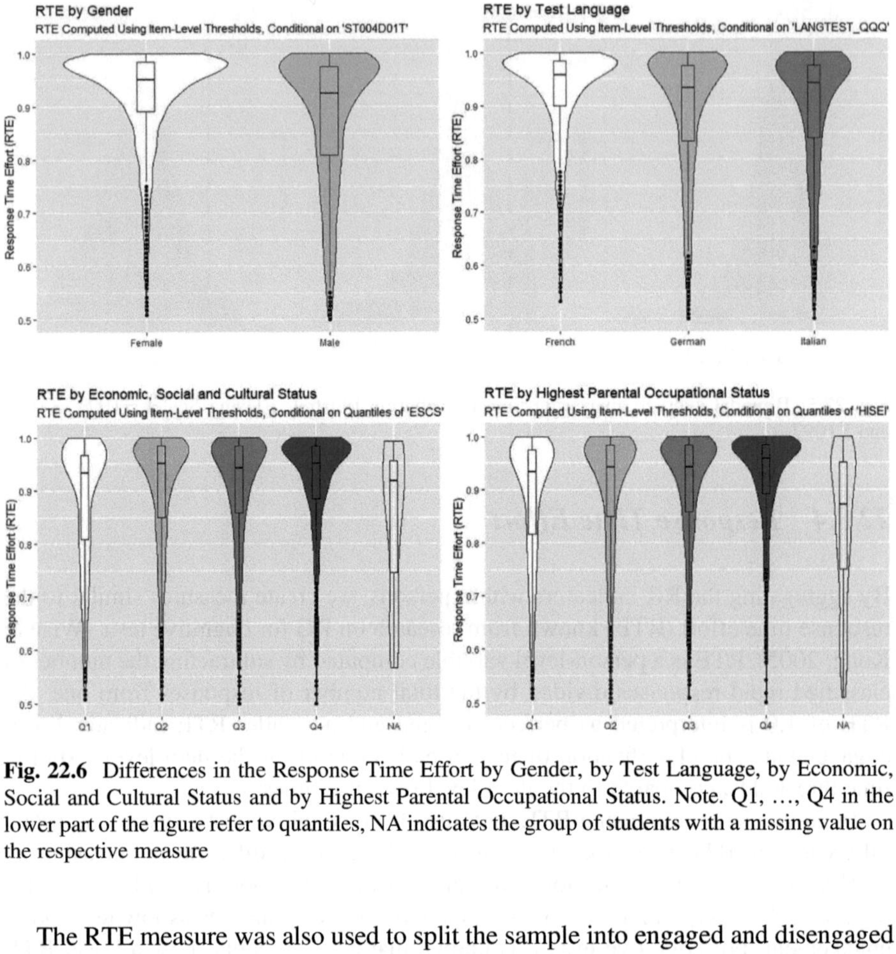

Fig. 22.6 Differences in the Response Time Effort by Gender, by Test Language, by Economic, Social and Cultural Status and by Highest Parental Occupational Status. Note. Q1, …, Q4 in the lower part of the figure refer to quantiles, NA indicates the group of students with a missing value on the respective measure

The RTE measure was also used to split the sample into engaged and disengaged students. For that purpose, we used the RTE measure estimated across all investigated scales and the cut-off values RTE < 0.8 (disengaged) and RTE $> = 0.8$ (engaged), which was chosen based on previous findings of the authors (Kroehne et al., 2019).

22.3.5 Application of the IRTree Modeling Approaches

As described earlier, the rating data of the selected questionnaire scales with a four-point Likert-type response format were decomposed into BPI to account for extreme responses (e-items) and trait-relevant responses (d-items). While e-items can be used for modeling ERS d-items are used to model the latent trait (DI). The resulting BPI data were modeled using multidimensional IRTree models (1D, 2D, 3D) as well as mixture IRTree models, all based on the 2PLM.

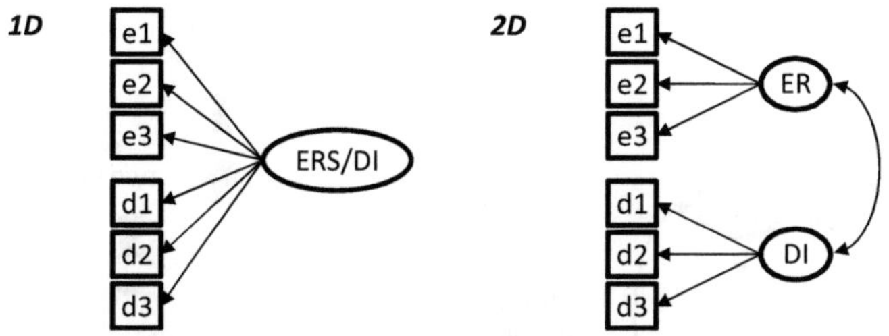

Fig. 22.7 1D and 2D Model to Test for the Presence of ERS in each Questionnaire Scale and Engagement Group (Engaged versus Disengaged or RR)

For overall model-data fit evaluation, we used the Akaike Information Criterion (AIC; Akaike, 1974) and the Bayesian Information Criterion (BIC; Schwarz, 1978). While the use of the AIC is recommended for conditions with small sample sizes, the BIC is more appropriate in conditions with large sample sizes (Dziak et al., 2012). Note that smaller AIC and BIC values indicate better fit.

IRTree models applied to known groups (RQ1). In a first step, the sample was divided into a group with engaged response behavior and a group with RR behavior by using the RTE indicator. For each group and each questionnaire scale, a 1D model (e-items and d-items are assigned to the same factor) and a 2D model (e-items and d-items are assigned to two separate factors) were estimated and the overall model fit was compared between these models. In addition, the latent correlation between the two factors in the 2D model is of interest. We would expect no high correlation of the factors that correspond to ERS and DI (see right part of Fig. 22.7) if ERS is present in the data and can be differentiated from DI.

Mixture IRTree models for examining latent classes (RQ2). In a second step, mixture IRTree models were applied to the total sample for each questionnaire scale to differentiate between latent classes of respondents with different response behaviors. 2D mixture IRTree models were applied to the BPI data with d-items and e-items being assigned to two separate factors (see Fig. 22.8). Groups are separated not based on observed RR but concerning the RS operationalized as class membership in the mixture IRTree model.

Mixture IRTree models with different number of latent classes were estimated and compared using overall model fit indices to select the best fitting model. Latent classes are then compared regarding their response pattern in the original Likert-type scale by looking at the response frequencies, and by the amount of RR behavior operationalized using time thresholds for the item-level response times and the AAT measure. We expect more RR behavior in classes that show ERS patterns.

Joint IRTree model for RR and ERS (RQ3). In a third step, we investigate the dimensionality of RR and ERS and introduce a new IRTree model that incorporates RR as RRP (r-items) in addition to DI (d-items) and ERS (e-items). For each scale, we compare a 3D model (all BPIs are assigned to different distinct factors) with two 2D models (2D^{a}: e-items and r-items are assigned to one and d-items are assigned to a second factor; 2D^{b}: r-items are assigned to one factor and d-items and e-items are

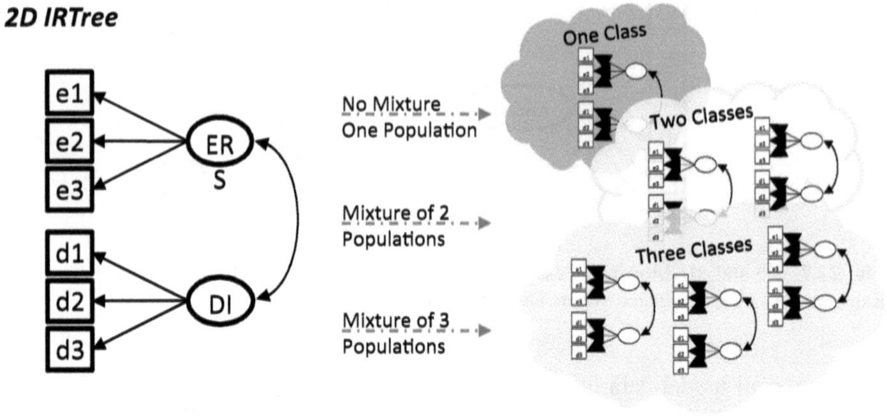

Fig. 22.8 2D Mixture IRTree Model for Differentiating between Latent Classes of Respondents with Different Response Behavior in each Questionnaire Scale

assigned to a second factor). Models are compared with regard to their overall model fit indices and correlations between the latent variables are examined. If the 2D[a] model shows the best fit, we could assume that RR and ERS are very similar behaviors. The 2D[b] model would be expected to fit the data best if response styles could not be identified when investigating RR simultaneously.

22.4 Results

The results regarding the three research questions will be presented using the scale ST094 as an example.

22.4.1 *RQ 1: ERS Behavior in Engaged Vs. Disengaged (RR) Groups (IRTree Models)*

Research question one addressed the correlation between responses in the direction of towards the trait (DI) and the extreme response style (ERS) measure. For that purpose, 2D models were compared to 1D models first. For the total sample as well as for sub-samples of engaged and disengaged students we found that the 2D model fits best (see Table 22.3 for the results regarding scale ST094). We found a similar result for all regular 4-point Likert-type scales but for the scale ST012 that was using an anchoring vignette.

The 2D IRTree models were used to estimate latent correlations between DI and ERS. Separate models were used to estimate the correlation for the sub-samples of

Table 22.3 Model fit for IRTree models illustrated using scale ST094

Model	Likelihood	Deviance	AIC	BIC	Model-based log penalty per item	Independence log penalty per item
All students						
2 D	−17354.0183	34708.03659	**34750.03659**	**34881.40353**	0.4526348	0.6435642
1 D	−21045.6106	42091.2212	42131.2212	42256.33257	0.5489205	0.6435642
Engaged students (RTE > = 0.8)						
2 D	−14437.66	28875.31997	**28917.31997**	**29044.71033**	0.4546435	0.6317425
1 D	−16857.1069	33714.21375	33754.21375	33875.5379	0.5308322	0.6317425
Disengaged students (RTE < 0.8)						
2 D	−2708.3411	5416.68226	**5458.68226**	**5553.14618**	0.4113519	0.6787948
1 D	−3293.4059	6586.81189	6626.81189	6716.77753	0.5002135	0.6787948

Note. Smallest AIC/BIC values printed in bold

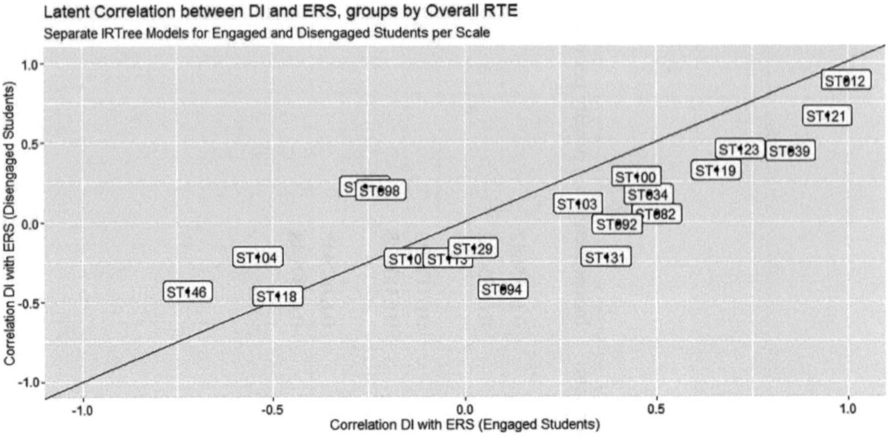

Fig. 22.9 Latent Correlation Between the Trait Measure (DI) and Extreme Response Style Measure (ERS) for Engaged and Disengaged Students

engaged and disengaged students for each scale. Results are presented in Fig. 22.9. The scatter plot shows for each investigated scale the latent correlation between DI and ERS for both sub-samples. On the x-axis the result from the sub-group of engaged students is plotted and on the y-axis the correlation estimated in the disengaged group is shown.

A consistent correlation pattern can be seen across all investigates scales: While there are some differences between the correlations estimated for disengaged and engaged students, the range and ranking of correlations between these groups is largely the same. Note that these are estimated correlations, not the true correlations, and the estimates have different error variances in subgroups of different sizes.

22.4.2 RQ 2: Amount of RR for Latent ERS Classes (Mixture IRTree Models)

Mixture IRTree models were estimated to identify classes of students with differences in ERS. For the example scale ST094, AIC and BIC point to either the three- or the four-class solution (see Table 22.4).

Inspecting the category frequencies reveals that for the three-class solution (the plot in the lower left part of Fig. 22.10) no clear ERS can be identified. In the three-class solution, extreme responses are observed only to some extent in class one. In the four-class solution (the plot in the lower right part of Fig. 22.10), extreme responses are more clearly grouped in class one (Category 0) and also to some extent in class two (Category 3).

Mixture modelling was applied to the indicators for DI and ERS, without taking the item-level response time, the average answering time or the indicators for rapid

Table 22.4 Model fit for the 2D mixture IRTree model illustrated using scale ST094

Model	Likelihood	Deviance	AIC	BIC	Model-based log penalty per item	Independence log penalty per item
1 class	−17354.0287	34708.05747	34750.05747	34881.42442	0.4526351	0.6435642
2 class	−16762.726	33525.45202	33611.45202	33880.44147	0.4372125	0.6435642
3 class	−16514.1319	33028.26382	33158.26382	**33564.87578**	0.4307285	0.6435642
4 class	−16452.4093	32904.81867	**33078.81867**	33623.05315	0.4291187	0.6435642

Note. Smallest AIC/BIC values printed in bold

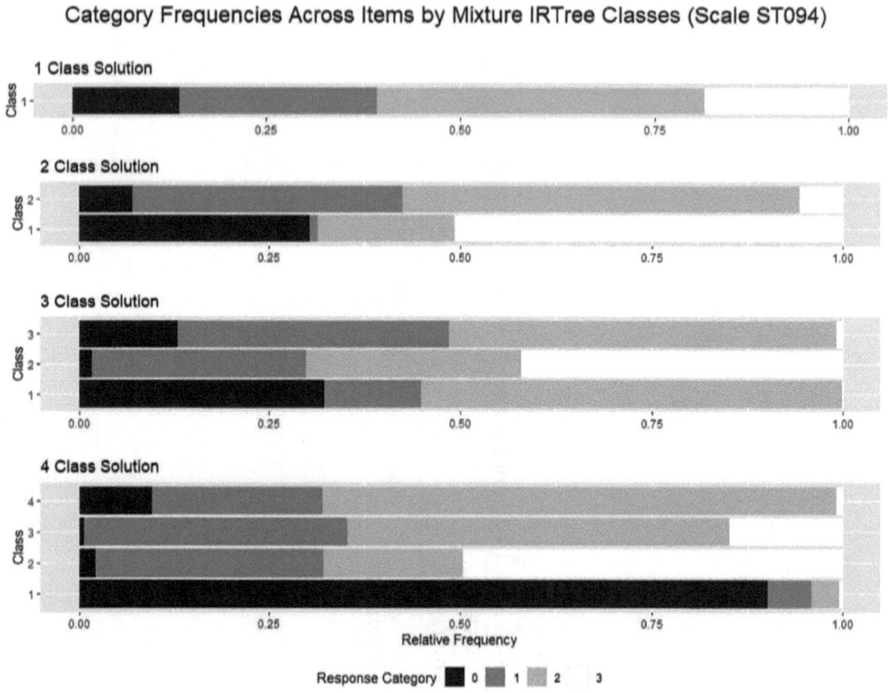

Fig. 22.10 Category Frequencies of Mixture IRTree Models Illustrated Using Scale ST094

responding into account. Hence, the solutions found for the three-class and four-class mixture model for the selected scale ST094 can be interpreted by comparing the differences between classes in response times and response time effort. The upper part of Fig. 22.11 shows the average answering time (AAT) and the first response time (FRT) by class membership for the three- and four-class solution (scale ST094). The difference between the classes becomes obvious only concerning the AAT with respect to the median. The classes that contain the extreme responses can be described as the classes with lower AAT.

The difference in the amount of RR for students with ERS can be seen for the four-class solution with respect to the engagement measure computed as RTE (see lower part of Fig. 22.11). Mixture modelling of DI and ERS seems to identify a class of respondents with ERS, and the respondents in this class show lower test-taking engagement in terms of RTE. Hence, differences in the average answering time are also reflected as differences between classes formed by the response style indicators (ERS and DI). No structural difference between the RTE computed by scale (using AAT) or by item (using item-level response times) can be observed.

The classes also differ with respect to the average number of rapid responses per scale. Regardless of a three- or four-class solution, both identify one class with notably more rapidly answered items (see Fig. 22.12 for the selected scale ST094).

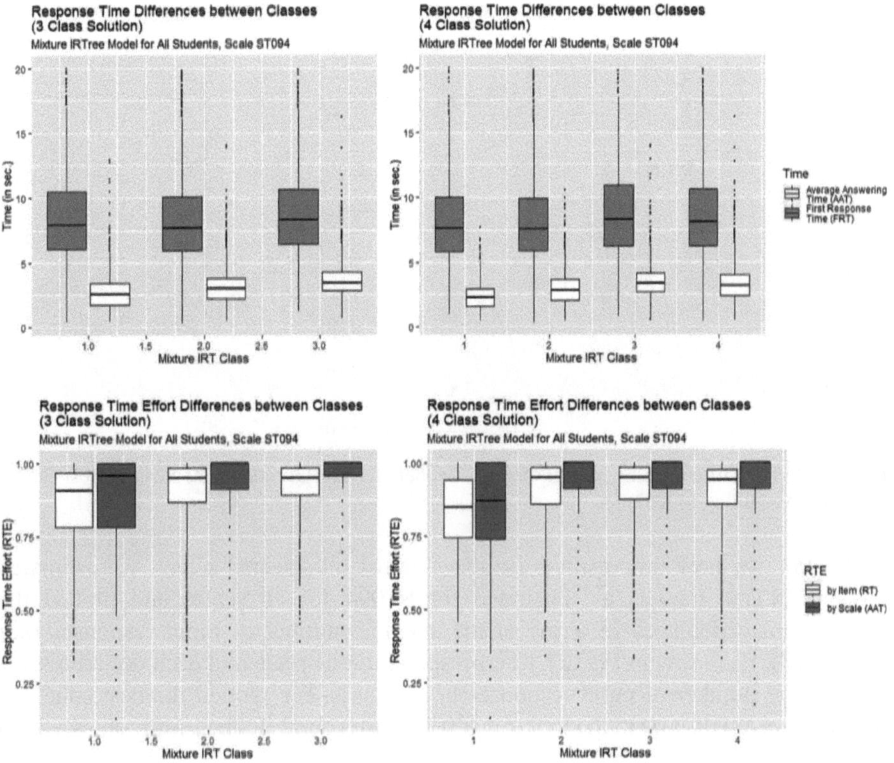

Fig. 22.11 Differences in the Response Time and Response Time Effort by Classes of the Mixture IRTree Model for the Two- and Three-Class Solutions for Scale ST094

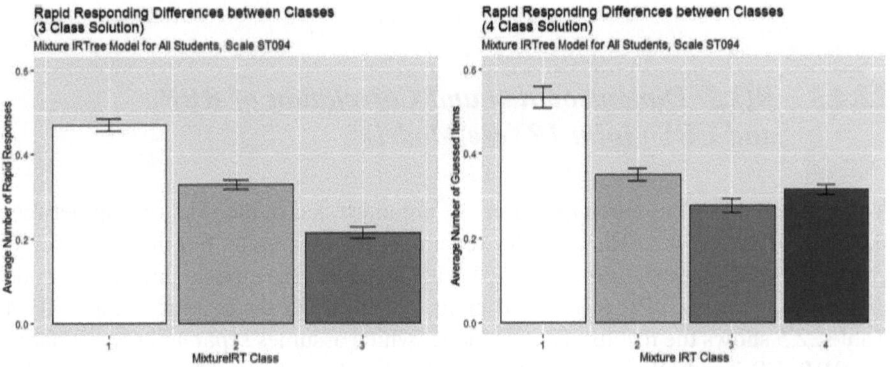

Fig. 22.12 RR Difference of Mixture IRTree Models for the Two- and Three-Class Solutions for Scale ST094

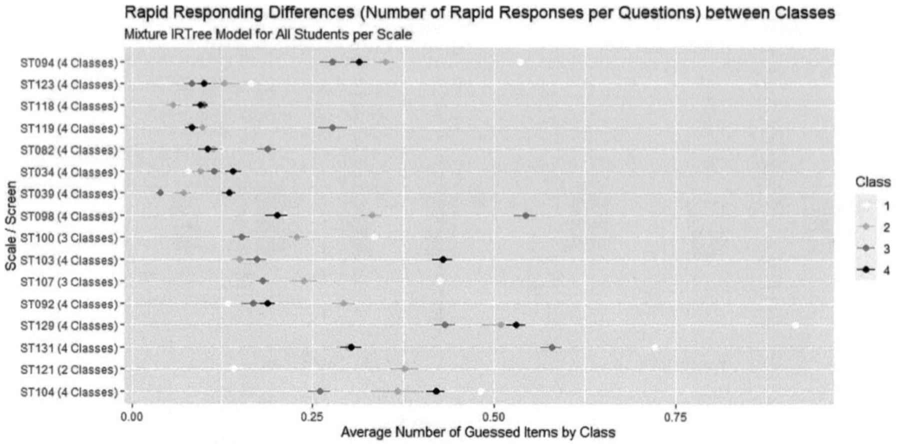

Fig. 22.13 RR Difference of Mixture IRTree Models for All Questionnaire Scales (Best Fitting Solutions According to BIC)

To describe how this result generalizes across all selected scales, we estimated mixture IRTree models as illustrated for ST094 for all scales and plotted the engagement differences in terms of the average number of rapid responses (see Fig. 22.13). Each row in Fig. 22.13 correspond to one scale and each dot represents the average number of rapid responses in one class. We plotted the best solutions with respect to BIC and in the first row ST094 corresponds to the results shown on in Fig. 22.12.

Inspecting the remaining scales shown in Fig. 22.13 reveals that for many but not all scales a similar picture can be observed. The latent classes identified with respect to item parameters of BPIs also differ with respect to number of rapid responses.

22.4.3 RQ 3: Dimensionality and Correlation of RRP and ERS (Joint IRTree Model)

To answer the third research question, we are interested in the relationship between the latent RRP and the ERS variables separately for each scale. For that purpose, we compared a 3D model (see Fig. 22.3) with different 2D models and a 1D model applied to the same BPIs (e-items, d-items, r-items). For the example scale ST094, Table 22.5 shows the results. The 3D model which assumes separate latent variables for RRP, ERS and DI shows the best model-data fit compared to all other models indicating that RRP and ERS are distinct phenomena that can be distinguished. However, the model comparison might lack power (see Pokropek et al., 2023) and results have to be interpreted with some care. The findings generalize to all investigated scales but ST012 as an exception.

As shown in Fig. 22.14, the RRP is only for some scales correlated with either ERS or DI, in the 3D model.

Table 22.5 Model fit for IRTree model with rapid responding propensity (RRP) illustrated using scale ST094

Model	Likelihood	Deviance	AIC	BIC	Model-based log penalty per item	Independence log penalty per item
3 D	−22925.1851	45850.37012	**45916.37012**	**46122.80389**	0.4001603	0.5475365
2 D[a]	−24023.2031	48046.40629	48108.40629	48302.32892	0.4193263	0.5475365
2 D[b]	−26621.092	53242.18391	53304.18391	53498.10654	0.4646726	0.5475365

[a] DI vs. RRP & ERS, [b] RRP vs. DI & ERS

Note. Smallest AIC/BIC values printed in bold

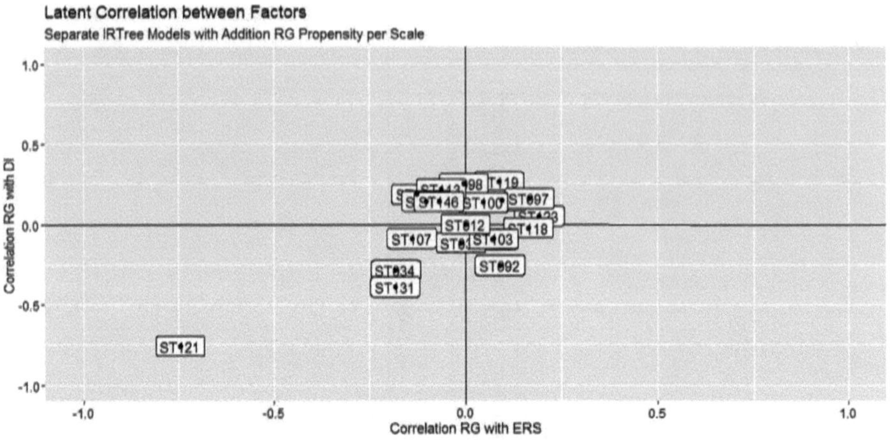

Fig. 22.14 Latent Correlations of Extreme Response Style (ERS) with Rapid Responding Propensity (RRP) and Responses in the Direction Towards the Trait (DI)

The weak latent correlations found for ERS and RRP confirm the result of the model comparison. ERS and RR in the selected questionnaire scales are empirically separable and not perfectly correlated.

22.5 Discussion

To our knowledge, this chapter presented the first empirical study that investigated the relationship of rapid responding (RR) and response style (in specific extreme response style, ERS) in questionnaire scales using IRTree models. RR is understood as the counterpart for survey questions to what is known as rapid guessing (RG) in cognitive tests. Aggregated across items within persons the amount of RR could be an indicator for engagement. By comparing groups with different engagement levels, we have shown that engagement affects the relationship of ERS and the trait (DI). We found hints that the correlation between ERS and DI seems stronger for engaged students indicating that engaged students show less ERS behavior as it cannot be completely separated from DI. At the same time, it seems like ERS and DI can be better separated for disengaged respondents with RR behavior as correlations between ERS and DI are smaller.

The mixture IRTree models for ERS and DI identified a class with higher average number of rapid responses and lower RTE for most questionnaire scales. That means, classes in mixture IRTree models can be validated with respect to engagement differences. The classes that contain respondents with more extreme responses seems to also contain respondents that show more RR behavior. We found no structural differences between using a global response time effort measure (RTE based on AAT) compared to either an item-level response time effort (RTE based on response times) or the number of rapid responses.

Across all respondents we found separate factors for ERS and DI. This can be interpreted as evidence that RS are present in the data. In a joint IRTree modeling approach, we also found a separate factor for the rapid responding propensity (RRP) that was only weakly correlated with ERS. Together with the other findings, this can be interpreted as evidence that ERS and RR overlap to some extent but are distinct phenomena that can be empirically separated.

This study comes with a couple of limitations. The mixture IRTree modeling approach for many scales requires more careful interpretation of the solutions. The superficial interpretation across scales was only conducted to report findings that, to some extent, generalize across scales. ERS was modelled using IRTree models, but other psychometric approaches such as mixture IRT modeling (e.g., Ulitzsch et al., 2021, 2023) might provide additional insights. Moreover, since investigating RR behavior in questionnaire scales is a rather new research field, some decisions (such as using 0.8 as RTE cut-off value to separate engaged versus disengaged students) could have impacted the findings. Therefore, the investigation of different cut-off values and replications are required, even with the same data (multiverse analysis). Finally, further studies should examine the impact of the item and scale position within the questionnaire on the relation between ERS and RR. Both can be expected to be more prone in scales administered towards the end of the questionnaire. It would be interesting to see whether RR and ERS show more overlap (similarities) at the end compared to the beginning of a questionnaire.

However, this chapter shows that both phenomena RR and ERS can be empirically distinguished but also overlap to some extent. Consequently, further research should develop additional models to investigate both phenomena simultaneously.

References

Akaike, H. (1974). A new look at the statistical model identification. *IEEE Transactions on Automatic Control, 19*, 716–723. https://doi.org/10.1109/TAC.1974.1100705

Austin, E. J., Deary, I. J., & Egan, V. (2006). Individual differences in response scale use: Mixed Rasch modelling of responses to NEO-FFI items. *Personality and Individual Differences, 40*(6), 1235–1245. https://doi.org/10.1016/j.paid.2005.10.018

Baumgartner, H., & Steenkamp, J.-B. E. M. (2001). Response styles in marketing research: A cross-national investigation. *Journal of Marketing Research, 38*(2), 143–156. https://10.1509/jmkr.38.2.143.18840

Birnbaum, A. (1968). Some latent trait models and their use in inferring an examinee's ability. In F. M. Lord & M. R. Novick (Eds.), *Statistical theories of mental test scores* (pp. 397–479). Addison-Wesley.

Böckenholt, U. (2012). Modeling multiple response processes in judgment and choice. *Psychological Methods, 17*(4), 665–678. https://doi.org/10.1037/a0028111

Böckenholt, U. (2019). Assessing item-feature effects with item response tree models. *British Journal of Mathematical and Statistical Psychology, 72*(3), 486–500. https://doi.org/10.1111/bmsp.12163

Böckenholt, U., & Meiser, T. (2017). Response style analysis with threshold and multi-process IRT models: A review and tutorial. *British Journal of Mathematical and Statistical Psychology, 70*(1), 159–181. https://doi.org/10.1111/bmsp.12086
Bolt, D. M., & Newton, J. R. (2011). Multiscale measurement of extreme response style. *Educational and Psychological Measurement, 71*(5), 814–833. https://doi.org/10.1177/0013164410388411
Buckley, J. (2009, June). *Cross-national response styles in international educational assessments: Evidence from PISA* 2006. NCES Conference on the Program for International Student Assessment: What we can learn from PISA. Retrieved from http://edsurveys.rti.org/PISA/
Curran, P. G. (2016). Methods for the detection of carelessly invalid responses in survey data. *Journal of Experimental Social Psychology, 66*, 4–19. https://doi.org/10.1016/j.jesp.2015.07.006
Chun, K. T., Campbell, J. B., & Yoo, J. H. (1974). Xtreme response style in cross-cultural research: A reminder. *Journal of Cross-Cultural Psychology, 5*(4), 465–480. https://doi.org/10.1177/002202217400500407
DeBoeck, P., & Partchev, I. (2012). IRTrees: Tree-based item response models of the GLMM family. *Journal of Statistical Software, 48*(1), 1–28. https://doi.org/10.18637/jss.v048.c01
De Jong, M. G., Steenkamp, J.-B. E. M., Fox, J.-P., & Baumgartner, H. (2008). Using item response theory to measure extreme response style in marketing research: A global investigation. *Journal of Marketing Research, 45*(1), 104–115. https://doi.org/10.1509/jmkr.45.1.104
DeMars, C. E., Bashkov, B. M., & Socha, A. B. (2013). The role of gender in test-taking motivation under low-stakes conditions. *Research and Practice in Assessment, 8*, 69–82. https://files.eric.ed.gov/fulltext/EJ1062839.pdf
Dïbek, M. I. (2019). Examination of the extreme response style of students using IRTree: The case of TIMSS 2015. *International Journal of Assessment Tools in Education, 6*(2), 300–313. https://doi.org/10.21449/ijate.534118
Dolnicar, S., & Grun, B. (2009). Response style contamination of student evaluation data. *Journal of Marketing Education, 31*(2), 160–172. https://doi.org/10.1177/0273475309335267
Dziak, J. J., Coffman, D. L., Lanza, S. T., & Li, R. (2012). *Sensitivity and specificity of information criteria* (Vol. 21, pp. 553–565). Methodology Center and Department of Statistics, Pennsylvania State University. https://doi.org/10.1093/bib/bbz016
Eid, M., & Rauber, M. (2000). Detecting measurement invariance in organizational surveys. *European Journal of Psychological Assessment, 16*(1), 20–30. https://doi.org/10.1027/1015-5759.16.1.20
Goldhammer, F., Martens, T., Christoph, G., & Lüdtke, O. (2016). *Test-taking engagement in PIAAC* (Vol. 133). OECD Publishing. https://doi.org/10.1787/19939019
Gollwitzer, M., Eid, M., & Jürgensen, R. (2005). Response styles in the assessment of anger expression. *Psychological Assessment, 17*(1), 56–69. https://doi.org/10.1037/1040-3590.17.1.56
Henninger, M., & Plieninger, H. (2020). Different styles, different times: How response times can inform our knowledge about the response process in rating scale measurement. *Assessment, 28*(5), 1–19. https://doi.org/10.1177/1073191119900003
Javaras, K. N., & Ripley, B. D. (2007). An "unfolding" latent variable model for Likert attitude data: Drawing inferences adjusted for response style. *Journal of the American Statistical Association, 102*, 454–463. https://doi.org/10.1198/016214506000000960
Jeon, M., & De Boeck, P. (2016). A generalized item response tree model for psychological assessments. *Behavior Research Methods, 48*(3), 1070–1085. https://doi.org/10.3758/s13428-015-0631-y
Khorramdel, L., & von Davier, M. (2014). Measuring response styles across the big five: A multiscale extension of an approach using multinomial processing trees. *Multivariate Behavioral Research, 49*(2), 161–177. https://doi.org/10.1080/00273171.2013.866536
Khorramdel, L., von Davier, M., & Pokropek, A. (2019). Combining mixture distribution and multidimensional IRTree models for the measurement of extreme response styles. *British*

Journal of Mathematical and Statistical Psychology, 72(3), 538–559. https://doi.org/10.1111/bmsp.12179

Khorramdel, L., von Davier, M., Bertling, J. P., Roberts, R. D., & Kyllonen, P. C. (2017). Recent IRT approaches to test and correct for response styles in PISA background questionnaire data: A feasibility study. *Psychological Test and Assessment Modeling, 59*(1), 71–92. https://www.psychologie-aktuell.com/fileadmin/download/ptam/1-2017_20170323/05_Khorramdel.pdf

Kim, N., & Bolt, D. M. (2020). A mixture IRTree model for extreme response style: Accounting for response process uncertainty. *Educational and Psychological Measurement, 81*(1), 131–154. https://doi.org/10.1177/2F0013164420913915

Köhler, C., Pohl, S., & Carstensen, C. H. (2015). Taking the missing propensity into account when estimating competence scores: Evaluation of item response theory models for nonignorable omissions. *Educational and Psychological Measurement, 75*(5), 850–874. https://doi.org/10.1177/0013164414561785

Kroehne, U., Buchholz, J., & Goldhammer, F. (2019, April). Detecting carelessly invalid effort guessing in item sets using average item-level response times. In *Paper presented at the annual meeting of the National Council on measurement in education (NCME)*. https://doi.org/10.5281/zenodo.10846003

Kroehne, U., & Goldhammer, F. (2018). How to conceptualize, represent, and analyze log data from technology-based assessments? A generic framework and an application to questionnaire items. *Behaviormetrika, 45*(2), 527–563. https://doi.org/10.1007/s41237-018-0063-y

Lechner, C. M., Partsch, M. V., Danner, D., & Rammstedt, B. (2019). Individual, situational, and cultural correlates of acquiescent responding: Towards a unified conceptual framework. *British Journal of Mathematical and Statistical Psychology, 72*(3), 426–446. https://doi.org/10.1111/bmsp.12164

Lee, Y.-H., & Jia, Y. (2014). Using response time to investigate students' test-taking behaviors in a NAEP computer-based study. *Large-Scale Assessments in Education, 2*(1), 8. https://doi.org/10.1186/s40536-014-0008-1

Mayerl, J. (2013). *Response latency measurement in surveys. Detecting strong attitudes and response effects*. Survey methods: Insights from the field. Retrieved from https://surveyinsights.org/?p=1063,https://doi.org/10.13094/SMIF-2013-00005

Meiser, T., & Machunsky, M. (2008). The personal structure of personal need for structure: A mixture-distribution Rasch analysis. *European Journal of Psychological Assessment, 24*(1), 27–34. https://doi.org/10.1027/1015-5759.24.1.27

Morren, M., Gelissen, J., & Vermunt, J. (2012). The impact of controlling for extreme responding on measurement equivalence in cross-cultural research. *Methodology, 8*(4), 159–170. https://doi.org/10.1027/1614-2241/a000048

Nunnally, J. (1967). *Psychometric theory*. McGraw-Hill.

Paulhus, D. L. (1991). Measurement and control of response bias. In J. P. Robinson, P. R. Shaver, & L. S. Wrightsman (Eds.), *Measures of personality and social psychological attitudes* (pp. 17–59). Academic.

Penk, C., & Richter, D. (2017). Change in test-taking motivation and its relationship to test performance in low-stakes assessments. *Educational Assessment, Evaluation and Accountability, 29*(1), 55–79. https://doi.org/10.1007/s11092-016-9248-7

Plieninger, H., & Meiser, T. (2014). Validity of multiprocess IRT models for separating content and response styles. *Educational and Psychological Measurement, 74*(5), 875–899. https://doi.org/10.1177/0013164413514998

Primi, R., Santos, D., De Fruyt, F., & John, O. P. (2019). Comparison of classical and modern methods for measuring and correcting for acquiescence. *British Journal of Mathematical and Statistical Psychology, 72*(3), 447–465. https://doi.org/10.1111/bmsp.12168

Pokropek, A., Khorramdel, L., & von Davier, M. (2023). Detecting and differentiating extreme and midpoint response styles in rating scales: Simulation study and empirical evidence. *Psychological Test and Assessment Modeling, 65*, 220–258. https://www.psychologie-aktuell.com/fileadmin/Redaktion/Journale/ptam_2023-2/PTAM__2-2023_3_kor_2te.pdf

Reckase, M. D. (2009). Multidimensional item response theory. In *Statistics for social and behavioral sciences*. Springer.

Rost, J. (2004). *Lehrbuch Testtheorie—Testkonstruktion [Textbook test theory—Test construction]* (2nd ed.). Huber.

Rost, J., Carstensen, C., & von Davier, M. (1999). Sind die big five Rasch-skalierbar? Eine reanalyse der NEO-FFI-Normierungsdaten [Are the big five Rasch scalable? A reanalysis of the NEO-FFI norm data]. *Diagnostica, 45*, 119–127.

Schnipke, D. L., & Scrams, D. J. (1997). Modeling item response times with a two-state mixture model: A new method of measuring speededness. *Journal of Educational Measurement, 34*(3), 213–232. https://doi.org/10.1111/j.1745-3984.1997.tb00516.x

Schwarz, G. (1978). Estimating the dimension of a model. *Annals of Statistics, 6*(2), 461–464. https://www.jstor.org/stable/2958889

Soland, J., Wise, S. L., & Gao, L. (2019). Identifying disengaged survey responses: New evidence using response time metadata. *Applied Measurement in Education, 32*(2), 151–165. https://doi.org/10.1080/08957347.2019.1577244

Ulitzsch, E., Pohl, S., Khorramdel, L., Kroehne, U., & von Davier, M. (2021). A response time based latent response mixture model for identifying and modeling careless and insufficient effort responding in survey data. *Psychometrika, 87*(2), 593–619. https://doi.org/10.1007/s11336-021-09817-7

Ulitzsch, E., Pohl, S., Khorramdel, L., Kroehne, U., & von Davier, M. (2023). Using response times for joint modeling of careless responding and attentive response styles. *Journal of Educational and Behavioral Statistics, 49*(2), 173–206. https://doi.org/10.3102/10769986231173607

Ulitzsch, E., von Davier, M., & Pohl, S. (2020). A hierarchical latent response model for inferences about examinee engagement in terms of guessing and item-level nonresponse. *British Journal of Mathematical and Statistical Psychology, 73*(S1), 83–112. https://doi.org/10.1111/bmsp.12188

Van Vaerenbergh, Y., & Thomas, T. D. (2013). Response styles in survey research: A literature review of antecedents, consequences, and remedies. *International Journal of Public Opinion Research, 25*(2), 195–217. https://doi.org/10.1093/ijpor/eds021

von Davier, M. (2005). *A general diagnostic model applied to language testing data*. (Research Report No. RR-05-16). Educational Testing Service.

von Davier, M. (2008). The mixture general diagnostic model. In G. R. Hancock & K. M. Samuelson (Eds.), *Advances in latent variable mixture models* (pp. 255–276). Information Age Publishing.

von Davier, M. (2010). Hierarchical mixtures of diagnostic models. *Psychological Test and Assessment Modeling, 52*, 8–28. Retrieved from http://www.psychologie-aktuell.com/fileadmin/download/ptam/1-2010/02_vonDavier.pdf

von Davier, M., & Khorramdel, L. (2013). Differentiating response styles and construct related responses: A new IRT approach using bifactor and second-order models. In R. E. Millsap, L. A. van der Ark, D. M. Bolt, & C. M. Woods (Eds.), *New developments in quantitative psychology: Presentations from the 77th annual psychometric society meeting* (pp. 463–488). Springer.

von Davier, M., Rost, R., & Carstensen, C. H. (2007). Introduction: Extending the Rasch model. In M. von Davier & C. H. Carstensen (Eds.), *Multivariate and mixture distribution Rasch models: Extensions and applications* (pp. 1–12). Springer.

Weijters, B., Geuens, M., & Schillewaert, N. (2010). The stability of individual response styles. *Psychological Methods, 15*(1), 96–110. https://doi.org/10.1037/a0018721

Weijters, B., Schillewaert, N., & Geuens, M. (2008). Assessing response styles across modes of data collection. *Journal of the Academy of Marketing Science, 36*, 409–422. https://doi.org/10.1007/s11747-007-0077-6

Wetzel, E., Böhnke, J. R., & Rose, N. (2016). A simulation study on methods of correcting for the effects of extreme response style. *Educational and Psychological Measurement, 76*(2), 304–324. https://doi.org/10.1177/0013164415591848

Wetzel, E., Carstensen, C. H., & Böhnke, J. R. (2013). Consistency of extreme response style and non-extreme response style across traits. *Journal of Research in Personality, 47*(2), 178–189. https://doi.org/10.1016/j.jrp.2012.10.010

Wise, S. L. (2017). Rapid-guessing behavior: Its identification, interpretation, and implications. *Educational Measurement: Issues and Practice, 36*(4), 52–61. https://doi.org/10.1111/emip.12165

Wise, S. L., & Kong, X. J. (2005). Response time effort: A new measure of examinee motivation in computer-based tests. *Applied Measurement in Education, 18*(2), 163–183. https://doi.org/10.1207/s15324818ame1802_2

Zehner, F., Eichmann, B., Deribo, T., Harrison, S., Bengs, D., Andersen, N., & Hahnel, C. (2021). Applying psychometric modeling to aid feature engineering in predictive log-data analytics: The naep edm competition. *Journal of Educational Data Mining, 13*(2), 80–107.

Chapter 23
Analyzing Test-Taking Behaviors Through Process Data Using the IRT Explanatory Model for Guessing

Artur Pokropek and Lale Khorramdel

Abstract Guessing behavior is a known problem in international large-scale assessments (ILSAs), and it can impact the assessment's reliability and validity by introducing random error and bias. Testing and accounting for such biases is essential in ensuring the accuracy and comparability of test scores and secondary analysis results. This chapter proposes a flexible item response theory (IRT) modeling approach for guessing behavior: the IRT Explanatory Model for Guessing (IRT-EMG). The IRT-EMG combines mixture models with response time and other process data in-formation (number of actions) from computer-based ILSAs. It presents an extension of the grade of membership response time model for detecting guessing behaviors proposed by Pokropek (*Journal of Educational and Behavioral Statistics* 41: 300–325, 2016) and is based on the grade of membership (GoM) model (Erosheva, 2002 *Grade of membership and latent structure models with application to disability survey data* (Doctoral dissertation). Carnegie Mellon University)) and the HYBRID model (Yamamoto, *HYBRID model of IRT and latent class models* (ETS Research Report RR-89-41). Educational Testing Service, 1989; Yamamoto & Everson, Modeling the effects of test length and test time on parameter estimation using the HYBRID model. In J. Rost (Ed.), Applications of latent trait and latent class models in the social sciences (pp. 89–98), Waxmann, 1997). Modeling the effects of test length and test time on parameter estimation using the HYBRID model. We show that IRT-EMG is a generalization of the classical 3PL-IRT model and that the 3PL-IRT model could be presented as a restricted case of the IRT-EMG. This chapter describes the features and advantages of the IRT-EMG model and illustrates its application on data from the Programme for the International Assessment of Adult Competencies (PIAAC). Results show how information about different characteristics of items, respondents, and process data

---

A. Pokropek (✉)
Educational Research Institute National Research Institute (IBE PIB), Warsaw, Poland

L. Khorramdel
Lynch School of Education and Human Development, TIMSS & PIRLS International Study Center at Boston College, Chestnut Hill, MA, USA

L. Khorramdel et al. (eds.), *Innovative Digital-Based International Large-Scale Assessments*, Methodology of Educational Measurement and Assessment,
https://doi.org/10.1007/978-3-031-90951-1_23

(response time, number of actions) might help detect different types of guessing behaviors and identify items that are particularly prone to guessing.

Considerations of response behavior and test-taking efforts in international large-scale assessments (ILSAs), such as the Programme for the International Assessment of Adult Competencies (PIAAC), are crucial. ILSAs present high stakes for participating countries while low stakes for the assessed students. In other words, the results of these studies are essential for decision-makers but not for the examined students, as they bear no immediate personal consequences. Behaviors like careless or insufficient effort responding (Curran, 2016; Huang et al., 2012; Huang et al., 2015), low test-taking effort (Penk et al., 2014), and guessing (Wise, 2017; Wise & Kong, 2005) can negatively impact the validity and reliability of test scores. Guessing behaviors can lead to significant biases in the estimation of item parameters and standard errors for ability estimates (Żółtak & Golonka, 2015). Moreover, guessing and related behaviors can decrease the comparability of scores across participating countries or subpopulations within countries (Burton, 2005; Pokropek, 2016; Rowley & Traub, 1977) as different populations tend to show different types and amounts of such behaviors.

This chapter deals with the problem of guessing behaviors and proposes a flexible modeling approach that allows for the differentiation between different guessing behaviors and effort responding. Guessing is a complex phenomenon that arises from the interaction between person, item, and test situation characteristics. Guessing may result from a lack of motivation to engage with an item or test, even if a respondent has the skill set to solve a given task (Wise, 2017; Wise & Kong, 2005) and by some of the authors is described as disengagement behavior (Ulitzsch et al., 2020). Guessing can also be a rational strategy of a motivated respondent who cannot solve the problem and uses guessing as a solution behavior (Chiu & Camilli, 2013; Hutchinson, 1991). Finally, guessing could be a rational strategy of a respondent who has no time to attempt to solve the problem (Schnipke & Scrams, 1997; Yamamoto & Everson, 1997).

In all cases, guessing behavior leads to the introduction of random error variance and bias (Rowley & Traub, 1977) and, as a consequence, to biased estimates of item parameters and abilities (Haladyna & Downing, 2004; O'Neil Jr. et al., 1995; Wise et al., 2010; Żółtak & Golonka, 2015). In this paper, we demonstrate a flexible modeling approach that addresses different types of guessing by accounting for both solution and rapid or disengagement guessing behavior but without the need to differentiate between them, as this is often not easy to do. Instead, they are treated as similar phenomena by broadly defining guessing as construct-irrelevant responses that are not directly related to the construct of interest. This definition can be compared to response styles in survey research (Paulhus, 1991; Rost, 2004; Van Vaerenbergh & Thomas, 2013) but is applied to guessing behaviors in cognitive tests.

The model we demonstrate to handle this broad definition is an extension of the IRT Response Time (IRT-RT) model proposed by Pokropek (2016), which is formulated to be much more flexible in detecting this generalized definition of guessing. We refer to this extended model as the IRT Explanatory Model for Guessing (IRT-EMG).

It easily incorporates a large variety of information that could be used to predict guessing, including response time (RT) and other process data (like the number of actions), item characteristics (types of items), and individual characteristics, such as background variables and ability. It allows the formulation of more complex relationships, including nonlinear relations between guessing and RT and abilities, and possible interactions between abilities and observed variables without a strong assumption on the guessing behavior's nature. Because the model can be formulated in the multilevel mixture modeling framework, the estimation of this model can be performed not only in the Bayesian framework but also using maximum likelihood estimation. Moreover, we illustrate that the IRT-EMG combines the 3PL model (Birnbaum, 1968) and a more complex mixture IRT models (Mislevy & Verhelst, 1990) by formally showing that it can be expressed as a generalization of both.

Using data from PIAAC, we will illustrate how information about different characteristics of items and respondents and process data might help detect different types of guessing behaviors, correct the ability estimates and identify items that are particularly prone to guessing.

23.1 Previous Approaches for Modeling and Handling Guessing

23.1.1 Solution-Based Guessing

As noted before, guessing can be a rational strategy of a motivated respondent who is not able to solve the problem or is not entirely sure about the correct answer. It is called ability-based or solution-based guessing behavior. This solution-based guessing has been addressed primarily in high-stakes test situations by introducing penalty scores for wrong answers (for example, in the SAT). In such approaches, test-takers are informed that for multiple-choice questions, some quantity (usually 1/k where k is the number of responses) will be subtracted from the final score if the answer is incorrect. It is assumed that this discourages test-takers if they are unsure about the correct answer (Chiu & Camilli, 2013).

Another approach is using the 3PL model for scaling, which implicitly adjusts for guessing (sometimes called pseudo guessing) by introducing a penalty to correct answers that are unlikely in terms of person and item parameters (Chiu & Camilli, 2013, p. 77). The model is used operationally in some of the ILSAs (e.g., in PIRLS and TIMSS). Hutchinson (1991) suggested that the 3PL model could be interpreted as a model composed of two processes: a solution process followed by a guessing process. A test-taker attempts to solve an item and resorts to guessing only if a correct response cannot be identified.

San Martin et al. (2006) proposed a solution-based guessing model expanding the 3PL model so that the probability of a correct guess depends on not only the item but also the ability of the individual, weighted with a general discrimination parameter. By so doing, the possibility that the individual uses his or her ability to differentiate among responses, while guessing, is also considered.

The work on efficient estimation procedures for this approach was conducted by several authors (Belov et al., 2024; Guo et al., 2021), but the substantial research on solution-based guessing is very limited. San Martin et al. (2006) found that this type of guessing depends on ability in reading tests but not on mathematics tests. To our knowledge, no further examination of solution-based guessing on empirical data was conducted, and, as far as we know, no work has been done on connecting solution-based guessing strategies to RT or other log data (which are intensely used in studies concerning other types of guessing).

23.1.2 Disengagement Guessing

It is assumed that motivated respondents take sufficient time to actively try solving an item (solution behavior), while respondents with lower motivation show faster responses and more omissions as they engage less with the test. Therefore, recent studies have used RT information obtained from computer-based assessments to measure rapid guessing and careless responding in cognitive (Goldhammer et al., 2016; Lee & Jia, 2014; Ulitzsch et al., 2020) and non-cognitive scales (Kroehne et al., 2019). Results of these studies show that rapid guessing and insufficient effort responding can indeed be related to shorter RTs and higher omission rates and can successfully be used to differentiate between engaged and disengaged students. Moreover, there is reason to assume that the number of actions per item can be helpful in differentiating between engagement and disengagement (Khorramdel et al., 2019). Different approaches exist for utilizing RT to account for engaged versus disengaged responses. Generally, RT-based scoring methods and model-based approaches can be distinguished.

RT-based scoring methods use RT thresholds where RTs below a certain threshold are considered to be rapid guessing. They either define a common threshold for all items representing the minimum amount of time needed to give an engaged response (Wise et al., 2004) or item-specific thresholds. Item-specific thresholds can be established by visually assessing bimodal RT distributions for a distinctive gap (Wise et al., 2009) or by assessing RT distributions jointly with the correct conditional proportion. The latter aims to identify an RT threshold at which accuracy exceeds what would be expected from random responding (Goldhammer et al., 2016; Guo et al., 2016; Lee & Jia, 2014). However, such techniques for identifying disengaged behavior might considerably vary depending on the utilized threshold methods (Goldhammer et al., 2016; Lee & Jia, 2014).

In mixture IRT models, different latent classes of examinees are assumed. Usually, two latent classes are considered: engaged or solution-based behavior (ability) and guessing behavior. Responses from the latent guessing class are assumed to contain no information about the examinees' abilities. Usually, mixture models make strong assumptions about the response process. Some models allow mixture proportions at the item level (the proportion of disengaged responses vary across items but is constant across examinees; Schnipke & Scrams, 1997), some models allow mixture proportions at the examinee level (the proportion of disengaged responses varies across examinees, but is constant across items; Cao & Stokes, 2008; Mislevy & Verhelst, 1990; Wang & Xu, 2015). RT can be included in a mixture of IRT modeling approaches (List et al., 2017; Meyer, 2010; Ulitzsch et al., 2020) based on the further development of the speed-accuracy model (van der Linden, 2007).

The RT-based scoring methods and the mixture modeling approaches usually assume that ability and guessing behavior are unrelated constructs. While RT-based scoring methods assume that responses identified as guessing are ignorable and code them as missing (Pokropek, 2016; Rios et al., 2017; Rubin, 1976), mixture modeling approaches model the probability of being engaged independently of the latent ability. However, some research suggests that disengagement and guessing behavior are related to ability (Boe et al., 2002; Braun et al., 2011; Goldhammer et al., 2016; Wise et al., 2009; Wise & DeMars, 2005) and that ignoring this relation can lead to biased ability estimates (Rios et al., 2017).

Most recent approaches propose a hierarchical latent response model for jointly identifying and modeling examinee disengagement and engagement (Nagy & Ulitzsch, 2022; Ulitzsch et al., 2020). Disengaged behavior is conceptualized in terms of both random guessing and omitting responses. This approach utilizes mixture models that identify disengagement at the item-by-examinee level by assuming different data-generating processes underlying item responses, omissions, and RTs associated with engaged and disengaged behavior, resulting in different distributions.

23.1.3 Guessing in Tests with Time Limits

Finally, we can distinguish a third type of guessing behavior that occurs when respondents have no time to attempt to solve an item. Guessing behavior due to time restrictions was mainly investigated using the HYBRID model (Yamamoto, 1989) and further extensions based on it (Bolt et al., 2002; Jin & Wang, 2014).

The HYBRID model was developed to handle test-taking behavior that occurs under conditions of test speediness in high-stakes tests and assumes two latent classes: an engaged or solution behavior class and a guessing class, where belonging to each class is defined by a threshold. This threshold is estimated for each respondent and determines the starting point of guessing behavior. The assumption is that respondents are switching from an engaged response to guessing if time pressure occurs, and once respondents start guessing, they continue this behavior until the end of the test.

Bolt et al. (2002) modified the HYBRID model by substituting the guessing class with a decline class where abilities are still related to the probability of a correct response, but higher item difficulties characterize the decline class. The model was first proposed as a mixture Rasch (1960) model but was further developed for the 2PL model (Cao & Stokes, 2008; Suh et al., 2012). Jin and Wang (2014) proposed an extension of the HYBRID model that is based on a multiclass mixture model. In their approach, a gradual decline in test performance is modeled. The last model is especially interesting but relies on many assumptions, including that the decrement function is the same for all test takers with the same switching point and that the decrement is minor when performance decline develops later in the test.

23.1.4 Limitations of the Previous Models

Already developed models have some limitations. Firstly, they concentrate on rapid guessing (e.g., Nagy & Ulitzsch, 2022; Pokropek, 2016) not accounting for solution-based guessing, which also might affect the reliability and validity of the scores. Second, almost all current guessing model applications rely on RT and abilities (e.g., Ulitzsch et al., 2020). Third, most models have rather strong assumptions about the nature of guessing, either assuming all examinees to be equally prone to show disengaged behavior (e.g., IRT 3PL model; Birnbaum, 1968) or that the proportion of disengaged responses is constant across items (e.g., IRT Mixture modeling; Mislevy & Verhelst, 1990). Only the most recent approaches (Nagy & Ulitzsch, 2022; Pokropek, 2016; Ulitzsch et al., 2020) allow for item-by-examinee interactions in guessing behaviors. The current study fills this gap by accounting for rapid and solution-based guessing and modeling these behaviors using different variables and process data such as RT and the number of actions.

23.2 The IRT-EMG Model

The IRT-EMG model proposed in this study is an extension of the mixture IRT model that was proposed by Pokropek (2016). The model can be seen as a combination of the Grade of Membership (GoM) Model (Erosheva, 2002) and the HYBRID model (Yamamoto, 1989; Yamamoto & Everson, 1997). It allows for the use of various process and log-data and other test-taker-level variables (such as gender, language spoken at home, etc.) in addition to response data for the estimation of latent classes with the aim to differentiate between different types of guessing behavior and effort responding. Moreover, the proposed approach can be seen as a generalization of the 3PL IRT model (Birnbaum, 1968).

In most measurement applications, the latent classes of latent class analysis (LCA) models are defined at the test-taker level, where Y_{ij} is the $i-$th measurement for individual j, and a categorical latent variable C_j is defined for each individual. The distribution of Y_{ij} is given by

$$P(Y_{ij}=1|C_j=c)=f_c(y_{ij}|\psi_c) \tag{23.1}$$

where $f_c(y_{ij}|\psi_c)$ is the class-specific response function with a vector of parameters ψ_c. Here, each individual is assumed to arise from a set that is a mixture of

C unobserved classes of unknown proportions π_c where the proportions of latent classes in each response are greater than 0, and their sum is 1. With such an approach, different latent classes could have different item parameters. For instance, one class could be defined as a solution behaviors response class (e.g., any simple-structure IRT model) and a second class as a guessing class where the probability of guessing is item-specific or constant across items (Mislevy & Verhelst, 1990). In such specifications, test-takers are assumed to belong to one class, either the solution behavior class or the guessing class, and they respond to all items according to their class belonging. This is similar to the HYBRID model, but the IRT-EMG model allows for more complex latent class specifications and a more flexible solution where respondents can switch their behavior (latent class membership) freely between items.

This could be formulated in terms of the GoM model, where each respondent can belong to both the engaged and the guessing class and show different behaviors for different items. To achieve this, the LCA formulation can be changed by defining latent classes based on a categorical latent variable that would be item specific similar to the GoM model (Erosheva, 2002):

$$P(Y_{ij}=1|C_{ij}=c)=f_c(y_{ij}|\psi_c) \tag{23.2}$$

The change between Eqs. (23.1) and (23.2) is small and based on changing the categorical latent variable C_j into C_{ij}, but the consequences are large. In this formulation, some responses might behave differently than others, but they will follow the model defined for one of the distinct classes. Each individual response is assumed to arise from a set that is a mixture of C unobserved classes of unknown proportions πc, assuming that the proportions of latent classes in each response are greater than 0 and their sum is 1. The model allows respondents to switch from one class to another freely.

As stated above, we assume two latent classes (1) solution behavior class and (2) guessing class. We assume that all observed variables are binary for simplicity, but a generalization for categorical data could also be applied. In the current application, the two-parameter logistic (2PL) model was chosen to represent the solution behavior class:

$$P(Y_{ij}=1|C_{ij}=1)=\frac{exp(a_i(\theta_j-\beta_i))}{1+exp(a_i(\theta_j-\beta_i))} \tag{23.3}$$

where θ_j is a continuous latent variable that reflects the ability of the respondent, a_i is the discrimination parameter (or loading), and β_i is the item difficulty parameter. The guessing class would be defined by:

$$P(Y_{ij}=1|C_{ij}=2)=1 \tag{23.4}$$

and

$$P(C_{ij}=2)=\pi_{ij} \tag{23.5}$$

where π_{ij} is the probability of belonging to a latent class that represents guessing. This might be interpreted as a response-specific guessing parameter for response i of test-taker j. This guessing parameter varies across all responses and between all items. This response-specific guessing parameter could be related to conditioning variables similarly to applications of LCA models (Dayton & Macready, 1988; Kim et al., 2016; Smit et al., 1999, 2000) but is also accounting for the multilevel structure of the data:

$$P(C_{ij}=2 \mid Z_{ij}, W_j, \theta_j)=\pi_{ij\mid zw\theta}=g(Z_{ij}, W_j, \theta_j, \gamma, \delta, \eta) \tag{23.6}$$

where $=\pi_{ij \mid zw\theta}$ is the proportion of the responses in the guessing class conditional on the item level variables (Z_{ij}; e.g., response time), respondent level variables (W_j; e.g., gender) and respondent-level latent variables (θ_j; abilities and their transformations such as squared terms).

The proposed model could also use interactions (including cross-level interactions and interactions with latent variables) not explicitly listed in the equation. Vectors γ, δ, η contain vectors of parameters that define the $g()$ function and refer to item-level variables, individual-level variables, and latent abilities (θ), respectively. Because $\pi_{ij \mid zw\theta}$ is defined on the [0,1] interval, the logistic function is a good choice for modeling the probability of guessing with explanatory variables:

$$\frac{\pi_{ij\mid zw\theta}}{1-\pi_{ij\mid zw\theta}}=a_{ij}+Z_{ij}\gamma+W_j\delta+\theta_j\eta+u_j \tag{23.7}$$

a_{ij} is the latent class prediction threshold that is the log odds of being in the guessing class versus the solution behavior class when all covariates are equal zero and u_j is a normally distributed person level error.

With this increased flexibility, the proposed approach can be used to model more complex relations between guessing and various variables, such as different process and log data (like RT and number of actions), different item characteristics (e.g., item types) or individual characteristics of respondents. The conditioning variables are at the heart of the proposed model because a correct prediction of latent class belonging might be difficult without additional information.

Conditioning variables make this task significantly more manageable, providing more information about class membership. With conditioning variables, the multinomial distribution of the class variable will be more concentrated around certain latent classes than the overall distribution (von Davier & Yamamoto, 2014, p. 110). There is, however, a caveat because the conditioning variables are "defining" the latent class solution. Therefore, model misspecifications might affect the estimation outcomes.

Interestingly the proposed model could be presented in a form that would make it similar to the IRT modeling convention. The probability of a correct response that partially belongs to two classes might be written as:

$$P(Y_{ij}=1|C_{ij}=1,2)=P(C_{ij}=2)*P(Y_{ij}=1|C_{ij}=2)+P(C_{ij}=1)*P(Y_{ij}=1|C_{ij}=1) \tag{23.8}$$

For the conditional probability formulation, we would write:

$$P(Y_{ij}=1|C_{ij}=1,2)=\pi_{ij\,|\,zw\theta}*P(Y_{ij}=1|C_{ij}=2)+(1-\pi_{ij\,|\,zw\theta})*P(Y_{ij}=1|C_{ij}=1) \tag{23.9}$$

and simplify Eqs. (23.3) and (23.4):

$$P(Y_{ij}=1|C_{ij}=1,2)=\pi_{ij\,|\,zw\theta}+(1-\pi_{ij\,|\,zw\theta})*\frac{exp(a_i(\theta_j-\beta_i))}{1+exp(a_i(\theta_j-\beta_i))} \tag{23.10}$$

The proposed IRT-EMG model could be expressed as a generalization of the 3PL model with a respondent-by-item varying conditional guessing parameter. It is also worth noticing that the 3PL model could be understood as a special case of the IRT-EMG model without conditional variables, and the probability of guessing is fixed to be the same for each item. Alternatively, the proposed IRT-EMG could be treated as a generalization of the 3PL-IRT model.

23.3 Application on PIAAC Data

23.3.1 Data

For a demonstration of the model, we use PIAAC data from three countries: Italy (ITA), Japan (JPN) and the United Kingdom (UK). Sample sizes are presented in Table 23.2. We chose those countries to represent higher (JPN), middle (UK), and lower (ITA) performing countries according to the PIAAC study. The data is available on the OECD PIAAC website (https://www.oecd.org/skills/piaac/).

PIAAC has two main components: a background questionnaire and a cognitive assessment of literacy, numeracy, and problem-solving in technology-rich environments (PSTRE). Both are administered to a target population of adults between the ages of 16 and 65. PIAAC was the first computer-based international large-scale assessment, although individuals who lacked familiarity with a computer (or a willingness to sit a test with a computer) were offered a paper-based version of the test.

The original PIAAC scale was concurrently estimated for data from the computer-based and paper-based assessments using a 2PL model. In this study, only items from the computer-based assessment were used. Data from respondents completing paper booklets were excluded from the analysis, as we wanted to utilize two different kinds of process data indicators that are only available for the computer-based assessment: response time (RT) and number of actions.

As Nagy and Ulitzsch (2022) pointed out, Pokropek's (2016) initial model implies that if a person responds more quickly to one question than another, they are more likely to be disengaged in this item. However, since questions can take different times to answer and engagement levels vary by question, this might not be a plausible assumption, even for rapid guessing. Nagy and Ulitzsch (2022) adjust the model to consider these differences by setting item-specific time thresholds. In the model discussed in this chapter, we introduce an even more straightforward approach to address this issue by utilizing standardized response time measures.

Because the RT variable is highly skewed, we used the inverse hyperbolic sine (IHS) transformation to standardize each variable separately. The number of actions variable was re-coded into a binary variable that takes the value of 1 if more than the minimum number of actions necessary to answer a question correctly was used and 0 if the number of actions was equal to or less than the minimum. The results of the transformations are presented in Fig. 23.1.

A U-shaped curve relation between guessing behavior and RT is expected. We assume that short response times indicate rapid guessing, while longer RTs indicate solution-based guessing and uncertainty of the correct answer. It is also expected that respondents who show solution-based guessing show a higher number of actions, as they are not sure of the correct answer and, therefore, change their responses. If only a few actions are taken, we would expect rapid guessing.

We also utilized the characteristics of the items for the analysis. For literacy items, we differentiated between "click items" (coded as 1) and text or numeric entry (coded as 0), and between "highlighting items" (coded as 1) and text or numeric entry (coded as 0). Numeracy PIAAC assessment does not have highlighting items. The reference category is text or numeric entry items. The rationale behind this categorization stems from the hypothesis that highlighting items, which often require selecting specific portions of text, could also be easily subject to guessing behaviors. Participants might randomly select parts of the text without engaging deeply with the item's content, especially under time pressure or if the item appears difficult at first glance. This guessing strategy could significantly distort the assessment's measurement of their true literacy skills. Similarly, for text or numeracy items that require entering a number or text, we also expect some level of guessing, particularly when the test-takers perceive the answer format as straightforward. For instance, if an item requires simply entering a number, test-takers might attempt to guess this number. This form of guessing might be less prevalent than clicking or highlighting items due to the necessity of generating an answer rather than selecting from given options. However, it remains a potential issue, especially in cases where the respondents might try to guess the answer without fully reading or understanding the item stimulus. Even with text entry, where crafting an answer requires more effort,

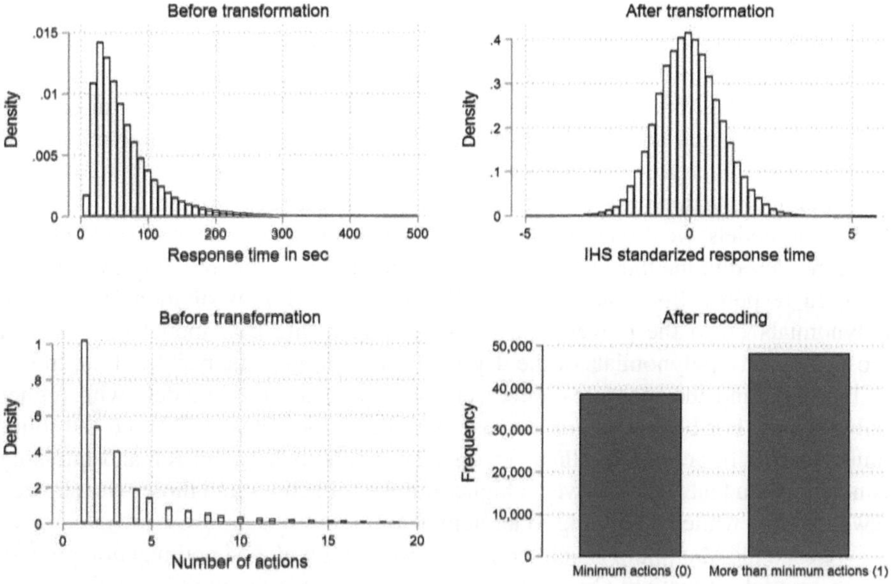

Fig. 23.1 Response time (RT) and number of actions before and after transformation, pooled data for all items

guessing could still occur if test-takers input common or generic responses in hopes of accidentally matching the correct answer.

Moreover, we used individual-level variables: gender (we used 1 for male and 0 for female), age (standardized), still in education (1 is used for respondents who are still in education, 0 is used for everyone else), years of completed education (standardized) and a binary indicator that identifies whether the native language is different from the language of the test.

For the presented analysis, we excluded all items that had less than 60 cases (i.e., the number of correct or incorrect responses for an item needs to be higher than 60).

23.3.2 Specification end Estimation of the Model IRT-EMG

We used the specification as described in Eqs. (23.2–23.10) to estimate the IRT-EMG model. The 2PL-IRT model defined the effort response class, while in the guessing class, the model assumes that responses were unrelated to the ability (similar to the HYBRID model). We used a set of conditional variables to predict the latent class membership. We used all observed variables described in the previous section and additional interaction terms for literacy. For numeracy, we have added additional latent abilities as conditioning variables with additional interaction terms with time. With this addition, we explicitly model the relation between guessing and ability. This interaction term allows to model the situation where the relation between RT and the propensity of guessing differ with different levels of abilities.

Models with latent abilities and interaction between abilities and response time for guessing class prediction for the domain literacy have problems with convergence, and we could not obtain stable and robust results even with a high number of random starts. Therefore, for literacy, we used simplified specifications that excluded ability and interaction terms, which brought no convergence problems.

For all models, we used third-degree polynomials for RT (time, root squared time and time raised to the third power). This was done to allow more complex relations between response time and guessing. We estimated dozens of models with the polynomials up to the power of five. According to the BIC model fit index, in most situations, polynomials of the degree of three provide the best fit of the model to the data, and the measures were very similar to those of models with higher polynomials. For both domains, we use the interaction between RT and the minimum number of actions to allow for different relations between RT and guessing concerning students who showed a higher number of actions and those who showed fewer actions while responding to an item.

We used the Maximum Likelihood (ML) estimator with EM optimization process for the model estimation. One of the biggest problems is the multi-modality of the likelihood when the algorithm only finds local modes. To avoid this problem, we used an algorithm that randomized the starting values for the optimization routine using the three-stage optimization proposed by Asparouhov and Muthèn (2019) and implemented it in Mplus version 8.4. An initial set of 800 random starting values was selected during the first stage estimation. Then, a partial optimization with 10 EM iterations for each starting value was completed. In the second stage, the continued estimation was based on 200 random starts with the best log-likelihood values obtained in the first stage. In the third stage, the first ten optimizations are completed until convergence. The algorithm continues with the estimation until the 75th iteration for the remaining optimizations. The optimization can continue only if a log-likelihood value is obtained at the 75th iteration mark, which is better than any of the top ten log-likelihoods at the 75th iteration mark to that point. This algorithm rejects underperforming optimizations and, thus, reduces the computational burden while preserving the efficiency (for details, see Asparouhov & Muthèn, 2019, p. 5). The user can set the number of iterations to be performed at each stage. The numbers 800, 200, and 10 were proven to be efficient and robust in many applications of the proposed model on PIAAC data but could be adjusted for other applications.

23.4 The IRT-EMG Model in Mplus

One of the advantages of the proposed model is that it can be implemented in different existing software. The software needs to allow for the estimation of multilevel mixture models for binary outcomes. One such software is Mplus (Muthèn & Muthèn, 1998–2017), a popular computer program for latent variable modeling. In Exhibit 23.1 we present the syntax that could be used to estimate the IRT-EMG model presented in this chapter. The syntax reflects the model specification for the numeracy model (which includes latent abilities and interactions between abilities and RT). The syntax in Exhibit 23.1 is simplified for illustration purposes to include only five items. Additional variables referring to the additional items should be added to the syntax to include more items. For example, in the case of ten items, we would have `resp ON item2* item3* item4* item5* item6* item7* item8* item9* item10*;` instead of `resp ON item2* item3* item4* item5*;`. Similarly, nine interactions (s2–s10) should be specified instead of four (s2–s5).

The data for the presented model should be prepared in a long format, like in the example presented in Table 23.1. In this setting, one record (row) is uniquely identified by respondent ID and item ID. The variable "response" defines whether the response to an item was correct (1) or incorrect (0). The "indicator variables"

```
Variable:
        Names are id resp time acs AGE FEMALE click
              item2 item3 item4 item5 time2 time3 acs1 acs2;
        MISSING are all (-9999) ;  ! Missing variables
        CATEGORICAL = resp;        ! Response variable is categorical
        CLUSTER = id ;             ! Cluster is defined by id
        WITHIN = item2-item5 time acs time2 time3 acs1 acs2 CW click  ;
        BETWEEN=AGE FEMALE ;       ! Setting within and between vars.
        CLASSES = CW(2);           ! Two latent classes are specified.

Analysis:
      TYPE=TWOLEVEL MIXTURE RANDOM;
      PROC=8; Integration=7; STARTS=800 200 10; ESTIMATOR = ML; ! Estimation settings

Define:
      time2 = time*time; time3 = time*time*time;  ! Define interactions
      acs1 = acs*time; acs2 = acs*time*time;      ! between observed variables

Model:
      %WITHIN%
              %OVERALL%
              abi_tim | f XWITH time; ! Define interactions between observed and latent
              s2 | f XWITH item2;     ! variables. Abi_tim - interaction between ability
              s3 | f XWITH item3;     ! and time s2-s5 - item slopes
              s4 | f XWITH item4;
              s5 | f XWITH item5
              resp ON item2-item5;
             resp ON s2-item5;

             cw#1 ON time time2 time3  ! Predictors of guessing on within level
                  acs acs1 acs2 click abi_tim;

              %CW#1%
              resp ON item2* item3* item4* item5*; ! Measurement model for response class
             resp ON s2* s3* s4* s5*;              ! item slopes and intercepts = IRT 2PL

             %CW#2%
             resp ON item2@0 item3@0 item4@0 item5@0;!Measurement model for guessing class

      %BETWEEN%
             %OVERALL%
             f BY resp;

            cw#1 ON AGE FEMALE;                    ! Predictors of guessing on individual
             f@1;                                    ! level

              %CW#1%
              f BY resp*; [ f@0 ]; [resp$1*];    ! Response class
              %CW#2%
              f BY resp*; [ f@0 ]; [resp$1@15]; ! Definition of correctly guessing
```

Exhibit 23.1 Mplus 8.11 syntax for IRT-EMG model estimation

Table 23.1 Example of dataset in long format

ID variables		Response	Indicator variables			Conditioning variables	
Respondent	Item	Correct (1) Incorrect (0)	Item2	Item3	(...)	RT	Female (1) Male (0)
1	Item1	1	0	0	(...)	1.234	1
1	Item2	1	1	0	(...)	3.455	1
1	Item3	0	0	1	(...)	4.533	1
(...)	(...)	(...)	(...)	(...)	(...)	(...)	(...)
2	Item1	1	0	0	(...)	2.111	0
2	Item2	0	1	0	(...)	1.021	0
2	Item3	0	0	1	(...)	0.939	0

define item indicators. We need indicators for all items except one to formulate the model properly. Such indicators take the value one if the records correspond to the item and zero if not. In our example, the first item is treated as a reference; therefore, a binary indicator is not necessary (all other indicators define the item without an indicator as a reference). Note that different ways of coding indicator variables exist (Hardy, 1993) and could be applied. Finally, in the last columns, "conditional variables" are specified both on the response level (e.g., RT) and on the respondent level (e.g., gender).

23.5 Results

Table 23.2 summarizes the basic statistics of the estimated IRT-EMG models. We estimated six models using samples from three countries (ITA, JPN, UK) for two domains (literacy, numeracy). In the third column, we present the percentage of correct responses. This could be treated as raw indicators of the ability levels in the countries. The next column is the entropy index, a classification measure representing the classification's precision. The entropy ranges from 0 to 1. Entropy with value 1 indicates clear differentiation of classes (all respondents have a probability equal to one of being in one class). Entropy with a value of 0 indicates that the probabilities of being assigned to a class are constant for all subjects (Celeux & Soromenho, 1996). According to Geiser (2013), values lower than 0.8 indicate problematic categorization and values above 0.80 indicate good classification accuracy. The following two columns estimated the proportions of guessing behaviors according to posterior probability and most likely memberships. The final columns report the number of responses and respondents that are used to estimate the model.

The entropy values range from 0.858 (literacy, JPN) to 0.705 (numeracy, ITA). Entropy is higher for literacy than numeracy. Values higher than 0.8, considered the threshold for accurate classification, were reached in half of the cases but, in all situations, were higher than 0.7. The results indicate that the proposed model effectively flagged guessing responses in some situations, although the performance,

Table 23.2 Summary of the six IRT-EMG models

Domain	Country	Percentage of correct responses	Entropy	Guessing class membership		Number of	
				Posterior probability	Most likely membership	Responses	Respondents
Literacy	Japan	67.5%	0.858	3.7%	1.7%	41,730	2183
	UK	60.0%	0.839	5.0%	1.9%	81,277	4336
	Italy	52.9%	0.754	8.1%	3.2%	46,898	2733
Numeracy	Japan	76.7%	0.844	7.8%	5.3%	42,987	2229
	UK	64.7%	0.720	14.0%	10.8%	83,584	4427
	Italy	62.8%	0.705	12.2%	5.9%	49,538	2765

as indicated by the entropy, was not universal and depends on the properties of the data.

The estimated proportion of guessing (according to the posterior probability indicator) ranges from 3.7% (literacy, JPN) to 14% (numeracy, ITA). In the presented study, the IRT-EMG model detects more guessing behavior in groups with higher proficiency than in groups with lower proficiency. However, the presented study only examines data from three countries. Hence, these results need to be interpreted with caution, and additional studies with more countries are needed.

Table 23.3 presents the results for the domain literacy of the part of the model that is used to predict guessing behavior as described in Eq. (23.6). Coefficients are provided in logistic and odds ratio metrics (in brackets). The intercepts reflect the propensity of guessing in each group (lowest in Japan, highest in Italy). Coefficients for RT, RT2, and RT3 describe the relation between the probability of guessing and RT. The third-degree polynomial is difficult to interpret; therefore, in Fig. 23.1, we present a graphical inspection of the relation between probability of guessing and predicted RT. In the upper panel of Fig. 23.1, the estimated probability of guessing is plotted separately against standardized RT for the three countries. The picture is consistent for all three countries, showing a U-shaped relationship. The probability of guessing is highest for very short RTs and lower for longer RTs (i.e., for responses with an RT slightly lower than the average RT), the probability of guessing increases and becomes relatively high for long RTs. We believe that this shape represents rapid guessing (for short RT) and solution-based guessing (for long RT).

Figure 23.2 also depicts the effect for when the number of actions a respondent took exceeds the number of minimum actions necessary to correctly solve an item and the interaction of these indicators with RT. The middle panel of Fig. 23.2 represents the relation between the estimated probability of guessing and RT for all responses where a minimum number or less than a minimum number of actions were taken to give a response. The lower panel represents the relation between the predicted probability of guessing and RT for responses where more than a minimum number of actions necessary for a correct response were taken. The results indicate that, for responses with a minimum number of actions (or fewer), practically only rapid guessing exists, while for more than the minimum number of actions needed for a correct response, two types of guessing behavior occur. This confirms our expectations. In solution-based guessing, respondents are unsure of the answer, so they might change their responses (showing somewhat experimental behavior). If only a few actions are taken, we would expect rapid guessing.

Parameters of variables "Click items" (in odds ratio metric JPN = 3.093; UK = 1.241; ITA = 3.737) and "Highlight items" (JPN = 2.268; UK = 1.640; ITA = 3.977) were included to describe the effect of item type on the probability of guessing behavior. The reference category for those two binary indicators is a text/numeric entry. The results are consistent for all countries and follow expectations (only results for click items for the UK are not statistically significant). Items where the respondent is asked to click on the answer or highlight parts of the text are much more prone to guessing than text entry items where some unique text or number must be entered.

Table 23.3 Literacy. Logistic regression coefficient from the model used to predict guessing behaviors (effort responding as reference). Coefficients are presented in logit metric; odds ratio are presented in brackets

	JPN	UK	ITA
	Logit (odds ratio)	Logit (odds ratio)	Logit (odds ratio)
Intercept (guessing class)	−5.961** (0.003)	−4.700** (0.009)	−4.603** (0.010)
Within level variables (response level)			
Response time	0.708** (2.030)	0.574** (1.775)	0.332** (1.420)
(Response time)^2	0.498** (1.645)	0.450** (1.568)	0.351** (1.455)
(Response time)^3	−0.107** (0.899)	−0.115** (0.891)	−0.093** (0.911)
More than minimum actions	1.283** (3.607)	1.362** (3.904)	0.947** (2.578)
Response time * (More than minimum actions)	0.226* (1.253)	0.136* (1.146)	0.012 (1.012)
(Response time)^2 * (More than minimum actions)	−0.159** (0.851)	−0.057* (0.924)	0.013 (1.013)
Click items (vs text/numeric entry)	1.129** (3.093)	0.216 (1.241)	1.214** (3.737)
Highlight items (vs text/numeric entry)	0.819* (2.268)	0.495** (1.640)	1.355** (3.977)
Between level variables (respondent level)			
Age (standardized)	−0.070 (0.932)	0.004 (1.004)	0.127* (1.127)
Gender (male reference category)	0.146 (1.147)	0.122 (1.130)	0.120 (1.080)
In education	0.329 (1.389)	−0.140 (0.869)	−0.615** (0.541)
Test language different from first language	0.320 (1.377)	0.202 (1.223)	−0.070 (0.932)
Years of completed education (standardized)	−0.273** (0.761)	−0.184** (0.832)	−0.629** (0.533)

Note: ** $p > 0.01$ * $p > 0.05$

Finally, the last five rows in Table 23.3 present the effects of individual-level variables (between-level variables) on guessing behavior. The strongest and most consistent effect is shown for the variable "years of completed education." For all investigated countries, the variable "years of completed education" was shown to be significantly and negatively related to the probability of guessing (in logistic metric: JPN = −0.273; UK = −0.184; ITA = −0.629). Moreover, in Italy, whether one is currently "in education" is significantly related to a lower probability of guessing (−0.615). All other predictors are small and insignificant ($p > 0.05$).

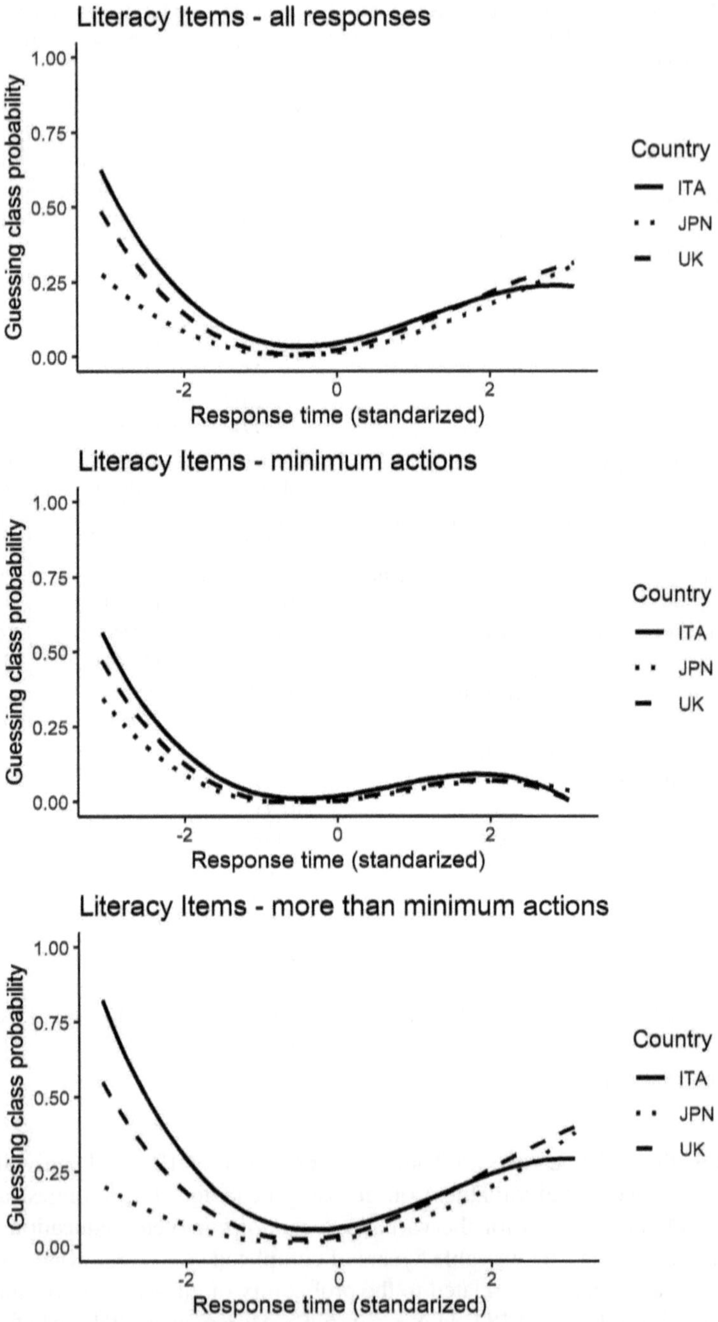

Fig. 23.2 Literacy items. Relation between estimated probability of guessing behaviors and RT for all responses (upper panel), responses with a minimum number of actions necessary for a correct response (middle panel), and responses with more than a minimum number of actions necessary for a correct response

Table 23.4 Numeracy. Logistic regression coefficient from the model used to predict guessing behaviors (effort responding class as reference). Coefficients are presented in logit metric; odds ratio are presented in brackets

	JPN	UK	ITA
	Logit (odds ratio)	Logit (odds ratio)	Logit (odds ratio)
Intercept	−3.188	−3.120	−3.120
Within level variables (response level)			
Response time	0.037 (1.038)	−0.184** (0.832)	−0.482** (0.618)
(Response time)^2	0.244** (1.276)	0.177** (1.194)	0.080* (1.083)
(Response time)^3	−0.034** (0.967)	−0.017** (0.983)	0.013 (1.013)
More than minimum actions	0.372* (1.451)	0.463** (1.589)	0.687** (1.988)
Response time * (More than minimum actions)	0.051 (1.052)	0.104 (1.110)	0.350** (1.419)
(Response time)^2 * (More than minimum actions)	−0.037 (0.964)	−0.006 (0.994)	−0.048 (0.953)
Click Items (vs text/numeric entry)	−0.221** (0.802)	−0.216** (0.806)	−0.157** (0.855)
Between level variables (respondent level)			
Ability	−0.974** (0.378)	−0.957** (0.384)	−0.695** (0.499)
Ability*Time	0.218** (1.244)	0.308** (1.361)	0.247** (1.280)
Age (standardized)	0.101* (1.106)	−0.015 (0.973)	−0.012 (0.985)
Gender (male reference category)	0.172 (1.188)	0.179 (1.196)	0.168 (1.184)
In education	0.038 (1.039)	0.025 (1.025)	−0.040 (0.961)
Test language different from first language	1.184 (3.267)	0.739** (2.094)	0.020 (1.020)
Years of completed education (standardized)	−0.394** (0.674)	−0.479** (0.619)	−0.502** (0.605)

Note: ** $p > 0.01$ * $p > 0.05$

Results for the numeracy domain are presented in Table 23.4 and Fig. 23.3. The model for numeracy differs slightly from the literacy model as it includes ability and interaction between ability and response time as a predictor for guessing behaviors. The results are consistent among countries; respondents with higher abilities, on average, exhibit guessing behaviors less than those with low abilities. According to the model, the predicted odds for guessing decrease significantly when the ability is higher by one standard deviation. More precisely, the odds are 0.378 lower in Japan, 0.384 in the UK, and 0.499 in Italy.

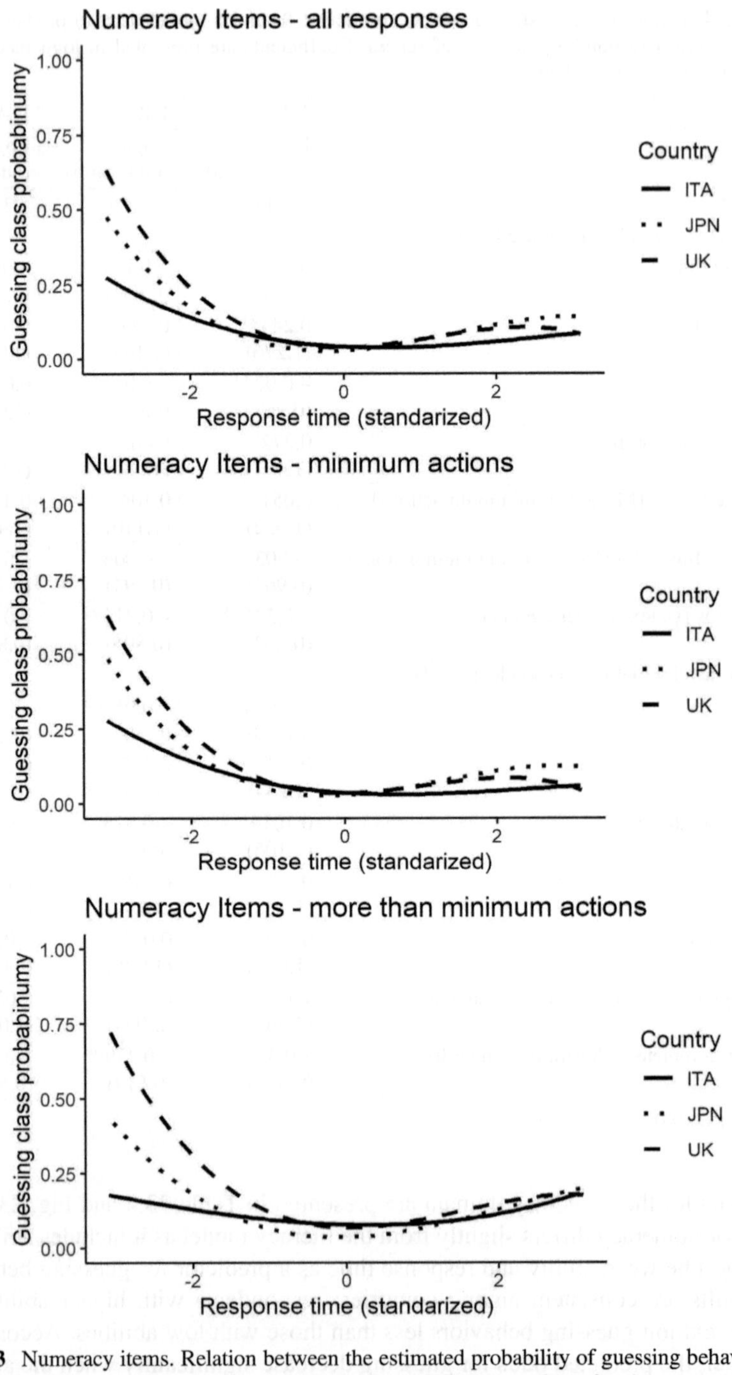

Fig. 23.3 Numeracy items. Relation between the estimated probability of guessing behaviors and RT for all responses (upper panel), responses with a minimum number of actions necessary for a correct response (middle panel), and responses with more than a minimum number of actions necessary for a correct response

Additionally, we observe significant and positive interaction with time: for long response times, the effect of abilities on probability of guessing is smaller. This suggests that the effect of abilities is higher for rapid guessing than for solution guessing. In other words, low-ability students exhibit rapid guessing behaviors more often than high-ability students.

The shape of the relations between RT and guessing in numeracy described by model parameters is depicted in Fig. 23.3. The results are similar to literacy, but the relations are less steep. RT for numeracy seems to be a less effective predictor of guessing behavior than literacy. However, respondents with RTs closer to the average RT also have a lower estimated probability of guessing here.

An unexpected result was observed for click items, where the odds of falling into the guessing class are lower than for text entry items. This challenges conventional assumptions about guessing behavior in assessment contexts. One possible explanation is the complexity of click items used in PIAAC, which often present test takers with more complex scenarios through which they need to navigate. Numeric entry items, in contrast, may be simpler and relatively straightforward as they only require to extract information from a text and enter a number into the response box. However, further investigation is needed to explore this finding.

All other results for numeracy are very similar to the results for literacy. The variable "more than the minimum of actions" is positively related to the probability of guessing. However, the interactions between RT and the minimum number of actions are significant only for Italy. Similar to literacy, only one respondent level-variable is significantly related to guessing among all countries: "years of completed education." The presented models predict that one standard deviation change in "completed years of education" would result in approximately 60% decrease in odds of guessing (JPN = 0.674; UK = 0.619; ITA = 0.605).

23.6 Discussion

The presented IRT Explanatory Model for Guessing (IRT-EMG) model is a generalized approach for modeling guessing behavior which combines IRT mixture modeling (e.g., HYBRID model) and classical 3PL modeling (Birnbaum, 1968; see also San Martin et al., 2006), and that allows the integration of new types of data from computer-based assessments such as process- and other para-data, which provide additional information about the response behavior. Moreover, it offers the advantage of modeling different guessing behaviors, including rapid and solution-based guessing. The model's capacity to integrate such diverse data types underscores its flexibility and the breadth of its applicability, enabling researchers to explore the multifaceted nature of guessing behaviors with unprecedented depth and precision.

The model shows its analytical capabilities and operationalization advantages, offering compatibility with widely used statistical software packages, such as Mplus, and possible estimation via maximum likelihood methods. Such accessibility allows a broader research community to explore the nuances of guessing behavior within large-scale assessments.

The IRT-EMG model's application on PIAAC data highlighted insights into guessing behaviors across literacy and numeracy assessments in Japan, the UK, and Italy. The model revealed variability in guessing probabilities across countries. Entropy indexes suggested adequate differentiation between guessing behaviors, with a notable U-shaped relationship between response time and guessing probability, indicating both rapid and solution-based guessing. Item types, such as "click" and "highlight" tasks, were more susceptible to guessing than text/numeric entry items. Additionally, respondent characteristics, particularly years of education, emerged as significant predictors of guessing, with higher education levels associated with lower guessing probabilities. This analysis underscores the complexity of guessing behaviors and the efficacy of the IRT-EMG model in capturing these nuances through detailed process data.

Despite these strengths, refining and optimizing the IRT-EMG model is ongoing. The complex landscape of modeling guessing behaviors demands computational rigor and a solid theoretical underpinning to guide the selection and interpretation of conditional variables critical to the model's structure. The delicate balance between model specificity and the risk of misspecification, particularly concerning polynomial degrees and the choice of conditional variables, presents a nuanced challenge. Selecting conditional variables for this model is not straightforward, and while model fit indices like the BIC or AIC measures can be used, they have known limitations and do not guarantee optimal variable selection (Royston & Sauerbrei, 2008). Exploring alternative methods or criteria for selecting conditional variables in this model would be beneficial. Moreover, the used software MPLUS offers limited capabilities for exhaustive model fit evaluation. Other statistical software might offer more comprehensive assessments of model fit. Furthermore, the computational intensity required for numerical integration underscores the need for ongoing advancements in algorithmic efficiency and model estimation techniques to streamline the model comparison process and enhance user experience.

Looking ahead, we envision a trajectory of continued evolution and refinement for the IRT-EMG model. Developing more robust estimation procedures and establishing clear, empirically validated criteria for variable selection stand as critical milestones on this path. The engagement with simulation studies will further elucidate the model's strengths and limitations, offering a scaffold for iterative improvement and adaptation. Additionally, the potential expansion of the model to accommodate ordered response data opens new avenues for exploration. There is also potential to enhance the accuracy of our model through the application of specific statistical techniques. For example, employing regularization or non-linear regression methods could lead to a better understanding of the impact of response time. Currently, response time is modeled as a combination of linear, quadratic, and cubic effects. Refining this through more advanced statistical methods could yield better predictions, but may require different software packages. Finally, while the model's generalization for ordered responses seems feasible, it requires additional assumptions and further work.

The IRT-EMG and similar models for analyzing guessing behavior hold substantial potential for application in the standardization and pilot phases of assessments. In the standardization phase of test development, the IRT-EMG model can be utilized to identify items particularly prone to guessing. This analysis can inform the selection or revision of test items. By identifying items that are susceptible to guessing, educators and test developers can make informed decisions about item revision or exclusion, thereby increasing the validity of the assessment.

Online learning platforms, characterized by their diverse user base and flexible learning environments, present unique challenges and opportunities for leveraging guessing models. Models that investigate guessing behaviors can be integrated into learning analytics systems to monitor and analyze student performance in real time. By identifying patterns of guessing behavior, the model can provide insights into student engagement, motivation, and learning strategies.

References

Asparouhov, T., & Muthèn, B. (2019). *Random starting values and multistage optimization.* https://www.statmodel.com/download/StartsUpdate.pdf

Belov, D. I., Lüdtke, O., & Ulitzsch, E. (2024). *Likelihood-free estimation of IRT models in small samples: A neural networks approach* [Preprint]. https://doi.org/10.31234/osf.io/w3cyq

Birnbaum, A. (1968). Some latent trait models and their use in inferring a student's ability. In F. M. Lord & M. R. Novick (Eds.), *Statistical theories of mental test scores*. Addison-Wesley.

Boe, E. E., May, H., & Boruch, R. F. (2002). *Student task persistence in the third international mathematics and science study: A major source of achievement differences at the national, classroom, and student levels* (Technical Report No. CRESP-RR-2002-TIMSS1). Center for Research and Evaluation in Social Policy, University of Pennsylvania.

Bolt, D. M., Cohen, A. S., & Wollack, J. A. (2002). Item parameter estimation under conditions of test speededness: Application of a mixture Rasch model with ordinal constraints. *Journal of Educational Measurement, 39*, 331–348. https://doi.org/10.1111/j.1745-3984.2002.tb01146.x

Braun, H., Kirsch, I., & Yamamoto, K. (2011). An experimental study of the effects of monetary incentives on performance on the 12th-grade NAEP reading assessment. *Teachers College Record, 113*(11), 2309–2344.

Burton, R. F. (2005). Multiple-choice and true/false tests: Myths and apprehensions. *Assessment & Evaluation in Higher Education, 30*, 65–72.

Cao, J., & Stokes, S. L. (2008). Bayesian IRT guessing models for partial guessing behaviors. *Psychometrika, 73*, 209–230. https://doi.org/10.1007/S11336-007-9045-9

Celeux, G., & Soromenho, G. (1996). An entropy criterion for assessing the number of clusters in a mixture model. *Journal of Classification, 13*, 195–212.

Chiu, T. W., & Camilli, G. (2013). Comment on 3PL IRT adjustment for guessing. *Applied Psychological Measurement, 37*(1), 76–86.

Curran, P. G. (2016). Methods for the detection of carelessly invalid responses in survey data. *Journal of Experimental Social Psychology, 66*, 4–19. https://doi.org/10.1016/j.jesp.2015.07.006

von Davier, M., & Yamamoto, K. (2014). Mixture-distribution and HYBRID Rasch models. *Multivariate and Mixture Distribution Rasch Models*, 99–115. https://link.springer.com/chapter/10.1007/978-0-387-49839-3_6

Dayton, C. M., & Macready, G. B. (1988). Concomitant-variable latent-class models. *Journal of the American Statistical Association, 83*(401), 173–178. https://doi.org/10.1080/01621459.1988.10478584

Erosheva, E. A. (2002). *Grade of membership and latent structure models with application to disability survey data* (Doctoral dissertation). Carnegie Mellon University. https://pdfs.semanticscholar.org/1fe4/64b6cae48d009697783bdbb72bcd4527608a.pdf

Geiser, C. (2013). *Methodology in the social sciences. Data analysis with Mplus*. Guilford Press.

Goldhammer, F., Martens, T., Christoph, G., & Lüdtke, O. (2016). *Test-taking engagement in PIAAC* (OECD Education Working Papers No. 133). OECD Publishing.

Guo, H., Rios, J. A., Haberman, S., Liu, O. L., Wang, J., & Paek, I. (2016). A new procedure for detection of students' rapid guessing responses using response time. *Applied Measurement in Education, 29*(3), 173–183. https://doi.org/10.1080/08957347.2016.1171766

Guo, S., Wu, T., Zheng, C., & Chen, Y. (2021). Bayesian Modal estimation for the one-parameter logistic ability-based guessing (1PL-AG) model. *Applied Psychological Measurement, 45*(3), 195–213.

Haladyna, T. M., & Downing, S. M. (2004). Construct-irrelevant variance in high-stakes testing. *Educational Measurement: Issues and Practice, 23*(1), 17–27. https://doi.org/10.1111/j.1745-3992.2004.tb00149.x

Hardy, M. A. (1993). *Regression with dummy variables* (Vol. 93). Sage.

Huang, J. L., Curran, P. G., Keeney, J., Poposki, E. M., & DeShon, R. P. (2012). Detecting and deterring insufficient effort responding to surveys. *Journal of Business and Psychology, 27*(1), 99–114.

Huang, J. L., Liu, M., & Bowling, N. A. (2015). Insufficient effort responding: Examining an insidious confound in survey data. *Journal of Applied Psychology, 100*(3), 828.

Hutchinson, T. (1991). *Ability, partial information, guessing: Statistical modeling applied to multiplechoice tests*. Rumsby Scientific.

Jin, K.-Y., & Wang, W.-C. (2014). Item response theory models for performance decline during testing. *Journal of Educational Measurement, 51*, 178–200. https://doi.org/10.1111/jedm.12041

Khorramdel, L., von Davier, M., & Pokropek, A. (2019). Combining mixture distribution and multidimensional IRTree models for the measurement of extreme response styles. *British Journal of Mathematical and Statistical Psychology, 72*, 538–559. https://doi.org/10.1111/bmsp.12179

Kim, M., Vermunt, J., Bakk, Z., Jaki, T., & Van Horn, M. L. (2016). Modeling predictors of latent classes in regression mixture models. *Structural Equation Modeling: A Multidisciplinary Journal, 23*(4), 601–614.

Kroehne, U., Buchholz, J., & Goldhammer, F. (2019). *Detecting carelessly invalid effort guessing in item sets using average item-level response times*. Paper presented at the Annual Meeting of the National Council on Measurement in Education (NCME), Toronto, Ontario, Canada, April 4–8, 2019.

Lee, Y.-H., & Jia, Y. (2014). Using response time to investigate students' test-taking behaviors in a NAEP computer-based study. *Large-Scale Assessments in Education, 2*(1). https://doi.org/10.1186/s40536-014-0008-1

van der Linden, W. J. (2007). A hierarchical framework for modeling speed and accuracy on test items. *Psychometrika, 72*(3), 287.

List, M. K., Robitzsch, A., Lüdtke, O., Köller, O., & Nagy, G. (2017). Performance decline in low-stakes educational assessments: Different mixture modeling approaches. *Large-scale Assessments in Education, 5*(1), 1–25.

Meyer, J. P. (2010). A mixture Rasch model with item response time components. *Applied Psychological Measurement, 34*(7), 521–538. https://doi.org/10.1177/0146621609355451

Mislevy, R. J., & Verhelst, N. (1990). Modeling item responses when different subjects employ different solution strategies. *Psychometrika, 55*(2), 195–215. https://doi.org/10.1007/BF02295283

Muthèn, L. K., & Muthèn, B. O. (1998–2017). *Mplus user's guide* (8th ed.). Muthèn & Muthèn.

Nagy, G., & Ulitzsch, E. (2022). A multilevel mixture IRT framework for modeling response times as predictors or indicators of response engagement in IRT models. *Educational and Psychological Measurement, 82*(5), 845–879.

O'Neil, H. F., Jr., Sugrue, B., & Baker, E. L. (1995). Effects of motivational interventions on the national assessment of educational progress mathematics performance. *Educational Assessment, 3*(2). https://doi.org/10.1207/s15326977ea0302_2

Paulhus, D. L. (1991). Measurement and control of response bias. In J. P. Robinson, P. R. Shaver, & L. S. Wrightsman (Eds.), *Measures of personality and social psychological attitudes* (pp. 17–59). Academic Press.

Penk, C., Pöhlmann, C., & Roppelt, A. (2014). The role of test-taking motivation for students' performance in low-stakes assessments: An investigation of school-track-specific differences. *Large-Scale Assessments in Education, 2*(5). https://doi.org/10.1186/s40536-014-0005-4

Pokropek, A. (2016). Grade of membership response time model for detecting guessing behaviors. *Journal of Educational and Behavioral Statistics, 41*(3), 300–325. https://doi.org/10.3102/1076998616636618

Rasch, G. (1960). *Probabilistic models for some intelligence and attainment tests*. Nielsen & Lydiche. (Expanded edition, Chicago, University of Chicago Press, 1980).

Rios, J. A., Guo, H., Mao, L., & Liu, O. L. (2017). Evaluating the impact of careless responding on aggregated-scores: To filter unmotivated examinees or not? *International Journal of Testing, 17*(1), 74–104. https://doi.org/10.1080/15305058.2016.1231193

Rost, J. (2004). *Lehrbuch Testtheorie—Testkonstruktion [Textbook test theory—test construction]* (2nd ed.). Huber.

Rowley, G., & Traub, R. E. (1977). Formula scoring, number-right scoring, and test-taking strategy. *Journal of Educational Measurement, 14*, 15–22.

Royston, P., & Sauerbrei, W. (2008). *Multivariable Model-Building: A pragmatic approach to regression analysis based on fractional polynomials for modelling continuous variables*. John Wiley and Sons, Ltd.

Rubin, D. B. (1976). Inference and missing data. *Biometrika, 63*(3), 581–592. https://doi.org/10.1093/biomet/63.3.581

San Martin, E., del Pino, G., & De Boeck, P. (2006). IRT models for ability-based guessing. *Applied Psychological Measurement, 30*(3). https://doi.org/10.1177/0146621605282773

Schnipke, D. L., & Scrams, D. J. (1997). Modeling item response times with a two-state mixture model: A new method of measuring speededness. *Journal of Educational Measurement, 34*(3), 213–232. https://doi.org/10.1111/j.1745-3984.1997.tb00516.x

Smit, J. A., Kelderman, H., & van der Flier, H. (1999). Collateral information and mixed Rasch models. *Methods of Psychological Research Online, 4*(3), 1–13. https://www.dgps.de/fachgruppen/methoden/mpr-online/issue8/art2/mrc.pdf

Smit, J. A., Kelderman, H., & van der Flier, H. (2000). The mixed Birnbaum model: Estimation using collateral information. *Methods of Psychological Research Online, 5*, 1–13. https://www.dgps.de/fachgruppen/methoden/mpr-online/issue12/art3/mbc.pd

Suh, Y., Cho, S.-J., & Wollack, J. A. (2012). A comparison of item calibration procedure in the presence of test speededness. *Journal of Educational Measurement, 49*, 285–311. https://doi.org/10.1111/j.1745-3984.2012.00176.x

Ulitzsch, E., von Davier, M., & Pohl, S. (2020). A hierarchical latent response model for inferences about examinee engagement in terms of guessing and item-level non-response. *British Journal of Mathematical and Statistical Psychology, 73*(1), 83–112. https://doi.org/10.1111/bmsp.12188

Van Vaerenbergh, Y., & Thomas, T. D. (2013). Response styles in survey research: A literature review of antecedents, consequences, and remedies. *International Journal of Public Opinion Research, 25*(2), 195–217. https://doi.org/10.1093/ijpor/eds021

Wang, C., & Xu, G. (2015). A mixture hierarchical model for response times and response accuracy. *British Journal of Mathematical and Statistical Psychology, 68*(3), 456–477. https://doi.org/10.1111/bmsp.12054

Wise, S. L. (2017). Rapid-guessing behavior: Its identification, interpretation, and implications. *Educational Measurement: Issues and Practice*. https://doi.org/10.1111/emip.12165

Wise, S. L., & DeMars, C. E. (2005). Low examinee effort in low-stakes assessment: Problems and potential solutions. *Educational Assessment, 10*(1), 1–17. https://doi.org/10.1207/s15326977ea1001_1

Wise, S. L., & Kong, X. J. (2005). Response time effort: A new measure of examinee motivation in computer-based tests. *Applied Measurement in Education, 18*, 163–183. https://doi.org/10.1207/s15324818ame1802_2

Wise, S. L., Kingsbury, G., Thomason, J., & Kong, X. (2004, April). *An investigation of motivation filtering in a statewide achievement testing program*. Paper presented at the Annual Meeting of the National Council on Measurement in Education, San Diego, CA.

Wise, S. L., Pastor, D. A., & Kong, X. J. (2009). Correlates of rapid-guessing behavior in low-stakes testing: Implications for test development and measurement practice. *Applied Measurement in Education, 22*(2), 185–205. https://doi.org/10.1080/08957340902754650

Wise, S. L., Ma, L., Kingsbury, G. G., Hauser, C. (2010). *An investigation of the relationship between time of testing and test-taking effort* [Conference Paper]. Paper presented at the annual meeting of the National Council on Measurement in Education (NCME). Denver, CO.

Yamamoto, K. (1989). *HYBRID model of IRT and latent class models* (ETS Research Report RR-89-41). Educational Testing Service.

Yamamoto, K., & Everson, H. (1997). Modeling the effects of test length and test time on parameter estimation using the HYBRID model. In J. Rost (Ed.), *Applications of latent trait and latent class models in the social sciences* (pp. 89–98). Waxmann.

Żółtak, T., & Golonka, G. (2015). Does guessing matter? Differences between ability estimates from 2PL and 3PL IRT models in case of guessing. *EDUKACJA Quarterly, 134*(3).

Part V
Software Tools for Analyzing Large-Scale Assessment Data

Chapter 24
The IEA's IDB Analyzer

Eugenio Gonzalez

Abstract The IEA's IDB Analyzer is an application created by the IEA to facilitate processing and analyses of international large-scale assessment databases. Originally developed in 2004 to be used with IEA's own data, it is currently configured to work with most international large-scale assessment databases, as well as a few national assessments. In this chapter we describe the main features and capabilities of the IEA's IDB Analyzer and present an illustrative example of how to use the two modules that comprise the IEA's IDB Analyzer. We present and describe an example output and provide a brief overview of the statistical techniques and procedures addressed by the application.

The IEA's International Database Analyzer (IEA's IDB Analyzer; IEA, 2021) is an application developed by the International Association for the Evaluation of Educational Achievement (IEA) that can be used to create SAS or SPSS syntax to combine and analyze data from IEA's large-scale assessments, as well as data from most major large-scale assessment surveys, including those conducted by the Organisation for Economic Co-operation and Development (OECD) and other international organizations. Originally designed for international large-scale assessments, it has also been configured to work with different databases from the US National Assessment of Educational Progress (NAEP).

At its core, the IEA's IDB Analyzer is a graphical user interface that reads the contents of SAS or SPSS files and creates and passes on syntax with parameters selected by the user to one or more of the macros that are bundled with the application. These macros can merge files from different sources (mainly schools, students, and their teachers), or set up analyses that conform to the technical specifications of the corresponding large-scale assessment. While the instructions

E. Gonzalez (✉)
TIMSS & PIRLS International Study Center at Boston College, Chestnut Hill, MA, USA
e-mail: gonzalee@bc.edu

L. Khorramdel et al. (eds.), *Innovative Digital-Based International Large-Scale Assessments*, Methodology of Educational Measurement and Assessment,
https://doi.org/10.1007/978-3-031-90951-1_24

created with the IEA's IDB Analyzer are configured to fit analysis specifications of the configured studies, these instructions can be modified to fit other studies.

The macros that the IEA's IDB Analyzer uses are written for SAS and SPSS.[1] The IEA's IDB Analyzer, as such, does not merge files or compute statistics but, rather, provides the user with an interface to select variables and analysis settings based on the specifications from a diverse portfolio of large-scale assessment programs, and creates the corresponding syntax. All data manipulation and analyses are done with SAS or SPSS, and therefore, the user needs to have one of these applications installed on their computer. Furthermore, the IEA's IDB Analyzer is a Windows based application, only available for systems using a Windows operating system.

24.1 About Large-Scale Assessments

Large-scale assessments have become ubiquitous in recent years among education experts and policy makers. Their importance is based on the amount and quality of information they provide in terms of what students and adults know and can do, usually labeled as skills, proficiency, or achievement levels, and the distribution of these among different groups of the populations of interest. The overarching goals of these large-scale assessments can be summarized in three main goals: determine the level and the distribution of proficiency in key domains for important subgroups of the population; better understand factors associated with the acquisition, development, maintenance, and loss of proficiency; and better understand the relationship of proficiency in these domains and other related outcomes. Therefore, to achieve these goals, it is imperative to provide analysis tools for researchers and policy makers that would help them extract and calculate statistical summaries of the data with confidence and follow the recommended specifications for the analysis.

The qualification of "large" in large-scale assessments has less to do with the number of persons assessed or the number of participants, but rather with the amount of information that is collected and the inferences that are to be made from that information. In this sense, large-scale assessments are set to evaluate extensively each of their target domains by including many items and to make inferences to relatively large populations, such as, for example, all 4th graders in the country. In some instances, large-scale assessments are also called group-score assessments, since their goal is not to provide individual level results or even for small groups, such as students attending a class or school, but rather to provide results for relatively large groups of the population based on the selection of representative samples.

[1] The IDB undergoes regular updates to incorporate compatibility with new studies and software. Since this chapter was first submitted for publication, there is a newer version of the IDB Analyzer that has been updated to work with R. For the latest version, functionality, and software compatibility of the IDB Analyzer, please visit www.iea.nl.

To achieve these goals, large-scale assessments rely on sampling methodology. Use of sampling methodology helps make it possible to administer relatively large sets of test items and survey questions, ensuring proper coverage of the domain, to relatively large samples of participants, without overburdening the participants and without making the costs of carrying out such an assessment prohibitive.

Large-scale assessments rely on what is called *multiple-matrix-sampling*, which in brief means that some members of the population (samples of individuals) are given subsets of the test items (samples of items). From the resulting sparse matrix of responses, we then need to be able make inferences to the entire population and about the entire domain covered by the items.

While using multiple-matrix-sampling designs make these assessments operationally feasible, they do require special computations that account for the different features of the designs. As such, there are three specific areas of the analysis that are affected and require special analytic treatment.

The first area is the relative contribution of each individual in the data file. Usually only a few people from the entire population are selected to participate in the assessment, and these might be selected at different rates depending on where or who they are, and once selected, some of them might not participate or need to be replaced, so in the end, we need to have an indication of how many people from the population are represented by each selected participant. In other words, what should be the relative contribution of each participant to the overall estimate? This is addressed by calculating what are called sampling weights and using them in the analysis. The underlying statistical theory for sample selection and the analysis of survey data using sampling weights can be found in Cochran (1977), Särndal et al. (1992), and Lohr (2010).

The second area that requires special analytic treatment is that of estimating and what is called sampling variance. *Sampling variance* deals with the question about how the results would vary had we followed the same random sampling procedure, but had ended up with a somewhat different, yet statistically equivalent, set of participants. In large-scale assessments, samples are selected using a series of strategies that help optimize the sample selection, the deployment of staff for data collection, and ensure the representation of different groups from the population for the purpose of reporting about them. Except for very limited cases where a simple random sample selection is possible to answer the design questions, samples in large-scale assessments are selected in different stages, using different stratification approaches, and ultimately in clusters where respondents are found. These clusters could be a class, school, or geographic location. Therefore, sampling variance in large-scale assessments is not that of a simple random sample. Calculating the sampling variance is addressed by replicating the analysis multiple times, using the sample selection and stratification information and the sample weights. The general approach is to systematically alter the overall contribution of selected elements of the sample to the estimate and summarize the results according to a specified formula. There are two main procedures currently used operationally in large-scale assessments to accomplish this—either Jackknife or Balanced repeated replication with Fay's variant. Each of these replication procedures have variations that are used and applicable under different assumptions. The IEA's IDB Analyzer macros can implement both procedures. The underlying statistical theory for sampling variance estimation procedures for survey data can be found in Efron (1982), Wolter (2007), Rust (1985), and Rust and Rao (1996).

The last area that requires special analytic treatment is that of the precision of the measurement of the domain of interest and the difference between groups in the population. In brief, since the assessments of a domain contain a subset of all possible items that would cover that domain, and each individual is only administered a subset of that subset, we end up with an imperfect estimate of each individual's proficiency level and an estimate of the uncertainty of the point estimate. Furthermore, when estimating proficiency levels and their uncertainty, background data are used to optimize these estimates to help compensate for the lack of information about a respondent. Similar to what is done to estimate sampling variance by creating replicate samples, large-scale assessments also include replicates of the proficiency estimates based on the point estimates and the uncertainty of these point estimates in their databases. These replicates are called plausible values. The underlying statistical theory for multiple imputation estimation procedures and estimating variance can be found in Little and Rubin (1983), Johnson and Rust (1992), Mislevy (1985, 1991), Mislevy et al. (1992), and Rubin (1987).

In addition to the references presented above for the underlying statistical theory, details on the specific implementation of these procedures can be found in the technical report for each of the large-scale assessments (OECD, 2020; Martin et al., 2017; Schulz et al., 2018, among others) that can be readily analyzed with the IEA's IDB Analyzer.

In summary, using large-scale assessments to obtain proper population estimates requires accounting for the differential contribution of the members of the sample and replicating the results numerous times to estimate the corresponding sampling and measurement variances. While the use of sampling weights is quite standard with most statistical procedures, implementing the corresponding replications to obtain the sampling and measurement errors and combine these results is not too common.

24.2 About the IEA's IDB Analyzer

The IEA's IDB Analyzer is a stand-alone application that creates SPSS and SAS syntax that can be used to combine files from across different countries and levels (student, teacher, school, etc.) and perform analyses with large-scale assessment databases. The user is presented with an interface where they can select from a variety of preconfigured study parameter settings. Based on this selection, the IEA's IDB Analyzer generates SPSS or SAS syntax that, when executed, accounts for information from the sampling design in the computation of sampling variance and handles the plausible values. The code generated by the IEA's IDB Analyzer enables the user to compute descriptive and inferential statistics and conduct statistical hypothesis testing among groups in the population without having to write any programming code.

The IEA's IDB Analyzer is licensed free of cost, not sold, and is for use only in accordance with the terms of the licensing agreement. The software is distributed for free, but the user is only authorized to use the SPSS and SAS macros in combination with the IEA's IDB Analyzer unless explicitly authorized by the IEA. The software and license need to be renewed every year, and an updated version is provided at the time of renewal.

The IEA's IDB Analyzer consists of two modules: the Merge Module and the Analysis Module. These two modules are integrated and available from a single interface.

24.2.1 The Merge Module

The Merge Module is used to create syntax to combine data files from different countries[2] and, when necessary, merge data files from different sources like student background questionnaires and achievement files, or student background files, with teacher or school level files. It also allows the user to select individual or sets of variables to create a smaller and more manageable dataset.

In the introduction to this chapter, we mentioned the analytic complexities because of the use of multiple matrix sampling designs, but international large-scale assessments typically have dozens of country participants, administering instruments to different actors in the educational system (mainly students, teachers, parents, and school administrators). At the time the IEA's IDB Analyzer was first conceived in early 2000, computing power and storage capacity was barely a fraction of what it is now. At that time, it was important, and to some extent necessary, to read from the data files only what was needed for a specific analysis to alleviate the storage and processing demands on the computer. The Merge Module served this function to allow the user to create syntax that would create a file with selected variables from a selection of country participants. Today, given computing and storage capacity, this function is not as necessary as it used to be, but it is still useful under certain circumstances. But the need to combine and merge files from different sources (students, teachers, and school principals) belonging to different countries, remains.

As is standard operating procedure of the IEA's IDB Analyzer, when running the Merge Module, the IEA's IDB Analyzer creates SPSS or SAS syntax that is used to merge and combines files specified by the user, keeping only the variables selected

[2]Throughout this chapter, the term "country" is used to refer to reporting jurisdictions in the corresponding study. These include countries per se, "states," "provinces," "benchmarking participants," "partner countries," etc. Each study has adopted its own convention for naming and identifying these entities, but in general these are identified by a unique value in the variables usually called IDCNTRY, CNT, CNTRYID, etc.

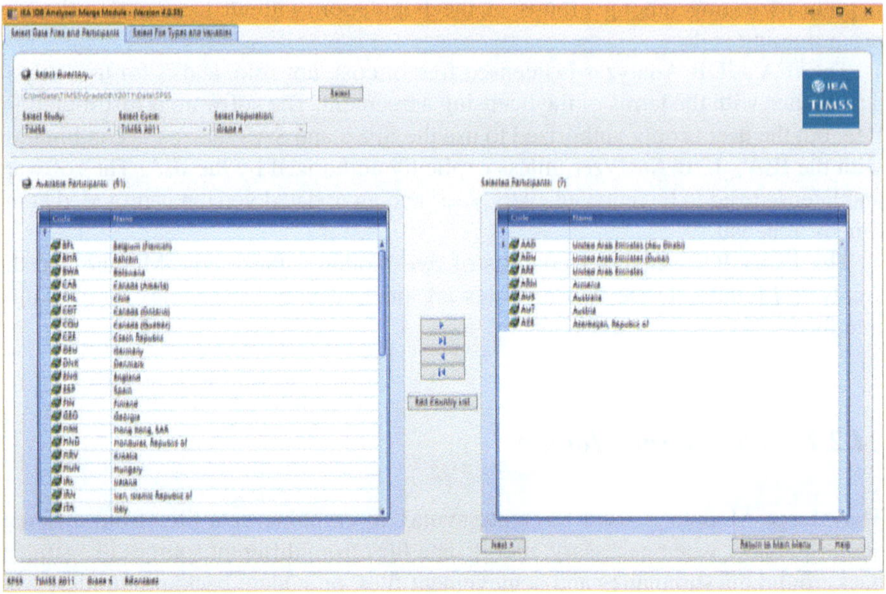

Fig. 24.1 First selection screen of the merge module

for the merge. The SPSS or SAS syntax is automatically saved to an SPSS or SAS syntax file and opened in an SPSS or SAS syntax window ready to be submitted for processing.

The data files created using the syntax created by the Merge Module can be processed either with the Analysis Module of the IEA's IDB Analyzer or by any other analysis software that accepts SPSS or SAS files as input.

As of the writing of this chapter, the Merge Module is only available for use with IEA publicly available databases and others in which the data are published separated by country (OECD-TALIS, OECD-PIAAC, IADB-PRIDI, etc.). The Merge Module is also only designed to combine data from the same assessment year or cycle. Combining data from across different years requires additional pre-processing given that, depending on the study, at times the same variable name is used for different variables, and the same variable receives different variable names in different years of the same study. Therefore, if a user wants to combine data from across two or more years of the same assessment, they will need to reconcile the different variables and variable names from each assessment cycle and ensure the same variable names are used consistently throughout.

The Merge Module uses two selection screens. In the first screen, shown in Fig. 24.1, the user indicates the location of the files, and selects the countries that will be included in the resulting merged file. Additionally, the user can edit the country names that will be displayed in the output. The countries selected are on the right panel, and the countries available for selection are in the left panel. Using

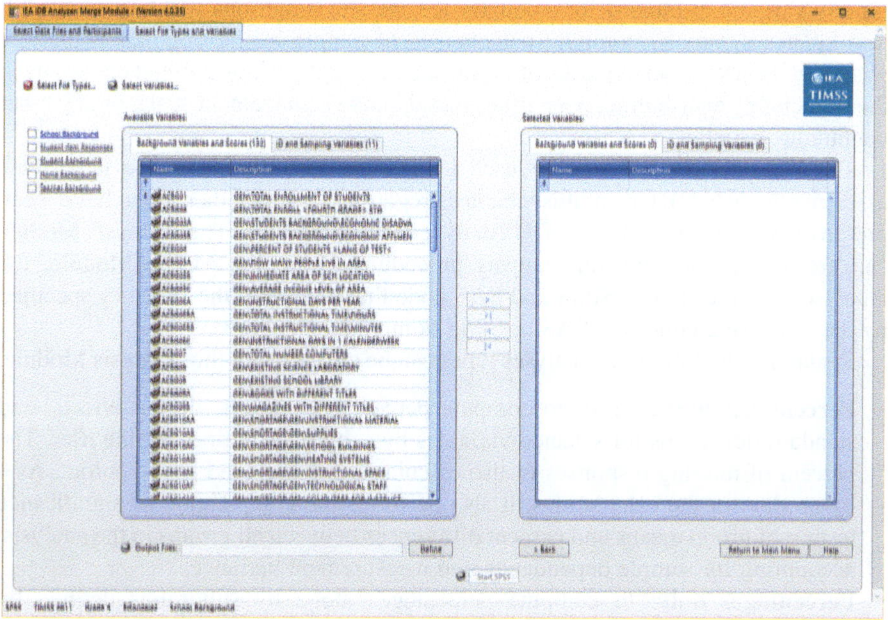

Fig. 24.2 Second selection screen of the merge module

the name of the files in the directory, the IEA's IDB Analyzer recognizes the study, cycle, and population.

In the second screen, shown in Fig. 24.2, the user selects the variables and file types that are going to be merged and specifies the name of the output file that will contain the variables and cases specified in the selection. In selecting the corresponding file types, the Merge Module will write the code to do the corresponding merge between the different file types. By default, the Merge Module selects all sampling related variables, as well as any available plausible values, in the corresponding files.

24.2.2 The Analysis Module

The Analysis Module of the IEA's IDB Analyzer allows the user to create syntax for the computation of a diverse set of descriptive and inferential statistics. The created syntax also incorporates instructions to create some graphics for the descriptive statistics, to write summary tables of the results in HTML format, and to export results to files that can be read with spreadsheet applications.

Specifically, the IEA's IDB Analyzer Analysis Module can be used to create SAS or SPSS syntax to compute means, percentages, standard deviations, Pearson and Spearman's rank-order correlations, and linear and logistic regression coefficients for any variable of interest for a country for specific subgroups within a country. It also creates syntax to compute percentages of people in the population that are within, at, or above benchmarks of performance or within user-defined cut points in the proficiency distribution, percentiles based on the achievement scale, or any other continuous variable.

The Analysis Module can be used to analyze data files from the large-scale assessments listed later in this chapter, regardless of whether they have been pre-processed using the IEA's IDB Analyzer Merge Module. The Analysis Module can create code for several analysis procedures. Like the Merge Module, the Analysis Module creates SPSS and SAS code that computes the statistics specified by the user using either the SAS or SPSS computation engine.

Syntax for the following analyses types can be created with the Analysis Module:

- **Percentages and Means**: to compute percentages, means, design effects, and standard deviations for selected variables by subgroups defined by the user. The percent of missing responses is also calculated and included in the output. As a by-product of the calculations, it also performs the computation of significance testing of group means and percent differences between all groups in the analysis, accounting for sample dependency and measurement variance.
- **Percentages only**: to compute percentages alone by subgroups defined by the user.
- **Linear Regression**: to compute linear regression coefficients for selected variables predicting a dependent variable by subgroups defined by the user. The IEA's IDB Analyzer has the capability of processing plausible values as dependent or independent variables in the equation. It also has the capability of creating contrast coded categorical variables using dummy or effect coding and including these in the equation.
- **Logistic Regression**: to compute logistic regression coefficients for selected variables predicting a dependent dichotomous variable, by subgroups defined by the user. The IEA's IDB Analyzer has the capability of processing plausible values as independent or dependent variables in the logistic regression equation. It also has the capability of contrast coding categorical variables and including them in the equation. Under the SAS version, the user can also run multinomial logistic regressions.
- **Benchmarks**: to compute the percentage of the population in two modes: (a) as the percent within a group meeting a set of user-specified performance or achievement benchmarks, or (b) as the percent within a proficiency level that belong to a specific group of the population. It also computes the percentages within a group meeting the benchmarks in two modes: cumulative, as the percentage of the population at or above given points in the distribution; or discrete, as the percentage of the population within given points of the distribution. It can also compute the mean of an analysis variable for those within a particular group defined by the cut points. In doing this, it also performs the computation of group mean differences and percent differences, as well as the

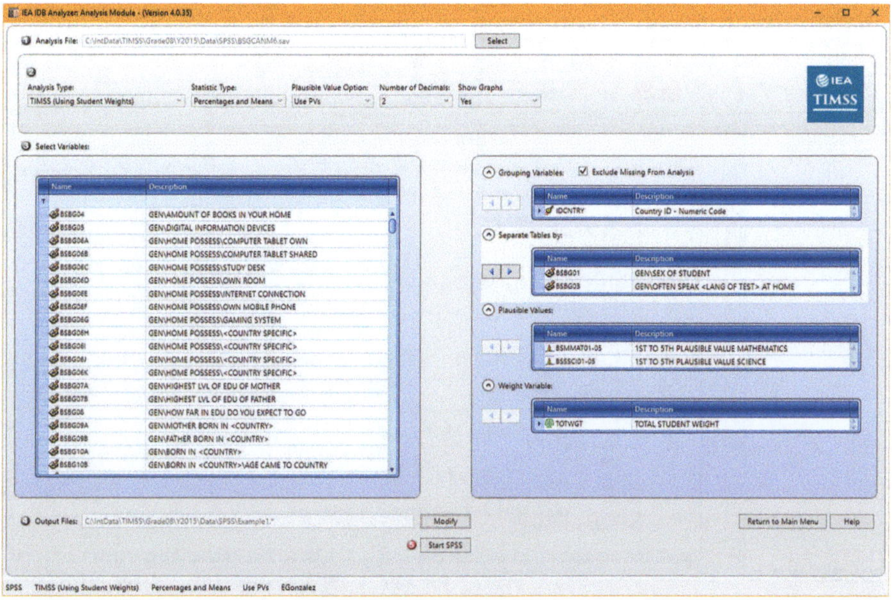

Fig. 24.3 Example setup for an analysis to compute percentages and means

corresponding significance testing, between groups, taking into account sample dependency.

- **Correlations**: to compute Pearson or Spearman's Rank Order correlation for selected variables by subgroups defined by the grouping variable(s). The IEA IDB Analyzer is capable of computing the correlation between sets of plausible values or between plausible values and any other variable.
- **Percentiles**: to compute the score points that separate a given proportion of the distribution of scores by subgroups defined by the grouping variable(s).

When creating the syntax to compute these statistics, the IEA's IDB Analyzer has the capability of using any continuous or categorical variable in the database, or it can make use of scores in the form of plausible values. When using plausible values, the IEA's IDB Analyzer generates code that accounts for the multiple imputation methodology in the calculation of the variance for statistics, as it applies to the corresponding study.

All procedures offered within the Analysis Module of the IEA's IDB Analyzer create syntax that make use of appropriate sampling weights and standard errors of the statistics that are computed according to the variance estimation procedure required by the design as it applies to the corresponding study.

In the figures above, we show an example setup and output for an analysis using the IEA's IDB Analyzer. In Fig. 24.3, we show the setup using the TIMSS 2015 Canadian data for eighth grade (BSGCANM6.sav). In this setup we will be computing percentages and means, by gender and by frequency the test language is

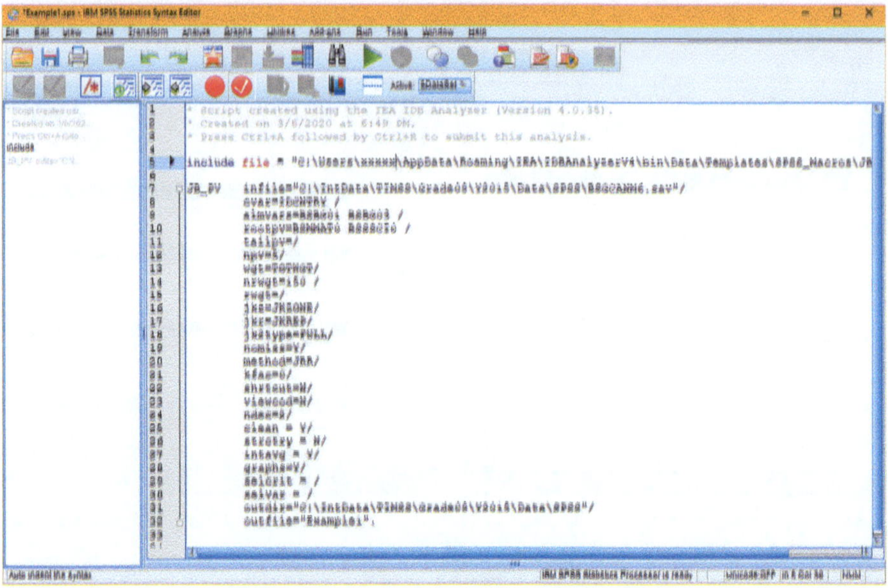

Fig. 24.4 Example syntax for an analysis to compute percentages and mean

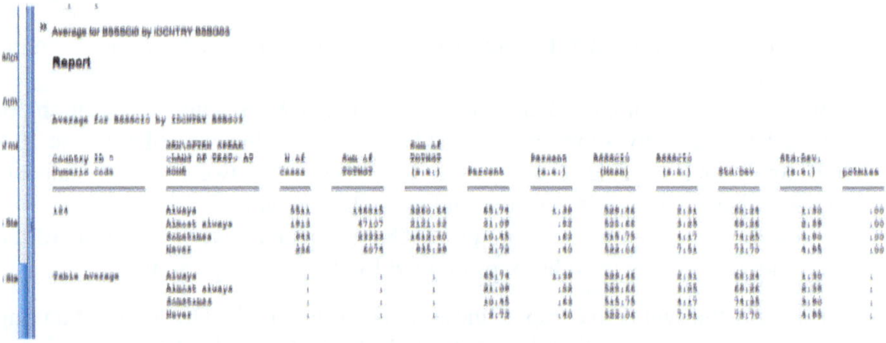

Fig. 24.5 Selected section of the output from the example analysis

spoken at home, for the mathematics and the science plausible values. The syntax and corresponding output will be written to a file called "Example 1."

In Fig. 24.4 we see the syntax created by the setup above. This syntax is in SPSS and the IEA's IDB Analyzer opens it automatically in an SPSS window. Should the user have selected to work with SAS, a similar syntax would have opened in a corresponding SAS window.

In Fig. 24.5 we see a subset of the output from the analysis. In this table we see the labels for the different groups, the number of cases, sum of the weights, average science achievement by subgroup, percentages, and standard deviations. When processed by SPSS, the syntax shown in Fig. 24.4 would produce four sets of results: mathematics by sex of students, mathematics by frequency the test language being spoken at home, and similar results for science. The output includes several graphics, and all tabular results are written to corresponding Excel files: one set of Excel files containing the descriptive statistics, and a second set results from significance testing results. When using SAS, the results are written to CSV files instead of Excel files.

In addition to the statistics of interest, such as means, percentages, and the corresponding errors, the IEA's IDB Analyzer also provides several statistics useful for evaluating the quality of the data in the output. These include the number of valid cases and percent or missing data for each variable by analysis group, weighted and unweighted descriptive statistics for the variables in the analysis, and the number of variance strata used in the calculation of the sampling variances.

As shown in Fig. 24.5, a line with a "Table Average" is included in the output. The "Table Average" presents the unweighted average of the countries in the table, and the corresponding standard error for that average computed from the countries that contribute to the table average. The table average α is computed as

$$\alpha = \frac{\sum_{c=1}^{C} \varepsilon_c}{C}$$

where, C is the number of countries, and ε_c is the statistic ε in country c. And its standard error is

$$SE_\alpha = \sqrt{\frac{\sum_{c=1}^{C} SE_{\varepsilon_c}^2}{C^2}}$$

where, C is the number of countries, and $SE_{\varepsilon_c}^2$ is the standard error for the statistic ε in country c.

24.3 Variance Estimation

The main advantage of using the IEA's IDB Analyzer to work with large-scale assessment data is its capability to create syntax in SAS or SPSS that takes into account the sample and assessment designs for the calculation of a wide range of statistics, as well as their corresponding standard errors, and calculate statistical significance. As such, the IEA's IDB Analyzer calculates the sampling variance for the statistics according to the predetermined sampling design, and when plausible values are involved, it computes the statistic of interest using all plausible values and calculates the corresponding imputation variance for the plausible values in the data. When necessary, it combines these two variance terms into an overall standard error and uses this standard error for the calculation of statistical significance.

24.3.1 Estimating Sampling Variance

For the estimation of sapling variance, the IEA's IDB Analyzer supports both, balanced repeated and jackknife approaches for sampling variance estimation, and provides the following options for the replication methods:

- Balances repeated replication (BRR) for designs with two primary sampling units per stratum
- Fay's BRR variant
- Jackknife for unstratified designs (JK1)
- Jackknife for designs with two primary sampling units per stratum (JK2).

Within the JK2 method, it can also handle two variants: the FULL-JK2, where both primary sampling units (PSUs) are dropped, one for each of the two replications, and the HALF-JK2, where only one of the PSUs is randomly dropped from a single replication.

It can also handle two variants when estimating sampling variance for plausible values. It can use the shortcut method, where the sampling variance is estimated using only the first plausible value in the analysis, or it can use all plausible values for the calculation of the sampling variance. The so-called shortcut method was popular in earlier days when computing power was substantially less than what it is today, but currently, most large-scale assessments do not use the shortcut methods operationally.

When JK2-type replicate weights are not included in the data file, the IEA's IDB Analyzer can create syntax to generate them on the fly based on sampling and stratification information provided in the data file.

When computing the sampling variance ($SE(s)_{\varepsilon}^{2}$), the IEA's IDB Analyzer uses the following general formula:

$$SE(s)_{\varepsilon}^{2} = f * \sum_{r=1}^{R} (\varepsilon_r - \varepsilon_0)^2$$

where,

ε_0 is any statistic of interest (percent, mean, variance, regression coefficient, etc.) computed using the full sample weight,

ε_r is any statistic of interest (percent, mean, variance, regression coefficient, etc.) computed using replicate weight r,

r is the number of replicates according to the sample and variance estimation design, and

f is a factor that takes on the following values, depending on the sample design, and replication method chosen:

- Under BRR: $f = \left(\frac{1}{R}\right)$
- Under BRR with Fay's variant: $f = \left(\frac{1}{R*(1-V)^2}\right)$, where V is the Fay factor used to create the replicate weights.

- Under JK1: $f = \left(\frac{R-1}{R}\right)$
- Under the HALF-JK2: $f = \left(\frac{1}{1}\right)$
- Under the FULL-JK2: $f = \left(\frac{1}{2}\right)$

When working with statistics that involve plausible values, using the shortcut method applies the formula above only to the first plausible value. When not using the shortcut method, the formula above is applied to each of the plausible values, and the sampling standard error for the statistic is calculated as follows:

$$SE(s)_{\varepsilon}^{2} = \frac{1}{P} * \sum_{p=1}^{P} SE(s)_{\varepsilon,p}^{2}$$

where,

P is the number of plausible values used for the analysis, and
$SE(s)_{\varepsilon,p}^{2}$ is the sampling variance calculated using each of the plausible values.

The standard error for a statistic that does not involve plausible values is simply the square root of its sampling variance, or $SE_{\varepsilon} = \sqrt{SE(s)_{\varepsilon}^{2}}$.

The macros used by the IEA's IDB Analyzer are highly parameterized, so a user could potentially use any combination of values for the number of replicates, number of plausible values, and resampling method, not to mention using the full or half JK2 approach, or the shortcut method in the case of working with plausible values. However, the IEA's IDB Analyzer uses only the values for these parameters as specified in the technical documentation of the studies for which it has been configured.

The underlying statistical theory for sampling variance estimation procedures for survey data are beyond the scope of this chapter, but they can be found in Efron (1982), Wolter (2007), Rust (1985), and Rust and Rao (1996), among many others. For a complete description and statistical formulas for the replication procedures available in the IEA's IDB Analyzer, please refer to the Wesvar 4.3 User Guide (Westat, 2007). Details on the calculations implemented with the macros in the IEA's IDB Analyzer are also provided by Gonzalez (2014).

24.3.2 Estimating Imputation Variance

When working with plausible values, it is necessary to calculate the imputation variance for a statistic, in addition to its sampling variance. For the estimation of the imputation variance ($SE(i)_{\varepsilon}^{2}$), the IEA's IDB Analyzer uses the procedure recommended by the technical documentation of the corresponding large-scale assessment. The imputation variance is computed by calculating the statistics of interest separately with each plausible value, and then calculating the variance of these statistics. The formula used is the following:

$$SE(i)_{\varepsilon}^{2} = \left(1 + \frac{1}{P}\right) \frac{\sum_{p=1}^{P} \left(\varepsilon_{0,p} - \overline{\varepsilon}_{0,P}\right)^{2}}{P-1}$$

where,

P is the number of plausible values used in the analysis
$\varepsilon_{0,\,p}$ is any statistic of interest (percent, mean, variance, regression coefficient, etc.) computed with plausible value p, and using the full sample weight, and
$\overline{\varepsilon}_{0,P}$ is the average of all available $\varepsilon_{0,\,p}$.

The IEA's IDB Analyzer can calculate the imputation variance with any number of plausible values provided. While each large-scale assessment has a specified number of plausible values used, the macros used by the IEA's IDB Analyzer can be instructed to use all of them, or fewer.

For the statistical underpinnings of the calculations of the imputation variance, please refer to the following sources: Little and Rubin (1987), Johnson and Rust (1992), Mislevy (1985), Mislevy (1991), Mislevy et al. (1992), and Rubin (1987). For the specific implementation and calculation of the plausible values, please refer to the technical report for the corresponding large-scale assessment. Details on the calculations implemented with the macros in the IEA's IDB Analyzer are also provided by Gonzalez (2014).

24.3.3 Working with Plausible Values

Working with plausible values requires additional steps in the analysis. In the previous section we mentioned how to calculate the imputation variance using the plausible values available in the data. But there are two additional steps to take when working with plausible values.

The first step is accomplished in-passing when calculating the imputation variance, and that is the calculation of the statistic of interest as it will be reported. This is $\overline{\varepsilon}_{0,P}$, or the average of the P estimates of ε_0, each calculated with each of the plausible values.

The second step is to combine the sampling and imputation variances to characterize the statistics reported or conduct significance testing. As mentioned earlier, when the statistic of interest does not involve the use of plausible values, we only calculate the sampling variance, and the standard error for the statistic is its square root. However, when working with plausible values we need to combine the sampling and imputation variances to obtain the standard error for a statistic, and therefore, $SE_{\varepsilon} = \sqrt{SE(s)_{\varepsilon}^{2} + SE(i)_{\varepsilon}^{2}}$. The macros accessed by the IEA's IDB Analyzer are properly configured to perform these calculations.

24.4 Configurations Available

The IEA's IDB Analyzer is highly configurable and has a settings file that can be modified to accommodate most existing large-scale assessments. There is a single version that works for all configured studies. As of the writing of this chapter, the IEA's IDB Analyzer has been configured to work with the studies listed in Table 24.1. Note that while all can be analyzed with the Analysis Module, the Merge Module has only been configured for the studies that publish their data separately by country. The list of studies and compatibility is updated every year. For the latest list of compatible studies, please download a copy of the IEA's IDB Analyzer, install it, and look for this information in the Help Manual. The Help Manual will also list updates to the analysis methods and procedures of the different studies, if any have occurred.

Table 24.1 Studies that can be analyzed with the IEA IDB analyzer (in alphabetical order)

Study/organization	Capability
ALLS/StatsCan & OECD: Adult Literacy and Life Skills Study	Analyze
CivEd/IEA: Civics Education Study	Merge and analyze
IALS/OECD: International Adult Literacy Study	Analyze
ICCS/IEA: International Civics and Citizenship Education Study	Merge and analyze
ICILS/IEA: International Computer and Information Literacy Study	Merge and analyze
NAEP/ U.S. National Assessment of Educational Progress	Analyze
PIAAC/OECD: Programme of International Assessment of Adult Competencies	Merge and analyze
PIRLS/IEA: Progress in International Literacy Study	Merge and analyze
PISA/OECD: Programme for International Student Assessment	Analyze
PRIDI/IADB: Project on Child Development Indicators	Merge and analyze
SITES/IEA: Second Information Technology in Education Study	Merge and analyze
TALIS/OECD: Teachers and Learning International Study	Merge and analyze
TEDS-M/IEA: Teacher Education Study	Merge and analyze
TIMSS/IEA: Trends in International Mathematics and Science Study	Merge and analyze

24.5 Additional Features

We have presented and described the main features of the IEA's IDB Analyzer, mainly to provide SAS and SPSS syntax to combine files from a large portfolio of large-scale assessments and to conduct the analysis of these data using the parameters and specifications as outlined in the corresponding technical reports for the study, but also in concordance with the underlying statistical theory. Before concluding, we want to provide an overview of additional functionality available in the IEA's IDB Analyzer.

24.5.1 *Selecting Cases for Analysis*

Any analyses done with the IEA's IDB Analyzer present the option to subset the data used for the analysis. By using the parameters "SELVAR" and "SELCRIT" in the macros, the user can conduct their analysis using a subset of the data. For example, by setting "SELVAR = YEAR" and "SELCRIT = (year = 1990)" a user can conduct the analysis with only those cases for which the variable YEAR equals 1990. When using a selection criterion, the program will still read through the entire file to search for the records that meet the selection criterion, so the initial processing speed will depend on the size of the file that contains all the cases. Alternatively, the user can create a smaller file with the records that meet a particular criterion and operate on that file.

24.5.2 *Output Files*

All statistical procedures carried out by the macros will create various output files with the results from the analysis and the syntax used to create the results. The syntax can later be used to replicate the analysis. Descriptions of the output files generated by SPSS and SAS are presented in Tables 24.2 and 24.3, respectively, below.

All output filenames use the name specified for the syntax file created by the IEA's IDB Analyzer and are written to the same directory where the syntax file is written. The output files can be distinguished by the extension used for the file and some additional text within the file name clarifying the contents of the file. The Help

Table 24.2 Output created by the IEA's IDB Analyzer when running with SPSS

Type	Generated by	Extension	Format	Open with
Syntax	IEA IDB Analyzer	*.sps	Text	SPSS or text editor
Results	SPSS	*.spv	SPSS Pivot Tables	SPSS
Results	SPSS	*.sav	SPSS Dataset	SPSS
Results	SPSS	*.xlsx	MS Excel	MS Excel

Table 24.3 Output created by the IEA's IDB Analyzer when running with SAS

Type	Generated by	Extension	Format	Open with
Syntax	IEA IDB Analyzer	*.sas	Text	SAS or other text editor
Results	SAS	*.lst	SAS Listing file	SAS or other text editor
Results	SAS	*.html	HTML	SAS or other HTML reader
Results	SAS	*.csv	Comma Separated variables	MS Excel or other CSV file reader
Results	SAS	*. sas7bdat	SAS Dataset	SAS
Analysis Steps	SAS	*.log	SAS log file	SAS or other text editor

Manual for the IEA's IDB Analyzer describes the file naming convention used for these files and their contents.

24.5.3 Analysis Settings

The IEA's IDB Analyzer works by using a set of macros with a particular set of parameters. This allows the macros to be highly configurable and flexible to accommodate different study configurations. The following are the parameter settings that pertain to the specifics of the analysis:

METHOD: This indicates the method used in the calculation of the sampling variance. Options are JK1 for jackknife1, JK2 for jackknife2, BRR for balanced repeated replicates, and PIAAC (the PIAAC study uses different variance estimation procedures for different countries). When option PIAAC is used, the program looks for the variables VEMETHOD, VENREPS, and VEFAYFAC to complete the analysis specifications.

KFAC The Fay factor to use in the calculation of the sampling variance when METHOD=BRR.

JK2TYPE When using METHOD = JK2, you have the option of using the FULL or HALF JK2 approach described earlier in the chapter.

SHRTCUT This indicates where to use all plausible values in the calculation of the sampling variance (SHRTCUT = N) or to take a shortcut and use only the first plausible value in the calculation of the sampling variance (SHRTCUT = Y).

WGT Name of the sampling weight variable used for the analysis.

RWGT Prefix of the replicate weight variables. If replicate weights are calculated on the fly, this entry is left blank. When replicate weights are included in the file, these are expected to be in the format RWGT1 to RWGT#, where # corresponds to the value of NRWGT below.

NPV The number of plausible values used for the analysis.

NRWGT The number of replicate weights used to calculate the sampling variance.

Further information about these parameters and the specific values for the different large-scale assessments are available in the Help Manual.

24.6 Future Directions

Some of the macros accessed by the IEA's IDB Analyzer were first created in 1994 for the operational processing of the TIMSS 1995 data and were published at that time as part of the User Guide for the TIMSS 1995 International Database (Gonzalez & Smith, 1997). Over time, the original macros have been significantly improved and adapted to keep up with the changing features and requirements of existing studies and the needs of the emerging ones. The interface accessing these macros, now known as the IEA's IDB Analyzer, was first created in 2004, and this interface has also undergone extensive adaptation to incorporate the many different large-scale assessment studies it serves, as well as the new features and analysis types it is capable of setting up. Most recently, in early 2021, the IEA's IDB Analyzer began to have the capability of creating code to calculate Spearman rank-order correlation coefficients, and extensive work is being undertaken to replicate the statistical functionality, now available using SAS or SPSS, to be available using R as well. In addition, there is constant work going on to extend the current functionality and to improve the quality and usefulness of the output produced by the underlying macros.

While certainly not all statistical analyses are possible with the IEA's IDB Analyzer, it does offer a solid and well-tested set of useful capabilities to conduct many of the analyses most users of large-scale assessment databases would want to conduct. Because of the way the underlying macros are written, and the way in which they can be accessed, they are also quite useful for batch processing of multiple requests for analysis and ideal for operational processing of data.

Looking towards the future, as resources allow, the IEA's IDB Analyzer will continue to be updated to incorporate analysis specifications for newer studies, as well as newer statistical and display functionality.

References

Cochran, W. G. (1977). *Sampling techniques* (3rd ed.). Wiley.

Efron, B. (1982). *The jackknife, the bootstrap, and other resampling plans* (CBMS-NSF Monographs, Vol. 38). Society of Industrial and Applied Mathematics.

Gonzalez, E. (2014). Calculating standard errors of sample statistics when using international large-scale assessment data. In R. Strietholt, W. Bos, J.-E. Gustafsson, & M. Rosén (Eds.), *Educational policy evaluation through international comparative assessments*. Münster.

Gonzalez, E. & Smith, T. A. (Eds.) (1997). User guide for the TIMSS International Database – Primary and middle school years. Boston College.
IEA. (2021). *Help manual for the IEA IDB analyzer* (Version 4.0). www.iea.nl/data.html
Johnson, E. G., & Rust, K. F. (1992). Population inferences and variance estimation for NAEP data. *Journal of Educational Statistics, 17*, 175–190. https://doi.org/10.2307/1165168
Little, R. J. A., & Rubin, D. B. (1983). On jointly estimating parameters and missing data. *American Statistician, 37*, 218–220. https://doi.org/10.1080/00031305.1983.10483106
Lohr, S. L. (2010). *Sampling: Design and analysis* (2nd ed.). Brooks/Cole.
Martin, M., Mullis, I., & Hooper, M. (Eds.). (2017). *Methods and procedures in TIMSS 2015.* Boston College. https://timssandpirls.bc.edu/publications/timss/2015-methods.html
Mislevy, R. J. (1985). Estimation of latent group effects. *Journal of the American Statistical Association, 80*(392), 993–997. https://doi.org/10.1080/01621459.1985.10478215
Mislevy, R. J. (1991). Randomization-based inference about latent variables from complex samples. *Psychometrika, 56*(2), 177–196. https://doi.org/10.1007/BF02294457
Mislevy, R. J., Beaton, A. E., Kaplan, B., & Sheehan, K. M. (1992). Estimating population characteristics from sparse matrix samples of item responses. *Journal of Educational Measurement, 29*, 133–161. https://doi.org/10.1111/j.1745-3984.1992.tb00371.x
OECD. (2020). *PISA 2018 technical report*. OECD Publishing. https://www.oecd.org/pisa/data/pisa2018technicalreport/
Rubin, D. B. (1987). *Multiple imputation for nonresponse in surveys*. Wiley.
Rust, K. (1985). Variance estimation for complex estimators in sample surveys. *Journal of Official Statistics, 1*(4), 381–397.
Rust, K., & Rao, J. N. K. (1996). Replication methods for analyzing complex survey data. *Statistical Methods in Medical Research: Special Issue on the Analysis of Complex Surveys, 5*, 283–310. https://doi.org/10.1177/096228029600500305
Särndal, C., Swensson, B., & Wretman, J. (1992). *Model assisted survey sampling*. Springer. https://www.springer.com/gp/book/9780387406206
Schulz, W., Carstens, R., Losito, B., & Fraillon, J. (Eds.). (2018). *ICCS 2016 technical report.* https://www.iea.nl/sites/default/files/2019-07/ICCS%202016_Technical%20Report_FINAL.pdf
Westat. (2007). *WesVar® 4.3 user's guide.*
Wolter, K. (2007). *Introduction to variance estimation* (2nd ed.). Springer. https://www.springer.com/gp/book/9780387329178

Chapter 25
The Data Explorer

John L. Barone and Alfred M. Rogers

Abstract In this chapter we chronicle the Educational Testing Service's (ETS) development and advancement of the family of web-based analysis applications known collectively as the Data Explorer (DX). The DX provides for secure primary and secondary analysis of National and International large-scale educational assessment (LSA) data on a public Web platform. It greatly simplifies the computationally intensive analyses that use these large, complex assessment databases, and provides reporting capabilities in the forms of tables and graphics. This chapter further provides an overview of the underlying Data Explorer Technologies (DXT™) that drive the application and how they have been leveraged to develop and deploy other quality control, analysis, and reporting methodologies. Although this chapter is directed to a non-technical audience, there are occasional digressions into certain technical issues for the curious reader and to give an idea of the complexities of the design and operation of the system.

25.1 Organization of This Chapter

We first trace the evolution of ETS-developed report delivery systems, from manually delivered PDF files to the National Assessment of Educational Progress (NAEP) Data Tool, the first Web-based application, through the currently deployed NAEP Data Explorer, and finally to the International Data Explorer, which is used in the reporting of results from the following studies: the Programme for International Student Assessment (PISA); the Programme for the International Assessment of Adult Competencies (PIAAC); Trends in International Mathematics and Science Study (TIMSS); Progress in International Reading Literacy Study (PIRLS); the Teaching and Learning International Survey (TALIS); and the International

---

J. L. Barone (✉) · A. M. Rogers
ETS, Princeton, NJ, USA

L. Khorramdel et al. (eds.), *Innovative Digital-Based International Large-Scale Assessments*, Methodology of Educational Measurement and Assessment,
https://doi.org/10.1007/978-3-031-90951-1_25

Computer and Information Literacy Survey (ICILS). Each step in this evolution represents the attainment of, or improvement upon, the following goals:

1. Produce rapid and reliable results for requested analyses on the appropriate set (s) of data
2. Provide access to all of the relevant data variables without exposing the underlying micro-data
3. Have flexibility in the types of statistical results to be tabulated or displayed graphically
4. Accurately apply corresponding statistical models in the production of results
5. Be appropriate for use by people with a wide range of interests, purposes, and technical sophistication

The overarching objectives of these goals are to make the data widely available and to free the user from needing in-depth knowledge of the assessment design and of the advanced statistical methodologies that it entails.

We next present examples of the input to and output from the current version of the DX as the external user experiences it. The remainder of the chapter describes how the DX is used internally at ETS, where we have direct access to the DX engine to perform quality control operations, as well as develop reports in support of the NAEP, PISA and PIAAC projects. The chapter concludes with a more technical overview of the DX architecture and technology and the planned development activities to further advance the capabilities and the functionality of the DX.

25.2 Large-Scale Assessment Analysis Overview

Large-scale group national and international assessment programs (LSA), such as NAEP, PISA, and PIAAC, are designed to provide valid and reliable data on respondent group and subgroup performance on a broad range of domain competencies (Johnson, 1992; Beaton & Barone, 2017). By their nature, LSA programs do not intend to provide individual scores to respondents or summary results to schools; rather, they offer results for relatively large groups of students (e.g., fourth-graders) and subgroups of those populations (e.g., male and female students, students attending public or private schools, etc.), as well as results regarding subject-matter achievement, instructional experiences, and school environment.

They accomplish this by using sampling methodologies at both the population and cognitive domain levels. A carefully designed respondent-sampling procedure allows the results of the assessment to be representative of the different targeted reporting groups. Usually, a complex multistage sampling methodology is employed to ensure appropriate population and sub-population coverage and to ensure that valid estimates and standard errors can be calculated for the derived estimates to the required degree of precision (Rust & Johnson, 1992). The sample sizes for these surveys are determined by a combination of reporting requirements and cost. A sample weight is assigned to each participant to account for differential selection probabilities and to indicate the optimal contribution of that participant to the overall estimate. In order to accurately represent the sampling error, several resampling methodologies are available for use in the sample design and the calculation of these errors, such as the jackknife or balanced repeated replication. To ensure that the results of these calculations can be reproduced consistently, a set of replicate weights (Hsieh et al., 2009) are usually generated for each case in the study, or they can be generated on the fly using the appropriate stratification information included in the data files.

Similarly, cognitive domain sampling allows for a breadth of content/framework coverage and ensures coverage over the range of abilities within a subject area. This domain sampling design also limits testing time to avoid imposing an excessive burden on respondents. Owing to the limited number of items that each respondent receives, there is substantial uncertainty in the estimates of individual abilities. This uncertainty is accounted for in the generation and use of plausible values (Mislevy et al., 1992; von Davier et al., 2009), which are drawn at random from a distribution of estimated ability for each respondent in the assessment. Plausible values are not point-value scores for individuals but are intermediary estimates to be used in conjunction with the sample weights for calculating statistics for groups of students.

A byproduct of the plausible value methodology that also makes analysis of the data more complex are the proficiency levels, which are ordered categories defined by ranges on the continuous proficiency scales (Xie & Zhang, 2017). The NAEP achievement levels (Bourque, 2009) are one example of proficiency levels in which students are categorized as *below basic*, *basic*, *proficient*, and *advanced*. Other examples are the proficiency levels in PIAAC (OECD, 2016) and PISA (OECD, 2017) and the benchmarks in TIMSS. Because each student receives more than one estimate of proficiency, a level category must be assigned to the respondent for each plausible value, and, therefore, a respondent could be assigned to different categories, depending on the plausible value used for the classification. As with the plausible values, any analysis involving these proficiency levels involves performing a separate analysis with each classification, then combining the results of these analyses in a prescribed way (Rutkowski et al., 2014).

Most assessment programs provide public-use or restricted-use data files to make the collected data available for secondary analysis. However, proper analysis of these LSA data files is difficult and, at times, unfeasible using standard procedures from statistical systems such as SAS and SPSS, as most procedures of these statistical systems typically assume the use of simple random samples. The correct analyses involve employing a complex analysis methodology that was first introduced in NAEP by ETS (Johnson, 1989). This methodology includes the use of plausible values in conjunction with replicate weights, which may entail the calculation of several hundred intermediate statistics just to derive a single statistic and its standard error. For example, each scale score in PISA has 10 plausible values and there are 80 replicate weights per student. In order to calculate a mean and standard error for a single group, 810 separate weighted means of plausible values have to be calculated, then reduced to the two desired quantities.

Applying the correct methodology obviously requires a substantial amount of user knowledge and computer power to perform analyses. When performing analyses across multiple cycles (years) and/or countries of a given study, there is the additional burden of dealing with multiple files with different variable names, separate analyses, and combining and displaying the results. Experience has shown that these tasks discourage many users and limit the amount of secondary analyses that can be performed. In addition, concerns over the confidentiality of the data and conditions imposed by the granting agencies restricts access to these data files, further limiting the number of users.

Addressing the problem that secondary users face in working with the complex LSA databases led to a continuing effort to increase accessibility and usability through the development of easy-to-use application software and web-based analysis and reporting tools. These tools, including the Data Explorer, allow technical and non-technical users to produce results and reports that adhere to strict statistical, reporting, and technical standards required to perform LSA analyses.

Two more recently developed analytic tools are: the International Database (IDB) Analyzer (Gonzalez, 2024), developed by IEA (International Association for the Evaluation of Educational Achievement) as a front end for conducting analyses in SPSS and SAS; and EdSurvey (Zhang et al., 2024), an R package developed by American Institutes for Research (AIR) under contract with NCES.

Before the advent of computer-based and on-the-fly delivery of results, ETS used its high-performance computing power to generate pre-determined tabular summaries of NAEP proficiencies for the nation and, later, by states. Central to the creation of these summaries was the ETS proprietary F4STAT Statistical System (Beaton, 1973; Beaton & Barone, 2017) employed in Fortran programs to produce summary data tables (SDT) These were made available in (large) binders of computer printout. Throughout the 1990s these print images were converted to PDF format and copied to CD-ROM and shipped to the client for review and generation of reports. Even in this reduced and pre-calculated format, it was still not always easy to locate the desired statistics, especially when looking across individual state summaries.

The first web-delivered NAEP results were simply the uploading of these PDF files to a Web server accompanied with a front-end interface to locate results by cycle, grade level, and reporting sample, and a table of contents indexed on the variables in the analysis to quickly access and present the results. This improved access to results still had limitations on the number of variables available, the variety of statistical results that could be reported, and the precision of those results, if further meta-analyses were desired (e.g. significance tests). The NAEP Data Tool was developed to address these limitations.

25.2.1 The NAEP Data Tool

The ETS NAEP Data Tool (NDT), deployed in 2001, was the first incarnation of a Web-based application that provided an interface for the specification of the analyses to be retrieved and to display the results, which could be extracted or used in further analyses, such as significance tests. The interface provided navigation through the cognitive domains (math, reading, etc.), reporting units (national, individual states), and cycles (assessment year); then, variables of interest could be selected and extracted, and static summary tables displayed.

This application drew on a relational database of pre-calculated results for its displays rather than static PDF files. In this way, analyses that went across cycles and/or reporting groups could be organized and displayed in a more coherent and logical format. A new class of programs was then developed by ETS to calculate all of the relevant statistics and their co-statistics (standard errors, degrees of freedom, etc.) for a larger set of reporting and contextual variables across all jurisdictions and to write them out in full precision to the appropriate table in the database. These tables were identified by cognitive domain and grade level, and each record in each table contained a cycle identifier, the reporting jurisdiction, variable name, response level, type of statistic, and the statistic with its co-statistics.

The NDT was the first LSA application developed at ETS to employ a three-tier architecture, consisting of:

1. A web-facing interface for the specification and collection of analysis requests and display of analysis results
2. A set of programs and procedures (known collectively as *middleware*) that operates between the interface and the database
3. A relational database for the storage of data/results and all other meta-data used in the operation of the middleware or the interface display

The details of the operation of this architecture will be explained later in this chapter.

Among the new statistics that were available in the NDT were pre-specified percentile values (at the 10th, 25th, 50th, 75th and 90th levels) and proficiency (achievement) levels. As described earlier, proficiency levels are assigned to students based on the plausible values. The NAEP achievement level statistics include a set of percentages for each discrete level, as well as cumulative "at or above" percentages for some levels.

As computing power and capacity continued to grow, the question was raised as to whether or not the NDT application could be redesigned to query the micro-data in real time to deliver results. This method was proven feasible and efficient; the NAEP Data Explorer was then created to use this method.

25.2.2 The NAEP Data Explorer

The specifications for the development of the NAEP Data Explorer (NDE) included several challenging requirements:

1. Provide for real-time access, navigation, and mining of the extensive NAEP database, requiring the management of large datasets across multiple cycles,

and provide proper integration and organization of the data in ways that facilitated analysis.
2. Develop optimal algorithms and data structures and efficient underlying computational engines.
3. Ensure the integrity and security of the underlying NAEP database, therefore protecting the identify of respondents from public disclosure.
4. Support the processing of up to hundreds of simultaneous analysis requests, from multiple users, accurately, and within a reasonable amount of time (<2 mins), given the complexity of the analysis.
5. Develop tabular and graphic representations of the results in the appropriate format (bar charts, line charts, etc.) that can be dynamically tailored to individual requirements.

The NDE application used the same three-tier design as developed for the NDT, but with the middleware taking on increased functionality, and the database growing significantly in size. The new middleware developed more complex instructions for the database to calculate and return large sets of intermediate statistics and then perform further calculations to reduce these statistics to the desired results. The new database performed the "heavy lifting" task of filtering, partitioning, and aggregating sums of variables, using standard database functions. The development team for the NDE consisted of three groups, each primarily focused on the design and implementation of each tier, but with overlapping expertise and responsibilities to ensure the deployment of an integrated product.

When it was first deployed in 2005, the NDE was still partitioned according to cognitive domain and age/grade cohort, as there were no requirements in the reporting for analyses to cross domains or cohorts. Over the course of time, several related Data Explorers were deployed for NAEP-related studies:

1. The NAEP Long Term Trend (LTT) study for age-based NAEP data collected from its inception in 1970 through 2012
2. The High School Transcript Study (HSTS) that linked high school seniors' transcript information with NAEP proficiency measures, which introduced additional measures to the set of scale scores that could be analyzed, including grade point averages and course credits
3. The National Indian Education Study (NIES), which oversampled Native American students and administered surveys to those students and their teachers and schools
4. A special Data Explorer deployed for use only by NCES to track student exclusion rates by state within assessment cycles

A new release of the NDE in 2018 introduced a new interface that permitted analyses to be conducted across subject domains and/or grade levels. In 2019, the report graphics were converted to a format that is more compatible with the several modern Web browsers that can be used to access the NDE application. Another enhancement was the incorporation of linking error into NAEP standard errors for comparisons across cycles. This last enhancement was necessitated by the program's change in assessment modes from paper to computer and the statistical linking adjustments required to make the results from the two different modes comparable (Mazzeo et al., 2024; Jewsbury et al., 2024, both of this volume).

25.2.3 The International Data Explorer

In the mid 2000s, the United States National Center for Education Statistics (NCES) provided funding for the development of a Data Explorer to produce results from the 2006 PISA study. Despite the close similarities between the PISA and NAEP designs and the analysis methods for their data, there were some slight differences that required modifications to the analytic engines and the rules and methods for displaying the results.

One of the major changes to the analytic engine involved a change in the calculation of estimates for the standard errors of the reported statistics. NAEP uses a standard jackknife method (JK2) for sampling variance estimation while PISA uses a modified balanced repeated replication method (BRR). For more detail, see *International Large-Scale Assessment Data: Issues in Secondary Analysis and Reporting* (Rutkowski et al., 2010).

The success of this initial effort enabled more funding for the addition of data from the other PISA cycles from 2000 to 2012 to this application. It also led to the development of Data Explorers for the TIMSS and PIRLS studies. One drawback to these Explorers was that the database design followed the NAEP model, in which each respondent is assessed in one and only one domain, and therefore, there was one dataset for each cognitive domain by cohort (age or grade). The interfaces for these Explorers were similarly partitioned, preventing the analysis across domains that would have been desirable for PISA and TIMSS.

In 2008, ETS was awarded the Organisation for Economic Co-operation and Development (OECD) contract for the PIAAC study and was required to develop and deploy a Data Explorer to be hosted on the OECD site. As a participant in this study, the United States, through NCES, requested the development of a parallel Data Explorer to accompany the other International Data Explorers identified above. A key difference between the NCES and OECD versions of the DX is that the OECD database is built from the restricted-use files (RUF) that are available to ETS as the contractor, while the NCES database is based on public-use files (PUF), to which NCES has access, but also includes US-specific variables that the OECD version could not include.

One significant difference between the PIAAC study and the other studies in its design is that each country was allowed to implement one of several sampling designs. All previous studies applied the same sampling design to all participating jurisdictions, allowing the Data Explorer to process a single dataset that could be grouped by jurisdiction and apply the same analytic process to each of those subgroups. The PIAAC design required the storage of each country's data into a separate dataset, each with its own set of parameters dictating how the sampling errors are to be determined. Furthermore, using the NDE database design required a separate dataset for each of the three cognitive domains that were scaled.

In 2012, ETS was awarded the PISA 2015 contract, which also required the development and deployment of a PISA Data Explorer to be hosted on the OECD site. The development of this Data Explorer presented several challenges, each of which represented a break from the NDE database structure and analysis methodology that had been used for all IDE's.

1. Because students responded to items from two or more cognitive domains, the interface should enable the selection of these domain scale scores from the same menu, rather than from a separate menu for each domain.
2. Some of the reports required collapsing of the proficiency levels in ways that were not possible with the NDE algorithms.
3. The combined number of international and national variables far exceeded the capacity of the database to contain them in a single table.
4. The specifications called for an early password-controlled version that participating countries could access before the official release in order to quality check the data and develop reports for their own country. Each country would have access to all international variables and their own national variables but no other country's national variables.

Further enhancements were made to the IDE database structure and the analysis engines. The new database design permitted the storage of all concurrently estimated plausible values and their associated proficiency level scores in a single table. It also allowed the linking of up to ten tables of variables with these plausible values to store up to 9000 variables for each case in a given dataset. Further, several tables were designed and added for the storage and maintenance of login ID's with their passwords and their linkages to the variables and jurisdictions that could be displayed for each ID. The analysis engine now properly handled proficiency levels, enabling the on-demand collapsing of the categories and exposing them to the full range of charting options within the IDE. These changes were first implemented in the release of the public PISA Data Explorer at the OECD in 2016 and were later propagated to the US-PISA, TIMSS, PIRLS, PIAAC, and US-PIAAC Data Explorers.

The International Data Explorers currently on the NCES site are:

1. Programme for International Student Assessment (PISA)
2. Trends in Mathematics and Science Study (TIMSS)
3. Programme for the International Assessment of Adult Competencies (PIAAC)
4. Teaching and Learning International Survey (TALIS)
5. Progress in International Reading Literacy Study (PIRLS)

The International Data Explorers currently on the OECD site are:

1. Programme for International Student Assessment (PISA)
2. Programme for the International Assessment of Adult Competencies (PIAAC)

The PISA, TIMSS, and PIRLS Data Explorers include data across multiple cycles that can be used to measure and report trends in achievement by targeted subgroups in each country. Although the PIAAC Data Explorer currently includes a single cycle of data for this study, it also includes data from earlier studies of Adult Literacy, of which the cognitive domains have been recast and rescaled to conform to the PIAAC reporting metrics. The responses to questionnaires in these surveys have also been linked to certain questions in PIAAC to support similar "trend" analyses.

For the PIAAC IDE on the OECD site, these studies include:

1. International Adult Literacy Study (IALS)
2. Adult Literacy and Life Skills Survey (ALL)

The PIAAC IDE on the NCES site includes the above studies as well as the following studies only conducted in the United States:

1. The National Adult Literacy Survey (NALS)
2. The National Assessment of Adult Literacy (NAAL)

25.3 Navigating the Interface

In this section, we will demonstrate how the user can build a report in the PISA Data Explorer. For this example, we look at three increasingly complex analyses using the reading results for Australia, Canada, and the United States in the PISA 2015 cycle:

1. What are the average reading scores by gender for each country?
2. What are the percentages in each reading proficiency level by country?
3. How do the distributions of reading proficiency differ by gender within country?

The URL for the application that produced the following set of images is https://pisadataexplorer.oecd.org/ide/idepisa/.

After opening the link, use the drop-down list beside **Subject** to locate the *Reading, Mathematics and Science* option, and click on it (Fig. 25.1).

The next screen presents three sets of choices that must be made before proceeding with the report:

1. The year(s) that is to be used in the report, listed across the top of the **Measure** and **Jurisdiction** lists
2. The available Measures or dependent variables that will be the basis of the average or percentile statistics
3. The Jurisdictions that are available for reporting, partitioned into International, OECD, Partners, and Adjudicated Sub-Region groupings

The grids formed by the years as columns and the measures and jurisdictions as rows contain one of two symbols to indicate whether or not data are available for the row and column combination: if data are available and if data are not available.

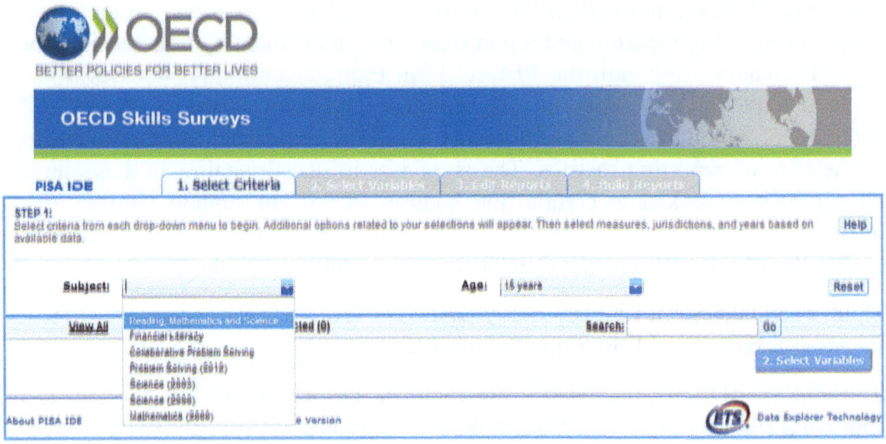

Fig. 25.1 Domain selection screen

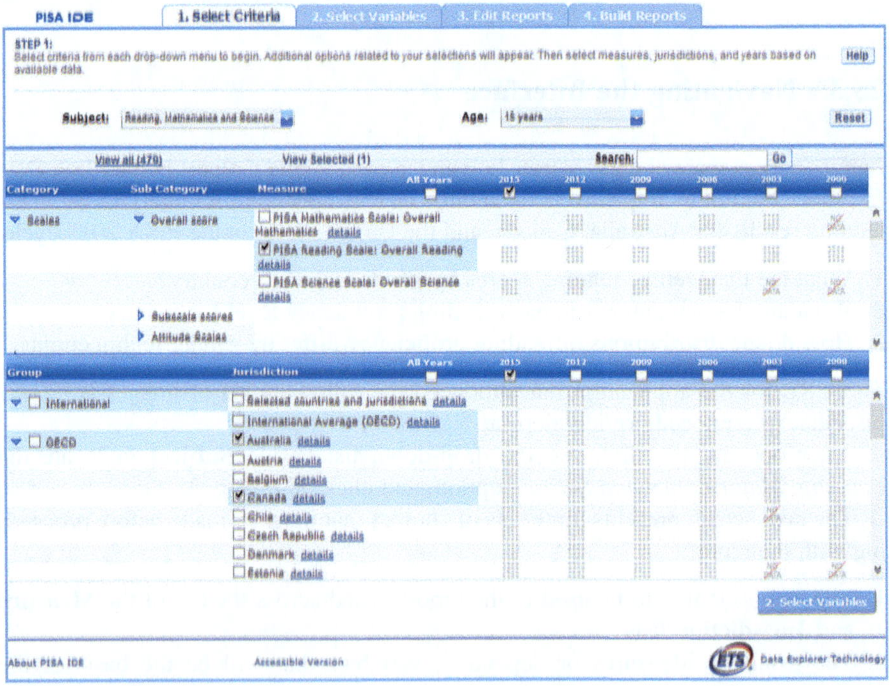

Fig. 25.2 Select criteria

From the **Measure** list, select "PISA Reading Scale: Overall Reading" and from the **Jurisdiction** list locate and use the scrolling feature to select Australia, Canada, and the United States. Then click on the **Select Variables** tab (top) or button (bottom) (Fig. 25.2).

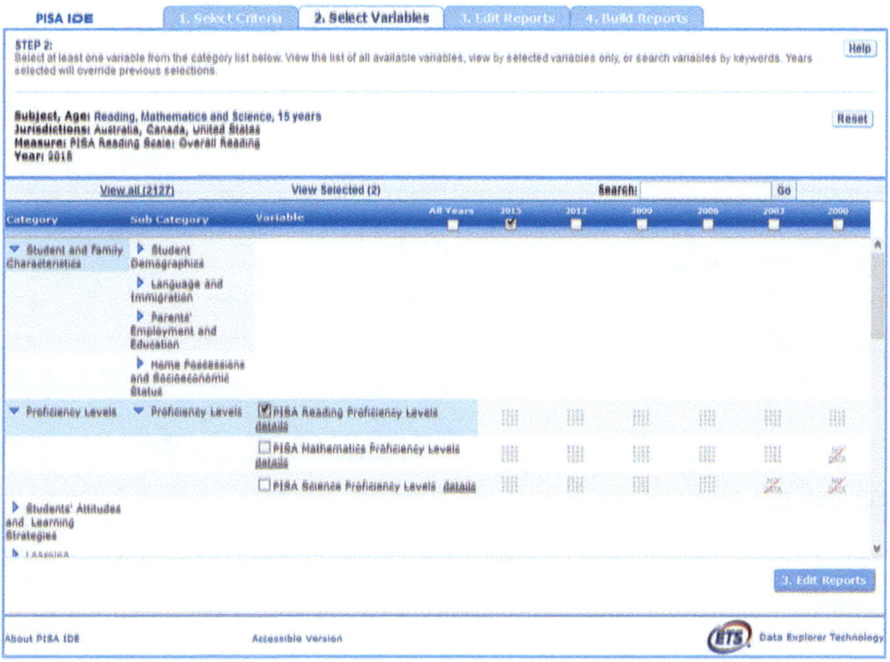

Fig. 25.3 Select variables

The *Select Variables* display contains a listing of all variables that are available for analysis. The categories and sub-categories for these variables are displayed in a format that can be collapsed or expanded by clicking on the symbol before each category label. For our example, first select the variable labeled as "Sex" under the *Student Demographics* category; then, collapse that category to identify and expand the *Proficiency Levels* category, and select the "PISA Reading Proficiency Levels" variable. Now click on the **Edit Reports** tab (top) or button (bottom) (Fig. 25.3).

The Edit Reports display allows the user to customize the report(s) based on the selections made to this point. The prominent part of the display is the listing of reports that are to be created as a result of the selections made in the previous displays. Notice that the DX automatically creates an extra report based on a cross-tabulation of the selected variables when two or three variables are selected (Fig. 25.4).

The first report will address the first research question without modification, as it will compute the average scale scores for males and females for each of the selected countries. The second report, however, will need modification in order to report percentages by proficiency level.

To modify this report, click on the **Edit** link within the Report 2 row. The Edit Report pop-up window will then be displayed. Under the Statistic column, de-select Averages, and select Percentages. Click on the **Done** button to save these changes (Fig. 25.5).

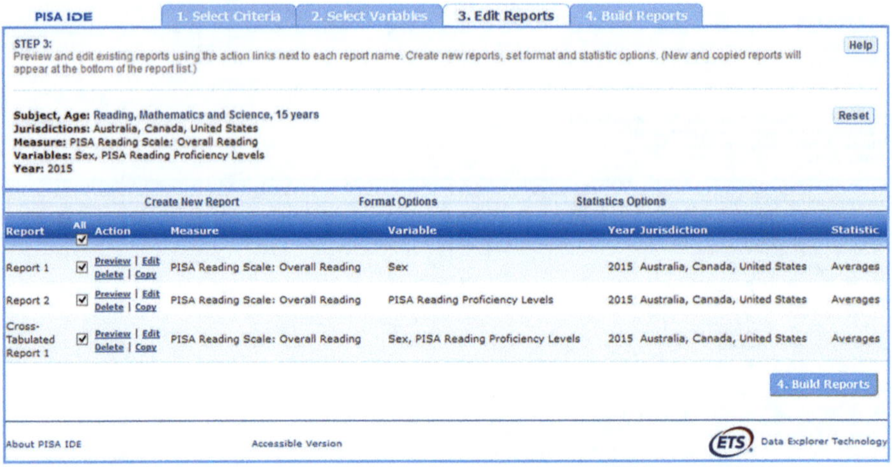

Fig. 25.4 Edit reports

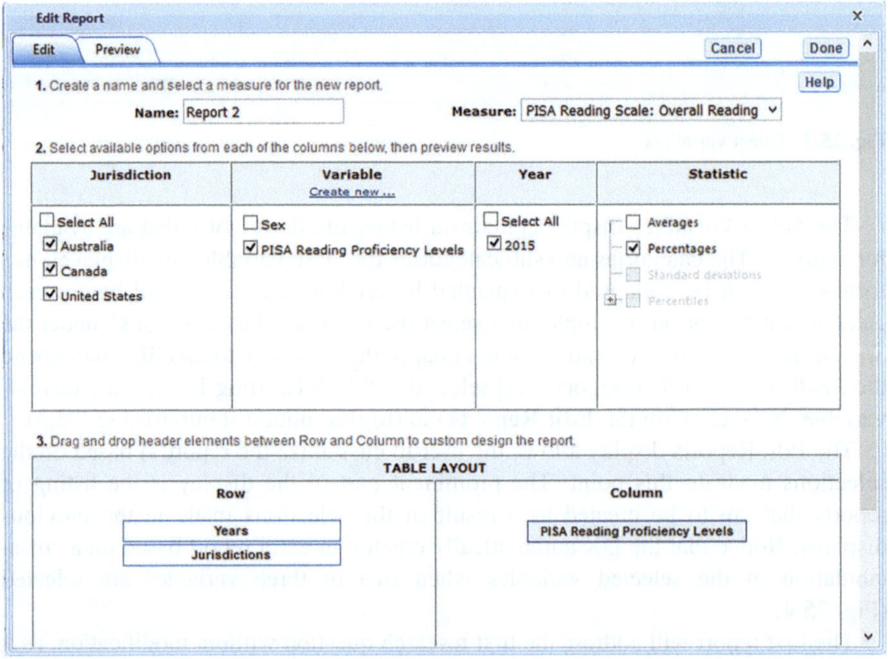

Fig. 25.5 Edit report 2

The third report will also require some editing, as it will not only require both averages and percentages, but its presentation must be arranged to address the third research question. Here is the Edit Report display as initially configured (Fig. 25.6).

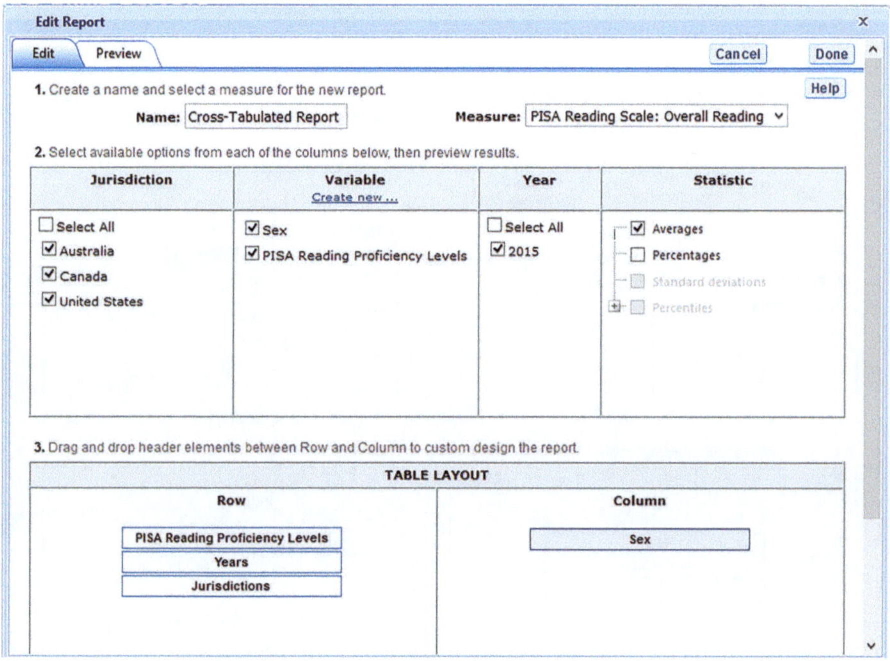

Fig. 25.6 Edit report 3 (Before)

The desired analysis requires that the percentages be reported by proficiency level, not sex, so those two variables must be "swapped' in the Table Layout section. This is accomplished by simply dragging the proficiency levels variable from the Row side of the section into the Column side, then dragging the Sex variable into the Row side, below the Years and Jurisdictions elements. This last move will result in the nesting of Males and Females within each country in the report. The display should now look like this (Fig. 25.7).

Clicking on the **Build Reports** tab (top) or button (bottom) will produce the results for Report 1 (Fig. 25.8).

Of special note are the buttons for producing charts of the results, tests of significance within groups or across countries, and exports of any or all of these report-related products into the common formats of Excel, Word, or HTML documents. The Gap Analysis is a more complex analysis that measures change in proficiency differences between targeted subgroups across cycles. The link labeled as **Link to this Page** will produce a URL that can be saved or shared with others to replicate the initial tabular results.

Use the **Select Report** drop down tool to locate and click on Report 2 to produce these results (Fig. 25.9).

Use the **Select Report** menu again and select the Cross-Tabulated Report 1 to produce the last results (Fig. 25.10).

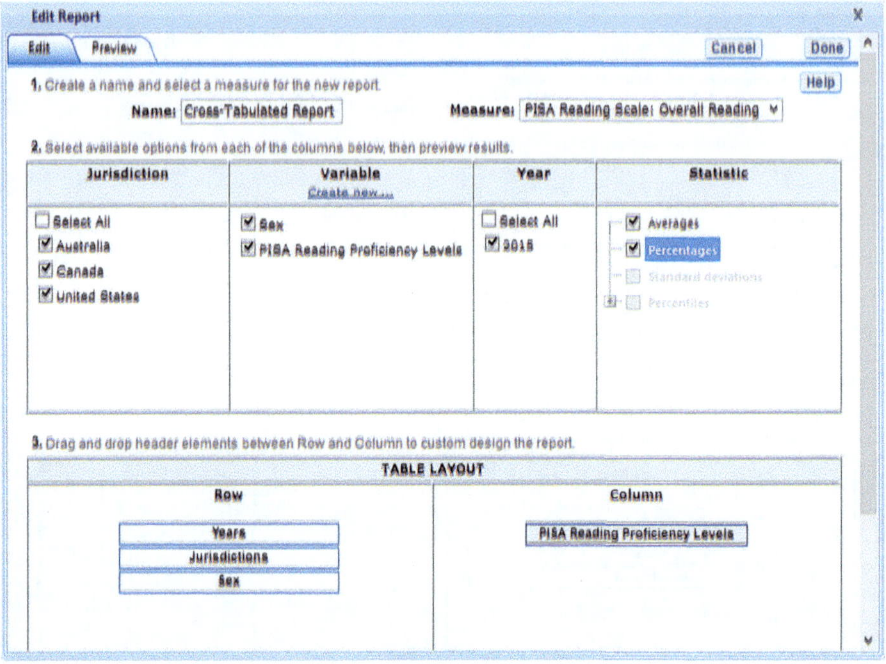

Fig. 25.7 Edit report 3 (After)

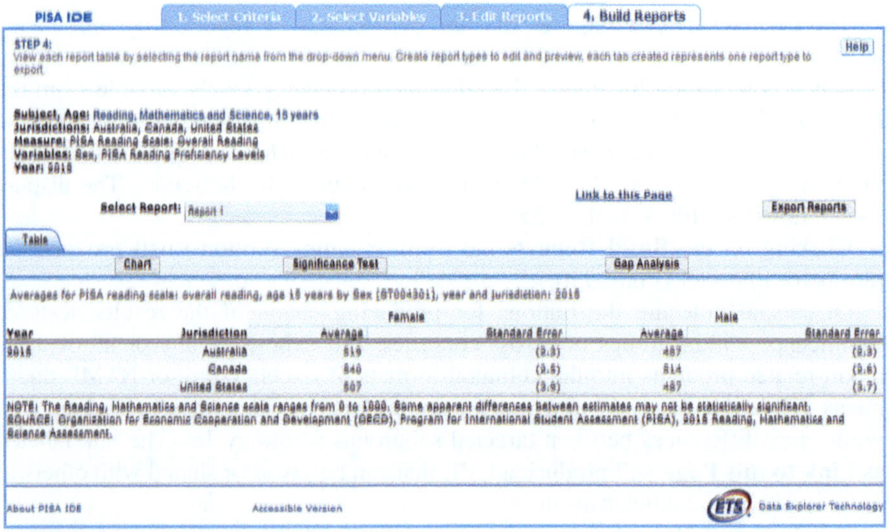

Fig. 25.8 Results for report 1

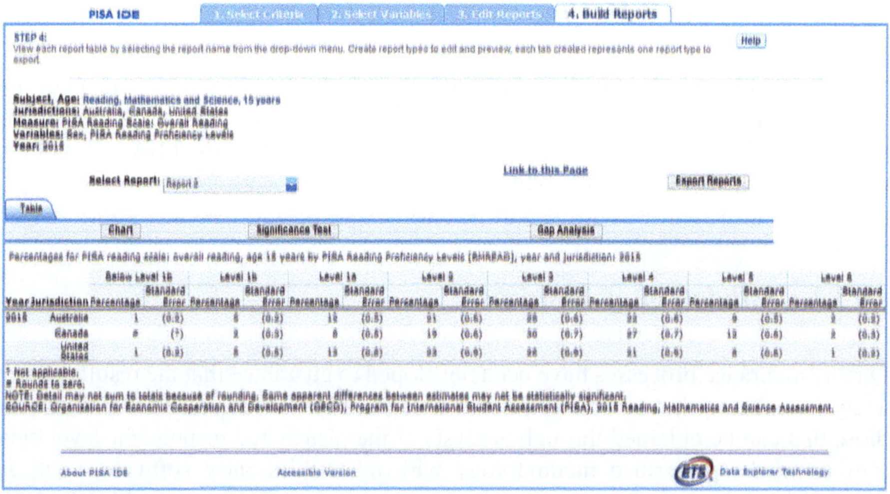

Fig. 25.9 Results for report 2

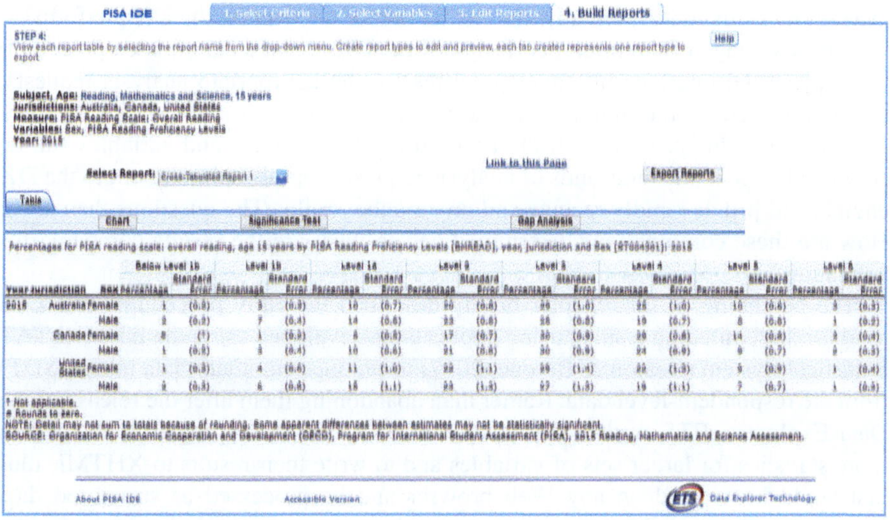

Fig. 25.10 Results for report 3

We can see here that within each country the percentages for Males are higher than those for Females at the low end of the proficiency levels and those for the Females are higher at the upper end of the levels, consistent with the findings in the first report where the averages for Females were higher than those for Males in each country.

25.3.1 Additional Documentation and Tutorials

The above examples only present the surface of the full set of navigation, analysis, and report generation capabilities of the DX. For additional, detailed DX documentation, as well as in-depth user tutorials, we direct the reader to the LSA web-portals provided by the NCES and the OECD.

25.3.2 Validation of Results

Quality assurance processes have been developed to guarantee that the results of any analysis in the myriad of Data Explorers are accurate. They are also consistent with those that can be obtained through analysis of the distributed respondent-level data files using the prescribed methodology with other third-party software, such as SPSS, STATA, or SAS. Given the large number of cases and variables in each data set and the complexity of the analysis design, how can we possibly perform adequate quality assurance procedures in the often very limited time between the availability of the data internally and the release of the data on the DX platform?

One advantage of the three-tier design of the DX is that an automated procedure can replace the interface by creating scripts for the appropriate analysis requests, submitting them to the middle tier, and processing the returned XML results data stream. Using this procedure, lists of measures, jurisdictions, and variables can be combined to generate thousands of analysis requests, rapidly submit them to the DX engine, and just as rapidly receive and process the results. The questions then arise: How are these combinations determined and to what are the results compared for validation?

The beginning of this chapter briefly described the early procedures that ETS used for disseminating results, using applications developed using the ETS F4STAT Statistical System (Beaton & Barone, 2017) to produce summary data tables (SDT) from the respondent-level data. Rather than abandoning them after the release of the Data Explorers, ETS further optimized and upgraded these programs to develop more statistics for larger sets of variables and to write their results to XHTML files that could be viewed on any Web browser and be processed as structured data documents. These XHTML result files would then serve as both the driver for the combinations of analyses to be performed, as well as the reference for comparison of the produced results.

The data validation process then consisted of running the SDT generation program with a set of parameters defining the cycle, measures, jurisdictions, and variables to be used in the analysis. The output XHTML file from this run is then submitted to the automated procedure (robot) that, for each analysis, would create a script defining the analysis and submit it to the appropriate DX engine. The received results would then be compared to those in the XHTML document and only produce a message if the results differed significantly.

The diagram in Fig. 25.11 summarizes the data validation process.

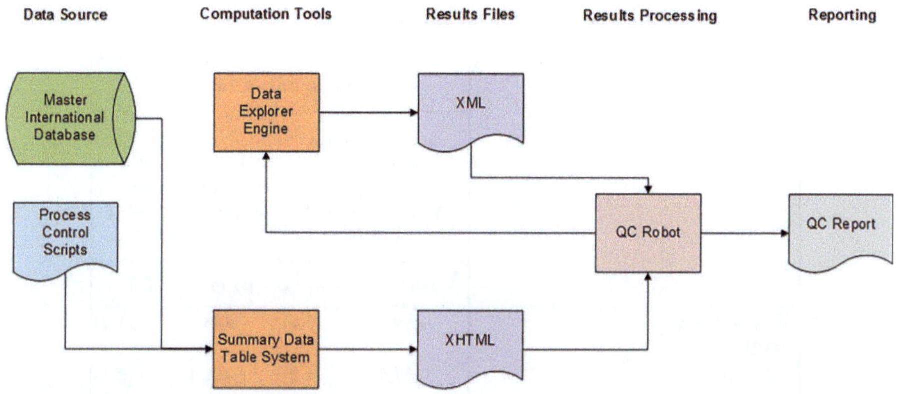

Fig. 25.11 Data validation process

25.3.3 Report Generation

In a similar manner to the validation process, ETS uses the Data Explorer engine to develop tables and graphics that are used in the publication of reports. The client (NCES or OECD) provides the shells for these tables in Excel files, including the text to be included in these tables as well as all formatting styles. ETS then converts the information for each table into a script specifying all of the parameters to be used by the DX in the analysis. Depending on the type of analysis requested, the scripts are submitted to one of three analytic processes:

1. Item analysis for the responses to the cognitive part of the survey
2. Data Explorer engine for the tabulations based on survey questions and/or derived variables
3. Special analysis for those requests that cannot be handled by the above analytics

The results from these analyses are saved until all analyses for a given report have been completed.

A design template document specifies the overall look of the report; a report structure document itemizes and orders the reports to be produced. When these documents have been finalized, the report generator procedure is launched, once for each type of desired publication. The economy of this approach is that the results only have to be calculated once and can be displayed in one or more ways across the publications, allowing the formats of the reports to be modified without necessitating a recalculation of the contained results (Fig. 25.12).

The diagram in Fig. 25.13 summarizes the dynamic reporting system.

Reading performance, by main language spoken at home

	Reading performance			
	Students who speak the language of assessment at home		Students who speak another language at home	
	Mean	S.E.	Mean	S.E.
OECD average	499.5	(0.5)	446.0	(1.8)
OECD				
Australia	507.0	(1.8)	485.4	(5.4)
Austria	498.3	(2.7)	427.8	(5.9)
Belgium	510.7	(2.3)	442.4	(5.9)
Canada	529.6	(2.3)	520.7	(3.5)
Chile	459.4	(2.6)	430.2	(14.4)
Czech Republic	490.3	(2.4)	440.5	(9.6)
Denmark	505.6	(2.5)	441.6	(6.2)
Estonia	522.6	(2.2)	473.3	(7.5)
Finland	531.0	(2.4)	461.4	(8.9)
France	507.9	(2.3)	430.8	(8.4)
Germany	523.9	(2.9)	446.4	(8.7)
Greece	471.4	(4.2)	404.8	(10.5)
Hungary	470.4	(2.7)	448.2	(13.1)
Iceland	486.6	(2.0)	408.6	(9.2)
Ireland	522.9	(2.5)	497.8	(5.6)
Israel	486.1	(3.7)	442.5	(7.6)
Italy	494.0	(2.6)	446.3	(5.2)
Japan	516.6	(3.1)	432.5	(38.6)
Korea, Republic of	518.1	(3.5)	‡	(†)
Latvia	492.5	(1.8)	449.8	(5.8)
Luxembourg	530.3	(4.3)	472.5	(1.5)
Mexico	426.0	(2.6)	352.2	(8.2)

Fig. 25.12 Excel table created by report generator

25.4 Architecture and Future Enhancements

Since its first operational deployment in 2005, the Data Explorer architecture has undergone several enhancements, some due to the emergence of new assistive technologies and others due to the particular features of the programs for which it has been deployed, including the demands of the clients for these programs. Each revision has delivered a more flexible, feature rich, and higher capacity version without sacrificing speed, precision, or efficiency. One immediate example of the continuing advancement of the DX is the recently released user interface, described earlier, that was deployed for the NAEP Data Explorer (NDE).

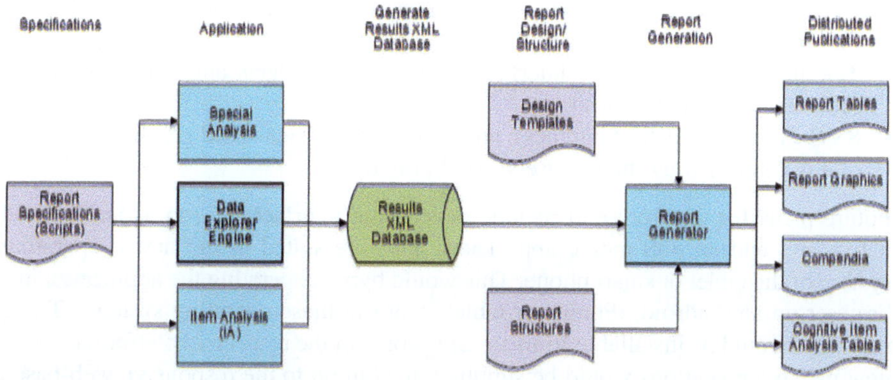

Fig. 25.13 Dynamic reporting system

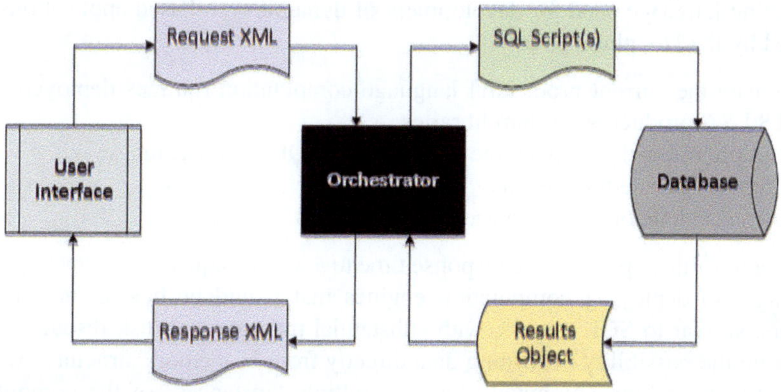

Fig. 25.14 Data explorer three-tier structure

A brief overview of the DX three-tiered architecture was described earlier in this chapter. The following section is presented for the reader interested in learning more about the underlying DX technology and a selection of proposed enhancements. The diagram in Fig. 25.14 presents an overview of the three-tiered architecture as described in the following text.

Currently, the front end is deployed as a Web-based interface, which has been configured to work with Internet Explorer, Firefox, Chrome, and Safari browsers. It employs custom designed controls and the following technologies:

- JavaScript: a browser-supported, web application development language that is used to make the DX dynamic. DX displays use responsive programming practices so that browser pages are rendered specific to the device making the request: desktop, tablet, or phone.
- HTML (Hyper Text Markup Language): used to make up the layout and structure for the DX web interface.

- CSS (Cascading Style Sheets): the language developers used to style the DX interface.
- Telerik: collection of user interface controls and widgets applied to assist DX browser interface development.
- SVG (Scalable Vector Graphics): an open standard, text-based graphics language that is supported by the browsers listed previously.

Future plans for the front end include developing interfaces to run as an Android native app and an iOS native app. These apps are suited to the native operating system of the tablet or smart phone. This would bypass operating the application in a browser on an Android, iPhone, or tablets that run these operating systems. These new apps would be installable from an "app store" of the respective platforms. These new native applications would be supplied in addition to the responsive web-based solutions already provided.

The middle tier (computation engine) is written in C#, an object-oriented programming language used for development of dynamic web-based applications, and is used by the DX platform to:

- Emulate the current procedural language computation routines deployed in the ETS LSA production system libraries
- Read micro-level data from the DX data tier SQL Server database
- Compute intermediate results
- Save cacheable objects to speed future requests

In order to further speed up the response time to analysis requests, the developers are looking into deploying computation engines that would be hosted on dedicated servers, similar to SQL servers, with substantial memory and disk resources. This opens up the possibility of reading data directly from in-memory structures without the need for caching computed objects. Multiple "instances" of the computation engine could then be launched for each database that the host machine can support. The end result would be services that depend on the amount of RAM installed and the speed of the processors rather the load placed on the hosted data sets. This would greatly reduce the database sizes by eliminating the "pre cache" that currently requires substantial computation and elapsed time to create and load before delivery of each product is made to a client.

The Data tier is a SQL server-relational database containing tables that conform to exact specifications, which enable the middle tier to operate efficiently and identically for each instance of the DX. A combination of stored procedures and SQL queries generated by the middle tier are used for retrieval of requested data objects required to analyze, display, and report user requests.

25.5 Conclusion

The number and diversity of Data Explorer applications since its inception in 2005 is a credit to the simplicity and flexibility of its three-tiered design. Each of these tiers have undergone several changes to meet new database, computational, and reporting requirements. ETS is committed to further improve and expand the application to new databases and a growing audience of government agencies, policy makers, foundations, the media, and the general public in the delivery of timely, accurate, and informative results from these large-scale surveys and assessments.

References

Beaton, A. E. (1973). F4STAT statistical system. In W. J. Kennedy (Ed.), *Proceedings of the computer science and statistics: Seventh annual symposium of the interface* (pp. 279–282). Iowa State University Press.

Beaton, A. E., & Barone, J. L. (2017). Large-scale group-score assessment. In R. E. Bennett & M. von Davier (Eds.), *Advancing human assessment: Methodological, psychological, and policy contributions of ETS* (pp. 233–284). Springer Open. https://doi.org/10.1007/978-3-319-58689-2_8

Bourque, M. L. (2009). *A history of NAEP achievement levels: Issues, implementation, and impact 1989–2009*. National Assessment Governing Board. https://files.eric.ed.gov/fulltext/ED509389.pdf

von Davier, M., Gonzalez, E., & Mislevy, R. J. (2009). What are plausible values and why are they useful? *IERI Monograph Series, 2*, 9–36. Princeton, NJ. https://www.ierinstitute.org/fileadmin/Documents/IERI_Monograph/IERI_Monograph_Volume_02_Chapter_01.pdf

Gonzalez, E. (2024). The IEA's IDB analyzer. In L. Khorramdel, M. von Davier, & K. Yamamoto (Eds.), *Innovative digital-based international large-scale assessments – Foundations, methodologies and quality assurance procedures*. Springer.

Hsieh, C., Xu, X., & von Davier, M. (2009). Variance estimation for NAEP data using a comprehensive resampling-based approach: An application of cognitive diagnostic models. *IERI Monograph Series, 2*, 161–174. Princeton, NJ. https://www.ierinstitute.org/fileadmin/Documents/IERI_Monograph/IERI_Monograph_Volume_02_Chapter_07.pdf

Jewsbury, P., Jia, Y., & Xi, N. (2024). Effects of mode transition on instruments and subpopulation performance in NAEP. In L. Khorramdel, M. von Davier, & K. Yamamoto (Eds.), *Innovative digital-based international large-scale assessments – Foundations, methodologies and quality assurance procedures*. Springer.

Johnson, E. G. (1989). Considerations and techniques for the analysis of NAEP data. *Journal of Educational Statistics, 14*(4), 303–334. https://doi.org/10.2307/1164942

Johnson, E. G. (1992). The design of the national assessment of educational progress. *Journal of Educational Measurement, 29*(2), 95–110. https://doi.org/10.1111/j.1745-3984.1992.tb00369.x

Mazzeo, J., Liu, B., Donoghue, J., & Xu, X. (2024). Approximate standard errors for NAEP results that incorporate linking error under the random group design. In L. Khorramdel, M. von Davier, & K. Yamamoto (Eds.), *Innovative digital-based international large-scale assessments – Foundations, methodologies and quality assurance procedures*. Springer.

Mislevy, R. J., Beaton, A. E., Kaplan, B., & Sheehan, K. M. (1992). Estimating population characteristics from sparse matrix samples of item responses. *Journal of Educational Measurement, 29*(2), 133–161. https://doi.org/10.1111/j.1745-3984.1992.tb00371.x

Organisation for Economic Co-operation and Development. (2016). *Technical report of the survey of adult skills (PIAAC)* (2nd ed.). OECD. https://www.oecd.org/skills/piaac/PIAAC_Technical_Report_2nd_Edition_Full_Report.pdf

Organisation for Economic Co-operation and Development. (2017). *PISA technical report*. OECD. https://www.oecd.org/pisa/data/2015-technical-report/

Rust, K. F., & Johnson, E. G. (1992). Sampling and weighting in the national assessment. *Journal of Educational and Behavioral Statistics, 17*(2), 111–129. https://doi.org/10.3102/10769986017002111

Rutkowski, L., Gonzalez, E., Joncas, M., & von Davier, M. (2010). International large-scale assessment data: Issues in secondary analysis and reporting. *Educational Researcher, 39*(2), 142–151. https://doi.org/10.3102/0013189X10363170

Rutkowski, L., von Davier, M., & Rutkowski, D. (Eds.). (2014). *Handbook of international large-scale assessments: Background, technical issues, and methods of data analysis*. CRC Press. https://doi.org/10.1111/jedm.12095

Xie, Q., & Zhang, T. (2017). *Analyses using achievement levels based on plausible values* (NAEP Working Paper #2017–02, NCES Data R Project Series #01). American Institutes for Research. https://www.air.org/sites/default/files/downloads/report/Analyses-Using-Achievement-Levels-Based-on-Plausible-Values-NAEP-April-2017.pdf

Zhang, T., Sikali, E., & Bailey, P. (2024). Analyzing large-scale assessment data in R: EdSurvey. In L. Khorramdel, M. von Davier, & K. Yamamoto (Eds.), *Innovative digital-based international large-scale assessments – Foundations, methodologies and quality assurance procedures*. Springer.

Chapter 26
Statistical Tools for Analyzing Data from Large-Scale Assessments in Education

Artur Pokropek and Tomasz Żółtak

Abstract Analysis of large-scale assessment in education (LSAE) datasets is challenging for many researchers because of the need to use specific statistical methods that properly account for complex sampling designs and multiple latent variable indicators, called plausible values (PVs). In this chapter, we discuss resampling methods that allow for the proper estimation of sampling variance and PVs methodology that allows for inferring the latent variables, in a short and accessible way. We show how to properly utilize those methods in the context of LSAEs presenting examples of typical LSAEs analyses. The examples show how to perform analyses in a convenient way using additional packages that are available for two popular statistical software programs: Stata and R. We present ***pisatools***, ***piaactools***, and ***repest*** packages from Stata and ***mitools***, ***survey***, and ***svyPVpm*** libraries from R. For both environments, analyses involving descriptive statistics, frequency tables, OLS regression, and logistic regression are presented.

26.1 Large-scale Assessments in Education and Their Specificity

Large-scale assessments in education (LSAE), such as the Programme for International Student Assessment (PISA), Progress in International Reading Literacy Study (PIRLS), Trends in International Mathematics and Science Study (TIMSS), and the Survey of Adult Skills (PIAAC) are designed to measure the average levels of competencies, skills, or knowledge in target populations. To achieve this goal, a group of respondents needs to solve a set of test items prepared to measure skills under investigation. To obtain the most accurate results, both the respondents and the

A. Pokropek (✉)
Educational Research Institute National Research Institute (IBE PIB), Warsaw, Poland

T. Żółtak
Institute of Philosophy and Sociology Polish Academy of Sciences, Warsaw, Poland

L. Khorramdel et al. (eds.), *Innovative Digital-Based International Large-Scale Assessments*, Methodology of Educational Measurement and Assessment,
https://doi.org/10.1007/978-3-031-90951-1_26

items need to be selected with care. For the former, random samples are drawn from the target populations to make sure that respondents are representative of those populations. For the latter, sets of test items are prepared by groups of experts who make sure that the contents of the measured domains (e.g., competence, skill, or area of knowledge) are well represented. Most LSAEs aim to measure very broad skills, like "reading abilities" or "numeracy skills." Measuring such broad domains requires testing many different types of tasks, so, usually, the item pool is sizeable, too large to be processed by a single respondent on a single day. To handle this problem, so-called missing-by-design plans are used in which not all items are solved by all respondents (see Pokropek, 2011 for an overview). Such designs, while very efficient, face two complications. First, they require large sample sizes, which are usually obtained by complex sampling methods. Second, they require complicated procedures for scaling cognitive outcomes and generating equivalent scores for respondents answering to different sets of items. Not only do these complications affect data collection and scaling procedures, which are conducted before substantial analyses, but they also determine the way in which secondary analyses must be conducted to properly account for sampling and measurement error. In this chapter, the consequences of sampling and scaling procedures used in LSAEs are briefly discussed, and proper techniques for handling such data are described, followed by examples that use proper statistical tools for analyzing data from LSAEs.

26.2 Sampling of Respondents and Accounting for Sampling Error

In the LSAEs, multi-stage and clustered sampling designs are used, in which the first stage of sampling comprises groups of respondents (e.g. schools or municipalities), and only in further steps are individual respondents or whole groups of individual respondents (classes, grades) sampled. Such designs allow for reducing the cost of data collection. Testing 30 students in one school at one time generally is much less expensive than testing 30 students spread across the country. Although such designs are usually less statistically efficient (i.e. they lead to less precise estimates) than simple random samples of the same size, they are still more efficient in terms of cost. Moreover, in some countries, multi-stage sampling is necessary because of the lack of a register that could be used as a sampling frame. Usually, it is much easier to obtain a list of all schools than a list of all students in a country. In addition, stratification, explicit or implicit, is used to increase statistical efficiency. In stratified samples, target populations are divided into strata during the sampling procedure (explicit) or in the phase of constructing weights (implicit). Typical strata are types of schools (in educational research) or geographical regions, and sampling from each strata is performed independently to guarantee proportional representation of them in a sample. Sometimes stratification is also used to oversample some small groups that are of special interest to researchers and policy makers to assure that characteristics of these groups can be estimated with reasonable precision. Specific sampling designs vary considerably between studies and sometimes, as in the case of PIAAC, also between different countries within one study (Mohadjer et al., 2016).

Features of multi-stage cluster stratified sampling design violate the assumption that sampling units are independent and identically distributed (i.i.d.) and, therefore, require special treatment to obtain correct point estimates and their standard errors. Two aspects must always be addressed in estimation using data that comes from studies based on complex sampling schemes: unequal weights of observations (i.e. each observation "represents" a different number of units in the population) and impact of the sampling scheme on the precision of estimates (i.e., magnitude of the estimator's variance). Proper sampling weights must be applied to calculate unbiased point estimates. Unbiased standard errors are obtained with the use of resampling techniques.

In the resampling approach, the estimator's variance is estimated by assessing variability of the parameter estimates in a set of subsamples, so called *replicates*, that are drawn from the set of respondents that took part in the study in a way that is somewhat similar to how they have been sampled from the population. Consequently, there are several different resampling techniques that are adequate for different sampling schemes; the most common in LSAEs include delete-one-jackknife (JK1), paired-jackknife (JK2), Balanced Repeated Replication (BRR), and its modification known as Fay's method (Rao & Shao, 1999; Shao & Tu, 1995). The first method can be applied only to sampling schemes that do not include stratification, and the other methods are adequate when different strata have been used. Table 26.1 describes sampling schemes and resampling techniques used in the most recognizable large-scale assessment studies.

To make use of replication techniques straightforward and replicable, a set of additional variables, called replicate weights, is often attached to a LSAE's datasets (with one variable corresponding to one replication) that are used to define which observations should be included and which should be excluded in each replication. These replicate weights are products of sampling weights (i.e. weights that allow to obtain correct point estimates) and a set of weights that describe resampling scheme. To obtain a standard error of an estimate, one has to perform an additional analysis as many times as there are different replicate weight variables provided in a dataset and, afterwards, to pool results using the formula:

$$SE_{samp.}\left(\widehat{\theta}\right) = \sqrt{f\sum_{r=1}^{R}\left(\widehat{\theta}_r - \widehat{\theta}_0\right)^2} \tag{26.1}$$

where:

$\widehat{\theta}_0$ is full sample estimate,

$\widehat{\theta}_r$ is estimate for r^th^ replicate,

Table 26.1 Sampling schemes and standard error estimation methods in large-scale international assessment studies

Study	Target population	Sampling scheme	Method of estimating standard errors	Number of replicates
PISA	15-years old students (age cohort)	Stratified two-stage sampling: 1. Shools 2. Individual students	BRR with Fay (0.5)	80
PIRLS, TIMSS	Students in a selected grade (typically fourth year of formal schooling)	Stratified two-stage sampling: 1. Schools 2. Classes	JK2*	150
PIAAC	Country residents between age 16 and 65	Stratified sampling, state dependent	JK1 or JK2	12–80
NAEP	Main: fourth-, eighth- and twelfth-grade students; Long-term trend: 9-, 13-, and 17-year-old students	Stratified three-stage sampling: 1. Region 2. Schools 3. Individual students	JK2	62

Sources: OECD (2017), Foy and LaRoche (2016, 2017), Mohadjer et al. (2016), and Johnson and Rust (1992)

BRR balanced repeated replication, *JK1* delete-one jackknife, *JK2* paired jackknife (with one replicate generated for each variance stratum), *JK2** paired jackknife with two replicates generated for each variance stratum (jackknife repeated replication)

$$f = \begin{cases} 1 & \text{for the paired jacknife (JK2)} \\ \frac{R-1}{R} & \text{for the delete} - \text{one jacknife (JK1)} \\ \frac{1}{R(1-k)^2} & \text{for the BRR or Fay's approach} \end{cases} \text{ is scaling factor,}$$

R is number of replicates,

k is weighing factor for Fay's method (0 in case of classical BRR).

This formula may be applied to compute correct standard errors for all variables in LSAE datasets, except ability estimates. The latter must be supplemented by additional treatment that accounts for scaling procedures of cognitive variables.

26.3 Scaling Cognitive Outcomes and Accounting for Measurement Error

The focus of most LSAEs is on the cognitive characteristics of surveyed respondents. These cognitive characteristics are conceptualized as latent variables that reflect respondents' abilities, skills, or knowledge. Latent variables are not directly observed but could be measured by observable indicators like responses to test items. Using the information about responses to the test and statistical modeling, it is possible to infer the characteristics of latent variables, but those estimates always will be inherently contaminated with measurement error. As a result, statistical properties of estimates for observed variables are not identical to those of latent traits. This especially affects the estimation of variances of abilities and their covariances with other variables that might lead to serious biases and threats the validity of the analysis.

Generally speaking, there are two strategies to mitigate the problem of measurement error. The first is to measure the latent variable very reliably, that is, using many indicators. The second is to use proper statistical tools to infer the latent trait, considering the problem of measurement error. The modern LSAEs use both strategies.

To cover more cognitive test items, modern LSAEs use a complex process of measuring abilities that contains missing-by-design plans and/or adaptive testing and/or different modes of administration. The bottom line of these approaches is that different groups of respondents (or, in extreme situations, each respondent) are assigned different sets of items. As a result, however, outcomes regarding some specific domains may be absent for some individuals. Inference about skills in these domains is performed in a similar way to imputation of missing data. That is, respective estimates cannot be obtained directly but, rather, are only able to be obtained using a model describing how other outcomes (that are known for these respondents) are related to these unknown ones. To achieve this goal and to properly account for measurement error, a combination of Item Response Theory (IRT) models, missing data techniques, and latent variable modeling are used. As a result, LSAEs provide not one variable as an indicator of the latent treat, but multiple indicators, called plausible values (PVs).

PVs are random draws from an empirically derived distribution of latent variables that is conditional on the IRT parameters of the measurement model, e.g., latent regression coefficients that carry the information on conditional variables, test taker outcomes (on test items), and characteristics (described by conditioning variables). While the IRT parameters define the measurement model and allow latent abilities to be estimated, covariates in the latent regression model increase the precision of estimates by providing additional information and allowing such random draws to be generated, which will result in unbiased results of analyses involving both latent ability and (some of the) variables used for conditioning (von Davier et al., 2009). As PVs are random draws, the random component provides a way to model the uncertainty associated with the estimate that is related to the measurement error

(Mislevy, 1991; Mislevy et al., 1992; Wu, 2005). Moreover, PVs have approximately the same conditional distribution as the latent trait being measured. The latent regression adjusts PVs to have conditional distribution on covariates (used in the generating process), the same as the latent distribution conditioned on covariates. Results obtained from the plausible values are, therefore, in most cases (Schofield et al., 2015) close to results that would be obtained using latent regression modelling—the most effective statistical tool that could be used to properly account for measurement error. Unlike latent variable modeling, PV methodology does not require specialized software nor high computational burden.

PVs are a special case of multiple imputations (MI) and the analysis that employs PVs should follow procedures known from MI. Each PV is used once in each analysis. For instance, for 10 generated PVs, 10 models are estimated; in each model one PV is used. The final estimates are obtained by aggregating the results. Point estimates are obtained by simply averaging over the estimates obtained from each PV. The standard errors are combined across P PVs using Rubin's (Little & Rubin, 1987) formula:

$$SE\left(\widehat{\theta}_P\right) = \sqrt{\frac{1}{P}\sum_{i=1}^{P} SE_{samp.}\left(\widehat{\theta}_i\right)^2 + \frac{P+1}{P(P-1)}\sum_{i=1}^{P}\left(\widehat{\theta}_{0,i} - \overline{\theta}_{0,P}\right)^2} \quad (26.2)$$

where:

P is the number of plausible values, $i = (1, \ldots, P)$;

$SE_{samp.}\left(\widehat{\theta}_i\right)$ is sampling error for the ith *plausible value* computed with Eq. (26.1);

$\widehat{\theta}_{0,i}$ is estimate using the final sample weight for the ith *plausible value*;

$\overline{\theta}_{0,P} = \frac{1}{P}\sum_{i=1}^{P}\widehat{\theta}_{0,i}$ is unweighted average (of the statistics computed for the individual *plausible value*) using whole sample and the final weight.

26.4 Combining Sampling and Measurement Error in LSAE

Obtaining correct estimates using PV methodology is not conceptually much more difficult in LSAE data sets than in other types of data sets. It requires that the same estimations be performed many times using different replicate weights and different PVs, and they need to be combined with simple aggregation rules that require basic arithmetic (Eqs. 26.1 and 26.2). Even though the procedure is technically not very sophisticated, handling many replications might be cumbersome. While using PISA data to obtain point estimates with correct standard errors involving PVs, one needs to perform 810 estimations using the same statistical model. First, ten estimates for each of the 10 PVs with the final weight must be computed to obtain correct point

Table 26.2 Statistical programs that allow for analyses on LSAE

Type	Name	Environment	Which LSAE	Reference
Basic computations: tables, means, regressions	IDB analyzer	SPSS, SAS/STAT	NAEP, PISA, PIAAC, TIMSS, PIRLS (…)	
	piaactools	Stata	PIAAC	Jakubowski and Pokropek (2019)
	pisatools	Stata	PISA	Jakubowski and Pokropek (2017)
	intsvy	R	TIMSS, PIRLS, PISA, ICILS, PIAAC	Caro and Biecek (2017)
	svyPVpack	R	PIAAC, (…)	
All purpose, allowing estimation of most statistical models	Pv	Stata	PISA, TIMSS, PIRLS, PIAAC, (…)	Macdonald (2008)
	repest	Stata	PISA, PIAAC, TALIS, ALL, IALS (…)	Avvisati and Keslair (2014)
	survey + mitools	R	All pourpose	Lumley (2010)

estimates. Then, the correct procedure calculates statistics with 80 replicate weights and repeats this for each plausible value, that is, 800 times. Finally results from each plausible value are aggregated using Eq. (26.2).

Popular statistical programs like SPSS, Stata, and SAS do not have built-in features that could automatically handle both PVs and replicate weights, thereby, providing an easy way to obtain proper estimates for LSAEs. Fortunately, many extensions were developed to facilitate the work with LSEAs. In the Table 26.2 we summarize the most popular tools.

In the next part of the chapter, we show the programs that allow easy and proper analysis of LSAE data involving PVs. We focus on Stata and R, showing basic computations for selected LSAEs. We chose Stata because this software has one of the easiest, but also very flexible, implementation of tools to facilitate working with PVs and replicate weights. R was selected because it is a powerful computing environment. As such, it may be more demanding, but R is easy to obtain, free, and uses an open-source statistical programing language.

26.5 Empirical Examples

26.5.1 Datasets

Datasets used in the analysis below are subsets of the officially released PISA 2015 student questionnaire data file and the PIAAC public use files. To decrease complexity and long outputs, examples were restricted to four countries: Germany, Korea, Poland, and the United States. In the PISA dataset, cases from all other countries were excluded, and the dataset was saved in the Stata 14 format. In the case of PIAAC, individual datasets from these four countries were merged into one and also saved in the Stata 14 format.

In PISA, the same replication approach is used within each country: Fay's method with k = 0.5 and 80 replicates. In PIAAC, however, three different replication schemes were used: paired jackknife with 80 replicates for Korea and Poland, paired jackknife with 45 replications for the United States, and a delete-one jackknife with 80 replicates in the case of Germany. Information regarding the replicate method is included in variables `vemethod` and `venreps` in the dataset.

Variables from the PISA 2015 dataset used in examples below are:

- `cntryid` – student's country
- `ST004D01T` – student's gender
- `PV1SCIE`, ... `PV10SCIE` – PV estimates of student's ability in science
- `cultposs`, `hedres`, `wealth` – weighted least squares (WLS) estimates of indexes describing the student's household; respectively: cultural possession index, educational resources index, and family wealth index
- `W_FSTUWT` – *sampling weight*
- `W_FSTURWT1`, ..., `W_FSTURWT80` – *replicate weights*

Variables from the PIAAC dataset used in examples below are:

- `cntryid` – respondent's country
- `GENDER_R` – respondent's gender
- `neet` – respondents status of being not employed and not participating in education for last 12 months
- `PVLIT1`, ..., `PVLIT10`, `PVNUM1`, ..., `PVNUM10`, `PVPSL1`, ..., `PVPSL10` – PV estimates of respondents abilities in literacy, numeracy and problem solving respectively
- `vemethod`, `venreps` – replicate method and number of *replicate weights* being used in a given country
- `SPFWT0` – *sampling weight*
- `SPFWT1`, ..., `SPFWT80` – *replicate weights*

26.5.2 Examples with Stata

Stata is popular general purpose statistical software. Although it has a graphical interface, most users use the command line to input commands and perform analyses. Stata has a very developed, intuitive syntax that is easy to learn and convenient to use; therefore, we use this language in the examples below.

26.5.2.1 Installation of Additional Packages

To start analyzing LSAE data in Stata, one needs to download additional user-written packages that will handle specific aspects of the LSAE for the user. Three packages will be used in the following examples:

- *pisatools* (Jakubowski & Pokropek, 2017) for performing analyses on PISA data
- *piaactools* (Jakubowski & Pokropek, 2019) for performing analyses on PIAAC data
- *repest* (Avvisati & Keslair, 2014) for general use

The first two packages are designed to work only with specific LSAEs. Their advantage is that they are very easy to use and provide clear HTML tables that compress information from many countries into an easy-to-use format. The `repest` package gives more flexibility and could be used with many different LSAEs, not only PISA and PIAAC, but it requires more effort to specify the models and format the output. There are two ways of installing the packages. First is built-in command, `ssc`, that will automatically download and install the required packages:

```
> ssc install pisatools
> ssc install piaactools
> ssc install repest
```

Alternatively, one could download the packages as ado files (files with extension .ado) from the authors' websites (or other repositories) and copy them to the "ado" folder on the local drive. For instance, the most recent version of the `pisatools` package can be found on the authors' webpage (Evidence Institute, 2017). To check where ado folders are located, the Stata command `adopath` can be used.

26.5.2.2 Setup and Data Import

After installing the additional packages, Stata is practically ready to work with LSAEs. The only thing left is to define the working directory. For convenience, we advise the reader to rename all variable names to lowercase. Stata syntax is case sensitive, and sometimes lower- and upper-case letters generate problems in user-defined programs. Changing the working directory, loading the data, and renaming variables can be achieved in three lines of code:

```
> cd <<<"Your path to the folder with the data">>>
> use PISA_LSEAbook.dta /* or different dataset */
> rename *, lower
```

26.5.2.3 Descriptive Statistics

Assuming that the PISA data are loaded to Stata, descriptive statistics can be obtained using the `pisastats` command from `pisatools` package:

```
> pisastats, pv(scie) stats(mean) ///
> cnt(ALL) save(ex_pisa_1) round(2) cycle(2015)
```

The `pv()` option defines which PVs are analyzed. In this example, science abilities are used by specifying `pv(scie)`, but option `pv()` accepts also other acronyms used to describe main scales or subscales of PISA (e.g. read, math, scie, etc). The `stats()` option defines the statistics to be computed. The expression "///" denotes a Stata line-continuation marker. To check the list of descriptive statistics supported by `pisastas`, please see `help(pisastats)`. Option `cnt()` defines whether the analysis is to be performed for all countries (country-by-country) or for selected countries only. To select a subset of countries, the three-letter ISCO codes of countries should be entered. For instance, to analyze only Poland and Korea, one needs to specify `cnt(POL KOR)`. In the present case, the data have already been subset, so "`ALL`" specifies the data what we want. The `save()` option defines the name of the HTML file where the results of the analysis are stored. Option `round()` defines the number of decimal places in the output tables, and option `cycle()` tells the package which year of PISA data collection is used in current analysis (e.g. 2000, 2003, etc.). The PISA cycles differ in the number of plausible values and the names of the core variables, so the last option needs to be specified to avoid errors. The detailed information about the command could be evoked by typing `help(pisastats)`.

The results of `pisastats` are stored in a HTML table. The table that was generated using the described command is presented as Table 26.3 without additional changes.

Table 26.3 Means for science achievement in PISA generated by `pisastats` command

	scie	
Country	Mean	S.E.
Germany	509.14	2.70
Korea	515.81	3.13
Poland	501.44	2.51
United States	496.24	3.18

Table 26.4 Descriptive statistics for science achievement generated by `pisastats` command

	scie									
Country	Min	S.E.	p25	S.E.	p50	S.E.	p75	S.E.	Max	S.E.
Germany	162.09	29.06	439.36	3.57	511.64	3.30	579.89	2.80	828.87	14.75
Korea	178.77	16.96	451.07	3.76	520.38	3.73	583.93	3.26	825.74	32.12
Poland	158.91	37.53	436.80	2.91	502.01	3.01	565.27	3.07	800.21	12.23
United States	172.55	22.00	424.94	3.68	495.24	3.82	567.49	3.89	833.28	28.57

Table 26.5 Means for numeracy proficiency in PIAAC generated by `piaacdes` command

	pvnum	
Country	Mean	S.E.
Germany	271.73	1.00
Korea	263.39	0.69
Poland	259.77	0.82
United States	252.84	1.17
Average	**261.93**	**0.47**

To compute more descriptive statistics, one needs to specify them using the `stats()` option. For computing minimum, maximum and selected quantiles, the names of the statistics should be entered into the options: `stats("min p25 p50 p75 max")`.

```
> pisastats, pv(scie) stats("min p25 p50 p75 max") ///
> cnt(ALL) save(ex_pisa_2) round(2) cycle(2015) ///
```

The resulting HTML table is presented in Table 26.4.

The easiest way to perform similar computations using the PIAAC data is to use the `piaacdes` command from the `piaactools` package. The syntax for computing the means of numeracy abilities defined by plausible values in `piaacdes` is as follows:

```
> piaacdes, pv(pvnum) stats(mean) round(2) ///
> save(ex_piaac_1) countryid(cntryid)
```

There is a small difference in the syntax of the two commands. In `piaacdes`, one needs to specify the variable that defines the countries in option `countryid`. Similar to `pisatools`, tables are generated in an HTML file. Table 26.5 shows the output of the described `piaacdes` command. The only difference is that `piaacdes` automatically adds the average for the analyzed groups. This average is the mean of the estimated statistics with equally weighted countries, which is rarely of interest.

To get more descriptive statistics, the option `stats()` needs to be specified. Percentiles could be obtained by specifying the additional option `centile()`. The full list of available descriptive statistics and additional options might be requested using help(piaacdes) command. To produce results similar to Table 26.4 using PIAAC data and piaacdes command options, stats(min max) and centile(25 50 75) should be specified:

```
> piaacdes, pv(pvnum) stats(min max) centile(25 50 75)///
> round(2) countryid(cntryid) save(ex_piaac_2)
```

We are not presenting the output of this command. The structure of results is very similar to Table 26.5 but without an additional row with the average results.

26.5.2.4 Frequency Tables

To compare distributions of achievements across groups in a way that is widely understandable, numeric scores are often translated into so-called proficiency levels. These levels are assigned by categorizing the numeric scores with respect to a set of thresholds that were selected in such a way that they describe important differences in the set of skills that someone has mastered. In PISA 2015, eight science proficiency levels were defined, which are "below 1", "1b", "1a", "2", "3", "4", "5" and "6", with the following thresholds separating these levels: 260.54, 334.94, 409.54, 484.14, 558.73, 633.33 and 707.93 (OECD, 2017, pp. 275–283). PIAAC uses six proficiency levels and the following thresholds: 175, 225, 275, 325 and 375 (OECD, 2019, pp. 587). In this example, cross-tabulation of proficiency levels by gender is discussed.

For the PIAAC data, this is very easy because piaactools package contains command piaactab, which essentially performs all the computations and data handling automatically. The input for computations is as follows:

```
> piaactab pvnum, over(gender_r) countryid(cntryid) ///
> round(2) save(ex_piaac_2)
```

This command automatically cuts the PVs for numeracy as indicated by pvnum and computes results by gender_r variable, which was specified by option over(). For further details of the command, see help(piaactab). Similar to other piaactools commands, piaactab also produces an HTML output file. Table 26.6 shows a part of the output returned by piaactab. We dropped columns representing higher levels of proficiency and the female sub-population to fit the table to the page, as the resulting output is rather large.

The situation for the PISA data is different because a command similar to piaactab for PISA has not yet been developed. Moreover, pisatools package has no command to compute frequencies. One solution is to recode data to binary categories and compute the means of the variables of interest. However, a more convenient way is to use general purpose command repest. The package is for

Table 26.6 Part of the frequency table of the proficiency level by gender using PIAAC data generated by `piaactab` command

	Male									
	Below Level 1		Level 1		Level 2		Level 3		(...)	
Country	%	S.E.	%	S.E.	%	S.E.	%	S.E.	(...)	(...)
Germany	3.38	0.44	11.44	0.80	28.63	1.08	37.55	1.20	(...)	(...)
Korea	3.31	0.37	12.98	0.77	37.66	1.20	37.28	1.31	(...)	(...)
Poland	6.79	0.56	17.05	0.90	35.36	1.20	30.71	1.34	(...)	(...)
United States	8.15	0.68	18.81	1.09	31.58	1.60	29.18	1.19	(...)	(...)
Average	**5.41**	**0.26**	**15.07**	**0.45**	**33.31**	**0.64**	**33.68**	**0.63**	**(...)**	**(...)**

general use, so it does not facilitate automatic recoding, and the proficiency levels need to be created by the user. In Stata, it could be easily done by `forvalues` loop accompanied by the `egen` command:

```
> forvalues i = 1(1)10 {
> egen pvlev_scie'i' = cut(pv'i'scie), ///
> at(-9999, 260.54, 334.94, 409.54, 484.14,558.73, ///
> 633.33, 707.93, 9999) label
> }
```

The `forvalues` loop generates a new categorical variable for each of the 10 PVs by cutting them using thresholds provided in option `at()`. Note that the list of values provided in `at()` also should contain a lower bound for the first level and an upper bound for the last level (here −9999 and 9999 respectively). For more details, see `help(forvalues)` and `help(egen)`. After this preliminary step, the `repest` command could be used in the following structure:

```
> repest PISA, estimate(freq pvlev_scie@) ///
> svyparms(NBpv(10)) by(cnt) over(st004d01t) ///
> outfile(ex_pisa_3)
```

Following the `repest` command, the name of the study needs to be specified. In the case above, it is PISA, and this string is sufficient to define the whole estimation scheme. The option `estimate(freq pvlev_scie@)` defines the estimation objectives; that is, the estimation of frequencies of variables `pvlev_scie@`, where @ indicates that this is a set of variables rather than one single variable. Option `svyparms(NBpv(10))` tells `repest` how many PVs should be used in computations (it is particularly important because early cycles of PISA used only five PVs, while starting from 2015, ten PVs were used). The "st004d01t" string in option `over()` is the name for variable that represents gender. Using option `outfile()`, the user defines the name of a Stata file where the results are to be stored. The `repest` command does not generate output tables but, rather, stores results in a Stata data frame. The data frame with results could be manipulated using Stata commands or easily converted to an MS Excel file where post processing of the results could be done manually. We are not presenting the initial output file provided by `repest`, as the results were already shown, and the output table is too big for presentation. For more `repest` options, see `help(repest)`.

26.5.2.5 Regression

Both `pisatools` and `piaactools` have specific commands for regression analysis. The following example estimates a regression model in PISA. In the model, science performance is predicted by student's gender and various dimensions of socio-economic status: culture possession index (`cultposs`), home educational resources (`hedres`), wealth indicator (`wealth`).

Before specifying the model, it is useful to recode the gender variable, st004d01t, into binary indicator:

```
> recode st004d01t (2=0), gen(female)
```

Then `pisareg` command could be used to estimate the model:

```
> pisareg scie female cultposs hedres wealth, ///
> cnt(ALL) save(ex_pisa_4) round(2) cycle(2015)
```

The syntax of the command is simple. The first string after the command defines the dependent variable (PVs are detected automatically so the prefix "pv" is not necessary), followed by the independent variables. All other options are similar to the previously presented commands from the `pisatools` package. To see additional options, see `help(pisareg)`.

Similar to other commands form `pisatools` package, `pisareg` generates an HTML table (see the example in Table 26.7). This table reports the results country-by country. It provides regression coefficients, standard errors, and R-squared of the models. By default, regression constants are not reported but could be required by adding option `cons`. Additionally, statistically significant results are bolded by the program.

Table 26.7 Regression coefficient table generated by `pisareg` command using PISA data

	Female		Cultposs		Hedres		Wealth		r2
Country	Coef.	S.E.	Coef.	S.E.	Coef.	S.E.	Coef.	S.E.	
Germany	**−15.49**	2.83	**19.96**	1.60	**11.95**	1.88	**9.14**	1.80	0.09
Korea	1.49	4.25	**18.71**	1.33	**14.78**	1.83	**−6.90**	3.50	0.09
Poland	**−11.38**	2.91	**30.66**	2.00	−0.71	2.00	**4.91**	2.43	0.10
United States	**−9.13**	3.06	**14.14**	1.40	**5.64**	1.35	**10.95**	1.59	0.07
Average	**−8.63**	1.66	**20.87**	0.80	**7.92**	0.89	**4.53**	1.22	0.09

Similar outputs could be obtained by `piaacreg` command from the `piaactools` package. In the example below, a model with numeracy ability that is predicted by gender, ICT resources at home (`icrhome`), and the index of readiness for learning (`readytolearn`) is specified.

```
> piaacreg female icthome readytolearn , pvdep(pvnum) ///
> countryid(cntryid) round(2) save(ex_piaac_4)
```

Syntax of the `piaacreg` command is slightly different from `pisareg`. Here, the dependent variable based on plausible values needs to be specified in the `pvdep ()` option without a suffix specifying the number of PVs. The output table is analogous to that presented by `pisareg`, therefore, we are not presenting it here.

26.5.2.6 Logistic Regression

Not only linear regression models can be estimated on the LSEA data. Another widely used regression model is logistic regression in which a binary variable is predicted by a set of covariates. Additional complications for specifying such models using standard software occur when one wants to use PVs as independent variables. Both `pisatools` and `piaactools` allow for easy specification and estimation of such models. The example uses the PISA data. The grade repetition variable (`repeat`) is a binary dependent variable and is predicted by gender, culture possession index (`cultposs`), and science abilities. For making results more readable before the analysis, we divide science PVs by 100 to achieve a scale with SD approximate equal to 1. Otherwise, the coefficients for PVs would be very small.

```
> forvalues i = 1(1)10 {
> replace pv'i'scie = pv'i'scie/100
> }

> pisacmd repeat female cultposs, cmd("logit") ///
> pvindep1(scie) cnt(ALL) save(ex_pisa_5) round(2) ///
> cycle(2015) pw r2(r2_p)
```

The syntax of `pisacmd` is similar to other commands from the `pisatools` package. The most important option is `cmd ()`, which defines which type of model is to be estimated (other regression models than logistic could be specified, e.g. probit, scobit, hetprobit etc.). The first string after the command specifies the dependent variable (`repeat`), while the following variables are predictors. To define PVs as an independent variable, `pvindep* ()`, is used. Asterisks stand for the number of dependent plausible value variables used in the analysis. One could decide to use more than one plausible value variable. For specifying second plausible values, dependent variable `pvindep2 ()` should be used etc. Additionally, the `pw` option is

Table 26.8 Logistic regression coefficient table generated by pisareg command using PISA data

	Scie		Female		Cultposs		r2_p
Country	Coef.	S.E.	Coef.	S.E.	Coef.	S.E.	
Germany	**−0.86**	0.05	**−0.29**	0.08	0.01	0.05	0.10
Korea	**−0.30**	0.07	−0.16	0.15	0.04	0.06	0.01
Poland	**−1.59**	0.14	**−1.05**	0.16	**−0.35**	0.12	0.24
United States	**−1.18**	0.07	**−0.52**	0.11	−0.07	0.06	0.16
Average	**−0.98**	0.04	**−0.50**	0.06	**−0.09**	0.04	0.13

required by pisacmd for models where Stata does not allow for analytic weights (pisacmd uses analytical weights by default to obtain replication point estimates, some of Stata commands, however, allow only for probability weights and this needs to be specified; as only point estimates are utilized in replications, this fact has no statistical implications). Finally r2() options define the type of the R-squared to be reported. Because logistic regression has no "classical" R-squared, McFadden's pseudo R-squared was specified using r2_p argument.

Like other pisatools commands, pisacmd provides output in an HTML file. The table generated by the command is presented as Table 26.8.

For estimating logistic regression in PIAAC, piaacreg can be used. In the following example, a NEET (not employed and not participating in education for last 12 months) indicator is the dependent variable, and gender, ICT resources at home (icthome), an index of readiness for learning (readytolearn), and numeracy (PV variable) are predictors.

```
> piaacreg neet icthome readytolearn female, ///
> pvindep1(pvnum) cmd("logit") countryid(cntryid) ///
> round(2) save(ex_piaac_5) r2(r2_p)
```

The syntax is similar to pisacmd, although pisacmd does not require the additional pw option. The output table is also analogous, so we are not presenting it to save the space.

26.5.3 Examples with R

We assume that readers have basic knowledge about the R language and working environment: They are aware of the R system's additional packages and are able to perform basic data manipulation operations on typical R data structures (vectors, lists and data frames). In a few cases, we use a more advanced data manipulation technique that is typical for the R language: function `lapply`.[1] Also, we want to remind readers about specific ways that most R functions compute statistical parameters and deal with missing values. If there is any missing value in a variable, "NA" code will be returned as a result, unless an additional argument `na.rm. = TRUE` was included in the function call.

26.5.3.1 Installation of Additional Packages

Before starting the analysis, some additional packages, not included in basic R installation, must be installed. These packages are:

- *haven* for data import
- *mitools* (Lumley, 2019a)
- *survey* (Lumley, 2019b)
- *svyPVpack* (Reif & Peterbauer, 2019) for performing analyses with complex sample designs and plausible values. Please note that the last package must be installed from the GitHub repository as it is not submited to CRAN.

26.5.3.2 R Session Setup and Data Import

Additional packages are not automatically available for use during a new R session. To use them, one needs to use the `library` command at the beginning of the R script:

```
> library(haven)
> library(survey)
> library(svyPVpack)
```

The next step is data import. This can be done with the function `read_dta` or `read_spss` from the `haven` package. Please note that variables in a data frame returned by these functions are mostly so-called *labelled* vectors, and this type of variable is not widely supported by R data analysis functions. Moreover, values declared as so-called *user missing values* in the Stata or SPSS files are, by default, converted to NA values during import into R, i.e., to an analogue of the *system missing value*.

```
> pisa <- read_dta("PISA_LSEAbook.dta")
> piaac <- read_dta("PIAAC_LSEAbook.dta")
```

[1] The `lapply` function takes a list and a function as arguments, It applies the function to each element of the list and returns a list of results.

Aforementioned labelled variables may be analyzed as numeric without performing any additional manipulation. However, if one wants to treat such variables as a categorical and make its value labels visible in the output of analysis, it must be converted to a typical R categorical variable format, called a factor. This can be achieved with the `as_factor` function from the `haven` package (not to be confused with the `as.factor` function from base R). The code below transforms variables describing each country in PISA and PIAAC datasets, respectively, to factors. The function `droplevels` allows the user to restrict the set of considered values of a variable to only those values that appear in the data.

```
> pisa$cntryid <- droplevels(as_factor(pisa$cntryid))
> piaac$cntryid <- droplevels(as_factor(piaac$cntryid))
```

26.5.3.3 Defining Survey Design Objects

Having data imported, one need to define so called *survey design objects*, i.e., R objects that store both data and the information on how to compute statistics and its standard errors using sampling weights and replicate weights. To define these objects, one should use `svrepdesign` function from the `survey` package:

- Argument `repweights` provides a regular expression that match the names of *replicate weights* ("^" at the beginning of the expression below stands for "starts with"). Alternatively this argument may be provided as a *character vector* of variables' names.
- Argument `weights` provides the name of the variable storing *sampling weights* (it is declared as a *formula*, so the name must be preceded with "~").
- Argument `data` specifies the *data frame* with previously imported dataset.
- Other parameters describe the type of replication scheme.

In the case of the PISA 2015 dataset, one must declare Fay's method with k = 0.5. There are ready-to-use replicate weights provided in the dataset (combining both sampling weights and replication scheme), so the argument `combined.weights` should be set to `TRUE`.

Setting argument `mse` to `TRUE` means that one wants to compute standard errors using deviations from the full sample estimate of a statistic (as in Eqs. 26.1 and 26.3), which is correct for LSEAs. It must be specified because the default behavior of the `survey` package is different: It computes deviations from the mean estimate across replications (i.e. deviations from $\frac{1}{R}\sum_{r=1}^{R}\widehat{\theta}_r$), which is not correct in analyzing LSEA data.

To define a survey design object for the PISA 2015 data, one may call:

```
> pisaSVY <- svrepdesign(repweights = "^W_FSTURWT",
+ weights = ~W_FSTUWT,
```

```
+ data = pisa,
+ type = "Fay",
+ rho = 0.5,
+ combined.weights = TRUE,
+ mse = TRUE)
> pisaSVY
Call: svrepdesign.default(repweights = "^W_FSTURWT", weights =
~W_FSTUWT,
 data = pisa, type = "Fay", rho = 0.5, combined.weights = TRUE,
 mse = TRUE)
Fay's variance method (rho= 0.5 ) with 80 replicates and MSE variances.
```

In the case of PIAAC, it is not possible to include all the countries into one survey design object, because of the different replication schemes used in different countries. Nevertheless, it helps to automate the computations if one defines a list of survey design object for each country. There are some preliminary data transformations that must be made beforehand:

- Changing names of replication methods that are stored in `vemethod` variable to these recognized by `svrepdesign`.
- Computing *scaling factor* (f in Eq. 26.1), as it must be provided explicitly to `svrepdesign` in the case of jackknife replication schemes.
- The PIAAC dataset must be split into a *list* of *data frames* with each element storing observations from a given country. Next, function `definePiaacDesign` is defined that will be used to define *survey design* object for a single country. Finally `lapply` function is called on the *list* to execute `definePiaacDesign` on data from each country separately.

```
> # changing name of replication method
> piaac$vemethod[piaac$vemethod %in% "JK2"] = "JKn"
> # calculating scaling factor
> piaac$scale <- ifelse(
+ piaac$vemethod %in% "JK1",
+ (piaac$venreps - 1) / piaac$venreps, 1)
> # splitting data by country
> piaacSVYs <- split(piaac, piaac$cntryid)
> # declaring function that defines survey design object
> # for a given country
> definePiaacDesign <- function(x) {
+ x$cntryid <- droplevels(x$cntryid)
+ svrepdesign(repweights = make.formula(
+ paste0("SPFWT", 1:x$venreps[1])),
+ weights = ~SPFWT0, data = x,
+ type = x$vemethod[1],
+ combined.weights = TRUE,
+ scale = x$scale[1], rscales = 1,
```

```
+ mse = TRUE)
+ }
> # applying this function to each element of the list
> piaacSVYs <- lapply(piaacSVYs, definePiaacDesign)
> piaacSVYs
$Germany
Call: svrepdesign.default(repweights = make.formula(paste0("SPFWT",
 1:x$venreps[1])), weights = ~SPFWT0, data = x, type = x$vemethod[1],
 combined.weights = TRUE, scale = x$scale[1], rscales = 1,
 mse = TRUE)
Unstratified cluster jacknife (JK1) with 80 replicates and MSE
variances.
(...)
$`United States`
Call: svrepdesign.default(repweights = make.formula(paste0("SPFWT",
 1:x$venreps[1])), weights = ~SPFWT0, data = x, type = x$vemethod[1],
 combined.weights = TRUE, scale = x$scale[1], rscales = 1,
 mse = TRUE)
Stratified cluster jackknife (JKn) with 45 replicates and MSE variances.
```

Please note that the argument `repweights` in the call above is provided as a formula, constructed with the `make.formula` function, given a vector containing names of variables SPFWT1-SPFWT80 (prepared with use of `paste0` function). This is necessary because variable SPFWT0 in the PIAAC dataset contains sampling weights, and one does not want this variable to be considered as a replicate weight, as it would be if regular expression "^SPFWT" were provided as `repweights`.

Use of such a list of survey design objects in the analysis is not supported by functions of neither `survey` nor `svyPVpm` packages. One can overcome this problem by defining a function that will perform the analysis on each element of the list separately and then combine the results:

Listing 26.1 Definition of the `piaacDoSvyPV` function

```
> piaacDoSvyPV <- function(SVYlist, expr) {
+ results <- sapply(SVYlist, function(.d, expr) {
+ eval(expr)
+ }, expr = expr)
+ if (is.list(results)) {
+ results <- as.data.frame(t(results))
+ results <- as.data.frame(lapply(results, unlist))
+ rownames(results) <- NULL
+ }
+ return(results)
+ }
```

The way this function works internally is somewhat complicated, and we do not discuss it here. Nevertheless we will describe how to use it using examples in the following sections.

26.5.3.4 Data Manipulation Using Survey Design Objects

If one wants to transform variables in a survey design object that has already been defined, it can be done using `transform` function. The first argument to this function is a survey design object and further ones are expressions describing transformations. For example, to create a new factor variable gender in the PISA survey design object by converting labelled variable ST004D01T that originally describes gender of respondents one may use code below.

```
> pisaSVY <- transform(
+ pisaSVY,
+ gender = droplevels(as_factor(ST004D01T)))
```

In the case of the PIAAC list of survey design objects, one needs to use `lapply` provide the list as first argument, the `transform` function (please note that this is not a function call) as a second argument and expressions describing transformations as further arguments (these expressions will be passed to each call of `transform`). The code below creates two factor variables, one describing gender and second an indicator describing being not in employment, education or training status.

```
> piaacSVYs <- lapply(
+ piaacSVYs, transform,
+ gender = droplevels(as_factor(GENDER_R)),
+ neet = droplevels(as_factor(neet)))
```

26.5.3.5 Descriptive Statistics

To get estimates of mean and standard deviation of an ability indicator, it is convenient to use the `svyPVpm` function from the `svyPVpm` package. The argument by specifies a grouping variable (it must be specified as a formula); argument `svydat` gives a survey design object, and with argument `pvs`, one should pass the names of variables containing "PV" to be analyzed. The argument `colN` specifies how the column describing grouping factor should be named in output. Code below estimates statistics of science ability indicator within each country using PISA 2015 data.

```
> svyPVpm(by = ~cntryid,
+ svydat = pisaSVY,
```

```
+ pvs = paste0("PV", 1:10, "SCIE"),
+ colN = TRUE)
cntryid N(...) cases (...) weights Prop(...) SE.Prop(...)
Germany 6504 741872 0.1432 0.00370
Korea 5581 569106 0.1098 0.00239
Poland 4478 345709 0.0667 0.00138
United States 5712 3524497 0.6802 0.00670
PVSCIE_mean PV(...)_mean:SE PV(...)_stddev PV(...)_stddev:SE
509 2.70 99.3 1.48
516 3.13 95.2 1.47
501 2.51 90.8 1.34
496 3.18 98.6 1.40
```

Function `svyPVpm` returns a data frame with columns describing the mean and standard deviation of proficiency indicators, corresponding standard errors, and additional statistics describing the size of each group: number of cases (rows), sum of sampling weights, and proportion with regards to the sum of sampling weights (there is also standard error of the latter).

Function `svyPVquantile` enables the estimation of quantiles. It has an analogous syntax to `svyPVpm`, with additional argument `quantiles` used to specify which quantiles should be reported. Please note that you may use values 0 and 100 in the quantile specification to get the minimum and maximum, respectively.

```
> svyPVquantile(by = ~cntryid,
+ svydat = pisaSVY,
+ pvs = paste0("PV", 1:10, "SCIE"),
+ quantile = c(0, 0.25, 0.5, 0.75, 100),
+ colN = TRUE)
cntryid N(...) cases (...) weights q0 q0.25 q0.5 q0.75
Germany 6504 741872 162 439 512 580
Korea 5581 569106 179 451 520 584
Poland 4478 345709 159 437 502 565
United States 5712 3524497 173 425 495 567
q100 se:q0 se:q0.25 se:q0.5 se:q0.75 se:q100
829 29.1 3.57 3.31 2.79 14.8
826 17.0 3.76 3.73 3.27 32.1
800 37.5 2.91 3.01 3.07 12.2
833 22.0 3.67 3.82 3.89 28.6
```

Both `svyPVpm` and `svyPVquantile` can be applied to the PIAAC list of a survey design object with the help of the `piaacDoSvyPV` function (see Listing 26.1). This list should be passed as the first argument, with the second argument being the call to `svyPVpm` or `svyPVquantile` enclosed with `quote()`. The last modification that one must make (compared to the syntax used for the PISA data) is substituting a survey design object passed with `svydat` argument with ".d" to specify that the elements of the list of survey design objects should be placed there (this is specific notation implemented in the `piaacDoSvyPV` function).

```
> piaacDoSvyPV(piaacSVYs,
+ quote(svyPVpm(by = ~cntryid,
+ svydat = .d,
+ pvs = paste0("PVNUM", 1:10),
+ colN = TRUE)))
cntryid N(...) cases (...) weights Percent SE.Percent
Germany 5379 5.29e+07 100 0
Korea 6651 3.45e+07 100 0
Poland 9366 2.67e+07 100 0
United States 4898 1.95e+08 100 0
PVNUM_mean PVNUM_mean.SE PVNUM_stddev PVNUM_stddev.SE
272 0.995 53.1 0.736
263 0.690 45.6 0.634
260 0.824 50.7 0.649
253 1.166 57.0 1.029
```

It should be noted that because estimation regarding each country is performed separately, column "Percent" does not show meaningful information.

Functions `svyPVpm` and `svyPVquantile` enable the analysis of PVs only. If one provides only one variable as the `pvs` argument, point estimates of statistics are computed correctly, but missing values are reported for the standard errors. Because of this, for regular variables, one should use `svymean`, `svyvar`, and `svyquantile` functions from the `survey` package called along with `svyby`.

26.5.3.6 Frequency Tables

In R, the function `svyPVlevel` from the `svyPVpack` package allows for constructing distributions of proficiency levels in a convenient way. The syntax of this function is similar to those discussed previously, but with two additional arguments: `CATDEF` is used to define thresholds and `levlab` to provide labels for the proficiency levels. One should note that the vector passed to `CATDEF` should also contain a lower bound for the first level and an upper bound for the last level (similar to the base R function `cut`). It is convenient to use `-Inf` and `Inf` (infinity), respectively, as these values. In the example below, the conditional distributions of proficiency levels in science crossed with gender, nested within countries, are constructed.

```
> pisaScienceLevels <- svyPVlevel(
+ by = ~cntryid + gender,
```

```
+ svydat = pisaSVY,
+ pvs = paste0("PV", 1:10, "SCIE"),
+ CATDEF = c(-Inf, 260.54, 334.94, 409.54, 484.14,
+ 558.73, 633.33, 707.93, Inf),
+ levlab = c("bel.1.1",
+ paste0("lev.", c("1b", "1a", 2:6))),
+ colN = TRUE)
> pisaScienceLevels
cntryid gender levels Prop(...) Prop(...)SE(...) cases(...) weights
Germany Female bel.1.1 0.00396 0.00194 13 1443
Germany Female lev.1b 0.03908 0.00537 118 14240
Germany Female lev.1a 0.13819 0.00900 420 50351
Germany Female lev.2 0.22536 0.01081 712 82114
Germany Female lev.3 0.28877 0.01063 932 105219
Germany Female lev.4 0.21723 0.00933 710 79151
Germany Female lev.5 0.07619 0.00612 254 27759
Germany Female lev.6 0.01122 0.00213 38 4086
 Korea Female bel.1.1 0.00179 0.00109 5 486
 Korea Female lev.1b 0.01948 0.00373 53 5297
```

PISA results are returned as a data frame with proportion, its standard error, the number of cases, and the sum of sampling weights reported for each combination of grouping factors and proficiency level. To convert this into a typical table, one may use the standard R `xtabs`, `addmargins`, and `ftable` functions.

The PIAAC data may be analyzed adopting the same approach as that described in the immediately preceding section on descriptive statistics. With six proficiency levels and their respective thresholds of 175, 225, 275, 325, and 375 (OECD, 2019, pp. 587), distributions of numeracy achievements, conditional on country and gender, may be obtained with the code below (for definition of the function `piaacDoSvyPV` see Listing 26.1). The additional argument `right` is set to `FALSE` to indicate that the constructed intervals should be opened on their right side (i.e. second proficiency level should contain value 176 but not contain value 227).

```
> piaacNumeracyLevels <- piaacDoSvyPV(piaacSVYs, quote(
+ svyPVlevel(by = ~cntryid + gender,
+ svydat = .d,
+ pvs = paste0("PVNUM", 1:10),
+ CATDEF = c(-Inf, 176, 227, 277, 327, 377,
+ Inf), right = FALSE,
+ levlab = c("<1.1", paste0("1.", 1:5)),
+ colN = TRUE)
+ ))
```

The returned object has the same format as in the case of previously described analysis regarding PISA data (due to data manipulation that is done internally within `piaacDoSvyPV`).

26.5.3.7 Regression

To estimate a regression model, one may use the `svyPVglm` function. Its first argument is a formula describing model specification. It is constructed in the same way as in the `lm` or `glm` functions with the only difference being a special notation used with regards to PVs. The name of the achievements indicator to be used within the formula is constructed by replacing the PV number in name of variables with ".." (two dots). Values that should replace this part of the name to construct names of actual variables with PVs are passed to the `svyPVglm` function as a character or numeric vector, which are supplied using the argument `placeholder`. Finally, the argument `design` should be provided to specify the survey design object to be used. By default, a linear regression model is estimated.

For example, to estimate a model using PISA data in which achievements in science are predicted by gender and indexes of household cultural possessions, educational resource possessions, and wealth, one should call:

```
> svyPVglm(PV..SCIE ~ gender+cultposs+hedres+wealth,
+ design = pisaSVY,
+ placeholder = 1:10)
$coef
 mean se t.value Pr.t
(Intercept) 497.910421 2.198496 226.477793 < 2.22e-16
genderMale 9.235097 2.231822 4.137918 8.6915e-05
cultposs 17.943495 1.041578 17.227214 < 2.22e-16
hedres 7.508880 1.072217 7.003133 7.3703e-10
wealth 6.675032 1.189463 5.611804 2.8603e-07

$mod.fit
 Working.2.logLR p
 1197.187 < 2.22e-16
```

It is important to note that the function `svyPVglm` does not return a regular R regression model object (like those returned by `lm`, glm, or `svyglm`). It is a list with the first element being a data frame containing a summary of model coefficients and the second element describing model fit in terms of so-called working log-likelihood. Because of this, one may not call regression-diagnosis functions like `residuals`, `predict`, `vcov`, etc. on the object returned by `svyPVglm`.

Moreover, unlike all the other functions in the `svyPVpack` package, `svyPVglm` performs estimation on the whole dataset without allowing one to specify a grouping variable. To estimate separate models on subsets of data, the syntax presented below must be used to subset the data and repeatedly perform the analysis. In case of PIAAC, this task is simpler because this data is already split by country (see the next section on logistic regression).

```
> # model only for Germany
> svyPVglm(PV..SCIE ~ gender+cultposs+hedres+wealth,
+ design = subset(pisaSVY,
+ cntryid == "Germany"),
+ placeholder = 1:10)
(...)
> # models for each country using PISA data
> lapply(setNames(levels(pisa$cntryid),
+ levels(pisa$cntryid)),
+ function(cntry) {
+ svyPVglm(PV..SCIE ~ gender+cultposs+hedres+wealth,
+ design = subset(pisaSVY,
+ cntryid == cntry),
+ placeholder = 1:10)})
$Germany
$Germany$coef
 mean se t.value Pr.t
(Intercept) 506.346923 2.461630 205.695765 < 2.22e-16
genderMale 15.486567 2.830498 5.471322 5.0953e-07
cultposs 19.955634 1.601828 12.458035 < 2.22e-16
hedres 11.949802 1.883511 6.344431 1.2988e-08
wealth 9.139444 1.799112 5.079974 2.4658e-06
```

26.5.3.8 Logistic Regression

One may use PVs as independent variables in the same way as dependent variables. For example, to estimate a model describing how chances of being a NEET (not employed and not participating in education for last 12 months) is related to cognitive skills and gender using PIAAC data, one may use the same function `svyPVglm`. To specify that this should be a logistic regression, the argument `family = quasibinomial` should be passed to the function `call` (using "quasibinomial" family instead of binomial suppresses useless warnings regarding non-integer number of successes and failures; these numbers are always non-integer when sampling weights are used).

```
> # model for Germany
> svyPVglm(grepl("NEET", neet) ~ icthome + gender +
+ readytolearn + PVNUM..,
```

```
+ design = piaacSVYs[[1]],
+ placeholder = 1:10,
+ family = quasibinomial)
(...)
> # models for each country using PIAAC data
> lapply(piaacSVYs, function(.d) {
+ svyPVglm(grepl("NEET", neet) ~ icthome + gender +
+ readytolearn + PVNUM..,
+ design = .d,
+ placeholder = 1:10,
+ family = quasibinomial)})
$Germany
$Germany$coef
 mean se t.value Pr.t
(Intercept) 1.1191673 0.3239291 3.454976 0.00088813
icthome -0.1066211 0.0666943 -1.598654 0.11388965
genderFemale 0.6260891 0.1277355 4.901450 4.9763e-06
readytolearn -0.1235414 0.0972209 -1.270729 0.20755456
PVNUM1 -0.0122612 0.0012419 -9.873222 1.9415e-15

$Germany$mod.fit
 Working.2.logLR p
 1173.158 < 2.22e-16
```

References

Avvisati, F., & Keslair, F. (2014). *REPEST: Stata module to run estimations with weighted replicate samples and plausible values* (Statistical Software Components (S457918), revised March 23, 2017). Boston College Department of Economics.

Caro, D., & Biecek, P. (2017). intsvy: An R package for analyzing international large-scale assessment data. *Journal of Statistical Software, 81*(7), 1–44. https://doi.org/10.18637/jss.v081.i07

von Davier, M., Gonzalez, E., & Mislevy, R. J. (2009). What are plausible values and why are they useful. In M. von Davier & D. Hastedt (Eds.), *IERI monograph series: Issues and methodologies in large-scale assessments* (Vol. 2, pp. 9–36). International Association for the Evaluation of Educational Achievement; Educational Testing Service.

Evidence Institute. (2017). PISATOOLS package to analyze PISA data in Stata. http://www.evidin.pl/en/data.

Foy, P., & LaRoche, S. (2016). Estimating standard errors in the TIMSS 2015 results. In M. Martin, I. Mullis, & M. Hooper (Eds.), *Methods and procedures in TIMSS 2015* (pp. 77–85). TIMSS & PIRLS International Study Center, Lynch School of Education, Boston College and International Association for the Evaluation of Educational Achievement (IEA). https://timssandpirls.bc.edu/publications/timss/2015-methods/T15-Methods-and-Procedures-TIMSS-2015.pdf

Foy, P., & LaRoche, S. (2017). Estimating standard errors in the PIRLS 2016 results. In M. Martin, I. Mullis, & M. Hooper (Eds.), *Methods and procedures in PIRLS 2016* (pp. 81–89). TIMSS & PIRLS International Study Center, Lynch School of Education, Boston College and International Association for the Evaluation of Educational Achievement (IEA). https://timssandpirls.bc.edu/publications/pirls/2016-methods/P16_Methods_and_Procedures.pdf

Jakubowski, M. & Pokropek, A. (2017), *PISATOOLS: Stata module to facilitate analysis of the data from the PISA OECD study* (Statistical Software Components (S457754), revised May 06, 2020). Boston College Department of Economics.

Jakubowski, M., & Pokropek, A. (2019). piaactools: A program for data analysis with PIAAC data. *The Stata Journal, 19*(1), 112–128. https://doi.org/10.1177/1536867X19830909

Johnson, E., & Rust, K. (1992). Population inferences and variance estimation for NAEP data. *Journal of Educational Statistics, 17*(2), 175–190. https://doi.org/10.2307/1165168

Little, R., & Rubin, D. (1987). *Statistical analysis with missing data.* John Wiley & Sons.

Lumley, T. (2010). *Complex surveys: A guide to analysis using R.* John Wiley.

Lumley, T. (2019a). *mitools: tools to perform analyses and combine results from multiple-imputation datasets.* R package version 2.4.

Lumley, T. (2024). *survey: Analysis of complex survey samples.* R package version 4.4.

Macdonald, K. (2008). *PV: Stata module to perform estimation with plausible values* (Technical report, Statistical Software Components (S456951), revised Feb 12, 2014). Boston College Department of Economics, .

Mislevy, R. J. (1991). Randomization-based inference about latent variables from complex samples. *Psychometrika, 56*, 177–196. https://doi.org/10.1007/BF02294457

Mislevy, R. J., Beaton, A., Kaplan, B., & Sheehan, K. (1992). Estimating population characteristics from sparse matrix samples of item response. *Journal of Educational Measurement, 29*(2), 133–161. https://doi.org/10.1111/j.1745-3984.1992.tb00371.x

Mohadjer, L., Krenzke, T., Van de Kerckhove, W., & Li, L. (2016). Survey weighting and variance estimation. In *Technical report of the survey of adult skills (PIAAC)* (2nd ed., pp. 395–428). OECD. http://www.oecd.org/skills/piaac/PIAAC_Technical_Report_2nd_Edition_Full_Report.pdf

OECD. (2017). *PISA 2015 technical report.* OECD Publishing. https://www.oecd.org/pisa/sitedocument/PISA-2015-technical-report-final.pdf

OECD. (2019). *Technical report of the survey of adult skills (PIAAC)* - (3rd edition), OECD. https://www.oecd.org/content/dam/oecd/en/about/programmes/edu/piaac/technicalreports/cycle-1/PIAAC_Technical_Report_2019.pdf.

Pokropek, A. (2011). Missing by design: Planned missing-data designs in social science. *ASK. Research & Methods, 20*, 81–105. http://hdl.handle.net/1811/69576

Rao, J. N. K., & Shao, J. (1999). Modified balanced repeated replication for complex survey data. *Biometrika, 86*(2), 403–415. https://doi.org/10.1093/biomet/86.2.403

Reif, M., & Peterbauer, J. (2024). *svyPVpack: A package for complex surveys including plausible values.* R package version 0.1-5. https://github.com/manuelreif/svyPVpack

Schofield, L. S., Junker, B., Taylor, L. J., & Black, D. A. (2015). Predictive inference using latent variables with covariates. *Psychometrika, 80*, 727–747. https://psycnet.apa.org/doi/10.1007/s11336-014-9415-z

Shao, J., & Tu, D. (1995). *The jackknife and the bootstrap.* Springer.

Wu, M. (2005). The role of plausible values in large-scale surveys. *Studies in Educational Evaluation, 31*(2–3), 114–128. https://doi.org/10.1016/j.stueduc.2005.05.005

Chapter 27
Software Tools for Analyzing Log Data

Ulf Kroehne and Frank Goldhammer

Abstract The analysis of log file data from computer-based large-scale assessments promises to provide new insights into test-taking processes by combining different perspectives and research disciplines. To gain such insights, the extraction of low-level features and the aggregation to process indicators is necessary. However, log file data generated by assessment software cannot be analyzed using standard educational research and psychometric modeling tools directly. Based on the typical preparation for results data, this chapter starts with a description of a workflow for log data preparation and the separation of feature extraction and derived process indicators. Five requirements for software tools for log file analyses are derived from the workflow: the need to handle the contextual dependency of log events (R1), openness regarding the developed workflow (R2), a design that enables reproducibility of log file analyses (R3), extensibility to allow extraction of new features (R4), and versatility to use data from different assessment platforms (R5). The requirements are then used to compare three different tools: the general R environment, a specific R package (LogFSM), and the PIAAC Log Data Analyzer. No gold standard yet exists concerning the level of detail at which log file data from large-scale assessments are made available to data users. Hence, the closing discussion summarizes how tools for the analyses of log data could further be developed and used by data providers and data users.

Cognitive measurement in large-scale assessment can be described as students (or test-takers) answering a series of test items designed to measure a particular construct. The test items have scoring rules developed beforehand that relate the *raw response* (i.e., the work product obtained from each) to a *scored response* (either dichotomous or polytomous). The scored item-level responses are aggregated in

---

U. Kroehne (✉) · F. Goldhammer
DIPF | Leibniz Institute for Research and Information in Education, Frankfurt am Main, Germany
e-mail: u.kroehne@dipf.de

L. Khorramdel et al. (eds.), *Innovative Digital-Based International Large-Scale Assessments*, Methodology of Educational Measurement and Assessment,
https://doi.org/10.1007/978-3-031-90951-1_27

some way to a *test score* or *ability estimate*, which is finally used to infer the students' knowledge, skill, or another cognitive attribute (Goldhammer et al., 2021). Large-scale assessment programs often deliver *Plausible Values* (PV) as an ability measure derived to simplify inferences at the population level. Test developers or researchers closely related to a particular assessment program have access to the work products at all levels of the processing pipeline (raw responses, scored responses, test score or ability estimates, and PVs), while researchers using the outcomes of a cognitive assessment for more complex analyses might have only access to PV or ability estimates, disseminated using national or international *research data centers*. Specific research questions can require access to item-level work products (i.e., raw or scored responses) or to the actual item material (i.e., test items, stimulus texts and instructions). Therefore, some assessment programs offer the possibility to request in-depth access for specifically described analyses. Access is then guarded with confidentiality agreements and contracts to protect the used items (including, for instance, link items in longitudinal studies or studies with repeated cycles). The underlying processes and steps of coding, scoring, and generation of aggregate measures are known and accessible described in the form of technical reports published by the data providers, even though access to item-level work products is limited for data users.

The transition from paper-based assessment to computer-based assessment in large-scale assessments (i.e., PIAAC 2012, PISA 2015, TIMSS 2019, and PIRLS 2021) provides the opportunity to collect not only the final work product (raw response), but also to allows the collection of so-called *log data*, that can include different events with event-related attributes and timestamps that originate from students' interactions with the computer-based assessment platform (e.g., clicked buttons, selected radio buttons or checkboxes, entered text, mouse-moves, etc.). These log data are used by data providers and test developers of cognitive assessments and are made more and more available in public use files to other researchers and data users (for PIAAC, see Goldhammer et al., 2020; OECD, 2019).

The examination of these log data from cognitive ability testing has gained increased attention due to the digitalization of measurements in applications, such as large-scale assessments (Goldhammer & Zehner, 2017). However, additional data management and preparation steps are required to fully unlock the potential of such log data and to allow a higher level of transparency, as it is known for the work products (i.e., the PV or ability estimates, disseminated in public use files). Operationally, log data analysis is not done manually but with software tools. Hence, as a starting point for describing software tools for log data analyses, we assume the following simple workflow shown in Table 27.1, which describes how so-called process indicators are generated from log files (Goldhammer et al., 2021).

Data from large-scale assessments are typically prepared and analyzed using software tools. Any description of tools for analyzing log data should begin by clarifying what the software must accomplish to be useful. Instead of providing a definition, we summarize essential requirements that have emerged in this research field so far. The requirements and tools discussed in this chapter address the first two steps of the workflow shown in Table 27.1. Comparable to the status quo of work products, where the data providers do the coding and scoring of responses and the

Table 27.1 Workflow and differentiation between pre-processing (Log data), low-level feature extraction, and analysis and validation

<table>
<tr><th>Phase</th><th>Purpose</th><th>Term</th><th>Responsibility</th></tr>
<tr><td rowspan="4">Pre-Processing</td><td>Preparing and parsing log data (cleaning)</td><td rowspan="10">Log Data
↓
Low-Level Features
↓
Process Indicators</td><td rowspan="10">Software Developer
↓
Test Devel-oper
↓
Data Provider
↓
Data User</td></tr>
<tr><td>Generation of datasets (transformation)</td></tr>
<tr><td>Checking and purifying log data (including anonymization)</td></tr>
<tr><td>Documentation of log data</td></tr>
<tr><td rowspan="4">Low-Level Feature Extraction</td><td>Verification of log data (sequences, expected pattern, contexts)</td></tr>
<tr><td>Theoretical derivation of indicators</td></tr>
<tr><td>Extraction of low-level features (e.g., actions and states)</td></tr>
<tr><td>Combination of low-level features to process indicators</td></tr>
<tr><td rowspan="2">Analysis and Validation</td><td>Psychometric modelling of indicators</td></tr>
<tr><td>Life cycle of process indicators validation</td></tr>
</table>

ability estimation, the first steps of analyzing log data also fall under the test developers' and data providers' responsibility. However, data users certainly require knowledge of the overall process to interpret and empirically validate the derived process indicators.

In the first phase, raw log data must be preprocessed to handle the heterogeneous data formats provided by different assessment platforms and create datasets suitable for dissemination to empirical researchers who are used to working with datasets in file formats, such as CSV, SPSS, or Stata. in order to handle the heterogeneous data provided by different assessment platforms and create datasets suitable for dissemination to empirical researchers who are used to working with data sets. Note that dissemination is important, since secondary data users are expected to contribute to the development, investigation, and in particular, to the validation of process indicators (life cycle of process indicators). However, using log data similarly to result data requires the anonymization and documentation of the raw log data, which also involves data security issues. Data security and anonymity not only concerns the handling of personal information, which may be contained in the log data (for instance, key-down events that allow reconstructing the text entered into text fields). Information about items and test instruments directly included in the log events or indirectly derivable must also be considered and removed to ensure that the item content remains protected.

Pre-processing is followed by the second phase, feature extraction, which aims to identify specific low-level features relevant to track, for instance, the solution behaviors of target persons during the assessment. This chapter distinguishes the following terms: *raw log data* as the events stored by an assessment platform, *low-level features* as the contextualized information, and derived *process indicators* representing properties of the test taking process extracted from the event stream using low-level features.

Large-scale assessment programs can disseminate raw log data, extracted features, and derived process indicators to data users, and different user groups might emerge in the future. However, since the theoretical foundation of process indicators might require incorporating specific low-level features, access to the raw log data will be necessary for particular research questions.

The third phase, labeled as *analysis and validation* of the process indicators, concludes the workflow outlined in Table 27.1. In this phase, the process indicators are used in psychometric models or to derive diagnostic information to add or contrast the result data. While phases one and two include the algorithmic and mechanical management of information, phase three involves the validation of statements and claims based on *process indicators* with either empirical or theoretical evidence.

Table 27.1 lists the primary purposes of each phase, which cannot, however, be considered exhaustive. For instance, decoding and transforming log data into a documented data format might be necessary before the data can be effectively anonymized by research data centers.

A discussion of tools for analyzing log data can benefit from incorporating general needs derived from open science considerations (replicating log data analyses using new software; e.g., Plesser, 2018) and the FAIR principle (Findability, Accessibility, Interoperability, and Reusability; e.g., Wilkinson et al., 2016). Hence, in this chapter, we describe two general representations of log data (i.e., a flat and sparse log data table, and a universal log format) that can be stored and documented similarly to the results data for (large-scale) educational assessments. If the data are documented using codebooks, events are described using metadata, and datasets are stored similarly to all other datasets, we believe log file data can align well with the general FAIR principle, making it possible to apply existing routines and procedures to this new type of data. Special attention is also given to the reproducibility of log data analyses using identical data. Specifically, we refer to the definition provided in the Companion Guidelines on Replication & Reproducibility in Education Research (National Science Foundation, 2018): "Reproducibility refers to the ability to achieve the same findings as another investigator using extant data from a prior study." The first step towards achieving this kind of reproducibility for analyses based on log data from large-scale assessments is the dissemination of log data (and indicators derived from log data) together with its documentation (see Table 27.1). For process indicators, a second step concerns the need to provide detailed and understandable descriptions of the operationalization of these indicators. This will be discussed as a requirement for each of the considered software tools.

In this chapter, we begin with a short description of the event data representing the source for typical log file analyses and identify five requirements for software tools to support the feature extraction phase of the workflow shown in Table 27.1. We will then briefly describe three different software environments and tools with respect to these requirements: the general R environment (and selected R packages), a specific R package for analyzing log file data using finite state machines

(LogFSM,[1] see Kroehne & Goldhammer, 2018) and the PIAAC Log Data Analyzer (PIAAC LDA; OECD, 2019; Goldhammer et al., 2020), a specific tool for the analysis of PIAAC 2012 log data (round 1). In the concluding discussion, the central findings of this examination of the tools and requirements are summarized, and some consequences for the life cycle of process indicators and future advances in log data analysis are described.

27.1 Pre-processing

We assume that educational researchers and policymakers are less familiar with event-based log data than typical results datasets disseminated as public or scientific use files. Therefore, we start with a brief description of this relatively new type of data and emphasize the specific characteristics of log events originating from technology-based assessments in more detail. First of all, it is worth mentioning that log data are generated and stored by assessment platforms in various technical data formats. Potential formats include the extensible markup language (XML), the JavaScript object notation (JSON), or in either relational or non-relational databases from which data can be exported in the form of XML, JSON, SQL statements, or comma-separated value (CSV) files. Specific formats developed for event stream data, such as the eXtensible Event Stream (XES, XES Working Group, 2016), are also used—for instance, in process mining research (e.g., Janssenswillen et al., 2019). Although both JSON and XML are meant to be human-readable, log data are often perceived as 'messy' data. Uncertainty and lack of transparency concerning log data's actual content make it difficult for data users to follow, e.g., which actions are necessary before using and published log data from large-scale assessments. In the past, log data were often created for technical purposes, such as debugging (i.e., to give software developers the possibility to trace and check the functioning of the assessment platform). Log data were also used by item developers, for instance, to learn from field trials about the required time to work on a particular task or to answer a set of questions. Assessment platforms might also use log data to restore the items' state after re-entry (for instance, due to back navigation) or crash recovery (i.e., continuing an interrupted assessment with as little data loss as possible). Collecting log data can put additional burden on the system performance of assessment platforms. For the various reasons described so far, and because no secondary use of log data was foreseen, it is often impossible to use log data directly for research purposes. Often no formal description of the data structure (e.g., XML schema definition [XSD] or JSON schema definition) exists, and only an arbitrary subset of possible log events are typically stored. Consequently, the selection and structure of log data are platform-specific (i.e., each assessment platform or implementation of a digital-based large-scale assessment collects a different set of log

[1] The R package can be obtained and installed from the website https://www.logfsm.com.

events and stores the data differently). Furthermore, raw files or database tables with log data cannot be opened or imported "as is" with standard software tools for substantive analyses.

27.1.1 Preparing and Parsing Log Data

Whether log data was collected for technical reasons or intentionally for research purposes, data preprocessing steps are necessary, and preparing log data is typically more challenging than preparing results data. Parsing is required if information must be extracted from log data within files or tables containing some amount of unstructured text. While text data is often parsed automatically (in the background), the manual and explicit parsing of log data can sometimes be necessary—for instance, from server (access-) logs created during assessment. Figure 27.1 presents a real data example from a large-scale digital reading assessment (PISA 2012), for which log data are provided by the OECD for research purposes. Although the data are disseminated in SPSS file format (one file for each item), the information is provided

Step 1: Syntax for reading log data for Item CR002Q03 using the R package haven	`library(haven)` `cr002q03 <-` `read_sav(file="CBA_cr002q03_logs12_SPSS.sav")`
Step 2: Show selected columns for a case with a particular ID	`View(cr002q03[cr002q03$StIDStd == "09047",` `c("event_number","time","event",` `"event_value","event_detail")])`

	time	event	event_value	event_detail
1	1066.2	START_ITEM	NULL	NULL
2	1077.5	click	noID	seraing_home.php?tok=&link=275
3	1110.3	click	noID	ccc_home.php?tok=&link=217
4	1119.8	click	noID	ccc_programme_by_date.php?tok=&link=16
5	1140.8	click	previousBtn	stimulus.php
6	1142.5	click	previousBtn	stimulus.php#
7	1146.3	click	era002q2o4	item2.php
8	1146.3	change	era002q2o4	NULL
9	1149.4	click	NextButtonPForm	unit.php?PROCESSURI=http%3A%2F%2F127.0.0.1...
10	1151.4	END_ITEM	NULL	NULL
11	1151.4	click	noID	unit.php?PROCESSURI=http%3A%2F%2F127.0.0.1...

Note. Log events for one particular test-taker interacting with one task in unit 'cr002' of the electronic reading assessment (data source https://www.oecd.org/pisa/pisaproducts/database-cbapisa2012.htm). Haven library (Wickham & Miller, 2020) version 2.1.3 supports reading the SPSS file in R.

Fig. 27.1 Screenshot showing log data provided by OECD for PISA 2012 digital reading assessment. Note. Log events for one particular test-taker interacting with one task in unit 'cr002' of the electronic reading assessment (data source https://www.oecd.org/pisa/pisaproducts/database-cbapisa2012.htm). Haven library (Wickham & Miller, 2020) version 2.1.3 supports reading the SPSS file in R in a way that requires explicit parsing. The log data include information about the page selected by test-takers in a simulated hyper-text reading environment. However, this information must be extracted from the provided string variable (see column '`event_detail`', which contains the page name for click events in rows 2, 3, and 4 in Fig. 27.1) before test-takers' (navigation) behavior can be reconstructed from the log data.

The data extract presented in Fig. 27.1 illustrates the need to disentangle encapsulated information. The data itself is provided as a regular dataset (i.e., a rectangular dataset following an event structure with each line corresponding to one event). However, since PISA 2012 was administered using web-based techniques, the platform logged the requested page as part of the so-called query strings (information after the question mark in column '`event_detail`' in Fig. 27.1). The name of the requested page must be pre-processed before it can be worked with directly, meaning for this example, the event details must be separated into separate variables. An additional task in the pre-processing step for log file data, beyond parsing text into structured data in order to disentangle the available information, is the harmonization of timestamps (e.g., so-called epoch values,[2] relative time in different units, and absolute timestamps in different formats with/without a time zone). Since detailed technical insights in the assessment platform are required, data providers should be responsible for preparing and parsing log data to make sure relevant information can be identified and used (see Table 27.1).

27.1.2 Generation of Cleaned Datasets

Log data can appear confusing at first glance. The notion of log data as messy, non-transparent data is to some extent due to the fact that special characters, line breaks, and tab stops are modified by the technical encoding of characters. For instance, when the assessment platform integrates data from different components or technologies, character encoding might be employed so that the syntactic rules of an outer layer (for instance, an XML document) are not violated by data in an inner layer (see Fig. 27.8 for a real data example). Such hurdles, trivial from a technical perspective, must be eliminated and cleaned up as part of the data preparation step if secondary data users are expected to work with the raw log event data. The generation of cleaned datasets requires data providers' internal knowledge of the assessment platform's functioning and implementation in order to transform log data into a clean form that can be described and documented for dissemination to secondary data users.

In the following paragraphs, we briefly describe log data characteristics to understand better the structure of datasets that software tools are expected to read,

[2]Milliseconds since January 1, 1970, for example, are a common convention for storing time information.

analyze, and process. Log data are event-based and composed of events of different types (i.e., data from a varying number of unique events are collected for each test-taker depending on the test-taker's behavior). Since each log event takes place at a certain point in time, the event structure also represents the temporal dimension of log data. *Timestamps* can be either absolute (with or without a date) or relative (e.g., fractions of a second since the first log event or since the start of a particular part of the assessment, such as a given unit or item). Beyond timestamps, events are also typically linked to a given test-taker, represented in the log data by a person identifier.[3] In addition, events are either related to a particular part of the instrument (e.g., item or unit) or the test level (e.g., login, logout, …). The *event type*, which defines the meaning of a particular log event, also determines which additional event-specific data are provided by a particular event. These event-specific data can be required (i.e., must be provided for each event of a particular type) or optional (i.e., can be missing for some instances).[4] Together, the event data of different types, with the associated required and optional event-specific data, constitute the *raw log data*. Raw log data can be transformed into different formats, and the selection of a certain format depends on various preferences and traditions. Two specific types of data representations are discussed here:

- **Flat and sparse log data table:** One combined table containing all event data, including the event-specific data located in columns, with missing values for event-specific data not provided for an instance of an event of a particular type (because event-specific data are either optional or not defined for this event type). The data shown in Fig. 27.1 would give an example for a *flat and sparse log data table* if the event-specific data from column '`event_detail`' were separated (i.e., parsed) and split into different columns.
- **Universal log format:** Multiple tables, each table containing data for one particular event type only, with event-specific data located in columns and missing values only for optional event-specific data not provided for a given instance of an event.

In both formats, the generated clean data table(s) contain a column for the event type, the timestamp (either relative or absolute), the person identifier (and additional mapping identifier if necessary, see footnote 4) and a string variable identifying the element (i.e., item, unit or string constant referring to the test level). The central difference between the two forms of representation lies in how the event-specific

[3] In large-scale assessments, test-takers are often nested within schools, which are nested within higher-order groups (such as states or countries), so the person identifier might also allow for mapping to additional identifiers (such as a country identifier).

[4] Event-specific data can fit into the schema of key-value pairs, in which the number of keys is known and only one value is provided for each key. Assessment platforms might also provide event-specific data with a data structure that requires repeating identical information multiple times. This kind of 'nesting' is not further considered here but can be incorporated in the two representation formats for log data presented here.

data is incorporated (identical column for event-specific values of the same name in the flat sparse log data table or event-specific tables in the universal log format).

The highest degree of structure is obtained with the universal log format, which divides the log data into different tables depending on the event type. Within each of these datasets, the number of columns is completely determined by the number of required and optional event-specific data. Each row in such a table with the data for a particular log event type represents a log event, and missing values can only occur for optional event-specific data. Combining the event-specific tables and sorting the resulting data records by person identifier and timestamp transforms the data into a flat and sparse log data table, with missing values for all columns representing event-specific data that are not shared by all event types.[5] The flat and sparse log data table is the format used internally by the LogFSM package (see below). In contrast, the universal log format, which simplifies the storage and handling of log data through procedures similar to those for results data, might be preferred by research data centers for exchanging and archiving log file data. The universal log format is particularly useful if analyses only require a subset of event types. Moreover, the data structure used in the universal log format also helps to identify documentation- and purification-related needs, which are described next.

27.1.3 Contextual Dependency and Verification of Log Events

The meaning of a single log event (e.g., clicking a button) can differ depending on which log events happen before (or after). We call this property the *contextual dependency* of log events. This context (i.e., the sequence in which events occur) must be taken into account in numerous log file analyses—for instance, on strategies when reading multiple documents (Hahnel et al., 2019) or in-depth technical verification of the correctness, integrity, and completeness of log data (Kroehne & Goldhammer, 2018)—requiring a kind of "memory" of information on previous log events.

As illustrated in Table 27.2 with a hypothetical unit consisting of two items, the interpretation of events can depend on the context (the current page in the example). Combining knowledge about the assessment platform and the context created by the possible sequences of log events makes it possible to verify (or technically validate) that a particular log event is allowed at a particular position in the event stream. Methods for the verification of log data, as well as methods for extracting indicators from log data like those discussed in this chapter, must deal with the contextual dependency of log events in order to ensure data quality and integrity, to derive

[5]Technical details and additional steps for nested event-specific data are not discussed here. Moreover, if different event types provide event-specific data with identical names but different data types, techniques such as string serialization might be necessary.

Table 27.2 Illustration of the contextual dependency of log events for a hypothetical unit with three pages

Hypothetical Screen Shot	Interpretation of events
Page 1: 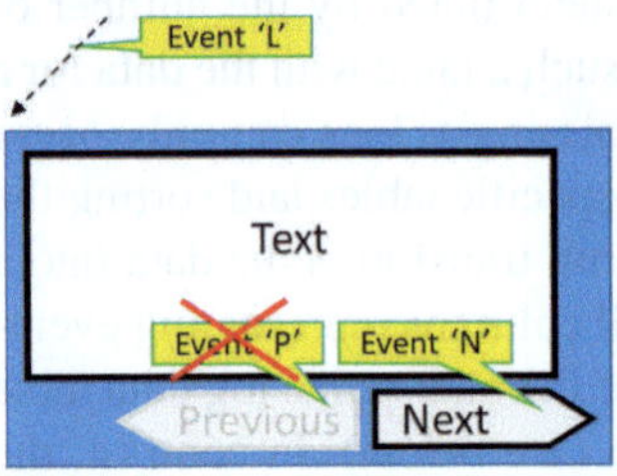	In this hypothetical example illustrating the contextual dependency of log events, we consider a simple unit with three pages. The first page contains a text, while the second and third pages contains the first and second items, respectively. Loading the unit is indicated by event 'L.' The pages contain two buttons, each of which triggers an event: The 'next' button triggers an event called 'N,' and the 'previous' button triggers an event called 'P.' If test-takers are not able to navigate to the previous page, event 'P' is invalid (as is the case, for instance, on the first page). Event 'N' identifies navigation to the next page.
Page 2: 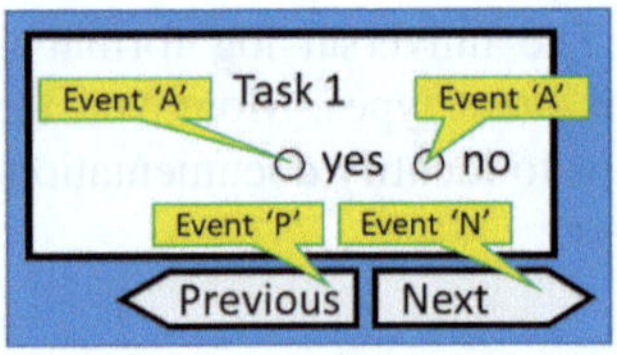	On the second page, both the 'N' event and 'P' event can occur. The 'P'event indicates back-navigation to the text page, and the 'N' event indicates navigation to the last page containing the second item. However, the interpretation of 'P' and 'N' for Page 2 is different from Page 1. Hence, in order to interpret the events 'N' and 'P' correctly, it is necessary to incorporate the context (i.e., the current page, which can be identified from a valid sequence of 'N' and 'P' events). Events of type 'A' indicate an answer change; 'A' events related to Item 1 are valid only on Page 2.
Page 3: 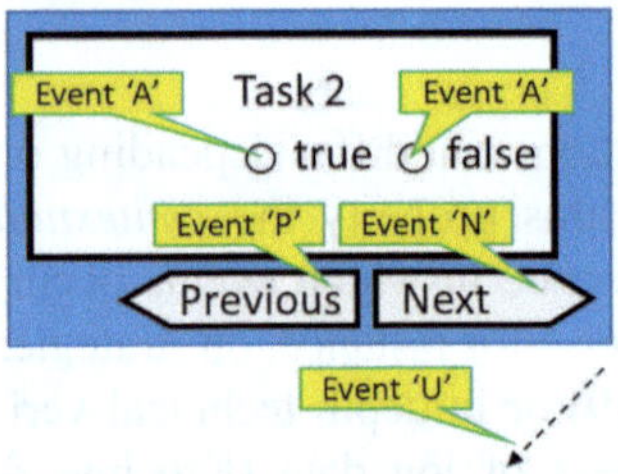	On the third page, the interpretation of events changes again. Events of type 'A' indicate an answer change for Item 2. Event 'P' now indicates back-navigation to the second page. In order to interpret event 'N', additional knowledge about the platform is required. For instance, this might indicate that the test-taker has finished the unit and navigated to the subsequent unit (or finished the entire test). In addition, the platform might provide an event identifying that the unit has been unloaded ('U'), which is only valid following the third page of the unit.

meaningful process indicators and, thereby, to fully unlock the potential of log data analysis. This, however, requires that the interactive items be documented in sufficient detail to provide an understanding of the test-taking and task processing context in which log events occur. Platform-specific knowledge about log events (for instance, the sequence in which different events must necessarily occur) is required to judge the correctness of the log data. Such data integrity checks can be understood as the first step towards ensuring the quality of log data.

27.1.4 Purification and Documentation of Log Data

Log data can contain unnecessary or redundant information. Hence, it should be mentioned that the pre-processing phase may also include a data purification step. If the log data collection was not primarily designed for subsequent advanced log data analysis, not all events or event-specific attributes are necessarily transferred into a log dataset. Documenting the removal of log information for particular event types or the deletion or modification of event-specific data increases the reproducibility of the data analyses.

A deep understanding of the interactive items' functioning is required to work with raw log event data, which are *contextualized*, requiring knowledge about and a profound experience of both the specific assessment instrument and its technical implementation. Possible approaches to providing information about log events within specific items include annotated screenshots (as illustrated in Table 27.2), screencasts of the items loaded in a web browser while the browsers inspection tool is running (e.g., Chrome's Developer Tools or Firefox's Firebug tool), and others (for instance, see below the documentation developed for the PIAAC items alongside the PIAAC Log Data Analyzer). There is currently a need to develop log data documentation strategies that balance the protection of item content while allowing log events to be interpreted within the interactive items, which is beyond the scope of this chapter.[6]

Assessment platforms typically only collect a subset of possible log events, and the names and values (including the structure) of the provided event-specific data are subject to the software implementation of a specific assessment platform (i.e., a standardization of event types and event-specific data has not yet been achieved, see Kroehne & Goldhammer, 2018). Accordingly, the different log event types must be described as part of a log data documentation. This encompasses describing the various tables for each event type provided in the universal log format. Additional documentation as *codebooks* is required to provide for each event type names, labels, and variable types for required or optional event-specific data. The log events providing string responses in event-specific attributes that necessitate additional anonymization steps need to be flagged in such codebooks. Similarly, log events containing data that could expose confidential information about the items or instruments require specific documentation to allow for proper pre-processing. Finally, identifiers or timestamps can be marked in the codebook where necessary to provide data centers with information about potential data preparation steps, such as de-anonymization checks.

[6] This issue is related to the extent to which raw responses embedded in log data contain information about sensitive item content, for instance, in terms of event-specific data or attribute names or disclose item content when combined with scored responses.

27.2 Feature Extraction

The (raw) log data discussed in the previous section comprise the source data for feature extraction after they have been pre-processed, cleaned, and purified. Low-level feature extraction is understood as the generation of new variables related to the test-taking process from the raw log data (i.e., the log events), which means applying rules to the observed events for each test-taker independently. Assessment platforms typically provide log events that do not necessarily correspond directly to actions with a unique interpretation. Hence, the interpretation of events can be further facilitated by considering their sequence, and, thus, the contextual dependency of log events (see above).

The number of possible characteristics and features is conceptually unlimited (see Kroehne & Goldhammer, 2018, for a taxonomy), and researchers are constantly adding to the total by suggesting new characteristics that might contain relevant information (for instance, the gaps between specific log events, see Eichmann et al., 2019). Table 27.3 presents one potential differentiation of features into individual process components (actions or states), sequences of process components (i.e., sequences of actions or states) and process indicators, which refer to properties of single process components or sequences of thereof (for instance, the presence or absence of particular action or state or sequences). We consider this distinction of extracted features into the three categories of action or state, sequence of actions or

Table 27.3 Classification of extracted features

What	Description
Action or *State*	Individual events or combinations of events designating certain *actions* or *states* are identified in the event stream. The distinction between states and actions is based on the idea that states have a certain duration (see Kroehne & Goldhammer, 2018), whereas for actions, only their occurrence is observed and the temporal duration can be neglected (similar to "verb clauses", see Hao & Mislevy, 2018; Mislevy et al., 2014; Mislevy, 2019).
Sequence of *Actions* or *States*	Actions or states can occur multiple times within the log event stream, gathered while a test-taker answers a particular task. The repeated occurrence of actions and states within the event stream makes it possible to derive a sequence of actions or sequence of states. The meaning of such sequences is based on the actions or states under consideration, and the length of sequences can vary between test-takers.
Process Indicator	Process indicators are understood as characteristics of the test-taking process that can be derived from the sequence of actions, sequence of states, or presence of a particular action or state (i.e., sub-sequences of length 1) or any other pattern created by actions and states. Process indicators can be stored in a wide format with one column per indicator. Process indicators can be qualitative, indicating whether an action or state was observed or a specified pattern occurred (dummy coding). Process indicators can also be quantitative, encoding, for instance, the number of occurrences of a particular action or the time required for a particular state or a sequence of states.

states, and process indicators as productive. While actions and states can be added to the raw log data (for instance, as an additional column in the sparse log data table format), process indicators can be merged with the results data (a fixed number of indicators for each item). Sequences of actions or states, meanwhile, can be represented either in a long format or as symbols in an n-gram representation model, inspired by methods developed for natural language processing (e.g., He & von Davier, 2016) and sequence mining techniques (e.g., Sukkarieh et al., 2012; Yamamoto et al., 2016; also applied by He et al., 2019).

Under the assumption that substantive analyses involve process indicators, linking the operationalization of process indicators to actions and states, and actions and states to log events can foster the transparency and reproducibility of log file analyses. In addition to automatic, data-driven feature extraction algorithm (e.g., action sequence auto-encoder, Tang et al., 2019), the following section provides a theory-driven perspective on low-level feature extraction. Although data users of large-scale assessments can construct their indicators only if they have access to raw log data, as in PIAAC (see below), the interpretation of process indicators requires a sound understanding of their construction based on feature extraction.

27.2.1 Theoretical Substantiation of Indicators

Theoretical arguments are necessary to justify the interpretation of process indicators derived from log data. General concepts, such as evidence-centered design (ECD; Almond et al., 2003), can be used to formulate validity arguments for process indicators, understood as characteristics of the test-taking process that can be derived from low-level features, such as sequences of actions or states. As shown in the hierarchical classification system presented in Table 27.3, process indicators can be understood as an aggregation or summary of specific information extracted from log events utilizing low-level features. Unfortunately, a complete list of all possible log events does not exist, and the completeness of log data is difficult to define in general (see Kroehne & Goldhammer, 2018). As highlighted by Bergner and von Davier (2019, p. 20), "emphasis on completeness should not come at the expense of clarifying the purpose of assessment and the nature of the assessment argument." Accordingly, following the ECD approach, the completeness needs to incorporate the goals and specific purposes of a log file analyses and the intended interpretation of process indicators (formulated as *state-completeness* by Kroehne & Goldhammer, 2018). Previous research has also not yet produced an exhaustive list of potential low-level features or process indicators. Hence, tools for analyzing log data cannot be described in terms of whether they implement all relevant indicators, even if all required log events are collected. Instead, the flexibility in defining low-level features will be considered as a requirement for potential software tools. Moreover, data analysis tools should support transparency regarding the relationship between log events and the extracted low-level features and derived process indicators.

Not all theoretically-derived process indicators created by test developers, data providers, or data users will necessarily work empirically in the expected or hypothesized way. Accordingly, a cumulative process of evidence accumulation is needed to develop valid indicators, following, for instance, the argument-based validation approach (Kane, 2013). The use and particularly the validation of indicators is conceptualized as an iterative research process, which can be described as an evolutionary development of process indicators (Goldhammer et al., 2021). For large-scale assessments, this might require disseminating the raw log events for secondary analyses to data users. Moreover, the reproducibility of feature extractions must be emphasized to facilitate the progressive development of knowledge about the validity of proposed indicators. As an illustrative example of the indicator development and validation process, we highlight the concept of rapid guessing. Suggested more than twenty years ago by Schnipke and Scrams (1997) as a measure of test speededness, a number of research papers showed and repeatedly validated that fast responses, identified using response time threshold or mixture models, are indicative of unmotivated test-taking behavior (see Wise, 2017, for a recent review). Item-level indicators of solution versus rapid guessing behavior have been investigated in large-scale assessments, such as PIAAC (e.g., Goldhammer et al., 2017) and NAEP (Lee & Jia, 2014). In future low-stakes assessments, these indicators (or the generalized response behavior index, Wise & Gao, 2017) might be computed directly by the assessment platform or even used to increase validity during the assessment (as shown by Wise et al., 2019). As a joint endeavor by the research community, investigating unmotivated test-taking processes using log file data and establishing corresponding indicators requires different researchers to apply the same operationalization of rapid guessing indicators. In particular, this requires transparency about the method used to derive the thresholds (see, for instance, Wise, 2019, for a detailed description) used to differentiate between solution and rapid guessing behavior. Goldhammer and Zehner (2017) illustrate the chain of inferences involved in converting log file data into a latent process of disengaged test-taking, and Goldhammer et al. (2021) systematize different validation strategies.

27.2.2 Extraction of Low-Level Features and Aggregation of Process Indicators

Having described the pre-processing of log data in the first section and emphasizing the need to relate process indicators to the assessment framework in the previous section, we will now identify five requirements of software tools for extracting features and process indicators. The first requirement for such tools refers to handling the *contextual dependency* of log events that is created by event sequences (see Table 27.2):

R1 (***contextual dependency***) The tool allows for the extraction of new information (low-level features), defined as prescriptive rules or algorithms for handling elementary raw log events, while taking into account that the same event can have different meanings depending on the context created by previous (and subsequent)[7] events. The extracted low-level features can be used to generate derived variables and process indicators.

In the simplest case, feature extraction is based on the presence of a single event or two consecutive events. For instance, the time difference between the event timestamps indicating the beginning of a task (for an example, see Fig. 27.1, event type START_ITEM) and its end (for an example, see Fig. 27.1, event type END_ITEM) is known as the 'time on task' indicator (e.g., Goldhammer et al., 2014). The platform might provide two events of different event types, as in this example. Depending on the event type, additional event-specific data might also be provided that can be used to derive additional low-level features. In more advanced cases, prescriptive rules can generalize to algorithms allowing for more sophisticated processing of multiple events, taking into account the context created by previous (or subsequent) events as well as additional event-specific data. The need for flexible prescriptive rules, also known as the algorithmic definition of indicators, becomes obvious when, for instance, the time after answering an item should be considered to derive a process indicator. Extracting such a feature requires selecting the last event of a particular type (e.g., the last event representing an answer change for a given item) before a second event occurs (e.g., the event representing navigation to the next item). Note that deciding that a given answer change event is the last answer change event represents a *look ahead* in the stream of events and is only possible after the student has finished working on a particular task (see footnote 8). Whether an extracted feature, such as the time after answering an item, provides meaningful insights into the test-taking process and leads to valid interpretations cannot be definitively judged without empirical evidence. Accordingly, we formulate an additional requirement regarding the integration of tools into workflows (such as the simple workflow shown in Table 27.1):

R2 (*openness*) Extracted indicators can be embedded in a workflow that distinguishes among the phases of pre-processing, feature extraction, and analysis and validation of extracted indicators.

R2 includes the possibility of disseminating process indicators in addition to or instead of elementary raw log data. Figure 27.2 provides an example of how the indicators computed by a specific tool that is available for data users (the PIAAC

[7]Taking into account not only previous events (i.e., involving some kind of memory) but also subsequent (future) events, as stipulated in R1, deserves special attention. While such a look ahead is possible when analyzing log data retrospectively, it is not available when log events are analyzed live during an ongoing assessment.

Step 1: Open a ZIP file with log file data from a particular country in the PIAAC Log Data Analyzer

Step 2: Select a task (e.g., CP32P001) and select an indicator (e.g., 'time on task'). Use the button 'Export Aggregated Variables' to request a data export and specify the name of the file to be created (e.g., "LDA_Export_Indicator.csv").

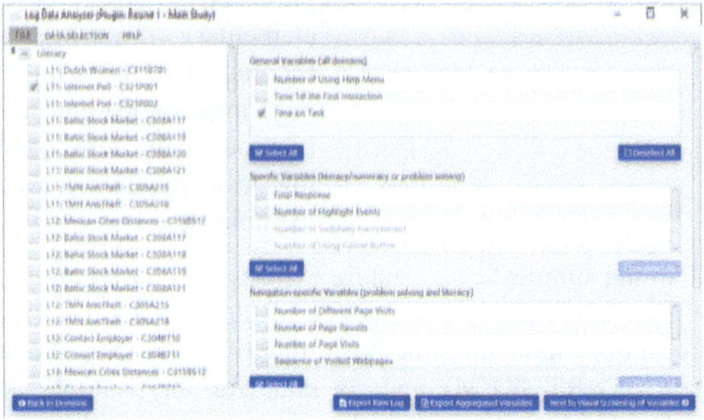

Step 3: Read the exported aggregated variable (i.e., the indicator) into R and plot histogram of the 'time on task' indicator

```
ind <- read.csv(file="LDA_Export_Indicator.csv",
           header = T, sep="\t")

hist(ind$L11_C321P001_time_on_task/1000,
   main="Time on Task 'C321P001'",
   br=seq(0,4000,2.5), xlim=c(0,200),
   xlab="Time on Task"
```

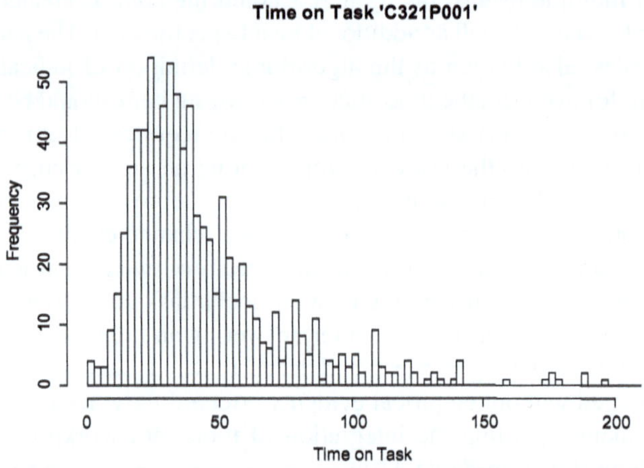

Note. Version 1.0 of the PIAAC Log Data Analyzer can be obtained from http://piaac-logdata.tba-hosting.de/download/.

Fig. 27.2 Example for requirement R2 (Openness) for the feature extraction phase of the PIAAC log data analyzer. Note. Version 1.0 of the PIAAC log data analyzer can be obtained from http://piaac-logdata.tba-hosting.de/download/

Log Data Analyzer, see below) in the feature extraction phase can be further investigated using R.

Moreover, to ensure that the extracted low-level features and process indicators can not only be shared, but that scientific knowledge about the functioning of the investigated indicators can be gathered, we identify the following additional requirement:

R3 (***reproducibility***) The tool stores the exact specification of the algorithms used for feature extraction in such a way that independent researchers can reproduce the creation of process indicators.

Conducting exactly the same feature extraction using identical data ensures reproducibility, a pre-requisite for scientific quality assurance. However, reproducibility does not necessarily imply the possibility of modifying and adjusting the algorithms used to extract features. This additional distinct feature is addressed in the following requirement:

R4 (***extensibility***) The tool enables the extraction of user-defined low-level features and process indicators based on newly developed algorithms or modifying the existing algorithms.

Requirement R4 does not explicitly limit the data to which a tool can be applied. Instead, the requirement is formulated so that it can also be fulfilled if only data from a particular dataset can be used, for instance, with tools provided for a specific assessment program. Hence, R4 refers to the need for software tools to formalize the algorithmic extraction of low-level features in a way that can be edited or modified by users to define new process indicators. Tools' scope, or applicability for the algorithmic extraction of indicators from data stemming from different studies, is addressed in the last requirement:

R5 (***versatility***) The tool can be used with log data gathered from different assessment platforms or various data sources.

The capability of applying algorithms to new log data (R5) might require that tools allow for the extraction of identical low-level feature or process indicators using different formulations of the algorithm, as log data from different sources and assessment platforms are likely to differ not only with respect to the types of events, but also the structure of event-specific data.

Together, the five requirements enable a conceptual classification of tools for analyzing log data. We propose that the term "tools for analyzing log data" be used when R1 is fulfilled. Since the reproducibility of analyses plays a central role in the validation of indicators, we also seek out tools that fulfill R2 and R3. Finally, with respect to scope, a distinction can be made between specific tools, which do not fulfill either R4 or R5, and general tools that are either extensible in terms of low-level features and process indicators (R4) or can be applied to different data (R5).

Solutions fulfilling both requirements, R4 (extensibility) and R5 (versatility), meaning that it is possible to define new algorithms for low-level feature extraction for custom log data, adding complexity.

27.2.3 Documentation of Indicators

As described above, process indicators are created using low-level features, extracted by applying prescriptive rules (i.e., algorithms) to the raw log data (event streams). Process indicators and low-level features can be created and disseminated by data providers. Process indicators documentation can gradually replace raw log data documentation if the computer-based instrument is not accessible for data users. Extracted features and process indicators can be understood as *decontextualized* regarding the specific assessment instrument and its implementation. The documentation of process indicators needs to be created and provided by data providers based on their particular knowledge about the platform required to extract low-level features. Low-level features or process indicators, similarly derived for multiple items, tasks, or even the whole instrument, can be more usable and self-evident for a larger group of data users and are more similar to the generated measures typically disseminated in public use files (such as PVs or ability estimates). If, however, as in the example of PIAAC, the raw log data are also disseminated, the extraction of low-level features and the aggregation of process indicators can also be done by data users, replicating results of previous research or investigating new features and process indicators. Accordingly, documenting the exact definition and operationalization of indicators is of particular importance. The documentation of indicators is closely related to the openness requirement (R2), as the dissemination of indicators to secondary data users can be considered a pre-requisite for the accumulation of knowledge about the functioning and validation of process indicators. The documentation of the process indicator construction is also related to the reproducibility of log file analysis (R3). As described, for instance, by Wilkinson et al. (2016), the FAIR principles apply not only to data (i.e., disseminated log file data for a particular large-scale assessment program) but also to the associated algorithms, tools, and workflows, ensuring transparency, reproducibility, and reusability. An evaluation of previous analyses of log data from large-scale assessments regarding the completeness of the documentation of the extracted indicators is beyond the scope of this chapter. However, in the following presentation of selected tools for analyzing log data, we will pay special attention to the question: To what extent are the tools particularly suitable for the documentation of the extraction of low-level features and the aggregation to process indicators?

In the remainder of the chapter, we will discuss the following tools with respect to the described requirements R1 to R5: the general purpose R environment with selected packages, a general tool for analyzing log data using finite state machines (LogFSM), and a specific tool for analyzing log data gathered in the first round of the PIAAC 2012 study (PIAAC Log Data Analyzer). Please note that this chapter was written in 2019 and last updated in 2021.

27.3 Analyzing Log File Data with R

Empirical research rests on the availability of tools for the analysis of datasets, such as those presented in introductory courses on quantitative (e.g., SPSS, Stata, or R) and qualitative (e.g., MaxQDA; Kuckartz & Raediker, 2019) research. Data users of large-scale assessments familiar with predefined and derived variables (e.g., PVs or the pre-computed *Total Time*) provided in public use files then use other software tools (e.g., MPlus, Muthèn & Muthèn, 2012; or the R package lavaan, Rosseel, 2012) for psychometric modeling. Concerning the workflow shown in Table 27.1, these tools are to be used for phase three (*analysis and validation*), while different tools are required for the phases one (pre-processing) and two (low-level feature extraction). While some statistical software environments are limited to datasets in one or multiple spreadsheet format(s), the R environment for statistical computing (R Core Team, 2020a) can handle both spreadsheet data and data not in spreadsheet format, including XML and JSON data.

27.3.1 Pre-Processing Using R

For importing XML and JSON data, the packages XML (Temple Lang, 2020), xml2 (Wickham, Hester, & Ooms, 2020b), and jsonlite (Ooms, 2014) are of great value. Moreover, packages such as DBI (R Special Interest Group on Databases, Wickham & Müller, 2019) allow access to various databases (through additional packages). Packages like haven (Wickham & Miller, 2020) and foreign (R Core Team, 2020b) provide support for datasets from other statistical software packages, such as SPSS and Stata, respectively. R supports the reading of data in the eXtensible Event Stream format (XES, XES Working Group, 2016) with the package xesreadR (Janssenswillen, 2019). As mentioned above, although *XML* and *JSON* formats are mostly human-readable, both file formats require developers to follow some rules in order to maintain syntactical correctness (such as avoiding un-encapsulated XML token characters < and > within attribute or node values). Various packages provide functions for decoding (and encoding) base64 strings (e.g., the jsonlite package) and URL strings (e.g., the urltools package, Keyes et al., 2019). Altogether, R fully meets requirement R5 (*versatility*).

27.3.2 Extraction of Indicators Using R

The R environment can be used to implement algorithms and to run the extraction of low-level features using rules and algorithms formulated in the R programming language (requirement R1). R packages, such as the *tidyverse* collection (Wickham et al., 2019), can simplify the implementation of algorithms to extract low-level

Step 1: Open a ZIP file with log file data from a particular country in the PIAAC Log Data Analyzer (see also Figure 27.2.).

Step 2: Export the raw log data using the button 'Export Raw Log' (file name "LDA_Export_Raw.txt")

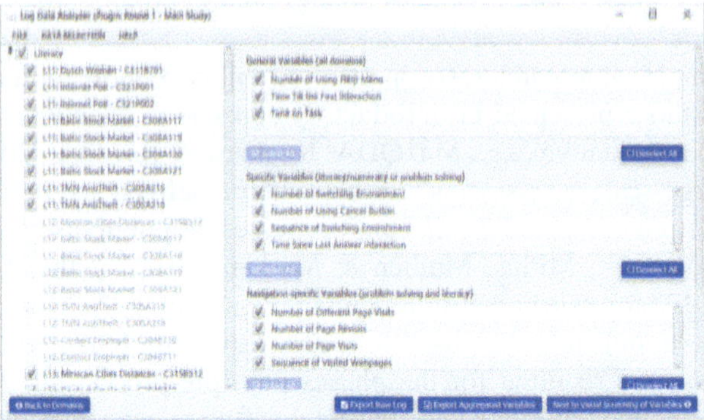

Step 3: Syntax for importing the raw log data into R

```
raw <- read.csv(file="LDA_Export_Raw.txt",
                header = T, sep="\t")
```

Step 4: Show selected columns for case with a particular ID

```
View(raw[raw$SEQID == 14 & raw$item_id == 1 &
    raw$booklet_id == 'L13',
      c("event_name", "event_type",
        "timestamp","event_description")])
```

	event_name	event_type	timestamp	event_description
3231	taoPIAAC	START	0	TEST_TIME=323
3232	stimulus	initialized	210	C315B512
3233	stimulus	cba_event	881	<cbaloggingmodel:CBAItemLogEntry xmlns:cbalogg...
3234	taoPIAAC	GET_HELP	28514	REQUEST
3235	stimulus	cba_event	131225	<cbaloggingmodel:ImageMapLogEntry xmlns:cbalo...
3236	taoPIAAC	NEXT_INQUIRY	132828	REQUEST
3237	taoPIAAC	NEXT_BUTTON	132828	id=nextinquiry_button
3238	taoPIAAC	BUTTON	132833	id=nextinquiry_button
3239	taoPIAAC	DOACTION	132833	action=tao_item.nextinquiry
3240	stimulus	itemScoreResult	137098	<cbascoringresultmm:itemScore xmi:version=2.0 x...
3241	stimulus	itemScore	137104	<cbascoringresultmm:itemScore xmi:version=2.0 x...
3242	stimulus	snapshotRef	137104	http://localhost:8080/cba-runtime/item?custom_ser...
3243	stimulus	END	137104	END
3244	taoPIAAC	NEXT_ITEM	137106	REQUEST
3245	taoPIAAC	END	137106	END

Fig. 27.3 Example of exporting raw log event data for PIAAC round 1 (with PIAAC log data analyzer)

features to some extent. For instance, the lead and lag operators (package dplyr; Wickham, François, et al., 2020a) make it possible to filter previous and subsequent events, facilitating the efficient handling of two-step sequences.

Figure 27.4 presents an example for creating an indicator (total number of help requests) based on the low-level feature (help request) from the raw log events as exported using the PIAAC Log Data Analyzer (see Figs. 27.2 and 27.3). As can be seen in the lower part of Fig. 27.4, there is a relationship between the number of visits to the help page within PIAAC items and test-takers' age. Older test-takers

Step 1: Create a data frame with an indicator for the total number of help page requests (based on raw log data from Figure 27.3)

```
library(dplyr)

help_indicator <- raw %>% group_by(SEQID) %>%
mutate(help_requested = sum(event_type== 'GET_HELP')) %>%
select(SEQID, help_requested) %>% unique()
```

Step 2: Read the public use file and merge the indicator and results data using the person identifier (SEQID)

```
library(foreign)

puf <- read.spss(file="data/prgusap1.sav",
        to.data.frame=T, use.missings=FALSE)

puf_help_indicator <- merge(puf, help_indicator, by="SEQID")
```

Step 3: Analyze indicator together with result variables, e.g., the relationship with age

```
tab <- table(puf_help_indicator$AGEG5LFS,
        puf_help_indicator$help_requested)

plot(tab[1:10,1:8],
   main="Total Number of Help Requests by Age")
```

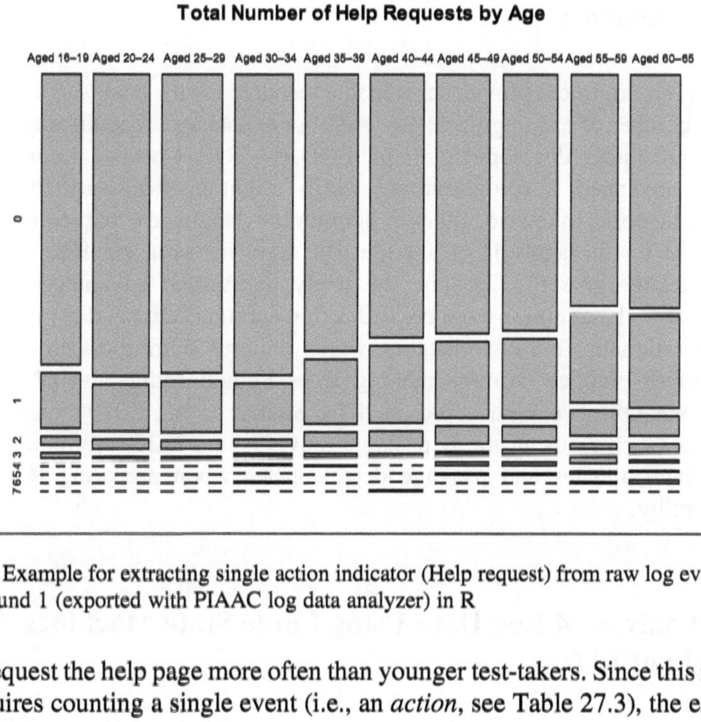

Fig. 27.4 Example for extracting single action indicator (Help request) from raw log event data in PIAAC round 1 (exported with PIAAC log data analyzer) in R

tend to request the help page more often than younger test-takers. Since this indicator only requires counting a single event (i.e., an *action*, see Table 27.3), the extraction of the indicator can be simplified to a single command—for instance, using the pipe operator provided by the dplyr package.

R allows pre-processing, low-level feature extraction, and statistical and psychometric modeling of process indicators in separate steps, with the possibility to adjust and adapt the process (requirement R2). Algorithms written in the R language can be extended or replaced to achieve *extensibility* (requirement R4), but programming skills are required to implement sophisticated algorithms for low-level feature extraction when there is no existing R package providing additional support. The strength of R becomes obvious in phase three of the sketched workflow. Due to the enormous number of already existing packages, various statistical and psychometrical modeling methods are available to jointly analyze the extracted features, process indicators, and outcome variables. A particular strength of R is the capability to read and process data of different formats, making R a versatile tool for the analysis of log data (requirement R5). However, as data pre-processing, feature extraction and aggregation to indicators must be implemented using the R language, the complete R source code is required to understand how process indicators have been operationalized. Requirement R3 (reproducibility) can be achieved for R-based log file analyses. However, due to different coding styles among researchers, substantial effort can be required to fully understand the R code used in previous research, if it is even shared.

27.3.3 Summary

The core R environment provides a valuable resource for the analysis of log file data given the number of existing packages. Additional packages, specifically for log file analyses, are under development (e.g., LOGAN; Reis Costa & Leoncio, 2019). However, advanced R (programming) skills might be necessary to implement custom indicators. Likewise, other programming languages, particularly Python, can be used for the analysis of log file data (see, for instance, Hao et al., 2016; Python package *glassPy*). Despite the many undisputed advantages of R (and Python), particular attention must be paid to the reproducibility and communicability of the operationalization of indicators, which requires sharing documented source code. Current practice requires making R or Python scripts available (see, for instance, the OSF workspace provided by Stadler et al., 2019), and additional approaches (such as using the R package renv; Ushey, 2020) are required to deal with package updates and package dependencies when preparing R source for reproducibility.

27.4 Analysis of Log Data Using Finite State Machines (LogFSM)

The R package LogFSM[8] aims to implement a general framework for analyzing log data using finite state machines described in Kroehne and Goldhammer (2018). The design of this tool is based on the assumption that the algorithmic decomposition of interaction processes (such as the test-taking process) is a general method that can be applied to define and operationalize different low-level features required for the

[8]LogFSM is available from https://github.com/kroehne/LogFSM.

definition of process indicators (as a general method for theory-driven feature extraction and construction of process indicators).[9]

27.4.1 Definition of States

To generate a specific semantic for analyzing log data, LogFSM requires that the researcher defines meaningful sections of the interaction between the test-taker and assessment platform, designated as *states*. States should be constructed on the basis of theory, for instance, by considering evidence-based arguments in the sense of ECD (Almond et al., 2003). After states have been defined, LogFSM can be used to actually extract the low-level features required for process indicators from log file data. By constructing indicators on the basis of states and *actions* (understood as designated events without a duration), the tool emphasizes the importance of conceptually relating the components of the test-taking process (i.e., states and actions) to the assessment framework. Instead of using (raw) log events directly, LogFSM operationalizes the process indicators with reference to properties of the reconstructed sequence of states and actions. One or multiple finite state machines can be used to reconstruct the test-taking process in terms of the considered states and actions. Thus, an additional layer of abstraction is included between the platform-specific, raw log events and the final indicators. This layer of low-level features seeks to strengthen the theoretical foundation of log file analyses and foster the applicability of the conceptualization of indicators to different data sources.

27.4.2 Implementation of Indicators

No pre-defined indicators are provided out of the box in the current version of LogFSM. Instead, the package allows for the use of a simple syntax language to define one (or multiple) finite state machines for the decomposition of the test-taking process in terms of rules triggered by the occurrence of specific raw log events. Given that the meaning of log events can change depending on their context (contextual dependency), state machines can process identical events differently when the state machines are in particular states. Moreover, as finite state machines are a mathematical computational model that can be used to represent a particular class of algorithms, LogFSM allows the algorithmic operationalization of low-level features (requirement R1), and the algorithms used for feature extraction can be extended (requirement R4).

[9]Note that finite state machines are typically used to implement state-based system behavior. For instance, the item authoring tool CBA ItemBuilder (Rölke, 2012) uses finite state machines for the implementation of complex or simulation-based items in computer-based assessment.

Indicators can be derived from raw log event data stored by a given assessment platform as an event stream if the necessary states can be distinguished on the basis of the observed log events. Given this property, described by Kroehne and Goldhammer (2018) as state-completeness, LogFSM can be used to implement user-defined low-level features as long as the required states can be identified by assigning one or multiple events to all transitions between states.

27.4.3 Reconstruction of the Test-Taking Process

When reconstructing the test-taking process for each test-taker, the finite state machine(s) begins in a specified start state. LogFSM then internally processes all log events for each test-taker serially, event by event. If a transition is defined for the finite state machine for a particular log event in the current state, the finite state machine accepts the event and moves on to the new state. Beyond accepting events, the finite state machine definition also makes it possible to specify events that should be ignored in particular states (i.e., events can be accepted without changing the machine's current state). After the last event is processed, the machine should reach one of potentially multiple end states. In relating the finite state machine definition to the log data provided to LogFSM, the event type and any event-specific data can be used as a trigger for a transition. The finite state machine rejects events if no transition is defined for events of a given event type (or generally for events with a specific combination of event-specific values) in a specific state. This feature can also be used to verify log data (concerning sequences, expected patterns, and contexts) regarding a finite state machine definition.

An example of the syntax used by LogFSM to define a finite state machine is presented in Fig. 27.7 for the PIAAC item "Internet Poll." The syntax composes transitions in the form "from-state" -> "to-state," which are triggered by events with particular event-specific values, as specified to the right of the @ sign. An event with matching event-specific values is accepted by the finite state machine (i.e., the rule is triggered) only if the finite state machine is in the specified "from-state." LogFSM applies finite state machines to the log data for each test-taker independently, starting with the designated start state for each test-taker. If the event data for a particular event (i.e., event name and event-specific attributes) fit the transition defined for the state machine in the start state, the event is accepted and the machine transitions from the prior state to the subsequent state (see line 2 in the augmented log data table, shown in Fig. 27.5).

Following this logic, LogFSM adds the states before and after processing each event to the *flat and sparse log data table* representation of the log file data (see columns "StateBefore_1" and "StateAfter_1" in Fig. 27.5 using the LogFSM-syntax shown in Fig. 27.6). LogFSM computes the time spent within each state using the relative or absolute timestamps provided with the raw log events. Finally, various summaries and statistics are created automatically from the reconstructed sequence of states for each test-taker. For instance, n-grams of observed state sequences and event sequences, state transition frequencies, the number of visits to each state, and

Step 1: Prepare the raw data exported by the PIAAC Log Data Analyzer for processing with LogFSM.

```
library(LogFSM)
myzipfilename <- paste0(getwd(),
    "/data/ZA6712_US.data_LDA_rawlog_txt.zip")
myprepared <- paste0(getwd(),
    "/data/temp__C321P001_C321P002.zip")
```

Only data from the two elements "C321P001" and "C321P002" (representing two items of a literacy unit) are processed in this example.

```
PrepareLogData(zipfilename = myzipfilename,
  outfilename = myprepared,
  elements = "C321P001;C321P002",
  datafiletype = "piaacldazip01a",
  flags = "RELATIVETIME")
```

Step 2: Define a LogFSM syntax as an R string variable.

```
Myfsmsyntax <- … # see Figure 6
```

Step 3: Run the decomposition (i.e., apply the FSM syntax to the prepared data).

```
out <- RunFSMSyntax(zipfilename = myprepared,
    datafilefilter = "*",
    fsmsyntax = myfsmsyntax,
    flags = "RELATIVETIME")
```

Request the augmented log data table for a particular test-taker with selected columns.

```
View(out$AugmentedLogDataTable %>%
    filter(!grepl("ItemScore",EventName)) %>%
    filter(PersonIdentifier == "ZA6712 16") %>%
    select(RelativeTime, EventName, StateBefore_1,
      StateAfter_1, Result_1,id,value))
```

	RelativeTime	EventName	StateBefore_1	StateAfter_1	Result_1	id	value
1	00:00:00.000	START	StartState	StartState	EventNotDefinedAsTrigger	NA	NA
2	00:00:00.278	initialized	StartState	C321P001	EventAccepted	NA	C321P001
3	00:00:02.337	CBAItemLogEntry	C321P001	C321P001	EventNotDefinedAsTrigger	NA	NA
4	00:00:35.308	SelectedInputFieldLogEntry	C321P001	C321P001	EventNotDefinedAsTrigger	[illegible]	NA
5	00:01:12.637	NEXT_INQUIRY	C321P001	C321P001	EventNotDefinedAsTrigger	NA	REQUEST
6	00:01:12.637	NEXT_BUTTON	C321P001	C321P001	EventNotDefinedAsTrigger	nextInquiry_button	NA
7	00:01:12.638	BUTTON	C321P001	C321P001	EventNotDefinedAsTrigger	nextInquiry_button	NA
8	00:01:12.638	DOACTION	C321P001	C321P001	EventNotDefinedAsTrigger	NA	NA
9	00:01:12.727	gotoFrame	C321P001	C321P002	EventAccepted	NA	C321P002
10	00:01:14.473	CBAItemLogEntry	C321P002	C321P002	EventNotDefinedAsTrigger	NA	NA
11	00:01:31.269	ButtonLogEntry	C321P002	votePageTask2	EventAccepted	L306F01B01	NA
12	00:01:34.085	ButtonLogEntry	votePageTask2	C321P002	EventAccepted	L306F02B01	NA
13	00:01:36.349	NEXT_INQUIRY	C321P002	C321P002	EventNotDefinedAsTrigger	NA	REQUEST
14	00:01:36.349	NEXT_BUTTON	C321P002	C321P002	EventNotDefinedAsTrigger	nextInquiry_button	NA
15	00:01:36.354	BUTTON	C321P002	C321P002	EventNotDefinedAsTrigger	nextInquiry_button	NA
16	00:01:36.354	DOACTION	C321P002	C321P002	EventNotDefinedAsTrigger	NA	NA
17	00:01:38.443	END	C321P002	EndState	EventAccepted	NA	END
18	00:01:38.445	NEXT_ITEM	EndState	EndState	EventNotDefinedAsTrigger	NA	REQUEST
19	00:01:38.445	END	EndState	EndState	EventNotPermitted	NA	END

Fig. 27.5 Example for using LogFSM to decompose the test-taking process using the state machine defined in Fig. 27.6 for the raw log data from PIAAC R1 (Exported with PIAAC log data analyzer)

time measures for single states and sequences are calculated. LogFSM generates various aggregate tables for all cases in the dataset as output alongside the reconstructed sequence of states. Accordingly, process indicators can be formulated as a function of the output of one or multiple finite state machines, using states and actions as low-level features in the additional layer of abstraction between the raw log events and extracted indicators.

LogFSM is provided as an R package enabling the creation of the required syntax to run finite state machines in R commands. The package provides access to the output data and, thus, also to the extracted low-level features, which can be further aggregated to process indicators and analyzed directly in the R environment (requirement R2). Using the package DiagrammeR, LogFSM can create state charts as graphical representations of the finite state machine(s) used to operationalize particular low-level features or indicators (see Fig. 27.7 for an example). These

LogFSM Syntax	Comment
myfsmsyntax <-"	Syntax is assigned to an R string variable.
Start: Startstate End: Endstate	Definition of start and end states.
Transitions: Startstate -> C321P001 @ EventName=initialized&value=C321P001	Unit start with an event named "initialized", and the attribute value must contain the item name ("C321P001").
Transitions: C321P001 -> VotePageTask1 @ EventName=ButtonLogEntry&id=L306F01B01 Transitions: C321P001 -> NonTargtePageTask1 @ EventName=EmbeddedLinkLogEntry Transitions: C321P001 -> ResultPageTask1 @ EventName=ButtonLogEntry&id=L306F01B02 Transitions: C321P001 -> C321P002 @ EventName=gotoFrame&value=C321P002	The name of the event to the right of the @ sign defines the state to which the machine transitions from state C23P001.
Transitions: C321P002 -> ResultPageTask2 @ EventName=ButtonLogEntry&id=L306F01B02 Transitions: C321P002 -> C321P001 @ EventName=gotoFrame&value=C321P001 Transitions: C321P002 -> NonTargtePageTask2 @ EventName=EmbeddedLinkLogEntry Transitions: C321P002 -> VotePageTask2 @ EventName=ButtonLogEntry&id=L306F01B01	Rules defined in this part apply only when the state machine is in state C321P002, since C321P002 is named as the "from-State" to the left of the "->".
Transitions: NonTargtePageTask1 -> C321P001 @ EventName=ButtonLogEntry;EmbeddedLinkLogEntry Transitions: NonTargtePageTask2 -> C321P002 @ EventName=ButtonLogEntry;EmbeddedLinkLogEntry Transitions: ResultPageTask1 -> C321P001 @ EventName=ButtonLogEntry;EmbeddedLinkLogEntry Transitions: ResultPageTask2 -> C321P002 @ EventName=ButtonLogEntry;EmbeddedLinkLogEntry	Events that can trigger the transition back to states C23P001 and C23P002 are listed, separated by ";".
Transitions: VotePageTask1 -> ResultPageTask1 @ EventName=ButtonLogEntry&id=L306F02B02 Transitions: VotePageTask1 -> C321P001 @ _ EventName=ButtonLogEntry&cbaBackButton_101_4606665852447718	Each line contains a single rule (_ marks rules continued on the next line).
Transitions: VotePageTask1 -> SubmitPageTask1 @ EventName=ButtonLogEntry&id=L306F02B01 Transitions: VotePageTask2 -> ResultPageTask2 @ EventName=ButtonLogEntry&id=L306F02B02 Transitions: VotePageTask2 -> C321P002 @ _ EventName=ButtonLogEntry&cbaBackButton_101_4606665852447718 Transitions: VotePageTask2 -> SubmitPageTask2 @ EventName=ButtonLogEntry&id=L306F02B01	Multiple conditions, e.g. EventName= ButtonLogEntry and id=L306F02B01, can be combined using the "&" sign.
Transitions: SubmitPageTask1 -> C321P001 @ EventName=ButtonLogEntry;EmbeddedLinkLogEntry Transitions: SubmitPageTask2 -> C321P002 @ EventName=ButtonLogEntry;EmbeddedLinkLogEntry	Note that the event ButtonLogEntry can trigger a transition to different to-states depending on the from-state.
Transitions: C321P001;C321P002;ResultPageTask1;ResultPageTask2;VotePageTask1; _ VotePageTask2;SubmitPageTask1;SubmitPageTask2 -> Endstate @ EventName=END"	For all defined states (C321P001, …, SubmitPageTask2), the END event results in a transition to the end state.

Fig. 27.6 Commented LogFSM-syntax for decomposing the test-taking process into task states (C321P001 and C321P002) and states related to task-specific pages (VotePageTask1, VotePageTask2, ResultPageTask1, ResultPageTask2, NonTargetPageTask1 and NonTargetPageTask2)

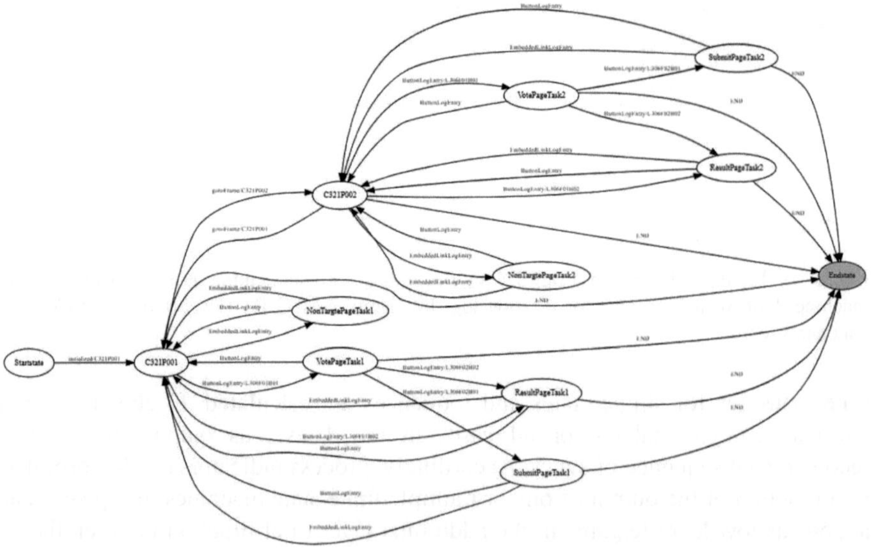

Note: Ellipses represent states and arrows represent transitions. Arrow labels refer to conditions (i.e., event with specific values) triggering a particular transition.

Fig. 27.7 UML state chart for the decomposition defined using the LogFSM—Syntax in Fig. 27.6. Note: Ellipses represent states and arrows represent transitions. *Arrow* labels refer to conditions (i.e., event with specific values) triggering a particular transition.Integration into R visualizations of the decomposition of the test-taking process should simplify the communication of low-level feature extraction and thus the reproducibility of log file analyses (requirement R3).

LogFSM can be applied to all log data that can be transformed into one of the two representations described above (flat and sparse log data table or universal log data format), fulfilling requirement R5. The R environment (including the packages mentioned above) can be used to read and pre-process custom log data before LogFSM is applied for feature extraction. LogFSM also supports different log file formats (for instance, log file data generated with the CBA ItemBuilder; Rölke, 2012), and can read raw log event data exported by the PIAAC LogData Analyzer.

27.4.4 Summary

The general framework described in Kroehne and Goldhammer (2018) uses finite state machines to conceptualize the decomposition of test-taking processes. Beyond the extraction of low-level features, the approach also allows, for instance, for the verification of the correctness of log data concerning the (expected) sequence. The validation of event streams depends on the existence of a complete finite state machine. Such a complete finite state machine definition incorporates existing knowledge about the functioning of the assessment platform and the computer-based instrument (i.e., it requires all expected transitions for a particular set of states). Log data are considered to be valid if a complete finite state machine rejects no event. A sound understanding of the meaning of individual raw log events from the particular assessment platform is required for use of both the general R environment and LogFSM. This, in turn, necessitates comprehensive documentation, which is provided, for instance, by the third tool described in the following subsection. With respect to the workflow outlined in Table 27.1, LogFSM can be used by data providers to extract low-level features that form the basis of process indicators. However, LogFSM can also be used by data users if the necessary log event data has been made available.

27.5 PIAAC Log Data Analyzer

The PIAAC Log Data Analyzer (PIAAC LDA; Goldhammer et al., 2020; OECD, 2019) was developed with funding from the OECD as a specific tool to make PIAAC log file data available and usable for the research community. No specific knowledge about the PIAAC log events is required to use the tool for the pre-processing and feature extraction phases of the workflow shown in Table 27.1. The PIAAC LDA can also be used solely for pre-processing the PIAAC log data. Here, the PIAAC LogDataAnalyzer enables the cleaned raw log event data to be exported so that it can then be used in combination with other approaches, such as LogFSM.

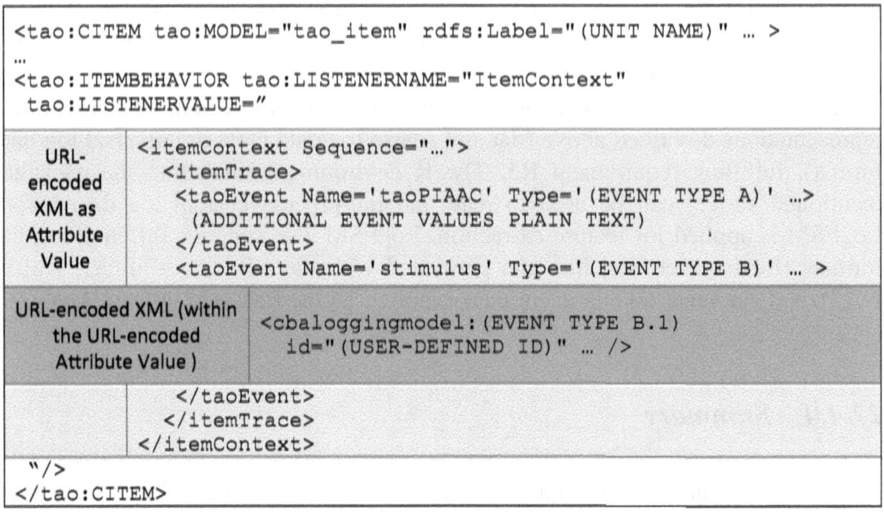

Fig. 27.8 Illustration of the nested layers of log data collected with PIAAC round 1

27.5.1 PIAAC Documentation

The PIAAC LDA includes detailed descriptions of all log events as well as detailed explanations of the algorithmic extraction process used to derive indicators and aggregated variables. The included program help pages are supplemented by complete documentation of the PIAAC items, which are provided in the form of interactive screenshots with information about the occurrence and meaning of all events and event-specific data.[10]

27.5.2 PIAAC Raw Log Files

The platform used in PIAAC Round 1 creates XML files containing the raw log event data. The XML files include multiple layers of encapsulation (i.e., XML nodes containing text that is itself encoded as XML) originating from the combination of different software components used in the PIAAC test delivery, including TAO (Jadoul et al., 2016) and the CBA ItemBuilder (Rölke, 2012). Figure 27.8 illustrates the nested raw data structure created by the assessment platform. The raw data in the RDF-style XML document produced by the TAO platform contains a node '`tao:CITEM`' for each unit, which has an attribute '`tao:LISTENERVALUE`' that includes the item context encoded as a URL. The URL-encoded attribute value is,

[10]The PIAAC log data documentation including interactive screenshots can be accessed via the hosting site URL (OECD, n.d.).

in fact, XML structured text that contains the log event data in a so-called '`itemTrace`' section, provided as a list of '`taoEvent`' nodes. Depending on the type of the '`taoEvent`' node, the event values either take the form of plain text or are included as another sub-component of URL-encoded text. URL encoding is used for the XML data originating from log events created in the sections of the PIAAC assessment implemented using the CBA ItemBuilder (see '`cbaloggingmodel`' in Fig. 27.8). The PIAAC LDA takes these technical details already into account when reading the log data. Without considering these details, the log data provided by the GESIS Data Archive (GESIS, 2021) cannot be processed. Hence, the log data disseminated from the PIAAC assessment require the PIAAC LDA to conduct the steps described in phase one of the workflow in Table 27.1. However, since the PIAAC LDA provides a user friendly interface, data users can run prepared parts of the pre-processing and low-level feature extraction.

The PIAAC LDA provides functionality for exporting raw log event data, which other tools can use to extract low-level features and to operationalize indicators not implemented in the PIAAC LDA (requirement R2). The nesting of the platform-specific data structures, illustrated for the PIAAC Round 1 data in Fig. 27.8, is removed when the PIAAC LDA exports data as flat and sparse log data table in CSV format, and the package LogFSM can read and transform this CSV format into formats, such as the universal log format.

27.5.3 Implemented Indicators

The PIAAC LDA provides direct access to various pre-defined aggregated variables (i.e., indicators that can be created in the feature extraction phase). Subject matter experts selected these indicators under the assumption that they are highly relevant for researchers interested in analyzing PIAAC log data. Table 27.4 provides a list of aggregated variables, here formulated with reference to the state decomposition described for LogFSM. Internally, the algorithms used to extract indicators from log events are implemented in Java. The implemented prescriptive rules and algorithms for low-level feature extraction consider the contextual dependency of log events (requirement R1). Openness, as described above (requirement R2), is achieved through the PIAAC LDA's ability to export generated variables (see Fig. 27.2), as well as the raw log event data (see above). Research using the PIAAC LDA can ensure reproducibility of pre-processing and feature extraction simply by providing a reference to the specific version of the PIAAC LDA (requirement R3).

The PIAAC LDA allows computing and exporting one or multiple indicators (as listed in Table 27.4) for either all or a subset of items from the PIAAC cognitive assessment. Exported aggregated variables and process indicators can be used for further analyses with the statistical software environment of the researcher's choice (requirement R2). Moreover, since no additions to the pre-defined list of aggregated variables can be made in the current version of the PIAAC LDA (requirement R4 is not fulfilled), interested researchers need to export raw log event data for further

Table 27.4 Indicators created by the PIAAC LDA, ordered according to state(s) of the test-taking process. Modified from Goldhammer et al. (2020)

Indicator/Aggregated Variable Created by the PIAAC LDA	Domain (s)	Property from the reconstructed sequence of states
Encompassing a single state corresponding to the complete task …		
Time on task	all	Total time on the state
Number of uses of help	all	Count of a specific event
Number of uses of 'Cancel' button	*PS*	Count of a specific event
Number of highlight events	*L*	Count of a specific event
Three states for each task (one prior to the first interaction, one after the last answer selection, and one in between) …		
Time till first interaction	all	Total time on the first state
Time since last answer interaction	*L / N*	Total time on the third state
State for each environment (in relevant tasks) …		
Number of environment switches	*PS*	Number of transitions
Sequence of environment switches	*PS*	Sequence of states (*n*-grams)
State for the email environment (in relevant tasks) …		
Number of created emails	*PS*	Count of specific events
State for each email (in relevant tasks)…		
Sequence of viewed emails	*PS*	Sequence of states (*n*-grams)
Number of email views	*PS*	Number of state visits
Number of different email views	*PS*	Number of unique state visits
Number of revisited Emails	*PS*	Number of state re-visits
States for each page (in relevant tasks) …		
Sequence of visited webpages	*L / PS*	Sequence of states (*n*-grams)
Time sequence of time spent on webpages	*L / PS*	Sequence of time on state for a state sequence (*n*-grams)
Number of different page visits	*L / PS*	Number of unique state visits
Number of page visits	*L / PS*	Number of state visits
Number of page revisits	*L / PS*	Number of state re-visits

Domain: *L* Literacy, *N* Numeracy, *PS* Problem solving

research when aiming at using the PIAAC log files with new indicators not foreseen by the aforementioned subject matter experts.

The published version of the PIAAC LDA can be used by data users to analyze data from Round 1 of the PIAAC 2012 study. Until plugins are developed and made available by the data providers to use the PIAAC LDA in combination with other data, requirement R5 is not fulfilled.

27.5.4 Summary

The primary advantage of the PIAAC LDA is to make extracted features and process indicators (labeled as aggregated variables) for a given PIAAC log file easily accessible to researchers and data users with only a couple of mouse clicks. By integrating the pre-processing and feature extraction phases into a user-friendly software, the PIAAC LDA creates a natural starting point for interested researchers new to the field of log file analyses. The practical usefulness of the PIAAC LDA is further enhanced by the accompanying documentation of both the indicators (within the tool) and items (via an additional website). After starting the process of analyzing log file data using the PIAAC LDA, both the process indicators and the prepared raw log event data itself can be exported for in-depth, advanced studies.

27.6 Discussion

Log data files created in computer-based large-scale assessments can be challenging to understand for researchers and stakeholders new to the field of log file analyses and learning analytics for various reasons. As we have shown, this is partly due to the technical nature of log files. In the past, event data were not primarily gathered for later scientific use, but rather for operational reasons, such as tracing and debugging technical (mis)behavior by assessment platforms. However, event data from computer-based large-scale assessments can be transformed into representations aligning with the infrastructure provided by research data centers (as illustrated with the universal log format) and disseminated to data users (as illustrated with the PIAAC LDA). The two described data formats for log data, the flat and sparse log data table containing all event data and the universal log format separating the event data according to the event type, are mutually transferable. Combined with documentation of the log event types and a ability to link the provided event-specific data to possible behavior within the interactive items, general and versatile tools can be developed for data users that allow low-level feature extraction as required for constructing process indicators based on log file data.

The operationalization of feature extractions typically involves multiple steps, meaning that prescriptive rules in the form of small algorithms are necessary to process log events accurately and automatically. These algorithms can be implemented directly in data analysis environments that allow for programming (e.g., R or Python). In this chapter, the R environment was presented as a general tool for analyzing log data. The algorithmic extraction of low-level features (i.e., actions and states) can also be performed with other programming languages, such as those used to implement the assessment platform itself (e.g., Java, PHP, or JavaScript). The LogFSM package was presented, which relies on finite state machines to implement an abstract representation of algorithms. Embedded in the R environment, LogFSM can be used to run finite state machines with empirical raw log event data in order to extract low-level features and to derive process indicators. Finally, the PIAAC LDA, implemented in Java, was presented as a user-friendly tool for the analysis of specific log file data for PIAAC R1 data users.

The availability of software tools for analyzing log data, either as specific tools for particular large-scale assessments or more general methods implementing analytical approaches, represents an important step towards providing new insights into the test-taking process for test developers, data providers and data users. Future standardization of data formats for log file data are expected to further increase the value of such tools. The generalizability of methods to analyze data from different sources is an important prerequisite to reduce the required resources to disseminate process indicators and low-level features. General purpose tools for analyzing log data can help to achieve this objective. Moreover, tools for analyzing log data can help to increase studies' reproducibility. However, the availability of data and analysis tools is not the only critical factor for further research into such indicators. Another key success factor will be the construction of validity arguments explaining why certain indicators allow conclusions to be drawn about the potential underlying (cognitive) processes (Goldhammer & Zehner, 2017; Goldhammer, et al., 2021). Hence, future development of tools should seek to create extensible and versatile methods and further develop methodological perspectives on log file analyses, which can in turn foster theory-driven research on process indicators.

A systematic approach to the analysis of log file data using specific tools also requires clarifying the status of indicators with respect to the dissemination of large-scale assessment data. We propose that the development, evaluation, and validation of process indicators is conceptualized as a development process. Access to the raw log event data is required to test new indicators and experiment with new low-level features of the test-taking process. However, at the end of the process, the validated process indicators can be either implemented directly in the assessment platform or created routinely as part of the data preparation process for large-scale assessment data, either as covariates or as additional score variables.

This chapter aimed to increase the transparency regarding the algorithmic low-level feature extraction and the aggregation to process indicators. Until now, the different assessment programs that introduced computer-based assessment do not follow a common strategy, whether process indicators, low-level features, or raw log event data are provided for data users. It is expected that data providers will find a balance between disseminating raw log event data, offering data users the flexibility to investigate custom process indicators, and disseminating extracted process indicators, making log data analyses easier. The combination of the PIAAC LDA and LogFSM presented in this chapter might serve as an example of how both goals can potentially be achieved with tools that make log data analyses more accessible for a larger group of data users.

References

Almond, R. G., Steinberg, L., & Mislevy, R. J. (2003). *A four-process architecture for assessment delivery, with conections to assessment design* [CSE Report 616]. http://cresst.org/wp-content/uploads/R616.pdf

Bergner, Y., & von Davier, A. A. (2019). Process data in NAEP: Past, present, and future. *Journal of Educational and Behavioral Statistics, 44*(6). https://doi.org/10.3102/1076998618784700

Eichmann, B., Goldhammer, F., Greiff, S., Pucite, L., & Naumann, J. (2019). The role of planning in complex problem solving. *Computer & Education, 128*, 1–12. https://doi.org/10.1016/j.compedu.2018.08.004

GESIS. (2021). *Programme for the International Assessment of Adult Competencies*. https://www.gesis.org/en/piaac/rdc

Goldhammer, F., & Zehner, F. (2017). What to make of and how to interpret process data. *Measurement: Interdisciplinary Research and Perspectives, 15*(3–4). https://doi.org/10.1080/15366367.2017.1411651

Goldhammer, F., Naumann, J., Stelter, A., Tóth, K., Rölke, H., & Klieme, E. (2014). The time on task effect in reading and problem solving is moderated by task difficulty and skill: Insights from a computer-based large-scale assessment. *Journal of Educational Psychology, 106*(3), 608–626. https://doi.org/10.1037/a0034716

Goldhammer, F., Martens, T., & Lüdtke, O. (2017). Conditioning factors of test-taking engagement in PIAAC: An exploratory IRT modelling approach considering person and item characteristics. *Large-Scale Assessments in Education, 5*(1), 18. https://doi.org/10.1186/s40536-017-0051-9

Goldhammer, F., Hahnel, C., & Kroehne, U. (2020). Analysing log file data from PIAAC. In D. B. Maehler & B. Rammstedt (Eds.), *Large-scale cognitive assessment* (pp. 239–269).

Goldhammer, F., Hahnel, C., Kroehne, U., & Zehner, F. (2021). From byproduct to design factor: On validating the interpretation of process indicators based on log data. Large-scale Assessments in Education, 9. https://doi.org/10.1186/s40536-021-00113-5

Hahnel, C., Kroehne, U., Goldhammer, F., Schoor, C., Mahlow, N., & Artelt, C. (2019). Validating process variables of sourcing in an assessment of multiple document comprehension. *British Journal of Educational Psychology, 89*(3), 524–537. https://doi.org/10.1111/bjep.12278

Hao, J., & Mislevy, R. J. (2018). The evidence trace file: A data structure for virtual performance assessments informed by data analytics and evidence-centered design. *ETS Research Report Series, 2018*(1), 1–16. https://doi.org/10.1002/ets2.12215

Hao, J., Smith, L., Mislevy, R., von Davier, A., & Bauer, M. (2016). Taming log files from game/simulation-based assessments: Data models and data analysis tools. *ETS Research Report Series, 1*, 1–17. https://doi.org/10.1002/ets2.12096

He, Q., & von Davier, M. (2016). Analyzing process data from problem-solving items with ngrams: Insights from a computer-based large-scale assessment. In Y. Rosen, S. Ferrara, & M. Mosharraf (Eds.), *Handbook of research on technology tools for real-world skill development* (pp. 749–776). Information Science Reference.

He, Q., Borgonovi, F., & Paccagnella, M. (2019). *Using process data to understand adults' problem-solving behaviour in the Programme for the International Assessment of Adult Competencies (PIAAC): Identifying generalised patterns across multiple tasks with sequence mining* (OECD Education Working Papers Nr. 205). https://doi.org/10.1787/650918f2-en

Jadoul, R., Plichart, P., Bogaerts, J., Henry, C., & Latour T. (2016). The Tao platform. *PIAAC technical report second Edition*. https://www.oecd.org/skills/piaac/PIAAC_Technical_Report_2nd_Edition_Full_Report.pdf

Janssenswillen, G. (2019). *XesreadR: Read and write XES files* [Manual]. https://CRAN.R-project.org/package=xesreadR

Janssenswillen, G., Depaire, B., Swennen, M., Jans, M., & Vanhoof, K. (2019). bupaR: Enabling reproducible business process analysis. *Knowledge-Based Systems, 163*, 927–930.

Kane, M. (2013). The argument-based approach to validation. *School Psychology Review, 42*(4), 448–457. https://doi.org/10.1080/02796015.2013.12087465

Keyes, O., Jacobs, J., Schmidt, D., Greenaway, M., Rudis, B., Pinto, A., Khezrzadeh, M., Meilstrup, P., Costello, A. M., Bezanson, J., Meilstrup, P., & Jiang, X. (2019). *urltools: Vectorised tools for URL handling and parsing* [Manual]. https://CRAN.R-project.org/package=urltools

Kroehne, U., & Goldhammer, F. (2018). How to conceptualize, represent, and analyze log data from technology-based assessments? A generic framework and an application to questionnaire items. *Behaviormetrika, 45*(2). https://doi.org/10.1007/s41237-018-0063-y

Kuckartz, U., & Raediker, S. (2019). *Analyzing qualitative data with MAXQDA: Text, audio, and video*. Springer.
Lee, Y.-H., & Jia, Y. (2014). Using response time to investigate students' test-taking behaviors in a NAEP computer based study. *Large-Scale Assessments in Education, 2*(1), 1–24. https://doi.org/10.1186/s40536-014-0008-1
Mislevy, R. J. (2019). On integrating psychometrics and learning analytics in complex assessments. In H. Jiao, R. W. Lissitz, & A. van Wie (Eds.), *Data analytics and psychometrics* (pp. 1–52). Information Age Publishing.
Mislevy, R. J., Oranje, A., Bauer, M. I., von Davier, A. A., Hao, J., Corrigan, S., Hoffman, E., DiCerbo, K. E., & John, M. (2014). *Psychometric considerations in game-based assessment*. GlassLab Research, Institute of Play.
Muthèn, L. K., & Muthèn, B. O. (2012). *MPlus: Statistical analysis with latent variables – User's guide*. http://www.statmodel.com/download/usersguide/Mplus%20user%20guide%20Ver_7_r3_web.pdf
National Science Foundation and Institute of Educational Science. (2018). *Companion guidelines on replication & Reproducibility in education research.* https://www.nsf.gov/pubs/2019/nsf19022/nsf19022.pdf
OECD. (2019). *Beyond proficiency: Using log files to understand respondent behaviour in the survey of adult skills.* OECD. https://doi.org/10.1787/0b1414ed-en
OECD. (n.d.). *OECD survey of adult skills (PIAAC) log data documentation*. DIPF/GESIS. https://piaac-logdata.tba-hosting.de/
Ooms, J. (2014). *The jsonlite package: A practical and consistent mapping between JSON data and r objects. arXiv:1403.2805* [stat.CO]. https://arxiv.org/abs/1403.2805
Plesser, H. E. (2018). Reproducibility vs. replicability: A brief history of a confused terminology. *Frontiers in Neuroinformatics, 11*, 76. https://doi.org/10.3389/fninf.2017.00076
R Core Team. (2020a). *R: A language and environment for statistical computing* [Manual]. https://www.R-project.org/
R Core Team. (2020b). *foreign: Read data stored by "minitab", "s", "SAS", "SPSS", "stata", "systat", "weka", "dBase", ...* [Manual]. https://CRAN.R-project.org/package=foreign
R Special Interest Group on Databases (R-SIG-DB), Wickham, H., & Müller, K. (2019). *DBI: R database interface* [Manual]. https://CRAN.R-project.org/package=DBI
Reis Costa, D., & Leoncio, W. (2019). *LOGAN: Log file analysis in international large-scale assessments* [Manual]. https://CRAN.R-project.org/package=LOGAN
Rölke, H. (2012). The itembuilder: A graphical authoring system for complex item development. In T. Bastiaens & G. Marks (Eds.), *Proceedings of E-Learn 2012--World Conference on E-Learning in Corporate, Government, Healthcare, and Higher Education 1* (pp. 344–353). Association for the Advancement of Computing in Education (AACE). https://www.learntechlib.org/primary/p/41614/
Rosseel, Y. (2012). Lavaan: An R package for structural equation modeling and more. Version 0.5–12 (BETA). *Journal of Statistical Software, 48*(2), 1–36. https://www.jstatsoft.org/article/view/v048i02
Schnipke, D. L., & Scrams, D. J. (1997). Modeling item response times with a two-state mixture model: A new method of measuring speededness. *Journal of Educational Measurement, 34*(3), 213–232. https://doi.org/10.1111/j.1745-3984.1997.tb00516.x
Stadler, M., Fischer, F., & Greiff, S. (2019). Taking a closer look: An exploratory analysis of successful and unsuccessful strategy use in complex problems. *Frontiers in Psychology, 10*, 777. https://doi.org/10.3389/fpsyg.2019.00777
Sukkarieh, J., von Davier, M. & Yamamoto, K. (2012). *From biology to education: Scoring and clustering multilingual text sequences and other sequential tasks* (ETS Research Report, ETS RR–12-25). https://doi.org/10.1002/j.2333-8504.2012.tb02307.x
Tang, X., Wang, T., Du, Y., & Dai, Y. (2019). Motor imagery EEG recognition with KNN-based smooth auto-encoder. *Artificial Intelligence in Medicine, 101*. https://doi.org/10.1016/j.artmed.2019.101747

Temple Lang, D. (2020). *XML: Tools for parsing and generating XML within r and s-plus* [Manual]. https://CRAN.R-project.org/package=XML

Ushey, K. (2020). *renv: Project environments* [Manual]. https://CRAN.R-project.org/package=renv

Wickham, H., & Miller, E. (2020). *haven: Import and export "SPSS", "stata" and "SAS" files* [Manual]. https://CRAN.R-project.org/package=haven

Wickham, H., Averick, M., Bryan, J., Chang, W., McGowan, L. D., François, R., Grolemund, G., Hayes, A., Henry, L., Hester, J., Kuhn, M., Pedersen, T. L., Miller, E., Bache, S. M., Müller, K., Ooms, J., Robinson, D., Seidel, D. P., Spinu, V., et al. (2019). Welcome to the tidyverse. *Journal of Open Source Software, 4*(43), 1686. https://doi.org/10.21105/joss.01686

Wickham, H., François, R., Henry, L., & Müller, K. (2020a). *dplyr: A grammar of data manipulation* [Manual]. https://CRAN.R-project.org/package=dplyr

Wickham, H., Hester, J., & Ooms, J. (2020b). *xml2: Parse XML* [Manual]. https://CRAN.R-project.org/package=xml2

Wilkinson, M. D., Dumontier, M., Aalbersberg, I. J., Appleton, G., Axton, M., Baak, A., & Mons, B. (2016). The FAIR guiding principles for scientific data management and stewardship. *Scientific Data, 3*(1). https://doi.org/10.1038/sdata.2016.18

Wise, S. L. (2017). Rapid-guessing behavior: Its identification, interpretation, and implications. *Educational Measurement: Issues and Practice, 36*(4), 52–61. https://doi.org/10.1111/emip.12165

Wise, S. L. (2019). An information-based approach to identifying rapid-guessing thresholds. *Applied Measurement in Education, 32*(4), 325–336. https://doi.org/10.1080/08957347.2019.1660350

Wise, S. L., & Gao, L. (2017). A general approach to measuring test-taking effort on computer-based tests. *Applied Measurement in Education, 30*(4), 343–354. https://doi.org/10.1080/08957347.2017.1353992

Wise, S. L., Kuhfeld, M. R., & Soland, J. (2019). The effects of effort monitoring with proctor notification on test-taking engagement, test performance, and validity. *Applied Measurement in Education, 32*(2), 183–192. https://doi.org/10.1080/08957347.2019.1577248

XES Working Group. (2016). *IEEE standard for eXtensible Event Stream (XES) for achieving interoperability in event logs and event streams* (IEEE Std 1849–2016). https://standards.ieee.org/standard/1849-2016.html

Yamamoto, K., von Davier, M., & Sukkarieh, J., (2016). *Systems and methods for evaluating multilingual text sequences* (Patent Serial No.: 13/848,837, issued July 27, 2016).

Chapter 28
Analyzing Large-Scale Assessment Data in R: EdSurvey

Ting Zhang, Emmanuel Sikali, and Paul Bailey

Abstract This chapter introduces EdSurvey, an R package for analyzing large-scale, complex survey response data with an assessment. The chapter begins with an overview of the package and its essential methodology for handling complex sampling designs and plausible values. Empirical examples illustrate how to use EdSurvey to perform data processing, exploration, manipulation, and some statistical analyses, including tables, gap analysis, and linear and logistic regressions.

Large-scale educational assessments—such as the National Assessment of Educational Progress (NAEP), Trends in International Mathematics and Science Study (TIMSS), and the Programme for International Student Assessment (PISA)—are valuable for studying various issues related to education and to inform policy. However, because of the scope and complexity of large-scale assessment data, researchers must normally use several software tools to (a) access the data, (b) perform data manipulation, and (c) analyze the data by considering its complexity in sampling and other special features.

EdSurvey is an R statistical package that serves as an extensible one-stop shop for analyzing large-scale education data. This package streamlines the entire process of large-scale data analysis from downloading, processing, exploration, and manipulation to conducting analyses. The intent is to make the process user oriented and efficient. This chapter gives an overview of the EdSurvey package, starting with the current state of the art and applications and moving on to the theoretical methods that deal with complex sampling designs and plausible values. Lastly, we present empirical examples and end the chapter with a discussion and outlook for future development.

T. Zhang (✉) · P. Bailey
American Institute for Research, Arlington, VA, USA
e-mail: tzhang@air.org

E. Sikali
National Center for Education Statistics, Washington, DC, USA

L. Khorramdel et al. (eds.), *Innovative Digital-Based International Large-Scale Assessments*, Methodology of Educational Measurement and Assessment,
https://doi.org/10.1007/978-3-031-90951-1_28

28.1 Current State of the Art and Applications

To assist large-scale assessment data analysis, the International Association for the Evaluation of Educational Achievement (IEA) provides free tools, such as the International Database Analyzer (IDB Analyzer), a stand-alone application that creates shareable code in SAS and SPSS for combining and analyzing data from most international large scale assessments, and even several large-scale national assessments, such as NAEP (Gonzalez, 2021). The U.S. National Center for Education Statistics (NCES) has traditionally supported the development of various tools—such as Electronic Code Books, NAEP Data Explorer, International Data Explorer, WesVar©,[1] and the AM software©[2]—for the research community to access NCES datasets with ease, perform data analysis, and test their hypotheses. WestVar and AM have the advantage of allowing users to manipulating data outside of commercially available software packages, such as SPSS, SAS, or STATA. However, most of their functionality has not been kept up to date to address researchers' needs of performing more advanced analyses and using recently developed procedures.

EdSurvey is a software package that one can use in the R environment; R is a programming language licensed under the GNU General Public License. The R software can be installed for free on computing devices that use the Windows, Linux, or Mac OS operating systems and servers.

The functionality that EdSurvey adds within the R computing environment is presented in this chapter. In addition, the R community of users and researchers have developed hundreds of programs to perform statistical analyses in bundles called *packages*. These packages are shared and available for free download in the comprehensive R archive network, usually called by its acronym, CRAN. Academia and the research community widely use R. Other packages in R will analyze large-scale data with complex sampling designs with plausible values, including survey (Lumley, 2004), lavaan.survey (Oberski, 2014), svyPVpack (Reif & Peterbauer, 2014), BIFIEsurvey (BIFIE et al., 2019), and intsvy (Caro & Biecek, 2017). Among these packages, some have limited data coverage (e.g., tailored to international large-scale assessments only), and some require user input about specifics of the survey design and plausible values before they can be used.

EdSurvey was developed to analyze all the large-scale assessments that the United States participates in under NCES, with the complex sampling design and plausible value methodology incorporated seamlessly. The advantages of EdSurvey include:

---

[1] WesVar is a software that handles multiply-imputed datasets and variance estimates are based on replicate weights.

[2] AM is a software for analyzing data from complex samples, especially large scale assessments.

- allowing for data manipulation inside and outside the package;
- enabling users to search the names and labels of variables, view frequencies and percentages of response categories of variables, and visualize the data;
- performing analysis operations with complex samples;
- performing analysis using plausible values;
- performing multilevel analyses and modeling;
- expanding to meet the needs of analysts by constantly adding new features; and
- incorporating user community contributions into EdSurvey (because users have access to the source code).

In the next section, we will overview the methodology that EdSurvey employs for calculating estimates and variances for large-scale assessment data with plausible values.

28.2 Statistical Methods Used in EdSurvey

Users new to large scale assessment data may be both unfamiliar with plausible values and estimating sampling variance, and they can find the methodology daunting. EdSurvey simplifies their interaction with the data by flattening the learning curve, for example, by representing scores as if they were a single variable, as users are used to thinking of them. To enable this, all EdSurvey analysis commands automatically take into account the specific design of the data file (e.g., for NAEP after 2015, twenty plausible values and a JK2 sampling design) and use that information to estimate sampling and imputation variance without the user having to further specify the design.

Bailey and Cohen (2018, 2019) described the methodology of EdSurvey for estimation procedures, including means, percentages, percentiles, achievement levels, regression analysis, and variance estimation with and without plausible values. EdSurvey follows a methodology similar to that used in large-scale assessments (e.g., NAEP and TIMSS) for estimation (Johnson & Rust, 1992; Mislevy et al., 1992). It offers three methods for variance estimation: (a) the jackknife replication methods (JK1, JK2, and JKn), (b) balanced repeated replication (BRR), and (c) the Taylor series approximation.

Comparing existing statistical tools for large-scale data, EdSurvey includes new functions and approaches for analyses and integrates with R packages to give additional tools to the user. The new functions include `cor.sdf` and `mixed.sdf`. The `cor.sdf` function estimates the weighted and unweighted Pearson, Spearman, polychoric, and polyserial correlation coefficients (see Bailey et al., 2018, for more details). The `mixed.sdf` function fits a mixed-effects model (also referred to as a multilevel or hierarchical linear model) to the data. For each mixed model level, weights (student and school) that are already available in large-scale assessment data can be applied. Although other mixed-effects model packages are available in R, EdSurvey is unique in that it implements methods for mixed models using weights at multiple levels; EdSurvey supports both linear and logit models. EdSurvey's implementation supports models with up to three levels (Bailey et al., 2020).

Because EdSurvey runs in the R environment, it inherits the ability to access the functionality of any R package in CRAN. For example, users can use the data frame generated from EdSurvey with the ggplot package, which is a plotting system that generates powerful data visualization displays with minimal code. Instructions and examples are in Liao et al. (2018).

28.2.1 Linking Error

As large-scale assessments have been moving to computer-based assessment, special variance arising from the mode change must be accounted for when comparing scores from computer-based assessments to scores from previous paper-based assessments. In NAEP assessments, the uncertainty from the mode change is calculated and referred to as a linking error. EdSurvey accounts for linking error between NAEP paper- and digitally-based assessments following the method by Mazzeo et al. (2021), in this volume. In addition, EdSurvey accounts for linking error caused by item parameter drift through time, as in PISA, but this is not discussed further in this chapter.

28.3 Empirical Examples Using Statistical Approaches

The EdSurvey package is designed as a one-stop shop for large scale assessments, with functions for downloading public use data, importing data, and manipulating and analyzing data. Recognizing that users come with varying levels of R skills and experience, the package provides flexibility when conducting survey analysis. It has built-in functions for manipulating the data; in addition, more experienced R users can manipulate their data using other R methods or supplementary packages. Each of these operations will be discussed in this section.

To facilitate the learning process, we recommend first becoming familiar with the EdSurvey commands before incorporating features from other packages. To find help and see examples for a specific function, the question mark function allows users to get help about a function. For example, at the R prompt, after installing and loading EdSurvey, a user can get help about the `downloadTIMSS` function by typing? `downloadTIMSS`.

In terms of data analysis procedures, we recommend the following steps in any analysis:

1. Install and load the package
2. Link to the data

3. Explore the data
4. Clean and manipulate the data
5. Run the analysis

These five steps are the focus of most of this chapter.

28.3.1 Install and Load EdSurvey

Users can install R from https://www.r-project.org/. In R, use the command `install.packages("EdSurvey")` to add EdSurvey in the packages that you can access while working in R. Then in every R session, the user must load the package by using the command `library(EdSurvey)`.

This chapter offers a brief demonstration of many EdSurvey functions. For more details on any function, a user can access extensive documentation, including examples, by entering a question mark and then the command name.

28.3.1.1 Obtain the Data

Data are available to the user either on a multimedia drive or for download from the Web. For downloadable data, EdSurvey provides a family of download functions for publicly available datasets. This type of function uses an Internet connection to download data from the host website. Users can specify the cycle(s) or assessment year(s) for the datasets that they wish to download. The following example shows how to download 2015 TIMSS datasets for all grades and jurisdictions, and the function connects to data files on timssandpirls.bc.edu.

```
> downloadTIMSS(years = 2015, root = "~/")
```

Sometimes, the host may update their repository addresses for certain datasets. EdSurvey will update the links accordingly. When data cannot be automatically downloaded but are available online due to a changed address or if the host data repository requires an account, the download functions or help files give instructions for users on how to download and extract the formatted data files.

Table 28.1 lists the download functions available in Version 2.7 of EdSurvey (the current version).

Other surveys are available only in a multimedia format; these surveys do not have a download function associated with them.

Table 28.1 EdSurvey download functions for publicly available datasets

Dataset	Built-in Functions
The Trends in International Mathematics and Science Study (TIMSS and TIMSS Advanced)	`downloadTIMSS, downloadTIMSSAdv`
Progress in International Reading Literacy Study (PIRLS and ePIRLS)	`downloadPIRLS, download_ePIRLS`
The Civic Education Study 1999 and the International Civic and Citizenship Study (CIVED and ICCS)	`downloadCivEDICCS`
The International Computer and Information Literacy Study (ICILS)	`downloadICILS`
The Programme for International Student Assessment (PISA)	`downloadPISA, downloadPISA_YAFS`
The Programme for the International Assessment of Adult Competencies (PIAAC)	`downloadPIAAC`
The Teaching and Learning International Survey (TALIS)	`downloadTALIS`
The Early Childhood Longitudinal Study, Kindergarten Class of 1998–99 and 2010–11 (ECLS-K 1998 and 2011)	`downloadECLS_K`
The Education Longitudinal Study of 2002 (ELS)	`downloadELS`
The High School Longitudinal Study of 2009 (HSLS)	`downloadHSLS`
The National Household Education Surveys Program (NHES)	`downloadNHES`
The School Survey on Crime and Safety (SSOCS)	`downloadSSOCS`

Note. Please carefully review the instructions when running the download function. Some data links have been updated. For instance, OECD datasets now require users to agree to a license before downloading

28.3.2 Link to the Data

With the data in hand, the user needs to open a connection to the data to use it. To avoid storing the data, it should be read into working memory; EdSurvey affords the option to selectively bring data into memory by reading rows and columns into memory on an as-needed basis. Some data files are exceptionally large and take up several gigabytes of space on a disk. It is unfathomable to load such data files in the memory to perform either data exploration or analysis. For example, in 2017, NAEP assessed more than 250,000 U.S. students at each of grades 4 and 8 for mathematics and reading. In each data file produced for these assessments, each student has more than 1500 variables. Analysts who load an entire grade and subject of this assessment into their computer will overwhelm their computer resources.

The EdSurvey `read` function bypasses this common constraint from large-scale data by using the LaF package (van der Laan, 2017) to quickly access gigantic data

Table 28.2 Read functions in EdSurvey

Dataset	Function
NAEP and NAEP High School Transcript Studies	readNAEP, readHSTS
TIMSS and TIMSS Advanced	readTIMSS, readTIMSSAdv
PIRLS and ePIRLS	readPIRLS, read_ePIRLS
CIVED 1999 and ICCS	readCivEDICCS
ICILS	readICILS
PISA	readPISA, readPISA_YAFS
PIAAC	readPIAAC
TALIS	readTALIS
ECLS-K: 1998 and 2011	readECLS_K1998, readECLS_K2011
ECLS-B	readECLS_B
ELS 2002	readELS
HSLS 2009	readHSLS
The Beginning Teacher Longitudinal Study (BTLS)	readBTLS
NHES	readNHES
SSOCS	readSSOCS

files. When running, the function creates an `edsurvey.data.frame` that stores only meta-level information about the specific data files processed. The meta-level information constitutes the location on the disk, the file format and layout, special values, and survey design attributes, such as plausible values, achievement levels, weights, the primary sampling unit, and stratum variables. EdSurvey uses these data for estimation and variance estimation in analytical functions.

For instance, to load the TIMSS 2015 Denmark data for the fourth graders and create an `edsurvey.data.frame`, select the pathway to the TIMSS 2015 data folder, and read in the Denmark ("dnk") grade 4 data:

```
> Dnk2015 <- readTIMSS("~/TIMSS/2015", countries = "dnk", gradeLvl = 4)
```

The EdSurvey `read` function opens a connection to the user-specified directory where the data file resides. The function creates an R control file to import data in ASCII, CSV, or SAV formats. Upon the first read-in, the EdSurvey package caches existing data being processed, which will speed up the read function in future sessions. For instance, the average first-time processing time for PISA data for all countries can be 10–15 min, but the next time this `read` function is called, the process will take only 5–10 s.

EdSurvey handles any necessary data merges automatically when reading. As an example, for international large-scale assessment data with both school- and student-level data available, school data will always be returned merged with student data, even if only school variables are requested. In some cases, teacher-level data are handled differently. For instance, in TIMSS, teacher data will be merged only if teacher variables are requested. Many students can be linked to many teachers, which varies widely between surveys and even countries.

Table 28.2 presents the family of EdSurvey `read` functions. Data process and merging details (and additional help) for each unique dataset are in the EdSurvey R help file, which can be traced using a question mark before the name of the function (e.g.,? `readNAEP`).

28.3.3 Data Exploration

To obtain survey design attributes and other general information regarding the data processed by the EdSurvey `read` function, one can print the `edsurvey.data.frame` object. We use the 2015 TIMSS grade 4 dataset for Denmark for illustration. All examples in this chapter will use this data file:

```
> Dnk2015 <- readTIMSS("~/TIMSS/2015", countries = "dnk", gradeLvl = 4)
> #show metadata about Grade 4 Denmark 2015 TIMSS survey
> Dnk2015
```

To pull more information about the survey design of the read-in data, use the commands in the following sections. Note that all data outputs shown in this chapter are only small segments of the overall output. The actual output could cover several pages.

28.3.3.1 Plausible Values

```
> showPlausibleValues(Dnk2015, verbose=TRUE)

There are 16 subject scale(s) or subscale(s) in this
 edsurvey.data.frame:
'mmat' subject scale or subscale with 5 plausible values (the default).
The plausible value variables are: 'asmmat01', 'asmmat02', 'asmmat03',
'asmmat04', and 'asmmat05'

'ssci' subject scale or subscale with 5 plausible values.
The plausible value variables are: 'asssci01', 'asssci02', 'asssci03',
'asssci04', and 'asssci05'
```

When running analyses with plausible values, one needs to supply only a short name for the set of plausible values, as found through this function, such as "`mmat`" and "`ssci`" in the TIMSS datasets. EdSurvey uses each plausible value in calculations and correctly accounts for the imputation variance from the plausible values following Mislevy et al. (1992).

28.3.3.2 Weights

Depending on the read-in data file, EdSurvey selects, by default, the most suitable weight for the analytical model specified and then extracts corresponding sampling attributes: replicate weights, strata, or primary sampling units (PSUs) information once the user chooses a variance estimation method (jackknife or Taylor series). A user can change the weight variable, and, accordingly, EdSurvey applies its associated survey attributes such as replicate weights. The weighting information can be obtained from the following:

```
> showWeights(Dnk2015, verbose=TRUE)

There are 4 full sample weights in this edsurvey.data.frame:
'totwgt' with 150 JK replicate weights (the default)
'tchwgt' with 150 JK replicate weights
'matwgt' with 150 JK replicate weights
'sciwgt' with 150 JK replicate weights
```

28.3.3.3 Other Survey Design Attributes

The functions `getStratumVar` and `getPSUVar` return, respectively, the stratum and PSU variables associated with the sample weight specified.

```
> getStratumVar(data = Dnk2015, weightVar = "totwgt")
[1] "jkzone"
> getPSUVar(data = Dnk2015, weightVar = "totwgt")
[1] "jkrep"
```

These functions are particularly useful for accessing the variables associated with weights in the Taylor series estimation.

28.3.3.4 Special Values

By default, EdSurvey shows the results from analyses after listwise deletion of cases that have special values. To see the levels that will be omitted by default, the user can print them with the following example code for the Dnk2015 data:

```
> getAttributes(Dnk2015, "omittedLevels")
```

Users wishing to see output with these values included can set `defaultConditions = FALSE` in data extraction or analysis.

EdSurvey offers a variety of ways to explore the data and variables. Some basic functions that work on a `data.frame`, such as `dim`, `nrow`, and `ncol`, also work on an `edsurvey.data.frame` for dimension checking.

A full codebook can be accessed using `showCodebook`, which reveals variable names, variable labels, and value labels of a survey dataset. This function pairs well with the `View` function to more easily explore the dataset:

```
> View(showCodebook(Dnk2015))
```

Because large-scale data can have several thousand variables, EdSurvey provides powerful searching functions to meet users' needs, including `searchSDF`, `levelsSDF`, and `summary2`. The `searchSDF` function returns variable names and labels meeting criteria specified by the user. For example, a search for variables with a name or label that contains "computer":

```
> searchSDF(string = "computer", data = Dnk2015)
 variableName Labels
1 asbg05a GEN\crHOME POSSESS\crCOMPUTER TABLET OWN
2 asbg05b GEN\crHOME POSSESS\crCOMPUTER TABLET SHARED
3 asbg10a GEN\crUSE COMPUTER TABLET FOR HOMEWORK\crHOME
```

By default, EdSurvey searches all data files read-in, including student, school, and teacher files, and allows users to search multiple strings simultaneously using the vertical pipe (|) "or" operator.

```
> searchSDF(string="computer|internet|phone", data=Dnk2015)
 variableName Labels
1 m051134 MONTHS PETER PAID LESS FOR PHONE (1)
2 asbg05a GEN\crHOME POSSESS\crCOMPUTER TABLET OWN
3 asbg05b GEN\crHOME POSSESS\crCOMPUTER TABLET SHAREDShow
```

Users can narrow down the search by specifying the data source (e.g., the student file or the school file) in the argument `fileFormat` or define a more precise search and, therefore, eliminate the variables that do not apply to a specific analysis.

Lastly, setting the `levels` argument to TRUE unfolds the levels and value labels of each variable. An alternative is to use `levelsSDF` with the variable name provided. For this example, we omit the output, describing it in a comment instead.

```
> #Search the keyword in the student file only
> searchSDF(string = "computer", data = Dnk2015, fileFormat = "student")
> #Provide multiple keywords
> searchSDF(string = c("computer","homework","school"), data =
Dnk2015, levels = TRUE)
> #Display levels of the variable "asbg10b"
```

```
> levelsSDF(varnames = "asbg10b", data = Dnk2015)
Levels for Variable 'asbg10b' (Lowest level first):
 1. EVERY DAY OR ALMOST EVERY DAY (n=840)
 2. ONCE OR TWICE A WEEK (n=1368)
 3. ONCE OR TWICE A MONTH (n=671)
 4. NEVER OR ALMOST NEVER (n=603)
 9. OMITTED OR INVALID* (n=122)
NOTE: * indicates an omitted level.
```

Once an analyst finds the variables of interest, the `summary2` function provides descriptive statistics of a given variable, such as response frequencies and response percentages. The `summary2` function can produce results with or without weighting. It calculates the weighted statistics by default, with standard errors accounted for in the survey design. If the specified variable is a set of plausible values, `summary2` accounts for both plausible values when weighting and pooling.

```
> summary2(Dnk2015, "mmat")
Estimates are weighted using weight variable 'totwgt'
 Variable N Weighted N Min. 1st Qu. Median Mean
1 mmat 3710 58472.65 264.8542 489.5665 542.1979 538.6533
 3rd Qu. Max. SD NA's Zero-weights
1 590.8396 793.3109 75.16361 0 0
```

To unweight the results, specify weightVar = NULL:

```
> summary2(Dnk2015, "asbg10b", weightVar = NULL)
Estimates are not weighted.
 asbg10b N Percent
1 (Missing) 106 2.857143
2 EVERY DAY OR ALMOST EVERY DAY 840 22.641509
3 ONCE OR TWICE A WEEK 1368 36.873315
4 ONCE OR TWICE A MONTH 671 18.086253
5 NEVER OR ALMOST NEVER 603 16.253369
6 OMITTED OR INVALID 122 3.288410
```

28.3.4 Manipulate the Data

As mentioned earlier, the user has the following options: Prepare and analyze data within the EdSurvey package functions, use `getData` to extract a `data.frame` to clean and edit with any R tool, and then use `rebindAttributes` to use EdSurvey functions to analyze the data.

28.3.4.1 Preparing Data in EdSurvey

EdSurvey offers three functions for data manipulation: `subset`, `recode.sdf`, and `rename.sdf`. The subset function selects only the rows that meet a specified condition. The following example subsets the data to male students only:

```
> Dnk2015A <- subset(Dnk2015, itsex %in% "MALE")
> summary2(Dnk2015A, "itsex")
```

Estimates are weighted using weight variable 'totwgt'

```
 itsex N Weighted N Weighted Percent Weighted Percent SE
1 MALE 1877 29616.63 100 0
```

The `recode.sdf` function collapses variable levels or changes value labels:

```
>#Collapses variable levels
>Dnk2015B1<- recode.sdf(Dnk2015, recode = list(asbg10b=list(from=c
("EVERY DAY OR ALMOST EVERY DAY",
+ "ONCE OR TWICE A WEEK"),
+ to=c("Frequently")),
+ asbg10b=list(from=c("ONCE OR TWICE A MONTH",
+ "NEVER OR ALMOST NEVER"),
+ to=c("Rarely"))))

>summary2(Dnk2015B1, "asbg10b")
Estimates are weighted using weight variable 'totwgt'
 asbg10b N Weighted N Weighted Percent Weighted Percent SE
1 (Missing) 106 1642.648 2.809259 0.3473685
2 OMITTED OR INVALID 122 1997.412 3.415976 0.3532134
3 Frequently 2208 35048.662 59.940269 2.0226713
4 Rarely 1274 19783.925 33.834496 1.9919606

> #Change value labels
> Dnk2015B <- recode.sdf(Dnk2015, recode =
+ list(itsex = list( from = c("MALE"), to = c("BOY")),
+ itsex = list( from = c("FEMALE"), to = c("GIRL"))))

> summary2(Dnk2015B, "itsex")
Estimates are weighted using weight variable 'totwgt'
 itsex N Weighted N Weighted Percent Weighted Percent SE
1 BOY 1877 29616.63 50.6504 0.7932687
2 GIRL 1833 28856.02 49.3496 0.7932687
Users can use the rename.sdf function to change a variable name:
```

```
> Dnk2015C <- rename.sdf(Dnk2015, c("itsex", "mmat"), c("gender",
+ "math_overall"))

> summary2(Dnk2015C, "gender")
Estimates are weighted using weight variable 'totwgt'
 gender N Weighted N Weighted Percent Weighted Percent SE
1 FEMALE 1833 28856.02 49.3496 0.7932687
2 MALE 1877 29616.63 50.6504 0.7932687

> summary2(Dnk2015C, "math_overall")
Estimates are weighted using weight variable 'totwgt'
 Variable N Weighted N Min. 1st Qu. Median Mean
1 math_overall 3710 58472.65 264.8542 489.5665 542.1979 538.6533
 3rd Qu. Max. SD NA's Zero-weights
1 590.8396 793.3109 75.16361 0 0
```

28.3.4.2 Extract a Data Frame

The `getData` function takes an `edsurvey.data.frame` and returns a `data.frame` with the chosen variables only. The `getData` function is also a data manipulation function for extracting a data frame with the variables of interest. This function gives users flexibility for efficiently managing computer resources to perform extensive variable editing. It accommodate those who have preferred methods of data manipulation using base R functions or specific R packages. We recommend importing only the necessary variables to optimize performance.

Second, design attributes needed for EdSurvey analytical functions may not be retained. Users can attach the attributes necessary to further use EdSurvey functions when calling `getData` by setting the `addAttributes` argument to TRUE. In addition, some data manipulation may incidentally remove the attributes from the data frame. When this happens, the user can again append the attributes to the data frame with the `rebindAttributes` function.

For example, one can use the base R `ifelse` function to conditionally recode the variable `asbm01e` (the student's response to "I like mathematics" on a scale from "Disagree a lot" to "Agree a lot") before running a regression:

```
> gddat2 <- getData(data = Dnk2015, varnames = c("itsex", "asbm01e",
"totwgt", "mmat"), addAttributes = FALSE, dropOmittedLevels = TRUE)
# recode "AGREE A LOT" and "AGREE A LITTLE"
> gddat2$likemath <- ifelse(gddat2$asbm01e %in% c("AGREE A LOT",
+ "AGREE A LITTLE"), "Agree", "Disagree")
> gddat3 <- rebindAttributes(gddat2, Dnk2015)
&gt; lm_c &lt;- lm.sdf(formula = mmat ~ itsex + likemath, data = gddat3)
> summary(lm_c)
```

```
Formula: mmat ~ itsex + likemath

Weight variable: 'totwgt'
Variance method: jackknife
JK replicates: 150
Plausible values: 5
jrrIMax: 1
full data n: 5897
n used: 3548

Coefficients:
 coef se t dof Pr(>|t|)
(Intercept) 541.7336 3.1568 171.6063 78.118 < 2.2e-16 ***
itsexMALE 5.9832 2.9356 2.0381 100.791 0.04415 *
likemathDisagree -32.3797 4.2874 -7.5522 127.490 7.241e-12 ***
---
Signif. codes: 0 '***' 0.001 '**' 0.01 '*' 0.05 '.' 0.1 ' ' 1

Multiple R-squared: 0.027
```

In the previous example, the `dropOmittedLevels` argument in `getData` was set as TRUE. In doing so, all cases with special values (in this example, "OMITTED OR INVALID" and missing) were listwise deleted. Changing that argument to FALSE will include all cases. For additional details on the features of the `getData` function and data manipulation, please refer to Lee, Bailey, and Emad (2018a) and Bailey and Zhang(2018b).

After cleaning and preparing the desired data, we are ready to move to the EdSurvey analytical functions. Next we highlight a few essential analytical functions: tables, linear and regression analyses, and gap analysis.

28.3.5 Run the Analysis in EdSurvey

When the data are ready, either as an `edsurvey.data.frame` or a `light.edsurvey.data.frame`, all the EdSurvey analytical functions can be used. In the following examples, we will employ the `edsurvey.data.frame` because it uses computer memory more efficiently.

28.3.5.1 Summary Tables

To create a table in EdSurvey, one should use `edsurveyTable`. This function returns a summary table with the frequency and percentage of each category, as well as the mean value of the outcome variable by predictor categorical variable(s).

The following example generates a table of mathematics mean scores ("mmat") by students' sex ("itsex") and computer at home ("asbg05a").

```
&gt; edsurveyTable(formula = mmat ~ itsex + asbg05a, data = Dnk2015,
+ varMethod = "jackknife", omittedLevels = TRUE)
Formula: mmat ~ itsex + asbg05a

Plausible values: 5
jrrIMax: 1
Weight variable: 'totwgt'
Variance method: jackknife
JK replicates: 150
full data n: 5897
n used: 3607

Summary Table:
 itsex asbg05a N WTD_N PCT SE(PCT) MEAN SE(MEAN)
 FEMALE YES 1514 23902.169 84.57526 0.9887521 536.0474 3.162143
 FEMALE NO 281 4359.251 15.42474 0.9887521 538.0483 6.307016
 MALE YES 1578 24947.719 87.14129 0.8254618 543.3249 2.795532
 MALE NO 234 3681.325 12.85871 0.8254618 538.4015 6.436956
```

The left side of the formula is usually a continuous variable or a subject scale (represented by a set of plausible values in NAEP and international large scale assessments), and the right side of the formula has categorical variables to tabulate. When the left side of the formula is omitted and, `returnMeans` is TRUE, the default subject scale is used, which can be revealed through the `showPlausibleValues` function. The order of the right side variables affects the output.

By default, the function returns estimates and standard errors with the survey design taken into account, following the methods described in Bailey and Cohen (2019). After running the analysis, the data are automatically removed from memory. This data removal occurs for all EdSurvey analytical functions.

In the previous example, each estimate was weighted with "totwgt" (the default, so it does not appear in the call) and pooled from the five plausible values composed of the mathematics scale, mmat. Variance is calculated using the jackknife method based on the TIMSS sampling design. When specifying the jackknife method in the command, EdSurvey automatically looks at the sampling design attributes associated with the data and calculates the variance using the proper approach. For example, EdSurvey uses the JK2 method (Johnson & Rust, 1992) for TIMSS and NAEP data analysis and the BRR method (Wolters, 2007) for OECD data such as PISA. By default, EdSurvey estimates variances based on the first plausible value of the set. Users can change the setting, for example, specifying `jrrIMax = Inf` to use all the plausible values for variance calculation. Specifying `varMethod = "Taylor"` leads EdSurvey to employ the Taylor series method. The Taylor series method is proper for NAEP data and some longitudinal datasets, but users should use it with caution for international assessment data because most international assessments, including TIMSS and PISA, were not designed to be analyzed with this approach. Special values (i.e., missing, "OMITTED OR INVALID" in this example) were listwise deleted by default in computation. Setting `dropOmittedLevels = FALSE` keeps these values.

28.3.5.2 Linear Regression

The `lm.sdf` function fits a linear model that properly accounts for a complex survey sampling design and plausible values. It allows the jackknife, BRR, or Taylor series method for variance estimation. Similar to the `edsurveyTable` command, the model is specified following the R `lm` convention through formula, such as `response ~ terms`, where response (dependent variables) is a continuous variable or a scale that consists of a set of plausible values. If the left side is left blank, the default subject scale will be used, which can be revealed through the `showPlausibleValues` function. The terms (independent variables) can be continuous or categorical variables, including plausible values. The following example predicts students' mathematics achievement scores ("mmat") with students' sex ("itsex"), computer at home ("asbg05a"), and the use of a computer for homework at home ("asbg10a").

```
&gt; lm1 &lt;- lm.sdf(mmat ~ itsex + asbg05a + asbg10a, Dnk2015,
+ varMethod = "jackknife")
> summary(lm1)
Formula: mmat ~ itsex + asbg05a + asbg10a

Weight variable: 'totwgt'
Variance method: jackknife
JK replicates: 150
Plausible values: 5
jrrIMax: 1
full data n: 5897
n used: 3519

Coefficients:
 coef se t dof
(Intercept) 519.72742 4.20182 123.69099 79.356
itsexMALE 7.82075 2.86993 2.72507 81.606
asbg05aNO -0.80213 4.17364 -0.19219 116.806
asbg10aONCE OR TWICE A WEEK 23.60695 3.60470 6.54893 133.237
asbg10aONCE OR TWICE A MONTH 36.49484 4.86415 7.50283 40.747
```

```
asbg10aNEVER OR ALMOST NEVER 15.12409 4.20319 3.59824 107.067
 Pr(>|t|)
(Intercept) < 2.2e-16 ***
itsexMALE 0.0078639 **
asbg05aNO 0.8479279
asbg10aONCE OR TWICE A WEEK 1.154e-09 ***
asbg10aONCE OR TWICE A MONTH 3.370e-09 ***
asbg10aNEVER OR ALMOST NEVER 0.0004867 ***
---
Signif. codes: 0 '***' 0.001 '**' 0.01 '*' 0.05 '.' 0.1 ' ' 1

Multiple R-squared: 0.0328
```

Following Bailey and Cohen (2019), the standard error of the coefficient was estimated with the jackknife replication method, but users can switch it to Taylor series when appropriate. The estimates were pooled from the five plausible values that consist of the "mmat" scale. Cases with special values were listwise deleted from the analysis. As with the `lm` package, dummy codes for categorical predictors do not need to be generated. By default, EdSurvey drops level 1 from the discrete predictor and treats it as the reference group. The reference level can be changed through the argument relevels. For example, in the previous model, the reference group is females. To change it to males, use the following:

```
&gt; lm2 &lt;- lm.sdf(mmat ~ itsex + asbg05a + asbg10a, Dnk2015,
+ relevels=list(itsex="MALE"))
> summary(lm2)
Formula: mmat ~ itsex + asbg05a + asbg10a

Weight variable: 'totwgt'
Variance method: jackknife
JK replicates: 150
Plausible values: 5
jrrIMax: 1
full data n: 5897
n used: 3519

Coefficients:
 coef se t dof
(Intercept) 527.54818 3.93642 134.01711 147.330
itsexFEMALE -7.82075 2.86993 -2.72507 81.606
asbg05aNO -0.80213 4.17364 -0.19219 116.806
asbg10aONCE OR TWICE A WEEK 23.60695 3.60470 6.54893 133.237
asbg10aONCE OR TWICE A MONTH 36.49484 4.86415 7.50283 40.747
asbg10aNEVER OR ALMOST NEVER 15.12409 4.20319 3.59824 107.067
```

```
Pr(>|t|)
(Intercept) < 2.2e-16 ***
itsexFEMALE 0.0078639 **
asbg05aNO 0.8479279
asbg10aONCE OR TWICE A WEEK 1.154e-09 ***
asbg10aONCE OR TWICE A MONTH 3.370e-09 ***
asbg10aNEVER OR ALMOST NEVER 0.0004867 ***
---
Signif. codes: 0 '***' 0.001 '**' 0.01 '*' 0.05 '.' 0.1 ' ' 1

Multiple R-squared: 0.0328
```

To return the standardized regression coefficient, add `src = TRUE` into the summary call to the regression model object, adding the `stdCoef` and `stdSE` columns to the output with standardized regression coefficients and their standard error:

```
> summary(lm2, src = TRUE)
Formula: mmat ~ itsex + asbg05a + asbg10a

Weight variable: 'totwgt'
Variance method: jackknife
JK replicates: 150
Plausible values: 5
jrrIMax: 1
full data n: 5897
n used: 3519

Coefficients:
 coef se t
(Intercept) 527.5481777 3.9364241 134.01711
itsexFEMALE -7.8207533 2.8699323 -2.72507
asbg05aNO -0.8021281 4.1736415 -0.19219
asbg10aONCE OR TWICE A WEEK 23.6069543 3.6047022 6.54893
asbg10aONCE OR TWICE A MONTH 36.4948379 4.8641453 7.50283
asbg10aNEVER OR ALMOST NEVER 15.1240936 4.2031895 3.59824
 dof Pr(>|t|) stdCoef stdSE
(Intercept) 147.330 0.0000e+00 NA NA
itsexFEMALE 81.606 7.8639e-03 -0.0524 0.01922 *
asbg05aNO 116.806 8.4793e-01 -0.0037 0.01942 *
asbg10aONCE OR TWICE A WEEK 133.237 1.1537e-09 0.1446 0.02208 *
asbg10aONCE OR TWICE A MONTH 40.747 3.3698e-09 0.1888 0.02517 *
asbg10aNEVER OR ALMOST NEVER 107.067 4.8669e-04 0.0820 0.02279 *
```

```
---
Signif. codes: 0 '***' 0.001 '**' 0.01 '*' 0.05 '.' 0.1 ' ' 1

Multiple R-squared: 0.0328
```

28.3.5.3 Binomial Regression

Binomial regression implements an estimator that correctly handles logical, left-side variables. It allows dichotomized variables, including plausible values, to be the response outcomes following Bailey and Cohen (2019). EdSurvey offers three binomial regression functions, namely `glm.sdf`, `logit.sdf`, and `probit.sdf`, for complex sampling design data with plausible values. `Glm.sdf` is an umbrella function that currently fits logit and probit models. Alternatively, users can choose `logit.sdf` or `probit.sdf` functions for binomial outcomes. This chapter provides examples for logit models, but most components also apply to the probit model. For more information about the probit models, please refer to the R help file.

```
&gt; logit1 &lt;- logit.sdf(I(mmat &gt;= 600) ~ itsex + asbg10a,
data=Dnk2015,
+ relevels=list(itsex="MALE"))
> summary(logit1)
Formula: I(mmat &gt;= 600) ~ itsex + asbg10a
Family: binomial (logit)

jrrIMax: 1
Weight variable: 'totwgt'
Variance method: jackknife
JK replicates: 150
full data n: 5897
n used: 3531

Coefficients:
 coef se t dof
(Intercept) -1.670253 0.187448 -8.910506 98.312
itsexFEMALE -0.281336 0.097308 -2.891205 146.165
asbg10aONCE OR TWICE A WEEK 0.649672 0.170091 3.819560 100.516
asbg10aONCE OR TWICE A MONTH 0.935835 0.185296 5.050497 117.436
asbg10aNEVER OR ALMOST NEVER 0.524762 0.224399 2.338525 127.232
 Pr(>|t|)
(Intercept) 2.730e-14 ***
itsexFEMALE 0.0044237 **
asbg10aONCE OR TWICE A WEEK 0.0002316 ***
```

```
asbg10aONCE OR TWICE A MONTH 1.632e-06 ***
asbg10aNEVER OR ALMOST NEVER 0.0209192 *
---
Signif. codes: 0 '***' 0.001 '**' 0.01 '*' 0.05 '.' 0.1 ' ' 1
```

The previous example models students whose mathematical scores are equal to or greater than 600 (about the 75th percentile), conditioning on their sex and use of a computer for homework at home. It implements the jackknife replication method by default, but users can choose the Taylor series method when appropriate. The dichotomization of the left-side variables is handled with the `I(·)` function, which makes the outcome variable a 1 when the condition is TRUE and a 0 when the condition is FALSE. The binomial functions contain similar features as in `lm.sdf`. For example, users can change the reference group of the categorical dependent variables using the relevels argument. A similar model can be run using `glm.sdf`.

```
&gt; logit2 &lt;- glm.sdf(I(mmat &gt;= 600) ~ itsex + asbg10a,
data=Dnk2015,
+ family=binomial(link="logit"), relevels=list(itsex="MALE"))
> summary(logit2)
Formula: I(mmat &gt;= 600) ~ itsex + asbg10a
Family: binomial (logit)

jrrIMax: 1
Weight variable: 'totwgt'
Variance method: jackknife
JK replicates: 150
full data n: 5897
n used: 3531

Coefficients:
 coef se t dof
(Intercept) -1.670253 0.187448 -8.910506 98.312
itsexFEMALE -0.281336 0.097308 -2.891205 146.165
asbg10aONCE OR TWICE A WEEK 0.649672 0.170091 3.819560 100.516
asbg10aONCE OR TWICE A MONTH 0.935835 0.185296 5.050497 117.436
asbg10aNEVER OR ALMOST NEVER 0.524762 0.224399 2.338525 127.232
 Pr(>|t|)
(Intercept) 2.730e-14 ***
itsexFEMALE 0.0044237 **
asbg10aONCE OR TWICE A WEEK 0.0002316 ***
asbg10aONCE OR TWICE A MONTH 1.632e-06 ***
asbg10aNEVER OR ALMOST NEVER 0.0209192 *
---
```

```
Signif. codes: 0 '***' 0.001 '**' 0.01 '*' 0.05 '.' 0.1 '' 1
```

To request the odds ratio, use the oddsRatio function, as shown in the following example:

```
> oddsRatio(logit1)
 OR 2.5% 97.5%
(Intercept) 0.1881995 0.1190555 0.2573435
itsexFEMALE 0.7547745 0.6108217 0.8987273
asbg10aONCE OR TWICE A WEEK 1.9149124 1.2765229 2.5533019
asbg10aONCE OR TWICE A MONTH 2.5493419 1.6234732 3.4752106
asbg10aNEVER OR ALMOST NEVER 1.6900566 0.9467335 2.4333796
```

For model fit statistics, EdSurvey provides the Wald test (Lishinski, 2018) to test estimated parameters in a model for data with plausible values. Because no well-established likelihood test exists for plausible values or residuals, the Wald test fills the role of the likelihood ratio test, the analysis of variance, and the *F*-test. The Wald test will test a single hypothesis or multiple hypotheses about multiple parameters simultaneously. In the following two examples, the first one tests whether the coefficient for level of "Once or twice a week" of asbg10a is equal to zero, whereas the second jointly tests all the coefficients corresponding to the levels of variable asbg10a.

```
> waldTest(model = logit1, coefficients = "asbg10aONCE OR TWICE A WEEK")
Wald test:
----------
H0:
asbg10aONCE OR TWICE A WEEK = 0

Chi-square test:
X2 = 14.6, df = 1, P(> X2) = 0.00013

F test:
W = 14.6, df1 = 1, df2 = 75, P(> W) = 0.00027
> # test all asbg10a coefficients simultaneously
> waldTest(model = logit1, coefficients = "asbg10a")
Wald test:
----------
H0:
asbg10aONCE OR TWICE A WEEK = 0
asbg10aONCE OR TWICE A MONTH = 0
asbg10aNEVER OR ALMOST NEVER = 0

Chi-square test:
X2 = 27.8, df = 3, P(> X2) = 4e-06
```

```
F test:
W = 9.0, df1 = 3, df2 = 73, P(> W) = 3.7e-05
```

28.3.5.4 Group Comparisons

The gap function in EdSurvey calculates the difference between two statistics for two groups in a population using the jackknife replication method. The function can perform the following comparisons with statistics as mean scores, achievement level percentages, percentiles, or student group percentages:

- Between groups by time point or jurisdiction
- Of the same group across time points or jurisdictions
- Between groups across time points or jurisdictions

A simple EdSurvey gap analysis is the comparison of scores between two groups within the same jurisdiction from one survey. In the following example, we compare the gender gap of Denmark fourth-grade students in 2015 TIMSS mathematics.

```
> gap(variable = 'mmat', data = Dnk2015, groupA=itsex %in% "FEMALE",
+ groupB = itsex %in% "MALE")

Call: gap(variable = "mmat", data = Dnk2015, groupA = itsex %in%
 "FEMALE", groupB = itsex %in% "MALE")

Labels:
 group definition nFullData nUsed
 A itsex %in% "FEMALE" 5897 1833
 B itsex %in% "MALE" 5897 1877

Percentage:
 pctA pctAse pctB pctBse diffAB covAB diffABse
 49.3496 0.7932687 50.6504 0.7932687 -1.300792 -0.6292753 1.586537
 diffABpValue dofAB
 0.414185 102.1871

Results:
 estimateA estimateAse estimateB estimateBse diffAB covAB
 535.5594 3.155563 541.6677 2.789448 -6.108257 4.938574
 diffABse diffABpValue dofAB
 2.803828 0.03201268 88.86473
```

The gap output contains three blocks: labels, percentage, and results. The first block, labels, shows the definition of groups A and B, along with a reminder of the full data *n* count (nFullData) and the *n* count of the number of individuals who are in the two subgroups with valid scores (nUsed). The second block, percentage, shows the percentage of individuals who fall into each category, with omitted levels removed. The third block, results, shows the estimated outcomes from the two groups listed in the columns estimateA and estimateB. The diffAB column shows the estimated difference between the two groups, and the diffABse column shows the standard error of the difference. The *t*-test significance results show in difABpValue with the degrees of freedom following. When sampling survey respondents through cluster sampling designs (e.g., a design involving sampling students from the same classrooms or schools), these respondents have more in common than randomly selected individuals. Therefore, EdSurvey calculates a covariance between groups from the same assessment sample (same assessment and year), which appears in the covAB column. Even when selecting respondents through simple random sampling, little harm occurs in estimating the covariance because it will be close to zero.

To support the comparisons across time points and jurisdictions, EdSurvey offers the `edsurvey.data.frame.list` function, which creates an `edsurvey.data.frame.list` object that combines a series of `edsurvey.data.frame` objects to reduce repetition in function calls. It is worth pointing out that the `edsurvey.data.frame.list` supports not only the gap function but also all EdSurvey data exploration (e.g., `searchSDF` and `summary2`) and analytical functions except for the mixed models. It can be particularly useful in analyzing large-scale international data collected continuously from numerous countries.

When performing gap analysis across time, it is necessary to check if variable information, such as variable name, value levels, or value labels changed across time. The analyst will have to recode them for consistency.

For example, in NAEP assessments, student race/ethnicity data from school records prior to 2011 have six mutually exclusive categories: White, Black, Hispanic, Asian/Pacific Islander, American Indian/Alaska Native, and Other or unclassified. Starting in 2011, the race/ethnicity variable expanded to seven categories with Asian students and Native Hawaiian/Other Pacific Islander becoming two separate categories. When comparing the results for racial/ethnic groups from or after 2011 to earlier assessment years, the data for Asian and Native Hawaiian/Other Pacific Islander students must be combined into a single Asian/Pacific Islander category, and their variable names must be harmonized.

When preparing an `edsurvey.data.frame.list`, it is necessary to ensure that variable information is consistent across datasets. We recommend using the functions `searchSDF` and `levelsSDF()`, which return variable names, value labels, and levels based on a character string.

In the following example, we show how to conduct analyses of gender gap in mathematics scores between 2007 and 2015, with the gender variable "itsex" in 2007 having different values label from the 2015 one. Before the gap analyses, we need to recode the data and then create the `edsurvey.data.frame.list` object.

```
#Read in 2007 Denmark data
> Dnk2007 <- readTIMSS("~/TIMSS/2007", countries="dnk", gradeLvl = 4)
> #Recode itsex
> Dnk2007 <- recode.sdf(Dnk2007, recode = list( itsex =
+ list(from = c("BOY"), to=c("MALE")), itsex=list(from=c("GIRL"),
+ to=c("FEMALE"))))
#Create an edsurvey.data.frame.list
#combining data from two years
TrendDnk <- edsurvey.data.frame.list(list(Dnk2007, Dnk2015))
#Gap analysis
> gap(variable = 'mmat', data = TrendDnk,
+ groupA=itsex %in% "FEMALE", groupB = itsex %in% "MALE")

gapList
Call: gap(variable = "mmat", data = TrendDnk, groupA = itsex %in%
 "FEMALE", groupB = itsex %in% "MALE")

Labels:
 group definition
 A itsex %in% "FEMALE"
 B itsex %in% "MALE"

Percentage:
 year pctA pctAse pctB pctBse diffAB covAB
 2007 51.11624 1.1761236 48.88376 1.1761236 2.232480 -1.3832667
 2015 49.34960 0.7932687 50.65040 0.7932687 -1.300792 -0.6292753
 diffABse diffABpValue dofAB diffAA covAA diffAAse
 2.352247 0.3453486 82.54044 NA NA NA
 1.586537 0.4141850 102.18706 1.766636 0 1.418641
 diffAApValue dofAA diffBB covBB diffBBse diffBBpValue
 NA NA NA NA NA NA
 0.2149666 149.6971 -1.766636 0 1.418641 0.2149666
 dofBB diffABAB covABAB diffABABse diffABABpValue dofABAB
 NA NA NA NA NA NA
 149.6971 3.533272 0 2.837282 0.2149666 149.6971

Results:
 year estimateA estimateAse estimateB estimateBse diffAB covAB
 2007 519.9165 2.860529 526.4408 3.186706 -6.524281 2.320667
 2015 535.5594 3.155563 541.6677 2.789448 -6.108257 4.938574
 diffABse diffABpValue dofAB diffAA covAA diffAAse
 3.700863 0.08304620 59.55705 NA NA NA
 2.803828 0.03201268 88.86473 -15.64292 0 4.259132
 diffAApValue dofAA diffBB covBB diffBBse diffBBpValue dofBB
 NA NA NA NA NA NA NA
```

```
0.0003354516 146.7124 -15.2269 0 4.235105 0.0004230123 171.8141
diffABAB covABAB diffABABse diffABABpValue dofABAB sameSurvey
NA NA NA NA NA NA
-0.4160244 0 4.643042 0.9287521 120.8613 FALSE
```

Similarly, a gap analysis across jurisdictions can use an `edsurvey.data.frame.list`. In the following example, we compare the gender gap in 2015 mathematics scores between Denmark and the United States.

```
#Read in the US 2015 data
> Usa2015 <- readTIMSS("~/TIMSS/2015", countries="usa", gradeLvl = 4)
> #Create an edsurvey.data.frame.list combining
> #data from two countries
> DnkUsa <- edsurvey.data.frame.list(list(Dnk2015, Usa2015))
> gap(variable = 'mmat', data = DnkUsa, groupA=itsex %in% "FEMALE",
+ groupB = itsex %in% "MALE")
gapList
Call: gap(variable = "mmat", data = DnkUsa, groupA = itsex %in%
 "FEMALE", groupB = itsex %in% "MALE")

Labels:
 group definition
 A itsex %in% "FEMALE"
 B itsex %in% "MALE"

Percentage:
 country pctA pctAse pctB pctBse diffAB
 Denmark 49.34960 0.7932687 50.65040 0.7932687 -1.300792
 United States 50.71408 0.5701627 49.28592 0.5701627 1.428164
 covAB diffABse diffABpValue dofAB diffAA covAA diffAAse
 -0.6292753 1.586537 0.4141850 102.1871 NA NA NA
 -0.3250855 1.140325 0.2127084 127.5956 -1.364478 0 0.9769139
 diffAApValue dofAA diffBB covBB diffBBse diffBBpValue
 NA NA NA NA NA NA
 0.1640956 193.6494 1.364478 0 0.9769139 0.1640956
 dofBB diffABAB covABAB diffABABse diffABABpValue dofABAB
 NA NA NA NA NA NA
 193.6494 -2.728956 0 1.953828 0.1640956 193.6494

Results:
 country estimateA estimateAse estimateB estimateBse diffAB
 Denmark 535.5594 3.155563 541.6677 2.789448 -6.108257
 United States 535.7482 2.312608 542.6617 2.501978 -6.913450
 covAB diffABse diffABpValue dofAB diffAA covAA diffAAse
```

```
4.938574 2.803828 0.032012676 88.86473 NA NA NA
4.068738 1.862948 0.000319999 114.68273 -0.1887894 0 3.912255
diffAApValue dofAA diffBB covBB diffBBse diffBBpValue
NA NA NA NA NA NA
0.9615723 160.5123 -0.9939828 0 3.747121 0.7910629
dofBB diffABAB covABAB diffABABse diffABABpValue dofABAB
NA NA NA NA NA NA
211.6759 0.8051934 0 3.366308 0.8112621 160.4194
sameSurvey
NA
FALSE
```

For more information about the full capabilities of the EdSurvey gap functions, refer to the help file or see Bailey and Zhang (2018b), Zhang and Bailey (2018), and Zhang et al. (2018).

28.4 Discussion and Outlook

EdSurvey was conceived to analyze all the data produced by the NCES, the statistical arm of the Institute of Education Sciences of the U.S. Department of Education. In this chapter, we focused solely on the functions for working with large-scale assessment data. We offered a comprehensive overview of the functions to download, read, manipulate, and analyze such data efficiently.

For users who are interested in advanced modeling, EdSurvey also provides multivariate regression, quantile regression, and multilevel mixed modeling (Bailey et al., 2018). Interested readers can find instructions about how to use these features, along with their methodology documents, on the EdSurvey website hosted by the American Institutes for Research. The scope of this chapter did not allow us to cover every EdSurvey function. Guidelines for other useful functions, including achievement level analysis, percentiles, and correlations, can be found through their R help files or related vignettes (e.g., Bailey and Zhang 2018b).

One distinctive feature of the EdSurvey R package is its extensibility to meet users' needs. The EdSurvey team will be releasing the direct estimation function (Cohen & Jiang, 1999) in Version 2.5 and multiple imputation and item response theory functions in the future. We are expanding EdSurvey's capacity to handle more large-scale datasets. In 2020, we plan to support most of the NCES longitudinal data as well as cross-sectional data, including, to name a few, the Beginning Teacher Longitudinal Study, Baccalaureate and Beyond, the Beginning Postsecondary Students Longitudinal Study, the School Survey on Crime and Safety, the National Household Education Surveys Program, and the High School Transcript Study.

References

Bailey, P., & Cohen, M. (2018). *Methods used for estimating percentiles in EdSurvey*. American Institutes for Research. https://www.air.org/sites/default/files/EdSurvey-Percentiles.pdf

Bailey, P., & Cohen, M. (2019). *Statistical methods used in EdSurvey*. American Institutes for Research. https://www.air.org/sites/default/files/EdSurvey-Statistics.pdf

Bailey, P., Emad, A., Zhang, T., Xie, Q., & Sikali, E. (2018). *Weighted and unweighted correlation methods for large-scale educational assessment: wCorr formulas* [AIR – NAEP Working Paper #2017–02, NCES Data R Project Series #02]. American Institutes for Research. https://www.air.org/sites/default/files/downloads/report/Weighted-and-Unweighted-Correlation-Methods-Large-Scale-Educational-Assessment-April-2018.pdf

Bailey, P., Kelley, C., Nguyen, T., & Huo, H. (2020). *WeMix: Weighted mixed-effects models using multilevel pseudo maximum likelihood estimation* [R package version 3.1.1]. https://CRAN.R-project.org/package=WeMix

BIFIE, Robitzsch, A., & Oberwimmer, K. (2019). *BIFIEsurvey: Tools for survey statistics in educational assessment* [R package version 3.3–12]. Federal Institute for Educational Research, Innovation and Development of the Austrian School System. https://CRAN.R-project.org/package=BIFIEsurvey

Caro, D. H., & Biecek, P. (2017). intsvy: An R package for analyzing international large-scale assessment data. *Journal of Statistical Software, 81*(7), 1–44. https://doi.org/10.18637/jss.v081.i07

Cohen, J. D., & Jiang, T. (1999). Comparison of partially measured latent traits across nominal subgroups. *Journal of the America Statistical Association, 94*(448), 1035–1044. https://doi.org/10.1080/01621459.1999.10473856

Gonzalez, E. (2021). The IDB analyzer. In L. Khorramdel, M. von Davier, & K. Yamamoto (Eds.), *Innovative computer-based international large-scale assessments – Foundations, methodologies and quality assurance procedures*. Springer.

Johnson, E. G., & Rust, K. F. (1992). Population inferences and variance estimation for NAEP data. *Journal of Educational Statistics, 17*(2), 175–190. https://doi.org/10.3102/10769986017002175

van der Laan, J. (2017). *LaF: Fast access to large ASCII files* [R package version 0.8.0]. https://CRAN.R-project.org/package=LaF

Lee, M., Bailey, P., & Emad, A. (2018a). *Using the getData function in EdSurvey*. Retrieved January 27, 2020, from https://www.air.org/sites/default/files/EdSurvey-getData.pdf

Bailey, P., & Zhang, T., (Eds.) (2024). EdSurvey user's guide. https://naep-research.airprojects.org/Portals/0/EdSurvey_A_Users_Guide/_book/index.html

Liao, Y., Bailey, P., & Zhang, T. (2018). *Exploratory data analysis on NCES data*. https://www.air.org/sites/default/files/EdSurvey-EDA.pdf

Lishinski, A. (2018). *Methods and overview of using EdSurvey for running Wald tests*. https://www.air.org/sites/default/files/EdSurvey-WaldTest.pdf

Lumley, T. (2004). Analysis of complex survey samples. *Journal of Statistical Software, 9*(8), 1–19. https://doi.org/10.18637/jss.v009.i08

Mazzeo, J., Liu, B., Donoghue, J. R., & Xu, X. (2021). Approximate standard errors for NAEP results that incorporate linking error under the random group design. Improved test designs and multistage adaptive testing in large-scale assessments. In L. Khorramdel, M. von Davier, & K. Yamamoto (Eds.), *Innovative computer-based international large-scale assessments – Foundations, methodologies and quality assurance procedures*. Springer.

Mislevy, R. J., Beaton, A., Kaplan, B. A., & Sheehan, K. (1992). Estimating population characteristics from sparse matrix samples of item responses. *Journal of Educational Measurement, 29*(2), 133–161. https://doi.org/10.1111/j.1745-3984.1992.tb00371.x

Oberski, D. (2014). lavaan.survey: An R package for complex survey analysis of structural equation models. *Journal of Statistical Software, 57*(1), 1–27. https://doi.org/10.18637/jss.v057.i01

Reif, M. & Peterbauer, J. (2014). *svyPVpack: A package for complex surveys including plausible values* [R package version 0.1–1]. https://github.com/manuelreif/svyPVpack

Wolters, K. M. (2007). *Introduction to variance estimation* (2nd ed.). Springer.

Zhang, T. & Bailey, P. (2018). *Using EdSurvey for trend analysis.* https://www.air.org/sites/default/files/EdSurvey-Trend.pdf

Zhang, T., Bailey, P., Lee, M., Cohen, M., & Yu, J. (2018). *Methods used for gap analysis in EdSurvey.* https://www.air.org/sites/default/files/EdSurvey-Gap.pdf

Part VI
The Future of International Large-Scale Assessments

Chapter 29
What Role Is There for International Large-Scale Assessments in an Age of Rapid Technological Change?

Francesco Avvisati and Francesca Borgonovi

Abstract In recent years, information and communication technology (ICT) has modified the world in which individuals work and live, learn, and socialise. As a result, international large-scale assessments have had to adapt in order to remain relevant and contribute evidence that reflects technological and social transformations. At the same time, international large-scale assessments have also influenced economic, political, social, and technological developments by inspiring policy making and contributing to the evidence base used to assess the relevance, efficiency, and effectiveness of existing policies related to skill development. This chapter reviews some of the key transformations that are likely to influence the demand for skills and skill measurement in the next 20 years, including technological innovations, population ageing, and international migration. Next, the chapter considers strategies, responses, and actions for the design, administration, and analysis of international large-scale assessments, which can ensure that they remain relevant in a changing world. In particular, the chapter identifies what the transformations imply for the assessment of foundation skills and for the assessment of new skills. Finally, the chapter reviews the challenge experts involved in the development, administration, and analysis of international large-scale assessments face with respect to building and maintaining trust in the information these assessments provide.

International large-scale assessments, which measure the knowledge and skills of entire cohorts of students or of the whole working-age population, play a key role for education and skills' policies: They reinforce public accountability, help monitor

---

F. Avvisati (✉)
Organisation for Economic Co-operation and Development, Paris, France
e-mail: francesco.avvisati@oecd.org

F. Borgonovi
Organisation for Economic Co-operation and Development, Paris, France
Social Research Institute, University College London, London, UK

L. Khorramdel et al. (eds.), *Innovative Digital-Based International Large-Scale Assessments*, Methodology of Educational Measurement and Assessment,
https://doi.org/10.1007/978-3-031-90951-1_29

global development goals, and facilitate international peer learning among policy-makers. The interdependence between assessments of skills and the contexts in which they occur has been recognised by many. For example, educational assessments are used to evaluate the effectiveness of schools and teachers (Agasisti & Longobardi, 2017; Lee et al., 1997; Palardy, 2008) and to assess whether students meet the expected learning standards;[1] but as a result, they influence what teachers do and what students learn. Educational assessments represent statements about what scientists, educators, policy-makers and parents want students to know and learn. In turn, the topics and methods of assessment inevitably end up being the focus of instruction. This reciprocal dependence implies that if assessments do not evolve at the same pace as changes in teaching and learning, then innovations in the latter become even more difficult to sustain (Pellegrino, 2013).

For international large-scale assessments, this interdependence extends beyond education systems. Their development, implementation, and interpretation are influenced by the economic, political, social and technological context in which they occur (OECD, 2016, 2019a). In turn, international large-scale assessments also influence economic, political, social and technological developments by inspiring policy making and contributing to the evidence base used to assess the relevance, efficiency and effectiveness of existing policies related to skill development (Breakspear, 2012; Hopfenbeck & Görgen, 2017). For example, in Germany, the education policy debate that arose from the publication of PISA 2000 results was very pronounced and led to the design and implementation of reform measures, including generating national standards and establishing further support for disadvantaged students, especially those from immigrant backgrounds (Ertl, 2006). Similarly, in response to the decline in PISA performance between PISA 2000 and 2003, the Japanese Ministry for Education reversed a contentious curriculum policy and enacted changes to national assessment practices (Takayama, 2008). If assessments do not evolve in order to remain attuned to changing circumstances, they risk not only losing relevance, but also holding back the necessary adaptations required for individuals and societies to thrive in these new circumstances.

In this chapter we will examine, in particular, the implications of technologically driven social and economic changes for two international large-scale assessments led by the Organisation for Economic Co-operation and Development (OECD): the Programme for International Student Assessment (PISA) and the OECD Survey of Adult Skills (PIAAC).

[1] Two prominent examples of the use of international assessment data to monitor goals are the inclusion of PISA indicators in the European Union's strategic framework for co-operation in education and training (known as ET 2020) (European Council, 2009), and more recently the inclusion of PIRLS, PISA and TIMSS indicators in the framework for monitoring progress towards the United Nations' Sustainable Development Goals (UNESCO Institute for Statistics, 2019).

29.1 PISA and PIAAC: Policy-Oriented Assessments

OECD-led assessments, such as PISA and PIAAC, are relatively young and build on the methodological and technical developments of earlier national and international assessments. Although the OECD-led assessments share many features with previous international assessments, they differ from other international assessments in a number of ways; chief among these is their strong policy orientation. Indeed, the OECD's role is to provide member countries and partner countries that participate in its assessment programmes with international comparative data that inform the decisions of governments and of other stakeholders about education and skills policies. The design of PISA and PIAAC developed in response to an increasing awareness of the role education can play in shaping a broad range of socio-economic outcomes and inequalities (Oreopoulos & Salvanes, 2011; Card, 2003) and to the resulting need for information about the acquisition and distribution of skills among policy makers, even among those not directly involved in the organisation of education services.

International institutions like the OECD provide countries not only with data, but also with a space for comparing themselves and collaborating with each other; for confronting ideas and engaging in peer learning in order to improve policies.

The two major OECD-led international assessments, PISA and PIAAC, are developed and implemented under the responsibility of education ministries and, in the case of PIAAC, also of ministries of labour. PISA and PIAAC boards involve representatives from all member countries and partner countries with associate status. Country representatives to these boards comprise a mix of government officials and staff of research and academic institutions. The boards determine policy priorities and oversee adherence to these priorities during implementation. This includes the setting of standards for data development, analysis and reporting.

Because of this unique governance structure, the development and implementation of these studies is highly sensitive to the needs of policy makers, and the work reflects the policy questions facing decision-makers involved in the development of skills in schools, training institutions and the labour market. The studies offer participants not only the opportunity to conduct international comparisons but also to pool resources to be able to answer long-term strategic questions that are often beyond the scope of day-to-day policy making in national administrations. At the same time, since the studies are steered by the OECD, the way in which such questions are framed, and solutions are sought are in line with the strategic orientations of the OECD, an organisation with the core mission of promoting multi-lateral co-operation among governments and developing global standards, including in the area of education. Given this strategic role, the OECD-led international large-scale assessments operate over time horizons that span decades and, as such, are importantly influenced by the anticipated medium and long-term technological, economic and social transformations but, at the same time, aim to influence such transformations by guiding the actions and decisions of governments and other stakeholders.

29.2 "A Brief History of the Future": What Are the Key Changes that Are Likely to Influence the Demand for Skills and for Skill Measurement in the Next 20 Years?

In the past thirty years, the policy priorities have evolved, but not to the extent of prompting significant redesign of large-scale assessment programmes. With the important exception of computer-based delivery, which begun with PIAAC and now has been used in all the major ILSAs, ILSAs administered in the last years of the 2010s resemble in administration conditions, assessment domains and assessment frameworks those developed in the 1990s. During the years ahead, it is possible that the landscape will change more decisively. Several trends—often referred to as "megatrends" in policy debates—will continue to drive change in societies and, in turn, will require changes in the design, implementation and analysis of international large-scale assessments. In particular, the pace of technological change is accelerating. Its interaction with other trends, such as ageing populations and increasing mobility of people across borders, has a significant impact on the way in which individuals develop and accumulate skills, how they are expected to use them in the labour market and in their everyday life, and what benefits such skills confer.

29.2.1 Technological Change and Its Impact on Life and Living

In recent years, information and communication technology (ICT) has modified the world in which individuals work and live, learn and socialise. This transformation has been particularly visible over the past decade. In 2009, about 15% of students in OECD countries, on average, reported that they did not have access to the Internet at home. By 2018, that proportion had shrunk to less than 5% (OECD, 2019a). The growth in access to online services is likely to be even larger than suggested by these percentages, which hide the exponential growth in the quality of Internet services and the explosion of mobile Internet services over the past decade. OECD statistics indicate, for example, that between 2009 and 2018, the number of mobile broadband subscriptions per capita increased by more than a factor of three across OECD countries, on average. At the end of 2018, there were more mobile broadband subscriptions than inhabitants, on average (109.7 per 100 inhabitants) (OECD, 2019b, p. Indicator 1.2 and 1.5). Increased connectivity means that children access and use ICT earlier than ever before—and that older cohorts are required to operate devices that are profoundly different from anything that they have known or have been trained for.

The rolling out of broadband infrastructures has meant an expansion in the bandwidth available for all types of services whose primary activity is the transfer of information. Greater availability of bandwidth, in turn, is driving many services to online platforms that can increasingly be accessed with mobile devices. These services, including not only traditional telecommunication, such as telephony, but also broadcast TV and radio, video and book publishing, as well as banking and money transfer services, can now be—and increasingly are—consumed "on the go" (OECD, 2013). As a result, new technologies have transformed not only what is required of people in the labour market, but also in their private lives.

Today's workplaces and daily lives already require individuals to solve mostly non-routine problems, because digital technologies can be used to perform routine manual or analytical tasks (Autor & Price, 2013; Spitz-Oener, 2006; Ikenaga & Kambayashi, 2010). The technological revolution that began in the last decades of the twentieth century has affected nearly every aspect of life in the twenty-first century: from how we "talk" with our family and friends, to how we shop, to how and where we work. Quicker and more efficient transportation and communication services have made it easier for people, goods, services and capital to move around the world, leading to the globalisation of economies. New means of communication and types of services have changed the way individuals interact with governments, service suppliers and each other. Web conferencing platforms (such as Zoom), business communication platforms (such as Slack), social media and e-government services mean that individuals can exchange messages cheaply and efficiently, citizens can request modules or make payments without queuing at the post office, and businesses can purchase goods rapidly and efficiently and effectively communicate with providers anywhere in the world. These social and economic transformations have, in turn, changed the demand for skills as well. While there are many factors responsible for these changes, digital technologies have profoundly altered what are considered to be the "key information-processing skills" that individuals need as economies and societies evolve in the twenty-first century (Greiff et al., 2017). As technology evolves, more and more routine tasks are in fact handled by machines with increasing levels of sophistication, while individuals are required to deal with unexpected and unfamiliar non-routine tasks (Autor et al., 2003; Spitz-Oener, 2006; Autor & Price, 2013; OECD, 2013).

As this trend continues, the capacity to manage information and solve problems using computers and digital technologies will become ever more a necessity. In most middle-income and high-income countries, ICT applications are increasingly fundamental to operate in the workplace, access social and welfare services, and be socially integrated. As ICT hardware, software and infrastructures change at a rapid pace, users must be able to adjust quickly to new ICT devices or applications that continually change in functionality.

Ever since the Industrial Revolution, fears of technology-induced job losses have been common in the public debate [for a short historical overview (ILO, 2016, 0.2-3). Despite such fears, employment in OECD countries has grown steadily over the past decades (OECD, 2019c). However, a number of authors have argued that the speed and intensity of technological progress is increasing and that the new wave of transformation may have more disruptive consequences for workers (Brynjolfsson & McAfee, 2011; Mokyr et al., 2015).

While worries of widespread technological unemployment may be overstated, the impact of digital transformation on the nature of work and the skills required is very real and already upon us (OECD, 2019c).

Many new, productive and rewarding forms of work and jobs are being created as part of the digital transformation. At the same time, many jobs have disappeared, and more are likely to go in the future (OECD, 2019d). Acknowledging that the impact of digitalisation on the workforce does not merely depend on the existence of new technologies, but also on the uptake of these innovations, OECD estimates suggest that some 14% of workers face a high risk that their tasks will be automated. Another 32% face major changes in the tasks required in their job and, consequently, the skills they would need to do their job (Nedelkoska & Quintini, 2018). These workers will need to significantly adapt to succeed in the new digitally enabled work environment.

Digital technologies and the resulting disruptions on job markets also bring important changes and opportunities to the way education is delivered. The accelerating nature of technological change and the potentially disruptive impact on the demand for basic skills of new technologies, such as artificial intelligence (Elliott, 2017), imply, at a minimum, the need to establish solid systems of lifelong learning and a re-definition of the role of schools. Skills acquired at the beginning of one's working life cannot be assumed to remain relevant over an increasingly long lifespan.

Technological innovations not only mean that individuals need to be able to operate digital technology efficiently to be productive and competitive in the labour market or be able to access and use services to go through their daily lives, but they are profoundly reshaping the way in which individuals access, analyse, and communicate information. The digital revolution has sharply changed the amount of information individuals can access and the cost of accessing such information. Customisation also allows individuals to access information anytime, anywhere. McLuhan's warning that "the medium is the message" (McLuhan, 1964) is as applicable to digital technologies as it was to the electric media that changed information exchange in the twentieth century. Technologies are not neutral information delivery platforms, they reshape the way in which individuals think and process information, and what information is available and when. Social media and other digital technologies provide a constant stream of information that users can tap from their smartphones and other connected devices. Such information is often made available without the editorial filters of traditional media and is pushed to connected users based on algorithms devised to maximise user engagement. This new information landscape demands the use of different strategies to sort, evaluate and prioritise information (Carr, 2011).

Today, the availability of information and services on demand poses, in our opinion, additional stress on individuals' capacity to persist on difficult tasks, to put in the time, effort and energy that is necessary to acquire deep knowledge and to avoid the temptation of spending time on digital entertainment contents. At the same time, the ubiquitous presence of digital social media has made it easier to disseminate information, with little cognitive or motivational capacity among many to separate information into facts, opinions and falsehoods. Algorithms that tailor what information reaches whom have been considered to be responsible for the segmentation of social groups into small niches that function as echo chambers reaffirming and solidifying individuals' prior beliefs with little or no opportunities for exposure to dissenting information or opinions (Pariser, 2011), though the role of technologies—as a cause, or as a mere amplifier of pre-existing tendencies—remains disputed (Davies, 2018; Bruns, 2019). This widens the gap between the "self" and the "other", making it difficult to establish common ground and foster social cohesion. At the same time, digital technologies have given a voice to some of the most marginalised and vulnerable groups in society that historically faced exclusion from participation in mainstream political debates, possibly because of their gender, immigration background, or because they are part of a visible or ethnic minority (Silverstone, 2001; Oiarzabal & Reips, 2012; Silverstone, 2000).

Moving forward, digitalisation and, in particular, the application of machine learning and artificial intelligence technologies, yield the promise of opening up yet unknown possibilities for humans. Such technologies will allow individuals to change their environment to a much higher degree than what was previously possible (for better, or for worse), and complement humans in the functions they perform. Like other key technological innovations in the past, such as the domestication of fire, the invention of writing systems, or the printing press, they will enhance capabilities and change the very nature of consciousness in the process (Harari, 2014).

Technological change has been a strong driver of societal change throughout human history, and the pace of technological change appears to be accelerating. Several other trends, many of which were related, initially, to technological advances, have acquired their own dynamics and can be expected to exert a strong influence on what the world will look like over the next 20–30 years. These include population ageing, international migration, climate change, urbanisation, increased inequality, and the fracturing of traditional information ecosystems. We turn now to two of these trends with direct implications on the demand and supply of skills: population ageing and international migration.

29.2.2 Population Ageing

The world's population is steadily ageing. OECD countries, in particular, are facing declining fertility rates and constant increases in life expectancy, which will inevitably lead to a steep increase in the share of elderly individuals in their populations. According to the most recent estimates, in 2025, around one in four individuals in OECD countries will be 65 or more and by 2050, people above 65 will constitute more than a quarter of the overall OECD population (25.3%), well above the current share of 16.4%. Similarly, the share of individuals above 80 years of age will more than double, both worldwide (from 2.0% to 4.8%) and in OECD countries (from 4.4% to 9.5%), while the share of people below 15 years old will experience a small decline (from 26.0% to 21.3% worldwide, and from 18% to 15.8% in OECD countries) (Paccagnella, 2016).

Population ageing is causing increasing strain on public finances, through higher healthcare costs and increases in public pension spending, notably in pay-as-you-go systems where a shrinking working population finances the pensions of a growing number of retirees (Gertler, 1999). Adjustments in pension entitlements and health insurance systems are being implemented in several countries. As a result, in many countries, individuals remain active in the labour market for longer. Some worry that as older cohorts remain active in the labour market, labour productivity may decline (Skirbekk, 2004). Indeed, evidence from the first cycle of PIAAC suggests that older cohorts tend to have lower levels of information processing skills than younger cohorts, although this may not generalise to other types of skills, and the age profile of skill proficiency varies considerably across countries (OECD, 2013; Paccagnella, 2016). Moreover, PIAAC results suggest that older cohorts are particularly more likely to report low levels of familiarity with digital technologies and demonstrate lower levels of proficiency in problem solving in technology rich environments (OECD, 2015).

Understanding how human capital is accumulated, and how it depreciates over time (Paccagnella, 2016), is crucial to address the challenges of population ageing, but also to improve the effectiveness of policies that aim to equip individuals with the skills needed at work and to thrive in society in the presence of rapid technological change.

29.2.3 International Migration

In 2015 and 2016, OECD countries recorded an unprecedented number of new immigrants. In particular, more than 5.3 million permanent entries of foreign citizens were registered by OECD countries in 2016 alone (OECD, 2018b). While geopolitical instability contributed to the large number of new arrivals registered in 2015 and 2016, immigrant entries in OECD countries have increased throughout recent decades—except during the most recent economic crisis (OECD, 2015)—and might be expected to intensify in the future. Several OECD countries have a long-term experience with integrating immigrants, but others are only beginning to experience immigration (OECD, 2018c).

The share of immigrants in OECD countries is not only growing; it is also becoming more diverse. For example, data from the OECD Database of Immigrants in OECD countries (DIOC) show that among the immigrant population in France, the share from the top five countries of origin declined from 57% in 2000 to less than 50% in 2010. Globally, migration from Asia is on the rise (OECD/EU, 2015, p. 69). In 2010, China was the second main source of highly-educated immigrants to Canada, the third source to Australia, and the fourth source to the United States and to New Zealand. In 1980, China did not appear among the main source countries of skilled immigrants to any of those countries (Brücker et al., 2016).

Increases in the number and diversity of immigrants will require that host communities upgrade their integration policies and programmes, to support their learning of the host country language and to ensure that the range of skills immigrants bring with them are effectively recognised in the host country labour market. The assessment and recognition of foreign qualifications improves the prospects of immigrants and ensures a better match between talent and opportunity, for the benefit of all (OECD, 2017); it also reduces the competition for low-skilled jobs, which may exacerbate wage inequality and aggravate unemployment.

International migration often contributes to urbanisation and takes place within the context of other demographic changes that vary widely among countries and regions. On one hand, centres of global poverty that often struggle with conflict, economic and political fragility tend to have higher fertility rates (World Bank, 2016), and will, therefore, likely account for an increasing share of migrants. On the other hand, countries with greater economic and political stability are preparing for the realities of a rapidly aging population, whereby the elderly outnumber those of working-age, and may use immigration to slow down this trend. Overall, migration is generally influenced by a variety of factors that "push" migrants away from their old home (be it a country or subnational region) or "pull" them toward a new location; these factors are motivated by global inequalities, climate change, demographic imbalances, geopolitical instability as well as search for better opportunities.

29.3 What Change Is Needed in Large-Scale Assessments?

Over the past century, formal learning institutions (mainly schools) have changed from serving the few to catering for the needs of the many, if not all. At the same time, they have become the main system to socialise children and youth and to licence professionals. School-based assessments, such as PISA, were established at a time where schools and classrooms were still the dominating channel and mode of learning (and they still are).

In the near future, schools and universities as we know them will remain important in providing space and opportunity for learning, but they may lose their quasi-monopoly and may have to interact and collaborate increasingly with many other channels and modes of learning (OECD, 2001, 2020). The family, community and the workplace have always contributed more to skills development than recognised by the credentials that determined, to a large extent, individuals' destinies on the labour market. In the coming years, technology increasingly will create new channels for human learning outside institutional frameworks, complementing and competing with the formal and traditional ones. Employers will create new modes of certifying skills, perhaps relying on distributed ledgers to avoid the need for a central gatekeeper and ensure mutual recognition and, therefore, portability of these credentials; as a result, training and informal learning at the workplace will acquire new value.

Anticipating the evolving economic, social, political and technological landscape and ensuring that large-scale assessments adapt to changing conditions poses a unique challenge. Because education fulfils several roles—including transmitting the heritage of the past and preparing young people for the future—education is ridden with tensions between tradition and innovation and does not react quickly to emerging needs.

In the Leopard, the historic novel that describes how Sicily's noblemen weathered the political changes brought by the Italian Risorgimento, Tancredi, the nephew of the prince of Salina says "For everything to stay the same, everything must change" (Tomasi di Lampedusa, 1957). The challenge for large-scale assessments is not to change "everything" but, rather, to adapt and stay relevant by striking the right balance between innovation and tradition. This can be achieved by recognising what elements used in past assessment administrations remain relevant in a changed environment, which elements lose relevance and need to be abandoned or modified, and which new elements need to be embedded in the assessments so that they maintain their relevance.

The technological and demographic changes detailed above determine a number of opportunities as well as challenges for international large-scale assessments. The strategies, responses and actions that can guide changes in the design, administration and analysis of international large-scale assessments so that they remain relevant can be classified in three phases: plan, absorb and adapt.

29.3.1 Plan

Planning starts with the recognition that technological and demographic transformations require changes for which adequate resources and decision-making mechanisms need to be put in place. Large-scale assessments may need to invest more than in the past in research and development, to be able to identify solutions to emerging needs. These will often correspond to solutions already pioneered in the context of national initiatives or small-scale research projects; still, resources need to be mobilised to identify how these solutions can be tested and brought to scale in international assessments involving different languages, cultural backgrounds, tight deadlines and standardised administration protocols.

Preparation may also involve the development of mid-term and long-term strategies and plans based on foresight exercises or other future-oriented tools and ensuring that different stakeholders are adequately involved in the process.

29.3.2 Absorb

Absorption refers to the short and medium-term changes in the design and instrumentation of international large-scale assessments required to keep satisfying existing standards of quality for official statistical activities. These standards include the need for relevance, accuracy, credibility, timeliness, and efficiency (value for money) (OECD, 2011).

Several other chapters in this volume have shown how technological advances can contribute to increase the validity (accuracy), reliability, and efficiency of assessment operations (e.g. through adaptive testing, automatic coding, closer monitoring of and adherence to standards, etc.; see Yamamoto et al., 2025a, 2025b, 2025c). Moving forward, it will be important to incorporate these advances in international large-scale assessment programmes, whenever this is not yet done. But integrating these new elements into the operations of large-scale assessments will not be sufficient to maintain relevance.

In order to satisfy the need for relevance, given technological and demographic transformations, international large-scale assessments will have to assess the set of skills that define an individual's capacity to keep learning and to adapt to future changes. This can be achieved by rethinking how foundation skills can and should be assessed, given technological changes, and by expanding the assessments to new skills—skills that are often considered hard to assess.

29.3.2.1 Rethinking the Assessments of Foundation Skills

The technological and demographic transformations described in previous sections require a reflection on the assessment of foundational skills—what it means to be proficient in reading, mathematics, and science, in the digital age. For example, contemporary science standards, such as the US Next Generation Science Standards or the German Bildungsstandards, emphasise that a central goal of science education is for students to be able to use their knowledge to make sense of the designed and natural world. In other words, students should be able to demonstrate knowledge-in-use by integrating disciplinary knowledge with scientific practices, such as modelling, using evidence-based explanation and argumentation, and designing investigations to test hypotheses, analyse results and construct explanations from data (Harris et al., 2016). This emphasis on doing science, rather than just on knowledge of and about science, has important implications, not only for the assessment content, but for its format as well. Assessment tasks will be expected to mimic closely authentic practice, from novice to expert level, and technology will be instrumental in achieving this. Similarly, while digitalisation has made sharing non-text-based sources of information, such as videos and images, easier, it has not necessarily done so at the expense of text-based information. Education systems are increasingly incorporating digital (reading) literacy into their programmes of instruction (Erstad, 2006), and the PISA 2018 reading assessment enabled to assess students' proficiency in processing digital texts (for example texts requiring scrolling and clicking on hyperlinks) and to evaluate the trustworthiness of information provided by examining sources and text features (OECD, 2019d).

A second aspect that needs to be considered is that low-stakes tests like PISA and PIAAC rely on the motivation of respondents to engage with the test. Traditionally, test-takers' engagement has been assumed, but there is increasing evidence that there are important differences in test-takers' engagement between countries and population subgroups (Borgonovi & Biecek, 2016; Eklöf, 2007; Chan et al., 1997). These differences may grow in the future, as some individuals may become used to the fast-paced and engaging nature of digital applications and online services and tools and, conversely, may be unable to put effort in less engaging tasks. In order to satisfy the need for validity and comparability, technological developments can be used to improve the experience for test takers, for example by developing tests that are more engaging for respondents both in terms of delivery mode (such as through the use of a stimulating computer platform) and test content (through the use of adaptive testing, so that the majority of the problems test takers are asked to solve are neither too easy nor too hard for them to solve). In the absence of such developments, the central argument of evidence-based design (Mislevy et al., 2003), that what we observe students say, do, or make in assessment tasks can be used to make inferences about what they know, can do, or have accomplished more generally, may falter.

29.3.2.2 Assessing New Skills

Both PISA and PIAAC have already begun developing assessments of emerging key competences: PISA 2015 assessed collaborative problem solving and PISA 2018, global competence, and an assessment of creative thinking is being developed for PISA 2022. PIAAC, in its first implementation, included an assessment of problem solving in technologically rich environments and will include an assessment of adaptive problem solving in the second implementation. The question is how to develop a coherent strategy for these "innovative assessments" and ensure that they cumulatively contribute to enhancing the programme's relevance.

Such a strategy must inevitably start with a realistic scenario about how future trends will affect the demand for people's skills. A growing body of research is dedicated to establishing frameworks of so-called twenty-first century competences (OECD, 2018a, 2005; Griffin et al., 2012; Center for Curriculum Redesign, 2019; National Research Council, 2012). Some of these frameworks identify the need to nurture learners' agency—i.e. to build the capacity and motivation for lifelong learning—as the most obvious implication for education of a world that requires constant adaptation and re-skilling (Center for Curriculum Redesign, 2019; OECD, 2018a). Adult and youth surveys alike will need to measure the extent to which learners have developed a sense of agency and the ability to actively design their own trajectories of learning, through combinations of formal, non-formal and informal learning.

For assessments such as PISA and PIAAC, which aspire to measure not mere knowledge or intellectual capacity, but knowledge-in-use and authentic competences, "contexts" are always a defining aspect of what is measured. Some groupings of so-called twenty-first century competences, like cognitive, intrapersonal and interpersonal competences (National Research Council, 2012), are relevant in most situations in which an individual is doing a task. Conversely, other groupings do not refer to completely distinct skills but rather to what individuals are able to do in specific contexts. For example, digital literacy competences are only relevant in a digital context; similarly, individuals can only engage their collaborative competences in an environment in which they interact with others or demonstrate civic or citizenship competences in a context that is relevant to civic issues. Rather than distinguishing between inherently different competences, these groupings might more appropriately be understood as describing the different contexts (i.e. digital, global/interconnected, collaborative) for which individuals need to be increasingly prepared and need to develop a specialised skillset (including cognitive, social, intrapersonal and self-regulatory/metacognitive components).

29.3.3 Adapt

Adaptation refers to the long-term changes in the design, instrumentation and analysis tools required to ensure that international large-scale assessments remain relevant given radically changed technological and social conditions. Such changes could involve, for example, significantly altering the structure and key parameters of the assessment programmes: not only what is assessed, but also who, where and how often.

As learning systems expand in time and space, blurring the limits between the formal and the informal, large-scale assessments focusing on youth will have to adapt, and, in order to remain relevant guides for policy, capture different forms of learning and their outcomes. The progressive erosion of schools as key players in the development of skills due to longer life expectancy coupled with rapid technological change may imply moving away from school-based assessments (at or around age 15) as a primary tool to evaluate the quality of education systems. On the other hand, the increased awareness that skills cannot be considered acquired once and forever, but that individuals go through phases of re-skilling and, sometimes, de-skilling, may increase the appetite to embed large-scale assessment programmes within longitudinal designs, which can capture crucial information on transitions and flows, in addition to the stocks measured by cross-sectional designs.

Another fundamental change may concern the frequency and periodicity of assessments. The administration in highly spaced time intervals at high levels of analysis, such as countries or large regions, may prove ill-suited to support the decision-making at the local level following technologically driven disruption or large-scale inflow of new arrivals because of international migration.

While PISA has expanded significantly in terms of global coverage (the number of participating education system increased from 32 in 2000 to more than 80 by 2018), it will be important to evaluate the opportunity for continued expansion. Global coverage can serve the important goal of establishing a global metric to monitor progress in the context of global initiatives, such as the Sustainable

Development Goals (SDGs). Rather than by rolling out the existing assessment programmes, their global reach may also be assured by integrating OECD-led initiatives within a system of skill monitoring and benchmarking, constituted by national, regional and other international assessment initiatives. Maximising complementarity through statistical linkages while differentiating the contextual information collected according to countries' conditions can ensure that the value of international assessments is maximised for local actors and stakeholders.

Finally, as statistical agencies are no longer the only organisations collecting data on individuals, the importance of data for many digital services also creates increased awareness, among citizens and consumers, of data ownership and privacy concerns. Will it still be possible to collect and publish individual data when the aim is to represent population distributions? Will response rates be affected?

29.4 The Trust Issue

Overall, trust is a ubiquitous ingredient in policymaking and implementation across many governance systems, including education, whether it concerns accountability mechanisms, capacity building or strategic thinking. The paradox of trust in institutions and policy making is that lack of trust often results in strong calls for transparency and accountability, yet putting in place systems that deliver on the need for accountability and transparency requires a degree of trust in those responsible for the design and implementation of such systems. Large-scale international assessments have played an important role in this respect and embody the tensions and paradox of trust. Large-scale assessments have, over the years, been used to measure the quality, equity and efficiency of national education systems worldwide. The shared decision-making governance structure, the involvement of international experts, including some experts from outside the OECD area in technical advisory groups, and the fact that the design, implementation and analyses was not under the direct influence of any one political party or any one decision-maker made large-scale assessments uniquely placed to provide evidence in a low-trust environment. Furthermore, technical standards, measurement frameworks, psychometric modelling and analysis procedures are all communicated publicly through various professional and research publications and are open to evaluation and criticism; they were largely inspired by national large-scale assessments, such as NAEP, and previous international surveys, such as TIMSS and IALS. However, in recent years, lack of trust has also been directed towards the community involved in the development of large-scale assessments, eroding the very foundations of how well large-scale assessments can perform their function.

Increased knowledge of international large-scale assessments and awareness of the role they play in policy-making has amplified the amount of interest in the information they provide and also in the level of scrutiny they are subject to, not only among the research community, but also the general public.

In the future, large-scale assessments will have to invest adequate resources to build trust among the stakeholders they serve, as well as the experts they rely on for their operations. This can be achieved by making technical materials more easily accessible by the research community and also by ensuring that information detailing the rationale guiding the design, implementation and analysis of the assessment is made available in a format suitable for being accessed by laypeople. Holding regular public consultations to ensure understanding and buy-in among stakeholders at key decision junctures over the evolution of international-large scale assessments may also ensure that they continue to serve the public interest.

29.5 Conclusion

All statistical tools "allow the discovery or creation of entities that support our descriptions of the world and the way we act on it" (Desrosières, 1998, p. 3). International large-scale assessments are, in this respect, a prime example of a statistical tool inspired by policy makers to support their decisions. But to continue to guide relevant decision making in a changing world, international large-scale assessments need to evolve as well. In this chapter we have presented the context in which the OECD established its largest assessment programmes, PIAAC and PISA, as well as the current trends which, in our view, will shape their future trajectories. In doing so, we highlight the role of technology, not only as a driver of methodological improvement (as illustrated in the remaining chapters of this volume), but also as a potentially disruptive force that may change (along with other so-called mega-trends) the economic value of particular skills and the institutional settings in which skills develop.

References

Agasisti, T., & Longobardi, S. (2017). Equality of educational opportunities, schools' characteristics and resilient students: An empirical study of EU-15 countries using OECD-PISA 2009 data. *Social Indicators Research, 134*(3), 917–953. https://doi.org/10.1007/s11205-016-1464-5

Autor, D. H., & Price, B. (2013). *The changing task composition of the US labor market: An update of Autor, Levy, and Murnane (2003)*. University of Minnesota. http://economics.mit.edu/files/9758

Autor, D., Levy, F., & Murnane, R. (2003). The skill content of recent technological change: An empirical exploration. *The Quarterly Journal of Economics, 118*(4), 1279–1333. https://doi.org/10.1162/003355303322552801

Borgonovi, F., & Biecek, P. (2016). An international comparison of students' ability to endure fatigue and maintain motivation during a low-stakes test. *Learning and Individual Differences, 49*, 128–137. https://doi.org/10.1016/j.lindif.2016.06.001

Breakspear, S. (2012). *The policy impact of PISA: An exploration of the normative effects of international benchmarking in school system performance* (OECD Education Working Papers, No. 71). OECD Publishing. https://doi.org/10.1787/5k9fdfqffr28-en

Brücker, H., Rother, N., Schupp, J., Babka Von Gostomski, C., Böhm, A., Fendel, T., et al. (2016). Forced migration, arrival in Germany, and first steps toward integration. *DIW Economic Bulletin, 6*(48), 541–556. Retrieved March 09, 2019, from https://www.diw.de/documents/publikationen/73/diw_01.c.548405.de/diw_econ_bull_2016-48-1.pdf

Bruns, A. (2019). Filter bubble. *Internet Policy Review, 8*(4). https://doi.org/10.14763/2019.4.1426

Brynjolfsson, E., & McAfee, A. (2011). *Race against the machine: How the digital revolution is accelerating innovation, driving productivity, and irreversibly transforming employment and the economy*. Digital Frontier Press.

Card, D. (2003). Estimating the return to schooling: Progress on some persistent econometric problems. *Econometrica, 69*(5), 1127–1160. https://doi.org/10.1111/1468-0262.00237

Carr, N. (2011). *The shallows. What the internet is doing to our brains*. W. W. Norton & Company.

Center for Curriculum Redesign. (2019). *Framework of competencies/subcompetencies*. Retrieved September 2, 2019, from https://curriculumredesign.org/framework/

Chan, D., Schmitt, N., DeShon, R., Clause, C., & Delbridge, K. (1997). Reactions to cognitive ability tests: The relationships between race, test performance, face validity perceptions, and test-taking motivation. *Journal of Applied Psychology, 82*(2), 300–310. https://doi.org/10.1037/0021-9010.82.2.300

Davies, H. (2018). Redefining filter bubbles as (Escapable) socio-technical recursion. *Sociological Research Online, 23*(3), 637–654. https://doi.org/10.1177/1360780418763824

Desrosières, A. (1998). *The politics of large numbers: A history of statistical reasoning*. Harvard University Press.

Eklöf, H. (2007). Test-taking motivation and mathematics performance in TIMSS 2003. *International Journal of Testing, 7*(3), 311–326. https://doi.org/10.1080/15305050701438074

Elliott, S. (2017). *Computers and the future of skill demand*. OECD Publishing, Paris. https://doi.org/10.1787/9789264284395-en

Erstad, O. (2006). A new direction?: Digital literacy, student participation and curriculum reform in Norway. *Education and Information Technologies, 11*, 415–429. https://doi.org/10.1007/s10639-006-9008-2

Ertl, H. (2006). Educational standards and the changing discourse on education: The reception and consequences of the PISA study in Germany. *Oxford Review of Education, 32*(5), 619–634. https://doi.org/10.1080/03054980600976320

European Council. (2009). Council conclusions of 12 May 2009 on a strategic framework for European cooperation in education and training ("ET 2020"). *Official Journal of the European Union*, pp. C 119/2–10. https://doi.org/10.3000/17252423.C_2009.119.eng

Gertler, M. (1999). Government debt and social security in a life-cycle economy. *Carnegie-Rochester Conference Series on Public Policy, 50*, 61–110. https://doi.org/10.1016/s0167-2231(99)00022-6

Greiff, S., et al. (2017). *Adaptive problem solving: Moving towards a new assessment domain in the second cycle of PIAAC* (OECD Education Working Papers, No. 156). OECD Publishing. https://doi.org/10.1787/90fde2f4-en

Griffin, P., McGaw, B., & Care, E. (Eds.). (2012). *Assessment and teaching of 21st century skills*. Springer. https://doi.org/10.1007/978-94-007-2324-5

Harari, Y. (2014). *Sapiens. A brief history of mankind*. Harper.

Harris, C. J., Krajcik, J. S., Pellegrino, J. W., & McElhaney, K. (2016). *Constructing assessment tasks that blend disciplinary core Ideas, crosscutting concepts, and science practices for classroom formative applications*. SRI International.

Hopfenbeck, T. N., & Görgen, K. (2017). The politics of PISA: The media, policy and public responses in Norway and England. *European Journal of Education, 52*(2), 192–205. https://doi.org/10.1111/ejed.12219

Ikenaga, T., & Kambayashi, R. (2010). *Long-term trends in the polarization of the Japanese labor Market: The increase of non-routine task Input and its valuation in the labor market*. Hitotsubashi University Institute of Economic Research Working Paper.

ILO. (2016). *Technological changes and work in the future: Making technology work for all*. ILO. https://www.ilo.org/global/topics/future-of-work/WCMS_534201/lang%2D%2Den/index.htm
Lee, V. E., Smith, J. B., & Croninger, R. G. (1997). How high school organization influences the equitable distribution of learning in mathematics and science. *Sociology of Education, 70*(2), 128–150. https://doi.org/10.2307/2673160
McLuhan, M. (1964). *Understanding media. The extension of man*. McGraw Hill.
Mislevy, R., Steinberg, L., & Almond, R. (2003). Focus article: On the structure of educational assessments. *Measurement: Interdisciplinary Research & Perspective, 1*(1), 3–62. https://doi.org/10.1207/s15366359mea0101_02
Mokyr, J., Vickers, C., & Ziebarth, N. (2015). The history of technological anxiety and the future of economic growth: Is this time different? *Journal of Economic Perspectives, 29*(3), 31–50. https://doi.org/10.1257/jep.29.3.31
National Research Council. (2012). *Education for life and work: Developing transferable knowledge and skills in the 21st century*. National Academies Press. https://doi.org/10.17226/13398
Nedelkoska, L., & Quintini, G. (2018). Automation, skills use and training. In *OECD social, employment and migration working papers*. OECD Publishing. https://doi.org/10.1787/2e2f4eea-en
OECD. (2001). What schools for the future? In *Schooling for tomorrow*. OECD Publishing. https://doi.org/10.1787/9789264195004-en
OECD. (2005). *Definition and selection of key competencies: Executive summary*. OECD Publishing.
OECD. (2011). *Quality framework for OECD statistical activities*. https://www.oecd.org/sdd/qualityframeworkforoecdstatisticalactivities.htm
OECD. (2013). *OECD skills outlook 2013: First results from the survey of adult skills*. OECD Publishing. https://doi.org/10.1787/9789264204256-en
OECD. (2015). *International migration outlook 2015*. https://doi.org/10.1787/migr_outlook-2015-en
OECD. (2016). Innovating education and educating for innovation: The power of digital technologies and skills. In *Educational research and innovation*. OECD Publishing. https://doi.org/10.1787/9789264265097-en
OECD. (2017). Making integration work: Assessment and recognition of foreign qualifications. In *Making integration work*. OECD Publishing. https://doi.org/10.1787/9789264278271-en
OECD. (2018a). *The future of education and skills: Education 2030*. OECD Publishing.
OECD. (2018b). *International migration outlook 2018*. https://doi.org/10.1787/migr_outlook-2018-en.
OECD. (2018c). The resilience of students with an immigrant background: Factors that shape well-being. In *OECD reviews of migrant education*. OECD Publishing. https://doi.org/10.1787/9789264292093-en
OECD. (2019a). *PISA 2018 results (Volume I): What students know and can do*. OECD Publishing. https://doi.org/10.1787/5f07c754-en
OECD. (2019b). *Broadband portal*. Retrieved July 10, 2019, from https://www.oecd.org/sti/broadband/broadband-statistics/
OECD. (2019c). The future of work: What do we know? In *OECD employment outlook 2019: The future of work*. OECD Publishing. https://doi.org/10.1787/ef00d169-en
OECD. (2019d). *PISA 2018 assessment and analytical framework*. OECD Publishing.
OECD. (2020). Back to the future of education: Four OECD scenarios for schooling. In *Educational Research and Innovation*. OECD Publishing. https://doi.org/10.1787/178ef527-en
OECD/EU. (2015). *Indicators of immigrant integration 2015: Settling in*. OECD Publishing/European Union. https://doi.org/10.1787/9789264234024-en
Oiarzabal, P., & Reips, U.-D. (2012). Migration and diaspora in the age of information and communication technologies. *Journal of Ethnic and Migration Studies, 38*(9), 1333–1338. https://doi.org/10.1080/1369183x.2012.698202

Oreopoulos, P., & Salvanes, K. G. (2011). Priceless: The nonpecuniary benefits of schooling. *Journal of Economic Perspectives, 25*(1), 159–184. https://doi.org/10.1257/jep.25.1.159

Paccagnella, M. (2016). *Age, ageing and skills: Results from the survey of adult skills* [OECD Education Working Papers, Vol. 2016]. OECD Publishing. https://doi.org/10.1787/5jm0q1n38lvc-en

Palardy, G. J. (2008). Differential school effects among low, middle, and high social class composition schools: A multiple group, multilevel latent growth curve analysis. *School Effectiveness and School Improvement, 19*(1), 21–49. https://doi.org/10.1080/09243450801936845

Pariser, E. (2011). *The filter bubble: How the new personalized web is changing what we read and how we think.* Penguin Books.

Pellegrino, J. (2013). Proficiency in science: Assessment challenges and opportunities. *Science, 340*(6130), 320–323. https://doi.org/10.1126/science.1232065

Silverstone, R. (2000). The sociology of mediation and communication. In C. Calhoun, C. Rojek, & B. Turner (Eds.), *The Sage handbook of sociology* (pp. 188–207). Sage Publications.

Silverstone, R. (2001). Finding a voice: Minorities, media and the global commons. *Emergences: Journal for the Study of Media & Composite Cultures, 11*(1), 13–27. https://doi.org/10.1080/10457220120044648

Skirbekk, V. (2004). Age and individual productivity: A literature survey. *Vienna Yearbook of Population Research, 2*, 133–153. http://epub.oeaw.ac.at/0xc1aa500d_0x00062025

Spitz-Oener, A. (2006). Technical change, job tasks, and rising educational demands: Looking outside the wage structure. *Journal of Labor Economics, 24*(2), 235–270. https://doi.org/10.1086/499972

Takayama, K. (2008). The politics of international league tables: PISA in Japan's achievement crisis debate. *Comparative Education, 44*(4), 387–407. https://doi.org/10.1080/03050060802481413

Tomasi di Lampedusa, G. (1957). *Il Gattopardo [The Leopard]*. Feltrinelli.

UNESCO Institute for Statistics. (2019). *Quick guide to education indicators for SDG 4.* UNESCO Institute for Statistics. Retrieved July 13, 2019, from https://unesdoc.unesco.org/ark:/48223/pf0000265396

World Bank. (2016). *Development goals in an era of demographic change*. International Bank for Reconstruction and Development/The World Bank.

Yamamoto, K., Ali, U., Robin, F., Kapur, L., & Kandathil, M. (2025a). Procedures to evaluate and improve data quality using timing and process data. In L. Khorramdel, M. von Davier, & K. Yamamoto (Eds.), *Innovative digital-based international large-scale assessments – Foundations, methodologies and quality assurance procedures*. Springer.

Yamamoto, K., Shin, H. J., He, Q., & von Davier, M. (2025b). Developing a machine assisted coding system for constructed response items in PISA. In L. Khorramdel, M. von Davier, & K. Yamamoto (Eds.), *Innovative digital-based international large-scale assessments – Foundations, methodologies and quality assurance procedures*. Springer.

Yamamoto, K., Shin, H. J., Robin, F., Khorramdel, L., & Halderman, L. (2025c). Improved test designs and multistage adaptive testing in large-scale assessments. In L. Khorramdel, M. von Davier, & K. Yamamoto (Eds.), *Innovative digital-based international large-scale assessments – Foundations, methodologies and quality assurance procedures*. Springer.

MIX
Papier aus verantwortungsvollen Quellen
Paper from responsible sources
FSC® C105338

If you have any concerns about our products,
you can contact us on
ProductSafety@springernature.com

In case Publisher is established outside the EU,
the EU authorized representative is:
Springer Nature Customer Service Center GmbH
Europaplatz 3, 69115 Heidelberg, Germany

Printed by Libri Plureos GmbH
in Hamburg, Germany